MW00843854

Ergebnisse der Mathematik und ihrer Grenzgebiete

Volume 44

3. Folge

A Series of Modern Surveys in Mathematics

Springer
Berlin
Heidelberg
New York
Barcelona
Hong Kong
London
Milan
Paris
Tokyo

Wolfgang Lück

L^2-Invariants: Theory and Applications to Geometry and K-Theory

 Springer

Wolfgang Lück
Mathematisches Institut
Universität Münster
Einsteinstraße 62
48149 Münster, Germany
e-mail: lueck@math.uni-muenster.de

Library of Congress Cataloging-in-Publication Data applied for

Die Deutsche Bibliothek – CIP-Einheitsaufnahme

Lück, Wolfgang:
L^2-invariants: theory and applications to geometry and K-theory / Wolfgang Lück. –
Berlin; Heidelberg; New York; Barcelona; Hong Kong; London; Milan; Paris; Tokyo: Springer, 2002

(Ergebnisse der Mathematik und ihrer Grenzgebiete ; Folge 3, Vol. 44)

ISBN 3-540-43566-2

Mathematics Subject Classification (2000):
57-99, 58J50, 46L99, 19A99

ISSN 0071-1136
ISBN 3-540-43566-2 Springer-Verlag Berlin Heidelberg New York

Springer-Verlag Berlin Heidelberg New York
a member of BertelsmannSpringer Science+Business Media GmbH

http://www.springer.de

© Springer-Verlag Berlin Heidelberg 2002
Printed in Germany

Typeset by the author using a Springer T$_E$X macro package.
Printed on acid-free paper   SPIN 10742230   44/3142LK - 5 4 3 2 1 0

Preface

There is the general principle to consider a classical invariant of a closed Riemannian manifold M and to define its analog for the universal covering $\widetilde{M}$ taking the action of the fundamental group $\pi = \pi_1(M)$ on $\widetilde{M}$ into account. Prominent examples are the Euler characteristic and the signature of M, which lead to Wall's finiteness obstruction and to all kinds of surgery obstructions such as the symmetric signature or higher signatures. The p-th L^2-*Betti number* $b_p^{(2)}(\widetilde{M})$ arises from this principle applied to the p-th Betti number $b_p(M)$. Some effort is necessary to define L^2-Betti numbers in the case where π is infinite. Typical problems for infinite π are that $\widetilde{M}$ is not compact and that the complex group ring $\mathbb{C}\pi$ is a complicated ring, in general not Noetherian. Therefore some new technical input is needed from operator theory, namely, the group von Neumann algebra and its trace. Analytically Atiyah defined L^2-Betti numbers in terms of the heat kernel on $\widetilde{M}$. There also is an equivalent combinatorial approach based on the cellular $\mathbb{C}\pi$-chain complex of $\widetilde{M}$. It is one of the main important and useful features of L^2-invariants that they can be defined both analytically and combinatorially. There are two further types of L^2-invariants. L^2-*torsion* generalizes the classical notion of Reidemeister torsion from finite to infinite π, whereas *Novikov-Shubin invariants* do not have a classical counterpart.

A very intriguing and important property of L^2-invariants is that they have relations to many other fields. From their construction it is clear that they have connections to operator theory, in particular to von Neumann algebras, and to the spectral theory of the Laplacian on $\widetilde{M}$. For instance Atiyah's motivation to consider L^2-Betti numbers was to establish his L^2-index theorem.

More suprising is the appearance of algebraic K-theory. In all examples where L^2-Betti numbers have been computed explicitly, the values turn out to be rational numbers whose denominators are linked to the orders of finite subgroups of π. This is very suprising in view of the actual definition of L^2-Betti numbers. This phenomenon is linked to questions in algebraic K-theory such as whether any finitely generated projective $\mathbb{C}\pi$-module M is obtained by induction from a finitely generated projective $\mathbb{C}H$-module for a finite subgroup $H \subset \pi$. This leads to the version of the so called *Atiyah Conjecture* that the L^2-Betti numbers are always integers if π is torsionfree.

It turns out that this conjecture implies the *Kaplansky Conjecture* that $\mathbb{C}\pi$ contains no non-trivial zero-divisors if π is torsionfree. For many groups π the Kaplansky Conjecture was not known until the Atiyah Conjecture was proved. We will investigate interactions between L^2-invariants and K-theory and applications of them in both directions throughout this book.

Next we explain a connection to geometry. Provided that M is aspherical, all computations lead to the result that $b_p^{(2)}(\widetilde{M}) = 0$ holds for $2p \neq \dim(M)$ and that $b_n^{(2)}(\widetilde{M}) = (-1)^n \cdot \chi(M)$ is true for the Euler characteristic $\chi(M)$ if $\dim(M) = 2n$ is even. In particular $(-1)^n \cdot \chi(M) \geq 0$ in the case $\dim(M) = 2n$, since each L^2-Betti number is larger or equal to zero by definition. This phenomenon seems to be typical and will be investigated in this book. Recall that M is aspherical if it carries a Riemannian metric with non-positive sectional curvature, but that the converse is not true. If $\dim(M) = 2n$ and M carries a Riemannian metric with negative sectional curvature, then all computations yield $b_n^{(2)}(\widetilde{M}) = (-1)^n \cdot \chi(M) > 0$. Hence L^2-Betti numbers are linked to the Hopf Conjecture which predicts $(-1)^n \cdot \chi(M) \geq 0$ if the $2n$-dimensional closed manifold M carries a Riemannian metric with non-positive sectional curvature, and $(-1)^n \cdot \chi(M) > 0$ if M carries a Riemannian metric with negative sectional curvature. Further connections between L^2-invariants and geometry and group theory will be presented in this book.

Why Study L^2-Invariants?

From the author's point of view there are certain criteria which decide whether a topic or an area in modern mathematics is worth studying or worth further development. Among them are the following:

- The topic has relations to other fields. There is a fruitful exchange of results and techniques with other areas which leads to solutions of problems and to innovations in both the topic of focus and other topics;
- There are some hard open problems which are challenging and promising. They create interesting activity and partial solutions and techniques for their proof already have applications to other problems;
- The topic is accessible with a reasonable amount of effort. In particular talented students are able to learn the basics of the topic within an appropriate period of time and while doing so get a broad basic education in mathematics.

The purpose of this book is to convince the reader that L^2-invariants do satisfy these criteria and to give a comprehensible and detailed approach to them which includes the most recent developments.

A User's Guide

We have tried to write this book in a way which enables the reader to pick out his favourite topic and to find the result she or he is interested in quickly and without being forced to go through other material. The various chapters are kept as independent of one another as possible. In the introduction of each chapter we state what input is needed from the previous chapters, which is in most cases not much, and how to browse through the chapter itself. It may also be worthwhile to go through the last section "Miscellaneous" in each chapter which contains some additional information. In general a first impression can be gained by just reading through the definitions and theorems themselves. Of course one can also read the book linearly.

Each chapter includes exercises. Some of them are easy, but some of them are rather difficult. Hints to their solutions can be found in Chapter 16. The exercises contain interesting additional material which could not be presented in detail in the text. The text contains some (mini) surveys about input from related material such as amenable groups, the Bass Conjecture, deficiency of groups, Isomorphism Conjectures in K-theory, 3-manifolds, Ore localization, residually finite groups, simplicial volume and bounded cohomology, symmetric spaces, unbounded operators, and von Neumann regular rings, which may be useful by themselves. (They are listed in the index under "survey". One can also find a list of all conjectures, questions and main theorems in the index.)

If one wants to run a seminar on the book, one should begin with Sections 1.1 and 1.2. Then one can continue depending on the own interest. For instance if one is algebraically oriented and not interested in the analysis, one may directly pass to Chapter 6, whereas an analyst may be interested in the rest of Chapter 1 and then pass to Chapter 2. Chapters 9, 10, 11, 12, 13 and 14 are independent of one another. One may directly approach these chapters and come back to the previous material when it is cited there.

We require that the reader is familiar with basic notions in topology (CW-complexes, chain complexes, homology, manifolds, differential forms, coverings), functional analysis (Hilbert spaces, bounded operators), differential geometry (Riemannian metric, sectional curvature) and algebra (groups, modules, elementary homological algebra).

Acknowledgements

I want to thank heartily the present and former members of the topology group in Münster who read through the manuscript and made a lot of useful comments, corrections and suggestions. These are Arthur Bartels, Clemens Bratzler, Clement de Seguins Pazzis, Eckehard Hess, Michael Joachim, Michel Matthey, David Meintrup, Morten Pohlers, Holger Reich, Juliane Sauer, Roman Sauer, Thomas Schick, Marco Schmidt, Roland Stamm, Marco Varisco, Julius Verrel and Christian Wegner.

I want to thank the Deutsche Forschungsgemeinschaft which has been and is financing the Sonderforschungsbereich 478 – "Geometrische Strukturen in der Mathematik" and the Graduiertenkolleg "Analytische Topologie und Metageometrie". These institutions made it possible to invite guests and run workshops on the topics of the book. This was very important for its writing. In particular I had fruitful discussions with and obtained a lot of information from Ulrich Bunke, Marc Burger, Dan Burghelea, Mike Davis, Jozef Dodziuk, Benno Eckmann, Michael Farber, Tom Farrell, Damien Gaboriau, Thomas Kappeler, Ian Leary, Peter Linnell, John Lott, Varghese Mathai, Guido Mislin, Nicolas Monod, Martin Olbrich, Pierre Pansu, Holger Reich, Thomas Schick, Michael Shubin, Alain Valette and Shmuel Weinberger.

I thank the Max-Planck Institute for Mathematics in Bonn for its hospitality during my stay in January and February 2001 while parts of this book were written.

Finally I want to express my deep gratitude to my wife Sibylle and our children Christian, Isabel, Tobias and Severina for everything.

Münster, February 2002   *Wolfgang Lück*

Contents

0. Introduction

0.1 What are L^2-Invariants?

There is the classical notion of the p-th Betti number $b_p(X)$ of a finite CW-complex X, for instance a closed manifold, which is the dimension of the complex vector space $H_p(X;\mathbb{C})$. Consider a G-covering $p\colon \overline{X} \to X$. If G is infinite, the p-th Betti number of $\overline{X}$ may be infinite and hence useless. Using some input from functional analysis involving Hilbert spaces, group von Neumann algebras and traces one can define the *p-th L^2-Betti number* $b_p^{(2)}(\overline{X};\mathcal{N}(G))$ of the total space $\overline{X}$ as the non-negative real number given by the von Neumann dimension of the (reduced) L^2-homology of $\overline{X}$. (Often we briefly write $b_p^{(2)}(\overline{X})$ if G is clear from the context.) If G is finite, $b_p^{(2)}(\overline{X}) = |G|^{-1} \cdot b_p(X)$ and we get nothing new. But L^2-Betti numbers carry new information and have interesting applications in the case where G is infinite. In general $b_p^{(2)}(\overline{X})$ of the total space $\overline{X}$ and $b_p(X)$ of the base space X have no relations except for the Euler-Poincaré formula, namely,

$$\chi(X) \;=\; \sum_{p\geq 0}(-1)^p \cdot b_p(X) \;=\; \sum_{p\geq 0}(-1)^p \cdot b_p^{(2)}(\overline{X}), \qquad (0.1)$$

where $\chi(X)$ is the Euler characteristic of X (see Section 0.6).

The notion of the classical Reidemeister torsion of $\overline{X}$ for finite groups G will be generalized to the notion of *L^2-torsion* $\rho^{(2)}(\overline{X}) \in \mathbb{R}$ in the case that G is infinite.

There is a third class of L^2-invariants, the *Novikov-Shubin invariants* $\alpha_p(\overline{X})$, which carry no information if G is finite.

All these types of L^2-invariants on the one hand have analytic definitions in terms of the heat kernel on $\overline{X}$, but on the other hand can be defined combinatorially in terms of the cellular $\mathbb{C}G$-chain complex of $\overline{X}$. These two approaches are equivalent. In the analytic context X must be a compact Riemannian manifold. For the combinatorial definition of L^2-Betti numbers and Novikov-Shubin invariants it suffices to require that the base space X is of finite type, i.e. each skeleton of X is finite, but X may be infinite-dimensional.

0.2 Some Applications of L^2-Invariants

In order to convince the reader about the potential of L^2-invariants we state some results which seem to have nothing to do with L^2-invariants but whose proofs — as we will see — use L^2-methods. The selection below consists of some easy to formulate examples and is not meant to represent the most important results about L^2-invariants. There are plenty of other very interesting and important theorems about L^2-invariants, a lot of which will be presented in this book. For simplicity we often will not state the most general formulations in this introduction. All notions appearing in the list of theorems below will be explained in the relevant chapters. The results below are due to Chang-Weinberger, Cheeger-Gromov, Cochran-Orr-Teichner, Dodziuk, Gaboriau, Gromov and Lück.

Theorem 0.2 (see Theorem 1.35 (2) and Corollary 6.75). *Let G be a group which contains a normal infinite amenable subgroup. Suppose that there is a finite CW-model for its classifying space BG. Then its Euler characteristic vanishes, i.e.*

$$\chi(G) := \chi(BG) = 0.$$

Theorem 0.3 (see Theorem 1.62 and Theorem 11.6). *Let M be a closed manifold of even dimension $2m$. Suppose that M is hyperbolic, or more generally, that its sectional curvature satisfies $-1 \le \sec(M) < -(1 - \frac{1}{m})^2$. Then*

$$(-1)^m \cdot \chi(M) > 0.$$

Theorem 0.4 (see Theorem 11.14 and Theorem 11.15). *Let M be a closed Kähler manifold of (real) dimension $2m$. Suppose that M is homotopy equivalent to a closed Riemannian manifold with negative sectional curvature. Then*

$$(-1)^m \cdot \chi(M) > 0.$$

Moreover, M is a projective algebraic variety and is Moishezon and Hodge.

Theorem 0.5 (see Theorem 7.25). *Let $1 \to H \to G \to K \to 1$ be an extension of infinite groups such that H is finitely generated and G is finitely presented. Then*

(1) The deficiency of G satisfies $\mathrm{def}(G) \le 1$;

(2) If M is a closed connected oriented 4-manifold with $\pi_1(M) \cong G$, then we get for its signature $\mathrm{sign}(M)$ and its Euler characteristic $\chi(M)$

$$|\mathrm{sign}(M)| \le \chi(M).$$

Theorem 0.6 (see Theorem 9.38). *Let $i\colon H \to G$ be the inclusion of a normal finite subgroup H into an arbitrary group G. Then the maps coming from i and the conjugation action of G on H*

$$\mathbb{Z} \otimes_{\mathbb{Z}G} \mathrm{Wh}(H) \to \mathrm{Wh}(G);$$
$$\mathrm{Wh}(H)^G \to \mathrm{Wh}(G)$$

have finite kernel, where Wh *denotes the Whitehead group.*

Theorem 0.7 (see Theorem 9.66). *Let* G *be a group and* $\mathbb{C}G$ *be its complex group ring. Let* $G_0(\mathbb{C}G)$ *be the Grothendieck group of finitely generated (not necessarily projective)* $\mathbb{C}G$-*modules. Then*

(1) If G *is amenable, the class* $[\mathbb{C}G] \in G_0(\mathbb{C}G)$ *is an element of infinite order;*

(2) If G *contains the free group* $\mathbb{Z} * \mathbb{Z}$ *of rank two, then* $[\mathbb{C}G] = 0$ *in* $G_0(\mathbb{C}G)$.

Theorem 0.8 (see Section 15.4). *There are non-slice knots in 3-space whose Casson-Gordon invariants are all trivial.*

Theorem 0.9 (see Section 7.5). *There are finitely generated groups which are quasi-isometric but not measurably equivalent.*

Theorem 0.10 (see Section 15.1). *Let* M^{4k+3} *be a closed oriented smooth manifold for* $k \geq 1$ *whose fundamental group has torsion. Then there are infinitely many smooth manifolds which are homotopy equivalent to* M *(and even simply and tangentially homotopy equivalent to* M*) but not homeomorphic to* M.

0.3 Some Open Problems Concerning L^2-Invariants

The following conjectures will be treated in detail in Section 2.5 and Chapters 10, 11, 12, 13 and 14. They have created a lot of activity. This book contains proofs of these conjectures in special cases which rely on general methods and give some structural insight or consist of explicit computations. Recall that a free G-CW-complex X is the same as the total space of a G-covering $X \to G\backslash X$ with a CW-complex $G\backslash X$ as base space, and that X is called finite or of finite type if the CW-complex $G\backslash X$ is finite or of finite type.

Conjecture 0.11 (Strong Atiyah Conjecture). *Let* X *be a free* G-CW-*complex of finite type. Denote by* $\frac{1}{|\mathcal{FIN}(G)|}\mathbb{Z}$ *the additive subgroup of* $\mathbb{R}$ *generated by the set of rational numbers* $|H|^{-1}$, *where* H *runs through the finite subgroups of* G. *Then we get for the* L^2-*Betti numbers of* X

$$b_p^{(2)}(X) \in \frac{1}{|\mathcal{FIN}(G)|}\mathbb{Z}.$$

In Subsection 10.1.4 we will explain that there are counterexamples to the strong Atiyah Conjecture 0.11 due to Grigorchuk and Żuk, but no counterexample is known to the author at the time of writing if one replaces $\frac{1}{|\mathcal{FIN}(G)|}\mathbb{Z}$

by $\mathbb{Q}$ or if one assumes that there is an upper bound for the orders of finite subgroups of G. The author is not aware of a counterexample to the following conjectures at the time of writing.

Conjecture 0.12. (Positivity and rationality of Novikov-Shubin invariants). *Let X be a free G-CW-complex of finite type. Then its Novikov-Shubin invariants $\alpha_p(X)$ are positive rational numbers unless they are ∞ or ∞^+.*

Conjecture 0.13 (Singer Conjecture). *Let M be an aspherical closed manifold. Then the L^2-Betti numbers of the universal covering $\widetilde{M}$ satisfy*

$$b_p^{(2)}(\widetilde{M}) \; = \; 0 \qquad \text{if } 2p \neq \dim(M)$$

and $(-1)^m \cdot \chi(M) \geq 0$ if $\dim(M) = 2m$ is even.

Let M be a closed connected Riemannian manifold with negative sectional curvature. Then

$$b_p^{(2)}(\widetilde{M}) \; \begin{cases} = 0 & \text{if } 2p \neq \dim(M); \\ > 0 & \text{if } 2p = \dim(M), \end{cases}$$

and $(-1)^m \cdot \chi(M) > 0$ if $\dim(M) = 2m$ is even.

Conjecture 0.14 (L^2-torsion for aspherical manifolds). *If M is an aspherical closed manifold of odd dimension $2m + 1$, then the L^2-torsion of its universal covering satisfies*

$$(-1)^m \cdot \rho^{(2)}(\widetilde{M}) \; \geq \; 0.$$

If M is a closed connected Riemannian manifold of odd dimension $2m + 1$ with negative sectional curvature, then

$$(-1)^m \cdot \rho^{(2)}(\widetilde{M}) \; > \; 0.$$

If M is an aspherical closed manifold whose fundamental group contains an amenable infinite normal subgroup, then

$$\rho^{(2)}(\widetilde{M}) \; = \; 0.$$

Conjecture 0.15 (Zero-in-the-spectrum Conjecture). *Let $\widetilde{M}$ be the universal covering of an aspherical closed Riemannian manifold M. Then for some $p \geq 0$ zero is in the spectrum of the minimal closure*

$$(\Delta_p)_{\min} \colon \; \mathrm{dom}\left((\Delta_p)_{\min}\right) \subset L^2\Omega^p(\widetilde{M}) \to L^2\Omega^p(\widetilde{M})$$

of the Laplacian acting on smooth p-forms on $\widetilde{M}$.

Conjecture 0.16 (Approximation Conjecture). *Let G be a group. Let $\{G_i \mid i \in I\}$ be an inverse system of normal subgroups of G directed by inclusion over the directed set I. Suppose that $\bigcap_{i \in I} G_i = \{1\}$. Let X be a* *free G-CW-complex of finite type. Then $G_i \backslash X$ is a free G/G_i-CW-complex of finite type and*

$$b_p^{(2)}(X; \mathcal{N}(G)) = \lim_{i \in I} b_p^{(2)}(G_i \backslash X; \mathcal{N}(G/G_i)).$$

Conjecture 0.17 (Simplicial volume and L^2-invariants). *Let M be an aspherical closed orientable manifold of dimension ≥ 1. Suppose that its simplicial volume $\|M\|$ vanishes. Then all the L^2-Betti numbers and the L^2-torsion of the universal covering $\widetilde{M}$ vanish, i.e.*

$$b_p^{(2)}(\widetilde{M}) = 0 \quad \text{for } p \geq 0;$$
$$\rho^{(2)}(\widetilde{M}) = 0.$$

0.4 L^2-Invariants and Heat Kernels

The *p-th L^2-Betti number* $b_p^{(2)}(\overline{M})$ of a G-covering $p \colon \overline{M} \to M$ of a closed Riemannian manifold M was first defined by Atiyah [9, page 71] in connection with his L^2-index theorem. By means of a Laplace transform, Atiyah's original definition agrees with the one given by the non-negative real number

$$b_p^{(2)}(\overline{M}) = \lim_{t \to \infty} \int_{\mathcal{F}} \mathrm{tr}_{\mathbb{C}}\left(e^{-t\Delta_p}(x, x)\right) d\mathrm{vol}. \tag{0.18}$$

Here $\mathcal{F}$ is a fundamental domain for the G-action on $\overline{M}$ and $e^{-t\Delta_p}(x, y)$ is the heat kernel on p-forms on $\overline{M}$. The p-th L^2-Betti number $b_p^{(2)}(\overline{M})$ measures the size of the kernel of the Laplacian acting on smooth p-forms on $\overline{M}$. If G is trivial, then $b_p^{(2)}(\overline{M})$ is the same as the ordinary Betti number $b_p(M)$ which is the real dimension of the p-th singular cohomology with real coefficients of M. One important consequence of the L^2-index theorem is the Euler-Poincaré formula (0.1) (see Theorem 1.35 (2)).

The *p-th Novikov-Shubin invariant* $\alpha_p^\Delta(\overline{M})$ measures how fast the expression $\int_{\mathcal{F}} \mathrm{tr}_{\mathbb{C}}(e^{-t\Delta_p}(x, x)) d\mathrm{vol}$ approaches its limit $b_p^{(2)}(\overline{M})$ for $t \to \infty$ (see (0.18)). The larger $\alpha_p^\Delta(\overline{M})$ is, the "thinner" is the spectrum of the p-th Laplacian on $\overline{M}$ at zero.

Notice that the L^2-Betti numbers and the Novikov-Shubin invariants are invariants of the large time asymptotics of the heat kernel and hence invariants of the global geometry, in contrast to invariants of the small time asymptotics, such as indices of operators, which are of local nature. For instance the Novikov-Shubin invariant associated to the Laplacian acting on 0-forms of the universal covering of a closed Riemannian manifold M is determined by group theoretic properties of the fundamental group $\pi_1(M)$ such as its growth rate or the question whether it is amenable (see Theorem 2.55 (5)).

In view of the definitions of the L^2-Betti numbers and Novikov-Shubin invariants, the strong Atiyah Conjecture 0.11 and the Conjecture 0.12 about the positivity and rationality of Novikov-Shubin invariants are very surprising. Some explanation for the strong Atiyah Conjecture 0.11 comes from connections with algebraic K-theory, whereas the only evidence for the Conjecture 0.12 about the positivity and rationality of Novikov-Shubin invariants is based on computations, and no conceptual reasons are known.

The third important L^2-invariant is the L^2-torsion $\rho^{(2)}(\overline{M})$ which was introduced by Carey-Mathai, Lott, Lück-Rothenberg, Mathai and Novikov-Shubin. It is only defined under a certain technical assumption, namely, that $\overline{M}$ is of determinant class. This condition is conjecturally always satisfied and we will suppress it in this discussion. If all L^2-Betti numbers of $\overline{M}$ vanish, the L^2-torsion $\rho^{(2)}(\overline{M})$ is independent of the Riemannian metric and depends only on the simple homotopy type. Actually, there is the conjecture that it depends only on the homotopy type (see Conjecture 3.94). Its analytic definition is complicated.

This analytic approach via the heat kernel is important in the following situations. One can compute the L^2-Betti numbers of the universal covering $\widetilde{M}$ of a closed Riemannian manifold M if M is hyperbolic (see Theorem 1.62), or, more generally, satisfies certain pinching conditions (see Theorem 11.4, Theorem 11.5 and Theorem 11.6). There are explicit computations of the L^2-Betti numbers, the Novikov-Shubin invariants and the L^2-torsion of the universal covering of a closed manifold M if M is a locally symmetric space (see Theorem 5.12 and Section 5.4). The proof of the Proportionality Principle 3.183 relies on the analytic description. The proofs of these facts do not have combinatorial counterparts.

0.5 L^2-Invariants and Cellular Chain Complexes

One important feature of all these L^2-invariants is that they can also be defined for a G-covering $p \colon \overline{X} \to X$ of a finite CW-complex X in terms of the cellular $\mathbb{Z}G$-chain complex $C_*(\overline{X})$. For L^2-Betti numbers and Novikov-Shubin invariants it suffices to require that X is of finite type. The associated L^2-chain complex $C_*^{(2)}(\overline{X})$ is defined by $l^2(G) \otimes_{\mathbb{Z}G} C_*(\overline{X})$. Each chain module $C_*^{(2)}(\overline{X})$ is a Hilbert space with isometric G-action of the special form $l^2(G)^n$, where $l^2(G)^n$ is the n-fold sum of the Hilbert space $l^2(G)$. Each differential $c_p^{(2)}$ is a bounded G-equivariant operator. The p-th L^2-homology $H_p^{(2)}(\overline{X})$ is defined to be the quotient of the kernel of $c_p^{(2)}$ by the closure of the image of $c_{p+1}^{(2)}$. Dividing out the closure of the image has the effect that $H_p^{(2)}(\overline{X})$ is again a Hilbert space with isometric G-action. It actually comes with the structure of a *finitely generated Hilbert $\mathcal{N}(G)$-module*, where $\mathcal{N}(G)$ denotes the *von*

Neumann algebra of the group G. This additional structure allows to define the *von Neumann dimension* of $H_p^{(2)}(\overline{X})$. Dodziuk has shown that this non-negative real number agrees with $b_p^{(2)}(\overline{X})$ as defined in (0.18) (see Theorem 1.59 and (1.60)). One can also read off the Novikov-Shubin invariants and the L^2-torsion from $C_*^{(2)}(\overline{X})$ by results of Efremov (see Theorem 2.68) and Burghelea-Friedlander-Kappeler-McDonald (see Theorem 3.149). The p-th Novikov-Shubin invariant $\alpha_p(\overline{X})$ measures the difference between the image of $c_p^{(2)}$ and the closure of the image of $c_p^{(2)}$.

The point of this cellular description is that it is much easier to handle and calculate than the analytic counterpart. For instance one can show homotopy invariance of L^2-Betti numbers, Novikov-Shubin invariants and L^2-torsion and prove some very useful formulas like sum formulas, product formulas, fibration formulas and so on using the combinatorial approach (see Theorem 1.35, Theorem 2.55, Theorem 3.93, Theorem 3.96 and Theorem 3.100). The combinatorial approach allows to show for an aspherical closed manifold M that all L^2-Betti numbers and the L^2-torsion of its universal covering vanish provided M carries a non-trivial S^1-action (see Theorem 3.105). There exists a combinatorial proof that all L^2-Betti numbers of the universal covering of a mapping torus of a self map of a CW-complex of finite type vanish (see Theorem 1.39). No analytic proofs or no simpler analytic proofs of these results are known to the author. The combination of the analytic and combinatorial methods yields a computation of the L^2-invariants of the universal covering of a compact 3-manifold provided Thurston's Geometrization Conjecture holds for the pieces appearing in the prime decomposition of M (see Theorem 4.1, Theorem 4.2 and Theorem 4.3).

For a kind of algorithmic computation of L^2-invariants based on the combinatorial approach we refer to Theorem 3.172.

The possibility to take both an analytic and a combinatorial point of view is one of the main reasons why L^2-invariants are so powerful.

0.6 L^2-Betti Numbers and Betti Numbers

Let $\widetilde{X} \to X$ be the universal covering of a connected CW-complex X of finite type. Then the L^2-Betti numbers $b_p^{(2)}(\widetilde{X})$ of $\widetilde{X}$ and the (classical) Betti numbers $b_p(X)$ share some basic properties such as homotopy invariance, the Euler-Poincaré formula, Poincaré duality, Morse inequalities, Künneth formulas and so on, just replace in the corresponding statement for the classical Betti numbers $b_p(X)$ by $b_p^{(2)}(\widetilde{X})$ everywhere (see Theorem 1.35). There is also an L^2-Hodge de Rham Theorem 1.59 which is one important input in the proof of Theorem 0.3.

But there are also differences. One important extra feature of the L^2-Betti numbers is that they are multiplicative under finite coverings in the following sense. If $p\colon Y \to X$ is a finite d-sheeted covering, then $b_p^{(2)}(\widetilde{Y}) = d \cdot b_p^{(2)}(\widetilde{X})$

(see Theorem 1.35 (9)). This implies for instance $b_p^{(2)}(\widetilde{S^1}) = 0$ for all $p \geq 0$ since there is a d-sheeted covering $S^1 \to S^1$ for $d \geq 2$. The corresponding statement is not true for the Betti numbers. This is one reason why L^2-Betti numbers more often tend to be zero than the classical Betti numbers. Often this is the key phenomenon for applications. Another reason for it is the fact that $b_0^{(2)}(\widetilde{X})$ is 0 if $\pi_1(X)$ is infinite and is $|\pi_1(X)|^{-1}$ if $\pi_1(X)$ is finite (see Theorem 1.35 (8)), whereas $b_0(X)$ is always 1.

If $\pi_1(X)$ is finite, then $b_p^{(2)}(\widetilde{X}) = |\pi_1(X)|^{-1} \cdot b_p(\widetilde{X})$. If $\pi_1(X)$ is infinite, the only general relation between the L^2-Betti numbers of $\widetilde{X}$ and the Betti numbers of X is the Euler-Poincaré formula (0.1). Given an integer $l \geq 1$ and a sequence $r_1, r_2, \ldots, r_l$ of non-negative rational numbers, we construct in Example 1.38 a group G such that BG is of finite type and

$$
\begin{aligned}
b_p^{(2)}(G) := b_p^{(2)}(EG) &= \begin{cases} r_p & \text{for } 1 \leq p \leq l; \\ 0 & \text{for } l+1 \leq p; \end{cases} \\
b_p(G) \quad := b_p(BG) \quad &= 0 \qquad \qquad \text{for } p \geq 1.
\end{aligned}
$$

On the other hand we can construct for any sequence $n_1, n_2, \ldots$ of non-negative integers a CW-complex X of finite type such that $b_p(X) = n_p$ and $b_p^{(2)}(\widetilde{X}) = 0$ hold for $p \geq 1$.

However, there is an asymptotic relation between the L^2-Betti numbers of $\widetilde{X}$ and the Betti numbers of X. Recall that the Betti numbers are not multiplicative. One may try to force multiplicativity of the Betti numbers by stabilizing under finite coverings as follows. Suppose that $\pi_1(X)$ possesses a nested sequence of normal subgroups of finite index

$$
\pi_1(X) = G_0 \supset G_1 \supset G_2 \supset G_3 \supset \ldots
$$

with $\cap_{i=0}^{\infty} G_i = \{1\}$. Then $G_i \backslash \widetilde{X}$ is a CW-complex of finite type and there is a $[G : G_i]$-sheeted covering $G_i \backslash \widetilde{X} \to X$. One may consider $\lim_{i \to \infty} \frac{b_p(G_i \backslash \widetilde{X})}{[G:G_i]}$. This expression is automatically multiplicative if the limit exists and is independent of the nested sequence. Actually it turns out that this is true and

$$
\lim_{i \to \infty} \frac{b_p(G_i \backslash \widetilde{X})}{[G : G_i]} = b_p^{(2)}(\widetilde{X}).
$$

This result is a special case of the Approximation Conjecture 0.16 which will be investigated in Chapter 13.

0.7 L^2-Invariants and Ring-Theory

A more algebraic approach will be presented in Chapter 6. It will enable us to define L^2-Betti numbers for arbitrary G-spaces and in particular for groups without any restrictions on BG. This allows to apply standard techniques of algebraic topology and homological algebra directly to L^2-Betti numbers. The idea is to view the group von Neumann algebra $\mathcal{N}(G)$ just as a ring forgetting the functional analysis and the topology. The von Neumann algebra $\mathcal{N}(G)$ has zero-divisors and is not Noetherian unless G is finite. This makes $\mathcal{N}(G)$ complicated as a ring. But it has one very nice property, it is *semihereditary*, i.e. any finitely generated submodule of a projective module is itself projective (see Theorem 6.5 and Theorem 6.7 (1)). This justifies the slogan that $\mathcal{N}(G)$ behaves like the ring $\mathbb{Z}$ if one ignores the facts that $\mathbb{Z}$ has no zero-divisors and is Noetherian. The main input for the ring-theoretic approach is the construction of a dimension function for arbitrary modules over the group von Neumann algebra $\mathcal{N}(G)$ (Theorem 6.7). It is uniquely characterized by the condition that it satisfies Additivity, Continuity and Cofinality and extends the classical dimension function for finitely generated projective modules which is defined in terms of the von Neumann trace of idempotents in $M_n(\mathcal{N}(G))$. One applies it to the $\mathcal{N}(G)$-modules $H_p(\mathcal{N}(G) \otimes_{\mathbb{Z}G} C_*^{\mathrm{sing}}(X))$ for a G-space X and gets an extension of the notion of L^2-Betti numbers to arbitrary G-spaces if one allows the value ∞. The second key result is that for amenable G the von Neumann algebra $\mathcal{N}(G)$ looks like a flat $\mathbb{C}G$-module from the point of view of dimension theory (see Theorem 6.37).

In Chapter 8 we introduce the algebra $\mathcal{U}(G)$ of operators affiliated to the group von Neumann algebra. From an algebraic point of view $\mathcal{U}(G)$ can be described as the Ore localization of $\mathcal{N}(G)$ with respect to the multiplicative set of non-zero divisors. The main ring theoretic property of $\mathcal{U}(G)$ is that it is *von Neumann regular* (see Theorem 8.22 (3)) which is a stronger property than to be semihereditary. The dimension theory of $\mathcal{N}(G)$ extends to $\mathcal{U}(G)$ (see Theorem 8.29). The relation of $\mathcal{U}(G)$ to $\mathcal{N}(G)$ is analogous to the relation of $\mathbb{Q}$ to $\mathbb{Z}$.

From the point of view of representation theory of finite groups the passage from $\mathbb{C}G$ to $\mathcal{N}(G)$ is the natural one for infinite groups. Namely, two finitely generated projective $\mathcal{N}(G)$-modules P and Q are $\mathcal{N}(G)$-isomorphic if and only if their center valued von Neumann dimensions $\dim_{\mathcal{N}(G)}^u(P)$ and $\dim_{\mathcal{N}(G)}^u(Q)$ agree (see Theorem 9.13). If G is finite, this reduces to the well-known theorem that two complex finite-dimensional G-representations are isomorphic if and only if they have the same character.

This algebraic approach may be preferred by algebraists who do not have much background in (functional) analysis.

Linnell's Theorem 10.19 says that the strong Atiyah Conjecture 0.11 is true for a class of groups $\mathcal{C}$ which contains all extensions of free groups with elementary amenable groups as quotients, provided that there is an upper bound on the orders of finite subgroups. Its proof is based on techniques from ring theory, in particular localization techniques, and from K-theory. The following square of inclusions of rings plays an important role as explained below

$$\begin{array}{ccc}
\mathbb{C}G & \longrightarrow & \mathcal{N}(G) \\
\downarrow & & \downarrow \\
\mathcal{D}(G) & \longrightarrow & \mathcal{U}(G)
\end{array} \qquad (0.19)$$

where $\mathcal{D}(G)$ denotes the *division closure* of $\mathbb{C}G$ in $\mathcal{U}(G)$.

0.8 L^2-Invariants and K-Theory

The strong Atiyah Conjecture 0.11 is related to K-theory in the following way. It is equivalent to the statement that for any finitely presented $\mathbb{C}G$-module M the generalized dimension $\dim_{\mathcal{N}(G)}(\mathcal{N}(G) \otimes_{\mathbb{C}G} M)$ (see Theorem 6.5 and Theorem 6.7 (1)) of the $\mathcal{N}(G)$-module $\mathcal{N}(G) \otimes_{\mathbb{C}G} M$ takes values in $\frac{1}{|\mathcal{FIN}(G)|}\mathbb{Z}$ (see Lemma 10.7). Notice that any non-negative real number occurs as $\dim_{\mathcal{N}(G)}(P)$ for a finitely generated projective $\mathcal{N}(G)$-module P, if G contains $\mathbb{Z}$ as subgroup (see Example 1.11, Theorem 6.24 (4) and Theorem 6.29 (2)). So the point is to understand the passage from $\mathbb{C}G$ to $\mathcal{N}(G)$, not only to investigate modules over $\mathcal{N}(G)$.

One may first consider the weaker statement that for any finitely generated *projective* $\mathbb{C}G$-module M the generalized dimension $\dim_{\mathcal{N}(G)}(\mathcal{N}(G) \otimes_{\mathbb{C}G} M)$ takes values in $\frac{1}{|\mathcal{FIN}(G)|}\mathbb{Z}$. This is equivalent to the statement that the composition $K_0(\mathbb{C}G) \xrightarrow{i} K_0(\mathcal{N}(G)) \xrightarrow{\dim_{\mathcal{N}(G)}} \mathbb{R}$ must have its image in $\frac{1}{|\mathcal{FIN}(G)|}\mathbb{Z}$, where i is the change of rings map. This is certainly true for the composition

$$\bigoplus_{\substack{H \subset G \\ |H| < \infty}} K_0(\mathbb{C}H) \xrightarrow{a} K_0(\mathbb{C}G) \xrightarrow{i} K_0(\mathcal{N}(G)) \xrightarrow{\dim_{\mathcal{N}(G)}} \mathbb{R}$$

where a is the sum of the various change of rings maps. The Isomorphism Conjecture 9.40 for $K_0(\mathbb{C}G)$ implies that a is surjective and hence that the image of $K_0(\mathbb{C}G) \xrightarrow{i} K_0(\mathcal{N}(G)) \xrightarrow{\dim_{\mathcal{N}(G)}} \mathbb{R}$ is contained in $\frac{1}{|\mathcal{FIN}(G)|}\mathbb{Z}$.

The proof of Linnell's Theorem 10.19 can be split into two parts, a ring-theoretic one and a K-theoretic one. Namely, one proves that any finitely presented $\mathbb{C}G$-module becomes finitely generated projective over the ring $\mathcal{D}(G)$ (see (0.19)) and that the composition

$$\bigoplus_{\substack{H \subset G \\ |H| < \infty}} K_0(\mathbb{C}H) \xrightarrow{a} K_0(\mathbb{C}G) \xrightarrow{j} K_0(\mathcal{D}(G))$$

for j the change of rings map is surjective (see Section 10.2). Then the claim follows from (0.19) and the facts that the change of rings homomorphism

$K_0(\mathcal{N}(G)) \to K_0(\mathcal{U}(G))$ is bijective (see Theorem 9.20 (1)) and that the dimension function $\dim_{\mathcal{N}(G)}$ for $\mathcal{N}(G)$ extends to a dimension function $\dim_{\mathcal{U}(G)}$ for $\mathcal{U}(G)$ satisfying $\dim_{\mathcal{U}(G)}(\mathcal{U}(G) \otimes_{\mathcal{N}(G)} M) = \dim_{\mathcal{N}(G)}(M)$ for any $\mathcal{N}(G)$-module M (see Theorem 8.29).

The extension of the dimension function to arbitrary modules has some applications to G-theory of $\mathbb{C}G$ as already mentioned in Theorem 0.7 (see Subsection 9.5.3). Computations of the middle K-theory and of the L-theory of von Neumann algebras and the associated algebras of affiliated operators are presented in Chapter 9. L^2-methods also lead to results about the Whitehead group $\mathrm{Wh}(G)$ (see Theorem 0.6) and some information about the Bass Conjecture (see Subsection 9.5.2). The question whether the L^2-torsion in the L^2-acyclic case is a homotopy invariant is equivalent to the question whether the map induced by the Fuglede-Kadison determinant $\mathrm{Wh}(G) \to \mathbb{R}$ is trivial (see Conjecture 3.94). This question is related to the Approximation Conjecture 0.16 by the Determinant Conjecture 13.2 (see Lemma 13.6 and Theorem 13.3 (1)). The Approximation Conjecture 0.16 also plays a role in proving that the class of groups for which the strong Atiyah Conjecture 0.11 is true is closed under direct and inverse limits (see Theorem 10.20).

0.9 L^2-Invariants and Aspherical Manifolds

Let M be an aspherical closed manifold, for instance a closed Riemannian manifold with non-positive sectional curvature. Then the Singer Conjecture 0.13, Conjecture 0.14 about L^2-torsion for aspherical manifolds and the zero-in-the-spectrum Conjecture 0.15 put some restrictions on the L^2-invariants of its universal covering. There are special cases where these conjectures have been proved by computations. For instance if M is a compact 3-manifold (see Chapter 4), a locally symmetric space (see Corollary 5.16) or carries a Riemannian metric whose sectional curvature satisfies certain pinching conditions (see Theorem 11.4, Theorem 11.5 and Theorem 11.6). They also have been proved under additional assumptions like the existence of a non-trivial S^1-action (see Theorem 3.105), the existence of the structure of a Kähler hyperbolic manifold (see Theorem 11.14) or the existence of a normal infinite (elementary) amenable subgroup of $\pi_1(X)$ (see Theorem 3.113 and Theorem 7.2). But it is still very mysterious why Poincaré duality together with asphericity may have such implications, or what kind of mechanism is responsible for these phenomenons. The status of Conjecture 0.17 about simplicial volume and L^2-invariants is similar. Conjectures 0.13, 0.14, 0.15 and 0.17 become false if one drops the condition that M is aspherical. Without this assumption it is easy to construct counterexamples to all but the zero-in-the-spectrum Conjecture 0.15. Counterexamples in the non-aspherical case to the zero-in-the-spectrum Conjecture 0.15 are presented by Farber-Weinberger [187] (see also [258]). We will deal with them in Section 12.3.

0.10 L^2-Invariants and Groups

L^2-Betti numbers $b_p^{(2)}(G)$ (and also Novikov-Shubin invariants $\alpha_p(G)$) can be defined for arbitrary (discrete) groups if one allows the value ∞. In Chapter 7 the L^2-Betti numbers of groups are investigated and in particular the question when they vanish is studied. The vanishing of all L^2-Betti numbers of G implies the vanishing of the L^2-*Euler characteristic* $\chi^{(2)}(G)$ of G. The notion of L^2-Euler characteristic agrees with the classical notion of Euler characteristic $\chi(BG)$ (or more generally the virtual Euler characteristic) if the latter is defined. Actually Theorem 0.2 is proved by showing that all L^2-Betti numbers of a group G vanish if G contains a normal infinite amenable subgroup. This example shows that it is important to extend the definition of L^2-Betti numbers from those groups for which BG is finite to arbitrary groups even if one may only be interested in groups with finite BG. Namely, if G has a finite model for BG, this does not mean that a normal subgroup $H \subset G$ has a model of finite type for BH. The vanishing of the first L^2-Betti number $b_1^{(2)}(G)$ has consequences for the deficiency of the group. The hard part of the proof of Theorem 0.5 is to show the vanishing of $b_1^{(2)}(G)$, then the claim follows by elementary considerations.

We show in Theorem 7.10 that all L^2-Betti numbers of Thompson's group F vanish. This is a necessary condition for F to be amenable. The group F cannot be elementary amenable and does not contain $\mathbb{Z} * \mathbb{Z}$ as subgroup but (at the time of writing) it is not known whether F is amenable or not.

In Section 7.4 a number $\rho^{(2)}(f) \in \mathbb{R}$ is associated to an automorphism $f \colon G \to G$ of a group G provided that BG has a finite model. One also needs the technical assumption of det ≥ 1-class which is conjecturally always true and proved for a large class of groups and will be suppressed in the following discussion. This invariant has nice properties such as the trace property $\rho^{(2)}(g \circ f) = \rho^{(2)}(f \circ g)$ and multiplicativity $\rho^{(2)}(f^n) = n \cdot \rho^{(2)}(f)$ and satisfies a sum formula $\rho^{(2)}(f_1 *_{f_0} f_2) = \rho^{(2)}(f_1) + \rho^{(2)}(f_2) - \rho^{(2)}(f_0)$ (see Theorem 7.27). If $f = \pi_1(g)$ for an automorphism $g \colon F \to F$ of a compact orientable 2-dimensional manifold F different from S^2, D^2 and T^2, then $\rho^{(2)}(f)$ is, up to a constant, the sum of the volumes of the hyperbolic pieces appearing in the Jaco-Shalen-Johannson-Thurston decomposition of the mapping torus of g along tori into Seifert pieces and hyperbolic pieces (see Theorem 7.28). If F is closed and g is irreducible, then $\rho^{(2)}(g) = 0$ if and only if g is periodic, and $\rho^{(2)}(g) \neq 0$ if and only if g is pseudo-Anosov.

In Section 7.5 the question is discussed whether or not the L^2-Betti numbers, Novikov-Shubin invariants and the L^2-torsion are quasi-isometry invariants or invariants of the measure equivalence class of a countable group G. Theorem 0.9 is one of the main applications of L^2-Betti numbers to measurable equivalence.

1. L^2-Betti Numbers

Introduction

In this chapter we introduce and study L^2-(co-)homology and L^2-Betti numbers for Hilbert chain complexes and for regular coverings of CW-complexes of finite type or of compact manifolds.

We follow the general strategy that a good invariant for a finite CW-complex X often has a refined version which is defined in terms of the universal covering $\widetilde{X}$ and the action of the fundamental group $\pi = \pi_1(X)$. For instance the Euler characteristic yields Wall's finiteness obstruction, and the signature yields all kinds of surgery obstructions under this passage from X to the π-space $\widetilde{X}$. The L^2-Betti numbers are derived from the Betti numbers in this way. Recall that the p-th Betti number $b_p(X)$ is defined by $\dim_{\mathbb{C}}(H_p(X;\mathbb{C}))$. In a naive approach one might try to define improved Betti numbers for a reasonable notion of $\dim_{\mathbb{C}\pi}$ by $\dim_{\mathbb{C}\pi}(H_p(\widetilde{X};\mathbb{C}))$, for instance for $\dim_{\mathbb{C}\pi}(H_p(\widetilde{X};\mathbb{C})) := \dim_{\mathbb{C}}(\mathbb{C} \otimes_{\mathbb{C}\pi} H_p(\widetilde{X};\mathbb{C}))$. The problem is that $\mathbb{C}\pi$ is in general not Noetherian and hence this number is not necessarily finite. The basic idea is to pass to the group von Neumann algebra $\mathcal{N}(\pi)$ and use its standard trace to define the notion of von Neumann dimension which is better behaved than $\dim_{\mathbb{C}\pi}$. Since one needs a Hilbert space setting, one completes the cellular $\mathbb{C}\pi$-chain complex of $\widetilde{X}$ to the cellular L^2-chain complex $C_*^{(2)}(\widetilde{X})$ and defines the L^2-homology $H_p^{(2)}(\widetilde{X})$ by the quotient of the kernel of the p-th differential by the closure of the image of the $(p+1)$-th differential. Then the p-th L^2-Betti number $b_p^{(2)}(\widetilde{X})$ of the π-space $\widetilde{X}$ is the von Neumann dimension of $H_p^{(2)}(\widetilde{X})$.

We will introduce the necessary input about von Neumann algebras and Hilbert modules in Section 1.1. The precise definitions of $C_*^{(2)}(\widetilde{X})$, $H_p^{(2)}(\widetilde{X})$ and $b_p^{(2)}(\widetilde{X})$ and their main properties are given in Section 1.2. The standard properties of Betti numbers such as homotopy invariance, the expression of the Euler characteristic as alternating sum of Betti numbers, Poincaré duality, Künneth formula, and Morse inequalities carry over to L^2-Betti numbers. L^2-Betti numbers have additional interesting new properties which Betti numbers do not have. For instance they are multiplicative under finite coverings. In general the values of $b_p(X)$ and $b_p^{(2)}(\widetilde{X})$ are not related. A priori L^2-Betti numbers can take any non-negative real number as value, but possibly they are always rational. L^2-Betti numbers tend to vanish more often than Betti numbers. For instance $b_p^{(2)}(\widetilde{X})$ vanishes for $p \geq 0$ if X is a mapping torus of a selfmap of a CW-complex of finite type, if X is an aspherical closed manifold with a non-trivial S^1-action or if X is a hyperbolic closed manifold of dimension n and $2p \neq n$.

In Section 1.3 we extend the classical relation of Betti numbers and the dimension of the space of harmonic forms given by the Hodge-de Rham decomposition and the de Rham isomorphism to the L^2-setting. Actually the original definition of L^2-Betti numbers of Atiyah in the context of his L^2-Index Theorem is analytic, namely, by an expression in the Schwartz kernels of the projections appearing in the spectral family of the Laplacian on the universal covering of a closed Riemannian manifold. The fact, proved in Section 1.4, that the analytic and the cellular version of the L^2-Betti numbers agree is one of the important features. It is also interesting to notice that some of the properties of the L^2-Betti numbers are proved analytically or topologically respectively, and often there are no topological or analytic respectively proofs available.

If one wants to get a quick impression of L^2-Betti numbers, one may ignore their definition and pass directly to Subsections 1.2.3 and 1.3.2.

For the remainder of this chapter G is a discrete group with the exception of Subsection 1.2.1. Manifolds are always smooth.

1.1 Group von Neumann Algebras and Hilbert Modules

In this section we deal with group von Neumann algebras and Hilbert modules. We defer the treatment of the more general notion of a finite von Neumann algebra to Section 9.1 since for the next chapters we will only need the concept of a group von Neumann algebra. We introduce the notions of von Neumann trace and von Neumann dimension and use them to define and study L^2-Betti numbers for Hilbert chain complexes.

1.1.1 Group von Neumann Algebras

Let G be a discrete group. Denote by $l^2(G)$ the Hilbert space of square-summable formal sums over G with complex coefficients. This is the same as the Hilbert space completion of the complex group ring $\mathbb{C}G$ with respect to the pre-Hilbert space structure for which G is an orthonormal basis. An element in $l^2(G)$ is represented by a formal sum $\sum_{g \in G} \lambda_g \cdot g$ for complex numbers λ_g such that $\sum_{g \in G} |\lambda_g|^2 < \infty$. The scalar product is defined by

$$\langle \sum_{g \in G} \lambda_g \cdot g, \sum_{g \in G} \mu_g \cdot g \rangle := \sum_{g \in G} \lambda_g \cdot \overline{\mu_g}.$$

Notice that left multiplication with elements in G induces an isometric G-action on $l^2(G)$. Given a Hilbert space H, denote by $\mathcal{B}(H)$ the C^*-algebra of bounded (linear) operators from H to itself, where the norm is the operator norm.

Definition 1.1 (Group von Neumann algebra). *The* group von Neumann algebra $\mathcal{N}(G)$ *of the group G is defined as the algebra of G-equivariant bounded operators from $l^2(G)$ to $l^2(G)$*

$$\mathcal{N}(G) := \mathcal{B}(l^2(G))^G.$$

An important feature of the group von Neumann algebra is its standard trace.

Definition 1.2 (Von Neumann trace). *The* von Neumann trace *on $\mathcal{N}(G)$ is defined by*

$$\mathrm{tr}_{\mathcal{N}(G)} \colon \mathcal{N}(G) \to \mathbb{C}, \qquad f \mapsto \langle f(e), e \rangle_{l^2(G)},$$

where $e \in G \subset l^2(G)$ is the unit element.

Example 1.3. If G is finite, then $\mathbb{C}G = l^2(G) = \mathcal{N}(G)$. The trace $\mathrm{tr}_{\mathcal{N}(G)}$ assigns to $\sum_{g \in G} \lambda_g \cdot g$ the coefficient λ_e of the unit element $e \in G$.

The next example is a key example. We recommend the reader to consider and check all the results for group von Neumann algebras in the following special case to get the right intuition.

Example 1.4. If G is $\mathbb{Z}^n$, there is the following model for the group von Neumann algebra $\mathcal{N}(\mathbb{Z}^n)$. Let $L^2(T^n)$ be the Hilbert space of equivalence classes of L^2-integrable complex-valued functions on the n-dimensional torus T^n, where two such functions are called equivalent if they differ only on a subset of measure zero. Define the Banach space $L^\infty(T^n)$ by equivalence classes of essentially bounded measurable functions $f \colon T^n \to \mathbb{C} \coprod \{\infty\}$, where essentially bounded means that there is a constant $C > 0$ such that the set $\{x \in T^n \mid |f(x)| \geq C\}$ has measure zero. An element $(k_1, \ldots, k_n)$ in $\mathbb{Z}^n$ acts isometrically on $L^2(T^n)$ by pointwise multiplication with the function $T^n \to \mathbb{C}$ which maps $(z_1, z_2, \ldots, z_n)$ to $z_1^{k_1} \cdot \ldots \cdot z_n^{k_n}$. Fourier transform yields an isometric $\mathbb{Z}^n$-equivariant isomorphism $l^2(\mathbb{Z}^n) \xrightarrow{\cong} L^2(T^n)$. Hence $\mathcal{N}(\mathbb{Z}^n) = \mathcal{B}(L^2(T^n))^{\mathbb{Z}^n}$. We obtain an isomorphism

$$L^\infty(T^n) \xrightarrow{\cong} \mathcal{N}(\mathbb{Z}^n)$$

by sending $f \in L^\infty(T^n)$ to the $\mathbb{Z}^n$-equivariant operator $M_f \colon L^2(T^n) \to L^2(T^n)$, $g \mapsto g \cdot f$ where $g \cdot f(x)$ is defined by $g(x) \cdot f(x)$. Under this identification the trace becomes

$$\mathrm{tr}_{\mathcal{N}(\mathbb{Z}^n)} \colon L^\infty(T^n) \to \mathbb{C}, \qquad f \mapsto \int_{T^n} f d\mu.$$

1.1.2 Hilbert Modules

Definition 1.5 (Hilbert module). *A Hilbert $\mathcal{N}(G)$-module V is a Hilbert space V together with a linear isometric G-action such that there exists a Hilbert space H and an isometric linear G-embedding of V into the tensor product of Hilbert spaces $H \otimes l^2(G)$ with the obvious G-action. A map of Hilbert $\mathcal{N}(G)$-modules $f \colon V \to W$ is a bounded G-equivariant operator. We call a Hilbert $\mathcal{N}(G)$-module V* finitely generated *if there is a non-negative integer n and a surjective map $\bigoplus_{i=1}^n l^2(G) \to V$ of Hilbert $\mathcal{N}(G)$-modules.*

The embedding of V into $H \otimes l^2(G)$ is not part of the structure, only its existence is required. We emphasize that a map of Hilbert $\mathcal{N}(G)$-modules is not required to be isometric. A Hilbert $\mathcal{N}(G)$-module V is finitely generated if and only if there is an isometric linear G-embedding of V into $\bigoplus_{i=1}^n l^2(G)$ for some non-negative integer n. If G is finite a finitely generated Hilbert $\mathcal{N}(G)$-module is the same as a finite dimensional unitary representation of G.

Definition 1.6 (Weak exactness). *We call a sequence of Hilbert $\mathcal{N}(G)$-modules $U \xrightarrow{i} V \xrightarrow{p} W$* weakly exact *at V if the kernel $\ker(p)$ of p and the closure $\mathrm{clos}(\mathrm{im}(i))$ of the image $\mathrm{im}(i)$ of i agree. A map of Hilbert $\mathcal{N}(G)$-modules $f \colon V \to W$ is a* weak isomorphism *if it is injective and has dense image.*

The following assertions are equivalent for a map $f \colon V \to W$ of Hilbert $\mathcal{N}(G)$-modules: (1) f is a weak isomorphism; (2) $0 \to V \xrightarrow{f} W \to 0$ is weakly exact; (3) f and f^* have dense image. An example of a weak isomorphism of $\mathcal{N}(G)$-modules which is not an isomorphism is $M_{z-1} \colon L^2(S^1) \to L^2(S^1)$ given by multiplication with $(z-1) \in L^\infty(S^1) = \mathcal{N}(\mathbb{Z})$, where $(z-1)$ denotes the function $S^1 \to \mathbb{C}, \quad z \mapsto (z-1)$ (see Example 1.4). A map $f \colon U \to V$ of Hilbert $\mathcal{N}(G)$-modules is an isomorphism if and only if it is bijective. This follows from the Inverse Mapping Theorem [434, Theorem III.11 on page 83] which ensures that f^{-1} is continuous for a bijective bounded operator of Hilbert spaces. If two Hilbert $\mathcal{N}(G)$-modules V and W are weakly isomorphic, then they are even isometrically isomorphic by Polar Decomposition [434, Theorem VI.10 on page 197]. Namely, if $f \colon V \to W$ is a weak isomorphism, the unitary part of its polar decomposition is an isometric isomorphism of Hilbert $\mathcal{N}(G)$-modules $V \to W$. More generally, if $0 \to U \to V \to W \to 0$ is a weakly exact sequence of Hilbert $\mathcal{N}(G)$-modules, then there is an isometric isomorphism of Hilbert $\mathcal{N}(G)$-modules $U \oplus W \to V$. Notice that a short exact sequence of Hilbert $\mathcal{N}(G)$-modules splits, but a weakly exact sequence does not split in general.

1.1.3 Dimension Theory

A bounded operator $f \colon H \to H$ of Hilbert spaces is called *positive* if $\langle f(v), v \rangle$ is a real number and ≥ 0 for all $v \in H$. This is equivalent to the condition that there is a bounded operator $g\colon H \to H$ with $f = g^*g$ since any positive operator f has a square root $f^{1/2}$, i.e. a positive operator with $f = f^{1/2} \circ f^{1/2}$ [434, Theorem VI.9 on page 196]. In particular a positive operator is selfadjoint. Given two bounded operators $f_1, f_2\colon H \to H$, we write

$$f_1 \leq f_2 \Leftrightarrow f_2 - f_1 \text{ is positive.} \tag{1.7}$$

Definition 1.8. *Let $f\colon V \to V$ be a positive endomorphism of a Hilbert $\mathcal{N}(G)$-module. Choose a Hilbert space H, a Hilbert basis $\{b_i \mid i \in I\}$ for H (sometimes also called complete orthonormal system, see [434, II.3]), a G-equivariant projection $\mathrm{pr}\colon H \otimes l^2(G) \to H \otimes l^2(G)$ and an isometric G-isomorphism $u\colon \mathrm{im}(\mathrm{pr}) \xrightarrow{\cong} V$. Let $\overline{f}\colon H \otimes l^2(G) \to H \otimes l^2(G)$ be the positive operator given by the composition*

$$\overline{f}\colon H \otimes l^2(G) \xrightarrow{\mathrm{pr}} \mathrm{im}(\mathrm{pr}) \xrightarrow{u} V \xrightarrow{f} V \xrightarrow{u^{-1}} \mathrm{im}(\mathrm{pr}) \hookrightarrow H \otimes l^2(G).$$

Define the von Neumann trace of $f\colon V \to V$ by

$$\mathrm{tr}_{\mathcal{N}(G)}(f) := \sum_{i \in I} \langle \overline{f}(b_i \otimes e), b_i \otimes e \rangle \qquad \in [0, \infty],$$

where $e \in G \subset l^2(G)$ is the unit element.

This definition is independent of the choices of H, $\{b_i \mid i \in I\}$, pr and u. At least we give the proof for the independence of the Hilbert basis. If $\{c_j \mid j \in J\}$ is a second Hilbert basis, it follows from the following calculation where all terms in the sums are non-negative and hence interchanging is allowed.

$$\sum_{i \in I} \langle \overline{f}(b_i \otimes e), b_i \otimes e \rangle = \sum_{i \in I} \|\overline{f}^{1/2}(b_i \otimes e)\|^2$$

$$= \sum_{i \in I} \sum_{j \in J} \sum_{g \in G} |\langle \overline{f}^{1/2}(b_i \otimes e), c_j \otimes g \rangle|^2$$

$$= \sum_{i \in I} \sum_{j \in J} \sum_{g \in G} |\langle b_i \otimes g^{-1}, \overline{f}^{1/2}(c_j \otimes e) \rangle|^2$$

$$= \sum_{j \in J} \sum_{i \in I} \sum_{g \in G} |\langle \overline{f}^{1/2}(c_j \otimes e), b_i \otimes g \rangle|^2$$

$$= \sum_{j \in J} \|\overline{f}^{1/2}(c_j \otimes e)\|^2$$

$$= \sum_{j \in J} \langle \overline{f}(c_j \otimes e), c_j \otimes e \rangle.$$

A *directed set* I is a non-empty set with a partial ordering $\leq$ such that for two elements i_0 and i_1 there exists an element i with $i_0 \leq i$ and $i_1 \leq i$. A *net* $(x_i)_{i \in I}$ in a topological space is a map from a directed set to the topological space. The net $(x_i)_{i \in I}$ *converges to* x if for any neighborhood U of x there is an index $i(U) \in I$ such that $x_i \in U$ for each $i \in I$ with $i(U) \leq i$. A net $(f_i)_{i \in I}$ in $\mathcal{B}(H)$ *converges strongly to* $f \in \mathcal{B}(H)$ if for any $v \in H$ the net $(f_i(v))_{i \in I}$ converges to $f(v)$ in H. A net $(f_i)_{i \in I}$ in $\mathcal{B}(H)$ *converges weakly to* $f \in \mathcal{B}(H)$ if for any $v, w \in H$ the net $(\langle f_i(v), w \rangle)_{i \in I}$ converges to $\langle f(v), w \rangle$ in $\mathbb{C}$. A net $(f_i)_{i \in I}$ in $\mathcal{B}(H)$ *converges ultra-weakly* if for any two sequences $(x_n)_{n \geq 0}$ and $(y_n)_{n \geq 0}$ of elements in H with $\sum_{n \geq 0} ||x_n||^2 < \infty$ and $\sum_{n \geq 0} ||y_n||^2 < \infty$ the net $\sum_{n \geq 0} |\langle f_i(x_n), y_n \rangle|$ converges to $\sum_{n \geq 0} |\langle f(x_n), y_n \rangle|$. Obviously norm-convergence implies both ultra-weak convergence and strong convergence, ultra-weak convergence implies weak convergence, strong convergence implies weak convergence. In general these implications cannot be reversed and there is no relation between ultra-weak convergence and strong convergence [144, I.3.1 and I.3.2].

Theorem 1.9 (Von Neumann trace). *Let U, V and W be Hilbert $\mathcal{N}(G)$-modules.*

(1) If $f, g \colon V \to V$ are positive endomorphisms, then

$$f \leq g \Rightarrow \mathrm{tr}_{\mathcal{N}(G)}(f) \leq \mathrm{tr}_{\mathcal{N}(G)}(g);$$

(2) If $(f_i)_{i \in I}$ is a directed system of positive endomorphisms $f_i \colon V \to V$, directed by the order relation $\leq$ for positive operators, and the net converges weakly to the endomorphism $f \colon V \to V$, then f is positive and

$$\mathrm{tr}_{\mathcal{N}(G)}(f) = \sup\{\mathrm{tr}_{\mathcal{N}(G)}(f_i) \mid i \in I\};$$

(3) We have for a positive endomorphism $f \colon V \to V$

$$\mathrm{tr}_{\mathcal{N}(G)}(f) = 0 \Leftrightarrow f = 0;$$

(4) We have for positive endomorphisms $f, g \colon V \to V$ and a real number $\lambda \geq 0$

$$\mathrm{tr}_{\mathcal{N}(G)}(f + \lambda \cdot g) = \mathrm{tr}_{\mathcal{N}(G)}(f) + \lambda \cdot \mathrm{tr}_{\mathcal{N}(G)}(g);$$

(5) If the following diagram of endomorphisms of Hilbert $\mathcal{N}(G)$-modules commutes, has exact rows and positive operators as vertical maps

$$
\begin{array}{ccccccccc}
0 & \longrightarrow & U & \overset{i}{\longrightarrow} & V & \overset{p}{\longrightarrow} & W & \longrightarrow & 0 \\
& & \downarrow{\scriptstyle f} & & \downarrow{\scriptstyle g} & & \downarrow{\scriptstyle h} & & \\
0 & \longrightarrow & U & \overset{i}{\longrightarrow} & V & \overset{p}{\longrightarrow} & W & \longrightarrow & 0
\end{array}
$$

then

$$\mathrm{tr}_{\mathcal{N}(G)}(g) = \mathrm{tr}_{\mathcal{N}(G)}(f) + \mathrm{tr}_{\mathcal{N}(G)}(h);$$

(6) We get for a map $f \colon V \to W$

$$\mathrm{tr}_{\mathcal{N}(G)}(f^* f) = \mathrm{tr}_{\mathcal{N}(G)}(f f^*);$$

(7) Let U_1 be a Hilbert $\mathcal{N}(G)$-module and let U_2 be a Hilbert $\mathcal{N}(H)$-module. Let $f_i \colon U_i \to U_i$ be a positive endomorphism for $i = 1, 2$. Then the Hilbert tensor product $f_1 \otimes f_2 \colon U_1 \otimes U_2 \to U_1 \otimes U_2$ is a positive endomorphism of Hilbert $\mathcal{N}(G \times H)$-modules and

$$\mathrm{tr}_{\mathcal{N}(G \times H)}(f_1 \otimes f_2) \;=\; \mathrm{tr}_{\mathcal{N}(G)}(f_1) \cdot \mathrm{tr}_{\mathcal{N}(H)}(f_2),$$

where we use the convention that $0 \cdot \infty = 0$ and $r \cdot \infty = \infty$ for $r \in (0, \infty]$;

(8) Let $H \subset G$ be a subgroup of finite index $[G : H]$. Let $f \colon V \to V$ be a positive endomorphism of a Hilbert $\mathcal{N}(G)$-module. Let $\mathrm{res}(V)$ be the restriction of V to $\mathcal{N}(H)$ which is a Hilbert $\mathcal{N}(H)$-module. If V is finitely generated, then $\mathrm{res}\, V$ is finitely generated. We have

$$\mathrm{tr}_{\mathcal{N}(H)}(\mathrm{res}(f)) = [G : H] \cdot \mathrm{tr}_{\mathcal{N}(G)}(f),$$

where we use the convention $[G : H] \cdot \infty = \infty$.

Proof. (1) follows directly from the definitions.

(2) Since each f_i is selfadjoint the same is true for f. We first show $f_i \le f$ for all $i \in I$. Fix $v \in V$ and $i \in I$. Given $\epsilon > 0$ there is an index $i(\epsilon)$ with $i \le i(\epsilon)$ and $\langle f_{i(\epsilon)}(v), v \rangle - \epsilon \le \langle f(v), v \rangle$. Since $f_i \le f_{i(\epsilon)}$, we get $\langle f_i(v), v \rangle - \epsilon \le \langle f(v), v \rangle$ for all $\epsilon > 0$. This implies $\langle f_i(v), v \rangle \le \langle f(v), v \rangle$ and hence $f_i \le f$. We conclude that f is positive and from (1)

$$\mathrm{tr}_{\mathcal{N}(G)}(f) \ge \sup\{\mathrm{tr}_{\mathcal{N}(G)}(f_i) \mid i \in I\}.$$

It remains to prove the reverse inequality. In the notation of Definition 1.8 the net $(\overline{f_i})_{i \in I}$ converges weakly to $\overline{f}$. Hence we can assume without loss of generality $V = H \otimes l^2(G)$.

Let $\{b_\lambda \mid \lambda \in \Lambda\}$ be a Hilbert basis for H. Fix $\epsilon > 0$. Choose a finite subset $\Lambda(\epsilon) \subset \Lambda$ such that

$$\mathrm{tr}_{\mathcal{N}(G)}(f) \le \epsilon/2 + \sum_{\lambda \in \Lambda(\epsilon)} \langle f(b_\lambda \otimes e), b_\lambda \otimes e \rangle.$$

Since the net $(f_i)_{i \in I}$ converges weakly to f and $\Lambda(\epsilon)$ is finite and I is directed, there is an index $i(\epsilon) \in I$ such that for all $\lambda \in \Lambda(\epsilon)$

$$\langle f(b_\lambda \otimes e), b_\lambda \otimes e \rangle \le \frac{\epsilon}{2 \cdot |\Lambda(\epsilon)|} + \langle f_{i(\epsilon)}(b_\lambda \otimes e), b_\lambda \otimes e \rangle$$

holds. We conclude

$$\text{tr}_{\mathcal{N}(G)}(f) \le \epsilon/2 + \sum_{\lambda \in \Lambda(\epsilon)} \langle f(b_\lambda \otimes e), b_\lambda \otimes e \rangle$$

$$\le \epsilon/2 + \epsilon/2 + \sum_{\lambda \in \Lambda(\epsilon)} \langle f_{i(\epsilon)}(b_\lambda \otimes e), b_\lambda \otimes e \rangle$$

$$\le \epsilon + \text{tr}_{\mathcal{N}(G)}(f_{i(\epsilon)})$$

$$\le \epsilon + \sup\{\text{tr}_{\mathcal{N}(G)}(f_i) \mid i \in I\}.$$

Since this holds for all $\epsilon > 0$, the claim follows.

(3) Suppose $\text{tr}_{\mathcal{N}(G)}(f) = 0$ for $f : V \to V$. Then we get for $\overline{f} : H \otimes l^2(G) \to H \otimes l^2(G)$ (using the notation of Definition 1.8)

$$0 = \sum_{i \in I} \langle \overline{f}(b_i \otimes e), b_i \otimes e \rangle = \sum_{i \in I} ||\overline{f}^{1/2}(b_i \otimes e)||^2.$$

This implies $\overline{f}^{1/2}(b_i \otimes e) = 0$ for all $i \in I$. Since $\overline{f}^{1/2}$ is G-equivariant, we get $\overline{f}^{1/2}(b_i \otimes g) = 0$ for all $i \in I$ and $g \in G$. This implies $\overline{f}^{1/2} = 0$ and hence $\overline{f} = 0$ and $f = 0$.

(4) follows directly from the definitions.

(5) By the Polar Decomposition and the Inverse Mapping Theorem we obtain a unitary isomorphism $U \oplus W \xrightarrow{\cong} V$ which induces together with the identity on U and W an isomorphism of the given exact sequence with the standard exact sequence $0 \to U \to U \oplus W \to W \to 0$. One easily checks for a positive operator

$$\begin{pmatrix} f & u \\ 0 & g \end{pmatrix} : U \oplus W \to U \oplus W$$

that

$$\text{tr}_{\mathcal{N}(G)} \begin{pmatrix} f & u \\ 0 & g \end{pmatrix} = \text{tr}_{\mathcal{N}(G)}(f) + \text{tr}_{\mathcal{N}(G)}(g)$$

holds. Hence it remains to show for a positive G-operator $h : V \to V$ and a unitary G-operator $v : V \to W$ that $\text{tr}_{\mathcal{N}(G)}(vhv^{-1}) = \text{tr}_{\mathcal{N}(G)}(h)$ holds. This follows from the proof that Definition 1.8 is independent of the various choices.

(6) Let $f = u|f|$ be the Polar Decomposition of f. We conclude

$$\text{tr}_{\mathcal{N}(G)}(ff^*) = \text{tr}_{\mathcal{N}(G)}(u|f|^2u^{-1}) = \text{tr}_{\mathcal{N}(G)}(|f|^2) = \text{tr}_{\mathcal{N}(G)}(f^*f).$$

(7) Obviously $l^2(G) \otimes l^2(H)$ is isometrically $G \times H$-isomorphic to $l^2(G \times H)$. Hence $U_1 \otimes U_2$ is a Hilbert $\mathcal{N}(G \times H)$-module and $f_1 \otimes f_2$ is a positive endomorphism. Because of assertion (5) it suffices to treat the case where $U_1 = H_1 \otimes l^2(G)$ and $U_2 = H_2 \otimes l^2(H)$ for appropriate Hilbert spaces H_1 and H_2. Then the claim follows from the obvious formula

$$\langle f_1 \otimes f_2(u_1 \otimes u_2), v_1 \otimes v_2 \rangle_{U_1 \otimes U_2} = \langle f_1(u_1), v_1 \rangle_{U_1} \cdot \langle f_2(u_2), v_2 \rangle_{U_2}$$

for $u_1, v_1 \in U_1$ and $u_2, v_2 \in U_2$.

(8) It suffices to treat the case $V = l^2(G)$. If $s \colon G/H \to G$ is a set-theoretic section of the projection $G \to G/H$, we obtain an isometric H-isomorphism

$$\bigoplus_{gH \in G/H} l^2(H) \xrightarrow{\cong} \operatorname{res} l^2(G), \quad \{u_{gH} \mid gH \in G/H\} \mapsto \sum_{gH \in G/H} u_{gH} \cdot s(gH).$$

This finishes the proof of Theorem 1.9. $\qquad\qquad\qquad\qquad\qquad\qquad\square$

Definition 1.10 (Von Neumann dimension). *Define the* von Neumann dimension *of a Hilbert $\mathcal{N}(G)$-module V*

$$\dim_{\mathcal{N}(G)}(V) := \operatorname{tr}_{\mathcal{N}(G)}(\operatorname{id} \colon V \to V) \qquad \in [0, \infty].$$

If G is finite, then $\dim_{\mathcal{N}(G)}(V)$ is $\frac{1}{|G|}$-times the complex dimension of the underlying complex vector space V. The next example shows that $\dim_{\mathcal{N}(G)}(V)$ can take any non-negative real number or ∞ as value.

Example 1.11. Let $X \subset T^n$ be any measurable set and $\chi_X \in L^\infty(T^n)$ be its characteristic function. Denote by $M_{\chi_X} \colon L^2(T^n) \to L^2(T^n)$ the $\mathbb{Z}^n$-equivariant unitary projection given by multiplication with χ_X. Its image V is a Hilbert $\mathcal{N}(\mathbb{Z}^n)$-module with $\dim_{\mathcal{N}(\mathbb{Z}^n)}(V) = \operatorname{vol}(X)$ (see Example 1.4).

Theorem 1.12 (von Neumann dimension). *(1) We have for a Hilbert $\mathcal{N}(G)$-module V*

$$V = 0 \Leftrightarrow \dim_{\mathcal{N}(G)}(V) = 0;$$

(2) If $0 \to U \to V \to W \to 0$ is a weakly exact sequence of Hilbert $\mathcal{N}(G)$-modules, then

$$\dim_{\mathcal{N}(G)}(U) + \dim_{\mathcal{N}(G)}(W) = \dim_{\mathcal{N}(G)}(V);$$

(3) Let $\{V_i \mid i \in I\}$ be a directed system of Hilbert $\mathcal{N}(G)$- submodules of V, directed by $\subset$. Then

$$\dim_{\mathcal{N}(G)}\left(\operatorname{clos}\left(\bigcup_{i \in I} V_i\right)\right) = \sup\{\dim_{\mathcal{N}(G)}(V_i) \mid i \in I\};$$

(4) Let V be an $\mathcal{N}(G)$-Hilbert module with $\dim_{\mathcal{N}(G)}(V) < \infty$ and $\{V_i \mid i \in I\}$ be a directed system of Hilbert $\mathcal{N}(G)$-submodules, directed by $\supset$. Then

$$\dim_{\mathcal{N}(G)}\left(\bigcap_{i \in I} V_i\right) = \inf\{\dim_{\mathcal{N}(G)}(V_i) \mid i \in I\};$$

(5) *Let U_1 be a Hilbert $\mathcal{N}(G)$-module and let U_2 be a Hilbert $\mathcal{N}(H)$-module. Then the Hilbert tensor product $U_1 \otimes U_2$ is a Hilbert $\mathcal{N}(G \times H)$-module and*

$$\dim_{\mathcal{N}(G \times H)}(U_1 \otimes U_2) = \dim_{\mathcal{N}(G)}(U_1) \cdot \dim_{\mathcal{N}(H)}(U_2),$$

where we use the convention that $0 \cdot \infty = 0$ and $r \cdot \infty = \infty$ for $r \in (0, \infty]$;

(6) *Let $H \subset G$ be a subgroup of finite index $[G : H]$. Let V be Hilbert $\mathcal{N}(G)$-module. Let $\mathrm{res}(V)$ be the restriction of V to $\mathcal{N}(H)$ which is a Hilbert $\mathcal{N}(H)$-module. If V is finitely generated, then $\mathrm{res}\, V$ is finitely generated. We have*

$$\dim_{\mathcal{N}(H)}(\mathrm{res}(V)) = [G : H] \cdot \dim_{\mathcal{N}(G)}(V),$$

where we use the convention $[G : H] \cdot \infty = \infty$.

Proof. (1) This follows from Theorem 1.9 (3).

(2) We conclude from the weak exactness and the Polar Decomposition that $U \oplus W$ is isometrically G-isomorphic to V. Now apply Theorem 1.9 (5).

(3) Let $\mathrm{pr} \colon V \to V$ and $\mathrm{pr}_i \colon V \to V$ be the projections onto $\mathrm{clos}\left(\bigcup_{i \in I} V_i\right)$ and V_i. Next we show that the net $(\mathrm{pr}_i)_{i \in I}$, directed by the order relation for positive operators, converges strongly (and hence in particular weakly) to pr. Given $v \in V$ and $\epsilon > 0$, there is an index $i(\epsilon) \in I$ and $v_{i(\epsilon)} \in V_{i(\epsilon)}$ with

$$|\mathrm{pr}(v) - v_{i(\epsilon)}| \leq \epsilon/2.$$

We conclude for $i \geq i(\epsilon)$

$$\begin{aligned}
|\mathrm{pr}(v) - \mathrm{pr}_i(v)| &\leq |\mathrm{pr}(v) - v_{i(\epsilon)}| + |v_{i(\epsilon)} - \mathrm{pr}_i(v)| \\
&= |\mathrm{pr}(v) - v_{i(\epsilon)}| + |\mathrm{pr}_i(v_{i(\epsilon)} - \mathrm{pr}(v))| \\
&\leq |\mathrm{pr}(v) - v_{i(\epsilon)}| + ||\,\mathrm{pr}_i\,|| \cdot |\mathrm{pr}(v) - v_{i(\epsilon)}| \\
&\leq \epsilon/2 + 1 \cdot \epsilon/2 \\
&= \epsilon.
\end{aligned}$$

We conclude from Theorem 1.9 (2)

$$\mathrm{tr}_{\mathcal{N}(G)}(\mathrm{pr}) = \sup\{\mathrm{tr}_{\mathcal{N}(G)}(\mathrm{pr}_i) \mid i \in I\}.$$

Since $\dim_{\mathcal{N}(G)}\left(\mathrm{clos}\left(\bigcup_{i \in I} V_i\right)\right) = \mathrm{tr}_{\mathcal{N}(G)}(\mathrm{pr})$ and $\dim_{\mathcal{N}(G)}(V_i) = \mathrm{tr}_{\mathcal{N}(G)}(\mathrm{pr}_i)$, the claim follows.

(4) We obtain a directed system $(V_i^{\perp})_{i \in I}$, directed by $\subset$. We have

$$\mathrm{clos}\left(\bigcup_{i \in I} V_i^{\perp}\right)^{\perp} = \bigcap_{i \in I} V_i;$$

$$\dim_{\mathcal{N}(G)}(V_i^{\perp}) = \dim_{\mathcal{N}(G)}(V) - \dim_{\mathcal{N}(G)}(V_i);$$

$$\dim_{\mathcal{N}(G)}\left(\mathrm{clos}\left(\bigcup_{i \in I} V_i^{\perp}\right)^{\perp}\right) = \dim_{\mathcal{N}(G)}(V) - \dim_{\mathcal{N}(G)}\left(\mathrm{clos}\left(\bigcup_{i \in I} V_i^{\perp}\right)\right).$$

Now the claim follows from (3).

(5) This follows from Theorem 1.9 (7).

(6) This follows from Theorem 1.9 (8).   □

We conclude from Theorem 1.12 (1) and (2)

Lemma 1.13. *Let* $f\colon U \to V$ *be a morphism of Hilbert* $\mathcal{N}(G)$-*modules whose von Neumann dimension is finite. Then the following statements are equivalent:*

(1) f is a weak isomorphism;
(2) The unitary part of the polar decomposition of f is an isomorphism;
(3) f^* is a weak isomorphism;
(4) f is injective and $\dim_{\mathcal{N}(G)}(U) = \dim_{\mathcal{N}(G)}(V)$;
(5) f has dense image and $\dim_{\mathcal{N}(G)}(U) = \dim_{\mathcal{N}(G)}(V)$.

Example 1.14. In this example we present a Hilbert $\mathcal{N}(\mathbb{Z})$-module U which is not finitely generated but has finite von Neumann dimension. In the sequel we use the identification of Example 1.4. For an interval $I \subset [0,1]$ let $\chi_I \in \mathcal{N}(\mathbb{Z}) = L^\infty(S^1)$ be the characteristic function of the subset $\{\exp(2\pi i t) \mid t \in I\}$. Define two Hilbert $\mathcal{N}(\mathbb{Z})$-modules by the orthogonal Hilbert sums

$$U = \bigoplus_{n=1}^{\infty} \mathrm{im}(\chi_{[0,2^{-n}]});$$

$$V = \bigoplus_{n=1}^{\infty} \mathrm{im}(\chi_{[1/(n+1),1/n]}),$$

where $\mathrm{im}(\chi_I)$ is the direct summand in $l^2(\mathbb{Z}) = L^2(S^1)$ given by the projection $\chi_I \in \mathcal{N}(\mathbb{Z})$. Theorem 1.12 and Example 1.11 imply

$$\dim_{\mathcal{N}(\mathbb{Z})}(U) \;=\; \dim_{\mathcal{N}(\mathbb{Z})}(V) \;=\; 1.$$

We want to show that U is not finitely generated. This is not obvious, for instance, V is isomorphic to the Hilbert $\mathcal{N}(\mathbb{Z})$-module $l^2(\mathbb{Z})$ and in particular finitely generated, although it is defined as an infinite Hilbert sum of non-trivial Hilbert $\mathcal{N}(\mathbb{Z})$-modules.

To show that U is not finitely generated, we use the center valued dimension function which assigns to a finitely generated Hilbert $\mathcal{N}(\mathbb{Z})$-module P an element

$$\dim^c_{\mathcal{N}(G)}(P) \in L^\infty(S^1) = \mathcal{N}(\mathbb{Z}).$$

Its definition is the same as the definition of $\dim_{\mathcal{N}(G)}(P)$ with the exception that one replaces the standard trace $\mathrm{tr}_{\mathcal{N}(\mathbb{Z})}\colon \mathcal{N}(\mathbb{Z}) \to \mathbb{C}$ by the identity $\mathrm{id}\colon \mathcal{N}(\mathbb{Z}) \to \mathcal{N}(\mathbb{Z})$. (More details will be given in Subsection 9.1.3 and 9.2.1.) Suppose that U is finitely generated. Then U is a direct summand in $l^2(\mathbb{Z})^k$ for some $k \geq 0$. Since $\bigoplus_{n=1}^{k+1} \mathrm{im}(\chi_{[0,2^{-n}]})$ is a direct summand in U and hence in $l^2(\mathbb{Z})^k$, we conclude

$$\sum_{n=1}^{k+1} \chi_{[0,2^{-n}]} = \dim_{\mathcal{N}(\mathbb{Z})}^c \left(\bigoplus_{n=1}^{k+1} \operatorname{im}(\chi_{[0,2^{-n}]}) \right) \le \dim_{\mathcal{N}(\mathbb{Z})}^c (l^2(\mathbb{Z})^k) = c_k,$$

where $c_k \colon S^1 \to \mathbb{C}$ is the constant function with value k. But this is a contradiction, since $\sum_{n=1}^{k+1} \chi_{[0,2^{-n}]}$ is equal to $k + 1$ on the subset $\{\exp(2\pi i t) \mid t \in [0, 2^{-k-1}]\}$ whose measure is different from zero.

1.1.4 Hilbert Chain Complexes

Definition 1.15 (Hilbert chain complex). *A Hilbert $\mathcal{N}(G)$-chain complex C_* is a sequence indexed by $p \in \mathbb{Z}$ of maps of Hilbert $\mathcal{N}(G)$-modules*

$$\ldots \xrightarrow{c_{p+2}} C_{p+1} \xrightarrow{c_{p+1}} C_p \xrightarrow{c_p} C_{p-1} \xrightarrow{c_{p-1}} \ldots$$

such that $c_p \circ c_{p+1} = 0$ for all $p \in \mathbb{Z}$. We call C_* positive if C_p vanishes for $p < 0$. We say it is finitely generated if each C_p is a finitely generated Hilbert $\mathcal{N}(G)$-module.

It is d-dimensional if C_p vanishes for $|p| > d$. It is finite dimensional if it is d-dimensional for some $d \in \mathbb{N}$. We call C_* finite if C_* is finitely generated and finite dimensional.

There are obvious notions of chain maps and chain homotopies.

Definition 1.16 (L^2-homology and L^2-Betti numbers). *Define the (reduced) p-th L^2-homology and the p-th L^2-Betti number of a Hilbert $\mathcal{N}(G)$-chain complex C_* by*

$$H_p^{(2)}(C_*) := \ker(c_p)/\operatorname{clos}(\operatorname{im}(c_{p+1}));$$
$$b_p^{(2)}(C_*) := \dim_{\mathcal{N}(G)}(H_p^{(2)}(C_*)).$$

Notice that we divide by the closure of the image and not by the image of c_{p+1}. This has the effect that $H_p^{(2)}(C_*)$ inherits the structure of a Hilbert space and a G-action from C_p. This is indeed a Hilbert $\mathcal{N}(G)$-structure on $H_p^{(2)}(C_*)$. Namely, $H_p^{(2)}(C_*)$ is isometrically G-isomorphic to $\ker(c_p) \cap \operatorname{im}(c_{p+1})^{\perp}$. The *Laplace operator* is defined by

$$\Delta_p = c_{p+1}c_{p+1}^* + c_p^* c_p \colon C_p \to C_p. \tag{1.17}$$

We have the following "baby"-version of the L^2-Hodge-de Rham Theorem 1.57.

Lemma 1.18. *Let C_* be a Hilbert chain complex. Then we get the orthogonal decomposition of Hilbert $\mathcal{N}(G)$-modules*

$$C_p = \ker(\Delta_p) \oplus \operatorname{clos}(\operatorname{im}(c_{p+1})) \oplus \operatorname{clos}(\operatorname{im}(c_p^*))$$

and the natural map

$$i\colon \ker(c_p) \cap \ker(c_{p+1}^*) = \ker(\Delta_p) \to H_p^{(2)}(C_*)$$

is an isometric G-isomorphism.

Proof. We have the obvious orthogonal decomposition

$$C_p \;=\; (\ker(c_p) \cap \operatorname{im}(c_{p+1})^\perp) \oplus \operatorname{clos}(\operatorname{im}(c_{p+1})) \oplus \ker(c_p)^\perp.$$

Since $\ker(c_p)^\perp = \operatorname{clos}(\operatorname{im}(c_p^*))$ and $\operatorname{im}(c_{p+1})^\perp = \ker(c_{p+1}^*)$, it remains to show

$$\ker(\Delta_p) \;=\; \ker(c_p) \cap \ker(c_{p+1}^*).$$

This follows from the calculation

$$\langle \Delta_p(v), v \rangle = |c_p(v)|^2 + |c_{p+1}^*(v)|^2. \qquad \square$$

One important tool in the theory of chain complexes is the long exact homology sequence. This does not go through directly, but in a weak sense under certain finiteness conditions. Let $0 \to C_* \xrightarrow{i_*} D_* \xrightarrow{p_*} E_* \to 0$ be an exact sequence of Hilbert $\mathcal{N}(G)$-chain complexes. The maps i_* and p_* induce maps on the homology groups denoted by $H_n^{(2)}(i_*)$ and $H_n^{(2)}(p_*)$. Next we want to define a natural *boundary operator*

$$\partial_n \colon H_n^{(2)}(E_*) \to H_{n-1}^{(2)}(C_*).$$

As a linear G-equivariant map it is defined in the usual way. Namely, let $z \in \ker(e_n)$ be a representative of $[z] \in H_n^{(2)}(E_*)$. Choose $y \in D_n$ and $x \in C_{n-1}$ with the properties that $p_n(y) = z$ and $i_{n-1}(x) = d_n(y)$. We can find such y since p_n is surjective and such x since $p_{n-1} \circ d_n(y) = e_n \circ p_n(y) = e_n(z) = 0$. We have $x \in \ker(c_{n-1})$ since i_{n-2} is injective and

$$i_{n-2} \circ c_{n-1}(x) = d_{n-1} \circ i_{n-1}(x) = d_{n-1} \circ d_n(y) = 0.$$

We define

$$\partial_n([z]) \;:=\; [x].$$

It is easy to check that this is independent of the choice of the representative z for $[z]$ and that this map is linear and G-equivariant. In order to prove that it is a bounded operator, we give a different description. Notice for the sequel that a bijective bounded operator of Hilbert spaces has a bounded inverse by the Inverse Mapping Theorem. Consider the following composition of maps of Hilbert $\mathcal{N}(G)$-modules.

$$\ker(e_n) \to E_n \xrightarrow{(p_n|_{\ker(p_n)^\perp})^{-1}} \ker(p_n)^\perp \xrightarrow{d_n|_{\ker(p_n)^\perp}} D_{n-1}$$

One easily checks that its image lies in $\ker(p_{n-1})$. Hence we can compose it with the isomorphism of Hilbert $\mathcal{N}(G)$-modules $i_{n-1}^{-1} \colon \ker(p_{n-1}) \to C_{n-1}$.

Again one easily checks that the image of this map lies in $\ker(c_{n-1})$ so that we have defined a map of Hilbert $\mathcal{N}(G)$-modules

$$\partial'_n \colon \ker(e_n) \to \ker(c_{n-1}).$$

Since it maps $\mathrm{im}(e_{n+1})$ to $\mathrm{im}(c_n)$, it induces a map on L^2-homology which is the desired map ∂_n.

Although the boundary operator can always be defined and one gets a long homology sequence, this long homology sequence does not have to be weakly exact as the following example shows.

Example 1.19. Let G be the trivial group. Then a Hilbert $\mathcal{N}(G)$-module is just a Hilbert space and a map of Hilbert $\mathcal{N}(G)$-modules is just a bounded operator. Denote by H the Hilbert space $\{(a_n \in \mathbb{C})_{n \in \mathbb{N}} \mid \sum |a_n|^2 < \infty\}$ with the inner product $\langle (a_n), (b_n) \rangle = \sum a_n \cdot \overline{b_n}$. Define a linear bounded operator $f \colon H \to H$ by sending (a_n) to $(1/n \cdot a_n)$. Obviously f is injective and has dense image. But f is not surjective because $u := (1/n)_n \in H$ cannot have a preimage. Let $\mathrm{pr} \colon H \to \mathrm{span}_{\mathbb{C}}(u)^{\perp}$ be the projection where $\mathrm{span}_{\mathbb{C}}(u)$ is the one-dimensional subspace generated by u. Then $\mathrm{pr} \circ f$ is injective with dense image. Hence f and $\mathrm{pr} \circ f$ are weak isomorphisms of Hilbert spaces but pr is not. View the following diagram as a short exact sequence of Hilbert chain complexes which are concentrated in dimensions 0 and 1

$$
\begin{array}{ccccccccc}
0 & \longrightarrow & 0 & \longrightarrow & H & \overset{\mathrm{id}}{\longrightarrow} & H & \longrightarrow & 0 \\
& & \downarrow & & {\scriptstyle f}\big\downarrow & & {\scriptstyle \mathrm{pr} \circ f}\big\downarrow & & \\
0 & \longrightarrow & \mathrm{span}_{\mathbb{C}}(u) & \longrightarrow & H & \overset{\mathrm{pr}}{\longrightarrow} & \mathrm{span}_{\mathbb{C}}(u)^{\perp} & \longrightarrow & 0
\end{array}
$$

All the L^2-homology groups are trivial except the zero-th L^2-homology group of the left Hilbert chain complex $0 \to \mathrm{span}_{\mathbb{C}}(u)$. Hence there cannot be a long weakly exact homology sequence.

Notice that in the example above the dimension of the Hilbert space H is infinite.

Definition 1.20. *A morphism $f \colon U \to V$ of Hilbert $\mathcal{N}(G)$-modules is called* Fredholm *if for some $\lambda > 0$ we have $\dim_{\mathcal{N}(G)}(\mathrm{im}(E_{\lambda}^{f^* f})) < \infty$, where $\{E_{\lambda}^{f^* f} \mid \lambda \in \mathbb{R}\}$ is the (right continuous) spectral family associated to the positive operator $f^* f$ (see Definition 1.68). A Hilbert $\mathcal{N}(G)$-chain complex C_* is called* Fredholm *at p if the induced morphism $\overline{c_p} \colon C_p / \mathrm{clos}(\mathrm{im}(c_{p+1})) \to C_{p-1}$ is* Fredholm. *We call C_** Fredholm *if it is Fredholm at p for all $p \in \mathbb{Z}$.*

For this chapter it suffices to know that a Hilbert $\mathcal{N}(G)$-chain complex C_*, for which C_p is finitely generated or satisfies $\dim_{\mathcal{N}(G)}(C_p) < \infty$, is automatically Fredholm at p. The next result is due to Cheeger and Gromov [105, Theorem 2.1].

Theorem 1.21 (Weakly exact long L^2-homology sequence). *Let $0 \to C_* \xrightarrow{i_*} D_* \xrightarrow{p_*} E_* \to 0$ be an exact sequence of Hilbert $\mathcal{N}(G)$-chain complexes which are Fredholm. Then the long homology sequence*

$$\cdots \xrightarrow{H_{n+1}^{(2)}(p_*)} H_{n+1}^{(2)}(E_*) \xrightarrow{\partial_{n+1}} H_n^{(2)}(C_*) \xrightarrow{H_n^{(2)}(i_*)} H_n^{(2)}(D_*)$$

$$\xrightarrow{H_n^{(2)}(p_*)} H_n^{(2)}(E_*) \xrightarrow{\partial_n} \cdots$$

is weakly exact.

Proof. We only prove weak exactness at $H_n^{(2)}(D_*)$, the other cases are similar. Since $p_* \circ i_* = 0$, we get $\operatorname{clos}(\operatorname{im}(H_n^{(2)}(i_*))) \subset \ker(H_n^{(2)}(p_*))$. It remains to prove equality. Let U be the orthogonal complement of $\operatorname{clos}(\operatorname{im}(H_n^{(2)}(i_*)))$ in $\ker(H_n^{(2)}(p_*))$. Let $V \subset \ker(d_{n+1}^*) \cap \ker(d_n)$ be the subspace corresponding to U under the isomorphism of Lemma 1.18

$$\ker(d_{n+1}^*) \cap \ker(d_n) \to H_n^{(2)}(D_*).$$

In view of Theorem 1.12 (1) it remains to show

$$\dim_{\mathcal{N}(G)}(V) = 0.$$

We have

$$p_n(V) \subset \operatorname{clos}(\operatorname{im}(e_{n+1}))$$

since $H_n^{(2)}(p_*)(U)$ is zero. The operator $e_{n+1} \circ e_{n+1}^* \colon E_n \to E_n$ is positive. Let $\{E_\lambda \mid \lambda \in \mathbb{R}\}$ be its (right continuous) spectral family (see Definition 1.68) of projections $E_\lambda \colon E_n \to E_n$. Since E_λ commutes with $e_{n+1} \circ e_{n+1}^*$, it sends $\operatorname{clos}(\operatorname{im}(e_{n+1} \circ e_{n+1}^*))$ to itself. We have

$$\operatorname{clos}(\operatorname{im}(e_{n+1} \circ e_{n+1}^*)) = \ker(e_{n+1} \circ e_{n+1}^*)^{\perp} = \ker(e_{n+1}^*)^{\perp} = \operatorname{clos}(\operatorname{im}(e_{n+1})),$$

where the second equality follows from $\langle e_{n+1} \circ e_{n+1}^*(v), v \rangle = |e_{n+1}^*(v)|^2$. This implies

$$E_\lambda \circ p_n(V) \subset \operatorname{clos}(\operatorname{im}(e_{n+1})).$$

Next we show that $E_\lambda \circ p_n$ is injective on V for $\lambda > 0$. Consider an element $v \in V$ with $E_\lambda \circ p_n(v) = 0$. We get $p_n(v) \in \operatorname{im}(E_\lambda)^{\perp}$. Notice that $e_{n+1} \circ e_{n+1}^*$ induces an invertible operator from $\operatorname{im}(E_\lambda)^{\perp}$ to itself for $\lambda > 0$, the inverse is given by

$$\int_\lambda^\infty \frac{1}{\mu} dE_\mu.$$

Hence we can find $w \in E_{n+1}$ satisfying

$$p_n(v) = e_{n+1}(w).$$

Since p_{n+1} is surjective and $\ker(p_n) = \operatorname{im}(i_n)$, we can find $x \in D_{n+1}$ and $y \in C_n$ satisfying

$$i_n(y) = d_{n+1}(x) + v.$$

This implies that $c_n(y) = 0$. Hence we get $[y] \in H_n^{(2)}(C)$ whose image under $H_n^{(2)}(i)$ is the element in U which corresponds to $v \in V$. Since U is orthogonal to $\operatorname{im}(H_n^{(2)}(i))$, we conclude $v = 0$. We have shown that $E_\lambda \circ p_n$ is injective on V for $\lambda > 0$. Now we conclude from Theorem 1.12 and the facts that E_λ is right continuous in λ and E_* is Fredholm by assumption

$$
\begin{aligned}
\dim_{\mathcal{N}(G)}(V) &= \dim_{\mathcal{N}(G)}\left(\operatorname{clos}(E_\lambda \circ p_n(V))\right) \\
&= \dim_{\mathcal{N}(G)}\left(\operatorname{clos}(E_\lambda \circ p_n(V)) \cap \operatorname{clos}(\operatorname{im}(e_{n+1}))\right) \\
&= \lim_{\lambda \to 0+} \dim_{\mathcal{N}(G)}\left(\operatorname{clos}(E_\lambda \circ p_n(V)) \cap \operatorname{clos}(\operatorname{im}(e_{n+1}))\right) \\
&= \dim_{\mathcal{N}(G)}\left(\bigcap_{\lambda>0} \operatorname{clos}(E_\lambda \circ p_n(V)) \cap \operatorname{clos}(\operatorname{im}(e_{n+1}))\right) \\
&\leq \dim_{\mathcal{N}(G)}\left(\bigcap_{\lambda>0} \operatorname{im}(E_\lambda) \cap \operatorname{clos}(\operatorname{im}(e_{n+1}))\right) \\
&= \dim_{\mathcal{N}(G)}\left(\operatorname{im}(E_0) \cap \operatorname{clos}(\operatorname{im}(e_{n+1}))\right) \\
&= \dim_{\mathcal{N}(G)}\left(\ker(e_{n+1} \circ e_{n+1}^*) \cap \operatorname{clos}(\operatorname{im}(e_{n+1}))\right) \\
&= \dim_{\mathcal{N}(G)}\left(\ker(e_{n+1}^*) \cap \operatorname{clos}(\operatorname{im}(e_{n+1}))\right) \\
&= \dim_{\mathcal{N}(G)}(0) \\
&= 0.
\end{aligned}
$$

This finishes the proof of Theorem 1.21.   □

Let G and H be groups. Let U be a Hilbert $\mathcal{N}(G)$-module and V be a Hilbert $\mathcal{N}(H)$-module. The Hilbert space tensor product $U \otimes V$ with the obvious $G \times H$-operation is a Hilbert $\mathcal{N}(G \times H)$-module because $l^2(G) \otimes l^2(H)$ is isometrically $G \times H$-isomorphic to $l^2(G \times H)$. Let C_* be a Hilbert $\mathcal{N}(G)$-chain complex and D_* be a Hilbert $\mathcal{N}(H)$-chain complex. Their tensor product is the Hilbert $\mathcal{N}(G \times H)$-chain complex $C_* \otimes D_*$ with n-th differential e_n which is given by

$$(C_* \otimes D_*)_n := \bigoplus_{i=0}^{n} C_i \otimes D_{n-i};$$

$$e_n|_{C_i \otimes D_{n-i}} := c_i \otimes \operatorname{id} + (-1)^i \cdot \operatorname{id} \otimes d_{n-i}.$$

Lemma 1.22. *Let C_* be a Hilbert $\mathcal{N}(G)$-chain complex and let D_* be a Hilbert $\mathcal{N}(H)$-chain complex. Then there is a natural isomorphism of Hilbert $\mathcal{N}(G \times H)$-modules for $n \geq 0$*

$$K_n: \bigoplus_{p+q=n} H_p^{(2)}(C_*) \otimes H_q^{(2)}(D_*) \xrightarrow{\cong} H_n^{(2)}(C_* \otimes D_*).$$

Proof. Given an element $u \in \ker(c_p)$ and $v \in \ker(d_q)$, define $K_n([u] \otimes [v])$ by the class of $u_p \otimes v_q \in \ker((c_* \otimes d_*)_n)$. Notice that the Laplacian Δ_n on $(C_* \otimes D_*)_n = \bigoplus_{p+q=n} C_p \otimes D_q$ is given by $\bigoplus_{p+q=n} \Delta_p^C \otimes \mathrm{id}_{D_q} + \mathrm{id}_{C_p} \otimes \Delta_q^D$, where Δ_*^C and Δ_*^D denote the Laplacians on C_* and D_*. To show that K_n is an isomorphism, it suffices to prove because of Lemma 1.18 that the kernel of $\Delta_p^C \otimes \mathrm{id}_{D_q} + \mathrm{id}_{C_p} \otimes \Delta_q^D$ is $\ker(\Delta_p^C) \otimes \ker(\Delta_q^D)$. Because of the orthogonal decompositions $C_p = \ker(\Delta_p^C) \oplus \ker(\Delta_p^C)^\perp$ and $D_q = \ker(\Delta_q^D) \oplus \ker(\Delta_q^D)^\perp$ it remains to show that the three induced positive endomorphisms

$$(\Delta_p^C)^\perp \otimes \mathrm{id} \colon \ker(\Delta_p^C)^\perp \otimes \ker(\Delta_q^D) \to \ker(\Delta_p^C)^\perp \otimes \ker(\Delta_q^D);$$
$$\mathrm{id} \otimes (\Delta_q^D)^\perp \colon \ker(\Delta_p^C) \otimes \ker(\Delta_q^D)^\perp \to \ker(\Delta_p^C) \otimes \ker(\Delta_q^D)^\perp;$$
$$(\Delta_p^C)^\perp \otimes \mathrm{id} + \mathrm{id} \otimes (\Delta_q^D)^\perp \colon \ker(\Delta_p^C)^\perp \otimes \ker(\Delta_q^D)^\perp \to \ker(\Delta_p^C)^\perp \otimes \ker(\Delta_q^D)^\perp$$

are injective. But this follows from the facts that a selfadjoint endomorphism is injective if and only if it has dense image, the sum of two injective positive operators is injective again and from the injectivity of $(\Delta_p^C)^\perp$ and $(\Delta_q^D)^\perp$. $\quad\square$

Of course everything in this section has an obvious analog for Hilbert $\mathcal{N}(G)$-cochain complexes.

1.1.5 Induction for Group von Neumann Algebras

Next we investigate how group von Neumann algebras and Hilbert modules behave under induction.

Let $i \colon H \to G$ be an injective group homomorphism. Let M be a Hilbert $\mathcal{N}(H)$-module. There is an obvious pre-Hilbert structure on $\mathbb{C}G \otimes_{\mathbb{C}H} M$ for which G acts by isometries since $\mathbb{C}G \otimes_{\mathbb{C}H} M$ as a complex vector space can be identified with $\bigoplus_{G/H} M$. Its Hilbert space completion is a Hilbert $\mathcal{N}(G)$-module and denoted by $i_* M$. A map of Hilbert $\mathcal{N}(H)$-modules $f \colon M \to N$ induces a map of Hilbert $\mathcal{N}(G)$-modules $i_* f \colon i_* M \to i_* N$. Thus we obtain a covariant functor from the category of Hilbert $\mathcal{N}(H)$-modules to the category of Hilbert $\mathcal{N}(G)$-modules. Obviously $i_* l^2(H) = l^2(G)$ and i_* is compatible with direct sums. If M is finitely generated, $i_* M$ is finitely generated.

Definition 1.23 (Induction). *Let $i \colon H \longrightarrow G$ be an injective group homomorphism. The Hilbert $\mathcal{N}(G)$-module $i_* M$ is called the* induction with i of M. *In particular i induces a ring homomorphism*

$$i \colon \mathcal{N}(H) \to \mathcal{N}(G).$$

Lemma 1.24. *Let $i \colon H \to G$ be an injective group homomorphism. Then*

(1) We have $\mathrm{tr}_{\mathcal{N}(H)} = \mathrm{tr}_{\mathcal{N}(G)} \circ i \colon \mathcal{N}(H) \to \mathbb{C}$;
(2) We get for a Hilbert $\mathcal{N}(H)$-module M

$$\dim_{\mathcal{N}(G)}(i_* M) = \dim_{\mathcal{N}(H)}(M);$$

(3) If $0 \to U \to V \to W \to 0$ is a sequence of Hilbert $\mathcal{N}(H)$-modules which is exact or weakly exact respectively, then the induced sequence $0 \to i_* U \to i_* V \to i_* W \to 0$ is a sequence of Hilbert $\mathcal{N}(G)$-modules which is exact or weakly exact respectively;

(4) Let C_* be a Hilbert $\mathcal{N}(H)$-chain complex. Then $i_* C_*$ is a Hilbert $\mathcal{N}(G)$-chain complex and for all $p \in \mathbb{Z}$

$$i_* H_p^{(2)}(C_*) = H_p^{(2)}(i_* C_*);$$
$$\dim_{\mathcal{N}(H)}(H_p^{(2)}(C_*)) = \dim_{\mathcal{N}(G)}(H_p^{(2)}(i_* C_*)).$$

Proof. (1) follows directly from the definitions.

(2) Let $\mathrm{pr}\colon V \otimes l^2(H) \to V \otimes l^2(H)$ be an H-equivariant unitary projection describing M for some Hilbert space V. Then $i_*(\mathrm{pr})$ describes $i_* M$.

(3) Since any exact sequence splits and i_* is compatible with direct sums, the claim follows for exact sequences. In order to treat the weakly exact case it suffices to show that for a weak isomorphism $f\colon U \to V$ of Hilbert $\mathcal{N}(H)$-modules the induced map $i_* f\colon i_* U \to i_* V$ is a weak isomorphism. By the Polar Decomposition we can assume without loss of generality that $U = V$ and f is positive. Since the kernel of a positive operator is the orthogonal complement of its image and $i_* f$ is positive it remains to show that the image of $i_* f$ is dense if the image of f is dense. This follows from the definition of $i_* f$ since $\mathbb{C}G$ is dense in $l^2(G)$.

(4) This follows from (2) and (3). $\square$

1.2 Cellular L^2-Betti Numbers

In this section we apply the material of Section 1.1 to regular coverings of CW-complexes of finite type (i.e. all skeleta are finite) and thus define and study cellular L^2-Betti numbers for them.

Since regular coverings of CW-complexes are special cases of G-CW-complexes and we will need the notion of a G-CW-complex in its full generality later, we collect some basic facts about G-CW-complexes. However, in order to read this and the next chapters one may skip subsection 1.2.1 and just keep in mind the following two facts: (1) A free finite G-CW-complex X or free G-CW-complex X of finite type respectively is the same as a G-space X such that the projection $X \to G\backslash X$ is a regular covering and $G\backslash X$ is a finite CW-complex or CW-complex of finite type respectively. (2) The cellular $\mathbb{Z}G$-chain complex of a free G-CW-complex is free and has a cellular $\mathbb{Z}G$-basis which is unique up to permutation and multiplication with trivial units $\pm g \in \mathbb{Z}G$ for $g \in G$.

1.2.1 Survey on G-CW-Complexes

In this subsection G can be any topological Hausdorff group. We recall some basic facts about G-CW-complexes. More informations and proofs can be found for instance in [495, Sections II.1 and II.2], [326, Sections 1 and 2], [365]. Throughout this book we will always work in the *category of compactly generated spaces* (see [482], [521, I.4]).

Definition 1.25 (G-CW-complex). *A G-CW-complex X is a G-space together with a G-invariant filtration*

$$\emptyset = X_{-1} \subset X_0 \subset X_1 \subset \ldots \subset X_n \subset \ldots \bigcup_{n \geq 0} X_n = X$$

such that X carries the colimit topology with respect to this filtration (i.e. a set $C \subset X$ is closed if and only if $C \cap X_n$ is closed in X_n for all $n \geq 0$) and X_n is obtained from X_{n-1} for each $n \geq 0$ by attaching equivariant n-dimensional cells, i.e. there exists a G-pushout

$$
\begin{array}{ccc}
\coprod_{i \in I_n} G/H_i \times S^{n-1} & \xrightarrow{\coprod_{i \in I_n} q_i} & X_{n-1} \\
\downarrow & & \downarrow \\
\coprod_{i \in I_n} G/H_i \times D^n & \xrightarrow{\coprod_{i \in I_n} Q_i} & X_n
\end{array}
$$

Provided that G is discrete, a G-CW-complex X is the same as a CW-complex X with G-action such that for any open cell $e \subset X$ and $g \in G$ with $ge \cap e \neq \emptyset$ left multiplication with g induces the identity on e.

The space X_n is called the *n-skeleton* of X. Notice that only the filtration by skeleta belongs to the G-CW-structure but not the G-pushouts, only their existence is required. An *equivariant open n-dimensional cell* is a G-component of $X_n - X_{n-1}$, i.e. the preimage of a path component of $G\backslash(X_n - X_{n-1})$. The closure of an equivariant open n-dimensional cell is called an *equivariant closed n-dimensional cell*. If one has chosen the G-pushouts in Definition 1.25, then the equivariant closed n-dimensional cells are just the G-subspaces $Q_i(G/H_i \times D^n)$.

If X is a G-CW-complex, then $G\backslash X$ is a CW-complex. If G is discrete or if G is a Lie group and $H \subset G$ is compact, then the H-fixed point set X^H inherits a WH-CW-complex structure. Here and in the sequel $NH = \{g \in G \mid gHg^{-1} = H\}$ is the *normalizer* of H in G and WH denotes the *Weyl group* NH/H of H in G. A G-space X is called *proper* if for each pair of points x and y in X there are open neighborhoods V_x of x and W_y of y in X such that the closure of the subset $\{g \in G \mid gV_x \cap W_y \neq \emptyset\}$ of G is compact. There are various slightly different definitions of proper G-space in the literature but for all these notions of properness a G-CW-complex X is proper if and only if all its isotropy groups are compact. In particular a free G-CW-complex is always proper. However, not every free G-space is proper. A G-space is called *cocompact* if $G\backslash X$ is compact. A G-CW-complex X is *finite* if X has only finitely many equivariant cells. A G-CW-complex is finite if and only if it is cocompact. A G-CW-complex X is *of finite type* if each n-skeleton is finite. It is called *of dimension $\leq n$* if $X = X_n$ and *finite dimensional* if it is of dimension $\leq n$ for some integer n. A free G-CW-complex X is the same as a regular covering $X \to Y$ of a CW-complex Y with G as group of deck transformations.

A G-map $f\colon X \to Y$ of G-CW-complexes is a G-homotopy equivalence if and only if for any subgroup $H \subset G$ which occurs as isotropy group of X or Y the induced map $f^H\colon X^H \to Y^H$ is a *weak homotopy equivalence*, i.e. induces a bijection on π_n for all base points and $n \geq 0$. A G-map of G-CW-complexes $f\colon X \to Y$ is *cellular* if $f(X_n) \subset Y_n$ holds for all $n \geq 0$. There is an equivariant version of the Cellular Approximation Theorem, namely, each G-map of G-CW-complexes is G-homotopic to a cellular one. If X is a G-CW-complex and Y is an H-CW-complex, then $X \times Y$ is a $G \times H$-CW-complex. Notice that one of the advantages of working in the category of compactly generated spaces is that this is true without any further assumptions on the topology of X or Y such as being locally compact.

Now suppose that G is discrete. The cellular $\mathbb{Z}G$-chain complex $C_*(X)$ of a G-CW-complex has as n-th chain group the singular homology $H_n(X_n, X_{n-1})$ and its n-th differential is the boundary homomorphism associated to the triple (X_n, X_{n-1}, X_{n-2}). If one has chosen a G-pushout as in Definition 1.25, then there is a preferred $\mathbb{Z}G$-isomorphism

$$\bigoplus_{i \in I_n} \mathbb{Z}[G/H_i] \xrightarrow{\cong} C_n(X). \tag{1.26}$$

If we choose a different G-pushout, we obtain another isomorphism, but the two differ only by the composition of an automorphism which permutes the summands appearing in the direct sum and an automorphism of the shape

$$\bigoplus_{i \in I_n} \mathbb{Z}[G/H_i] \xrightarrow{\bigoplus_{i \in I_n} \epsilon_i \cdot r_{g_i}} \bigoplus_{i \in I_n} \mathbb{Z}[G/H_i] \tag{1.27}$$

where $g_i \in G$, $\epsilon_i \in \{\pm 1\}$ and $\epsilon_i \cdot r_{g_i}$ sends gH_i to $\epsilon_i \cdot gg_iH_i$. In particular we obtain for a free G-CW-complex X a cellular $\mathbb{Z}G$-basis B_n for $C_n(X)$ which is unique up to permutation and multiplication with trivial units in $\mathbb{Z}G$, i.e. elements of the shape $\pm g \in \mathbb{Z}G$ for $g \in G$.

If G is a Lie group and M is a (smooth) proper G-manifold, then an equivariant smooth triangulation induces a G-CW-structure on M. For *equivariant smooth triangulations* we refer to [271], [272], [273]. There are obvious notions of pairs of G-CW-complexes (X, A) and of a relative G-CW-complex (X, A). In the latter case replace $X_{-1} = \emptyset$ by $X_{-1} = A$ for an arbitrary (compactly generated) G-space A.

There are some special G-CW-complexes. A *family of subgroups* $\mathcal{F}$ is a non-empty set of subgroups of G which is closed under conjugation and taking finite intersections. Examples are the trivial family $\mathcal{TR}$ consisting of the trivial subgroup and the family $\mathcal{FIN}$ of finite subgroups.

Definition 1.28. *A classifying space $E(G, \mathcal{F})$ for $\mathcal{F}$ is a G-CW-complex such that $E(G, \mathcal{F})^H$ is contractible for $H \in \mathcal{F}$ and all isotropy groups of $E(G, \mathcal{F})$ belong to $\mathcal{F}$.*

If $\mathcal{F}$ is the family $\mathcal{TR}$ or $\mathcal{FIN}$, we abbreviate

$$EG = E(G, \mathcal{TR});$$
$$BG = G \backslash EG;$$
$$\underline{E}G = E(G, \mathcal{FIN}).$$

The existence of $E(G, \mathcal{F})$ and proofs of their main property, namely, that for any G-CW-complex X whose isotropy groups belong to $\mathcal{F}$ there is up to G-homotopy precisely one G-map from X to $E(G, \mathcal{F})$ and thus that two such classifying spaces are G-homotopy equivalent, are presented in [493],[495, I.6]. A functorial "bar-type" construction is given in [128, section 7]. Notice that $G \to EG \to BG$ is a model for the universal G-principal bundle. If L is a Lie group with finitely many components and $K \subset L$ is a maximal compact subgroup, then for any discrete subgroup $G \subset L$ we get a model for $E(G; \mathcal{FIN})$ by L/K (see [1, Corollary 4.14]). The Rips complex of a word-hyperbolic group G is also a model for $E(G; \mathcal{FIN})$ (see [370]). The space $E(G; \mathcal{FIN})$ for a discrete group G is also called the *classifying space for proper G-spaces* since for any numerably proper G-space X there is up to G-homotopy precisely one G-map from X to $E(G, \mathcal{FIN})$. For more information about $E(G, \mathcal{FIN})$ we refer for instance to [27], [128, section 7], [495, section I.6], [336] and [342].

1.2.2 The Cellular L^2-Chain Complex

Definition 1.29. *Let X be a free G-CW-complex of finite type. Define its cellular L^2-chain complex and its cellular L^2-cochain complex by*

$$C_*^{(2)}(X) := l^2(G) \otimes_{\mathbb{Z}G} C_*(X);$$
$$C_{(2)}^*(X) := \hom_{\mathbb{Z}G}(C_*(X), l^2(G)),$$

where $C_*(X)$ is the cellular $\mathbb{Z}G$-chain complex.

If we fix a cellular basis for $C_n(X)$ we obtain explicit isomorphisms

$$C_n^{(2)}(X) \cong C_{(2)}^n(X) \cong \bigoplus_{i=1}^{k} l^2(G)$$

for some non-negative integer k and this induces the structure of a finitely generated Hilbert $\mathcal{N}(G)$-module on $C_n^{(2)}(X)$ and $C_{(2)}^n(X)$. One easily checks that the choice of cellular basis does not affect this structure. Hence the cellular L^2-chain complex and the cellular L^2-cochain complex come with preferred structures of finitely generated Hilbert $\mathcal{N}(G)$-chain complexes, provided that X is a free G-CW-complex of finite type.

Definition 1.30 (L^2-homology and L^2-Betti numbers). *Let X be a free G-CW-complex of finite type. Define its (reduced) p-th L^2-homology, (reduced) p-th L^2-cohomology and p-th L^2-Betti number by the corresponding notions of the cellular L^2-(co)chain complexes (see Definition 1.16)*

$$H_p^{(2)}(X;\mathcal{N}(G)) := H_p^{(2)}(C_*^{(2)}(X));$$
$$H_{(2)}^p(X;\mathcal{N}(G)) := H_{(2)}^p(C_{(2)}^*(X));$$
$$b_p^{(2)}(X;\mathcal{N}(G)) := b_p^{(2)}(C_*^{(2)}(X)).$$

If the group G and its action are clear from the context, we omit $\mathcal{N}(G)$ in the notation above. For instance for a connected CW-complex Y of finite type we denote by $\widetilde{Y}$ its universal covering, G is understood to be $\pi_1(Y)$ and we abbreviate $b_p^{(2)}(\widetilde{Y}) = b_p^{(2)}(\widetilde{Y};\mathcal{N}(\pi_1(Y)))$.

If M is a cocompact free proper G-manifold, define its p-th L^2-Betti number by the corresponding notion for any equivariant smooth triangulation.

Remark 1.31. The Hilbert $\mathcal{N}(G)$-modules $H_p^{(2)}(X)$ and $H_{(2)}^p(X)$ are isometrically G-isomorphic because of Lemma 1.18 since the p-th Laplace operator of the chain complex and the cochain complex are the same. In particular the cohomological and homological L^2-Betti numbers are the same.

Since we will prove in Theorem 1.35 (1) that $b_p^{(2)}(X)$ depends only on the G-homotopy type of X and two equivariant smooth triangulations of a cocompact free proper G-manifold M are G-homotopy equivalent, the definition of the p-th L^2-Betti number for M makes sense.

All these notions extend in the obvious way to pairs or more generally to relative free G-CW-complexes of finite type (X, A).

Example 1.32. If G is finite and X is a free G-CW-complex of finite type, then $b_p^{(2)}(X)$ is the classical p-th Betti number of X multiplied with $\frac{1}{|G|}$.

Let $f_*: C_* \to D_*$ be a chain map of Hilbert chain complexes or chain complexes of modules over a ring. Define $\mathrm{cyl}_*(f_*)$ to be the chain complex with n-th differential

$$C_{n-1} \oplus C_n \oplus D_n \xrightarrow{\begin{pmatrix} -c_{n-1} & 0 & 0 \\ -\mathrm{id} & c_n & 0 \\ f_{n-1} & 0 & d_n \end{pmatrix}} C_{n-2} \oplus C_{n-1} \oplus D_{n-1}.$$

Define $\mathrm{cone}_*(f_*)$ to be the quotient of $\mathrm{cyl}_*(f_*)$ by the obvious copy of C_*. Hence the n-th differential of $\mathrm{cone}_*(f_*)$ is

$$C_{n-1} \oplus D_n \xrightarrow{\begin{pmatrix} -c_{n-1} & 0 \\ f_{n-1} & d_n \end{pmatrix}} C_{n-2} \oplus D_{n-1}.$$

Given a chain complex C_*, define ΣC_* to be the quotient of $\mathrm{cone}_*(\mathrm{id}_{C_*})$ by the obvious copy of D_*, i.e. the chain complex with n-th differential

$$C_{n-1} \xrightarrow{-c_{n-1}} C_{n-2}.$$

Definition 1.33. *We call* $\mathrm{cyl}_*(f_*)$ *the* mapping cylinder, $\mathrm{cone}_*(f_*)$ *the* mapping cone *of the chain map* f_* *and* ΣC_* *the* suspension *of the chain complex* C_*.

Notice that with these definitions the cellular chain complex of a mapping cylinder of a cellular map $f\colon X \to Y$ of G-CW-complexes is the mapping cylinder of the induced chain map $C_*(f)\colon C_*(X) \to C_*(Y)$. Analogous statements holds for the mapping cone and the suspension (relative to the cone and suspension points). For the next lemma see also [115, section 5], [152], [169] and [331, Example 4.3].

Lemma 1.34. *(1) Let* C_* *be a free* $\mathbb{C}[\mathbb{Z}^n]$-*chain complex of finite type with some basis. Denote by* $\mathbb{C}[\mathbb{Z}^n]^{(0)}$ *the quotient field of the integral domain* $\mathbb{C}[\mathbb{Z}^n]$. *Then*

$$b_p^{(2)}\left(l^2(\mathbb{Z}^n) \otimes_{\mathbb{C}[\mathbb{Z}^n]} C_*\right) = \dim_{\mathbb{C}[\mathbb{Z}^n]^{(0)}}\left(\mathbb{C}[\mathbb{Z}^n]^{(0)} \otimes_{\mathbb{C}[\mathbb{Z}^n]} H_p(C_*)\right).$$

(2) Let X *be a free* $\mathbb{Z}^n$-CW-*complex of finite type. Then*

$$b_p^{(2)}(X) = \dim_{\mathbb{C}[\mathbb{Z}^n]^{(0)}}\left(\mathbb{C}[\mathbb{Z}^n]^{(0)} \otimes_{\mathbb{Z}[\mathbb{Z}^n]} H_p(X)\right).$$

Proof. (1) We abbreviate

$$C_*^{(0)} := \mathbb{C}[\mathbb{Z}^n]^{(0)} \otimes_{\mathbb{C}[\mathbb{Z}^n]} C_*;$$
$$C_*^{(2)} := l^2(\mathbb{Z}^n) \otimes_{\mathbb{C}[\mathbb{Z}^n]} C_*.$$

We have to show

$$\dim_{\mathbb{C}[\mathbb{Z}^n]^{(0)}}\left(\mathbb{C}[\mathbb{Z}^n]^{(0)} \otimes_{\mathbb{C}[\mathbb{Z}^n]} H_p(C_*)\right) = \dim_{\mathcal{N}(\mathbb{Z}^n)}\left(H_p^{(2)}(C_*^{(2)})\right).$$

In the sequel we can assume without loss of generality that C_* is finite dimensional. We first treat the case where $C_*^{(0)}$ has trivial homology. Then we can find a $\mathbb{C}[\mathbb{Z}^n]^{(0)}$-chain contraction γ'_* for $C_*^{(0)}$. Choose an element $u \in \mathbb{C}[\mathbb{Z}^n]$ and $\mathbb{C}[\mathbb{Z}^n]$-homomorphisms $\gamma_p\colon C_p \to C_{p+1}$ such that $u \neq 0$ and

$l_u \circ \gamma'_p = (\gamma_p)^{(0)}$ holds for all p where l_u is multiplication with u. Then γ_* is a chain homotopy of $\mathbb{C}[\mathbb{Z}^n]$-chain maps $l_u \simeq 0 \colon C_* \to C_*$. This induces a chain homotopy of chain maps of finite Hilbert $\mathcal{N}(\mathbb{Z}^n)$-chain complexes $l_u \simeq 0 \colon C_*^{(2)} \to C_*^{(2)}$. Hence multiplication with u induces the zero map on the L^2-homology of $C_*^{(2)}$. This is only possible if the L^2-homology is trivial and hence all L^2-Betti numbers of $C_*^{(2)}$ vanish because $l_u \colon l^2(\mathbb{Z}^n)^k \to l^2(\mathbb{Z}^n)^k$ is injective for any non-negative integer k (use the Fourier transform in Example 1.4).

Next we treat the general case. Put

$$b_p = \dim_{\mathbb{C}[\mathbb{Z}^n]^{(0)}} \left(\mathbb{C}[\mathbb{Z}^n]^{(0)} \otimes_{\mathbb{C}[\mathbb{Z}^n]} H_p(C_*) \right).$$

Notice that $\mathbb{C}[\mathbb{Z}^n]^{(0)}$ is flat over $\mathbb{C}[\mathbb{Z}^n]$. There is a $\mathbb{C}[\mathbb{Z}^n]^{(0)}$-isomorphism

$$\bigoplus_{i=1}^{b_p} \mathbb{C}[\mathbb{Z}^n]^{(0)} \to H_p(C_*^{(0)}) = \mathbb{C}[\mathbb{Z}^n]^{(0)} \otimes_{\mathbb{C}[\mathbb{Z}^n]} H_p(C_*).$$

By composing it with a map given by multiplication with a suitable element in $\mathbb{C}[\mathbb{Z}^n]$ one can construct a $\mathbb{C}[\mathbb{Z}^n]$-map

$$i_p \colon \bigoplus_{i=1}^{b_p} \mathbb{C}[\mathbb{Z}^n] \to H_p(C_*)$$

such that $(i_p)^{(0)}$ is a $\mathbb{C}[\mathbb{Z}^n]^{(0)}$-isomorphism. Let D_* be the finite free $\mathbb{C}[\mathbb{Z}^n]$-chain complex whose p-th chain module is $D_p = \bigoplus_{i=1}^{b_p} \mathbb{C}[\mathbb{Z}^n]$ and whose differentials are all trivial. Choose a $\mathbb{C}[\mathbb{Z}^n]$-chain map $j_* \colon D_* \to C_*$ which induces on the p-th homology the map i_p. Let $\mathrm{cone}_*(j_*)$ be its mapping cone. There is a canonical exact sequence of $\mathbb{C}[\mathbb{Z}^n]$-chain complexes $0 \to C_* \to \mathrm{cone}_*(j_*) \to \Sigma D_* \to 0$. Since it is split-exact in each dimension, it remains exact under the passage from C_* to $C_*^{(0)}$ or $C_*^{(2)}$. We conclude from the long exact homology sequence that $\mathrm{cone}_*(j_*)^{(0)}$ is acyclic since the boundary operator can be identified with the map induced by $j_*^{(0)}$. Hence the L^2-homology of $\mathrm{cone}_*(j_*)^{(2)}$ is trivial by the first step. We conclude from the long weakly exact L^2-homology sequence (see Theorem 1.21) and additivity of the von Neumann dimension (see Theorem 1.12 (2))

$$b_p^{(2)}(C_*^{(2)}) = b_p^{(2)}(D_*^{(2)}) = b_p = \dim_{\mathbb{C}[\mathbb{Z}^n]^{(0)}} \left(\mathbb{C}[\mathbb{Z}^n]^{(0)} \otimes_{\mathbb{C}[\mathbb{Z}^n]} H_p(C_*) \right).$$

(2) This follows from (1) applied to $C_*(X) \otimes_{\mathbb{Z}} \mathbb{C}$. $\qquad\square$

Notice that Example 1.32 and Lemma 1.34 (2) imply that the L^2-Betti numbers of a free G-complex of finite type are always rational if G is finite and integral if G is $\mathbb{Z}^n$. We will investigate the question, what values these can take for general G, in detail in Chapter 10.

1.2.3 Basic Properties of Cellular L^2-Betti Numbers

Theorem 1.35 (L^2-Betti numbers). *(1) Homotopy invariance*

Let $f\colon X \to Y$ be a G-map of free G-CW-complexes of finite type. If the map induced on homology with complex coefficients $H_p(f;\mathbb{C})\colon H_p(X;\mathbb{C}) \to H_p(Y;\mathbb{C})$ is bijective for $p \le d-1$ and surjective for $p = d$, then

$$b_p^{(2)}(X) = b_p^{(2)}(Y) \qquad \text{for } p < d;$$
$$b_d^{(2)}(X) \ge b_d^{(2)}(Y).$$

In particular we get for all $p \ge 0$ if f is a weak homotopy equivalence

$$b_p^{(2)}(X) = b_p^{(2)}(Y);$$

(2) Euler-Poincaré formula

Let X be a free finite G-CW-complex. Let $\chi(G\backslash X)$ be the Euler characteristic of the finite CW-complex $G\backslash X$, i.e.

$$\chi(G\backslash X) := \sum_{p\ge 0}(-1)^p \cdot \beta_p(G\backslash X) \qquad \in \mathbb{Z},$$

where $\beta_p(G\backslash X)$ is the number of p-cells of $G\backslash X$. Then

$$\chi(G\backslash X) = \sum_{p\ge 0}(-1)^p \cdot b_p^{(2)}(X);$$

(3) Poincaré duality

Let M be a cocompact free proper G-manifold of dimension n which is orientable. Then

$$b_p^{(2)}(M) = b_{n-p}^{(2)}(M, \partial M);$$

(4) Künneth formula

Let X be a free G-CW-complex of finite type and Y be a free H-CW-complex of finite type. Then $X \times Y$ is a free $G \times H$-CW-complex of finite type and we get for all $n \ge 0$

$$b_n^{(2)}(X \times Y) = \sum_{p+q=n} b_p^{(2)}(X) \cdot b_q^{(2)}(Y);$$

(5) Wedges

Let $X_1, X_2, \ldots, X_r$ be connected (pointed) CW-complexes of finite type and $X = \bigvee_{i=1}^r X_i$ be their wedge. Then

$$b_1^{(2)}(\widetilde{X}) - b_0^{(2)}(\widetilde{X}) = r - 1 + \sum_{j=1}^r \left(b_1^{(2)}(\widetilde{X_j}) - b_0^{(2)}(\widetilde{X_j}) \right);$$
$$b_p^{(2)}(\widetilde{X}) = \sum_{j=1}^r b_p^{(2)}(\widetilde{X_j}) \qquad \text{for } 2 \le p;$$

(6) Connected sums

Let M_1, M_2, ..., M_r be compact connected m-dimensional manifolds with $m \geq 3$. Let M be their connected sum $M_1 \# \ldots \# M_r$. Then

$$b_1^{(2)}(\widetilde{M}) - b_0^{(2)}(\widetilde{M}) = r - 1 + \sum_{j=1}^{r} \left(b_1^{(2)}(\widetilde{M_j}) - b_0^{(2)}(\widetilde{M_j}) \right) ;$$

$$b_p^{(2)}(\widetilde{M}) = \sum_{j=1}^{r} b_p^{(2)}(\widetilde{M_j}) \qquad \text{for } 2 \leq p \leq m - 2;$$

(7) Morse inequalities

Let X be a free G-CW-complex of finite type. Let $\beta_p(G \backslash X)$ be the number of p-cells in $G \backslash X$. Then we get for $n \geq 0$

$$\sum_{p=0}^{n} (-1)^{n-p} \cdot b_p^{(2)}(X) \leq \sum_{p=0}^{n} (-1)^{n-p} \cdot \beta_p(G \backslash X);$$

(8) Zero-th L^2-Betti number

Let X be a connected free G-CW-complex of finite type. Then

$$b_0^{(2)}(X) = \frac{1}{|G|},$$

where $\frac{1}{|G|}$ is to be understood to be zero if the order $|G|$ of G is infinite;

(9) Restriction

Let X be a free G-CW-complex of finite type and let $H \subset G$ be a subgroup of finite index $[G : H]$. Let $\mathrm{res}_G^H X$ be the H-space obtained from X by restricting the G-action to an H-action. This is a free H-CW-complex of finite type. Then we get for $p \geq 0$

$$[G : H] \cdot b_p^{(2)}(X; \mathcal{N}(G)) = b_p^{(2)}(\mathrm{res}_G^H X; \mathcal{N}(H));$$

(10) Induction

Let H be a subgroup of G and let X be a free H-CW-complex of finite type. Then $G \times_H X$ is a G-CW-complex of finite type and

$$b_p^{(2)}(G \times_H X; \mathcal{N}(G)) = b_p^{(2)}(X; \mathcal{N}(H)).$$

Proof. (1) We can assume that f is cellular. Let $C_*(f; \mathbb{C})$ be the map induced on the cellular $\mathbb{C}G$-chain complexes. The homology $H_p(\mathrm{cone}_*(C_*(f; \mathbb{C})))$ is trivial for $p \leq d$. Since $\mathrm{cone}_*(C_*(f; \mathbb{C}))$ is a finitely generated free $\mathbb{C}G$-chain complex, it is $\mathbb{C}G$-chain homotopy equivalent to a finitely generated free $\mathbb{C}G$-chain complex which is trivial in dimensions $\leq d$. Hence $\mathrm{cone}_*(C_*^{(2)}(f) : C_*^{(2)}(X) \to C_*^{(2)}(Y))$ is chain homotopy equivalent to a Hilbert $\mathcal{N}(G)$-chain complex which is trivial in dimensions $\leq d$ and in particular its L^2-homology is trivial in dimensions $\leq d$. Now the claim follows from the long weakly exact L^2-homology sequence associated to $C_*^{(2)}(f)$ (see Theorem 1.21) and additivity of the von Neumann dimension (see Theorem 1.12 (2)).

(2) Analogous to the classical situation where $\chi(X)$ is expressed in terms of (ordinary) Betti numbers the claim follows from the fact that the von Neumann dimension is additive (see Theorem 1.12 (2)).

(3) There is a subgroup $G_0 \subset G$ of finite index which acts orientation preserving on M. Since $b_p^{(2)}(\operatorname{res}_G^{G_0}(M); \mathcal{N}(G_0)) = [G : G_0] \cdot b_p^{(2)}(M; \mathcal{N}(G))$ follows from assertion (9) and similarly for $(M, \partial M)$, we can assume without loss of generality that $G = G_0$, i.e. $G \backslash M$ is orientable. Then the *Poincaré $\mathbb{Z}G$-chain homotopy equivalence* [510, Theorem 2.1 on page 23]

$$\cap [G \backslash M]\colon C^{n-*}(M, \partial M) \to C_*(M)$$

induces a homotopy equivalence of finitely generated Hilbert $\mathcal{N}(G)$-chain complexes $C_{(2)}^{n-*}(M, \partial M) \to C_*^{(2)}(M)$. Now the assertion follows because $\dim(H_p(C_{(2)}^{n-*}(M, \partial M))) = b_{n-p}^{(2)}(M; \partial M)$ holds by Lemma 1.18.

(4) The obvious isomorphism of $\mathbb{Z}[G \times H]$-chain complexes $C_*(X) \otimes_{\mathbb{Z}} C_*(Y) \to C_*(X \times Y)$ induces an isomorphism of Hilbert $\mathcal{N}(G \times H)$-chain complexes $C_*^{(2)}(X) \otimes C_*^{(2)}(Y) \to C_*^{(2)}(X \times Y)$. Now apply Theorem 1.12 (5) and Lemma 1.22. (see also [530, Corollary 2.36 on page 181]).

(5) We may assume without loss of generality that $r = 2$. We obtain an exact sequence of Hilbert $\mathcal{N}(\pi)$-chain complexes for $\pi = \pi_1(X_1 \vee X_2) = \pi_1(X_1) * \pi_1(X_2)$ and $i_k\colon \pi_1(X_k) \to \pi$ and $i_0\colon \{1\} \to \pi$ the obvious inclusions

$$0 \to (i_0)_* C_*^{(2)}(\{*\}) \to (i_1)_* C_*^{(2)}(\widetilde{X_1}) \oplus (i_2)_* C_*^{(2)}(\widetilde{X_2}) \to C_*^{(2)}(\widetilde{X_1 \vee X_2}) \to 0.$$

Notice that $(i_0)_* C_*^{(2)}(\{*\})$ is concentrated in dimension zero. Now the claim follows from Theorem 1.12 (2), Theorem 1.21 and Lemma 1.24 (2).

(6) We may assume without loss of generality that $r = 2$. The connected sum $M_1 \# M_2$ is obtained by glueing $M_1 \backslash \operatorname{int}(D^m)$ and $M_2 \backslash \operatorname{int}(D^m)$ together along ∂D^m. Since $\partial D^m \to D^m$ is $(m-1)$-connected the inclusion

$$M_1 \backslash \operatorname{int}(D^m) \cup_{\partial D^m} M_2 \backslash \operatorname{int}(D^m) \to M_1 \cup_{D^m} M_2$$

is *d-connected* for $d = m - 1$, i.e. induces an isomorphism on π_n for $n \leq d-1$ and an epimorphism on π_d for all base points. Obviously $M_1 \cup_{D^m} M_2$ is homotopy equivalent to the wedge $M_1 \vee M_2$. Because of assertion (1) it suffices to prove the claims for $M_1 \vee M_2$ which has already been done.

(7) analogous to (2).

(8) If G is finite, the claim follows from Example 1.32. It remains to show $b_0^{(2)}(X) = 0$ if G is infinite. Since $X \to G \backslash X$ is a covering with a path-connected total space, there is a group epimorphism $\pi_1(G \backslash X) \to G$. Since

$G\backslash X$ is of finite type, G is finitely generated. Let S be a finite set of generators. Let D_* be the $\mathbb{Z}G$-chain complex which is concentrated in dimension 0 and 1 and has as first differential

$$\bigoplus_{s \in S} \mathbb{Z}G \xrightarrow{\oplus_{s \in S} r_{s-1}} \mathbb{Z}G,$$

where r_{s-1} is right multiplication with $s - 1$. There is a $\mathbb{Z}G$-chain map $f_* \colon D_* \to C_*(X)$ which is 0-connected. Hence it suffices because of the argument in the proof of (1) to show $b_0^{(2)}(l^2(G) \otimes_{\mathbb{Z}G} D_*) = 0$. We have already seen that this is the same as the zero-th L^2-Betti number of the associated Hilbert $\mathcal{N}(G)$-cochain complex $\hom_{\mathbb{Z}G}(D_*, l^2(G))$ which is the dimension of $l^2(G)^G$. Since $l^2(G)^G = 0$ for infinite G, the claim follows.

(9) This follows from Lemma 1.9 (8).

(10) This follows from Lemma 1.24 (4) and $i_*(C_*^{(2)}(X)) = C_*^{(2)}(G \times_H X)$ for the inclusion $i \colon H \to G$. $\qquad\square$

Example 1.36. We give the values of the L^2-Betti numbers for the universal coverings of all compact connected 1- and 2-manifolds.

In dimension 1 there are only S^1 and the unit interval I. We get from Theorem 1.35 (9) or Lemma 1.34 (2) that $b_p^{(2)}(\widetilde{S^1}) = 0$ for all $p \geq 0$. As I is contractible, we have $b_0^{(2)}(\tilde{I}) = b_0(I) = 1$ and $b_p^{(2)}(\tilde{I}) = 0$ for $p \geq 1$.

A manifold is called *closed* if it is compact and has no boundary. Let F_g^d be the orientable closed surface of genus g with d embedded 2-disks removed. (As any nonorientable compact surface is finitely-covered by an orientable surface, Theorem 1.35 (9) shows that it is enough to handle the orientable case.) From Theorem 1.35 (2), (3) and (8) and the fact that a compact surface with boundary is homotopy equivalent to a bouquet of circles, we conclude

$$b_0^{(2)}(\widetilde{F_g^d}) = \begin{cases} 1 & \text{if } g = 0, d = 0, 1 \\ 0 & \text{otherwise} \end{cases};$$

$$b_1^{(2)}(\widetilde{F_g^d}) = \begin{cases} 0 & \text{if } g = 0, d = 0, 1 \\ d + 2 \cdot (g - 1) & \text{otherwise} \end{cases};$$

$$b_2^{(2)}(\widetilde{F_g^d}) = \begin{cases} 1 & \text{if } g = 0, d = 0 \\ 0 & \text{otherwise} \end{cases}.$$

Of course $b_p^{(2)}(\widetilde{F_g^d}) = 0$ for $p \geq 3$.

Example 1.37. Let $X \to Y$ be a finite covering with d sheets of connected CW-complexes of finite type. Then Theorem 1.35 (9) implies

$$b_p^{(2)}(\tilde{X}) = d \cdot b_p^{(2)}(\tilde{Y}).$$

In particular we get for a connected CW-complex X of finite type for which there is a selfcovering $X \to X$ with d sheets for some integer $d \geq 2$ that $b_p^{(2)}(\tilde{X}) = 0$ for all $p \geq 0$.

Example 1.38. The following examples show that in general there are hardly any relations between the ordinary Betti numbers $b_p(X)$ and the L^2-Betti numbers $b_p^{(2)}(\widetilde{X})$ for a connected CW-complex X of finite type.

Given a group G such that BG is of finite type, define its p-th L^2-Betti number and its p-th Betti number by

$$b_p^{(2)}(G) := b_p^{(2)}(EG; \mathcal{N}(G));$$
$$b_p(G) := b_p(BG).$$

We get from Theorem 1.35 (4), (5) and (9) for $r \geq 2$ and non-trivial groups $G_1, G_2, \ldots, G_r$ whose classifying spaces BG_i are of finite type

$$b_1^{(2)}(*_{i=1}^r G_i) = r - 1 + \sum_{i=1}^r \left(b_1^{(2)}(G_i) - \frac{1}{|G_i|} \right);$$

$$b_0^{(2)}(*_{i=1}^r G_i) = 0;$$

$$b_p^{(2)}(*_{i=1}^r G_i) = \sum_{i=1}^r b_p^{(2)}(G_i) \qquad \text{for } p \geq 2;$$

$$b_p(*_{i=1}^r G_i) = \sum_{i=1}^r b_p(G_i) \qquad \text{for } p \geq 1;$$

$$b_p^{(2)}(G_1 \times G_2) = \sum_{i=0}^p b_i^{(2)}(G_1) \cdot b_{p-i}^{(2)}(G_2);$$

$$b_p(G_1 \times G_2) = \sum_{i=0}^p b_i(G_1) \cdot b_{p-i}(G_2);$$

$$b_0^{(2)}(\mathbb{Z}/n) = \frac{1}{n};$$

$$b_p^{(2)}(\mathbb{Z}/n) = 0 \qquad \text{for } p \geq 1;$$

$$b_p(\mathbb{Z}/n) = 0 \qquad \text{for } p \geq 1.$$

From this one easily verifies for any integers $m \geq 0$, $n \geq 1$ and $i \geq 1$ that for the group

$$G_i(m, n) = \mathbb{Z}/n \times \left(*_{k=1}^{2m+2} \mathbb{Z}/2 \right) \times \left(\prod_{j=1}^{i-1} *_{l=1}^4 \mathbb{Z}/2 \right)$$

its classifying space $BG_i(m, n)$ is of finite type and

$$b_i^{(2)}(G_i(m, n)) = \frac{m}{n};$$

$$b_p^{(2)}(G_i(m, n)) = 0 \qquad \text{for } p \neq i;$$

$$b_p(G_i(m, n)) = 0 \qquad \text{for } p \geq 1.$$

Given an integer $l \geq 1$ and a sequence $r_1, r_2, \ldots, r_l$ of non-negative rational numbers, we can construct a group G such that BG is of finite type and

$$b_p^{(2)}(G) = \begin{cases} r_p & \text{for } 1 \leq p \leq l; \\ 0 & \text{for } l+1 \leq p; \end{cases}$$
$$b_p(G) = 0 \qquad \text{for } p \geq 1$$

holds as follows. For $l = 1$ we have already done this, so assume $l \geq 2$ in the sequel. Choose integers $n \geq 1$ and $k \geq l$ with $r_1 = \frac{k-2}{n}$. Fix for $i = 2, 3 \ldots, k$ integers $m_i \geq 0$ and $n_i \geq 1$ such that $\frac{m_i}{n \cdot n_i} = r_i$ holds for $1 \leq i \leq l$ and $m_i = 0$ holds for $i > l$. Put

$$G = \mathbb{Z}/n \times *_{i=2}^k G_i(m_i, n_i).$$

On the other hand we can construct for any sequence $n_1, n_2, \ldots$ of non-negative integers a CW-complex X of finite type such that $b_p(X) = n_p$ and $b_p^{(2)}(\widetilde{X}) = 0$ holds for $p \geq 1$, namely take

$$X = B(\mathbb{Z}/2 * \mathbb{Z}/2) \times \bigvee_{p=1}^{\infty} \left(\bigvee_{i=1}^{n_p} S^p \right).$$

Let $f \colon X \to X$ be a selfmap. Its *mapping torus* T_f is obtained from the cylinder $X \times [0, 1]$ by glueing the bottom of the cylinder $X \times [0, 1]$ to the top by the identification $(x, 1) = (f(x), 0)$. There is a canonical map $p \colon T_f \to S^1$ which sends (x, t) to $\exp(2\pi i t)$. It induces a canonical epimorphism $\pi_1(T_f) \to \mathbb{Z} = \pi_1(S^1)$ if X is path-connected.

The next theorem will be generalized in Theorem 6.63 and will play a key role in the proof of the vanishing of all L^2-Betti numbers of Thompson's group F in Subsection 7.1.2 (see also [329, Theorem 2.1], [331, Theorem 0.8]). It has been conjectured in [237, page 229] for an aspherical closed manifold M which fibers over the circle S^1.

Theorem 1.39 (Vanishing of L^2-Betti numbers of mapping tori).
Let $f \colon X \to X$ be a cellular selfmap of a connected CW-complex X of finite type and $\pi_1(T_f) \xrightarrow{\phi} G \xrightarrow{\psi} \mathbb{Z}$ be a factorization of the canonical epimorphism into epimorphisms ϕ and ψ. Let $\overline{T_f}$ be the covering of T_f associated to ϕ which is a free G-CW-complex of finite type. Then we get for all $p \geq 0$

$$b_p^{(2)}(\overline{T_f}) = 0.$$

Proof. Fix integers $p \geq 0$ and $n \geq 1$. Let $G_n \subset G$ be the preimage of the subgroup $n \cdot \mathbb{Z} \subset \mathbb{Z}$ under $\psi \colon G \to \mathbb{Z}$. This is a subgroup of index n. We get from Theorem 1.35 (9)

$$b_p^{(2)}(\overline{T_f}) = \frac{1}{n} \cdot b_p^{(2)}(\operatorname{res}_G^{G_n} \overline{T_f}).$$

There is a homotopy equivalence $h\colon T_{f^n} \to G_n\backslash\overline{T_f}$, where f^n is the n-fold composition of f. Let $\overline{T_{f^n}}$ be the G_n-space obtained by the pullback of $\overline{T_f} \to G_n\backslash\overline{T_f}$ with h. Then h induces a G_n-homotopy equivalence $\overline{h}\colon \overline{T_{f^n}} \to \operatorname{res}_G^{G_n} \overline{T_f}$. From homotopy invariance of the L^2-Betti numbers of Theorem 1.35 (1) we conclude

$$b_p^{(2)}(\operatorname{res}_G^{G_n} \overline{T_f}) = b_p^{(2)}(\overline{T_{f^n}}).$$

Let β_p be the number of p-cells in X. Then T_{f^n} has a CW-structure with $\beta_p + \beta_{p-1}$ cells of dimension p. Hence the von Neumann dimension of the cellular Hilbert $\mathcal{N}(G_n)$-chain module $C_p(\overline{T_{f^n}})$ is $\beta_p + \beta_{p-1}$. This implies by the additivity of the von Neumann dimension of Theorem 1.12 (2) that $b_p^{(2)}(\overline{T_{f^n}}) \le \beta_p + \beta_{p-1}$. We have shown

$$0 \le b_p^{(2)}(\overline{T_f}) \le \frac{\beta_p + \beta_{p-1}}{n}.$$

Since $\beta_p + \beta_{p-1}$ is independent of n, the claim follows by taking the limit for $n \to \infty$. □

A map $f\colon E \to B$ is called a *fibration* if it has the homotopy lifting property, i.e. for any homotopy $h\colon X \times [0,1] \to B$ and map $f\colon X \to E$ with $p \circ f = h_0$ there is a homotopy $H\colon X \times [0,1] \to E$ with $H_0 = f$ and $p \circ H = h$ (see [488, Definition 4.2 on page 53]). This is a weaker notion than the notion of a (locally trivial) fiber bundle with typical fiber F [269, chapter 4, section 5]. For instance the fiber F of a fibration is only well-defined up to homotopy equivalence, the one of a fiber bundle up to homeomorphism. The notion of a fibration is more general and more flexible than that of a fiber bundle. For example a group extension $1 \to \Delta \to \Gamma \to \pi \to 1$ such that $B\pi$ and $B\Delta$ are finite CW-complexes yields a fibration $B\Gamma \to B\pi$ with fiber $B\Delta$ such that $B\Gamma$ has the homotopy type of a finite CW-complex but in general one cannot expect the existence of a fiber bundle $B\Delta \to B\Gamma \to B\pi$ of finite CW-complexes.

If $f\colon F \to F$ is a homotopy equivalence, T_f is homotopy equivalent to the total space of a fibration over S^1 with fiber F. Conversely, the total space of such a fibration is homotopy equivalent to the mapping torus of the selfhomotopy equivalence of F given by the fiber transport with a generator of $\pi_1(S^1)$. Therefore L^2-Betti numbers are obstructions for a closed manifold to fiber over the circle S^1. This problem has been treated by Farrell [189]. The same idea of proof as in Theorem 1.39 yields Novikov-type inequalities for Morse 1-forms [185].

Theorem 1.40 (L^2-Betti numbers and S^1-actions). *Let X be a connected S^1-CW-complex of finite type, for instance a connected compact manifold with smooth S^1-action. Suppose that for one orbit S^1/H (and hence for all orbits) the inclusion into X induces a map on π_1 with infinite image. (In particular the S^1-action has no fixed points.) Let $\tilde{X}$ be the universal covering of X with the canonical $\pi_1(X)$-action. Then we get for all $p \ge 0$*

$$b_p^{(2)}(\widetilde{X}) = 0.$$

Proof. We show for any S^1-CW-complex Y of finite type together with an S^1-map $f\colon Y \to X$ that all the L^2-Betti numbers $b_p^{(2)}(f^*\widetilde{X}; \mathcal{N}(\pi))$ are trivial where $f^*\widetilde{X}$ is the pullback of the universal covering of X and $\pi = \pi_1(X)$. Since the p-th L^2-Betti number only depends on the $(p+1)$-skeleton, we can assume without loss of generality that Y is a finite S^1-CW-complex. We use induction over the dimension n of Y. The beginning $n = -1$ is trivial, the induction step from $n-1$ to $n \geq 0$ is done as follows.

Choose an equivariant S^1-pushout with $\dim(Z) = n-1$.

$$
\begin{array}{ccc}
\coprod_{i \in I} S^1/H_i \times S^{n-1} & \xrightarrow{\coprod_{i \in I} q_i} & Z \\
\downarrow & & \downarrow{\scriptstyle j} \\
\coprod_{i \in I} S^1/H_i \times D^n & \xrightarrow{\coprod_{i \in I} Q_i} & Y
\end{array}
$$

This induces a π-equivariant pushout

$$
\begin{array}{ccc}
\coprod_{i \in I} q_i^* j^* f^* \widetilde{X} & \longrightarrow & j^* f^* \widetilde{X} \\
\downarrow & & \downarrow \\
\coprod_{i \in I} Q_i^* f^* \widetilde{X} & \longrightarrow & f^* \widetilde{X}
\end{array}
$$

It induces a short exact sequence of finitely generated Hilbert $\mathcal{N}(\pi)$-chain complexes

$$0 \to C_*^{(2)}(j^* f^* \widetilde{X}) \to C_*^{(2)}(f^* \widetilde{X}) \to \bigoplus_{i \in I} C_*^{(2)}(Q_i^* f^* \widetilde{X}, q_i^* j^* f^* \widetilde{X}) \to 0.$$

Because of Theorem 1.21 and Theorem 1.12 (2) it suffices to prove

$$\dim_{\mathcal{N}(\pi)}\left(H_p^{(2)}(j^* f^* \widetilde{X}; \mathcal{N}(\pi))\right) = 0;$$

$$\dim_{\mathcal{N}(\pi)}\left(H_p^{(2)}(C_*(Q_i^* f^* \widetilde{X}, q_i^* j^* f^* \widetilde{X}))\right) = 0.$$

This follows for the first equation by the induction hypothesis. The pair $(Q_i^* f^* \widetilde{X}, q_i^* j^* f^* \widetilde{X})$ is π-homeomorphic to $\pi \times_{\mathbb{Z}} \widetilde{S^1} \times (D^n, S^{n-1})$ for an appropriate subgroup $\mathbb{Z} \subset \pi$ since S^1/H_i is homeomorphic to S^1 and the inclusion of an orbit into X induces an injection on the fundamental groups by assumption. Hence $C_*(Q_i^* f^* \widetilde{X}, q_i^* j^* f^* \widetilde{X})$ is $\mathbb{Z}\pi$-isomorphic to the n-fold suspension of $C_*(\pi \times_{\mathbb{Z}} \widetilde{S^1})$ for a subgroup $\mathbb{Z} \subset \pi$. We conclude from Theorem 1.35 (10)

$$\dim_{\mathcal{N}(\pi)}\left(H_p^{(2)}(C_*(Q_i^* f^* \widetilde{X}, q_i^* j^* f^* \widetilde{X}))\right) = b_{p-n}^{(2)}(\widetilde{S^1}).$$

We conclude $b_p^{(2)}(\widetilde{S^1}) = 0$ for $p \geq 0$ from Theorem 1.35 (9) or Lemma 1.34 (2).   □

Theorem 1.40 will be generalized in Theorem 6.65 and a more general version of Lemma 1.41 below will be proved in Lemma 6.66.

Lemma 1.41. *Let $F \to E \to B$ be a fibration of connected CW-complexes of finite type such that the inclusion induces an injection $\pi_1(F) \to \pi_1(E)$. Suppose that the L^2-Betti numbers $b_p^{(2)}(\widetilde{F})$ of the universal covering $\widetilde{F}$ (with the canonical $\pi_1(F)$-action) vanish for all $p \geq 0$. Then the same is true for the universal covering $\widetilde{E}$ (with the canonical $\pi_1(E)$-action) of E.*

1.2.4 L^2-Betti Numbers and Aspherical Spaces

A CW-complex X is *aspherical* if it is connected and its universal covering is contractible. This is equivalent to the condition that $\pi_n(X, x)$ is trivial for $n \neq 1$ and $x \in X$. The universal covering $\widetilde{X} \to X$ of an aspherical CW-complex X with fundamental group π is a model for the universal principal π-bundle $E\pi \to B\pi$. Given a group π, an aspherical CW-complex X with base point $x \in X$ together with an isomorphism $\pi_1(X, x) \xrightarrow{\cong} \pi$ is called an *Eilenberg-MacLane space of type* $(\pi, 1)$. Examples for aspherical CW-complexes are closed Riemannian manifolds with non-positive sectional curvature since their universal coverings are diffeomorphic to $\mathbb{R}^n$ by the Hadamard-Cartan Theorem [218, Theorem 3.87 on page 134].

Assertion (1) of the next Lemma 1.42 is a variation of [124, Lemma 5.1 on page 242] and assertion (2) appears in [232, page 95], where the result is attributed to unpublished work of Sullivan.

Lemma 1.42. *(1) Let X be a connected finite dimensional S^1-CW-complex with empty fixed point set X^{S^1}. Suppose that the universal covering $\widetilde{X}$ of X satisfies $H^*(\widetilde{X}; \mathbb{Q}) = H^*(\{*\}; \mathbb{Q})$. Then the inclusion of any orbit into X induces an injection on the fundamental group.*

(2) Let M be a connected closed oriented n-dimensional manifold with fundamental group π and fundamental class $[M] \in H_n(M; \mathbb{Z})$. Let $f: M \to B\pi$ be the classifying map of the universal covering $\widetilde{M} \to M$. Suppose that M carries a non-trivial (smooth) S^1-action with $M^{S^1} \neq \emptyset$. Then the image of the fundamental class in the homology of $B\pi$ with rational coefficients $f_*[M] \in H_n(B\pi; \mathbb{Q})$ is trivial.

Proof. (1) We first show for any fixed point free S^1-CW-complex Z that the canonical projection $p: ES^1 \times_{S^1} Z \to S^1 \backslash Z$ is a rational cohomology equivalence where $S^1 \to ES^1 \to BS^1$ is a model for the universal principal S^1-bundle. Recall that ES^1 is a free S^1-CW-complex which is (non-equivariantly) contractible. It suffices to prove the claim for each k-skeleton of Z because of the Five-Lemma and the short exact sequence due to Milnor [521, Theorem XIII.1.3 on page 605] for Z

$$0 \to \lim_{k \to \infty}^1 H^{n-1}(Z_k; \mathbb{Q}) \to H^n(Z; \mathbb{Q}) \to \lim_{k \to \infty} H^n(Z_k; \mathbb{Q}) \to 0$$

and the corresponding one for $ES^1 \times_{S^1} Z$. We use induction over k. The beginning $k = -1$ is trivial, the induction step from $k-1$ to $k \geq 0$ is done as follows. Suppose that the k-skeleton Z_k is obtained from the $(k-1)$-skeleton Z_{k-1} by attaching equivariant cells

$$\begin{array}{ccc}
\coprod_{i \in I} S^1/H_i \times S^{k-1} & \longrightarrow & Z_{k-1} \\
\downarrow & & \downarrow \\
\coprod_{i \in I} S^1/H_i \times D^k & \longrightarrow & Z_k
\end{array}$$

where $H_i \subset S^1$ is a finite subgroup for each $i \in I$. Then $ES^1 \times_{S^1} Z_k$ is the pushout

$$\begin{array}{ccc}
\coprod_{i \in I} ES^1 \times_{S^1} (S^1/H_i \times S^{k-1}) & \longrightarrow & ES^1 \times_{S^1} Z_{k-1} \\
\downarrow & & \downarrow \\
\coprod_{i \in I} ES^1 \times_{S^1} (S^1/H_i \times D^k) & \longrightarrow & ES^1 \times_{S^1} Z_k
\end{array}$$

and $S^1 \backslash Z_k$ is the pushout:

$$\begin{array}{ccc}
\coprod_{i \in I} S^{k-1} & \longrightarrow & S^1 \backslash Z_{k-1} \\
\downarrow & & \downarrow \\
\coprod_{i \in I} D^k & \longrightarrow & S^1 \backslash Z_k
\end{array}$$

The projections $ES^1 \times_{S^1} Y \to S^1 \backslash Y$ for $Y = Z_{k-1}, \coprod_{i \in I} S^1/H_i \times S^{k-1}$ and $\coprod_{i \in I} S^1/H_i \times D^k$ are rational cohomology equivalences by the induction hypothesis and because $BH_i \to \{*\}$ is one for each $i \in I$. By a Mayer-Vietoris argument the projection for $Y = Z_k$ is also a rational cohomology equivalence.

Let $\widetilde{X}$ be the universal covering of X and $\pi = \pi_1(X)$. There is an extension of Lie groups $1 \to \pi \to L \to S^1 \to 1$ and an action of L on $\widetilde{X}$ such that the L-action extends the π-action on $\widetilde{X}$ and covers the S^1-action on X. Moreover, there is an L-CW-structure on $\widetilde{X}$ such that the induced S^1-CW-structure on $X = \pi \backslash \widetilde{X}$ is the given one [326, Theorem 8.1 on page 138, Lemma 8.9 on page 140]. The basic idea of this construction is to define L as the group of pairs (f, g), where g is an element in S^1 and $f \colon \widetilde{X} \to \widetilde{X}$ is map which covers the map $l_g \colon X \to X$ given by multiplication with g.

The map ∂ in the exact sequence of homotopy groups associated to the regular π-covering $L \to S^1$

$$\{1\} \to \pi_1(L) \to \pi_1(S^1) \xrightarrow{\partial} \pi \to \pi_0(L) \to \{1\}$$

agrees with $\mathrm{ev}_* \colon \pi_1(S^1) \to \pi_1(X)$ induced by evaluating the S^1-action at the base point. We want to show that this map is injective. Suppose it is not true. Then $\pi_1(L)$ is an infinite cyclic group. As L is a 1-dimensional Lie group, its component of the identity is just S^1. Obviously each isotropy group under the L-action on $\widetilde{X}$ is finite as the S^1-action on X has only finite isotropy groups. Hence $\widetilde{X}$ carries a fixed point free S^1-action. The Gysin sequence [521, Theorem VII.5.12 on page 356] of $S^1 \to ES^1 \times \widetilde{X} \to ES^1 \times_{S^1} \widetilde{X}$ looks like:

$$\ldots \to H^p(ES^1 \times_{S^1} \widetilde{X}; \mathbb{Q}) \to H^{p+2}(ES^1 \times_{S^1} \widetilde{X}; \mathbb{Q}) \to$$
$$H^{p+2}(ES^1 \times \widetilde{X}; \mathbb{Q}) \to H^{p+1}(ES^1 \times_{S^1} \widetilde{X}; \mathbb{Q}) \to \ldots$$

We have $H^p(ES^1 \times \widetilde{X}; \mathbb{Q}) = H^p(\widetilde{X}; \mathbb{Q}) = 0$ for $p \neq 0$ by assumption. Hence we get $H^{2p}(ES^1 \times_{S^1} \widetilde{X}; \mathbb{Q}) \neq \{0\}$ for $p \geq 0$. As the projection $ES^1 \times_{S^1} \widetilde{X} \to S^1 \backslash \widetilde{X}$ induces an isomorphism on rational cohomology, $S^1 \backslash \widetilde{X}$ has non-trivial rational cohomology in all even dimensions. Since X has a finite dimensional S^1-CW-structure, $S^1 \backslash \widetilde{X}$ has a finite dimensional CW-structure, a contradiction. This proves assertion (1).

(2) Let $\rho \colon S^1 \times M \to M$ be the given S^1-operation and let $p \colon \widetilde{M} \to M$ be the universal covering. Choose $x \in M^{S^1}$ with a lift $\widetilde{x} \in \widetilde{M}$. Notice that the map induced by evaluation $\pi_1(S^1, e) \to \pi = \pi_1(M, x)$ is trivial. Hence there is precisely one map $\widetilde{\rho} \colon S^1 \times \widetilde{M} \to \widetilde{M}$ which satisfies $p \circ \widetilde{\rho} = \rho \circ (\mathrm{id}_{S^1} \times p)$ and $\widetilde{\rho}(e, \widetilde{x}) = \widetilde{x}$. One easily checks that $\widetilde{\rho}$ defines an S^1-action on $\widetilde{M}$ which commutes with the canonical $\pi = \pi_1(M, x)$-action on $\widetilde{M}$ with respect to $\widetilde{x}$ and covers the S^1-action on M. In particular $S^1 \backslash \widetilde{M}$ inherits a π-action and the projection $\widetilde{\mathrm{pr}} \colon \widetilde{M} \to S^1 \backslash \widetilde{M}$ is π-equivariant.

Next we want to show for $\widetilde{y} \in \widetilde{M}$ that the isotropy group under the π-action of $S^1 \widetilde{y} \in S^1 \backslash \widetilde{M}$ is finite. Let u be the projection to M of any path $\widetilde{u}$ joining $\widetilde{x}$ and $\widetilde{y}$ in $\widetilde{M}$. Then $w \cdot \widetilde{y}$ for $w \in \pi$ is by definition $\widetilde{u^- * w * u}(1)$ for any path $u^- * w * u$ with initial point $\widetilde{y}$ whose projection under p is the loop $u^- * w * u$ in M. Suppose that $w \cdot S^1 \widetilde{y} = S^1 \widetilde{y}$. Then we can find $z \in S^1$ with $z \cdot \widetilde{y} = w \cdot \widetilde{y}$. Let v be any path in S^1 from e to z. Put $y = p(\widetilde{y})$. Let $v\widetilde{y}$ and vy be the paths obtained from v and the S^1-actions. The paths $\widetilde{u^- * w * u}$ and $v\widetilde{y}$ are homotopic relative endpoints. This implies $z \in S^1_y$ for S^1_y the isotropy group of $y \in M$ under the S^1-action and that vy and $u^- * w * u$ are homotopic relative endpoints. Hence the isotropy group of $S^1 \widetilde{y}$ under the $\pi = \pi_1(M, x)$ action can be identified with the image of $\pi_1(S^1/S^1_y, eS^1_y) \to \pi_1(M, y) \xrightarrow{c_u} \pi = \pi_1(M, x)$ where c_u is given by conjugation with u. The composition of this map with the obvious map $\pi_1(S^1, e) \to \pi_1(S^1/S^1_y, eS^1_y)$ is trivial since the obvious map $\pi_1(S^1, e) \to \pi_1(M, x)$ is trivial because of $x \in M^{S^1}$. Hence all isotropy subgroups of the π-action on $S^1 \backslash \widetilde{M}$ are finite.

We have introduced $\underline{E}\pi = E(\pi; \mathcal{FIN})$ in Definition 1.28. Choose π-maps $\widetilde{f} \colon \widetilde{M} \to \underline{E}\pi$, $\widetilde{g} \colon E\pi \to \underline{E}\pi$ and $\widetilde{h} \colon S^1 \backslash \widetilde{M} \to \underline{E}\pi$. Then $\widetilde{g} \circ \widetilde{f}$ and $\widetilde{h} \circ \widetilde{\mathrm{pr}}$ are

π-homotopy equivalent. By passing to the quotients under the π-action we obtain an up to homotopy commutative diagram

$$
\begin{array}{ccc}
M & \xrightarrow{\ \text{pr}\ } & S^1 \backslash M \\
{\scriptstyle f}\downarrow & & \downarrow{\scriptstyle h} \\
B\pi & \xrightarrow{\ g\ } & \pi \backslash \underline{E}\pi
\end{array}
$$

where $f\colon M \to B\pi$ is a classifying map. Hence it suffices to show that $H_n(S^1 \backslash M; \mathbb{Z}) = 0$ holds and that $g_*\colon H_p(B\pi; \mathbb{Q}) \to H_p(\pi \backslash \underline{E}\pi; \mathbb{Q})$ is an isomorphism for each $p \geq 0$.

Since M is connected and closed and $M \neq M^{S^1}$ by assumption and each component C of M^{S^1} is a connected closed submanifold [64, Corollary VI.2.4 on page 308], the dimension of each component C is $\leq n - 1$. Since M is obtained from M^{S^1} by attaching cells of the type $S^1/H \times D^k$ for finite $H \subset S^1$ and $k \leq n - 1$, the dimension of the CW-complex $S^1 \backslash M$ is $\leq n - 1$ and hence $H_n(S^1 \backslash M; \mathbb{Z}) = 0$.

Hence it remains to prove that $H_*(g; \mathbb{Q})\colon H_*(B\pi; \mathbb{Q}) \to H_p(\pi \backslash \underline{E}\pi; \mathbb{Q})$ is bijective for all $p \geq 0$. Notice that $C_p(\underline{E}G) \otimes_{\mathbb{Z}} \mathbb{Q}$ is projective over $\mathbb{Q}G$ for each $p \geq 0$ since it is a direct sum of $\mathbb{Q}G$-modules of the shape $\mathbb{Q}[G/H]$ for finite $H \subset G$. The canonical map $EG \to \underline{E}G$ induces a homology equivalence of projective $\mathbb{Q}G$-chain complexes $C_*(EG) \otimes_{\mathbb{Z}} \mathbb{Q} \to C_*(\underline{E}G) \otimes_{\mathbb{Z}} \mathbb{Q}$. We conclude that this chain map is a $\mathbb{Q}G$-chain homotopy equivalence. Hence it induces a chain homotopy equivalence $C_*(EG) \otimes_{\mathbb{Z}G} \mathbb{Q} \to C_*(\underline{E}G) \otimes_{\mathbb{Z}G} \mathbb{Q}$ which induces on homology $H_*(g; \mathbb{Q})$. $\square$

Corollary 1.43. *Let M be an aspherical closed manifold with non-trivial S^1-action. Then the action has no fixed points and the inclusion of any orbit into X induces an injection on the fundamental groups. All L^2-Betti numbers $b_p^{(2)}(\widetilde{M})$ are trivial and $\chi(M) = 0$.*

Proof. This follows from Theorem 1.35 (2), Theorem 1.40 and Lemma 1.42 provided M is orientable. Recall that the *orientation covering* is the double covering $\overline{M} = S\Lambda^{\dim(M)} TM \to M$ and has the property that its total space $\overline{M}$ is orientable and it is trivial if and only if M is orientable. The non-orientable case then follows from Theorem 1.35 (9) since $\overline{M}$ is orientable and satisfies the same hypothesis as M. $\square$

We will show in Corollary 3.111 that under the conditions of Corollary 1.43 all Novikov-Shubin invariants of $\widetilde{M}$ are less or equal to 1 and that the L^2-torsion of $\widetilde{M}$ is trivial.

The next result will be a special case of a more general result, namely Theorem 7.2 (1) and (2)

Theorem 1.44 (L^2-Betti numbers and aspherical CW-complexes).
Let X be an aspherical CW-complex of finite type. Suppose that $\pi_1(X)$ contains an amenable infinite normal subgroup. Then $b_p^{(2)}(\widetilde{X}) = 0$ for all $p \geq 0$.

1.3 Analytic L^2-Betti Numbers

In this section we state the L^2-Hodge-de Rham Theorem 1.59 and give an analytic interpretation of the L^2-Betti numbers.

1.3.1 The Classical Hodge-de Rham Theorem

In this subsection we recall the classical Hodge-de Rham Theorem since we need it and the necessary notations later and it motivates the L^2-version.

Let M be a manifold without boundary (which is not necessarily compact). Denote by $\Omega^p(M)$ the *space of smooth p-forms* on M, i.e. the space of smooth sections of the bundle $\mathrm{Alt}^p(\mathbb{C} \otimes_{\mathbb{R}} TM)$ of alternating p-linear $\mathbb{C}$-forms associated to the complexification $\mathbb{C} \otimes_{\mathbb{R}} TM$ of the tangent bundle TM. Hence a p-form ω assigns to each $x \in M$ an alternating p-linear $\mathbb{C}$-form $\omega_x \colon \mathbb{C} \otimes_{\mathbb{R}} T_x M \times \ldots \times \mathbb{C} \otimes_{\mathbb{R}} T_x M \to \mathbb{C}$ on the complexified tangent space $\mathbb{C} \otimes_{\mathbb{R}} T_x M$. This is the same as an alternating p-linear $\mathbb{R}$-form $\omega_x \colon T_x M \times \ldots \times T_x M \to \mathrm{res}^{\mathbb{R}}_{\mathbb{C}} \mathbb{C}$. Notice that $\Omega^p(M)$ inherits the structure of a complex vector space. One can identify $\Omega^p(M)$ with the space of alternating p-forms on the $C^\infty(M)$-module $C^\infty(\mathbb{C} \otimes_{\mathbb{R}} TM)$, where $C^\infty(M) = \Omega^0(M)$ is the $\mathbb{C}$-algebra of smooth $\mathbb{C}$-valued functions on M and $C^\infty(\mathbb{C} \otimes_{\mathbb{R}} TM)$ is the $C^\infty(M)$-module of smooth sections of the complexified tangent bundle $\mathbb{C} \otimes_{\mathbb{R}} TM$. Denote by

$$\wedge \colon \Omega^p(M) \otimes \Omega^q(M) \to \Omega^{p+q}(M) \tag{1.45}$$

the $\wedge$-*product* of smooth forms. Recall that the $\wedge$-product is defined by the corresponding notion for finite dimensional vector spaces applied fiberwise. Denote by

$$d^p \colon \Omega^p(M) \to \Omega^{p+1}(M) \tag{1.46}$$

the p-*exterior differential*. Recall that the exterior differentials are uniquely determined by the following properties:

(1) d^p is $\mathbb{C}$-linear;
(2) $(d^0 f)(X) = Xf$ for $f \in \Omega^0(M) = C^\infty(M)$, X a vector field on M and Xf the derivative of f along X;
(3) $d^{p+q}(\omega \wedge \eta) = (d^p \omega) \wedge \eta + (-1)^p \omega \wedge d^q \eta$ for $\omega \in \Omega^p(M)$ and $\eta \in \Omega^q(M)$;
(4) $d^{p+1} \circ d^p = 0$.

The cohomology of the *de Rham cochain complex*

$$\ldots \xrightarrow{d^{p-2}} \Omega^{p-1}(M) \xrightarrow{d^{p-1}} \Omega^p(M) \xrightarrow{d^p} \Omega^{p+1}(M) \xrightarrow{d^{p+1}} \ldots$$

is the *de Rham cohomology* of M and denoted by $H^p_{\mathrm{dR}}(M)$.

Let $C^{\mathrm{sing}}_*(M)$ be the $\mathbb{Z}$-chain complex given by singular simplices $\Delta_p \to M$ which are continuous. Let $C^{\mathrm{sing},C^\infty}_*(M)$ be the $\mathbb{Z}$-chain complex given by singular simplices $\Delta_p \to M$ which are smooth. (Here Δ_p is the standard p-simplex and not the Laplace operator). We denote by $C^*_{\text{sing}}(M;\mathbb{C})$ and $C^*_{\text{sing},C^\infty}(M;\mathbb{C})$ the corresponding cochain complexes $\hom_{\mathbb{Z}}(C^{\text{sing}}_*(M),\mathbb{C})$ and $\hom_{\mathbb{Z}}(C^{\text{sing},C^\infty}_*(M),\mathbb{C})$. Let $H^*_{\text{sing}}(M;\mathbb{C}) := H^*(C^*_{\text{sing}}(M;\mathbb{C}))$ be the singular cohomology of M. The obvious inclusion induces a cochain map

$$i^* : C^*_{\text{sing}}(M;\mathbb{C}) \to C^*_{\text{sing},C^\infty}(M;\mathbb{C}).$$

There is a well-defined cochain map

$$I^* : \Omega^*(M) \to C^*_{\text{sing},C^\infty}(M;\mathbb{C})$$

which sends $\omega \in \Omega^p(M)$ to the cochain $I^p(\omega) : C^{\text{sing},C^\infty}_*(M) \to \mathbb{C}$ which maps $\sigma : \Delta_p \to M$ to $\int_{\Delta_p} \sigma^*\omega$. The proof of the de Rham Theorem 1.47 can be found for instance in [63, Section V.9.], [134], [156, Theorem 1.5 on page 11 and Theorem 2.4 on page 20], [357, Theorem A.31 on page 413].

Theorem 1.47 (De Rham Isomorphism Theorem). *Let M be a manifold. Then the cochain maps i^* and I^* above induce a natural isomorphisms for $p \geq 0$*

$$A^p : H^p_{\text{dR}}(M) \to H^p_{\text{sing}}(M;\mathbb{C}).$$

They are compatible with the multiplicative structures given by the $\wedge$-product and the $\cup$-product.

Now suppose that M comes with a Riemannian metric and an orientation. Let n be the dimension of M. Denote by

$$*^p : \Omega^p(M) \to \Omega^{n-p}(M) \tag{1.48}$$

the *Hodge star-operator* which is defined by the corresponding notion for oriented finite dimensional Hermitian vector spaces applied fiberwise. It is uniquely characterized by the property

$$\int_M \omega \wedge *^p\eta = \int_M \langle \omega_x, \eta_x \rangle_{\text{Alt}^p(T_xM)} \, d\text{vol}, \tag{1.49}$$

where ω and η are p-forms and $\langle \omega_x, \eta_x \rangle_{\text{Alt}^p(T_xM)}$ is the inner product on $\text{Alt}^p(T_xM)$ which is induced by the inner product on T_xM given by the Riemannian metric.

Define the *adjoint of the exterior differential*

$$\delta^p = (-1)^{np+n+1} *^{n-p+1} \circ d^{n-p} \circ *^p : \Omega^p(M) \to \Omega^{p-1}(M). \tag{1.50}$$

Notice that in the definition of δ^p the Hodge star-operator appears twice and the definition is local. Hence we can define δ^p without using an orientation of

M, only the Riemannian metric is needed. This is also true for the *Laplace operator* which is defined by

$$\Delta_p = d^{p-1} \circ \delta^p + \delta^{p+1} \circ d^p \colon \Omega^p(M) \to \Omega^p(M). \tag{1.51}$$

Let $\Omega^p_c(M) \subset \Omega^p(M)$ be the *space of smooth p-forms with compact support*. There is the following inner product and norm on it

$$\langle \omega, \eta \rangle_{L^2} := \int_M \omega \wedge *^p \eta \;=\; \int_M \langle \omega_x, \eta_x \rangle_{\mathrm{Alt}^p(T_x M)} \, d\mathrm{vol}; \tag{1.52}$$

$$\|\omega\|_{L^2} := \sqrt{\langle \omega, \omega \rangle_{L^2}}. \tag{1.53}$$

Notice that d^p and δ^p are formally adjoint in the sense that we have for $\omega \in \Omega^p_c(M)$ and $\eta \in \Omega^{p+1}_c(M)$

$$\langle d^p(\omega), \eta \rangle_{L^2} = \langle \omega, \delta^{p+1}(\eta) \rangle_{L^2}. \tag{1.54}$$

Let $L^2\Omega^p(M)$ be the Hilbert space completion of $\Omega^p_c(M)$. Define the *space of L^2-integrable harmonic smooth p-forms*

$$\mathcal{H}^p_{(2)}(M) := \{\omega \in \Omega^p(M) \mid \Delta_p(\omega) = 0, \int_M \omega \wedge *\omega < \infty\}. \tag{1.55}$$

Recall that a Riemannian manifold M is *complete* if each path component of M equipped with the metric induced by the Riemannian metric is a complete metric space. By the Hopf-Rinow Theorem the following statements are equivalent provided that M has no boundary: (1) M is complete, (2) the exponential map is defined for any point $x \in M$ everywhere on $T_x M$, (3) any geodesic of M can be extended to a geodesic defined on $\mathbb{R}$ (see [218, page 94 and 95]). Completeness enters in a crucial way, namely, it will allow us to integrate by parts [217].

Lemma 1.56. *Let M be a complete Riemannian manifold. Let $\omega \in \Omega^p(M)$ and $\eta \in \Omega^{p+1}(M)$ be smooth forms such that ω, $d^p\omega$, η and $\delta^{p+1}\eta$ are square-integrable. Then*

$$\langle d^p\omega, \eta \rangle_{L^2} \;=\; \langle \omega, \delta^{p+1}\eta \rangle_{L^2}.$$

Proof. Completeness ensures the existence of a sequence $f_n \colon M \to [0,1]$ of smooth functions with compact support such that M is the union of the compact sets $\{x \in M \mid f_n(x) = 1\}$ and $\|df_n\|_\infty := \sup\{\|(df_n)_x\|_x \mid x \in M\} < \frac{1}{n}$ holds. With the help of the sequence $(f_n)_{n \geq 1}$ one can reduce the claim to the easy case, where ω and η have compact support. $\square$

Theorem 1.57 (Hodge-de Rham Theorem). *Let M be a complete Riemannian manifold without boundary. Then we obtain an orthogonal decomposition, the so called* Hodge-de Rham decomposition

$$L^2\Omega^p(M) = \mathcal{H}^p_{(2)}(M) \oplus \mathrm{clos}(d^{p-1}(\Omega^{p-1}_c(M))) \oplus \mathrm{clos}(\delta^{p+1}(\Omega^{p+1}_c(M))).$$

1.3.2 Analytic Definition of L^2-Betti Numbers

Suppose for a moment that M is a Riemannian manifold which is closed and that we require no group action. Then in view of Theorem 1.57 the space $\mathcal{H}^p(M) = \{\omega \in \Omega^p(M) \mid \Delta_p(\omega) = 0\}$ is isomorphic to the singular cohomology $H^p_{\text{sing}}(M; \mathbb{C})$. In particular $\mathcal{H}^p(M)$ is a finite dimensional $\mathbb{C}$-vector space and we can define the *analytic p-th Betti number* by its dimension. Of course the analytic p-th Betti number agrees with the classical p-th Betti number which is defined by the dimension of $H^p_{\text{sing}}(M; \mathbb{C})$. The analytic p-th Betti number can be interpreted in terms of the heat kernel $e^{-t\Delta_p}(x, y)$, namely by the following expression [220, 1.6.52 on page 56]

$$b_p(M) = \lim_{t \to \infty} \int_M \text{tr}_{\mathbb{C}}(e^{-t\Delta_p}(x, x)) \, d\text{vol}. \tag{1.58}$$

Here (and elsewhere) $e^{-t\Delta_p}(x, y)$ denotes the *heat kernel*. This is a smooth section of the bundle $\hom(p_1^* \text{Alt}^p(TM), p_2^* \text{Alt}^p(TM))$ over $M \times M$ for $p_k \colon M \times M \to M$ the projection to the k-th factor. It is uniquely characterized by the property that for $\omega \in L^2\Omega^p(M)$ and $e^{-t\Delta_p}$ the operator obtained from Δ_p by spectral calculus (see (1.64))

$$e^{-t\Delta_p}(\omega)(x) = \int_M e^{-t\Delta_p}(x, y)(\omega_y) \, d\text{vol}_y$$

holds. The real number $\text{tr}_{\mathbb{C}}(e^{-t\Delta_p}(x, x))$ is the trace of the endomorphism of finite dimensional vector spaces $e^{-t\Delta_p}(x, x) \colon \text{Alt}^p(T_xM) \to \text{Alt}^p(T_xM)$.

We want to generalize these results and notions to the L^2-setting. From now on M is a cocompact free proper G-manifold without boundary and with G-invariant Riemannian metric. Recall that an *equivariant smooth triangulation* K of M consists of a simplicial complex K with simplicial G-action such that for each open simplex σ and $g \in G$ with $g\sigma \cap \sigma \neq \emptyset$ left multiplication with g induces the identity on σ, together with a G-homeomorphism $f \colon |K| \xrightarrow{\cong} M$ such that f restricted to each closed simplex is a smooth immersion of the simplex into M. This is the same as a lift of a smooth triangulation of $G\backslash M$ to M. Notice that K is a free cocompact G-CW-complex. In the sequel we will not distinguish between K and its realization $|K|$.

The key result, which is due to Dodziuk [145] and whose proof we will present in Section 1.4, is

Theorem 1.59 (L^2-Hodge-de Rham Theorem). *Let M be a cocompact free proper G-manifold with G-invariant Riemannian metric and let K be an equivariant smooth triangulation of M. Suppose that M has no boundary. Let $\mathcal{H}^p_{(2)}(M)$ be the space of L^2-integrable harmonic smooth p-forms on M introduced in (1.55). Then integration defines an isomorphism of finitely generated Hilbert $\mathcal{N}(G)$-modules*

$$\mathcal{H}^p_{(2)}(M) \xrightarrow{\cong} H^p_{(2)}(K).$$

This allows us to define the *analytic p-th L^2-Betti number* by the von Neumann dimension of the finitely generated Hilbert $\mathcal{N}(G)$-module $\mathcal{H}^p_{(2)}(M)$. Obviously this is the same as the cellular p-th L^2-Betti number. Moreover, this can be interpreted in terms of the heat kernel by the following expression [9, Proposition 4.16 on page 63] (or Theorem 3.136 (1))

$$b_p^{(2)}(M) = \lim_{t\to\infty} \int_{\mathcal{F}} \mathrm{tr}_{\mathbb{C}}(e^{-t\Delta_p}(x,x))\, d\mathrm{vol}. \tag{1.60}$$

Here $\mathcal{F}$ is a *fundamental domain* for the G-action, i.e. an open subset $\mathcal{F} \subset M$ such that M is the union $\bigcup_{g\in G} g \cdot \mathrm{clos}(\mathcal{F})$ and $g\mathcal{F} \cap \mathcal{F} \neq \emptyset \Rightarrow g = 1$ and the topological boundary of $\mathcal{F}$ is a set of measure zero. If one fixes a triangulation of $G\backslash M$ and fixes for each open simplex σ of dimension $n = \dim(M)$ a lift $\overline{\sigma} \subset M$, then the union of the lifts $\overline{\sigma}$ for all n-simplices σ in $G\backslash M$ is a fundamental domain.

Recall that a manifold is called *hyperbolic* if it is equipped with a Riemannian metric whose sectional curvature is constant -1. This is equivalent to the statement that M comes with a Riemannian metric such that the universal covering $\widetilde{M}$ with the induced Riemannian metric is isometrically diffeomorphic to the *(real) hyperbolic space* $\mathbb{H}^n$ for $n = \dim(M)$ [218, Theorem 3.82 on page 131]. Recall that the *Poincaré model* for $\mathbb{H}^n$ is the open unit disk $\{x \in \mathbb{R}^n \mid |x| < 1\}$ with the Riemannian metric which is given at a point $x \in \mathbb{R}^n, |x| < 1$ by

$$T_x M \times T_x M \to \mathbb{R}, \quad (v,w) \mapsto \frac{4}{(1-|x|^2)^2} \cdot \langle v, w \rangle_{\mathrm{Eucl}}, \tag{1.61}$$

where $\langle v, w \rangle_{\mathrm{Eucl}}$ is the standard Euclidean metric on $T_x\mathbb{R}^n = \mathbb{R}^n$.

Theorem 1.62. *Let M be a hyperbolic closed Riemannian manifold of dimension n. Then*

$$b_p^{(2)}(\widetilde{M}) \begin{cases} = 0 & \text{if } 2p \neq n \\ > 0 & \text{if } 2p = n \end{cases}.$$

If n is even, then

$$(-1)^{n/2} \cdot \chi(M) > 0.$$

Proof. The statement about the Euler characteristic for M follows from the statement about the L^2-Betti numbers from Theorem 1.35 (2). Since the von Neumann dimension is faithful by Theorem 1.12 (1) it suffices to show, because of Theorem 1.59, that

$$\mathcal{H}^p_{(2)}(\mathbb{H}^n) \begin{cases} = 0 & \text{if } 2p \neq n \\ \neq 0 & \text{if } 2p = n \end{cases}.$$

This is done in [146] (even for all rotationally symmetric Riemannian manifolds). The idea of the proof is summarized as follows.

An element $\omega \in \mathcal{H}^p_{(2)}(\mathbb{H}^n)$ is written (outside 0) in polar coordinates

$$\omega = a(r, \theta) \wedge dr + b(r, \theta) \wedge d\theta$$

for $(p-1)$-forms $a(r, \theta)$ and $b(r, \theta)$ on S^{n-1} in the variable θ which depend on a parameter $r \in (0, \infty)$. Then the conditions $d^p\omega = 0$, $\delta^p(\omega) = 0$ and $\int_{\mathbb{H}^n} \omega \wedge *\omega < \infty$ are explicitly rewritten in terms of a and b and the function $f(r)$ for which the Riemannian metric with respect to the polar coordinates looks like $ds^2 = dr^2 + f(r)^2 d\theta^2$. The function f satisfies $f(0) = 0$, $f'(0) = 1$, $f(r) > 0$ for $r > 0$, and $\int_1^\infty f(r)^{-1} dr < \infty$. Then a further calculation shows that $\mathcal{H}^p_{(2)}(\mathbb{H}^n) \neq 0$ implies

$$\int_1^\infty f^{n-2p-1}(r) \, dr < \infty.$$

Since the Hodge star operator yields an isomorphism $\mathcal{H}^{n-p}_{(2)}(\mathbb{H}^n) = \mathcal{H}^p_{(2)}(\mathbb{H}^n)$, $\mathcal{H}^p_{(2)}(\mathbb{H}^n) \neq 0$ implies that both $\int_1^\infty f^{n-2p-1}(r) dr$ and $\int_1^\infty f^{-n+2p-1}(r) dr$ are finite. If $2p = n$, these two integrals are equal and finite. However, if $2p \neq n$, then the exponents $(n - 2p - 1)$ and $(-n + 2p - 1)$ have different signs and one of the integrals has to diverge. This shows $\mathcal{H}^p_{(2)}(\mathbb{H}^n) = 0$ for $2p \neq n$. One can show explicitly that $\mathcal{H}^p_{(2)}(\mathbb{H}^n)$ is an infinite dimensional vector space if $2p = n$. $\square$

We will extend Theorem 1.62 in Theorem 5.12 (1), where the computation of the L^2-Betti numbers of the universal covering $\widetilde{M}$ of a closed Riemannian manifold M is given provided that $\widetilde{M}$ is a symmetric space of non-compact type.

1.4 Comparison of Analytic and Cellular L^2-Betti Numbers

In this section we give the proof of the L^2-Hodge-de Rham Theorem 1.59.

1.4.1 Survey on Unbounded Operators and Spectral Families

We have to explain some facts about unbounded densely defined operators on Hilbert spaces and their domains, about spectral families and about the Spectral Theorem. For a general discussion we refer for instance to [434, Section VIII.1 - 3], [421, Chapter 5].

Let H be a Hilbert space and $T \colon \text{dom}(T) \to H$ be a (not necessarily bounded) linear operator defined on a dense linear subspace $\text{dom}(T)$ which is called *(initial) domain*. We call T *closed* if its graph $\text{gr}(T) := \{(u, T(u)) \mid u \in \text{dom}(T)\} \subset H \times H$ is closed. We say that $S \colon \text{dom}(S) \to H$ is an *extension* of T and write $T \subset S$ if $\text{dom}(T) \subset \text{dom}(S)$ and $S(u) = T(u)$

holds for all $u \in \mathrm{dom}(T)$. We write $T = S$ if $\mathrm{dom}(T) = \mathrm{dom}(S)$ and $S(u) = T(u)$ holds for all $u \in \mathrm{dom}(T)$. We call T *closable* if and only if T has a closed extension. Since the intersection of an arbitrary family of closed sets is closed again, a closable unbounded densely defined operator T has a unique *minimal closure*, also called *minimal closed extension* , i.e. a closed operator $T_{\min} \colon \mathrm{dom}(T_{\min}) \to H$ with $T \subset T_{\min}$ such that $T_{\min} \subset S$ holds for any closed extension S of T. Explicitly $\mathrm{dom}(T_{\min})$ consists of elements $u \in H$ for which there exists a sequence $(u_n)_{n \geq 0}$ in $\mathrm{dom}(T)$ and an element v in H satisfying $\lim_{n \to \infty} u_n = u$ and $\lim_{n \to \infty} T(u_n) = v$. Then v is uniquely determined by this property and we put $T_{\min}(u) = v$. Equivalently, $\mathrm{dom}(T_{\min})$ is the Hilbert space completion of $\mathrm{dom}(T)$ with respect to the inner product

$$\langle u, v \rangle_{\mathrm{gr}} := \langle u, v \rangle_H + \langle T(u), T(v) \rangle_H. \tag{1.63}$$

If not stated otherwise we always use the minimal closed extension as the closed extension of a closable unbounded densely defined linear operator.

The *adjoint* of T is the operator $T^* \colon \mathrm{dom}(T^*) \to H$ whose domain consists of elements $v \in H$ for which there is an element u in H such that $\langle u', u \rangle = \langle T(u'), v \rangle$ holds for all $u' \in \mathrm{dom}(T)$. Then u is uniquely determined by this property and we put $T^*(v) = u$. Notice that T^* may not have a dense domain in general. Its domain is dense if and only if T is closable. If T is closable, then $T_{\min}^* = T^*$ and $T_{\min} = (T^*)^*$. We call T *symmetric* if $T \subset T^*$ and *selfadjoint* if $T = T^*$. Any selfadjoint operator is necessarily closed and symmetric. A bounded operator $T \colon H \to H$ is always closed and is selfadjoint if and only if it is symmetric. We call T *essentially selfadjoint* if $T_{\min}$ is selfadjoint.

A densely defined unbounded operator $T \colon \mathrm{dom}(f) \subset H \to H$ for a Hilbert space H is called *positive* if T is selfadjoint and the real number $\langle T(v), v \rangle$ is ≥ 0 for all $v \in \mathrm{dom}(T)$. Notice that the Polar Decomposition $T = US$ also exists for densely defined unbounded operators $T \colon \mathrm{dom}(T) \subset H_1 \to H_2$ [434, Theorem VIII.32 on page 297]. Here U and S are uniquely determined by the properties that U is a *partial isometry*, i.e. U^*U is a projection, S is a positive selfadjoint operator and $\ker(U) = \ker(T) = \ker(S)$. The operator S is constructed by $\sqrt{S^*S}$ in the sense of functional calculus as explained below. We have $U^*US = S$, $U^*T = S$ and $U^*UT = T$.

Let H be a Hilbert space and $T \colon \mathrm{dom}(T) \subset H \to H$ be a densely defined closed operator with domain $\mathrm{dom}(T)$. Define its *resolvent set* $\mathrm{resolv}(T)$ as the set of complex numbers $\lambda \in \mathbb{C}$ for which there is a bounded operator $S \colon H \to H$ whose image is $\mathrm{dom}(T)$ and which satisfies $(T - \lambda \cdot \mathrm{id}) \circ S = \mathrm{id}_H$ and $S \circ (T - \lambda \cdot \mathrm{id}) = \mathrm{id}_{\mathrm{dom}(T)}$. Define the *spectrum* $\mathrm{spec}(T)$ to be the complement of $\mathrm{resolv}(T) \subset \mathbb{C}$. The spectrum of T is always a closed subset of $\mathbb{C}$. If T is selfadjoint, then $\mathrm{spec}(T) \subset \mathbb{R}$. If T is positive, then $\mathrm{spec}(T) \subset [0, \infty)$. Notice the elementary facts that $0 \notin \mathrm{spec}(T)$ implies injectivity of T, but that there are everywhere defined bounded closed injective positive operators T with $0 \in \mathrm{spec}(T)$.

We call a family $\{E_\lambda \mid \lambda \in \mathbb{R}\}$ of orthogonal projections $E_\lambda \colon H \to H$ a *spectral family* if it satisfies for $x \in H$ and $\lambda, \mu \in \mathbb{R}$

$$\lim_{\lambda \to -\infty} E_\lambda(x) = 0;$$

$$\lim_{\lambda \to \infty} E_\lambda(x) = x;$$

$$\lim_{\lambda \to \mu+} E_\lambda(x) = E_\mu(x);$$

$$E_\lambda E_\mu = E_\mu E_\lambda = E_{\min\{\lambda,\mu\}}.$$

Denote by $d\langle u, E_\lambda(u) \rangle$ the Borel measure on $\mathbb{R}$ which assigns to the half open interval $(\lambda, \mu]$ the measure $\langle u, E_\mu(u) \rangle - \langle u, E_\lambda(u) \rangle$. If $g \colon \mathbb{R} \to \mathbb{R}$ is a Borel function, we obtain a selfadjoint operator on H

$$\int_{-\infty}^{\infty} g(\lambda) \, dE_\lambda \tag{1.64}$$

with dense domain

$$\operatorname{dom} \left(\int_{-\infty}^{\infty} g(\lambda) dE_\lambda \right) := \left\{ u \in H \ \bigg| \ \int_{-\infty}^{\infty} g(\lambda)^2 \, d\langle u, E_\lambda(u) \rangle < \infty \right\}.$$

It is uniquely determined by the property that for all $u \in \operatorname{dom} \left(\int_{-\infty}^{\infty} g(\lambda) dE_\lambda \right)$

$$\left\langle u, \left(\int_{-\infty}^{\infty} g(\lambda) dE_\lambda \right)(u) \right\rangle = \int_{-\infty}^{\infty} g(\lambda) d\langle u, E_\lambda(u) \rangle. \tag{1.65}$$

If g is a complex valued Borel function on $\mathbb{R}$, one defines $\int_{-\infty}^{\infty} g(\lambda) dE_\lambda$ using the Polar Decomposition $g = \Re(g) + i \cdot \Im(g)$.

Any selfadjoint operator $T \colon \operatorname{dom}(T) \to H$ determines a spectral family $\{E_\lambda^T \mid \lambda \in \mathbb{R}\}$. It is uniquely determined by the property

$$T = \int_{-\infty}^{\infty} \lambda \, dE_\lambda^T. \tag{1.66}$$

We often abbreviate for a Borel function $g \colon \mathbb{R} \to \mathbb{R}$

$$g(T) := \int_{-\infty}^{\infty} g(\lambda) \, dE_\lambda^T. \tag{1.67}$$

We have $E_\lambda^T = \chi_{(-\infty,\lambda]}(T) = \int_{-\infty}^{\infty} \chi_{(-\infty,\lambda]} dE_\lambda^T$ for $\chi_{(-\infty,\lambda]}$ the characteristic function of $(-\infty, \lambda]$.

Definition 1.68 (Spectral family). *We call the spectral family associated to a selfadjoint operator T by (1.66) the* spectral family associated to T. *The spectral family associated to an essentially selfadjoint operator T is the spectral family associated to $T_{\min}$.*

Notation 1.69. *Given two unbounded operators $S, T\colon H_1 \to H_2$ with domains $\mathrm{dom}(S)$ and $\mathrm{dom}(T)$ and complex numbers λ, μ, define the unbounded operator $\lambda \cdot S + \mu \cdot T\colon H_1 \to H_2$ with domain $\mathrm{dom}(\lambda \cdot S + \mu \cdot T) = \mathrm{dom}(S) \cap \mathrm{dom}(T)$ by $(\lambda \cdot S + \mu \cdot T)(x) = \lambda \cdot S(x) + \mu \cdot T(x)$ for $x \in \mathrm{dom}(S) \cap \mathrm{dom}(T)$.*

Given two unbounded operators $S\colon \mathrm{dom}(S) \subset H_1 \to H_2$ and $T\colon \mathrm{dom}(T) \subset H_2 \to H_3$, define the unbounded operator $T \circ S$ with domain $\mathrm{dom}(T \circ S) = S^{-1}(\mathrm{dom}(T))$ by $T \circ S(x) := T(S(x))$ for $x \in S^{-1}(\mathrm{dom}(T))$.

Given a selfadjoint operator T the functional calculus $g \mapsto g(T)$ is an essential homomorphism of $\mathbb{C}$-algebras in the following sense. Given Borel functions g_1 and g_2 and complex numbers λ_1 and λ_2, the minimal closure of $\lambda_1 \cdot g_1(T) + \lambda_2 \cdot g_2(T)$ is $(\lambda_1 \cdot g_1 + \lambda_2 \cdot g_2)(T)$ and the minimal closure of $g_2(T) \circ g_1(T)$ is $(g_1 \cdot g_2)(T)$, where $(\lambda_1 \cdot g_1 + \lambda_2 \cdot g_2)$ and $(g_1 \cdot g_2)$ are the obvious Borel functions given by pointwise addition and multiplication.

Let E and F be Hermitian vector bundles over a complete Riemannian manifold without boundary. Let $D\colon C_c^\infty(E) \to C_c^\infty(F)$ be an elliptic differential operator where $C_c^\infty(E)$ is the space of smooth sections with compact support. Our main examples are $d^p\colon \Omega_c^p(M) \to \Omega_c^{p+1}(M)$, $\delta^p\colon \Omega_c^p(M) \to \Omega_c^{p-1}(M)$ and $\Delta_p\colon \Omega_c^p(M) \to \Omega_c^p(M)$. Notice that there is a *formally adjoint operator* $D^t\colon C_c^\infty(F) \to C_c^\infty(E)$ which is uniquely determined by the property that $\langle D(u), v \rangle_{L^2} = \langle u, D^t(v) \rangle_{L^2}$ holds for all $u, v \in C_c^\infty(E)$. It is again an elliptic differential operator. For instance, δ^{p+1} is the formal adjoint of d^p. The minimal closure $D_{\min}$ of $D\colon C_c^\infty(E) \to L^2C^\infty(F)$ has been defined above where $L^2C^\infty(F)$ is the Hilbert space completion of $C_c^\infty(F)$. The *maximal closure* $D_{\max}$ of D is defined by the adjoint of $(D^t)_{\min}$. Indeed, for any closure $\overline{D}$ of $D\colon C_c^\infty(E) \to L^2C^\infty(F)$ we have $D_{\min} \subset \overline{D} \subset D_{\max}$. One can also describe $\mathrm{dom}(D_{\max})$ as the space of $u \in L^2C^\infty(E)$ for which the distribution $D(u)$ actually lies in $L^2C^\infty(F)$.

Lemma 1.70. *(1) Let M be a complete Riemannian manifold without boundary. Then the Laplacian $\Delta_p\colon \Omega_c^p(M) \to L^2\Omega^p(M)$ is essentially selfadjoint. The minimal and maximal closures of $d^p\colon \Omega_c^p(M) \to L^2\Omega^{p+1}(M)$, $\delta^p\colon \Omega_c^p(M) \to L^2\Omega^{p-1}(M)$ and $\Delta_p\colon \Omega_c^p(M) \to L^2\Omega_c^p(M)$ agree;*

(2) Let M be a cocompact G-manifold without boundary and with G-invariant Riemannian metric and let E and F be G-vector bundles with G-invariant Hermitian metrics. Let $D\colon C_c^\infty(E) \to C_c^\infty(F)$ be a G-equivariant elliptic operator. Then the minimal closure and the maximal closure of $D\colon C_c^\infty(E) \to L^2C^\infty(F)$ agree.

Proof. (1) The Laplacian is essentially selfadjoint by [109, Section 3],[216]. The claims about the equality of the minimal and maximal domains follow then from [73, Lemma 3.8 on page 113]

(2) is proved in [9, Proposition 3.1 on page 53].   □

Notice that Lemma 1.70 (1) allows us to talk about the spectral family associated to the Laplace operator $\Delta_p\colon \Omega_c^p(M) \to L^2\Omega^p(M)$ on a complete Riemannian manifold without boundary.

1.4.2 L^2-Hodge-de Rham Theorem

In this subsection we explain parts of the proof of the L^2-Hodge-de Rham-Theorem which we have already stated in Theorem 1.59. For more details we refer to the original proof in [145] (see also [147]).

Definition 1.71 (L^2-de Rham cohomology). *Let M be a complete Riemannian manifold without boundary. Put*

$$Z^p(M) := \ker\left(\left(d^p \colon \Omega_c^p(M) \to L^2\Omega^{p+1}(M)\right)_{\min}\right);$$
$$B^p(M) := \operatorname{im}\left(\left(d^{p-1} \colon \Omega_c^{p-1}(M) \to L^2\Omega^p(M)\right)_{\min}\right).$$

Define the unreduced L^2-de Rham cohomology *by*

$$H^p_{(2),\mathrm{unr}}(M) := Z^p(M)/B^p(M)$$

and the (reduced) L^2-de Rham cohomology *by the Hilbert space*

$$H^p_{(2)}(M) := Z^p(M)/\operatorname{clos}(B^p(M)).$$

Notice that this definition makes sense since $Z^p(M) \subset L^2\Omega^p(M)$ is closed and $B^p(M)$ lies in $Z^p(M)$. Because of Lemma 1.70 (1) we get the same if we use the minimal closures instead of the maximal closures. The difference of the unreduced and reduced L^2-cohomology is measured by the Novikov-Shubin invariants which we will introduce in Chapter 2.

Lemma 1.72. *Let M be a complete Riemannian manifold without boundary. The inclusion of $\mathcal{H}^p_{(2)}(M)$ into $L^2\Omega^p(M)$ induces an isometric isomorphism*

$$\mathcal{H}^p_{(2)}(M) \xrightarrow{\cong} H^p_{(2)}(M).$$

Proof. The following inequalities can easily be derived from Lemma 1.56

$$\operatorname{clos}(B^p(M)) = \operatorname{clos}(d^{p-1}(\Omega_c^{p-1}(M)));$$
$$Z^p(M) = \operatorname{clos}(\delta^{p+1}(\Omega_c^{p+1}(M)))^\perp.$$

Now apply Theorem 1.57. □

Let M be a cocompact free proper G-manifold with G-invariant Riemannian metric. Suppose that M has no boundary. Let $H^k\Omega^p(M)$ for a nonnegative integer k be the k-th *Sobolev space of p-forms* on M, i.e. the Hilbert space completion of $\Omega_c^p(M)$ with respect to the inner product or norm

$$\langle \omega, \eta \rangle_k = \langle (1+\Delta_p)^{k/2}\omega, (1+\Delta_p)^{k/2}\eta \rangle_{L^2} = \langle \omega, (1+\Delta_p)^k\eta \rangle_{L^2},$$
$$\|\omega\|_k = \sqrt{\|(1+\Delta_p)^{k/2}\omega\|_{L^2}} = \sqrt{\langle \omega, (1+\Delta_p)^k\omega \rangle_{L^2}}$$

using the L^2-inner product (1.52) or L^2-norm of (1.53). In particular we get $H^0\Omega^p(M) = L^2\Omega^p(M)$. One can identify $H^{2k}\Omega^p(M)$ with $\{\omega \in L^2\Omega^p(M) \mid$

$(1+\Delta_p)^k\omega \in L^2\Omega^p(M)\}$, where $(1+\Delta_p)^k\omega$ is to be understood in the sense of distributions by Lemma 1.70 (2). The operator $(1+\Delta_p)^{k/2}\colon \Omega_c^p(M) \to \Omega_c^p(M)$ induces for $k \leq l$ a G-equivariant isometric isomorphism

$$(1+\Delta_p)^{k/2}\colon H^l\Omega^p(M) \xrightarrow{\cong} H^{l-k}\Omega^p(M) \tag{1.73}$$

by the following argument. Obviously $(1+\Delta_p)^{k/2}$ is well-defined and isometric. In particular its image is closed. It remains to show that it is surjective. For this purpose it suffices to show that the orthogonal complement of the image of $(1+\Delta_p)^k\colon H^{2k}\Omega^p(M) \to L^2\Omega^p(M)$ is trivial. Since this is the subspace of those elements $\omega \in L^2\Omega^p(M)$ for which $\langle\omega,(1+\Delta_p)^k(\eta)\rangle_{L^2} = 0$ holds for all $\eta \in \Omega_c^p(M)$, ω lies in the kernel of $((1+\Delta_p)^k)_{\max}$. From Lemma 1.70 (2) we get $((1+\Delta_p)^k)_{\max} = ((1+\Delta_p)^k)_{\min}$. Hence there is a sequence $(\omega_n)_{n\geq 0}$ of elements in $\Omega_c^p(M)$ which converges in $L^2\Omega^p(M)$ to ω and for which $(1+\Delta_p)^k(\omega_n)$ converges in $L^2\Omega^p(M)$ to zero. Since $||\omega_n||_{L^2} \leq ||(1+\Delta_p)^k(\omega_n)||_{L^2}$ holds, we get $\omega = 0$.

The definition of $H^k\Omega^p(\mathbb{R}^n)$ corresponds to the usual definition of Sobolev space in the literature (see for instance [473, §7 in Chapter I]). In [145], only Sobolev spaces $H^{2k}\Omega^p(M)$ are considered and denoted by H^k.

Notice that using a fundamental domain of the G-action one obtains a G-equivariant isometric isomorphism from $L^2\Omega^p(M)$ to the tensor product of Hilbert spaces $l^2(G) \otimes L^2\Omega^p(G\backslash M)$ where G acts on $l^2(G)$ by left multiplication and on $L^2\Omega^p(G\backslash M)$ trivially [9, 4.1 on page 57 and page 65]. This implies that $H^k\Omega^p(M)$ is a Hilbert $\mathcal{N}(G)$-module in the sense of Definition 1.5 for all $k, p \geq 0$. The exterior differential induces a G-equivariant bounded operator, denoted in the same way, $d^p\colon H^{k+1}\Omega^p(M) \to H^k\Omega^{p+1}(M)$ for all $k, p \geq 0$. (To get bounded operators as differentials is the main reasons why we have introduced the Sobolev spaces.) Thus we obtain for $l \geq n$ with $n = \dim(M)$ a Hilbert $\mathcal{N}(G)$-cochain complex $H^{l-*}\Omega^*(M)$ in the sense of Definition 1.15

$$\ldots \to 0 \to H^l\Omega^0(M) \xrightarrow{d^0} H^{l-1}\Omega^1(M) \xrightarrow{d^1} H^{l-2}\Omega^2(M) \xrightarrow{d^2} \ldots$$
$$\xrightarrow{d^{n-1}} H^{l-n}\Omega^n(M) \to 0 \to \ldots \tag{1.74}$$

We have introduced its (reduced) L^2-cohomology $H_{(2)}^p(H^{l-*}\Omega^*(M))$ in Definition 1.16. The next lemma will show that it is independent of $l \geq \dim(M)$.

Lemma 1.75. *Let* $l \geq \dim(M)$. *Then the obvious inclusion of* $\mathcal{H}_{(2)}^p(M)$ *with the* L^2-*norm into* $H^{l-p}\Omega^p(M)$ *is a* G-*equivariant isometric embedding and induces a* G-*equivariant isometric isomorphism*

$$\mathcal{H}_{(2)}^p(M) \xrightarrow{\cong} H_{(2)}^p(H^{l-*}\Omega^*(M)).$$

Proof. Consider $\omega \in \mathcal{H}_{(2)}^p(M)$. Obviously $||\omega||_k = ||\omega||_0 < \infty$ for all $k \geq 0$. This implies that $d\omega$, $\delta d\omega$, $\delta\omega$ and $d\delta\omega$ are square-integrable. We conclude from Lemma 1.56

$$\langle \omega, \omega \rangle_0 \;=\; \langle (1 + \Delta_p)\omega, \omega \rangle_0 \;=\; \langle \omega, \omega \rangle_0 + \langle d^p\omega, d^p\omega \rangle_0 + \langle \delta^p\omega, \delta^p\omega \rangle_0.$$

This implies $d\omega = \delta\omega = 0$. For $\eta \in H^{l-p+1}\Omega^{p-1}(M)$ we get from Lemma 1.56

$$\langle \omega, d^{p-1}\eta \rangle_{l-p} \;=\; \langle \omega, (1 + \Delta_p)^{l-p} d^{p-1}\eta \rangle_0 \;=\; \langle \delta^p\omega, (1 + \Delta_{p-1})^{l-p}\eta \rangle_0 \;=\; 0.$$

This shows that $\mathcal{H}^p_{(2)}(M)$ lies in $\ker(d^p)$ and is orthogonal to $\mathrm{clos}(\mathrm{im}(d^{p-1}))$ in $H^{l-p}\Omega^p(M)$. It remains to show that the orthogonal complement of $\mathcal{H}^p_{(2)}(M)$ in $\ker(d^p)$ is contained in $\mathrm{clos}(\mathrm{im}(d^{p-1}))$.

Given $\mu \in \ker(d^p)$, we can decompose it orthogonally as $\mu = \omega + \eta$ for $\omega \in \mathcal{H}^p_{(2)}(M)$ and η in the orthogonal complement of $\mathcal{H}^p_{(2)}(M)$ in $\ker(d^p) \subset H^{l-p}\Omega^p(M)$. Put $\eta' = (1 + \Delta_p)^{(l-p)/2}(\eta)$. Then $d^p_{\min}\eta' = 0$ and for every $\nu \in \mathcal{H}^p_{(2)}(M)$ we get

$$
\begin{aligned}
\langle \eta', \nu \rangle_0 &= \langle (1 + \Delta_p)^{(l-p)/2}\eta, \nu \rangle_0 \\
&= \langle (1 + \Delta_p)^{(l-p)/2}\eta, (1 + \Delta_p)^{(l-p)/2}\nu \rangle_0 \\
&= \langle \eta, \nu \rangle_{l-p} \\
&= 0.
\end{aligned}
$$

From Theorem 1.57, we obtain a sequence of elements $\eta_n \in \Omega^{p-1}_c(M)$ such that $\eta' = \lim_{n\to\infty} d^{p-1}\eta_n$ with respect to the L^2-norm. Now we derive from 1.73 in $H^{l-p}\Omega^p(M)$

$$\eta \;=\; \left((1 + \Delta_p)^{(l-p)/2} \right)^{-1} \eta' \;=\; \lim_{n\to\infty} d^{p-1} \left((1 + \Delta_{p-1})^{(l-p)/2} \right)^{-1} \eta_n,$$

where we think of η_n as an element in $H^1\Omega^{p-1}(M)$. Hence η belongs to $\mathrm{clos}(\mathrm{im}(d^{p-1}))$ in $H^{l-p}\Omega^p(M)$. $\square$

Fix an equivariant smooth triangulation K of M. In particular K is a free cocompact G-CW-complex and we have introduced its cellular L^2-cochain complex $C^*_{(2)}(K)$ in Definition 1.29. Next we give a different model for it in terms of the *cellular cochain complex* $C^*(K) = \hom_{\mathbb{Z}}(C_*(K), \mathbb{C})$. Fix for any simplex in K some (arbitrary) orientation. Then we obtain a basis for $C_p(K)$ for each $p \geq 0$ and an element in $C^p(K)$ is the same as a function f from the set $S_p(K)$ of p-dimensional simplices of K to $\mathbb{C}$. We call f square-summable if $\sum_{\sigma \in S_n(K)} |f(\sigma)|^2 < \infty$. Let $l^2 C^p(K) \subset C^p(K)$ be the subspace of square-summable cochains. One easily checks that we obtain a subcomplex $l^2 C^*(K) \subset C^*(K)$, the *cochain complex of square-summable cochains*. It becomes a finite Hilbert $\mathcal{N}(G)$-chain complex with respect to the inner product $\langle f, g \rangle_{L^2} := \sum_{\sigma \in S_n(K)} f(\sigma)\overline{g(\sigma)}$.

Lemma 1.76. *There is a natural isometric G-chain isomorphism*

$$f^* \colon C^*_{(2)}(K) \xrightarrow{\;\cong\;} l^2 C^*(K).$$

Proof. The map f^p sends $u \in \hom_{\mathbb{Z}G}(C_p(K), l^2(G))$ to the element $f^p(u) \in \text{map}(S_p(K), \mathbb{C})$ which assigns to a simplex σ the coefficient λ_e of the unit element $e \in G$ in $u(\sigma) = \sum_{g \in G} \lambda_g \cdot g$. $\qquad \square$

If $\omega \in H^k \Omega^p(M)$ for $k > \dim(M)/2 + 1$, then ω is a C^1-form by the Sobolev inequality. In particular we can integrate ω over any oriented p-simplex of K and thus obtain an element in $C^p(K)$. It turns out that this element actually lies in $l^2 C^p(K)$. By this construction and the isomorphism appearing in Lemma 1.76 we obtain for large enough $l > 0$ a cochain map of Hilbert $\mathcal{N}(G)$-cochain complexes, the *L^2-de Rham cochain map*

$$A^*: H^{l-*}\Omega^*(M) \to C^*_{(2)}(K). \qquad (1.77)$$

Notice that $C^*_{(2)}(K)$, $l^2 C^p(K)$, f^* and A^* are independent of the choices of orientations of the simplices. A change of the orientation of a simplex σ affects $\int_\sigma \omega$ and the basis for $C_p(K)$ in a way which cancels out.

Next we define a right inverse of A^* as follows. The construction is due to Whitney [522, VII.11 on page 226]. Let $\{U_\sigma\}_{\sigma \in S_0(K)}$ be the open covering given by the open stars of the 0-simplices. Recall that the *closed star* $\text{st}(\sigma)$ of a simplex σ consists of all simplices τ which are faces of some simplex τ' which has σ as face. The *open star* is the interior of the closed star. Obviously $gU_\sigma = U_{g\sigma}$ for $\sigma \in S_0(K)$ and $g \in G$. Choose a *G-invariant subordinate smooth partition* $\{e_\sigma\}_{\sigma \in S_0(K)}$ of unity, i.e. smooth function $e_\sigma: M \to [0,1]$ with support in U_σ such that $e_{g\sigma} \circ l_g = e_\sigma$ for $l_g: M \to M$ left multiplication with g holds for all $\sigma \in S_0(K)$ and $g \in G$ and that the (locally finite) sum $\sum_{\sigma \in S_0(K)} e_\sigma$ is constant 1. Given a p-simplex τ with vertices $\sigma_0, \sigma_1, \ldots, \sigma_p$, let $c_\tau \in l^2 C^p(M)$ be the characteristic function associated to τ and define a smooth p-form $W(c_\tau)$ with support in the star of τ by

$$W(c_\tau) := p! \sum_{i=0}^{p} (-1)^i e_{\sigma_i} d^0 e_{\sigma_0} \wedge \ldots \wedge d^0 e_{\sigma_{i-1}} \wedge d^0 e_{\sigma_{i+1}} \wedge \ldots \wedge d^0 e_{\sigma_p}$$

if $p > 0$ and by $W(c_\tau) = e_{\sigma_0}$ if $p = 0$. Using the isomorphism appearing in Lemma 1.76 we obtain a well-defined cochain map of Hilbert $\mathcal{N}(G)$-cochain complexes, the so called *Whitney map*.

$$W^*: C^*_{(2)}(K) \to H^{l-*}\Omega^*(M). \qquad (1.78)$$

Again W^* does not depend on the choice of orientations of the simplices. It satisfies

$$A^* \circ W^* = \text{id}. \qquad (1.79)$$

In particular we see that the map induced by A^* on L^2-cohomology

$$H^*_{(2)}(A^*): H^*_{(2)}(H^{l-*}\Omega^*(M)) \to H^*_{(2)}(C^*_{(2)}(K)) \qquad (1.80)$$

is surjective. Next we explain why it is injective. This is not done by constructing a homotopy between $W^* \circ A^*$ and id as it can be done for $G \backslash K$ and $G \backslash M$ since the construction on the quotients is not local and cannot be lifted to K and M. Instead one modifies the definition of W, namely one uses the partition given by barycentric coordinate functions e_σ for a vertex σ instead of the smooth partition before. Then the image of this new Whitney map consists of L^2-integrable forms and yields a bounded operator $W^p \colon C^p_{(2)}(K) \to L^2 \Omega^p(M)$ (see Lemma 2.79), but the image does not lie in $H^l \Omega^p(M)$ for all $l \geq 0$ anymore, since the barycentric coordinate functions are not smooth and hence $W(c_\tau)$ is not smooth for a p-simplex τ. (We will see in the proof of Lemma 2.79 that $W(c_\tau)$ is at least continuous.) The advantage of this modified Whitney map lies in the following result [145, Lemma 3.9 on page 164].

Lemma 1.81. *Let $\omega \in H^k \Omega^p(M)$ be a fixed element and $k > \dim(M)/2 + 1$. For every $\epsilon > 0$ we can find an equivariant smooth triangulation $K(\epsilon)$ which is a subdivision of K such that*

$$||\omega - W^p_{K(\epsilon)} \circ A^p_{K(\epsilon)}(\omega)||_{L^2} < \epsilon$$

holds for the L^2-de Rham map $A^p_{K(\epsilon)}$ and the modified Whitney map $W^p_{K(\epsilon)}$ associated to $K(\epsilon)$.   □

This Lemma 1.81 is the key ingredient in the proof of injectivity of the map (1.80) as we explain next. Because of Lemma 1.72 and Lemma 1.75 it suffices to show for an element $\omega \in \mathcal{H}^p_{(2)}(M)$ with $H^*_{(2)}(A^*)([\omega]) = 0$ that its class $[\omega] \in H^p_{(2)}(M)$ vanishes.

Fix $\epsilon > 0$. Lemma 1.81 implies for an appropriate subdivision $K(\epsilon)$

$$||\omega - W^p_{K(\epsilon)} \circ A^p_{K(\epsilon)}(\omega)||_{L^2} < \epsilon/2.$$

There is a cochain map $l^2 C^*(K) \to l^2 C^*(K(\epsilon))$ which induces an isomorphism on L^2-cohomology and is compatible with the de Rham cochain maps for K and $K(\epsilon)$. We conclude $H^*_{(2)}(A^*_{K(\epsilon)})([\omega]) = 0$. Hence we can find a cochain $u \in l^2 C^{p-1}(K(\epsilon))$ with

$$||A^p_{K(\epsilon)}(\omega) - c^{p-1}_{K(\epsilon)}(u)||_{L^2} < \frac{\epsilon}{2 \cdot ||W^p_{K(\epsilon)}||}.$$

This implies

$$||\omega - W^p_{K(\epsilon)} \circ c^{p-1}_{K(\epsilon)}(u)||_{L^2}$$
$$\leq ||\omega - W^p_{K(\epsilon)} \circ A^p_{K(\epsilon)}(\omega)||_{L^2} + ||W^p_{K(\epsilon)} \circ A^p_{K(\epsilon)}(\omega) - W^p_{K(\epsilon)} \circ c^{p-1}_{K(\epsilon)}(u)||_{L^2}$$
$$\leq ||\omega - W^p_{K(\epsilon)} \circ A^p_{K(\epsilon)}(\omega)||_{L^2} + ||W^p_{K(\epsilon)}|| \cdot ||A^p_{K(\epsilon)}(\omega) - c^{p-1}_{K(\epsilon)}(u)||_{L^2}$$
$$\leq \epsilon/2 + ||W^p_{K(\epsilon)}|| \cdot \frac{\epsilon}{2 \cdot ||W^p_{K(\epsilon)}||}$$
$$= \epsilon.$$

Now one checks that the image of $W^p_{K(\epsilon)}$ is contained in the domain of $\left(d^p \colon \Omega^p_c(M) \to L^2\Omega^{p+1}(M)\right)_{\max}$ and that $d^{p-1}_{\max} \circ W^{p-1}_{K(\epsilon)} = W^p_{K(\epsilon)} \circ c^{p-1}_{K(\epsilon)}$. Hence for $\epsilon > 0$ there is v in the domain of $\left(d^p \colon \Omega^p_c(M) \to L^2\Omega^{p+1}(M)\right)_{\max}$ such that $\|\omega - d^{p-1}_{\max}(v)\|_{L^2} < \epsilon$ holds. Hence the class represented by ω in the reduced L^2-de Rham cohomology $H^p_{(2)}(M)$ is trivial. This shows that the map appearing in (1.80) is bijective. Now the map appearing in Theorem 1.59 is bijective since it is the composition of the bijective maps appearing in Lemma 1.75 and (1.80). This finishes the proof of Theorem 1.59.   □

1.5 L^2-Betti Numbers of Manifolds with Boundary

We briefly discuss the case of a manifold with boundary. A useful discussion what smooth means for a function or a section of a bundle on a manifold with boundary is given in [467]. Lemma 1.56 says that for a complete Riemannian manifold with boundary and for smooth forms $\omega \in \Omega^p(M)$ and $\eta \in \Omega^{p+1}(M)$ such that ω, $d^p\omega$, η and $\delta^{p+1}\eta$ are square-integrable, we get

$$\langle d^p\omega, \eta\rangle_{L^2} - \langle\omega, \delta^{p+1}\eta\rangle_{L^2} = \int_{\partial M} (\omega \wedge *^{p+1}\eta)|_{\partial M}. \tag{1.82}$$

The classical de Rham isomorphism is an isomorphism for any manifold regardless whether it has a boundary or not (see Theorem 1.47).

Suppose that the boundary ∂M of the complete Riemannian manifold M is the disjoint union of $\partial_0 M$ and $\partial_1 M$ where we allow that $\partial_0 M$, $\partial_1 M$ or both are empty. A form $\omega \in \Omega^p(M)$ satisfies *Dirichlet boundary conditions* on $\partial_0 M$ if $\omega|_{\partial_0 M} = 0$ holds. It satisfies *Neumann boundary conditions* on $\partial_1 M$ if $*^p\omega|_{\partial_1 M} = 0$ holds. Define

$$\Omega^p_2(M, \partial_0 M) := \{\omega \in \Omega^p_c(M) \mid \omega|_{\partial_0 M} = 0, (\delta^p\omega)|_{\partial_0 M} = 0,$$
$$(*^p\omega)|_{\partial_1 M} = 0, (*^{p+1}d^p\omega)|_{\partial_1 M} = 0\}; \tag{1.83}$$
$$\Omega^p_1(M, \partial_0 M) := \{\omega \in \Omega^p_c(M) \mid \omega|_{\partial_0 M} = 0, (*^p\omega)|_{\partial_1 M} = 0\}; \tag{1.84}$$
$$\Omega^p_d(M, \partial_0 M) := \{\omega \in \Omega^p_c(M) \mid \omega|_{\partial_0 M} = 0\}; \tag{1.85}$$
$$\Omega^p_\delta(M, \partial_0 M) := \{\omega \in \Omega^p_c(M) \mid (*^p\omega)|_{\partial_1 M} = 0\}. \tag{1.86}$$

Define the *space of L^2-integrable harmonic smooth p-forms satisfying boundary conditions*

$$\mathcal{H}^p_{(2)}(M, \partial_0 M) := \{\omega \in \Omega^p_2(M, \partial_0 M) \mid \Delta_p(\omega) = 0, \int_M \omega \wedge *\omega < \infty\}. \tag{1.87}$$

We have the following version of the Hodge-de Rham Theorem 1.57.

Theorem 1.88. (Hodge-de Rham Theorem for manifolds with boundary). *Let M be a complete Riemannian manifold whose boundary ∂M*

is the disjoint union $\partial_0 M$ *and* $\partial_1 M$. *Then we have the so called* Hodge-de Rham decomposition

$$L^2 \Omega^p(M) = \mathcal{H}^p_{(2)}(M, \partial_0 M) \oplus \text{clos}(d^{p-1}(\Omega_d^{p-1}(M, \partial_0 M))) \oplus$$
$$\text{clos}(\delta^{p+1}(\Omega_\delta^{p+1}(M, \partial_0 M))).$$

We have the following version of the L^2-Hodge-de Rham Theorem 1.59.

Theorem 1.89. (L^2-Hodge-de Rham Theorem for manifolds with boundary.) *Let M be a cocompact free proper G-manifold with G-invariant Riemannian metric whose boundary ∂M is the disjoint union of G-spaces $\partial_0 M$ and $\partial_1 M$. Let $(K; \partial_0 K, \partial_1 K)$ be an equivariant smooth triangulation of the triad $(M; \partial_0 M, \partial_1 M)$. Then integration defines an isomorphism of finitely generated Hilbert $\mathcal{N}(G)$-modules*

$$\mathcal{H}^p_{(2)}(M, \partial_0 M) \xrightarrow{\cong} H^p_{(2)}(K, \partial_0 K).$$

The detailed proof of Theorem 1.88 and of Theorem 1.89 in the more general context of Riemannian manifolds with bounded geometry can be found in [459, Theorem 5.10, Theorem 8.2 and Corollary 8.15] (see also [322, Theorem 5.13]). Theorem 1.88 remains true if one replaces $\Omega_d^{p-1}(M, \partial_0 M)$ by $\Omega_1^{p-1}(M, \partial_0 M)$ and $\Omega_\delta^{p+1}(M, \partial_0 M)$ by $\Omega_1^{p+1}(M, \partial_0 M)$.

The expression (1.60) in terms of the heat kernel on M for the analytic L^2-Betti numbers which are defined by

$$b_p^{(2)}(M, \partial_0 M) := \dim_{\mathcal{N}(G)}(\mathcal{H}^p_{(2)}(M, \partial_0 M)) \qquad (1.90)$$

carries over to the situation studied in Theorem 1.89 if one uses $\Omega_2(M, \partial_0 M)$ as initial domain for Δ_p.

There is also the following version of Lemma 1.72 proved in [459, Theorem 6.2]. Define the (reduced) L^2-cohomology $H^p_{(2)}(M, \partial_0 M)$ as in Definition 1.71 but now using $d^p: \Omega_d^p(M, \partial_0 M) \to L^2 \Omega^{p+1}(M)$ instead of $d^p: \Omega_c^p(M) \to L^2 \Omega^{p+1}(M)$.

Lemma 1.91. *Let M be a complete Riemannian manifold whose boundary ∂M is the disjoint union of $\partial_0 M$ and $\partial_1 M$. Then the inclusion of $\mathcal{H}^p_{(2)}(M, \partial_0 M)$ into $L^2 \Omega_p(M)$ induces an isomorphism*

$$\mathcal{H}^p_{(2)}(M, \partial_0 M) \xrightarrow{\cong} H^p_{(2)}(M, \partial_0 M).$$

Notice that we do not require here (nor later in the context of Novikov-Shubin invariants) that the metric is a product near the boundary. However, this will be crucial when we deal with torsion and L^2-torsion in Remark 3.162.

1.6 Miscellaneous

The following result is sometimes useful since it gives a criterion for the reduced L^2-cohomology to be non-trivial.

Lemma 1.92. *Let M be a complete Riemannian manifold without boundary. Denote by $i \colon H_c^p(M; \mathbb{C}) \to H^p(M; \mathbb{C})$ the natural map from the cohomology with compact support to the cohomology of M. Then $\mathrm{im}(i)$ injects into the reduced L^2-cohomology $H_{(2)}^p(M)$.*

Proof. Let $k \colon H_c^p(M; \mathbb{C}) \to H_{(2)}^p(M)$ be the canonical map. Choose any linear map $s \colon \mathrm{im}(i) \to H_c^p(M; \mathbb{C})$ with $i \circ s = \mathrm{id}$. We want to show that $k \circ s$ is injective. It suffices to prove for a smooth closed p-form ω with compact support satisfying $k([\omega]) = 0$ that $i([\omega]) = 0$ holds.

Since $k([\omega]) = 0$, there is a sequence $(\eta_n)_{n \geq 0}$ in $\Omega_c^{p-1}(M)$ such that $\omega = \lim_{n \to \infty} d^{p-1}(\eta_n)$ with respect to the L^2-norm holds. Suppose $i([\omega]) \neq 0$. Let $p \colon \overline{M} \to M$ be the orientation covering of M. This is a double covering with orientable total space and the induced map $p^* \colon H^p(M; \mathbb{C}) \to H^p(\overline{M}; \mathbb{C})$ is injective. Hence $p^* \omega$ represents a non-zero element in $H^p(\overline{M}; \mathbb{C})$. Put $m = \dim(M)$. By Poincaré duality [57, (5.4) on page 44 and Remark I.5.7 on page 46] there is a smooth $(m - p)$-form μ with compact support on $\overline{M}$ such that $d^{m-p}\mu = 0$ and $\int_{\overline{M}} p^* \omega \wedge \mu$ is different from zero. This yields the contradiction

$$\int_{\overline{M}} p^* \omega \wedge \mu \;=\; \lim_{n \to \infty} \int_{\overline{M}} d^{p-1} p^* \eta_n \wedge \mu \;=\; \lim_{n \to \infty} \int_{\overline{M}} d^{m-1}(p^* \eta_n \wedge \mu) \;=\; 0. \;\square$$

Cellular L^2-Betti numbers and L^2-(co-)homology can also be defined for a cocompact proper G-CW-complex X. Recall that the isotropy group of each point of a free G-CW-complex is trivial, whereas it is finite for a proper G-CW-complex. The point is that the cellular L^2-chain complex $C_*^{(2)}(X) = l^2(G) \otimes_{\mathbb{Z}G} C_*(X)$ is still defined as Hilbert $\mathcal{N}(G)$-chain complex for a proper G-CW-complex. After a choice of a characteristic map $(Q_i^n, q_i^n) \colon G/H_i \times (D^n, S^{n-1}) \to (X_n, X_{n-1})$ for each element of the set I_n of equivariant n-dimensional cells we obtain explicit isomorphisms

$$C_n(X) \cong \bigoplus_{i \in I_n} \mathbb{Z}[G/H_i];$$

$$C_n^{(2)}(X) \cong \bigoplus_{i \in I_n} l^2(G/H_i),$$

and we put on $C_n^{(2)}(X)$ the induced Hilbert $\mathcal{N}(G)$-module structure. This is independent of the choice of characteristic maps since for a different choice the two identifications of $C_n(X)$ differ by a direct sum of automorphisms of the shape $\epsilon_i \cdot l_{g_i} \colon \mathbb{Z}[G/H_i] \to \mathbb{Z}[G/H_i]$ for some $\epsilon_i \in \{\pm 1\}$ and $g_i \in G$.

Theorem 1.35 (1), (3), (4), (8), (9) and (10) remain true if one replaces free by proper. The basic observation is that $C_*^{(2)}(X)$ can be written as

$l^2(G) \otimes_{\mathbb{C}G} (\mathbb{C} \otimes_{\mathbb{Z}} C_*(X))$ and $\mathbb{C} \otimes_{\mathbb{Z}} C_n(X)$ is a finitely generated projective $\mathbb{C}[G]$-module for all n. Theorem 1.35 (2) becomes

$$\sum_{p \geq 0} (-1)^p \cdot b_p^{(2)}(X) = \sum_{p \geq 0} (-1)^p \cdot \sum_{i \in I_p} |H_i|^{-1} \qquad (1.93)$$

since $\dim_{\mathcal{N}(G)}(l^2(G/H_i)) = |H_i|^{-1}$. The space $\mathcal{H}_{(2)}^p(M)$ of harmonic L^2-forms of a cocompact proper G-manifold M with G-invariant Riemannian metric inherits the structure of a Hilbert $\mathcal{N}(G)$-module and Theorem 1.59 remains true without the assumption that M is free.

We will give a combinatorial approach to the L^2-Betti numbers in Section 3.7 which is useful for concrete calculations.

We will explain a proportionality principle for L^2-Betti numbers in Theorem 3.183, Theorem 7.34 and Corollary 7.37.

We will compute the values of L^2-Betti numbers of the universal coverings of compact 3-manifolds in Theorem 4.1 and of locally symmetric spaces in Theorem 5.12 and (5.15).

A more algebraic treatment of Hilbert modules, dimension functions and L^2-Betti numbers and a definition of L^2-Betti numbers for arbitrary topological spaces with an arbitrary G-action will be given in Chapter 6. This will apply in particular to EG for any (discrete) group G.

We will discuss the behaviour of L^2-Betti numbers of groups under quasi-isometry and measure equivalence in Section 7.5.

Lemma 1.94. *Let B be the set of real numbers r, for which there exists a CW-complex X of finite type and an integer $p \geq 0$ with $b_p^{(2)}(\widetilde{X}) = r$. Then B is countable and contains $\{r \mid r \in \mathbb{Q}, r \geq 0\}$.*

Proof. We get $\{r \mid r \in \mathbb{Q}, r \geq 0\} \subset B$ from Example 1.38. Let B' be the set of real numbers r, for which there exists a finitely presented group G, positive integers m and n and a matrix $A \in M(m, n, \mathbb{Z}G)$ satisfying

$$r = \dim_{\mathcal{N}(G)} \left(\ker \left(r_A^{(2)} : l^2(G)^m \to l^2(G)^n \right) \right),$$

where $r_A^{(2)}$ is given by right multiplication with A. The fundamental group π of a CW-complex of finite type X is finitely presented and

$$b_p^{(2)}(\widetilde{X}) = \dim_{\mathcal{N}(\pi)}(\ker(c_p^{(2)})) + \dim_{\mathcal{N}(\pi)}(\ker(c_{p+1}^{(2)})) - n_{p+1}$$

follows from Additivity (see Theorem 1.12 (2)), where n_{p+1} is the number of $(p+1)$-cells of X. Hence it suffices to show that B' is countable. Since there are only countably many finite presentations and $\mathbb{Z}G$ is countable for any finitely presented and hence countable group G, Lemma 1.94 follows. $\square$

In Chapter 10 we deal with the question how large the set B is. At the time of writing no counterexample to the statement $B = \{q \in \mathbb{Q} \mid q \geq 0\}$ is known to the author.

We will deal with the Singer Conjecture 11.1 that the L^2-Betti numbers of the universal covering of an aspherical closed manifold vanish outside the middle dimension in Chapter 11.

We will relate the L^2-Betti numbers of the universal covering of an aspherical closed manifold to its simplicial and its minimal volume (see Conjecture 14.1 and Subsection 14.2.6).

Question 1.95. (Vanishing of L^2-Betti numbers of the base and the total space of a fibration).

Let $F \to E \xrightarrow{p} B$ be a fibration of connected finite CW-complexes. Suppose that all L^2-Betti numbers of $\widetilde{B}$ are trivial. Does this imply that all the L^2-Betti numbers of $\widetilde{E}$ are trivial?

The answer is yes if $B = S^1$ by Theorem 1.39. The answer is also known to be affirmative if one of the following conditions is satisfied [61]: i.) B is an S^1-CW-complex such that the inclusion of one (and hence all) orbits induces a map on the fundamental groups with infinite image. ii.) The map $\pi_1(p) \colon \pi_1(E) \to \pi_1(B)$ is bijective and $\pi_1(B)$ is virtually poly-cyclic. iii.) The map $\pi_1(p) \colon \pi_1(E) \to \pi_1(B)$ is bijective and $\pi_1(B)$ operates trivially on $H_*(F)$. If B is aspherical, more information will be given in Theorem 7.4.

Using the center-valued trace (see Theorem 9.5) one can define a center-valued von Neumann dimension and thus elements $b_p^u(X; \mathcal{N}(G))$, which take values in the center $\mathcal{Z}(\mathcal{N}(G))$ of the von Neumann algebra $\mathcal{N}(G)$. Essentially we get for any conjugacy class (g), which contains only finitely many elements, a number. The value at (1) is the L^2-Betti number. Lott [321] defines analytically delocalized L^2-Betti numbers for the universal covering $\widetilde{M}$ of a closed Riemannian manifold M, which assigns to any conjugacy class a number. The invariant is presently only defined under certain technical assumptions. For example, the p-th delocalized Betti number is well-defined if G is virtually abelian, or if G is virtually nilpotent or Gromov hyperbolic and there is a gap away from zero in the spectrum of Δ_p [321, Proposition 6].

Elek [171] defines an invariant $b_E^p(\widetilde{X})$ for a finite CW-complex X replacing $C_{(2)}^*(\widetilde{X})$ by $\hom_{\mathbb{Z}}(C_*(\widetilde{X}); \mathbb{F}_2)$ for $\mathbb{F}_2$ the field of two elements and replacing the von Neumann dimension by the topological entropy of linear subshifts, provided that G is amenable. It has properties similar to the L^2-Betti numbers.

Finally we give some further references. Survey articles on L^2-cohomology and L^2-Betti numbers are [164], [168], [237, section 8], [332], [338], [341], [360] and [411]. More information about Morse inequalities in the L^2-context can be found for instance in [182], [363], [364], [401], [400], [475].

Exercises

1.1. Let $H \subset G$ be a subgroup. Show that the Hilbert space $l^2(G/H)$ with the obvious G-action is a Hilbert $\mathcal{N}(G)$-module if and only if H is finite, and that in this case its von Neumann dimension is $|H|^{-1}$.

1.2. Show that Lemma 1.9 (5) remains true if one only requires that the rows are weakly exact instead of exact, provided that all Hilbert $\mathcal{N}(G)$-modules have finite dimension.

1.3. Show by an example that Theorem 1.12 (4) becomes false if one drops the condition $\dim_{\mathcal{N}(G)}(V) < \infty$.

1.4. Consider the following diagram of Hilbert $\mathcal{N}(G)$-modules of finite dimension with weakly exact rows

$$
\begin{array}{ccccccccc}
U_1 & \xrightarrow{u_1} & U_2 & \xrightarrow{u_2} & U_3 & \xrightarrow{u_3} & U_4 & \xrightarrow{u_4} & U_5 \\
\downarrow{f_1} & & \downarrow{f_2} & & \downarrow{f_3} & & \downarrow{f_4} & & \downarrow{f_5} \\
V_1 & \xrightarrow{v_1} & V_2 & \xrightarrow{v_2} & V_3 & \xrightarrow{v_3} & V_4 & \xrightarrow{v_4} & V_5
\end{array}
$$

Suppose that f_1 has dense image, f_2 and f_4 are weak isomorphisms and f_5 is injective. Show that then f_3 is a weak isomorphism.

1.5. The group von Neumann algebra $\mathcal{N}(G)$ carries an involution of rings given by taking adjoints of bounded G-equivariant operators $l^2(G) \to l^2(G)$. Show that the ring homomorphism $i\colon \mathcal{N}(H) \to \mathcal{N}(G)$ given by induction (see Definition 1.23) is compatible with these involutions.

1.6. Find as many different proofs as possible for the fact that all the L^2-Betti numbers of the universal covering of the n-dimensional torus T^n for $n \geq 1$ vanish.

1.7. Let X be an aspherical CW-complex of finite type. Suppose that its fundamental group contains a normal infinite cyclic subgroup. Show that then all L^2-Betti numbers of $\tilde{X}$ vanish.

1.8. Show that a symmetric densely defined linear operator $T\colon \mathrm{dom}(T) \to H$ is selfadjoint if and only if the range of $T \pm i \cdot \mathrm{id}$ is H and that it is essentially selfadjoint if and only if the range of $T \pm i \cdot \mathrm{id}$ in H is dense.

1.9. Let $f\colon H \to H$ be a linear map of finite dimensional Hilbert spaces which is symmetric. Show that the projection E_λ^f of the associated spectral family is the sum $\bigoplus_{\mu \leq \lambda} \mathrm{pr}_\mu$, where pr_μ is the projection onto the eigenspace of $\mu \in \mathbb{R}$. Let $u\colon H \to \mathbb{C}^n$ be a unitary operator such that $ufu^{-1}\colon \mathbb{C}^n \to \mathbb{C}^n$ is given by a diagonal matrix D with real entries. Given a Borel function $g\colon \mathbb{R} \to \mathbb{C}$, let $g(D)$ be the diagonal matrix obtained from D by applying g to the diagonal entries. Show that $\int_{-\infty}^{\infty} g(\lambda)dE_\lambda$ in the sense of (1.64) is $u^{-1}g(D)u$.

1.10.   Let M be a complete Riemannian manifold without boundary. Show that we obtain a Hilbert cochain complex (in the sense of Definition 1.15 for $G = 1$) by

$$\ldots \xrightarrow{d^{p-2}_{\min}} \operatorname{dom}(d^{p-1}_{\min}) \xrightarrow{d^{p-1}_{\min}} \operatorname{dom}(d^{p}_{\min}) \xrightarrow{d^{p}_{\min}} \ldots$$

if we equip $\operatorname{dom}(d^{p}_{\min})$ with the Hilbert structure given by (1.63). Show that its unreduced and reduced L^2-cohomology in the sense of Definition 1.16 agrees with the notions of Definition 1.71.

1.11.   Show by giving an example that Lemma 1.56 does not hold without the hypothesis that M is complete.

1.12.   Show for any free $\mathbb{Z}$-CW-complex X of finite type that $b^{(2)}_p(X) \le b_p(\mathbb{Z}\backslash X)$ holds.

1.13.   Let M be an orientable closed 4-manifold with $\pi_1(M) = \mathbb{Z}/2 * \mathbb{Z}/2$. Show

$$b^{(2)}_2(\widetilde{M}) \;=\; \chi(M) \;=\; 2 + b_2(M).$$

1.14.   Fix an integer $n \ge 1$. Show that for a sequence $r^{(2)}_1, r^{(2)}_2, \ldots, r^{(2)}_n$ of non-negative rational numbers and a sequence $r_1, r_2, \ldots, r_n$ of non-negative integers there is a finite $(n+1)$-dimensional CW-complex X with infinite fundamental group satisfying $b^{(2)}_p(\widetilde{X}) = r^{(2)}_p$ and $b_p(X) = r_p$ for $p = 1, \ldots, n$.

2. Novikov-Shubin Invariants

Introduction

In this chapter we introduce and study Novikov-Shubin invariants for Hilbert chain complexes and for regular coverings of CW-complexes of finite type or of compact manifolds.

We will associate to the Laplacian Δ_p of a cocompact free proper G-manifold M with G-invariant Riemannian metric its spectral density function $F_p^\Delta(M) \colon [0, \infty) \to [0, \infty)$ which assigns to λ the von Neumann dimension of the image of E_λ, where $\{E_\lambda \mid \lambda \geq 0\}$ is the spectral family of Δ_p. We rediscover the p-th L^2-Betti number $b_p^{(2)}(M)$ by $F_p^\Delta(M)(0)$. If G is finite $F_p^\Delta(M)$ measures how many eigenvalues $\leq \lambda$, counted with multiplicity, Δ_p has, and this is the right intuition also for infinite G. $F_p^\Delta(M)$ contains a lot of information about the spectrum of Δ_p.

The spectral density function is an example of a so called density function $F \colon [0, \infty) \to [0, \infty]$, i.e. F is monotone non-decreasing and right-continuous. Two density functions F and G are called dilatationally equivalent if there are $C > 0$ and $\epsilon > 0$ such that $G(C^{-1} \cdot \lambda) \leq F(\lambda) \leq G(C \cdot \lambda)$ holds for all $\lambda \in [0, \epsilon]$. The spectral density function itself depends on the Riemannian structure. However, it turns out that its dilatational equivalence class only depends on the G-homotopy type of M. Hence it becomes a very interesting object because it is defined in terms of analytic data but is also a topological invariant. The Novikov-Shubin invariant of a density function is defined by

$$\alpha(F) = \liminf_{\lambda \to 0^+} \frac{\ln(F(\lambda) - F(0))}{\ln(\lambda)} \in [0, \infty],$$

provided that $F(\lambda) > F(0)$ holds for all $\lambda > 0$. Otherwise, one puts formally $\alpha(F) = \infty^+$. It measures how fast $F(\lambda)$ approaches $F(0)$ for $\lambda \to 0^+$ and takes the value ∞^+ if and only if there is a gap in the spectrum at zero. It only depends on the dilatational equivalence class of F. Define the Novikov-Shubin invariant $\alpha_p^\Delta(M)$ of M by the one of $F_p^\Delta(M)$. Novikov-Shubin invariants were originally defined in [400],[401]. In some sense Novikov-Shubin invariants measure how "thin" the spectrum of Δ_p at zero is. We will also introduce $\alpha_p(M)$ which is defined in terms of the differential d^{p-1}. It measures the difference between the unreduced and the reduced L^2-de Rham cohomology, the unreduced and the reduced one agree in dimension p if and only if $\alpha_p(M) = \infty^+$. We have $\alpha_p^\Delta(M) = 1/2 \cdot \min\{\alpha_p(M), \alpha_{p+1}(M)\}$.

Via the Laplace transform the dilatational equivalence class of the spectral density function and the Novikov-Shubin invariants are invariants of the asymptotic behaviour of the heat kernel for large times. Such invariants reflect the global geometry but are in general very hard to study and to compute. Notice that index theory yields invariants of the asymptotic behaviour of the heat kernel for small times and hence local invariants in contrast to invariants of the large time asymptotics which cannot be expressed locally.

We will introduce and study spectral density functions and Novikov-Shubin invariants for Hilbert chain complexes and possibly unbounded operators in Section 2.1. In particular we will show that the spectral density function is a homotopy invariant and satisfies a subadditivity-relation for short exact sequences. In Section 2.2 we apply this to the cellular L^2-chain complex and thus obtain the cellular spectral density function and Novikov-Shubin invariants of a free G-CW-complex of finite type. We prove their main properties such as homotopy invariance, Poincaré duality, product formula and express the first Novikov-Shubin invariant in terms of the growth rate of the fundamental group. We give an explicit formula in the case $G = \mathbb{Z}$. In Section 2.3 we introduce their analytic counterparts and show in Section 2.4 that the analytic and the cellular spectral density functions are dilatationally equivalent and hence the analytic and cellular Novikov-Shubin invariants agree [167], [240]. In Section 2.5 we discuss the conjecture that the Novikov-Shubin invariants are either rational positive numbers, ∞ or ∞^+. We briefly treat the case of a manifold with boundary in Section 2.6.

To get a quick overview about Novikov-Shubin invariants one should read through Sections 2.2, 2.3 and 2.5. The material of this chapter is rather independent of the following chapters which can be read without studying this chapter beforehand.

2.1 Spectral Density Functions

In this section we introduce and study the spectral density function associated to morphisms of Hilbert modules and to Hilbert chain complexes.

2.1.1 Spectral Density Functions of Morphisms

We next deal with the spectral density function of a map of Hilbert modules. We treat the more general case of an unbounded densely defined operator since this setting will be needed when we deal with the analytic case. If one only wants to investigate the cellular version it suffices to treat bounded operators.

Definition 2.1 (Spectral density function). *Let U and V be Hilbert $\mathcal{N}(G)$-modules. Let $f\colon \mathrm{dom}(f) \subset U \to V$ be a G-equivariant closed densely defined operator. For $\lambda \geq 0$ we define $\mathcal{L}(f, \lambda)$ as the set of Hilbert $\mathcal{N}(G)$-submodules $L \subset U$ such that $L \subset \mathrm{dom}(f)$ and $||f(x)|| \leq \lambda \cdot ||x||$ holds for all $x \in L$. Define the spectral density function of f*

$$F(f)\colon [0, \infty) \to [0, \infty], \qquad \lambda \mapsto \sup\{\dim_{\mathcal{N}(G)}(L) \mid L \in \mathcal{L}(f, \lambda)\}.$$

Define the L^2-Betti number of f by

$$b^{(2)}(f) := \dim_{\mathcal{N}(G)}(\ker(f)) = F(f)(0).$$

We call f Fredholm if there is $\epsilon > 0$ such that $F(f)(\epsilon) < \infty$.

We will conclude from Lemma 2.3 that the definition of Fredholm for a morphism of Hilbert $\mathcal{N}(G)$-modules of Definition 1.20 and Definition 2.1 agree.

Notice that $\ker(f) \subset U$ is closed and hence a Hilbert $\mathcal{N}(G)$-module and its von Neumann dimension is defined. The meaning of the spectral density function becomes more evident if one expresses it in terms of the spectrum of f^*f, as we explain next.

Lemma 2.2. *Let $f\colon \mathrm{dom}(f) \subset H_1 \to H_2$ be a closed densely defined operator of Hilbert spaces H_1 and H_2. Define*

$$\mathrm{dom}(f^*f) := \{x \in H_1 \mid x \in \mathrm{dom}(f), f(x) \in \mathrm{dom}(f^*)\}$$

and thus an operator $f^*f\colon \mathrm{dom}(f^*f) \subset H_1 \to H_1$. Then

(1) The subspace $\mathrm{dom}(f^*f)$ is dense and f^*f is a selfadjoint operator. Moreover, $\mathrm{dom}(f^*f)$ is a core for $\mathrm{dom}(f)$, i.e. $\mathrm{dom}(f^*f) \subset \mathrm{dom}(f)$ and for any $x \in \mathrm{dom}(f)$ there is a sequence $x_n \in \mathrm{dom}(f^*f)$ such that $\lim_{n\to\infty} x_n = x$ and $\lim_{n\to\infty} f(x_n) = f(x)$ holds;

(2) Let $\lambda \geq 0$ and $x \in \mathrm{dom}(f)$. Then

$$\begin{aligned} ||f(x)|| &> \lambda \cdot ||x|| && \text{if } E^{f^*f}_{\lambda^2}(x) = 0,\ x \neq 0; \\ ||f(x)|| &\leq \lambda \cdot ||x|| && \text{if } E^{f^*f}_{\lambda^2}(x) = x, \end{aligned}$$

where $\{E^{f^*f}_{\lambda} \mid \lambda \in \mathbb{R}\}$ denotes the spectral family of the selfadjoint operator f^*f.

Proof. (1) see [287, Theorem V.3.24 on page 275].

(2) Suppose that $E^{f^*f}_{\lambda^2}(x) = 0$ and $x \neq 0$. Choose a sequence $x_n \in \mathrm{dom}(f^*f)$ such that $\lim_{n\to\infty} x_n = x$ and $\lim_{n\to\infty} f(x_n) = f(x)$ holds. Since $x \neq 0$, and $0 = E^{f^*f}_{\lambda^2}(x) = \lim_{\mu\to\lambda^2+} E^{f^*f}_{\mu}(x)$ holds, we can find $\epsilon > 0$ with the property

$$\langle E^{f^*f}_{\lambda^2+\epsilon}(x), x \rangle < \frac{1}{2}\langle x, x \rangle.$$

We get from the definition of the spectral family

$$\|f(x_n)\|^2 = \langle f^* f(x_n), x_n \rangle$$
$$= \int_0^\infty \mu \, d\langle E_\mu^{f^*f}(x_n), x_n \rangle$$
$$\geq \int_{(\lambda^2, \infty)} \mu \, d\langle E_\mu^{f^*f}(x_n), x_n \rangle$$
$$= \int_{(\lambda^2, \lambda^2+\epsilon]} \mu \, d\langle E_\mu^{f^*f}(x_n), x_n \rangle + \int_{(\lambda^2+\epsilon, \infty)} \mu \, d\langle E_\mu^{f^*f}(x_n), x_n \rangle$$
$$\geq \lambda^2 \cdot (\langle E_{\lambda^2+\epsilon}^{f^*f}(x_n), x_n \rangle - \langle E_{\lambda^2}^{f^*f}(x_n), x_n \rangle)$$
$$\quad + (\lambda^2 + \epsilon) \cdot (\langle x_n, x_n \rangle - \langle E_{\lambda^2+\epsilon}^{f^*f}(x_n), x_n \rangle)$$
$$= (\lambda^2 + \epsilon) \cdot \langle x_n, x_n \rangle - \epsilon \cdot \langle E_{\lambda^2+\epsilon}^{f^*f}(x_n), x_n \rangle - \lambda^2 \cdot \langle E_{\lambda^2}^{f^*f}(x_n), x_n \rangle.$$

Taking the limit $n \to \infty$ yields

$$\|f(x)\|^2 \geq (\lambda^2 + \epsilon) \cdot \langle x, x \rangle - \epsilon \cdot \langle E_{\lambda^2+\epsilon}^{f^*f}(x), x \rangle - \lambda^2 \cdot \langle E_{\lambda^2}^{f^*f}(x), x \rangle$$
$$\geq (\lambda^2 + \epsilon) \cdot \|x\|^2 - \frac{\epsilon}{2} \cdot \|x\|^2 - 0$$
$$= (\lambda^2 + \frac{\epsilon}{2}) \cdot \|x\|^2$$
$$> \lambda^2 \cdot \|x\|^2$$

and hence the desired inequality $\|f(x)\| > \lambda \cdot \|x\|$. The easier analogous proof of $\|f(x)\| \leq \lambda \cdot \|x\|$ for $E_{\lambda^2}^{f^*f}(x) = x$ is left to the reader. $\qquad \square$

Lemma 2.3. *Let U and V be Hilbert $\mathcal{N}(G)$-modules. Let $f \colon \mathrm{dom}(f) \subset U \to V$ be a G-equivariant closed densely defined operator. Then for $\lambda \in \mathbb{R}$ the spectral projection $E_{\lambda^2}^{f^*f}$ is G-equivariant and*

$$F(f)(\lambda) = \dim_{\mathcal{N}(G)}(\mathrm{im}(E_{\lambda^2}^{f^*f})).$$

Proof. As f^*f commutes with the unitary G-action, the same is true for its spectral projections. Since $\mathrm{im}(E_{\lambda^2}^{f^*f}) \in \mathcal{L}(f, \lambda)$ for $\lambda \geq 0$ holds by Lemma 2.2 (2), we have $\dim_{\mathcal{N}(G)}(\mathrm{im}(E_{\lambda^2}^{f^*f})) \leq F(f)(\lambda)$. From Lemma 2.2 (2) and Theorem 1.12 (2) we conclude that $E_{\lambda^2}^{f^*f}|_L \colon L \to \mathrm{im}(E_{\lambda^2}^{f^*f})$ is injective and hence $\dim_{\mathcal{N}(G)}(L) \leq \dim_{\mathcal{N}(G)}(\mathrm{im}(E_{\lambda^2}^{f^*f}))$ for $L \in \mathcal{L}(f, \lambda)$. $\qquad \square$

Lemma 2.4. *Let U and V be Hilbert $\mathcal{N}(G)$-modules. Let $f \colon \mathrm{dom}(f) \subset U \to V$ be a G-equivariant closed densely defined operator. Let $f = u|f|$ be its polar decomposition into a partial isometry u and a positive operator $|f|$ with $\mathrm{dom}(f) = \mathrm{dom}(|f|)$. Suppose that f is Fredholm and $b^{(2)}(f^*)$ is finite. Then $|f|$ and f^* are Fredholm,*

$$b^{(2)}(f) = b^{(2)}(|f|)$$

and

$$F(f)(\lambda) - b^{(2)}(f) = F(|f|)(\lambda) - b^{(2)}(|f|) = F(f^*)(\lambda) - b^{(2)}(f^*).$$

Proof. We can assume without loss of generality that f and f^* are injective, otherwise pass to the induced operator $\ker(f)^\perp \to \mathrm{clos}(\mathrm{im}(f))$. Obviously $F(u|f|)(\lambda) = F(|f|)(\lambda) = F(|f|u^{-1})(\lambda)$ and $f^* = |f|u^{-1}$ where $\mathrm{dom}(u|f|) = \mathrm{dom}(|f|)$ and $\mathrm{dom}(|f|u^{-1}) = u(\mathrm{dom}(f))$. $\qquad\square$

The intuitive meaning of the spectral density function becomes clear from the next example, namely, $F(f)(\lambda)$ counts with multiplicity the eigenvalues μ of $|f|$ satisfying $\mu \le \lambda$.

Example 2.5. Let G be finite and $f \colon U \to V$ be a map of finitely generated Hilbert $\mathcal{N}(G)$-modules, i.e. of finite dimensional unitary G-representations. Then $F(f)$ is the right-continuous step function whose value at λ is the sum of the complex dimensions of the eigenspaces of f^*f for eigenvalues $\mu \le \lambda^2$ divided by the order of G, or, equivalently, the sum of the complex dimensions of the eigenspaces of $|f|$ for eigenvalues $\mu \le \lambda$ divided by the order of G.

Example 2.6. Let $G = \mathbb{Z}^n$. In the sequel we use the notation and the identification $\mathcal{N}(\mathbb{Z}^n) = L^\infty(T^n)$ of Example 1.4. For $f \in L^\infty(T^n)$ the spectral density function $F(M_f)$ of $M_f \colon L^2(T^n) \to L^2(T^n)$ sends λ to the volume of the set $\{z \in T^n \mid |f(z)| \le \lambda\}$ (see Example 1.11).

Definition 2.7. *We say that a function* $F \colon [0, \infty) \to [0, \infty]$ *is a* density function *if* F *is monotone non-decreasing and right-continuous. If* F *and* G *are two density functions, we write* $F \preceq G$ *if there are* $C > 0$ *and* $\epsilon > 0$ *such that* $F(\lambda) \le G(C \cdot \lambda)$ *holds for all* $\lambda \in [0, \epsilon]$. *We say that* F *and* G *are* dilatationally equivalent *(in signs* $F \simeq G$*) if* $F \preceq G$ *and* $G \preceq F$. *We say that* F *is* Fredholm *if there exists* $\lambda > 0$ *such that* $F(\lambda) < \infty$.

Of course, the spectral density function $F(f)$ is a density function, and f is Fredholm if and only if the density function $F(f)$ is Fredholm. Recall that for a function $h \colon (0, \infty) \to [0, \infty)$ its *limit inferior* for $\lambda \to 0+$ is defined by

$$\liminf_{\lambda \to 0+} h(\lambda) := \sup \left\{\inf\{h(\lambda) \mid 0 < \lambda \le \mu\} \mid 0 < \mu\right\} \qquad \in [0, \infty].$$

Definition 2.8 (Novikov-Shubin invariants). *Let* F *be a Fredholm density function. The* L^2-Betti number *of* F *is*

$$b^{(2)}(F) := F(0).$$

Its Novikov-Shubin invariant *is*

$$\alpha(F) := \liminf_{\lambda \to 0+} \frac{\ln(F(\lambda) - F(0))}{\ln(\lambda)} \qquad \in [0, \infty],$$

provided that $F(\lambda) > b^{(2)}(F)$ holds for all $\lambda > 0$. Otherwise, we put $\alpha(F) := \infty^+$. If f is a Fredholm morphism of Hilbert $\mathcal{N}(G)$-modules, we write

$$\alpha(f) = \alpha(F(f)).$$

Here ∞^+ is a new formal symbol which should not be confused with ∞. We have $\alpha(F) = \infty^+$ if and only if there is an $\epsilon > 0$ such that $F(\epsilon) = b^{(2)}(F)$. Notice that $b^{(2)}(f) = b^{(2)}(F(f))$ in the notation of Definition 2.1 and Definition 2.8. The Novikov-Shubin invariant of a spectral density function F measures how fast $F(\lambda)$ approaches $F(0)$ for $\lambda \to 0+$. It is independent of the actual value $F(0)$ and depends only on the germ of $F(\lambda)$ at zero. The Novikov-Shubin invariant of a Fredholm map $f\colon U \to V$ of Hilbert $\mathcal{N}(G)$-modules measures how "thick" the spectrum of $|f|$ (or f^*f) is at zero. It is ∞^+ if and only if the spectrum of $|f|$ (or f^*f) has a gap at zero. In particular $\alpha(f) = \infty^+$ always holds if G is finite. This shows that the Novikov-Shubin invariants are only interesting for infinite G.

Example 2.9. In this example we show that any possible value can occur as Novikov-Shubin invariant of a map of Hilbert $\mathcal{N}(G)$-modules and in particular of a spectral density function. Define spectral density functions F_t for $t \in [0, \infty] \coprod \{\infty^+\}$ by $F_t(0) = 0$ and for $\lambda > 0$ by

$$F_0(\lambda) = \begin{cases} 0 & \lambda = 0; \\ \frac{-1}{\ln(\lambda)} & 0 < \lambda < e^{-1}; \\ 1 & e^{-1} < \lambda; \end{cases}$$

$$F_t(\lambda) = \lambda^t \qquad t \in (0, \infty);$$
$$F_\infty(\lambda) = \exp(-\lambda^{-1});$$
$$F_{\infty^+}(\lambda) = 0.$$

Then one easily checks for $t \in [0, \infty] \coprod \{\infty^+\}$

$$\alpha(F_t) = t.$$

Using Example 2.6 it is not hard to construct for $t \in [0, \infty] \coprod \{\infty^+\}$ maps $f_t\colon l^2(\mathbb{Z}) \to l^2(\mathbb{Z})$ of Hilbert $\mathcal{N}(\mathbb{Z})$-modules such that $F(f_t)(\lambda) = F_t(\lambda)$ for small λ and hence $\alpha(f_t) = t$ holds.

Notation 2.10. *Define an ordering on $[0, \infty] \coprod \{\infty^+\}$ by the standard ordering on $\mathbb{R}$ along with $r < \infty < \infty^+$ for all $r \in \mathbb{R}$. For all $\alpha, \beta \in [0, \infty] \coprod \{\infty^+\}$ we define*

$$\frac{1}{\alpha} \le \frac{1}{\beta} \Leftrightarrow \alpha \ge \beta.$$

Given $\alpha, \beta \in [0, \infty] \coprod \{\infty^+\}$, we give meaning to γ in the expression

$$\frac{1}{\alpha} + \frac{1}{\beta} = \frac{1}{\gamma}$$

as follows: If $\alpha, \beta \in (0, \infty)$, let γ be the real number for which this arithmetic expression of real numbers is true. If $\alpha = 0$ or $\beta = 0$, put $\gamma = 0$. If $\alpha \in (0, \infty)$ and $\beta \in \{\infty, \infty^+\}$, put γ to be α. If $\beta \in (0, \infty)$ and $\alpha \in \{\infty, \infty^+\}$, put γ to be β. If α and β belong to $\{\infty, \infty^+\}$ and are not both ∞^+, put $\gamma = \infty$. If both α and β are ∞^+, put $\gamma = \infty^+$. For $r, \alpha \in [0, \infty] \coprod \{\infty^+\}$ we define

$$r \cdot \alpha \in [0, \infty] \coprod \{\infty^+\}$$

as follows. Given $r \in [0, \infty)$ and $\alpha \in [0, \infty)$, we define $r \cdot \alpha \in [0, \infty)$ to be the ordinary product of real numbers. For $r \in (0, \infty)$ we put $r \cdot \infty = \infty \cdot r = \infty$. Put $0 \cdot \infty = \infty \cdot 0 = 0$. Define $\infty^+ \cdot \alpha = \alpha \cdot \infty^+ = \infty^+$ for $\alpha \in [0, \infty] \coprod \{\infty^+\}$.

For example,

$$\frac{1}{\infty} + \frac{1}{\pi} = \frac{1}{\pi}, \quad \frac{1}{\infty^+} + \frac{1}{\pi} = \frac{1}{\pi}, \quad \frac{1}{\infty} + \frac{1}{\infty^+} = \frac{1}{\infty}, \quad \frac{1}{\infty^+} + \frac{1}{\infty^+} = \frac{1}{\infty^+},$$

$$\frac{1}{\alpha} \leq \frac{1}{\infty} + \frac{1}{4} + \frac{1}{2} \Leftrightarrow \alpha \geq 4/3, \quad \frac{1}{\alpha} \leq \frac{1}{\infty} + \frac{1}{\infty^+} + \frac{1}{\infty} \Leftrightarrow \alpha \geq \infty \text{ and } 5 \cdot \infty = \infty.$$

Lemma 2.11. Let F and F' be density functions and $f \colon U \to V$ be a morphism of $\mathcal{N}(G)$-Hilbert modules. Assume that F' is Fredholm. Then

(1) If $F \preceq F'$, then F is Fredholm and $b^{(2)}(F) \leq b^{(2)}(F')$;
(2) If $F \preceq F'$ and $b^{(2)}(F) = b^{(2)}(F')$, then $\alpha(F) \geq \alpha(F')$;
(3) If $F \simeq F'$, then $b^{(2)}(F) = b^{(2)}(F')$ and $\alpha(F) = \alpha(F')$;
(4) $\alpha(F(\lambda^r)) = r \cdot \alpha(F(\lambda))$ for $r \in (0, \infty)$;
(5) $\alpha(F) = \alpha(F - b^{(2)}(F))$;
(6) $b^{(2)}(f) = \dim_{\mathcal{N}(G)}(\ker(f^* f)) = \dim_{\mathcal{N}(G)}(\ker(f))$;
(7) If f is zero and $\dim_{\mathcal{N}(G)}(U) < \infty$, then f is Fredholm and $\alpha(f) = \infty^+$;
(8) Suppose that $\dim_{\mathcal{N}(G)}(U) = \dim_{\mathcal{N}(G)}(V) < \infty$ or that $U = V$ and f is selfadjoint. Then $f \colon U \to V$ is an isomorphism if and only if f is Fredholm, $b^{(2)}(f) = 0$ and $\alpha(f) = \infty^+$;
(9) Assume that $i \colon V \to V'$ is injective with closed image and $p \colon U' \to U$ is surjective with $\dim_{\mathcal{N}(G)}(\ker(p)) < \infty$. Then f is Fredholm if and only if $i \circ f \circ p$ is Fredholm, and in this case $\alpha(i \circ f \circ p) = \alpha(f)$;
(10) If F and F' are Fredholm then $\alpha(F + F') = \min\{\alpha(F), \alpha(F')\}$;
(11) $F(f^* f)(\lambda^2) = F(f)(\lambda)$, $b^{(2)}(f^* f) = b^{(2)}(f)$ and $\alpha(f^* f) = 1/2 \cdot \alpha(f)$;
(12) If $F(0) = F'(0)$, then $\alpha(\max\{F, F'\}) = \min\{\alpha(F), \alpha(F')\}$. $\square$

Proof. (1) to (7) are easy and left to the reader.

(8) The map $f^* f$ of Hilbert $\mathcal{N}(G)$-modules is invertible if and only if there is $\epsilon > 0$ such that $E_\lambda^{f^* f}$ is zero for $\lambda < \epsilon$. If such $\epsilon > 0$ exists, the inverse is given by $\int_{[\epsilon, \|f^* f\|]} \lambda^{-1} dE_\lambda^{f^* f}$ (see (1.64)). Provided that f is selfadjoint, $f^* f = f^2$ and $f^* f$ is invertible if and only if f is invertible. Provided that $\dim_{\mathcal{N}(G)}(U) = \dim_{\mathcal{N}(G)}(V) < \infty$, Theorem 1.12 (1) and (2) imply that f is invertible if and only if $f^* f$ is invertible.

(9) The Inverse Mapping Theorem implies $F(i \circ f \circ p)(\lambda) \simeq F(f)(\lambda) - \dim_{\mathcal{N}(G)}(\ker(p))$.

(10) Since $b^{(2)}(F + F') = b^{(2)}(F) + b^{(2)}(F')$, we may assume without loss of generality by assertion (5) that $b^{(2)}(F) = b^{(2)}(F') = b^{(2)}(F + F') = 0$. As $F, F' \leq F + F'$, assertion (2) implies that $\alpha(F + F') \leq \min\{\alpha(F), \alpha(F')\}$. To verify the reverse inequality, we may assume without loss of generality that $\alpha(F) \leq \alpha(F')$. The cases $\alpha(F) = 0$ and $\alpha(F) = \infty^+$ are trivial, and so we assume that $0 < \alpha(F) \leq \infty$. Consider any real number α satisfying $0 < \alpha < \alpha(F)$. Then there exists a constant $K > 0$ such that for small positive λ we have $F(\lambda), F'(\lambda) \leq K\lambda^{\alpha}$, and so $F(\lambda) + F'(\lambda) \leq 2K \cdot \lambda^{\alpha}$, implying that $\alpha \leq \alpha(F + F')$. The assertion follows.

The elementary proof of the other assertions is left to the reader. $\qquad \square$

In the rest of this Subsection 2.1.1 we state and prove some basic properties of spectral density functions and Novikov-Shubin invariants for compositions and exact sequences which will be needed for the proofs of the main results in Subsection 2.1.2.

Lemma 2.12. *Let $f\colon U \to V$ be a map of Hilbert $\mathcal{N}(G)$-modules which is Fredholm and a weak isomorphism. Let $L \subset V$ be a Hilbert $\mathcal{N}(G)$-submodule. Then f restricts to a weak isomorphism from $f^{-1}(L)$ to L and*

$$\dim_{\mathcal{N}(G)}(L) = \dim_{\mathcal{N}(G)}(f^{-1}(L)).$$

Proof. From the Polar Decomposition of f, we can assume without loss of generality that $U = V$ and f is positive. Obviously the restriction of f to $f^{-1}(L)$ is injective. It remains to show that $f(f^{-1}(L))$ is dense in L because then f induces a weak isomorphism from $f^{-1}(L)$ to L and the claim about the dimension follows from Theorem 1.12 (2).

Fix an orthogonal decomposition $L = \mathrm{clos}(f(f^{-1}(L))) \oplus M$, where M is an Hilbert $\mathcal{N}(G)$-submodule of L. It remains to prove $\dim_{\mathcal{N}(G)}(M) = 0$ because then $M = 0$ follows from Theorem 1.12 (1). As $f(f^{-1}(M)) \subset M$ and $f(f^{-1}(M)) \subset f(f^{-1}(L))$, we get $f(f^{-1}(M)) = 0$ and therefore $M \cap \mathrm{im}(f) = 0$. For $\lambda > 0$ restriction defines a map $E^f_{\lambda}|_M \colon M \to E^f_{\lambda}(U)$. If $m \in \ker(E^f_{\lambda}|_M)$ then the Spectral Theorem shows that $m \in \mathrm{im}(f)$, a preimage is given by $\left(\int_{\lambda}^{||f||} \mu^{-1} dE^f_{\mu} \right)(m)$. Thus $\ker(E^f_{\lambda}|_M) = 0$. Theorem 1.12 (2) shows $\dim_{\mathcal{N}(G)}(M) \leq \dim_{\mathcal{N}(G)}(E^f_{\lambda}(U))$. As f is injective and Fredholm and E_{λ} is right-continuous in λ, Theorem 1.12 (4) implies $\lim_{\lambda \to 0+} \dim_{\mathcal{N}(G)}(E^f_{\lambda}(U)) = \dim_{\mathcal{N}(G)}(E^f_0(U)) = 0$. Thus $\dim_{\mathcal{N}(G)}(M) = 0$. $\qquad \square$

Lemma 2.13. *Let $f\colon U \to V$ and $g\colon V \to W$ be morphisms of Hilbert $\mathcal{N}(G)$-modules. Then*

(1) $F(f)(\lambda) \leq F(gf)(||g|| \cdot \lambda)$;
(2) $F(g)(\lambda) \leq F(gf)(||f|| \cdot \lambda)$ if f is Fredholm and has dense image;
(3) $F(gf)(\lambda) \leq F(g)(\lambda^{1-r}) + F(f)(\lambda^r)$ for all $r \in (0,1)$.

Proof. (1) Consider $L \in \mathcal{L}(f, \lambda)$. For all $x \in L$, $|gf(x)| \leq ||g|| \cdot |f(x)| \leq ||g|| \cdot \lambda \cdot |x|$. This implies that $L \in \mathcal{L}(gf, ||g|| \cdot \lambda)$, and the claim follows.

(2) Consider $L \in \mathcal{L}(g, \lambda)$. For all $x \in f^{-1}(L)$, we have $|gf(x)| \leq \lambda \cdot |f(x)| \leq \lambda \cdot ||f|| \cdot |x|$, implying $f^{-1}(L) \in \mathcal{L}(gf, ||f|| \cdot \lambda)$. Hence it remains to show $\dim_{\mathcal{N}(G)}(L) \leq \dim_{\mathcal{N}(G)}(f^{-1}(L))$. Let $p: U \to U/\ker f$ be the projection and let $\overline{f}: U/\ker(f) \to V$ be the map induced by f. Clearly $\overline{f}$ is also Fredholm. Since p is surjective and $\overline{f}$ is a weak isomorphism, Theorem 1.12 (2) and Lemma 2.12 imply that $\dim_{\mathcal{N}(G)}\left(f^{-1}(L)\right) \geq \dim_{\mathcal{N}(G)}\left(p(f^{-1}(L))\right) = \dim_{\mathcal{N}(G)}\left(\overline{f}^{-1}(L)\right) = \dim_{\mathcal{N}(G)}(L)$.

(3) Consider $L \in \mathcal{L}(gf, \lambda)$. Let L_0 be the kernel of $E_{\lambda^{2r}}^{f^*f}|_L$. We have a weakly exact sequence $0 \to L_0 \to L \to \text{clos}(E_{\lambda^{2r}}^{f^*f}(L)) \to 0$. From Lemma 2.2 (2) we get that $|f(x)| > \lambda^r \cdot |x|$ for all nonzero $x \in L_0$. In particular, $f|_{L_0}: L_0 \to \text{clos}(f(L_0))$ is a weak isomorphism, and so Theorem 1.12 (2) implies that $\dim_{\mathcal{N}(G)}(L_0) = \dim_{\mathcal{N}(G)}(\text{clos}(f(L_0)))$. For $x \in L_0$ we have

$$|gf(x)| \leq \lambda \cdot |x| \leq \frac{\lambda}{\lambda^r} \cdot |f(x)| = \lambda^{1-r} \cdot |f(x)|.$$

Hence $\text{clos}(f(L_0)) \in \mathcal{L}(g, \lambda^{1-r})$. This shows that $\dim_{\mathcal{N}(G)}(L_0) \leq F(g)(\lambda^{1-r})$. From Theorem 1.12 and Lemma 2.3 we conclude

$$\dim_{\mathcal{N}(G)}\left(\text{clos}(E_{\lambda^{2r}}^{f^*f}(L))\right) \leq \dim_{\mathcal{N}(G)}\left(\text{im}(E_{\lambda^{2r}}^{f^*f})\right) = F(f)(\lambda^r);$$
$$\dim_{\mathcal{N}(G)}(L) = \dim_{\mathcal{N}(G)}(L_0) + \dim_{\mathcal{N}(G)}\left(\text{clos}(E_{\lambda^r}^{f^*f}(L))\right).$$

This implies that $\dim_{\mathcal{N}(G)}(L) \leq F(g)(\lambda^{1-r}) + F(f)(\lambda^r)$. □

Lemma 2.14. *Let $f: U \to V$ and $g: V \to W$ be morphisms of Hilbert $\mathcal{N}(G)$-modules.*

(1) If f and g are Fredholm, then the composition gf is Fredholm. If f and g are Fredholm and $\ker(g) \subset \text{clos}(\text{im}(f))$ then

$$\frac{1}{\alpha(gf)} \leq \frac{1}{\alpha(f)} + \frac{1}{\alpha(g)};$$

(2) If gf is Fredholm, then f is Fredholm. If gf is Fredholm and $\ker(g) \cap \text{im}(f) = 0$, then

$$\alpha(f) \geq \alpha(gf).$$

If gf is Fredholm and f has dense image, then g is Fredholm and

$$\alpha(g) \geq \alpha(gf).$$

Proof. We can assume without loss of generality that f is injective by Lemma 2.11 (9), otherwise replace f by the induced map $U/\ker(f) \to V$.

(1) We conclude from Lemma 2.13 (3) that gf is Fredholm. Now assume $\ker(g) \subset \text{clos}(\text{im}(f))$. As $f \colon U \to \text{clos}(\text{im}(f))$ is a weak isomorphism, Lemma 2.12 implies that $b^{(2)}(gf) = b^{(2)}(g) = b^{(2)}(f) + b^{(2)}(g)$. From Lemma 2.13 (3) we conclude for $0 < r < 1$

$$F(gf, \lambda) - b^{(2)}(gf) \le F(f, \lambda^r) - b^{(2)}(f) + F(g, \lambda^{1-r}) - b^{(2)}(g).$$

Assertions (2), (4), (5) and (10) of Lemma 2.11 show

$$\alpha(gf) \ge \min\{r \cdot \alpha(f), (1-r) \cdot \alpha(g)\}.$$

We only need to consider the case $\alpha(f), \alpha(g) \in (0, \infty)$, the other cases being now obvious. Taking inverses gives

$$\frac{1}{\alpha(gf)} \le \max\left\{\frac{1}{r \cdot \alpha(f)}, \frac{1}{(1-r) \cdot \alpha(g)}\right\}.$$

Since $\frac{1}{r \cdot \alpha(f)}$ (resp. $\frac{1}{(1-r) \cdot \alpha(g)}$) is a strictly monotone decreasing (resp. increasing) function in r, the maximum on the right side, viewed as a function of r, obtains its minimum precisely when the two functions of r have the same value. One easily checks that this is the case if and only if $r = \frac{\alpha(g)}{\alpha(f) + \alpha(g)}$, and the claim follows.

(2) This follows from Lemma 2.13 (1) and (2) and Lemma 2.11 (2), (5) and (9) since $\ker(g) \cap \text{im}(f) = 0 \Rightarrow b^{(2)}(gf) = b^{(2)}(f)$ and $b^{(2)}(gf) \le b^{(2)}(g)$ if f is injective. $\qquad\square$

Lemma 2.15. *Let* $u \colon U \to U'$ *and* $v \colon V \to V'$ *be morphisms of Hilbert* $\mathcal{N}(G)$-*modules and let*

$$
\begin{array}{ccccccccc}
0 & \longrightarrow & U_1 & \longrightarrow & U_0 & \longrightarrow & U_2 & \longrightarrow & 0 \\
& & \downarrow{\scriptstyle f_1} & & \downarrow{\scriptstyle f_0} & & \downarrow{\scriptstyle f_2} & & \\
0 & \longrightarrow & V_1 & \longrightarrow & V_0 & \longrightarrow & V_2 & \longrightarrow & 0
\end{array}
$$

be a commutative diagram of maps of Hilbert $\mathcal{N}(G)$-*modules whose rows are exact. Then*

(1) $u \oplus v$ *is Fredholm if and only if both* u *and* v *are Fredholm. In this case,*

$$\alpha(u \oplus v) = \min\{\alpha(u), \alpha(v)\};$$

(2) If f_1 *and* f_2 *are Fredholm, then* f_0 *is Fredholm. If* f_1 *and* f_2 *are Fredholm and* f_1 *has dense image or* f_2 *is injective, then*

$$\frac{1}{\alpha(f_0)} \le \frac{1}{\alpha(f_1)} + \frac{1}{\alpha(f_2)};$$

(3) If f_0 is Fredholm, then f_1 is Fredholm. If f_0 is Fredholm and f_2 is injective, then $\alpha(f_1) \geq \alpha(f_0)$. If f_0 is Fredholm and f_1 has dense image, then f_2 is Fredholm and $\alpha(f_2) \geq \alpha(f_0)$.

Proof. (1) This follows from Lemma 2.11 (10) using $F(u \oplus v, \lambda) = F(u, \lambda) + F(v, \lambda)$.

(2) Since an exact sequence of Hilbert $\mathcal{N}(G)$-modules always splits by the Open Mapping Theorem we can assume without loss of generality by Lemma 2.11 (9) that $U_0 = U_1 \oplus U_2$ and $V_0 = V_1 \oplus V_2$ and f_0 has the shape

$$f_0 = \begin{pmatrix} f_1 & f_3 \\ 0 & f_2 \end{pmatrix}$$

for a morphism $f_3 \colon U_2 \to V_1$. We have $f_0 = gf$, where $g = \begin{pmatrix} 1 & f_3 \\ 0 & f_2 \end{pmatrix}$ and $f = \begin{pmatrix} f_1 & 0 \\ 0 & 1 \end{pmatrix}$. Since f_1 is Fredholm, f is Fredholm and $\alpha(f_1) = \alpha(f)$ by assertion (1). Since we can write

$$g = \begin{pmatrix} 1 & 0 \\ 0 & f_2 \end{pmatrix} \cdot \begin{pmatrix} 1 & f_3 \\ 0 & 1 \end{pmatrix}$$

we conclude from Lemma 2.11 (9) and assertion (1) that g is Fredholm and $\alpha(g) = \alpha(f_2)$. If f_2 is injective then g is injective, and if f_1 has dense image then f has dense image. In both cases we have $\ker(g) \subset \operatorname{clos}(\operatorname{im}(f))$. Now apply Lemma 2.14 (1).

(3) follows analogously using Lemma 2.14 (2).   □

2.1.2 Spectral Density Functions of Hilbert Chain Complexes

Definition 2.16 (Spectral density function). *Let C_* be a Hilbert $\mathcal{N}(G)$-chain complex. Define its p-th spectral density function to be the spectral density function (see Definition 2.1) of its p-th differential c_p restricted to $\operatorname{im}(c_{p+1})^{\perp}$*

$$F_p(C_*) := F\left(c_p|_{\operatorname{im}(c_{p+1})^{\perp}} \colon \operatorname{im}(c_{p+1})^{\perp} \to C_{p-1}\right).$$

Suppose that C_* is Fredholm, i.e. for $p \in \mathbb{Z}$ there exists $\lambda_p > 0$ with $F_p(\lambda_p) < \infty$ (see Definition 1.20 and Lemma 2.3). Define its p-th Novikov-Shubin invariant be the Novikov-Shubin invariant (see Definition 2.8) of $F_p(C_*)$,

$$\alpha_p(C_*) := \alpha(F_p(C_*)).$$

Recall that $\Delta_p \colon C_p \to C_p$ is the Laplacian of C_* (see (1.17)). Put

$$F_p^{\Delta}(C_*) := F(\Delta_p);$$
$$\alpha_p^{\Delta}(C_*) := \alpha(\Delta_p).$$

Define for a Hilbert $\mathcal{N}(G)$-cochain complex C^* which is Fredholm

$$F_p(C^*) := F\left(c^p|_{\mathrm{im}(c^{p-1})^\perp} : \mathrm{im}(c^{p-1})^\perp \to C^{p+1}\right);$$
$$\alpha_p(C^*) := \alpha(F_p(C^*)).$$

The *dual* $(C_*)^*$ of a Hilbert chain complex C_* is the Hilbert cochain complex whose p-th cochain module is C_p and whose p-th codifferential $c^p :=$ $c^*_{p+1} : C_p \to C_{p+1}$ is the adjoint of c_{p+1}.

Lemma 2.17. *Let C_* and D_* be Hilbert chain complexes. Then*

(1) If C_* is Fredholm, then Δ_p is Fredholm and

$$\alpha_p^\Delta(C_*) = 1/2 \cdot \min\{\alpha_p(C_*), \alpha_{p+1}(C_*)\};$$

(2) Suppose that C_* is Fredholm. Then $(C_*)^*$ is Fredholm and

$$F_{p+1}(C_*) = F_p((C_*)^*);$$
$$b_p^{(2)}(C_*) = b_p^{(2)}((C_*)^*);$$
$$\alpha_{p+1}(C_*) = \alpha_p((C_*)^*);$$

(3) $C_* \oplus D_*$ is Fredholm if and only if both C_* and D_* are Fredholm. In this case

$$F_p(C_* \oplus D_*) = F_p(C_*) + F_p(D_*);$$
$$\alpha_p(C_* \oplus D_*) = \min\{\alpha_p(C_*), \alpha_p(D_*)\}.$$

Proof. (1) Lemma 1.18 and Lemma 2.15 (1) imply

$$F(\Delta_p) - b^{(2)}(\Delta_p) = F(c_p^* c_p : \ker(c_p)^\perp \to \ker(c_p)^\perp)$$
$$+ F(c_{p+1}c_{p+1}^* : \ker(c_{p+1}^*)^\perp \to \ker(c_{p+1}^*)^\perp);$$
$$\alpha(\Delta_p) = \min\{\alpha(c_p^* c_p : \ker(c_p)^\perp \to \ker(c_p)^\perp),$$
$$\alpha(c_{p+1}c_{p+1}^* : \ker(c_{p+1}^*)^\perp \to \ker(c_{p+1}^*)^\perp)\}.$$

We conclude from Lemma 2.11 (5), (9) and (11)

$$1/2 \cdot \alpha_p(C_*) = \alpha(c_p^* c_p : \ker(c_p)^\perp \to \ker(c_p)^\perp);$$
$$1/2 \cdot \alpha_{p+1}(C_*) = \alpha(c_{p+1}c_{p+1}^* : \ker(c_{p+1}^*)^\perp \to \ker(c_{p+1}^*)^\perp).$$

(2) This follows from Lemma 1.18 and Lemma 2.4.

(3) This follows from Lemma 2.15 (1). $\square$

We recall that a Hilbert $\mathcal{N}(G)$-chain complex C_* is said to be *contractible* if there exists a *chain contraction* γ_*, i.e. a collection of morphisms $\gamma_p : C_p \to C_{p+1}$ for $p \in \mathbb{Z}$ such that $\gamma_{p-1}c_p + c_{p+1}\gamma_p = \mathrm{id}$ for all p.

Lemma 2.18. *The following assertions are equivalent for a Hilbert $\mathcal{N}(G)$-chain complex C_*:*

(1) C_* is contractible;
(2) Δ_p is invertible for all p;
(3) C_* is Fredholm, $b_p^{(2)}(C_*) = 0$ and $\alpha_p(C_*) = \infty^+$ for all p.

Proof. (1) $\Rightarrow$ (3) We can construct morphisms $\overline{c_p}: C_p/\operatorname{clos}(\operatorname{im}(c_{p+1})) \to C_{p-1}$ and $\overline{\gamma_{p-1}}: C_{p-1} \to C_p/\operatorname{clos}(\operatorname{im}(c_{p+1}))$, using c_p and γ_{p-1}, such that $\overline{\gamma_{p-1}} \circ \overline{c_p} = \operatorname{id}$. Hence $\overline{c_p}$ induces an invertible operator onto its image.

Lemma 2.11 (8) and (9) imply that $\overline{c_p}$ is Fredholm and hence C_* is Fredholm at p, $b_p^{(2)}(C_*) = b^{(2)}(\overline{c_p}) = 0$ and $\alpha_p(C_*) = \alpha(\overline{c_p}) = \infty^+$.

(3) $\Rightarrow$ (2) From Lemma 1.18 and Lemma 2.17 (1) we conclude that Δ_p is Fredholm, $b^{(2)}(\Delta_p) = 0$ and $\alpha(\Delta_p) = \infty^+$ for all p. Now apply Lemma 2.11 (8).

(2) $\Rightarrow$ (1) Suppose that Δ_p is invertible for all p. Then γ_* with $\gamma_p := \Delta_{p+1}^{-1} \circ c_{p+1}^*$ is a chain contraction of C_*. $\square$

Next we reprove the homotopy invariance of the Novikov-Shubin invariants [240, Proposition 4.1].

Theorem 2.19. *If $f_*: C_* \to D_*$ is a chain homotopy equivalence of Hilbert $\mathcal{N}(G)$-chain complexes, then for all $p \in \mathbb{Z}$ we have*

$$F_p(C_*) \simeq F_p(D_*).$$

In particular C_* is Fredholm at p if and only if D_* is Fredholm at p. In this case

$$\alpha_p(C_*) = \alpha_p(D_*).$$

Proof. There are exact sequences of chain complexes $0 \to C_* \to \operatorname{cyl}_*(f_*) \to \operatorname{cone}_*(f_*) \to 0$ and $0 \to D_* \to \operatorname{cyl}_*(f_*) \to \operatorname{cone}_*(C_*) \to 0$ with $\operatorname{cone}_*(f_*)$ and $\operatorname{cone}_*(C_*)$ being contractible. We obtain chain isomorphisms $C_* \oplus \operatorname{cone}_*(f_*) \xrightarrow{\cong} \operatorname{cyl}_*(f_*)$ and $D_* \oplus \operatorname{cone}_*(C_*) \xrightarrow{\cong} \operatorname{cyl}_*(f_*)$ by the following general construction for an exact sequence $0 \to C_* \xrightarrow{j_*} D_* \xrightarrow{q_*} E_* \to 0$ with contractible E_*: Choose a chain contraction ϵ_* for E_* and for each p a morphism $t_p: E_p \to D_p$ such that $q_p \circ t_p = \operatorname{id}$. Put

$$s_p = d_{p+1} \circ t_{p+1} \circ \epsilon_p + t_p \circ \epsilon_{p-1} \circ e_p.$$

This defines a chain map $s_*: E_* \to D_*$ such that $q_* \circ s_* = \operatorname{id}$. Define a chain map $u_*: D_* \to C_*$ by requiring that for $x \in D_p$ its image $u_p(x)$ is the unique $y \in C_p$ such that $x = s_p q_p(x) + j_p(y)$. Then $j_* + s_*$ is a chain isomorphism $C_* \oplus E_* \to D_*$ with inverse $u_* \oplus q_*$. Since $C_* \oplus \operatorname{cone}_*(f_*)$ and $D_* \oplus \operatorname{cone}_*(C_*)$ are isomorphic and $\operatorname{cone}_*(f_*)$ and $\operatorname{cone}_*(C_*)$ are contractible, Lemma 2.11 (9), Lemma 2.17 (3) and Lemma 2.18 imply $F_p(C_*) \simeq F_p(D_*)$. Now the other statements follow from Lemma 2.11 (3). $\square$

The next result is taken from [322, Theorem 2.3 on page 27].

Theorem 2.20. *Let* $0 \to C_* \xrightarrow{j_*} D_* \xrightarrow{q_*} E_* \to 0$ *be an exact sequence of Hilbert* $\mathcal{N}(G)$-*chain complexes. Suppose that two of the chain complexes are Fredholm. Then all three chain complexes are Fredholm and we have*

$$\frac{1}{\alpha_p(D_*)} \leq \frac{1}{\alpha_p(C_*)} + \frac{1}{\alpha_p(E_*)} + \frac{1}{\alpha(\delta_p)};$$

$$\frac{1}{\alpha_p(E_*)} \leq \frac{1}{\alpha_{p-1}(C_*)} + \frac{1}{\alpha_p(D_*)} + \frac{1}{\alpha(H_{p-1}(j_*))};$$

$$\frac{1}{\alpha_p(C_*)} \leq \frac{1}{\alpha_p(D_*)} + \frac{1}{\alpha_{p+1}(E_*)} + \frac{1}{\alpha(H_p(q_*))},$$

where $\delta_p \colon H_p^{(2)}(E_*) \to H_{p-1}^{(2)}(C_*)$ *is the boundary operator in the long weakly exact homology sequence (see Theorem 1.21).*

Proof. We first treat the case where C_* and E_* are Fredholm. Then D_* is Fredholm by Lemma 2.15 (2). We now show that

$$\frac{1}{\alpha_p(D_*)} \leq \frac{1}{\alpha_p(C_*)} + \frac{1}{\alpha_p(E_*)} + \frac{1}{\alpha(\delta_p)}. \tag{2.21}$$

The given exact sequence $0 \to C_* \xrightarrow{j_*} D_* \xrightarrow{q_*} E_* \to 0$ induces the following commutative diagram with exact rows, where $\overline{q_p}$, $\overline{d_p}$ and $\overline{e_p}$ are the canonical homomorphisms induced by q_p, d_p and e_p, and i is the inclusion

$$
\begin{array}{ccccccccc}
0 & \longrightarrow & \ker \overline{q_p} & \xrightarrow{\ i\ } & D_p/\operatorname{clos}(\operatorname{im}(d_{p+1})) & \xrightarrow{\overline{q_p}} & E_p/\ker(e_p) & \longrightarrow & 0 \\
 & & \partial_p \downarrow & & \overline{d_p} \downarrow & & \overline{e_p} \downarrow & & \\
0 & \longrightarrow & C_{p-1} & \xrightarrow[j_{p-1}]{} & D_{p-1} & \xrightarrow[q_{p-1}]{} & E_{p-1} & \longrightarrow & 0
\end{array}
\tag{2.22}
$$

We define ∂_p in the above diagram as follows. Let $x \in \ker(e_p q_p)$ represent $[x] \in \ker(\overline{q_p})$. Then $d_p(x) = j_{p-1}(y)$ for a unique $y \in C_{p-1}$. We put $\partial_p([x]) = y$. (In fact, $y \in \ker(c_{p-1})$.)

Next we construct a sequence which we will show to be weakly exact

$$C_p \xrightarrow{\overline{j_p}} \ker(\overline{q_p}) \xrightarrow{\widehat{q_p}} H_p^{(2)}(E_*) \to 0 \tag{2.23}$$

The map $\overline{j_p}$ is induced by j_p in the obvious way. To define $\widehat{q_p}$, consider an $x \in D_p$ whose class $[x] \in D_p/\operatorname{clos}(\operatorname{im}(d_{p+1}))$ lies in $\ker(\overline{q_p})$. Then $q_p(x)$ is in the kernel of e_p and determines a class $[q_p(x)]$ in $H_p^{(2)}(E_*)$. Define $\widehat{q_p}([x])$ to be $[q_p(x)]$. One easily checks that $\widehat{q_p} \circ \overline{j_p}$ is zero and $\widehat{q_p}$ is surjective. We will show that $\ker(\widehat{q_p})$ is contained in $\operatorname{clos}(\operatorname{im}(\overline{j_p}))$. Consider $[x] \in \ker(\widehat{q_p})$ with representative $x \in D_p$. Since $[q_p(x)] \in H_p^{(2)}(E_*)$ is zero, there is a sequence $(y_n)_{n \geq 1}$ in E_{p+1} such that in E_p

$$\lim_{n \to \infty} (q_p(x) - e_{p+1}(y_n)) = 0.$$

As q_{p+1} is surjective, there is a sequence $\{u_n\}_{n \geq 1}$ in D_{p+1} such that $y_n = q_{p+1}(u_n)$. Thus

$$\lim_{n \to \infty} q_p (x - d_{p+1}(u_n)) = 0.$$

We write $x - d_{p+1}(u_n) = j_p(w_n) + r_n$, where $w_n \in C_p$ and $r_n \in \mathrm{im}(j_p)^{\perp}$. Then we obtain $\lim_{n \to \infty} q_p(r_n) = 0$. As the restriction of q_p to $\mathrm{im}(j_p)^{\perp}$ is an isomorphism, we conclude $\lim_{n \to \infty} r_n = 0$. Thus

$$x = \lim_{n \to \infty} (j_p(w_n) + d_{p+1}(u_n)),$$

and hence in $D_p/\operatorname{clos}(\mathrm{im}(d_{p+1}))$

$$[x] = \lim_{n \to \infty} \overline{j}_p(w_n).$$

This finishes the proof of weak exactness of (2.23).

Next, we construct a commutative diagram with exact rows

$$
\begin{array}{ccccccccc}
0 & \longrightarrow & \ker(\widehat{q}_p) & \xrightarrow{\; l_1 \;} & \ker(\overline{q}_p) & \xrightarrow{\; \widehat{q}_p \;} & H_p^{(2)}(E_*) & \longrightarrow & 0 \\
 & & \Big\downarrow{\overline{\partial}_p} & & \Big\downarrow{\partial_p} & & \Big\downarrow{\delta_p} & & \quad (2.24) \\
0 & \longrightarrow & \operatorname{clos}(\mathrm{im}(c_p)) & \xrightarrow[\; l_2 \;]{} & \ker(c_{p-1}) & \xrightarrow[\;\mathrm{pr}\;]{} & H_{p-1}^{(2)}(C_*) & \longrightarrow & 0
\end{array}
$$

The maps l_1 and l_2 are the canonical inclusions and the map pr is the canonical projection. The map $\overline{\partial}_p$ is induced by ∂_p. One easily verifies that the diagram commutes and has exact rows.

Let $\widetilde{\overline{j}}_p \colon C_p \to \ker(\widehat{q}_p)$ be the morphism induced by $\overline{j}_p$ whose image is dense by the weak exactness of (2.23). One easily checks that $\overline{\partial}_p \circ \widetilde{\overline{j}}_p = c_p$. As c_p is Fredholm by assumption, Lemma 2.14 (2) implies that $\overline{\partial}_p$ is Fredholm and

$$\alpha(\overline{\partial}_p) \geq \alpha_p(C_*). \qquad (2.25)$$

As E_* is Fredholm and hence $H_p^{(2)}(E_*)$ is finite dimensional, δ_p is Fredholm. As $\overline{\partial}_p$ has dense image, Lemma 2.15 (2) applied to (2.24) shows that ∂_p is Fredholm and

$$\frac{1}{\alpha(\partial_p)} \leq \frac{1}{\alpha(\overline{\partial}_p)} + \frac{1}{\alpha(\delta_p)}. \qquad (2.26)$$

From Lemma 2.11 (9) $\overline{e}_p$ is Fredholm and

$$\alpha(\overline{e}_p) = \alpha_p(E). \qquad (2.27)$$

As $\overline{e_p}$ is injective, Lemma 2.15 (2) applied to (2.22) shows

$$\frac{1}{\alpha_p(D_*)} \le \frac{1}{\alpha(\partial_p)} + \frac{1}{\alpha(\overline{e_p})}. \tag{2.28}$$

Now (2.21) follows from (2.25), (2.26), (2.27) and (2.28).

Next we show under the assumption that D_* and C_* are Fredholm that E_* is Fredholm and that

$$\frac{1}{\alpha_p(E_*)} \le \frac{1}{\alpha_{p-1}(C_*)} + \frac{1}{\alpha_p(D_*)} + \frac{1}{\alpha(H^{(2)}_{p-1}(j_*))}. \tag{2.29}$$

There are an exact sequence $0 \to D_* \to \mathrm{cyl}_*(q_*) \to \mathrm{cone}_*(q_*) \to 0$ and chain homotopy equivalences $E_* \to \mathrm{cyl}_*(q_*)$ and $\Sigma C_* \to \mathrm{cone}_*(q_*)$. We conclude from Theorem 2.19 and (2.21) that E_* is Fredholm and (2.29) is true since the connecting map from $H^{(2)}_p(\mathrm{cone}_*(q_*)) \to H^{(2)}_{p-1}(D_*)$ agrees under the identification $H^{(2)}_p(\mathrm{cone}_*(q_*)) = H^{(2)}_p(\Sigma C_*) = H^{(2)}_{p-1}(C_*)$ with the map $H^{(2)}_{p-1}(j_*) \colon H^{(2)}_{p-1}(C_*) \to H^{(2)}_{p-1}(D_*)$.

Analogously one shows under the assumption that D_* and E_* are Fredholm that C_* is Fredholm and

$$\frac{1}{\alpha_p(C_*)} \le \frac{1}{\alpha_p(D_*)} + \frac{1}{\alpha_{p+1}(E_*)} + \frac{1}{\alpha(H^{(2)}_p(q_*))}. \tag{2.30}$$

This finishes the proof of Theorem 2.20.   □

2.1.3 Product Formula for Novikov-Shubin Invariants

In this subsection we deal with the Novikov-Shubin invariants of a tensor product of Hilbert chain complexes. We will only consider positive chain complexes C_*, i.e. $C_n = 0$ for $n < 0$.

Lemma 2.31. *Let G and H be discrete groups. Let $f \colon U \to U$ and $g \colon V \to V$ be positive maps of Hilbert $\mathcal{N}(G)$-modules and $\mathcal{N}(H)$-modules. Then $f \otimes \mathrm{id} + \mathrm{id} \otimes g \colon U \otimes V \to U \otimes V$ is a positive map of Hilbert $\mathcal{N}(G \times H)$-modules and*

$$F(f \otimes \mathrm{id} + \mathrm{id} \otimes g) \simeq F(f) \cdot F(g),$$

where $F \cdot G$ is defined in terms of Notation 2.10.

Proof. Notice for the sequel that $\dim_{\mathcal{N}(G \times H)}(U \otimes V) = \dim_{\mathcal{N}(G)}(U) \cdot \dim_{\mathcal{N}(H)}(V)$ (see Theorem 1.12 (5)). For $x \in \mathrm{im}(E^f_{\lambda/2})$ and $y \in \mathrm{im}(E^g_{\lambda/2})$ we compute using Lemma 2.2 (2)

$$||(f \otimes \mathrm{id} + \mathrm{id} \otimes g)(x \otimes y)|| \leq ||(f \otimes \mathrm{id})(x \otimes y)|| + ||(\mathrm{id} \otimes g)(x \otimes y)||$$
$$= ||f(x)|| \cdot ||y|| + ||x|| \cdot ||g(y)||$$
$$\leq \lambda/2 \cdot ||x|| \cdot ||y|| + \lambda/2 \cdot ||x|| \cdot ||y||$$
$$= \lambda \cdot ||x \otimes y||.$$

This shows $\mathrm{im}(E^f_{\lambda/2}) \otimes \mathrm{im}(E^g_{\lambda/2}) \in \mathcal{L}(f \otimes \mathrm{id} + \mathrm{id} \otimes g, \lambda)$. Lemma 2.3 implies

$$F(f)(\lambda/2) \cdot F(g)(\lambda/2) \leq F(f \otimes \mathrm{id} + \mathrm{id} \otimes g)(\lambda). \tag{2.32}$$

Since $0 \leq f \otimes \mathrm{id} \leq f \otimes \mathrm{id} + \mathrm{id} \otimes g$ and $f \otimes \mathrm{id}$ and $f \otimes \mathrm{id} + \mathrm{id} \otimes g$ commute, we have $E^{f \otimes \mathrm{id} + \mathrm{id} \otimes g}_\lambda \leq E^{f \otimes \mathrm{id}}_\lambda$. Analogously we get $E^{f \otimes \mathrm{id} + \mathrm{id} \otimes g}_\lambda \leq E^{\mathrm{id} \otimes g}_\lambda$. This implies

$$\mathrm{im}(E^{f \otimes \mathrm{id} + \mathrm{id} \otimes g}_\lambda) \subset \mathrm{im}(E^{f \otimes \mathrm{id}}_\lambda) \cap \mathrm{im}(E^{\mathrm{id} \otimes g}_\lambda) = \mathrm{im}(E^f_\lambda) \otimes \mathrm{im}(E^g_\lambda).$$

Hence we get from Theorem 1.12 (5)

$$F(f)(\lambda) \cdot F(g)(\lambda) \geq F(f \otimes \mathrm{id} + \mathrm{id} \otimes g)(\lambda). \tag{2.33}$$

Now Lemma 2.31 follows from (2.32) and (2.33). $\qquad\square$

Notation 2.34. *Given a spectral density function* $F \colon [0, \infty) \to [0, \infty]$, *define*

$$F^\perp \colon [0, \infty) \to [0, \infty]$$

by $F^\perp(\lambda) = F(\lambda) - F(0)$.
   Define δ_r *for* $r \in [0, \infty]$ *by* $\delta_0 := \infty^+$ *and by* $\delta_r := 1$ *for* $r \neq 0$.

Notice that $\alpha(F) = \alpha(F^\perp)$ by definition. Recall Notation 2.10.

Lemma 2.35. *Let* G *and* H *be groups. Let* C_* *be a Hilbert* $\mathcal{N}(G)$-*chain complex and* D_* *be a Hilbert* $\mathcal{N}(H)$-*chain complex. Suppose that both* C_* *and* D_* *are Fredholm and positive. Then the tensor product of Hilbert chain complexes* $C_* \otimes D_*$ *is a Fredholm Hilbert* $\mathcal{N}(G \times H)$-*chain complex and*

(1) We have

$$F^\Delta_n(C_* \otimes D_*) \simeq \sum_{i=0}^n F^\Delta_i(C_*) \cdot F^\Delta_{n-i}(D_*)$$

and

$$\alpha^\Delta_n(C_* \otimes D_*) = \min_{i=0,1,\dots,n} \{ \alpha\left(F^\Delta_i(C_*)^\perp \cdot F^\Delta_{n-i}(D_*)^\perp \right),$$
$$\delta_{b_i^{(2)}(C_*)} \cdot \alpha^\Delta_{n-i}(D_*), \alpha^\Delta_i(C_*) \cdot \delta_{b_{n-i}^{(2)}(D_*)} \};$$

(2) We have

$$F_n(C_* \otimes D_*)^\perp \succeq \sum_{i=0}^{n} F_i(C_*)^\perp \cdot F_{n-i}(D_*)^\perp + b_i^{(2)}(C_*) \cdot F_{n-i}(D_*)^\perp$$

$$+ F_i(C_*)^\perp \cdot b_{n-i}^{(2)}(D_*); \qquad (2.36)$$

$$F_n(C_* \otimes D_*)^\perp \succeq \sum_{i=0}^{n} F_{i+1}(C_*)^\perp \cdot F_{n-i}(D_*)^\perp + b_i^{(2)}(C_*) \cdot F_{n-i}(D_*)^\perp$$

$$+ F_{i+1}(C_*)^\perp \cdot b_{n-1-i}^{(2)}(D_*); \qquad (2.37)$$

$$F_n(C_* \otimes D_*)^\perp \preceq \sum_{i=0}^{n} F_{i+1}(C_*)^\perp \cdot F_{n-i}(D_*)^\perp + F_i(C_*)^\perp \cdot F_{n+1-i}(D_*)^\perp$$

$$+ F_i(C_*)^\perp \cdot F_{n-i}(D_*)^\perp + b_i^{(2)}(C_*) \cdot F_{n-i}(D_*)^\perp$$

$$+ F_i(C_*)^\perp \cdot b_{n-i}^{(2)}(D_*) \qquad (2.38)$$

and

$$\alpha_n(C_* \otimes D_*) = \min_{i=0,1,\ldots,n} \{\alpha\left(F_{i+1}(C_*)^\perp \cdot F_{n-i}(D_*)^\perp\right),$$

$$\alpha\left(F_i(C_*)^\perp \cdot F_{n-i}(D_*)^\perp\right), \delta_{b_i^{(2)}(C_*)} \cdot \alpha_{n-i}(D_*),$$

$$\alpha_i(C_*) \cdot \delta_{b_{n-i}^{(2)}(D_*)}\}. \qquad (2.39)$$

Proof. (1) The Laplace operator $\Delta_n^{C_* \otimes D_*} : (C_* \otimes D_*)_n \to (C_* \otimes D_*)_n$ is the orthogonal sum

$$\bigoplus_{i=0}^{n} \Delta_i^{C_*} \otimes \mathrm{id} + \mathrm{id} \otimes \Delta_{n-i}^{D_*} : \bigoplus_{i=0}^{n} C_i \otimes D_{n-i} \to \bigoplus_{i=0}^{n} C_i \otimes D_{n-i}.$$

Now apply Lemma 2.11 (10) and Lemma 2.31.

(2) Let $e_n : (C_* \otimes D_*)_n \to (C_* \otimes D_*)_{n-1}$ be the n-th differential of the $\mathcal{N}(G \times H)$-chain complex $C_* \otimes D_*$. Consider the maps of Hilbert modules

$$e_{n\perp} := e_n|_{\ker(e_{n+1}^*)} : \ker(e_{n+1}^*) \to (C_* \otimes D_*)_{n-1};$$

$$c_{i\perp} := c_i|_{\ker(c_{i+1}^*)} : \ker(c_{i+1}^*) \to C_{i-1};$$

$$d_{n-i\perp} := d_{n-i}|_{\ker(d_{n-i+1}^*)} : \ker(d_{n-i+1}^*) \to D_{n-1-i}.$$

The following diagram commutes

$$\begin{array}{ccc}
\ker(e_{n+1}^*) & \xrightarrow{\;(e_{n\perp})^* e_{n\perp}\;} & \ker(e_{n+1}^*) \\
\uparrow & & \uparrow \\
\bigoplus_{i=0}^{n} \ker(c_{i+1}^*) \otimes \ker(d_{n-i+1}^*) & \xrightarrow{\quad u \quad} & \bigoplus_{i=0}^{n} \ker(c_{i+1}^*) \otimes \ker(d_{n-i+1}^*)
\end{array}$$

where u is $\bigoplus_{i=0}^{n}(c_{i\perp})^*c_{i\perp} \otimes \mathrm{id} + \mathrm{id} \otimes (d_{n-i\perp})^*d_{n-i\perp}$ and the vertical maps are the canonical inclusions. Since $(e_{n\perp})^*e_{n\perp}$ is selfadjoint, it splits as the orthogonal sum of $\bigoplus_{i=0}^{n}(c_{i\perp})^*c_{i\perp} \otimes \mathrm{id} + \mathrm{id} \otimes (d_{n-i\perp})^*d_{n-i\perp}$ and some endomorphism of the orthogonal complement of $\bigoplus_{i=0}^{n}\ker(c_{i+1}^*) \otimes \ker(d_{n-i+1}^*)$ in $\ker(e_{n+1}^*)$. Now Lemma 2.11 (11), Lemma 2.15 (1) and Lemma 2.31 imply

$$
\begin{aligned}
F_n(C_* \otimes D_*)(\lambda) &= F_n(e_{n\perp})(\lambda) \\
&= F_n((e_{n\perp})^*e_{n\perp})(\sqrt{\lambda}) \\
&\geq \sum_{i=0}^{n} F\left((c_{i\perp})^*c_{i\perp} \otimes \mathrm{id} + \mathrm{id} \otimes (d_{n-i\perp})^*d_{n-i\perp}\right)(\sqrt{\lambda}) \\
&\simeq \sum_{i=0}^{n} F((c_{i\perp})^*c_{i\perp})(\sqrt{\lambda}) \cdot F((d_{n-i\perp})^*d_{n-i\perp})(\sqrt{\lambda}) \\
&= \sum_{i=0}^{n} F(c_{i\perp})(\lambda) \cdot F(d_{n-i\perp})(\lambda) \\
&= \sum_{i=0}^{n} F_i(C_*)(\lambda) \cdot F_{n-i}(D_*)(\lambda).
\end{aligned}
$$

Now equation (2.36) follows since $b_i^{(2)}(C_*) = F_i(C_*)(0)$ and analogously for D_*.

Analogously we prove (2.37). We define maps of Hilbert modules

$$
\begin{aligned}
e_{n\perp}^* &:= e_n^*|_{\ker(e_{n-1})}\colon \ker(e_{n-1}) \to (C \otimes D)_n; \\
c_{i+1\perp}^* &:= c_{i+1}^*|_{\ker(c_i)}\colon \ker(c_i) \to C_{i+1}; \\
d_{n-i\perp}^* &:= d_{n-i}^*|_{\ker(d_{n-i-1})}\colon \ker(d_{n-i-1}) \to D_{n-i}.
\end{aligned}
$$

We obtain a commutative diagram with inclusions as vertical arrows

$$
\begin{array}{ccc}
\ker(e_{n-1}) & \xrightarrow{(e_{n\perp}^*)^*e_{n\perp}^*} & \ker(e_{n-1}) \\
\uparrow & & \uparrow \\
\bigoplus_{i=0}^{n}\ker(c_i) \otimes \ker(d_{n-i-1}) & \xrightarrow{\quad v \quad} & \bigoplus_{i=0}^{n}\ker(c_i) \otimes \ker(d_{n-i-1})
\end{array}
$$

where v is given by $\bigoplus_{i=0}^{n}(c_{i+1\perp}^*)^*c_{i+1\perp}^* \otimes \mathrm{id} + \mathrm{id} \otimes (d_{n-i\perp}^*)^*d_{n-i\perp}^*$. We conclude (2.37), where we use additionally the conclusion of Lemma 1.18

$$
\begin{aligned}
b^{(2)}(c_{i+1\perp}^*) &= b_i^{(2)}(C_*); \\
b^{(2)}(d_{n-i\perp}^*) &= b_{n-1-i}^{(2)}(D_*); \\
b^{(2)}(e_{n\perp}^*) &= b_{n-1}^{(2)}(E_*)
\end{aligned}
$$

and the conclusion of Lemma 2.4

$$F(c_{i+1\perp}^*)^\perp = F(c_{i+1\perp})^\perp;$$
$$F(d_{n-i\perp}^*)^\perp = F(d_{n-i\perp})^\perp;$$
$$F(e_{n\perp}^*)^\perp = F(e_{n\perp})^\perp.$$

Next we prove (2.38). We have the orthogonal decomposition

$$C_i = \ker(c_i) \oplus \ker(c_i)^\perp;$$
$$D_{n-i} = \ker(d_{n-i}) \oplus \ker(d_{n-i})^\perp;$$
$$C_i \otimes D_{n-i} = \ker(c_i) \otimes \ker(d_{n-i}) \oplus \ker(c_i) \otimes \ker(d_{n-i})^\perp \oplus$$
$$\ker(c_i)^\perp \otimes \ker(d_{n-i}) \oplus \ker(c_i)^\perp \otimes \ker(d_{n-i})^\perp.$$

The summand $\ker(c_i) \otimes \ker(d_{n-i})$ lies in the kernel of the n-th-differential e_n of $C_* \otimes D_*$. Hence $\ker(e_n)^\perp$ lies in the direct sum of the other three summands. Define maps by restricting the obvious maps

$$c_{i+1\perp}^* : \ker(c_i) \to C_{i+1};$$
$$d_{n-i\perp} : \ker(d_{n-i})^\perp \to D_{n-1-i};$$
$$d_{n+1-i\perp}^* : \ker(d_{n-i}) \to D_{n+1-i};$$
$$c_{i\perp} : \ker(c_i)^\perp \to C_{i-1}.$$

Consider the following three maps

$$(c_{i+1\perp}^*)^* c_{i+1\perp}^* \otimes \mathrm{id} + \mathrm{id} \otimes (d_{n-i\perp})^* d_{n-i\perp} : \ \ker(c_i) \otimes \ker(d_{n-i})^\perp \to$$
$$\ker(c_i) \otimes \ker(d_{n-i})^\perp;$$

$$(c_{i\perp})^* c_{i\perp} \otimes \mathrm{id} + \mathrm{id} \otimes (d_{n+1-i\perp}^*)^* d_{n+1-i\perp}^* : \ \ker(c_i)^\perp \otimes \ker(d_{n-i}) \to$$
$$\ker(c_i)^\perp \otimes \ker(d_{n-i});$$

$$(c_{i\perp})^* c_{i\perp} \otimes \mathrm{id} + \mathrm{id} \otimes (d_{n-i\perp})^* d_{n-i\perp} : \ \ker(c_i)^\perp \otimes \ker(d_{n-i})^\perp \to$$
$$\ker(c_i)^\perp \otimes \ker(d_{n-i})^\perp.$$

The orthogonal sum of these three selfadjoint maps is the orthogonal sum of the map

$$(e_{n\perp})^* e_{n\perp} : \ker(e_n)^\perp \to \ker(e_n)^\perp$$

and of an endomorphism of the orthogonal complement of $\ker(e_n)^\perp$ in $\big(\ker(c_i) \otimes \ker(d_{n-i})^\perp\big) \oplus \big(\ker(c_i)^\perp \otimes \ker(d_{n-i})\big) \oplus \big(\ker(c_i)^\perp \otimes \ker(d_{n-i})^\perp\big)$. Now Lemma 2.4, Lemma 2.11 (11), Lemma 2.15 (1) and Lemma 2.31 imply (2.38) by the following calculation:

$$F_n(C_* \otimes D_*)^\perp(\lambda) = F(e_{n\perp})(\lambda)$$
$$= F((e_{n\perp})^* e_{n\perp})(\sqrt{\lambda})$$
$$\leq \sum_{i=0}^{n} F((c_{i+1\perp}^*)^* c_{i+1\perp}^* \otimes \mathrm{id} + \mathrm{id} \otimes (d_{n-i\perp})^* d_{n-i\perp})(\sqrt{\lambda})$$
$$+ F((c_{i\perp})^* c_{i\perp} \otimes \mathrm{id} + \mathrm{id} \otimes (d_{n+1-i\perp}^*)^* d_{n+1-i\perp}^*)(\sqrt{\lambda})$$
$$+ F((c_{i\perp})^* c_{i\perp} \otimes \mathrm{id} + \mathrm{id} \otimes (d_{n-i\perp})^* d_{n-i\perp})(\sqrt{\lambda})$$
$$\simeq \sum_{i=0}^{n} F(c_{i+1\perp}^*)^* c_{i+1\perp}^*)(\sqrt{\lambda}) \cdot F((d_{n-i\perp})^* d_{n-i\perp})(\sqrt{\lambda})$$
$$+ F((c_{i\perp})^* c_{i\perp})(\sqrt{\lambda}) \cdot F((d_{n+1-i\perp}^*)^* d_{n+1-i\perp}^*)(\sqrt{\lambda})$$
$$+ F((c_{i\perp})^* c_{i\perp})(\sqrt{\lambda}) \cdot F((d_{n-i\perp})^* d_{n-i\perp})(\sqrt{\lambda})$$
$$= \sum_{i=0}^{n} F(c_{i+1\perp}^*)(\lambda) \cdot F(d_{n-i\perp})(\lambda)$$
$$+ F(c_{i\perp})(\lambda) \cdot F(d_{n+1-i\perp}^*)(\lambda) + F(c_{i\perp})(\lambda) \cdot F(d_{n-i\perp})(\lambda)$$
$$= \sum_{i=0}^{n} (F_{i+1}(C_*)^\perp(\lambda) + b_i^{(2)}(C_*)) \cdot F_{n-i}(D_*)^\perp(\lambda)$$
$$+ F_i(C_*)^\perp(\lambda) \cdot (F_{n+1-i}(D_*)^\perp(\lambda) + b_{n-i}^{(2)}(D_*))$$
$$+ F_i(C_*)^\perp(\lambda) \cdot F_{n-i}(D_*)^\perp(\lambda).$$

Finally (2.39) follows from (2.36), (2.37), and (2.38) using Lemma 2.11 (2) and (10). $\square$

Example 2.40. Let F and G be spectral density functions with $F(0) = G(0) = 0$. It is not hard to check that then

$$\alpha(F) + \alpha(G) \leq \alpha(F \cdot G)$$

is true with respect to Notation 2.10. However, the other inequality is not true in general as the following example shows. Define (continuous) density functions $F, G \colon [0, \infty) \to [0, \infty]$ by

$$
\begin{aligned}
F(\lambda) &= 2^{2^{2n+1}} \lambda^3 & \lambda &\in [2^{-2^{2n+1}}, 2^{-2^{2n}}]; \\
G(\lambda) &= 2^{-2^{2n+1}} & \lambda &\in [2^{-2^{2n+1}}, 2^{-2^{2n}}]; \\
F(\lambda) &= 2^{-2^{2(n+1)}} & \lambda &\in [2^{-2^{2(n+1)}}, 2^{-2^{2n+1}}]; \\
G(\lambda) &= 2^{2^{2(n+1)}} \lambda^3 & \lambda &\in [2^{-2^{2(n+1)}}, 2^{-2^{2n+1}}]; \\
F(0) &= 0; \\
G(0) &= 0; \\
F(\lambda) &= 1/2 & \lambda &\geq 1/2; \\
G(\lambda) &= 1/4 & \lambda &\geq 1/2.
\end{aligned}
$$

One easily checks $F(\lambda) \leq \lambda$ and $G(\lambda) \leq \lambda$ for $\lambda \geq 0$. Since

$$F(2^{-2^{2(n+1)}}) = 2^{-2^{2(n+1)}};$$

$$\lim_{n \to \infty} 2^{-2^{2(n+1)}} = 0;$$

$$G(2^{-2^{2n+1}}) = 2^{-2^{2n+1}};$$

$$\lim_{n \to \infty} 2^{-2^{2n+1}} = 0,$$

we get $\alpha(F) = \alpha(G) = 1$ and $\alpha(F) + \alpha(G) = 2$. On the other hand $F(\lambda) \cdot G(\lambda) = \lambda^3$ and hence $\alpha(F \cdot G) = 3$.

Definition 2.41. *A Fredholm spectral density function has the* limit *property if either $F(\lambda) = F(0)$ for some $\lambda > 0$ or if $F(\lambda) > F(0)$ for $\lambda > 0$ and the limit*

$$\lim_{\lambda \to 0+} \frac{\ln(F(\lambda) - F(0))}{\ln(\lambda)}$$

exists. A Fredholm Hilbert chain complex C_* has the limit *property if $F_p(C_*)$ has the limit property for all $p \in \mathbb{Z}$. A G-CW-complex X of finite type has the* limit *property if its cellular L^2-chain complex has the limit property.*

Remark 2.42. To the author's knowledge there is no example of a G-CW-complex X of finite type which does not have the limit property. See also the discussion in [240, page 381].

Lemma 2.43. *Let $F, G \colon [0, \infty) \to [0, \infty]$ be density functions which are Fredholm and have the limit property. Then $F \cdot G$ is a density function which is Fredholm and has the limit property and*

$$\alpha(F \cdot G) = \min\{\alpha(F) + \alpha(G), \delta_{b^{(2)}(F)} \cdot \alpha(G), \alpha(F) \cdot \delta_{b^{(2)}(G)}\}.$$

Proof. Since $F(\lambda) = F^\perp(\lambda) + b^{(2)}(F)$, it suffices to treat the case $F(0) = G(0) = 0$ because of Lemma 2.11 (10). Since $\ln(F(\lambda) \cdot G(\lambda)) = \ln(F(\lambda)) + \ln(G(\lambda))$ holds and $\lim_{\lambda \to 0+}$ is compatible with $+$ (in contrast to $\liminf_{\lambda \to 0+}$), Lemma 2.43 follows. $\square$

We conclude from Lemma 2.35 and Lemma 2.43

Corollary 2.44. *Let G and H be groups. Let C_* be a Hilbert $\mathcal{N}(G)$-chain complex and D_* be a Hilbert $\mathcal{N}(H)$-chain complex. Suppose that both C_* and D_* are positive, Fredholm and have the limit property. Then the tensor product of Hilbert chain complexes $C_* \otimes D_*$ is a Fredholm Hilbert $\mathcal{N}(G \times H)$-chain complex which has the limit property and we get*

$$\alpha_n^\Delta(C_* \otimes D_*) = \min_{i=0,1,\ldots,n} \{\alpha_i^\Delta(C_*) + \alpha_{n-i}^\Delta(D_*),$$

$$\delta_{b_i^{(2)}(C_*)} \cdot \alpha_{n-i}^\Delta(D_*), \alpha_i^\Delta(C_*) \cdot \delta_{b_{n-i}^{(2)}(D_*)}\}.$$

and

$$\alpha_n(C_* \otimes D_*) = \min_{i=0,1,\ldots,n} \{\alpha_{i+1}(C_*) + \alpha_{n-i}(D_*), \alpha_i(C_*) + \alpha_{n-i}(D_*),$$

$$\delta_{b_i^{(2)}(C_*)} \cdot \alpha_{n-i}(D_*), \alpha_i(C_*) \cdot \delta_{b_{n-i}^{(2)}(D_*)}\}.$$

2.1.4 The Laplacian in Dimension Zero

In this subsection G is a finitely generated group and S denotes a finite set of generators. We want to study the Novikov-Shubin invariant of the following operator

$$c_S \colon \bigoplus_{s \in S} l^2(G) \xrightarrow{\oplus_{s \in S} r_{s-1}} l^2(G),$$

where r_{s-1} is right multiplication with $s - 1 \in \mathbb{Z}[G]$. This is motivated by the following result.

Lemma 2.45. *Let G be a finitely generated group and let X be a connected free G-CW-complex of finite type. Then for any finite set S of generators of G we have*

$$\alpha_1(X) = \alpha(c_S).$$

Proof. The *Cayley graph* of G is the following connected one-dimensional free G-CW-complex. Its 0-skeleton is G. For each element $s \in S$ we attach a free equivariant G-cell $G \times D^1$ by the attaching map $G \times S^0 \to G$ which sends $(g, -1)$ to g and $(g, 1)$ to gs. Since X and the Cayley graph C are connected we an choose a G-map $f \colon X_1 \to C$, where X_1 is 1-skeleton of X. Theorem 2.55 (1) implies that $\alpha_1(X) = \alpha_1(X_1)$ agrees with $\alpha_1(C) = \alpha(c_S)$. $\qquad\square$

If S is a finite set of generators for the group G, let $b_S(n)$ be the number of elements in G which can be written as a word in n letters of $S \cup S^{-1} \cup \{1\}$. The group G has *polynomial growth of degree not greater than d* if there is C with $b_S(n) \le Cn^d$ for all $n \ge 1$. This property is a property of G and independent of the choice of the finite set S of generators. We say that G has *polynomial growth* if it has polynomial growth of degree not greater than d for some $d > 0$. A finitely generated group G is *nilpotent* if G possesses a finite *lower central series*

$$G = G_1 \supset G_2 \supset \ldots \supset G_s = \{1\} \qquad\qquad G_{k+1} = [G, G_k].$$

If (P) is a property of groups, a group G is called *virtually (P)* if G contains a subgroup $H \subset G$ of finite index such that H has property (P). Hence the notions of *virtually finitely generated abelian*, *virtually free* and *virtually nilpotent* are clear. In particular a group is virtually trivial if and only if it is finite. Let $\overline{G}$ be virtually nilpotent. Let $G \subset \overline{G}$ be a subgroup of finite index which is nilpotent and has the lower central series $G = G_1 \supset G_2 \supset \ldots \supset G_s = \{1\}$. Let d_i be the rank of the finitely generated abelian group G_i/G_{i+1} and let d be the integer $\sum_{i \ge 1} i d_i$. Then for any finite set S of generators of $\overline{G}$ there is a constant $C > 0$ such that $C^{-1} n^d \le b_S(n) \le Cn^d$ holds for any $n \ge 1$ and in particular $\overline{G}$ has polynomial growth precisely of degree d [22, page 607 and Theorem 2 on page 608]. A famous result of Gromov [231] says that a finitely generated group has polynomial growth if and only if it is virtually nilpotent. The notion of an amenable group will be reviewed in Subsection 6.4.1.

Lemma 2.46. *(1) $\alpha(c_S) = \infty^+$ if and only if G is non-amenable or finite;*
(2) $\alpha(c_S) < \infty$ if and only if G is infinite virtually nilpotent. If G is virtually nilpotent, $\alpha(c_S)$ is precisely the degree of the growth rate of G;
(3) $\alpha(c_S) = \infty$ if and only if G is amenable and not virtually nilpotent.

Proof. Because of Lemma 2.45 $\alpha(c_S)$ is independent of the choice of the finite set S of generators. If G is finite, then obviously $\alpha(c_S) = \infty^+$. Hence we can assume in the sequel that G is infinite and that S is symmetric, i.e. $s \in S$ implies $s^{-1} \in S$. Define

$$P \colon l^2(G) \xrightarrow{\sum_{s \in S} \frac{1}{|S|} \cdot r_s} l^2(G).$$

Then $\mathrm{id} - P = \frac{1}{2 \cdot |S|} c_S \circ c_S^*$. As G is infinite, $\dim_{\mathcal{N}(G)}(l^2(G) / \mathrm{clos}(\mathrm{im}(c_S))) = 0$. This has been shown in the proof of Theorem 1.35 (8). Hence the kernel of c_S^* is trivial. The spectrum of the selfadjoint operator P is contained in $[-1,1]$ and we conclude from Lemma 2.3 and Lemma 2.4

$$\mathrm{tr}_{\mathcal{N}(G)}(\chi_{[1-\lambda,1]}(P)) = F(c_S)(\sqrt{2|S|\lambda}) - b^{(2)}(c_S), \qquad (2.47)$$

where $\chi_{[1-\lambda,1]}$ is the characteristic function of $[1 - \lambda, 1]$.

(1) From (2.47) $\alpha(c_S)$ has the value ∞^+ if and only if the spectrum of the operator P does not contain 1. Since this operator is convolution with a probability distribution whose support contains S, namely

$$G \to [0, 1], \qquad \gamma \mapsto \begin{cases} |S|^{-1}, \gamma \in S \\ 0, \quad \gamma \notin S \end{cases},$$

the spectrum of P contains 1 if and only if G is amenable by a result of Kesten [290, page 150], [524, Theorem 3.2 on page 7].

(2) The recurrency probability of the natural random walk on G is defined by

$$p(n) = \mathrm{tr}_{\mathcal{N}(G)}(P^n).$$

We will use the following result due to Varopoulos [502], which is also explained in [524, Theorem 6.5 and Theorem 6.6 on page 24]. (The assumption below that n even is needed since the period of the natural random walk on the Cayley graph with respect to symmetric set S of generators is 1 or 2.)

Theorem 2.48. *The finitely generated group G has polynomial growth precisely of degree $2a$ if and only if there is a constant $C > 0$ such that $C^{-1} n^{-a} \le p(n) \le C n^{-a}$ holds for all even positive integers n. If the finitely generated group G does not have polynomial growth, then there is for each $a > 0$ a constant $C(a) > 0$ such that $p(n) \le C(a) n^{-a}$ holds for all $n \ge 1$.*

In the sequel let n be an even positive integer. Notice that then P^n is positive. We have

$$(1-\lambda)^n \chi_{[1-\lambda,1]}(P) \leq P^n \leq (1-\lambda)^n \chi_{[0,1-\lambda]}(P) + \chi_{[1-\lambda,1]}(P).$$

This implies because of $\|\chi_{[0,1-\lambda]}(P)\| \leq 1$

$$(1-\lambda)^n \operatorname{tr}_{\mathcal{N}(G)}(\chi_{[1-\lambda,1]}(P)) \leq p(n) \leq (1-\lambda)^n + \operatorname{tr}_{\mathcal{N}(G)}(\chi_{[1-\lambda,1]}(P)).$$

In the sequel we consider $\lambda \in (0, 1/4)$ only. We conclude from (2.47)

$$\frac{\ln(F(c_S)(\sqrt{2|S|\lambda}) - b^{(2)}(c_S))}{\ln(\lambda)} \geq \frac{\ln(p(n))}{\ln(\lambda)} - n \cdot \frac{\ln(1-\lambda)}{\ln(\lambda)}; \qquad (2.49)$$

$$\frac{\ln(F(c_S)(\sqrt{2|S|\lambda}) - b^{(2)}(c_S))}{\ln(\lambda)} \leq \frac{\ln(p(n) - (1-\lambda)^n)}{\ln(\lambda)}. \qquad (2.50)$$

Next we show for $a > 0$

$$\alpha(c_s) \geq 2a \qquad \text{if } p(n) \leq Cn^{-a} \text{ for } n \geq 1, n \text{ even}; \qquad (2.51)$$

$$\alpha(c_s) \leq 2a \qquad \text{if } p(n) \geq Cn^{-a} \text{ for } n \geq 1, n \text{ even}, \qquad (2.52)$$

where $C > 0$ is some constant independent of n. Suppose $p(n) \leq Cn^{-a}$ for all even positive integers n. Let n be the largest even integer which satisfies $n \leq \lambda^{-1}$. Then $n \geq 2$ since $\lambda \in (0, 1/4)$. We estimate using (2.49)

$$\frac{\ln(F(c_S)(\sqrt{2|S|\lambda}) - b^{(2)}(c_S))}{\ln(\lambda)} \geq \frac{\ln(Cn^{-a})}{\ln(\lambda)} - n \cdot \frac{\ln(1-\lambda)}{\ln(\lambda)}$$

$$= a + \frac{\ln(C) - a \ln(\lambda \cdot n)}{\ln(\lambda)} - n \cdot \frac{\ln(1-\lambda)}{\ln(\lambda)}$$

$$\geq a + \frac{\ln(C) - a \ln(\lambda \cdot n)}{\ln(\lambda)} - \frac{\ln(1-\lambda)}{\lambda \ln(\lambda)}.$$

Since $|\ln(C) - a \ln(\lambda \cdot n)|$ is bounded by $|\ln(C)| + a \ln(2)$ and l'Hospital's rule implies $\lim_{\lambda \to 0+} \frac{\ln(1-\lambda)}{\lambda \ln(\lambda)} = 0$, we conclude (2.51) from Lemma 2.11 (4).

Suppose $p(n) \geq Cn^{-a}$ for all even positive integers n. Fix $\epsilon > 0$. Put $\mu = 3^{-a-1}$. Let $[\lambda^{-(1+\epsilon)}]$ be the largest integer which is less or equal to $\lambda^{-(1+\epsilon)}$. Then we get for all $\lambda \in (0, 1)$ and $k \in \{1, 2\}$

$$2\mu \leq (([\lambda^{-(1+\epsilon)}] + k)\lambda^{1+\epsilon})^{-a} \qquad (2.53)$$

since $1 \leq ([\lambda^{-(1+\epsilon)}] + k)\lambda^{1+\epsilon} \leq 3$ holds for $\lambda \in (0, 1)$ and $k \in \{1, 2\}$. From l'Hospital's rule we get

$$\lim_{\lambda \to 0+} \frac{(\ln(C\mu) + a(1+\epsilon)\ln(\lambda)) \cdot \lambda^{1+\epsilon}}{\ln(1-\lambda)} = 0.$$

Hence we can choose $\lambda_0 \in (0, 1)$ such that for $0 < \lambda < \lambda_0$

$$\frac{(\ln(C\mu) + a(1 + \epsilon)\ln(\lambda)) \cdot \lambda^{1+\epsilon}}{\ln(1 - \lambda)} \le \frac{1}{2}.$$

Put $n = [\lambda^{-(1+\epsilon)}] + k$, where we choose $k \in \{1, 2\}$ such that n is even. Since $(1 - \lambda) \le 1$ and $\lambda^{-(1+\epsilon)}/2 \le [\lambda^{-(1+\epsilon)}]$ we get for $0 < \lambda < \lambda_0$

$$(1 - \lambda)^n \le (1 - \lambda)^{[\lambda^{-(1+\epsilon)}]} \le C\mu \cdot \lambda^{a(1+\epsilon)}.$$

Equation (2.53) implies $0 < C\mu \le C(([\lambda^{-(1+\epsilon)}] + k)\lambda^{1+\epsilon})^{-a} - C\mu$. We conclude from (2.50) for $\lambda \in (0, \lambda_0)$

$$\frac{\ln(F(c_S)(\sqrt{2|S|\lambda}) - b^{(2)}(c_S))}{\ln(\lambda)}$$

$$\le \frac{\ln(C \cdot n^{-a} - C\mu \cdot \lambda^{a(1+\epsilon)})}{\ln(\lambda)}$$

$$= a(1 + \epsilon) + \frac{\ln(C(([\lambda^{-(1+\epsilon)}] + k)\lambda^{1+\epsilon})^{-a} - C\mu)}{\ln(\lambda)}$$

$$\le a(1 + \epsilon) + \frac{\ln(C\mu)}{\ln(\lambda)}.$$

Lemma 2.11 (4) implies $\alpha(c_S) \le 2a(1 + \epsilon)$. Since this is true for all $\epsilon > 0$, (2.52) follows. Now assertion (2) follows from (2.51), (2.52) and Theorem 2.48.

(3) This follows from (1) and (2). This finishes the proof of Lemma 2.46.   □

2.2 Cellular Novikov-Shubin Invariants

In this section we apply the invariants of Subsection 2.1.2 to the cellular L^2-chain complex and thus define Novikov-Shubin invariants for free G-CW-complexes of finite type. We will describe and prove their main properties.

Definition 2.54 (Novikov-Shubin invariants). *Let X be a free G-CW-complex of finite type. Define its cellular p-th spectral density function and its cellular p-th Novikov-Shubin invariant by the corresponding notions (see Definition 2.16) of the cellular L^2-chain complex $C_*^{(2)}(X)$ of X (see Definition 1.29).*

$$F_p(X; \mathcal{N}(G)) := F_p(C_*^{(2)}(X));$$
$$\alpha_p(X; \mathcal{N}(G)) := \alpha_p(C_*^{(2)}(X));$$
$$\alpha_p^\Delta(X; \mathcal{N}(G)) := \alpha_p^\Delta(C_*^{(2)}(X)).$$

If the group G and its action are clear from the context, we omit $\mathcal{N}(G)$ in the notation above.

If M is a cocompact free proper G-manifold, define its p-th cellular spectral density function and its cellular p-th Novikov-Shubin invariant by the corresponding notion for some equivariant smooth triangulation.

Since the Novikov-Shubin invariant will turn out to be homotopy invariant, the definition of the p-th Novikov-Shubin invariant for a cocompact free proper G-manifold is independent of the choice of equivariant smooth triangulation. This is also true for the dilatational equivalence class of the spectral density function. All these definitions extend in the obvious way to pairs.

Theorem 2.55 (Novikov-Shubin invariants).

(1) Homotopy invariance

Let $f\colon X \to Y$ be a G-map of free G-CW-complexes of finite type. Suppose that the map induced on homology with complex coefficients $H_p(f;\mathbb{C})\colon H_p(X;\mathbb{C}) \to H_p(Y;\mathbb{C})$ is an isomorphism for $p \le d - 1$. Then

$$F_p(X) \simeq F_p(Y) \qquad \text{for } p \le d;$$
$$\alpha_p(X) = \alpha_p(Y) \qquad \text{for } p \le d.$$

In particular, if f is a weak homotopy equivalence, we get for all $p \ge 0$

$$F_p(X) \simeq F_p(Y);$$
$$\alpha_p(X) = \alpha_p(Y);$$

(2) Poincaré duality

Let M be a cocompact free proper G-manifold of dimension n which is orientable. Then

$$F_p(M) \simeq F_{n+1-p}(M, \partial M);$$
$$\alpha_p(M) = \alpha_{n+1-p}(M, \partial M);$$

(3) Product formula

Let X be a free G-CW-complex of finite type and let Y be a free H-CW-complex of finite type. Suppose that both X and Y have the limit property (see Definition 2.41). Then $X \times Y$ has the limit property and $\alpha_p(X \times Y)$ is the minimum in $[0, \infty] \coprod \{\infty^+\}$ of the union of the following four sets

$$\{\alpha_{i+1}(X) + \alpha_{p-i}(Y) \mid i = 0, 1, \ldots, (p-1)\};$$
$$\{\alpha_i(X) + \alpha_{p-i}(Y) \mid i = 1, \ldots, (p-1)\};$$
$$\{\alpha_{p-i}(Y) \mid i = 0, 1, \ldots, (p-1), b_i^{(2)}(X) \neq 0\};$$
$$\{\alpha_i(X) \mid i = 1, 2, \ldots, p, b_{p-i}^{(2)}(Y) \neq 0\};$$

(4) Connected sums

Let M_1, M_2, ..., M_r be compact connected m-dimensional manifolds, with $m \geq 3$. Let M be their connected sum $M_1 \# \ldots \# M_r$. Then

$$\alpha_p(\widetilde{M}) = \min\left\{\alpha_p(\widetilde{M_j}) \mid 1 \leq j \leq r\right\} \qquad \text{for } 2 \leq p \leq m-1.$$

If $\pi_1(M_i)$ is trivial for all i except for $i = i_0$, then $\alpha_1(\widetilde{M}) = \alpha_1(\widetilde{M_{i_0}})$. If $\pi_1(M_i)$ is trivial for all i except for $i \in \{i_0, i_1\}$, $i_0 \neq i_1$ and $\pi_1(M_{i_0}) = \pi_1(M_{i_1}) = \mathbb{Z}/2$, then $\alpha_1(\widetilde{M}) = 1$. In all other cases $\alpha_1(\widetilde{M}) = \infty^+$;

(5) First Novikov-Shubin invariant

Let X be a connected free G-CW-complex of finite type. Then G is finitely generated and

(a) $\alpha_1(X)$ is finite if and only if G is infinite and virtually nilpotent. In this case $\alpha_1(X)$ is the growth rate of G;

(b) $\alpha_1(X)$ is ∞^+ if and only if G is finite or non-amenable;

(c) $\alpha_1(X)$ is ∞ if and only if G is amenable and not virtually nilpotent;

(6) Restriction

Let X be a free G-CW-complex of finite type and let $H \subset G$ be a subgroup of finite index $[G : H]$. Let $\text{res}_G^H X$ be the H-space obtained from X by restricting the G-action to an H-action. Then this is a free H-CW-complex of finite type and we get for $p \geq 0$

$$F_p(X; \mathcal{N}(G)) = \frac{1}{[G : H]} \cdot F_p(\text{res}_G^H X; \mathcal{N}(H));$$

$$\alpha_p(X; \mathcal{N}(G)) = \alpha_p(\text{res}_G^H X; \mathcal{N}(H));$$

(7) Induction

Let H be a subgroup of G and let X be a free H-CW-complex of finite type. Then $G \times_H X$ is a free G-CW-complex of finite type and

$$F_p(G \times_H X; \mathcal{N}(G)) = F_p(X; \mathcal{N}(H));$$

$$\alpha_p(G \times_H X; \mathcal{N}(G)) = \alpha_p(X; \mathcal{N}(H)).$$

Proof. (1) We can assume without loss of generality that f is cellular. We have the canonical exact sequence of finitely generated free $\mathbb{C}G$-chain complexes $0 \to C_*(X) \to \text{cyl}(C_*(f)) \to \text{cone}(C_*(f)) \to 0$, where $C_*(f)$ is the $\mathbb{C}G$-chain map induced by f on the cellular $\mathbb{C}G$-chain complexes. Let C_*, D_* and E_* be the d-dimensional finitely generated $\mathbb{C}G$-chain complexes which are obtained from $C_*(X)$, $\text{cyl}(C_*(f))$ and $\text{cone}(C_*(f))$ by truncating everything in dimension $d+1$ and higher. We get an exact sequence of finitely generated free $\mathbb{C}G$-chain complexes $0 \to C_* \xrightarrow{i_*} D_* \xrightarrow{p_*} E_* \to 0$. Since the canonical inclusion $C_*(Y) \to \text{cyl}(C_*(f))$ is a $\mathbb{C}G$-chain homotopy equivalence, it induces a chain homotopy equivalence of finitely generated Hilbert $\mathcal{N}(G)$-chain complexes $C_*^{(2)}(Y) \to l^2(G) \otimes_{\mathbb{C}G} \text{cyl}(C_*(f))$. We conclude from Theorem 2.19

that $F_p(l^2(G) \otimes_{\mathbb{C}G} C_*) \simeq F_p(X)$ and $F_p(l^2(G) \otimes_{\mathbb{C}G} D_*) \simeq F_p(Y)$ holds for $p \leq d$. Hence it remains to show

$$F_p(l^2(G) \otimes_{\mathbb{C}G} C_*) \simeq F_p(l^2(G) \otimes_{\mathbb{C}G} D_*) \qquad \text{for } p \leq d.$$

The map $H_p(i_*) \colon H_p(C_*) \to H_p(D_*)$ is bijective for $p \leq d-1$ since $H_p(f; \mathbb{C})$ is bijective for $p \leq d-1$. This implies that $H_p(E_*) = 0$ for $p \leq d-1$. Let P be the kernel of $e_d \colon E_d \to E_{d-1}$. Since $0 \to P \to E_d \to E_{d-1} \to E_{d-2} \to \ldots \to E_0 \to 0$ is an exact $\mathbb{C}G$-sequence and each E_i is finitely generated free, there is a finitely generated free $\mathbb{C}G$-module F' such that $F := P \oplus F'$ is finitely generated free and P is a direct summand in E_d. For a $\mathbb{C}G$-module W let $d[W]_*$ be the $\mathbb{C}G$-chain complex concentrated in dimension d with W as d-th chain module. Let D'_* be the preimage of $d[F]_* = d[P]_* \oplus d[F']_* \subset E_* \oplus d[F']_*$ under $p_* \oplus \text{id}_{d[F']_*} \colon D_* \oplus d[F']_* \to E_* \oplus d[F']_*$. We obtain a commutative diagram of $\mathbb{C}G$-chain complexes with exact rows

$$
\begin{array}{ccccccccc}
0 & \longrightarrow & C_* & \xrightarrow{i_*} & D_* \oplus d[F']_* & \xrightarrow{p_* \oplus \text{id}_{d[F']_*}} & E_* \oplus d[F']_* & \longrightarrow & 0 \\
& & \text{id} \uparrow & & j_* \uparrow & & k_* \uparrow & & \\
0 & \longrightarrow & C_* & \xrightarrow{i'_*} & D'_* & \xrightarrow{p'_*} & d[F]_* & \longrightarrow & 0
\end{array}
$$

where j_* and k_* are the inclusions. Since id_{C_*} and k_* are $\mathbb{C}G$-homotopy equivalences, $j_* \colon D'_* \to D_* \oplus d[F']_*$ is a $\mathbb{C}G$-chain homotopy equivalence. Theorem 2.19 and Lemma 2.11 (9) imply for $p \leq d$

$$F_p(l^2(G) \otimes_{\mathbb{C}G} D'_*) \simeq F_p(l^2(G) \otimes_{\mathbb{C}G} (D_* \oplus d[F']_*)) \simeq F_p(l^2(G) \otimes_{\mathbb{C}G} D_*).$$

Hence it remains to show

$$F_p(l^2(G) \otimes_{\mathbb{C}G} C_*) \simeq F_p(l^2(G) \otimes_{\mathbb{C}G} D'_*) \qquad \text{for } p \leq d.$$

There is a $\mathbb{C}G$-chain complex C'_* such that $C'_p = C_p$ and $c'_p = c_p$ for $p \neq d$, $C'_d = C_d \oplus F$, $c'_d = c_d \oplus u \colon C_d \oplus F \to C_{d-1}$ for some $\mathbb{C}G$-map $u \colon F \to C_{d-1}$ and a chain isomorphism $g_* \colon C'_* \xrightarrow{\cong} D'_*$ such that g_* composed with the obvious inclusion $l_* \colon C_* \to C'_*$ is i'_*. Hence $F_p(l^2(G) \otimes_{\mathbb{C}G} C'_*) \simeq F_p(l^2(G) \otimes_{\mathbb{C}G} D'_*)$ for $p \leq d$ and $H_{d-1}(l_*) \colon H_{d-1}(C_*) \to H_{d-1}(C'_*)$ is bijective. It remains to show

$$F\left(\text{id}_{l^2(G)} \otimes_{\mathbb{C}G}(c_d \oplus u) \colon l^2(G) \otimes_{\mathbb{C}G} (C_d \oplus F) \to l^2(G) \otimes_{\mathbb{C}G} C_{d-1}\right)$$
$$\simeq F\left(\text{id}_{l^2(G)} \otimes_{\mathbb{C}G} c_d \colon l^2(G) \otimes_{\mathbb{C}G} C_d \to l^2(G) \otimes_{\mathbb{C}G} C_{d-1}\right). \quad (2.56)$$

Since $H_{d-1}(l_*)$ is bijective, $\text{im}(c_d) = \text{im}(c'_d)$. This shows $\text{im}(u) \subset \text{im}(c_d)$. As F is finitely generated free, we can find a $\mathbb{C}G$-map $v \colon F \to C_d$ with $u = c_d \circ v$. The map $\text{id}_{C_d} \oplus v \colon C_d \oplus F \to C_d$ is split surjective. Hence the map

$$\text{id}_{l^2(G)} \otimes_{\mathbb{C}G}(\text{id}_{C_d} \oplus v) \colon l^2(G) \otimes_{\mathbb{C}G} (C_d \oplus F) \to l^2(G) \otimes_{\mathbb{C}G} C_d$$

is a surjective map of finitely generated Hilbert $\mathcal{N}(G)$-modules whose composition with $\mathrm{id}_{l^2(G)} \otimes_{CG} c_d \colon l^2(G) \otimes_{CG} C_d \to l^2(G) \otimes_{CG} C_{d-1}$ is $\mathrm{id}_{l^2(G)} \otimes_{CG} (c_d \oplus u) \colon l^2(G) \otimes_{CG} (C_d \oplus F) \to l^2(G) \otimes_{CG} C_{d-1}$. Now (2.56) and thus assertion (1) follow from Lemma 2.11 (9).

(2) is proved analogously to Theorem 1.35 (3) using Lemma 2.17 (2) and Theorem 2.19.

(3) This follows from Corollary 2.44 since the cellular chain complex of $X \times Y$ is the tensor product of the one of X and the one of Y.

(4) This is proved analogously to Theorem 1.35 (6) for $2 \le p \le m - 1$. The claim for α_1 follows from assertion (5) and assertion (6) since a free product $G_1 * G_2$ of non-trivial groups is amenable if and only if $G_1 = G_2 = \mathbb{Z}/2$, the group $\mathbb{Z}/2 * \mathbb{Z}/2$ contains $\mathbb{Z}$ as a subgroup of finite index and $\alpha_1(\widetilde{S^1}) = 1$ by Lemma 2.58.

(5) Since X is connected and $X \to G\backslash X$ is a regular covering, there is a short exact sequence $1 \to \pi_1(X) \to \pi_1(G\backslash X) \to G \to 1$. Since $G\backslash X$ has finite 1-skeleton by assumption, $\pi_1(G\backslash X)$ and hence G are finitely generated. Now we can apply Lemma 2.45 and Lemma 2.46.

(6) This follows from Theorem 1.12 (6).

(7) This follows from Lemma 1.24 (2), since for each morphism $f \colon U \to V$ of Hilbert $\mathcal{N}(H)$-modules $i_* E_\lambda^{f^* f} = E_\lambda^{(i_* f)^* i_* f}$ and hence

$$F_{\mathcal{N}(H)}(f) = F_{\mathcal{N}(G)}(i_* f) \tag{2.57}$$

holds.   □

A more conceptual proof for Theorem 2.55 (1) can be given after we have developed some theory in Section 6.7. We will see that $F_p(X) \simeq F_p(Y)$ for $p \le d$ holds if $H_p^G(f; \mathcal{N}(G)) \colon H_p^G(X; \mathcal{N}(G)) \to H_p^G(Y; \mathcal{N}(G))$ is bijective for $p \le d - 1$. This will only use that the Novikov-Shubin invariants of homotopy equivalent $\mathcal{N}(G)$-Hilbert chain complexes agree (see Theorem 2.19). Here $H_p^G(X; \mathcal{N}(G))$ is the homology of the chain complex of modules over the ring $\mathcal{N}(G)$ given by $\mathcal{N}(G) \otimes_{CG} C_*(X)$, where only the ring structure of $\mathcal{N}(G)$ enters. The bijectivity of $H_p^G(f; \mathcal{N}(G))$ for $p \le d - 1$ follows from the universal coefficients spectral sequence [518, Theorem 5.6.4 on page 143] which converges to $H_{p+q}^G(X; \mathcal{N}(G))$ and has as E^2-term $E_{p,q}^2 = \mathrm{Tor}_p^{CG}(\mathcal{N}(G), H_q(X; \mathbb{C}))$ since by assumption $H_p(f; \mathbb{C}) \colon H_p(X; \mathbb{C}) \to H_p(Y; \mathbb{C})$ is bijective for $p \le d - 1$.

We will see in Example 3.110 that the condition "orientable" in Theorem 2.55 (2) is necessary.

Lemma 2.58. *Let C_* be a free $\mathbb{C}[\mathbb{Z}]$-chain complex of finite type. Since $\mathbb{C}[\mathbb{Z}]$ is a principal ideal domain [15, Proposition V.5.8 on page 151 and Corollary V.8.7 on page 162], we can write*

$$H_p(C_*) = \mathbb{C}[\mathbb{Z}]^{n_p} \oplus \left(\bigoplus_{i_p=1}^{s_p} \mathbb{C}[\mathbb{Z}]/((z - a_{p,i_p})^{r_{p,i_p}}) \right)$$

for $a_{p,i_p} \in \mathbb{C}$ and $n_p, s_p, r_{p,i_p} \in \mathbb{Z}$ with $n_p, s_p \geq 0$ and $r_{p,i_p} \geq 1$, where z is a fixed generator of $\mathbb{Z}$.

Then $l^2(\mathbb{Z}) \otimes_{\mathbb{C}[\mathbb{Z}]} C_*$ has the limit property and

$$b_p^{(2)}(l^2(\mathbb{Z}) \otimes_{\mathbb{C}[\mathbb{Z}]} C_*) = n_p.$$

If $s_p \geq 1$ and $\{i_p = 1, 2 \ldots, s_p, |a_{p,i_p}| = 1\} \neq \emptyset$, then

$$\alpha_{p+1}(l^2(\mathbb{Z}) \otimes_{\mathbb{C}[\mathbb{Z}]} C_*) = \min\{\frac{1}{r_{p,i_p}} \mid i_p = 1, 2 \ldots, s_p, |a_{p,i_p}| = 1\},$$

otherwise

$$\alpha_{p+1}(l^2(\mathbb{Z}) \otimes_{\mathbb{C}[\mathbb{Z}]} C_*) = \infty^+.$$

Proof. The statement about the L^2-Betti numbers has already been proved in Lemma 1.34.

Let $P(n_p)_*$ be the chain complex concentrated in dimension 0 whose 0-th module is $\mathbb{C}[\mathbb{Z}]^{n_p}$. Let $Q(a_{p,i_p}, r_{p,i_p})_*$ be the $\mathbb{C}[\mathbb{Z}]$-chain complex concentrated in dimensions 0 and 1 whose first differential $M_{(z-a_{p,i_p})^{r_{p,i_p}}} : \mathbb{C}[\mathbb{Z}] \to \mathbb{C}[\mathbb{Z}]$ is multiplication with $(z - a_{p,i_p})^{r_{p,i_p}}$. Notice that its homology is trivial except in dimension 0 where it is given by $\mathbb{C}[\mathbb{Z}]/((z - a_i)^{r_{p,i_p}})$. One easily constructs a $\mathbb{C}[\mathbb{Z}]$-chain map

$$f_*: \bigoplus_{p\geq 0} \Sigma^p \left(P(n_p)_* \oplus \bigoplus_{i_p=1}^{s_p} Q(a_{p,i_p}, r_{p,i_p})_* \right) \to C_*$$

which induces an isomorphism on homology and is therefore a $\mathbb{C}[\mathbb{Z}]$-chain equivalence. Because of Lemma 2.17 (3) and Theorem 2.19 it suffices to show for $a \in \mathbb{C}, r \in \mathbb{Z}, r \geq 1$

$$\alpha_1(Q(a,r)_*) = \begin{cases} \frac{1}{r} & \text{if } |a| = 1 \\ \infty^+ & \text{if } |a| \neq 1 \end{cases}.$$

Because of Example 2.6 we get

$$F_1(Q(a,r)_*)(\lambda) = \text{vol}\{z \in S^1 \mid |(z - a)^r| \leq \lambda\}.$$

If $|a| \neq 1$, then $F_1(Q(a,r)_*)(\lambda) = 0$ for $0 \leq \lambda < |1 - |a||^r$ and hence $\alpha_1(Q(a,r)_*) = \infty^+$. If $|a| = 1$ we conclude $\alpha_1(Q(a,r)_*) = \frac{1}{r}$ from

$$\text{vol}\{z \in S^1 \mid |(z - a)^r| \leq \lambda\} = \text{vol}\{\cos(\phi) + i\sin(\phi) \mid |2 - 2\cos(\phi)|^{r/2} \leq \lambda\};$$

$$\lim_{\phi \to 0} \frac{2 - 2\cos(\phi)}{\phi^2} = 1.$$

This finishes the proof of Lemma 2.58.   □

Example 2.59. For $G = \mathbb{Z}$ we conclude from Lemma 2.58 that $\alpha_1(\widetilde{S^1}) = 1$ and $\alpha_p(\widetilde{S^1}) = \infty^+$ for $p \geq 2$ and that $\widetilde{S^1}$ has the limit property. We get for the n-torus T^n from Theorem 2.55 (3) that $\widetilde{T^n}$ has the limit property and

$$\alpha_p(\widetilde{T^n}) = \begin{cases} n & 1 \leq p \leq n \\ \infty^+ & \text{otherwise} \end{cases}.$$

Example 2.60. Let t_1 and t_2 be the generators of $\mathbb{Z}^2$ and let $f \colon l^2(\mathbb{Z}^2) \to l^2(\mathbb{Z}^2)$ be given by right multiplication with $(t_1 - 1)(t_2 - 1)$. Then we get from Example 2.6 for small $\lambda > 0$

$$\begin{aligned}
F(f)(\lambda) &= \mathrm{vol}\{(z_1, z_2) \in T^2 \mid |z_1 - 1| \cdot |z_2 - 1| \leq \lambda\} \\
&= \mathrm{vol}\{(u_1, u_2) \in [-\pi, \pi] \times [-\pi, \pi] \mid \\
&\qquad\qquad |2 - 2\cos(u_1)|^{1/2} \cdot |2 - 2\cos(u_1)|^{1/2} \leq \lambda\} \\
&\simeq \mathrm{vol}\{(u_1, u_2) \in [-\pi, \pi] \times [-\pi, \pi] \mid |u_1||u_2| \leq \lambda\} \\
&= 4 \cdot \left(\int_{\lambda/\pi}^{\pi} \frac{\lambda}{u} du + \frac{\lambda}{\pi} \pi \right) \\
&= 4 \cdot (\lambda \ln(\pi) - \lambda \ln(\lambda/\pi) + \lambda) \\
&= 4\lambda \cdot (-\ln(\lambda) + 2\ln(\pi) + 1) \\
&\simeq -\ln(\lambda) \cdot \lambda.
\end{aligned}$$

This shows that $\alpha(f) = 1$ and that $F(f)$ is not dilatationally equivalent to λ.

Theorem 2.61 (Novikov-Shubin invariants and S^1-actions). *Let X be a connected S^1-CW-complex of finite type. Suppose that for one orbit S^1/H (and hence for all orbits) the inclusion into X induces a map on π_1 with infinite image. (In particular the S^1-action has no fixed points.) Let $\widetilde{X}$ be the universal covering of X with the canonical $\pi_1(X)$-action. Then we get for all $p \geq 0$*

$$b_p^{(2)}(\widetilde{X}) = 0;$$

$$\alpha_p(\widetilde{X}) \geq 1.$$

Proof. We show for any S^1-CW-complex Y of finite type together with a S^1-map $f \colon Y \to X$ that $b_p^{(2)}(f^*\widetilde{X}; \mathcal{N}(\pi)) = 0$ and $\alpha_p(f^*\widetilde{X}; \mathcal{N}(\pi)) \geq 1$ for all $p \geq 0$, where $f^*\widetilde{X}$ is the pullback of the universal covering of X and $\pi = \pi_1(X)$. The statement about the L^2-Betti numbers has already been proved in Theorem 1.40. The proof of the statement about the Novikov-Shubin invariants is analogous using Theorem 2.20, Theorem 2.55 (7) and the conclusion from Lemma 2.58 that

$$\alpha_p(\widetilde{S^1}) = \begin{cases} 1 & \text{if } p = 1; \\ \infty^+ & \text{otherwise}. \end{cases} \qquad\qquad \square$$

The notion of elementary amenable group will be explained in Definition 6.34. The next result is taken from [260, Corollary 2 on page 240]. Notice that a group H, which has a finite-dimensional model for BH, is torsionfree.

Lemma 2.62. *Let $1 \to H \to G \to Q \to 1$ be an extension of groups such that H is elementary amenable and BG has a finite-dimensional model. Then G contains a normal torsionfree abelian subgroup.*

Theorem 2.63. (Novikov Shubin invariants and aspherical CW-complexes). *Let X be an aspherical finite CW-complex. Suppose that its fundamental group contains an elementary amenable infinite normal subgroup H. Then*

$$b_p^{(2)}(\widetilde{X}) = 0 \quad \text{for } p \geq 0;$$
$$\alpha_p(\widetilde{X}) \geq 1 \quad \text{for } p \geq 1.$$

Proof. The claim for the L^2-Betti numbers is a special case of Theorem 1.44. Since X is a finite-dimensional model for $B\pi_1(X)$, we can assume without loss of generality by Lemma 2.62 that H is torsionfree abelian. In the case $H = \mathbb{Z}^n$, the claim follows from [343, Theorem 3.9 (6) on page 174]. In the general case, one has to notice that any finitely generated subgroup of H is isomorphic to $\mathbb{Z}^n$ for some $n \leq \dim(X)$. Now the claim follows from [343, Theorem 3.7 on page 172]. More details can also be found in [515, section 4.5]. $\square$

Theorem 2.63 still makes sense and still is true (by the same proof) if one replaces finite by finite-dimensional and uses the extension of the definition of Novikov-Shubin invariant to arbitrary spaces in [343].

2.3 Analytic Novikov-Shubin Invariants

In this section we introduce the analytic version of spectral density functions and Novikov-Shubin invariants.

Definition 2.64 (Analytic spectral density function). *Let M be a co-compact free proper G-manifold without boundary and with G-invariant Riemannian metric. Let $d_{\min}^p$ and $(\Delta_p)_{\min}$ be the minimal closures of the densely defined operators $d^p \colon \Omega_c^p(M) \to L^2\Omega^{p+1}(M)$ and $\Delta_p \colon \Omega_c^p(M) \to L^2\Omega^p(M)$. Let $d_{\min}^{p\perp} \colon \operatorname{dom}(d_{\min}^p) \cap \operatorname{im}(d_{\min}^{p-1})^\perp \to \operatorname{im}(\delta_{\min}^{p+2})^\perp$ be the operator induced by $d_{\min}^p$. Define the* analytic p-th spectral density function *of M by*

$$F_p(M) := F(d_{\min}^{p\perp}) \colon [0, \infty) \to [0, \infty)$$

and define

$$F_p^\Delta(M) := F((\Delta_p)_{\min}) \colon [0, \infty) \to [0, \infty).$$

Define the analytic *p-th Novikov-Shubin invariant of M by*

$$\alpha_p(M) := \alpha(F_{p-1}(M)).$$

Put

$$\alpha_p^{\Delta}(M) := \alpha(F_p^{\Delta}(M)).$$

Notice that here we define $\alpha_p(M)$ to be $\alpha(F_{p-1}(M))$ and not to be $\alpha(F_p(M))$ as in Definition 2.16 and Definition 2.54 because here we are dealing with cochain complexes whereas in the cellular context we use chain complexes and we want to show later that both definitions agree (see Lemma 2.17 (2)).

Notation 2.65. *Denote by $(\delta^{p+1}d^p)_{\min}^{\perp}$ and $(d^{p-1}\delta^p)_{\min}^{\perp}$ the minimal closure of the operators*

$$\delta^{p+1}d^p : \Omega_c^p(M) \cap \operatorname{im}(d_{\min}^{p-1})^{\perp} \to \operatorname{im}(d_{\min}^{p-1})^{\perp};$$
$$d^{p-1}\delta^p : \Omega_c^p(M) \cap \operatorname{im}(\delta_{\min}^{p+1})^{\perp} \to \operatorname{im}(\delta_{\min}^{p+1})^{\perp}.$$

Let $d_{\min}^{p\perp} : \operatorname{dom}(d_{\min}^p) \cap \operatorname{im}(d_{\min}^{p-1})^{\perp} \to \operatorname{im}(\delta_{\min}^{p+2})^{\perp}$ and $\delta_{\min}^{p\perp} : \operatorname{dom}(\delta_{\min}^p) \cap \operatorname{im}(\delta_{\min}^{p+1})^{\perp} \to \operatorname{im}(d_{\min}^{p-2})^{\perp}$ be the closed densely defined operators induced by $d_{\min}^p$ and $\delta_{\min}^p$. Denote by $E_\lambda^{\Delta_p}(x,y)$ the smooth Schwartz kernel for the projection $E_\lambda^{\Delta_p}$ appearing in the spectral family of the selfadjoint operator $(\Delta_p)_{\min}$.

The existence of the smooth Schwartz kernel $E_\lambda^{\Delta_p}(x,y)$ is for instance proved in [9, Proposition 2.4]. It is a smooth section of the vector bundle $\hom(p_1^* \operatorname{Alt}^p(TM), p_2^* \operatorname{Alt}^p(TM))$ over $M \times M$ for $p_k : M \times M \to M$ the projection to the k-th factor. It is uniquely characterized by the property that for $\omega \in L^2\Omega^p(M)$

$$E_\lambda^{\Delta_p}(\omega)(x) = \int_M E_\lambda^{\Delta_p}(x,y)(\omega_y)\, d\mathrm{vol}_y$$

holds.

Lemma 2.66. *Let M be a cocompact free proper G-manifold without boundary and with G-invariant Riemannian metric. Then*

(1) The values of the functions $F_p(M)(\lambda)$ and $F_p^{\Delta}(M)(\lambda)$ are finite for all $\lambda \in \mathbb{R}$. In particular $F_p(M)$ and $F_p^{\Delta}(M)$ are Fredholm;
(2) We have

$$F_p^{\Delta}(M)(\lambda^2) - b_p^{(2)}(M) = (F_{p-1}(M)(\lambda) - b_{p-1}^{(2)}(M))$$
$$+ (F_p(M)(\lambda) - b_p^{(2)}(M));$$
$$\alpha_p^{\Delta}(M) = 1/2 \cdot \min\{\alpha_p(M), \alpha_{p+1}(M)\};$$

(3) We have

$$F_p^\Delta(M)(\lambda) = \int_{\mathcal{F}} \operatorname{tr}(E_\lambda^{\Delta^p}(x,x))\, dvol_x,$$

where $\mathcal{F}$ is a fundamental domain of the G-action;

(4) If $F_p(H^{l-*}\Omega^*(M))(\lambda)$ is the p-th spectral density function of the Hilbert $\mathcal{N}(G)$-cochain complex $H^{l-*}\Omega^*(M)$ (see (1.74)), then

$$F_p(M) \simeq F_p(H^{l-*}\Omega^*(M)).$$

Proof. We get from [9, Proposition 4.16 on page 63]

$$\operatorname{tr}_{\mathcal{N}(G)}(E_\lambda^{\Delta_p}) = \int_{\mathcal{F}} \operatorname{tr}(E_\lambda^{\Delta_p}(x,x))\, dvol_x.$$

This and $F_p^\Delta(M)(\lambda)) = \operatorname{tr}_{\mathcal{N}(G)}(E_\lambda^{\Delta_p})$ imply that $F_p^\Delta(M)(\lambda))$ is finite for all $\lambda \in \mathbb{R}$ and satisfies

$$F_p^\Delta(M)(\lambda) = \int_{\mathcal{F}} \operatorname{tr}(E_\lambda^{\Delta_p}(x,x))\, dvol_x. \tag{2.67}$$

One easily checks the following equalities of densely defined operators using Lemma 1.70 (1)

$$(\Delta_p)_{\min}|_{\operatorname{im}(d_{\min}^{p-1})^\perp} = (\delta^{p+1}d^p)_{\min}^\perp;$$
$$(\Delta_p)_{\min}|_{\operatorname{im}(\delta_{\min}^{p+1})^\perp} = (d^{p-1}\delta^p)_{\min}^\perp;$$
$$\delta_{\min}^{p\perp} = (d_{\min}^{(p-1)\perp})^*;$$
$$(\delta^{p+1}d^p)_{\min}^\perp = (d_{\min}^{p\perp})^* d_{\min}^{p\perp};$$
$$(d^{p-1}\delta^p)_{\min}^\perp = (\delta_{\min}^{p\perp})^* \delta_{\min}^{p\perp}.$$

We conclude from Theorem 1.57

$$\ker((\Delta_p)_{\min}) = \ker(d_{\min}^{p\perp}) = \ker((\delta^{p+1}d^p)_{\min}^\perp) = \ker((d^{p-1}\delta^p)_{\min}^\perp)$$
$$= \ker(\delta_{\min}^{p\perp}) = \mathcal{H}_{(2)}^p(M).$$

We conclude from Theorem 1.57, Lemma 2.3 and Lemma 2.4

$$F_p^\Delta(M)(\lambda) + \dim_{\mathcal{N}(G)}(\mathcal{H}_{(2)}^p(M)) = F((\delta^{p+1}d^p)_{\min}^\perp)(\lambda) + F((d^{p-1}\delta^p)_{\min}^\perp)(\lambda);$$
$$F((\delta^{p+1}d^p)_{\min}^\perp)(\lambda^2) = F_p(M)(\lambda);$$
$$F((d^{p-1}\delta^p)_{\min}^\perp)(\lambda^2) - b_p^{(2)}(M) = F_{p-1}(M)(\lambda) - b_{p-1}^{(2)}(M).$$

This together with (2.67) proves assertions (1), (2) and (3). We will prove (4) in the next Section 2.4. □

The following result is due to Efremov [167]. We will prove it in the next Section 2.4.

Theorem 2.68. (Analytic and combinatorial Novikov-Shubin invariants). *Let M be a cocompact free proper G-manifold without boundary and with G-invariant Riemannian metric. Then the cellular and the analytic spectral density functions (see Definition 2.54 and Definition 2.64) are dilatationally equivalent and the cellular and analytic Novikov-Shubin invariants agree in each dimension p.*

Example 2.69. The Novikov-Shubin invariants of the universal covering $\widetilde{M}$ of a closed hyperbolic manifold of dimension n have been computed [316, Proposition 46 on page 499] using the analytic approach, namely

$$\alpha_p(\widetilde{M}) = \begin{cases} 1 & \text{if } n \text{ is odd and } 2p = n \pm 1 \\ \infty^+ & \text{otherwise} \end{cases} .$$

Moreover, the spectral density function has the limit property.

Example 2.70. In the notation of Example 1.36 we get

$$\alpha_p(\widetilde{F_g^d}) = \begin{cases} 2 & \text{if } g = 1, d = 0, p = 1, 2 \\ 1 & \text{if } g = 0, d = 2, p = 1 \\ \infty^+ & \text{otherwise} \end{cases} .$$

This follows from Theorem 2.55 (1) and (5), Example 2.59, Example 2.69 and the facts that F_g^d is homotopy equivalent to $\bigvee_{i=1}^{2g+d-1} S^1$ for $d \geq 1$, F_g^0 is hyperbolic for $g \geq 2$, $F_1^0 = T^2$, $\pi_1(F_0^0) = 1$ and a free group of rank ≥ 2 is not amenable.

2.4 Comparison of Analytic and Cellular Novikov-Shubin Invariants

In this section we give the proofs of Lemma 2.66 (4) and of Theorem 2.68. They will follow from Lemma 2.71, Lemma 2.72 and Lemma 2.80.

Lemma 2.71. *For any equivariant smooth triangulation K we have*

$$F_p(K) \preceq F_p(H^{l-*}\Omega^*(M)).$$

Proof. We have constructed cochain maps of Hilbert $\mathcal{N}(G)$-modules in (1.77) and (1.78)

$$A^* : H^{l-*}\Omega^*(M) \to C_{(2)}^*(K);$$
$$W^* : C_{(2)}^*(K) \to H^{l-*}\Omega^*(M)$$

such that $A^* \circ W^* = \mathrm{id}$ holds (see (1.79)). Now apply Lemma 2.17 (3).   □

Lemma 2.72. *We have*

$$F_p(H^{l-*}\Omega^*(M))) \preceq F_p(M).$$

Proof. The following diagram commutes and the vertical maps are given by the isometric isomorphisms of Hilbert $\mathcal{N}(G)$-modules of (1.73)

$$
\begin{array}{ccc}
H^{l-p}\Omega^p(M) & \xrightarrow{d^p_{H^{l-p}}} & H^{l-p-1}\Omega^{p+1}(M) \\
{\scriptstyle (1+\Delta_p)^{(l-p-1)/2}}\Big\downarrow & & {\scriptstyle (1+\Delta_p)^{(l-p-1)/2}}\Big\downarrow \\
H^1\Omega^p(M) & \xrightarrow{d^p_{H^1}} & L^2\Omega^{p+1}(M)
\end{array}
$$

where $d^p_{H^{l-p}}\colon H^{l-p}\Omega^p(M) \to H^{l-p-1}\Omega^{p+1}(M)$ is the bounded operator which is induced by $d^p\colon \Omega^p_c(M) \to \Omega^{p+1}_c(M)$. Hence it remains to show

$$F(d^{p\perp}_{H^1}) \preceq F(d^{p\perp}_{\min}),\tag{2.73}$$

where $d^{p\perp}_{H^1}\colon \operatorname{im}(d^{p-1}_{H^2})^\perp \to L^2\Omega^{p+1}(M)$ is the operator which is induced by $d^p_{H^1}\colon H^1\Omega^p(M) \to L^2\Omega^{p+1}(M)$.

For $\omega \in \Omega^{p-1}_c(M)$ and $\eta \in \operatorname{im}(d^{p-1}_{H^2})^\perp$ we conclude using the isomorphism (1.73) and partial integration (see Lemma 1.56)

$$
\begin{aligned}
\langle d^{p-1}(\omega), \eta\rangle_{L^2} &= \langle (1+\Delta_p)\circ d^{p-1}_{H^2}\circ(1+\Delta_p)^{-1}(\omega), \eta\rangle_{L^2} \\
&= \langle (1+\Delta_p)^{1/2}\circ d^{p-1}_{H^2}\circ(1+\Delta_p)^{-1}(\omega), (1+\Delta_p)^{1/2}(\eta)\rangle_{L^2} \\
&= \langle d^{p-1}_{H^2}((1+\Delta_p)^{-1}(\omega)), \eta\rangle_{H^1} \\
&= 0.
\end{aligned}\tag{2.74}
$$

One easily checks that for $\omega \in \Omega^p_c(M)$ and hence for all $\omega \in H^1\Omega^p(M)$

$$||\omega||^2_1 = ||\omega||^2_{L^2} + ||d^p\omega||^2_{L^2} + ||\delta^p\omega||^2_{L^2}\tag{2.75}$$

holds. Now (2.74) and (2.75) imply that we obtain a well-defined injective morphism of Hilbert $\mathcal{N}(G)$-modules induced by the inclusion $H^1\Omega^p(M) \to L^2\Omega^p(M)$

$$j\colon \operatorname{im}(d^{p-1}_{H^2})^\perp \to \operatorname{im}(d^{p-1}_{\min})^\perp.$$

Consider $0 \le \lambda < 1$ and $L \in \mathcal{L}(d^{p\perp}_{H^1}, \lambda)$. For $\omega \in L$ we have $j(w) \in \operatorname{im}(d^{p-1}_{\min})^\perp$ and hence $\delta^p(\omega) = 0$ and we get from (2.75)

$$||d^p(\omega)||^2_{L^2} \le \lambda^2 \cdot ||\omega||^2_1 = \lambda^2 \cdot ||\omega||^2_{L^2} + \lambda^2 \cdot ||d^p(\omega)||^2_{L^2}.$$

This implies

$$||d^p(\omega)||_{L^2} \le \frac{\lambda}{\sqrt{1-\lambda^2}} \cdot ||\omega||^2_{L^2}.$$

We conclude from Lemma 2.2 (2) that the composition

$$E^{(d^{p\perp}_{\min})^* d^{p\perp}_{\min}}_{\lambda^2/(1-\lambda^2)} \circ j \colon \operatorname{im}(d^{p-1}_{H^2})^{\perp} \to \operatorname{im}(E^{(d^{p\perp}_{\min})^* d^{p\perp}_{\min}}_{\lambda^2/(1-\lambda^2)})$$

is injective on L. This implies using Lemma 2.3

$$\dim_{\mathcal{N}(G)}(L) \leq \dim_{\mathcal{N}(G)}(\operatorname{im}(E^{(d^{p\perp}_{\min})^* d^{p\perp}_{\min}}_{\lambda^2/(1-\lambda^2)})) = F_p(M)(\lambda/\sqrt{1-\lambda^2})$$

and hence

$$F_p(H^{l-*}\Omega^*(M))(\lambda) \leq F_p(M)(\lambda/\sqrt{1-\lambda^2}).$$

This finishes the proof of Lemma 2.72.   □

The *mesh* of a triangulation is defined by

$$\operatorname{mesh}(K) := \sup\{d(p,q) \mid p,q \text{ vertices of a 1-simplex}\},$$

where $d(p,q)$ is the metric on M induced by the Riemannian metric. The *fullness* of a triangulation is defined by

$$\operatorname{full}(K) := \inf\left\{ \frac{\operatorname{vol}(\sigma)}{\dim(M)\operatorname{mesh}(K)} \,\middle|\, \sigma \in S_{\dim(M)}(K) \right\}.$$

The next result is taken from [145, page 165]. Its proof is based on [150, Proposition 2.4 on page 8]. Notice that from now on W^* is given in terms of barycentric coordinate functions.

Lemma 2.76. *Let M be a cocompact free proper G-manifold without boundary and with G-invariant Riemannian metric. Fix $\theta > 0$, $k > \dim(M)/2 + 1$ and an equivariant smooth triangulation K. Then there is a constant $C > 0$ such that for any equivariant barycentric subdivision K' of M with fullness $\operatorname{full}(K') \geq \theta$ and any p-form $\omega \in H^k\Omega^p(M)$*

$$\|\omega - W^p_{K'} \circ A^p_{K'}(\omega)\|_{L^2} \leq C \cdot \operatorname{mesh}(K') \cdot \|\omega\|_k$$

holds.

Fix $\epsilon > 0$. Lemma 2.2 (2) implies for $\omega \in \operatorname{im}(E^{(d^{p\perp}_{\min})^* d^{p\perp}_{\min}}_{\epsilon^2})$

$$\begin{aligned}
\|\omega\|_k^2 &= \langle \omega, (1+\Delta_p)^k(\omega) \rangle_{L^2} \\
&= \langle \omega, (1+(d^{p\perp}_{\min})^* d^{p\perp}_{\min})^k(\omega) \rangle_{L^2} \\
&\leq (1+\epsilon^{2k}) \cdot \|\omega\|_{L^2}^2.
\end{aligned}$$

Hence the inclusion of $\operatorname{im}(E^{(d^{p\perp}_{\min})^* d^{p\perp}_{\min}}_{\epsilon^2})$ into $L^2\Omega^p(M)$ induces a bounded G-equivariant operator

$$i_\epsilon \colon \operatorname{im}(E^{(d^{p\perp}_{\min})^* d^{p\perp}_{\min}}_{\epsilon^2}) \to H^{l-p}\Omega^p(M). \tag{2.77}$$

Lemma 2.78. *Fix $k > \dim(M)/2 + 1$ and $\epsilon > 0$. Then there is an equivariant smooth triangulation K and a constant $C_1 > 0$ such that for all $\omega \in \mathrm{im}(E_{\epsilon^2}^{(d_{\min}^{p\perp})^* d_{\min}^{p\perp}})$*

$$||\omega||_{L^2} \leq C_1 \cdot ||\operatorname{pr} \circ A^p \circ i_\epsilon(\omega)||_{L^2}$$

holds, where $\operatorname{pr}: l^2 C^p(K) \to l^2 C^p(K)$ *is the orthogonal projection onto* $\mathrm{im}\left(c^{p-1}: l^2 C^{p-1}(K) \to l^2 C^p(K)\right)^\perp$.

Proof. Given $\theta > 0$ and an equivariant smooth triangulation of M, we can find an equivariant subdivision whose fullness is bounded from below by θ and whose mesh is arbitrary small [522]. Hence we can find by Lemma 2.76 an equivariant smooth triangulation K and a constant $0 < C_0 < 1$ such that for any p-form $\eta \in H^k \Omega^p(M)$

$$||\eta - W^p \circ A^p(\eta)||_{L^2} \leq C_0 \cdot ||\eta||_{L^2}$$

holds. Let $\operatorname{pr}': L^2 \Omega^p(M) \to L^2 \Omega^p(M)$ be the orthogonal projection onto $\mathrm{im}(d_{\min}^{p-1})^\perp$. Recall that W^p sends $\mathrm{im}(c^{p-1})$ to $\mathrm{im}(d_{\min}^{p-1})$ and hence $\operatorname{pr}' \circ W^p \circ \operatorname{pr} = \operatorname{pr}' \circ W^p$. Now we estimate for $\omega \in \mathrm{im}(E_{\epsilon^2}^{(d_{\min}^{p\perp})^* d_{\min}^{p\perp}})$ using $\operatorname{pr}'(\omega) = \omega$

$$\begin{aligned}
||\omega||_{L^2} &\leq ||\operatorname{pr}' \circ W^p \circ A^p \circ i_\epsilon(\omega)||_{L^2} + ||\omega - \operatorname{pr}' \circ W^p \circ A^p \circ i_\epsilon(\omega)||_{L^2} \\
&= ||\operatorname{pr}' \circ W^p \circ \operatorname{pr} \circ A^p \circ i_\epsilon(\omega)||_{L^2} + ||\omega - \operatorname{pr}' \circ W^p \circ A^p \circ i_\epsilon(\omega)||_{L^2} \\
&\leq ||W^p \circ \operatorname{pr} \circ A^p \circ i_\epsilon(\omega)||_{L^2} + ||\operatorname{pr}'(\omega - W^p \circ A^p \circ i_\epsilon(\omega))||_{L^2} \\
&\leq ||W^p|| \cdot ||\operatorname{pr} \circ A^p \circ i_\epsilon(\omega)||_{L^2} + ||\omega - W^p \circ A^p \circ i_\epsilon(\omega)||_{L^2} \\
&\leq ||W^p|| \cdot ||\operatorname{pr} \circ A^p \circ i_\epsilon(\omega)||_{L^2} + C_0 \cdot ||\omega||_{L^2}.
\end{aligned}$$

If we put $C_1 = \max\{\frac{||W^p||}{1-C_0} \mid p = 0, 1, \ldots, \dim(M)\}$, the claim follows. $\square$

Lemma 2.79. *Given an equivariant smooth triangulation K, there is a constant C_2 such that for all $u \in l^2 C^p(M)$*

$$||u||_{L^2} \leq C_2 \cdot ||W_K^p(u)||_{L^2}.$$

Proof. For an element $u \in l^2 C^p(M)$ the element $W^p(u) \in L^2 \Omega^p(M)$ is smooth outside the $(\dim(M) - 1)$-skeleton. The p-form $W^p(u)$, which is a priori defined in the sense of distributions, restricted to the interior of a $\dim(M)$-simplex σ is smooth and has a unique smooth extension to σ itself which we denote in the sequel by $W(u)|_\sigma$. If τ is a p-dimensional face of both the $\dim(M)$-simplices σ_0 and σ_1, then $i_0^* W(u)|_{\sigma_0} = i_1^* W(u)|_{\sigma_1}$ for the inclusions $i_k: \tau \to \sigma_k$ for $k = 0, 1$. Hence we can define for any p-simplex τ a smooth p-form $W(u)|_\tau$ by $i^* W(u)|_\sigma$ for any $\dim(M)$-simplex σ which contains τ as face and $i: \tau \to \sigma$ the inclusion. In particular $W(c_\tau)$ which is a priori given in a distributional sense is a continuous (not necessarily smooth) p-form. Since G acts cocompactly, there is a constant $K > 0$ such that $||W(c_\tau)||_{L^2} \leq K$ holds for all p-simplices τ and hence $W(u) \in L^2\Omega^p(M)$ for all $u \in l^2C^p(M)$. If τ and σ are p-simplices, then we get for the characteristic functions $c_\tau, c_\sigma \in l^2C^p(K)$ of σ and τ

$$W(c_\tau)|_\sigma = 0 \qquad \text{if } \tau \neq \sigma;$$

$$\int_\sigma W(c_\sigma) = 1.$$

Recall that G acts freely and cocompactly. Hence there are numbers $D > 0$ and $S > 0$ such that for any p-simplex σ

$$\int_\sigma ||W(c_\sigma)||_x^2 \, d\mathrm{vol}_x^\sigma \geq 2 \cdot D;$$

$$|\{\tau \mid \tau \in \mathrm{st}(\sigma)\}| \leq S;$$

$$|\{\tau \mid \sigma \in \mathrm{st}(\tau)\}| \leq S,$$

where τ runs through all simplices.

Recall that the support of $W(c_\tau)$ lies in the star $\mathrm{st}(\tau)$ of τ. For any p-simplex σ we can choose an open neighborhood $U(\sigma)$ of the interior $\mathrm{int}(\sigma)$ which is obtained by thickening the interior of σ into the $\dim(M)$-simplices having σ as faces such that

$$||W(c_\tau)||_x^2 \leq \frac{D}{2 \cdot (4S - 2)S \, \mathrm{vol}(\sigma)} \qquad \text{if } \tau \neq \sigma, x \in U(\sigma);$$

$$g \cdot U(\sigma) = U(g \cdot \sigma);$$

$$U(\sigma) \cap U(\tau) = \emptyset \qquad \text{for } \sigma \neq \tau;$$

$$W(c_\tau)(x) = 0 \qquad \text{if } x \in U(\sigma) \text{ and } \sigma \notin \mathrm{st}(\tau).$$

holds. There is a number $\delta > 0$ such that possibly by shrinking $U(\sigma)$ to something which is up to small error a product neighbourhood $\mathrm{int}(\sigma) \times (-\frac{\delta}{2D}, \frac{\delta}{2D})$ we can additionally achieve that for all $g \in G$ and p-simplices σ and τ

$$\int_{U(\sigma)} ||W^p(c_\tau))||_x^2 \, d\mathrm{vol}_x \begin{cases} \geq \delta & \tau = \sigma \\ \leq \frac{\delta}{(4S-2)S} & \tau \neq \sigma \end{cases}.$$

holds. Then we get for $u = \sum_\sigma u_\sigma \cdot c_\sigma$ in $l^2C^p(M)$ using the inequality $||\sum_{i=1}^r a_i||^2 \leq (2r - 1) \cdot \sum_{i=1}^r ||a_i||^2$.

$$\|u\|_{L^2}^2 = \sum_\sigma |u_\sigma|^2$$

$$\leq \frac{1}{\delta} \cdot \sum_\sigma |u_\sigma|^2 \cdot \int_{U(\sigma)} \|W(c_\sigma)\|_x^2 \, d\mathrm{vol}_x$$

$$\leq \frac{1}{\delta} \cdot \sum_\sigma \int_{U(\sigma)} \|u_\sigma \cdot W(c_\sigma)\|_x^2 \, d\mathrm{vol}_x$$

$$\leq \frac{1}{\delta} \cdot \sum_\sigma \int_{U(\sigma)} \left(\|\sum_\tau u_\tau \cdot W(c_\tau)\|_x^2 + \|\sum_{\tau \neq \sigma} u_\tau \cdot W(c_\tau)\|_x^2 \right) d\mathrm{vol}_x$$

$$\leq \frac{1}{\delta} \cdot \int_{\coprod_\sigma U(\sigma)} \|\sum_\tau u_\tau \cdot W(c_\tau)\|_x^2 \, d\mathrm{vol}_x \; +$$

$$\frac{1}{\delta} \cdot \sum_\sigma \int_{U(\sigma)} \| \sum_{\tau \neq \sigma, \sigma \in \mathrm{st}(\tau)} u_\tau \cdot W(c_\tau)\|_x^2 \, d\mathrm{vol}_x$$

$$\leq \frac{1}{\delta} \cdot \int_M \|\sum_\tau u_\tau \cdot W(c_\tau)\|_x^2 \, d\mathrm{vol}_x \; +$$

$$\frac{1}{\delta} \cdot \sum_\sigma \int_{U(\sigma)} (2S - 1) \cdot \sum_{\tau \neq \sigma, \sigma \in \mathrm{st}(\tau)} \|u_\tau \cdot W(c_\tau)\|_x^2 \, d\mathrm{vol}_x$$

$$\leq \frac{1}{\delta} \cdot \|W(u)\|_{L^2} + \frac{2S - 1}{\delta} \cdot \sum_\sigma \sum_{\tau \neq \sigma, \sigma \in \mathrm{st}(\tau)} |u_\tau|^2 \cdot \int_{U(\sigma)} \|W(c_\tau)\|_x^2 \, d\mathrm{vol}_x$$

$$\leq \frac{1}{\delta} \cdot \|W(u)\|_{L^2} + \frac{2S - 1}{\delta} \cdot \sum_\sigma \sum_{\tau \neq \sigma, \sigma \in \mathrm{st}(\tau)} |u_\tau|^2 \cdot \frac{\delta}{(4S - 2)S}$$

$$\leq \frac{1}{\delta} \cdot \|W(u)\|_{L^2} + \frac{2S - 1}{\delta} \cdot S \cdot \sum_\tau |u_\tau|^2 \cdot \frac{\delta}{(4S - 2)S}$$

$$\leq \frac{1}{\delta} \cdot \|W(u)\|_{L^2} + \frac{1}{2} \cdot \sum_\tau |u_\tau|^2$$

$$\leq \frac{1}{\delta} \cdot \|W(u)\|_{L^2} + \frac{1}{2} \cdot \|u\|_{L^2}^2.$$

This implies

$$\|u\|_{L^2} \leq \frac{2}{\delta} \cdot \|W(u)\|_{L^2}$$

and hence Lemma 2.79 is proved.   □

Lemma 2.80. *For any equivariant smooth triangulation K we have*

$$F_p(M) \preceq F_p(K).$$

Proof. Because of Theorem 2.55 (1) it suffices to show the claim for one equivariant smooth triangulation. Let K be the equivariant smooth triangulation and $C_1 > 0$ be the constant appearing in Lemma 2.78. Recall that

$\mathrm{pr} \colon l^2 C^p(K) \to l^2 C^p(K)$ is the orthogonal projection onto $\mathrm{im}(c^{p-1})^{\perp}$ and $\mathrm{pr}' \colon L^2 \Omega^p(M) \to L^2 \Omega^p(M)$ is the orthogonal projection onto $\mathrm{im}(d^{p-1}_{\min})^{\perp}$. Fix $\epsilon > 0$. Next we show that there is a constant $C_3 > 0$ such that for all $\lambda \le \epsilon$ and $\omega \in \mathrm{im}(E^{(d^{p\perp}_{\min})^* d^{p\perp}_{\min}}_{\lambda^2})$

$$||c^p \circ \mathrm{pr} \circ A^p \circ i_\epsilon(\omega)||_{L^2} \le C_3 \cdot \lambda \cdot ||\mathrm{pr} \circ A^p \circ i_\epsilon(\omega)||_{L^2} \qquad (2.81)$$

holds. We estimate using Lemma 2.2 (2), Lemma 2.78 and Lemma 2.79

$$
\begin{aligned}
||c^p \circ \mathrm{pr} \circ A^p \circ i_\epsilon(\omega)||_{L^2} \\
&= ||c^p \circ A^p \circ i_\epsilon(\omega)||_{L^2} \\
&\le C_2 \cdot ||W^{p+1} \circ c^p \circ A^p \circ i_\epsilon(\omega)||_{L^2} \\
&= C_2 \cdot ||W^{p+1} \circ A^{p+1} \circ d^p \circ i_\epsilon(\omega)||_{L^2} \\
&= C_2 \cdot ||W^{p+1} \circ A^{p+1} \circ i_\epsilon \circ d^{p\perp}_{\min}(\omega)||_{L^2} \\
&\le C_2 \cdot ||W^{p+1} \circ A^{p+1} \circ i_\epsilon|| \cdot ||d^{p\perp}_{\min}(\omega)||_{L^2} \\
&\le C_2 \cdot ||W^{p+1} \circ A^{p+1} \circ i_\epsilon|| \cdot \lambda \cdot ||\omega||_{L^2} \\
&\le C_2 \cdot ||W^{p+1} \circ A^{p+1} \circ i_\epsilon|| \cdot \lambda \cdot C_1 \cdot ||\mathrm{pr} \circ A^p \circ i_\epsilon(\omega)||_{L^2}.
\end{aligned}
$$

If we put $C_3 := C_1 \cdot C_2 \cdot ||W^{p+1} \circ A^{p+1} \circ i_\epsilon||$, then (2.81) follows.

We conclude from (2.81) that $\mathrm{clos}(\mathrm{pr} \circ A^p \circ i_\epsilon(\mathrm{im}(E^{(d^{p\perp}_{\min})^* d^{p\perp}_{\min}}_{\lambda^2})))$ belongs to $\mathcal{L}(c^p, C_3 \lambda)$. Since $\mathrm{pr} \circ A^p \circ i_\epsilon$ is injective by Lemma 2.78 we get from Theorem 1.12 (2) and Lemma 2.2 (2) for $\lambda \le \epsilon$

$$
\begin{aligned}
F_p(M)(\lambda) &= \dim_{\mathcal{N}(G)} \left(\mathrm{im}(E^{(d^{p\perp}_{\min})^* d^{p\perp}_{\min}}_{\lambda^2}) \right) \\
&= \dim_{\mathcal{N}(G)} \left(\mathrm{clos}(\mathrm{pr} \circ A^p \circ i_\epsilon(\mathrm{im}(E^{(d^{p\perp}_{\min})^* d^{p\perp}_{\min}}_{\lambda^2}))) \right) \\
&\le F_p(K)(C_3 \lambda).
\end{aligned}
$$

This finishes the proof of Lemma 2.80.   □

2.5 On the Positivity and Rationality of the Novikov-Shubin Invariants

The following conjecture is taken from [322, Conjecture 7.1 on page 56].

Conjecture 2.82. (Positivity and rationality of Novikov-Shubin invariants). *Let G be a group. Then for any free G-CW-complex X of finite type its Novikov-Shubin invariants $\alpha_p(X)$ are positive rational numbers unless they are ∞ or ∞^+.*

This conjecture is equivalent to the statement that for any matrix $A \in M(m, n, \mathbb{Z}G)$ the Novikov-Shubin invariant of the induced morphism of finitely generated Hilbert $\mathcal{N}(G)$-modules $l^2(G)^m \to l^2(G)^n$ is a positive rational number, ∞ or ∞^+. It is also equivalent to the version where X is any cocompact free proper G-manifold without boundary and with G-invariant Riemannian metric. The proof of these equivalent formulations is analogous to the proof of Lemma 10.5.

Here is some evidence for Conjecture 2.82. Unfortunately, all the evidence comes from computations, no convincing conceptual reason is known. Conjecture 2.82 has been proved for $G = \mathbb{Z}$ in Lemma 2.58. Conjecture 2.82 is true for virtually abelian G by [316, Proposition 39 on page 494]. (The author of [316] informs us that his proof of this statement is correct when $G = \mathbb{Z}$ but has a gap when $G = \mathbb{Z}^k$ for $k > 1$. The nature of the gap is described in [321, page 16]. The proof in this case can be completed by the same basic method used in [316]. Moreover, the value ∞ does not occur for $G = \mathbb{Z}^k$.) D. Voiculescu informs us that Conjecture 2.82 is also true for a free group G. Details of the proof will appear in the Ph. D.thesis of Roman Sauer [456]. If Conjecture 2.82 is true for the free G-CW-complex X of finite type and for the cocompact free H-CW-complex Y of finite type and both X and Y have the limit property (see Definition 2.41), then Conjecture 2.82 holds for the $G \times H$-CW-complex $X \times Y$ by Theorem 2.55 (3). In all examples, where Novikov-Shubin invariants can be computed explicitly, the result confirms Conjecture 2.82. For any finitely generated group G it is true for $\alpha_1(X)$ by Theorem 2.55 (5). For 3-manifolds we refer to Theorem 4.2, for Heisenberg groups to Theorem 2.85 and for symmetric spaces to Theorem 5.12 (2).

Here is further evidence if one replaces positive rational number by positive number in Conjecture 2.82. If X is a finite aspherical CW-complex such that its fundamental group contains an elementary amenable infinite normal subgroup, then $\alpha_p(\widetilde{X}) \geq 1$ holds for $p \geq 1$ (see Theorem 2.63). We get $\alpha_p(\widetilde{X}) \geq 1$ for all $p \geq 1$ if X is a connected S^1-CW-complex of finite type such that for one orbit S^1/H the inclusion into X induces a map on π_1 with infinite image by Theorem 2.61. Let $p\colon E \to B$ be a fibration such that B is a connected finite CW-complex and the fiber F has the homotopy type of a connected finite CW-complex. Suppose that the inclusion of F into E induces an injection on the fundamental groups and that $b_p^{(2)}(\widetilde{F}) = 0$ for all $p \geq 0$. Then $\alpha_p(\widetilde{E}) > 0$ for all $p \geq 1$ if $\alpha_p(\widetilde{F}) > 0$ for all $p \geq 1$ (see Theorem 3.100 and Remark 3.184). A similar statement for pushouts follows from Theorem 3.96 (2) and Remark 3.184.

Conjecture 2.82 is related to the question whether a cocompact free G-CW-complex is of determinant class (see Theorem 3.14 (4)) and to Theorem 3.28. The question about determinant class arises in the construction of L^2-torsion in Subsection 3.3.1 and has a positive answer for the groups appearing in the class $\mathcal{G}$ as explained in Chapter 13. Some evidence for Conjecture 2.82

comes from Theorem 13.7, where an estimate of the shape $F_p(\lambda) \leq \frac{C}{-\ln(\lambda)}$ is proved for small λ.

We do not know of an example of an aspherical closed manifold M such that $\alpha_p(\widetilde{M}) < 1$ holds for some $p \geq 0$.

2.6 Novikov-Shubin Invariants of Manifolds with Boundary

In this section we briefly discuss the case of a manifold with boundary. Let M be a cocompact free proper G-manifold with G-invariant Riemannian metric. Suppose that the boundary ∂M of the complete Riemannian manifold M is the disjoint union $\partial_0 M$ and $\partial_1 M$ where we allow that $\partial_0 M$, $\partial_1 M$ or both are empty. Denote by $d^p_{\min} \colon L^2\Omega^p(M) \to L^2\Omega^{p+1}(M)$ the minimal closure of the operator $d^p \colon \Omega^p_c(M, \partial_0 M) \to L^2\Omega^{p+1}(M)$, where $\Omega^p_c(M, \partial_0 M)$ is the subspace of $\Omega^p_c(M)$ consisting of those p-forms ω with compact support whose restriction to $\partial_0 M$ vanishes. Then $\mathrm{im}(d^{p-1}_{\min})$ is contained in the domain of $d^p_{\min}$ and we obtain a closed densely defined operator $d^{p\perp}_{\min} \colon \mathrm{im}(d^{p-1}_{\min})^\perp \to L^2\Omega^p(M)$. Define the *p-th analytic spectral density function* and the *p-th Novikov-Shubin invariant* of $(M, \partial_0 M)$ by

$$F_p(M, \partial_0 M) := F(d^{p\perp}_{\min});$$
$$\alpha_p(M, \partial_0 M) := \alpha(F_{p-1}(\partial_0 M)).$$

This is consistent with [240, Definition 3.1 on page 387]. Let $(K, \partial_0 K, \partial_1 K) \to (M, \partial_0 M, \partial_1 M)$ be any equivariant smooth triangulation. Then we have the following version of Theorem 2.68

$$F_p(M, \partial_0 M) \simeq F_p(K, \partial_0 K); \tag{2.83}$$
$$\alpha_p(M, \partial_0 M) = \alpha_p(K, \partial_0 K). \tag{2.84}$$

Analogously the notions and results for the Laplacian carry over to $(M, \partial_0 M)$ if we use as initial domain the space $\Omega^p_2(M, \partial_0 M)$ (see (1.83)).

We briefly explain the idea of the proof of $F_p(M, \partial_0 M) \simeq F_p(K, \partial_0 K)$. Let g_0 and g_1 be two G-invariant Riemannian metrics on M. Since M is cocompact the L^2-norms on $\Omega^p_c(M)$ with respect to g_0 and g_1 are equivalent. This implies that we obtain a commutative square with invertible bounded G-equivariant operators induced by the identity on $\Omega^p_c(M)$ as vertical arrows

$$
\begin{array}{ccc}
L^2_{g_0}\Omega^p(M) & \xrightarrow{\;d^p_{\min}\;} & L^2_{g_0}\Omega^{p+1}(M) \\[2pt]
\cong \big\downarrow & & \cong \big\downarrow \\[2pt]
L^2_{g_1}\Omega^p(M) & \xrightarrow{\;d^p_{\min}\;} & L^2_{g_1}\Omega^{p+1}(M)
\end{array}
$$

This implies that the spectral density functions of $(M, \partial_0 M)$ with respect to g_0 and g_1 are dilationally equivalent. Hence we can assume without loss of generality that the G-invariant Riemannian metric on M is a product near the boundary and hence induces a G-invariant Riemannian metric on $M \cup_{\partial_1 M} M$. Let $\tau\colon M \cup_{\partial_1 M} M \to M \cup_{\partial_1 M} M$ be the isometric involution given by flipping the two copies of M inside $M \cup_{\partial_1 M} M$. It induces an isometric involution on all spaces $L^2 \Omega^p(M)$ and the operators $d^p_{\min}$ commute with this involution. Hence the whole picture decomposes orthogonally into a $+$-part where this involution is the identity and a $-$-part where this involution is $-\operatorname{id}$. So we can define $(d^p_{\min})^\pm\colon L^2 \Omega^p(M \cup_{\partial_1 M} M)^\pm \to L^2 \Omega^{p+1}(M \cup_{\partial_1 M} M)^\pm$ and $F_p(M \cup_{\partial_1 M} M)^\pm$ and analogously $F_p(K \cup_{\partial_1 K} K)^\pm$. One checks that $(d^p_{\min})^+\colon L^2 \Omega^p(M \cup_{\partial_1 M} M)^+ \to L^2 \Omega^{p+1}(M \cup_{\partial_1 M} M)^+$ is the same as $d^p_{\min}\colon L^2 \Omega^p(M) \to L^2 \Omega^{p+1}(M)$ for the pair $(M, \partial_0 M)$ and for $-$ instead of $+$ one gets it for the pair $(M, \partial_1 M)$. In particular

$$F_p(M \cup_{\partial_1 M} M)^+ = F_p(M, \partial_0 M);$$
$$F_p(M \cup_{\partial_1 M} M)^- = F_p(M, \partial_1 M).$$

Similarly one gets

$$F_p(K \cup_{\partial_1 K} K)^+ = F_p(K, \partial_0 K);$$
$$F_p(K \cup_{\partial_1 K} K)^- = F_p(K, \partial_1 K).$$

Put $\partial_1 M = \partial M$. Then $M \cup_{\partial M} M$ has no boundary. The proof of Theorem 2.68 that $F_p(M \cup_{\partial M} M) = F_p(K \cup_{\partial K} K)$ can be easily modified to show

$$F_p(M \cup_{\partial M} M)^+ = F_p(K \cup_{\partial K} K)^+;$$
$$F_p(M \cup_{\partial M} M)^- = F_p(K \cup_{\partial K} K)^-,$$

because everything is compatible with the involution and $||\omega||_k^2 = ||\omega^+||_k^2 + ||\omega^-||_k^2$ holds for all $\omega \in H^k \Omega^p(M)$. Hence the claim is true in the case $\partial_1 M = \partial M$. Now repeating this doubling trick allows to conclude the general case by inspecting $M \cup_{\partial_1 M} M$.

2.7 Miscellaneous

Let M be a closed Riemannian manifold. The analytic Laplace operator $\Delta_p\colon L^2 \Omega^p(\widetilde{M}) \to L^2 \Omega^p(\widetilde{M})$ on the universal covering of M has zero in its spectrum if and only if $b_p^{(2)}(\widetilde{M}) \neq 0$ or $\alpha_p^\Delta(\widetilde{M}) \neq \infty^+$. In Chapter 12 we will deal with the conjecture that for an aspherical closed Riemannian manifold M there is at least one $p \geq 0$ such that Δ_p has zero in the spectrum. We conclude from Theorem 1.35 (8), Theorem 2.55 (5b) and Theorem 2.68 the result of Brooks [68] that for a closed Riemannian manifold M the Laplacian $\Delta_0\colon L^2 \Omega^0(\widetilde{M}) \to L^2 \Omega^0(\widetilde{M})$ acting on functions on the universal covering has zero in its spectrum if and only if $\pi_1(M)$ is amenable. Moreover, the result of Brooks extends to the case where M is compact and has a boundary if one uses Neumann boundary conditions on $\widetilde{M}$.

We have interpreted the L^2-Betti numbers in terms of the heat kernel (see (1.60)). We will do this also for the Novikov-Shubin invariants $\alpha_p^{\Delta}(M)$ in Theorem 3.136 (3) and (4).

The notions of spectral density function and Novikov-Shubin invariants make also sense for proper G-CW-complexes of finite type because in this setting the cellular L^2-chain complex is still defined as a Hilbert $\mathcal{N}(G)$-chain complex as explained in Section 1.6. Theorem 2.55 (1), (2), (3), (5), (6) and (7) remain true word by word for proper G-CW-complexes of finite type. The L^2-de Rham complex is also defined for a cocompact proper G-manifold with G-invariant Riemannian metric. Hence the analytic versions of spectral density function and Novikov-Shubin invariants are still well-defined in this context and Theorem 2.68 remains true if one drops the condition free.

We will explain a proportionality principle for Novikov-Shubin invariants in Theorem 3.183.

A combinatorial approach to the Novikov-Shubin invariants which is useful for concrete calculations will be given in Section 3.7.

We will give further computations of Novikov-Shubin invariants for universal coverings of compact 3-manifolds in Theorem 4.2 and for universal coverings of closed locally symmetric spaces in Theorem 5.12 (2).

We will discuss the behaviour of L^2-Betti numbers and Novikov-Shubin invariants of groups under quasi-isometry and measure equivalence in Section 7.5).

Finally we mention the following computation due to Rumin [450, Corollary 7.15 on page 449]. Notice that the invariants computed there corresponds in our notation to $\alpha_p^{\Delta}(\widetilde{M})$ and one can easily deduce the values of $\alpha_p(\widetilde{M})$ from the relation $2 \cdot \alpha_p^{\Delta}(\widetilde{M}) = \min\{\alpha_p(\widetilde{M}), \alpha_{p+1}(\widetilde{M})\}$.

Theorem 2.85. *Let M be a closed Riemannian manifold whose universal covering $\widetilde{M}$ is the Heisenberg group H^{2n+1}. Then M is a non-trivial S^1-bundle over a torus and*

$$
\alpha_p(\widetilde{M}) = \begin{cases} n+1 & \text{if } p = n+1; \\ 2 \cdot (n+1) & \text{if } 1 \leq p \leq \dim(M), p \neq n+1; \\ \infty^+ & \text{otherwise.} \end{cases}
$$

In [343] the notion of Novikov-Shubin invariants for free G-CW-complexes of finite type is extended to arbitrary G-spaces. In particular $\alpha_p(G)$ can be defined by $\alpha_p(EG)$ for any group G.

Exercises

2.1. Let $f\colon U \to V$ and $g\colon V \to W$ be morphisms of Hilbert $\mathcal{N}(G)$-modules. Show: i.) if f and g are weak isomorphisms, then $g \circ f$ is a weak isomorphism and ii.) if two of the morphisms f, g and $g \circ f$ are weak isomorphisms and Fredholm, then all three are weak isomorphisms and Fredholm. Find a counterexample to assertion ii.) if one replaces "weak isomorphisms and Fredholm" by "weak isomorphisms".

2.2. Let G be a group which contains an element of infinite order. Then for any integer $n \geq 1$ there is an element $u \in \mathbb{C}G$ such that $\alpha(r_u) = 1/n$ and $b^{(2)}(r_u) = 0$ holds for the map $r_u\colon l^2(G) \to l^2(G)$ given by right multiplication with u. Moreover, if we additionally assume that G is finitely presented we can find for any sequence $\alpha_3, \alpha_4, \ldots$ of elements $\alpha_p \in \{1/n \mid n \in \mathbb{Z}, n \geq 1\} \coprod \{\infty^+\}$ a connected CW-complex X of finite type with $\pi_1(X) = G$ such that $b_p^{(2)}(\tilde{X}) = 0$ and $\alpha_p(\tilde{X}) = \alpha_p$ holds for $p \geq 3$.

2.3. Let G be locally finite, i.e. any finitely generated subgroup is finite. Let $A \in M(m, n, \mathbb{C}G)$ be any matrix. It induces a map of Hilbert $\mathcal{N}(G)$-modules $r_A\colon \bigoplus_{i=1}^{m} l^2(G) \to \bigoplus_{i=1}^{n} l^2(G)$. Then $\alpha(r_A) = \infty^+$. If X is any free G-CW-complex of finite type, then $\alpha_p(X) = \infty^+$ for all $p \geq 1$.

2.4. Show that there is no connected CW-complex X of finite type with $\alpha_1(\tilde{X}) = 1$ and $\alpha_2(\tilde{X}) \neq \infty^+$.

2.5. Show for two connected closed 4-dimensional manifolds M and N with isomorphic fundamental groups that $\alpha_p(\widetilde{M}) = \alpha_p(\widetilde{N})$ holds for all $p \geq 1$. Show the analogous statement if M and N are connected compact 3-manifolds possibly with boundary.

2.6. Let $S^1 \to E \to B$ be a principal S^1-bundle of CW-complexes of finite type such that B is simply connected. Show that either $\alpha_p(\tilde{E}) = 1$ or $\alpha_p(\tilde{E}) = \infty^+$ for $p \geq 1$ holds.

2.7. Show that $l^2(\mathbb{Z}^n)$ viewed as $\mathbb{C}[\mathbb{Z}^n]$-module is flat if and only if $n \leq 1$.

2.8. Let M be a closed hyperbolic manifold. Compute the L^2-Betti numbers and the Novikov-Shubin invariants of the universal covering of $T^n \times M$.

2.9. Let M and N be two compact manifolds (possibly with boundary) whose dimension is less or equal to 2. Compute the L^2-Betti numbers and Novikov-Shubin invariants of $\widetilde{M \times N}$.

2.10. Let X and Z be connected CW-complexes of finite type. Suppose that $\pi_1(Z)$ is finite and $\tilde{X}$ has the limit property. Show

$$\alpha_p(\widetilde{X \times Z}) = \min\{\alpha_i(\tilde{X}) \mid i = 1, \ldots, p, \, H_{p-i}(\tilde{Z}; \mathbb{C}) \neq 0\};$$

$$\alpha_p(\widetilde{X \times T^n}) = \begin{cases} \min\{n + \alpha_i(\tilde{X}) \mid i = p - n, \ldots, p\} & \text{if } b_i^{(2)}(\tilde{X}) = 0 \text{ for} \\ & \qquad p - n \leq i \leq p - 1; \\ n & \text{otherwise.} \end{cases}$$

2.11. Let $F \to E \to B$ be a fibration of connected CW-complexes such that the inclusion induces an injection $\pi_1(F) \to \pi_1(E)$. Suppose that $b_p^{(2)}(\widetilde{F}) = 0$ and $\alpha_p(\widetilde{F}) > 0$ holds for all $p \geq 0$. Show that then the same is true for $\widetilde{E}$.

2.12.   Let A be the set of real numbers r, for which there exists a CW-complex X of finite type and an integer $p \geq 0$ with $\alpha_p(\widetilde{X}) = r$. Then A is countable and contains $\{r \mid r \in \mathbb{Q}, r \geq 0\}$.

3. L^2-Torsion

Introduction

In this chapter we introduce and study L^2-torsion for Hilbert chain complexes and for regular coverings of finite CW-complexes or of compact manifolds.

There are various notions of torsion invariants, such as Reidemeister torsion, Whitehead torsion and analytic Ray-Singer torsion, which have been intensively studied since the twenties and have remained in the focus of attention. They will be reviewed in Section 3.1. L^2-torsion is the L^2-analog of Reidemeister torsion as L^2-Betti numbers are the L^2-analogs of the classical Betti numbers. The situation for the classical Reidemeister torsion is best if the homology vanishes, and the same is true in the L^2-context. Therefore we will consider in this introduction only the case of the universal covering $\widetilde{X}$ of a finite CW-complex X for which all L^2-Betti numbers are trivial. In many interesting geometric situations this assumption will be satisfied.

For the construction of Reidemeister torsion the notion of a determinant is crucial and we will define and investigate its L^2-version, the Fuglede-Kadison determinant, in Section 3.2. In the classical context of Reidemeister torsion the vanishing of the homology implies that the chain complexes under consideration are contractible. In the L^2-context the vanishing of the L^2-Betti numbers does not imply that the Hilbert chain complex under consideration is contractible, the Novikov-Shubin invariants of Chapter 2 measure the difference. Notice that the zero-in-the-spectrum Conjecture 12.1 says in the aspherical case that the cellular Hilbert $\mathcal{N}(\pi_1(X))$-chain complex of $\widetilde{X}$ is never contractible so that it is too restrictive to demand contractibility. This forces us to deal with Fuglede-Kadison determinants for weak automorphisms and with the problem whether $\widetilde{X}$ is of determinant class, which means that the Fuglede-Kadison determinant of each differential is different from zero and hence the L^2-torsion is defined as a real number. The universal covering $\widetilde{X}$ is of determinant class, provided that all its Novikov-Shubin invariants are positive or that $\pi_1(X)$ belongs to the class $\mathcal{G}$, which will be investigated in Subsection 13.1.3. There is the Conjecture 3.94 that $\widetilde{X}$ is always of determinant class. We will deal with it in Chapter 13.

In Section 3.3 we introduce L^2-torsion for finite Hilbert $\mathcal{N}(G)$-chain complexes and define $\rho^{(2)}(\widetilde{X})$ in terms of these invariants applied to the cellular

L^2-chain complex in Section 3.4. It turns out that the L^2-torsion $\rho^{(2)}(\widetilde{X})$ has the same formal properties as the classical Euler characteristic $\chi(X)$ if one ensures that relevant maps induce injections on the fundamental groups. For instance, $\rho^{(2)}(\widetilde{X})$ depends only on the simple homotopy type of X and, provided that a certain map in K-theory is always trivial, it depends on the homotopy type of X only. There are sum formulas, product formulas and fibration formulas and Poincaré duality holds. For a self map $f: X \to X$ inducing an isomorphism on $\pi_1(X)$ one can compute $\rho^{(2)}(\widetilde{T_f})$ of the mapping torus T_f in terms of the map induced on the L^2-homology of $\widetilde{X}$ by f. If X is an aspherical finite CW-complex such that $\pi_1(X)$ contains an elementary amenable infinite normal subgroup and is of det $\geq$ 1-class, then $\rho^{(2)}(\widetilde{X})$ vanishes. There are explicit formulas in terms of $H_p(\widetilde{X}; \mathbb{C})$ if $\pi_1(X) = \mathbb{Z}$.

In Section 3.5 we introduce the analytic version of $\rho^{(2)}(\widetilde{M})$ for a closed Riemannian manifold M in terms of the Laplace operator on differential forms on $\widetilde{M}$ following Lott and Mathai. The analytic and topological version agree by a deep result of Burghelea, Friedlander, Kappeler and McDonald. The L^2-torsion $\rho^{(2)}(\widetilde{M})$ of an odd-dimensional closed hyperbolic manifold is up to a dimension constant, which is computable and different from zero, the volume of M. Manifolds with boundary are briefly discussed in Section 3.6.

In Section 3.7 we give a combinatorial approach for the computation of L^2-Betti numbers and L^2-torsion. Namely, we give an algorithm to produce monotone decreasing sequences of rational numbers which converge to the L^2-Betti numbers and the Fuglede-Kadison determinants respectively. It is easier to compute than for instance the analytic versions which are defined in terms of the heat kernel (and meromorphic extensions). In practice this yields numerical upper bounds for the L^2-Betti numbers and the Fuglede-Kadison determinants. The speed of convergence is $\approx n^{-\alpha}$, where α is the relevant Novikov-Shubin invariant.

In order to get a quick overview one should read through Sections 3.1, (only if the reader is not familiar with the classical concept of torsion invariants) 3.4 and 3.5 and skip the very technical Sections 3.2 and 3.3.

3.1 Survey on Torsion Invariants

In this section we give a brief review of torsion invariants in order to motivate the definition of L^2-torsion.

3.1.1 Whitehead Groups

Let R be an associative ring with unit. Denote by $GL(n, R)$ the group of invertible (n, n)-matrices with entries in R. Define the group $GL(R)$ by the colimit of the system indexed by the natural numbers $\ldots \subset GL(n, R) \subset$

$GL(n+1, R) \subset \ldots$ where the inclusion $GL(n, R)$ to $GL(n+1, R)$ is given by stabilization

$$A \mapsto \begin{pmatrix} A & 0 \\ 0 & 1 \end{pmatrix}.$$

Define $K_1(R)$ by the abelianization $GL(R)/[GL(R), GL(R)]$ of $GL(R)$. Define the *Whitehead group* $\mathrm{Wh}(G)$ of a group G to be the cokernel of the map $G \times \{\pm 1\} \to K_1(\mathbb{Z}G)$ which sends $(g, \pm 1)$ to the class of the invertible $(1, 1)$-matrix $(\pm g)$. This will be the group where Whitehead torsion will take its values in.

The Whitehead group $\mathrm{Wh}(G)$ is known to be trivial if G is the free abelian group $\mathbb{Z}^n$ of rank n [25] or the free group $*_{i=1}^n \mathbb{Z}$ of rank n [481]. There is the conjecture that it vanishes for any torsionfree group. This has been proved by Farrell and Jones [190], [191], [193], [194], [196] for a large class of groups. This class contains any subgroup $G \subset G'$ where G' is a discrete cocompact subgroup of a Lie group with finitely many path components and any group G which is the fundamental group of a non-positively curved closed Riemannian manifold or of a complete pinched negatively curved Riemannian manifold. The Whitehead group satisfies $\mathrm{Wh}(G * H) = \mathrm{Wh}(G) \oplus \mathrm{Wh}(H)$ [481].

If G is finite, then $\mathrm{Wh}(G)$ is very well understood (see [406]). Namely, $\mathrm{Wh}(G)$ is finitely generated, its rank as abelian group is the number of conjugacy classes of unordered pairs $\{g, g^{-1}\}$ in G minus the number of conjugacy classes of cyclic subgroups and its torsion subgroup is isomorphic to the kernel $SK_1(G)$ of the change of coefficient homomorphism $K_1(\mathbb{Z}G) \to K_1(\mathbb{Q}G)$. For a finite cyclic group G the Whitehead group $\mathrm{Wh}(G)$ is torsionfree. For instance the Whitehead group $\mathrm{Wh}(\mathbb{Z}/p)$ of a cyclic group of order p for an odd prime p is the free abelian group of rank $(p-3)/2$ and $\mathrm{Wh}(\mathbb{Z}/2) = 0$. The Whitehead group of the symmetric group S_n is trivial. The Whitehead group of $\mathbb{Z}^2 \times \mathbb{Z}/4$ is not finitely generated as abelian group.

3.1.2 Whitehead Torsion

Let R be an associative ring with unit. Let C_* be a based free finite R-chain complex, where *based free* means that each chain module C_p is equipped with an (ordered) basis. Suppose that C_* is acyclic. Choose a chain contraction $\gamma_* \colon C_* \to C_{*+1}$. In the sequel we write :

$$C_{\mathrm{odd}} := \bigoplus_{n \in \mathbb{Z}} C_{2n+1};$$

$$C_{\mathrm{ev}} := \bigoplus_{n \in \mathbb{Z}} C_{2n}.$$

Then we obtain an isomorphism $(c + \gamma)_{\mathrm{odd}} \colon C_{\mathrm{odd}} \to C_{\mathrm{ev}}$. Since we have a basis for the source and target of this isomorphism, it determines an invertible matrix and hence an element

$$\rho(C_*) \in K_1(R). \tag{3.1}$$

Let $f\colon X \to Y$ be a G-homotopy equivalence of finite free G-CW-complexes. It induces a chain homotopy equivalence $C_*(f)\colon C_*(X) \to C_*(Y)$ of the cellular $\mathbb{Z}G$-chain complexes. Let $\mathrm{cone}_*(C_*(f))$ be its mapping cone which is a contractible finite free $\mathbb{Z}G$-chain complex. The G-CW-complex-structure determines a cellular $\mathbb{Z}G$-basis which is not quite unique, one may permute the basis elements or multiply a basis element with an element of the form $\pm g \in \mathbb{Z}G$. Hence we can define the *Whitehead torsion*

$$\tau(f) \in \mathrm{Wh}(G) \tag{3.2}$$

by the image of the element $\rho(\mathrm{cone}_*(C_*(f)))$ defined in (3.1) under the canonical projection $K_1(\mathbb{Z}G) \to \mathrm{Wh}(G)$. The Whitehead torsion $\tau(f)$ depends only on the G-homotopy class of f and satisfies $\tau(g \circ f) = \tau(f) + \tau(g)$ for G-homotopy equivalences $f\colon X \to Y$ and $g\colon Y \to Z$ of finite free G-CW-complexes [116, (22.4)]. There are sum and product formulas for Whitehead torsion [116, (23.1) and (23.2)].

Given a homotopy equivalence $f\colon X \to Y$ of connected finite CW-complexes, we can pick a lift $\widetilde{f}\colon \widetilde{X} \to \widetilde{Y}$ to the universal coverings and obtain an element $\tau(\widetilde{f}) \in \mathrm{Wh}(\pi_1(Y))$. The vanishing of this element has a specific meaning, namely it is zero if and only if f is a *simple homotopy equivalence*, i.e. up to homotopy f can be written as a finite sequence of combinatorial moves, so called elementary collapses and elementary expansions [116, (22.2)]. If f is a homeomorphism, then it is a simple homotopy equivalence, or equivalently $\tau(\widetilde{f}) = 0$, by a result of Chapman [99], [100]. A map between finite polyhedra is a simple homotopy equivalence if and only if there are regular neighbourhoods $i_X\colon X \to N_X$ and $i_Y\colon Y \to N_Y$ in high dimensional Euclidean space and a homeomorphism $g\colon N_X \to N_Y$ such that $g \circ i_X$ and $i_Y \circ f$ are homotopic [445, Chapter 3]. If Y is not connected, we define $\mathrm{Wh}(\pi_1(Y)) := \bigoplus_{C \in \pi_0(Y)} \mathrm{Wh}(\pi_1(C))$ and $\tau(f)$ by $\{\tau(f|_{f^{-1}(C)}\colon f^{-1}(C) \to C) \mid C \in \pi_0(Y)\}$.

The main importance of Whitehead torsion lies in the s-Cobordism Theorem we will explain next. A $(n+1)$-dimensional *h-cobordism* is a compact $(n+1)$-dimensional manifold W whose boundary is the disjoint union $\partial W = \partial_0 W \coprod \partial_1 W$ such that the inclusions $i_k\colon \partial_k W \to W$ for $k = 0,1$ are homotopy equivalences. Given a closed n-dimensional manifold M, an *h-cobordism over M* is a $(n+1)$-dimensional h-cobordism W together with a diffeomorphism $f\colon M \to \partial_0 W$. Two such h-cobordisms (W,f) and (W',f') over M are diffeomorphic if there is a diffeomorphism $F\colon W \to W'$ with $F \circ f = f'$. The next result is due to Barden, Mazur, Stallings (for its proof see for instance [288], [337, Section 1]), its topological version was proved by Kirby and Siebenmann [292, Essay II].

Theorem 3.3 (S-Cobordism Theorem). *Let M be a closed smooth n-dimensional manifold. Suppose $n \geq 5$. Then the map from the set of* *diffeomorphism classes of h-cobordisms (W, f) over M to the Whitehead group* $\mathrm{Wh}(\pi_1(M))$ *which sends the class of* (W, f) *to the image of the Whitehead torsion* $\tau(\widetilde{i_0}: \widetilde{\partial_0 W} \to \widetilde{W})$ *under the inverse of the isomorphism* $\mathrm{Wh}(\pi_1(M)) \to \mathrm{Wh}(\pi_1(W))$ *induced by* $i_0 \circ f$, *is a bijection. In particular an h-cobordism* (W, f) *over* M *is trivial, i.e. diffeomorphic to the trivial h-cobordism* $(M \times [0, 1], i_0)$, *if and only if* $\tau(\widetilde{i}: \widetilde{\partial_0 W} \to \widetilde{W}) = 0$.

Notice that the Whitehead group of the trivial group vanishes. Hence any h-cobordism over M is trivial if M is simply connected. This implies the *Poincaré Conjecture* in dimensions ≥ 5 which says that a closed n-dimensional manifold is homeomorphic to S^n if it is homotopy equivalent to S^n. The s-Cobordism Theorem is known to be false in dimension $n = 4$ [151] but it is still true for "good" fundamental groups in the topological category by results of Freedman [202], [203]. This implies that the Poincaré Conjecture is true also in dimension 4. Counterexamples to the s-Cobordism Theorem in dimension $n = 3$ are constructed by Cappell and Shaneson [90]. The Poincaré Conjecture in dimension 3 is open at the time of writing.

The s-Cobordism Theorem is one key ingredient in surgery theory which is designed to classify manifolds up to diffeomorphism. For instance, in order to show that two closed manifolds M and N are diffeomorphic, the strategy is to construct a cobordism W' (with appropriate bundle data) between M and N, then to modify W' to an h-cobordism W over M via surgery on the interior of W' such that the Whitehead torsion is trivial, and finally to apply the s-Cobordism Theorem to conclude that W is diffeomorphic to $M \times [0, 1]$ and in particular its two ends M and N are diffeomorphic. More information about Whitehead torsion can be found for instance in [116], [375]. Generalizations like bounded, controlled, equivariant or stratified versions can be found for instance in [4], [270], [326], [424], [425], [426], [483], [519].

3.1.3 Reidemeister Torsion

Let X be a finite free G-CW-complex. Let V be an orthogonal finite dimensional G-representation. Suppose that $H_p^G(X; V) := H_p(V \otimes_{\mathbb{Z}G} C_*(X))$ vanishes for all $p \geq 0$. After a choice of a cellular $\mathbb{Z}G$-basis, we obtain an isomorphism $\bigoplus_{i=1}^{l_p} V \xrightarrow{\cong} V \otimes_{\mathbb{Z}G} C_p(X)$. Now choose any orthonormal basis of V and equip $V \otimes_{\mathbb{Z}G} C_*(X)$ with the induced basis. With these choices we obtain a well-defined element $\rho(V \otimes_{\mathbb{Z}G} C_*(X)) \in K_1(\mathbb{R})$ (see (3.1)). The determinant induces an isomorphism $\det_{\mathbb{R}}: K_1(\mathbb{R}) \xrightarrow{\cong} \mathbb{R}^{\mathrm{inv}}$. The *Reidemeister torsion* of X with coefficients in V is defined to be the real number

$$\rho(X; V) := \ln\left(\left|\det_{\mathbb{R}}(\rho(V \otimes_{\mathbb{Z}G} C_*(X)))\right|\right). \tag{3.4}$$

It is independent of the choices of a cellular basis for X and an orthonormal basis for V. If $f: X' \to X$ is a G-homeomorphism, then $\rho(X'; V) = \rho(X; V)$.

Let V and W be two orthogonal $\mathbb{Z}/n$-representations for some $n \geq 3$ such that $\mathbb{Z}/n$ acts freely on the unit spheres SV and SW. If U is any orthogonal representation with trivial fixed point set, then one computes $H_p^{\mathbb{Z}/n}(SV;U) = H_p^{\mathbb{Z}/n}(SW;U) = 0$ for all $p \geq 0$ and hence $\rho(SV;U)$ and $\rho(SW;U)$ are defined. Suppose that SV and SW are $\mathbb{Z}/n$-homeomorphic. Since then $\rho(SV;U) = \rho(SW;U)$ holds for any such U, one can compute using Franz' Lemma [201] that V and W are isomorphic as orthogonal $\mathbb{Z}/n$-representations. Reidemeister torsion was the first invariant in algebraic topology which is not a homotopy invariant. Namely, for suitable choices of V and W, the associated *lens spaces* $L(V) := (\mathbb{Z}/n)\backslash SV$ and $L(W)$ are homotopy equivalent but not homeomorphic [436], [116, chapter V]. The difference of the diffeomorphism type is detected by $\rho(SV;U)$ for suitable U. On the other hand Reidemeister torsion can be used to prove rigidity. Namely, one can show using Reidemeister torsion that $L(V)$ and $L(W)$ are homeomorphic if and only if they are isometrically diffeomorphic with respect to the Riemannian metric induced by the orthogonal structure on V and W. Lens spaces with this Riemannian metric have constant positive sectional curvature. A closed Riemannian manifold with constant positive sectional curvature and cyclic fundamental group is isometrically diffeomorphic to a lens space after possibly rescaling the Riemannian metric with a constant [525]. The result above for free representations is generalized by De Rham's Theorem [133] (see also [323, Proposition 3.2 on page 478], [327, page 317], [443, section 4]) as follows. It says for a finite group G and two orthogonal G-representations V and W whose unit spheres SV and SW are G-diffeomorphic that V and W are isomorphic as orthogonal G-representations. This remains true if one replaces G-diffeomorphic by G-homeomorphic provided that G has odd order (see [265], [353]), but not for any finite group G (see [89], [91], [244] and [245]).

The Alexander polynomial of a knot can be interpreted as a kind of Reidemeister torsion of the canonical infinite cyclic covering of the knot complement (see [374], [497]). Reidemeister torsion appears naturally in surgery theory [352]. Counterexamples to the (polyhedral) Hauptvermutung that two homeomorphic simplicial complexes are already PL-homeomorphic are given by Milnor [373] (see also [431]) and detected by Reidemeister torsion. Seiberg-Witten invariants for 3-manifolds are essentially given by torsion invariants [498].

Definition (3.4) can be extended to the case where $H_p^G(X;V)$ is not trivial, provided a Hilbert space structure is specified on each $H_p^G(X;V)$. Namely, there is up to chain homotopy precisely one chain map $i_*: H_*^G(X;V) \to V \otimes_{\mathbb{Z}G} C_*(X)$ which induces the identity on homology where we consider $H_*^G(X;V)$ as a chain complex with trivial differentials. Its mapping cone $\mathrm{cone}_*(i_*)$ is an acyclic $\mathbb{R}$-chain complex whose chain modules are Hilbert spaces. As above choose for each chain module an orthonormal basis and define using (3.4)

$$\rho(X; V) := \ln\left(|\det_{\mathbb{R}}\left(\rho(\mathrm{cone}_*(i_*))\right)|\right). \tag{3.5}$$

The real number $\rho(X; V)$ is independent of the choice of i_* and orthonormal basis (see also [375, page 365]). There is one preferred Hilbert structure on $H_*^G(X; V)$ which is induced by the one on $V \otimes_{\mathbb{Z}G} C_*(X)$. However, with this choice $\rho(X; V)$ is not invariant under barycentric subdivision if X is a free cocompact simplicial G-complex with non-trivial $H_*^G(X; V)$ so that we do not get a useful invariant for manifolds (see [345, Example 5.1 on page 240]).

The situation is better in the presence of a G-invariant Riemannian metric on the cocompact free proper G-manifold M. Let $f: X \to M$ be any equivariant smooth triangulation of M. Then we can use a variant of the Hodge-de Rham isomorphism $\mathcal{H}^p(M; V) \xrightarrow{\cong} H_G^p(X; f^*V)$ (see Theorem 1.57), the natural isomorphism $H_G^p(X; f^*V)^* \xrightarrow{\cong} H_p^G(X; f^*V)$ and the Hilbert space structure on $\mathcal{H}^p(M; V)$ coming from the Riemannian metric to put a preferred Hilbert space structure on $H_p^G(X; f^*V)$. With respect to this preferred Hilbert space structure on $H_p^G(X; f^*V)$ we define the *topological Reidemeister torsion*

$$\rho_{\mathrm{top}}(M; V) := \rho(X; f^*V), \tag{3.6}$$

where $\rho(X; f^*V)$ was defined in (3.5). The topological Reidemeister torsion is independent of the choice of (X, f) and is invariant under isometric G-diffeomorphisms. If $H^p(M; V)$ is trivial for all $p \geq 0$, then we are back in the situation of (3.4) and $\rho_{\mathrm{top}}(M; V)$ depends only on the G-diffeomorphism type of M but not any more on the G-invariant Riemannian metric.

Ray-Singer [432] defined the analytic counterpart of topological Reidemeister torsion using a regularization of the zeta-function as follows. The first observation is that one can compute $\rho(X; V)$, which was introduced in (3.5) with respect to the Hilbert structure on $H_*^G(X; V)$ induced by the one on $V \otimes_{\mathbb{Z}G} C_*(X)$, in terms of the cellular Laplace operator $\Delta_p: V \otimes_{\mathbb{Z}G} C_p(X) \to V \otimes_{\mathbb{Z}G} C_p(X)$ by

$$\rho(C_*) = -\frac{1}{2} \cdot \sum_{p \in \mathbb{Z}} (-1)^p \cdot p \cdot \ln(\det_{\mathbb{R}}(\Delta_p^\perp)), \tag{3.7}$$

where $\Delta_p^\perp: \ker(\Delta_p)^\perp \to \ker(\Delta_p)^\perp$ is the positive automorphism of finite dimensional Hilbert spaces induced by Δ_p (cf. Lemma 3.30). Let M be a cocompact free proper G-manifold with G-invariant Riemannian metric. Let $\Delta_p: \Omega^p(M; V) \to \Omega^p(M; V)$ be the Laplace operator acting on smooth p-forms on M with coefficients in the orthogonal (finite dimensional) G-representation V. The Laplacian above on M is an essentially selfadjoint operator with discrete spectrum since M is cocompact. One wants to use the expression (3.7) also for the analytic Laplace operator and has to take into account that it is defined on infinite-dimensional spaces and hence $\ln(\det_{\mathbb{R}}(\Delta_p^\perp))$ does not make sense a priori. This is done as follows. The *zeta-function* is defined by

$$\zeta_p(s) := \sum_{\lambda > 0} \lambda^{-s}, \qquad (3.8)$$

where λ runs through the positive eigenvalues of Δ_p listed with multiplicity. Since the eigenvalues grow fast enough, the zeta-function is holomorphic for $\Re(s) > \dim(M)/2$. Moreover it has a meromorphic extension to $\mathbb{C}$ with no pole in 0 [468]. So its derivative for $s = 0$ is defined. The *analytic Reidemeister torsion or Ray-Singer torsion* of M is defined by [432, Definition 1.6 on page 149]

$$\rho_{\mathrm{an}}(M; V) := \frac{1}{2} \cdot \sum_{p \geq 0} (-1)^p \cdot p \cdot \frac{d}{ds} \zeta_p(s)|_{s=0}. \qquad (3.9)$$

The basic idea is that $\frac{d}{ds}\zeta_p(s)|_{s=0}$ is a generalization of the (logarithm of) the ordinary determinant $\det_{\mathbb{R}}$. Namely, if $f : V \to V$ is a positive linear automorphism of the finite-dimensional real vector space V and the positive real numbers $\lambda_1, \lambda_2, \ldots, \lambda_r$ are the eigenvalues of f listed with multiplicity, then we get

$$\frac{d}{ds} \zeta_p(s)|_{s=0} = \frac{d}{ds} \sum_{i=1}^{r} \lambda_i^{-s} \bigg|_{s=0}$$

$$= \sum_{i=1}^{r} \left(-\ln(\lambda_i) \cdot \lambda_i^{-s} \right)\big|_{s=0}$$

$$= -\ln \left(\prod_{i=1}^{r} \lambda_i \right)$$

$$= -\ln \left(\det_{\mathbb{R}}(f) \right).$$

Ray and Singer conjectured that the analytic and topological Reidemeister torsion agree. This conjecture was proved independently by Cheeger [103] and Müller [390]. Manifolds with boundary and manifolds with symmetries, sum (= glueing) formulas and fibration formulas are treated in [74], [82], [126], [127], [323], [327], [348], [503], [504], [505]. Non-orthogonal coefficient systems are studied in [49], [50], [80], [393]. Further references are [42], [44], [43], [45], [46], [47], [48], [62], [83], [136], [181], [198], [206], [207], [221], [295], [296], [317], [387], [423], [433], [506].

3.2 Fuglede-Kadison Determinant

In this section we extend the notion of the Fuglede-Kadison determinant for invertible morphisms of finite dimensional Hilbert $\mathcal{N}(G)$-modules [208] to arbitrary morphisms and study its main properties. It is not enough to consider only invertible morphisms because then we could construct L^2-torsion only for finite free G-CW-complexes whose L^2-Betti numbers are all trivial and whose Novikov-Shubin invariants are all ∞^+. But this condition is in view of the zero-in-the-spectrum Conjecture 12.1 much too restrictive.

Let $f\colon U \to V$ be a morphism of finite dimensional Hilbert $\mathcal{N}(G)$-modules, i.e. $\dim_{\mathcal{N}(G)}(U), \dim_{\mathcal{N}(G)}(V) < \infty$. Recall that $\{E_\lambda^{f^*f} \mid \lambda \in \mathbb{R}\}$ is the spectral family of the positive operator f^*f and that the spectral density function $F = F(f)\colon [0, \infty) \to [0, \infty)$ sends λ to $\dim_{\mathcal{N}(G)}(\operatorname{im}(E_{\lambda^2}^{f^*f}))$ (see Lemma 2.3). Recall that F is a monotone non-decreasing right-continuous function. Denote by dF the measure on the Borel σ-algebra on $\mathbb{R}$ which is uniquely determined by its values on the half open intervals $(a, b]$ for $a < b$

$$dF((a, b]) = F(b) - F(a). \tag{3.10}$$

Notice that the measure of the one point set $\{a\}$ is $\lim_{x \to 0+} F(a) - F(a - x)$ and is zero if and only if F is left-continuous in a. We will use here and in the sequel the convention that $\int_a^b, \int_{a+}^b, \int_a^\infty$ and $\int_{a+}^\infty$ respectively means integration over the interval $[a, b]$, $(a, b]$, $[a, \infty)$ and (a, ∞) respectively.

Definition 3.11 (Fuglede-Kadison determinant). *Let $f\colon U \to V$ be a morphism of finite dimensional Hilbert $\mathcal{N}(G)$-modules with spectral density function $F = F(f)$. Define its* (generalized) *Fuglede-Kadison determinant*

$$\operatorname{det}_{\mathcal{N}(G)}(f) \in [0, \infty)$$

by $\operatorname{det}_{\mathcal{N}(G)}(f) := \exp\left(\int_{0+}^\infty \ln(\lambda)\, dF\right)$ *if* $\int_{0+}^\infty \ln(\lambda)\, dF > -\infty$ *and by* $\operatorname{det}_{\mathcal{N}(G)}(f) := 0$ *if* $\int_{0+}^\infty \ln(\lambda)\, dF = -\infty$. *We call f of determinant class if and only if* $\int_{0+}^\infty \ln(\lambda)\, dF > -\infty$. *Often we omit $\mathcal{N}(G)$ from the notation.*

Notice that for $\int_{0+}^\infty \ln(\lambda)\, dF$ there is only a problem of convergence at zero where $\ln(\lambda)$ goes to $-\infty$ but not at ∞ because we have $F(\lambda) = F(||f||_\infty)$ for all $\lambda \geq ||f||_\infty$ and hence

$$\int_{0+}^\infty \ln(\lambda)\, dF = \int_{0+}^a \ln(\lambda)\, dF \qquad \text{for } a \geq ||f||_\infty.$$

The notion of determinant class and the first investigations of its basic properties are due to Burghelea-Friedlander-Kappeler-McDonald [84, Definition 4.1 on page 800, Definition 5.7 on page 817].

If $0\colon U \to V$ is the zero homomorphism for finite-dimensional Hilbert $\mathcal{N}(G)$-modules U and V, then $\det(0\colon U \to V) = 1$ by definition.

Example 3.12. Let G be finite and $f\colon U \to V$ be a morphism of finite dimensional Hilbert $\mathcal{N}(G)$-modules, i.e. a linear G-equivariant map of finite-dimensional unitary G-representations. Let $\lambda_1, \lambda_2, \ldots, \lambda_r$ be the positive eigenvalues of the positive map f^*f. Then we conclude from Example 2.5

$$\det(f) = \left(\prod_{i=1}^r \lambda_i\right)^{\frac{1}{2\cdot|G|}}.$$

If f is an isomorphism $\det(f)$ is the $|G|$-th root of the classical determinant of the positive automorphism of complex vector spaces $|f|\colon U \to U$.

Example 3.13. Let $G = \mathbb{Z}^n$. In the sequel we use the notation and the identification $\mathcal{N}(\mathbb{Z}^n) = L^\infty(T^n)$ of Example 1.4. We conclude from Example 2.6 for $f \in L^\infty(T^n)$

$$\det\left(M_f\colon L^2(T^n) \to L^2(T^n)\right) = \exp\left(\int_{T^n} \ln(|f(z)|) \cdot \chi_{\{u \in S^1 | f(u) \neq 0\}} \ dvol_z\right)$$

using the convention $\exp(-\infty) = 0$.

The next result says that the main important properties of the classical determinant of endomorphisms of finite-dimensional complex vector spaces carry over to the (generalized) Fuglede-Kadison determinant. But the proof in our context is of course more complicated than the one in the classical case.

Theorem 3.14 (Kadison-Fuglede determinant). *(1) Let $f\colon U \to V$ and $g\colon V \to W$ be morphisms of finite dimensional Hilbert $\mathcal{N}(G)$-modules such that f has dense image and g is injective. Then*

$$\det(g \circ f) \ = \ \det(f) \cdot \det(g);$$

(2) Let $f_1\colon U_1 \to V_1$, $f_2\colon U_2 \to V_2$ and $f_3\colon U_2 \to V_1$ be morphisms of finite dimensional Hilbert $\mathcal{N}(G)$-modules such that f_1 has dense image and f_2 is injective. Then

$$\det\begin{pmatrix} f_1 & f_3 \\ 0 & f_2 \end{pmatrix} \ = \ \det(f_1) \cdot \det(f_2);$$

(3) Let $f\colon U \to V$ be a morphism of finite dimensional Hilbert $\mathcal{N}(G)$-modules. Then

$$\det(f) = \det(f^*);$$

(4) If the Novikov-Shubin invariant of the morphism $f\colon U \to V$ of finite dimensional Hilbert $\mathcal{N}(G)$-modules satisfies $\alpha(f) > 0$, then f is of determinant class;

(5) Let $H \subset G$ be a subgroup of finite index $[G:H]$. Let $\mathrm{res}\, f\colon \mathrm{res}\, U \to \mathrm{res}\, V$ be the morphism of finite dimensional Hilbert $\mathcal{N}(H)$-modules obtained from f by restriction. Then

$$\det\nolimits_{\mathcal{N}(H)}(\mathrm{res}\, f) = \det\nolimits_{\mathcal{N}(G)}(f)^{[G:H]};$$

(6) Let $i\colon H \to G$ be an injective group homomorphism and let $f\colon U \to V$ be a morphism of finite dimensional Hilbert $\mathcal{N}(H)$-modules. Then

$$\det\nolimits_{\mathcal{N}(G)}(i_* f) = \det\nolimits_{\mathcal{N}(H)}(f).$$

Before we can give the proof of Theorem 3.14, we need

Lemma 3.15. Let $f\colon U \to V$ be a morphism of finite dimensional Hilbert $\mathcal{N}(G)$-modules and let $F\colon [0, \infty) \to [0, \infty)$ be a density function. Then

(1) We have for $0 < \epsilon < a$

$$\int_{\epsilon+}^{a} \ln(\lambda) \, dF = -\int_{\epsilon}^{a} \frac{1}{\lambda} \cdot (F(\lambda) - F(0)) \, d\lambda$$
$$+ \ln(a) \cdot (F(a) - F(0)) - \ln(\epsilon) \cdot (F(\epsilon) - F(0));$$

$$\int_{0+}^{a} \ln(\lambda) \, dF = \lim_{\epsilon \to 0+} \int_{\epsilon+}^{a} \ln(\lambda) \, dF;$$

$$\int_{0+}^{a} \frac{1}{\lambda} \cdot (F(\lambda) - F(0)) \, d\lambda = \lim_{\epsilon \to 0+} \int_{\epsilon}^{a} \frac{1}{\lambda} \cdot (F(\lambda) - F(0)) \, d\lambda.$$

We have $\int_{0+}^{a} \ln(\lambda) \, dF > -\infty$ if and only if $\int_{0+}^{a} \frac{1}{\lambda} \cdot (F(\lambda) - F(0)) \, d\lambda < \infty$, and in this case

$$\lim_{\lambda \to 0+} \ln(\lambda) \cdot (F(\lambda) - F(0)) = 0;$$

$$\int_{0+}^{a} \ln(\lambda) \, dF = -\int_{0+}^{a} \frac{1}{\lambda} \cdot (F(\lambda) - F(0)) \, d\lambda$$
$$+ \ln(a) \cdot (F(a) - F(0));$$

(2) If f is invertible, we get

$$\det(f) = \exp\left(\frac{1}{2} \cdot \operatorname{tr}(\ln(f^* f))\right);$$

(3) If $f^\perp \colon \ker(f)^\perp \to \operatorname{clos}(\operatorname{im}(f))$ is the weak isomorphism induced by f we get
$$\det(f) = \det(f^\perp);$$

(4) $\det(f) = \det(f^*) = \sqrt{\det(f^* f)} = \sqrt{\det(f f^*)};$

(5) If $f\colon U \to U$ is an injective positive operator, then

$$\lim_{\epsilon \to 0+} \det(f + \epsilon \cdot \operatorname{id}_U) = \det(f);$$

(6) If $f \leq g$ for injective positive morphisms $f, g\colon U \to U$, then

$$\det(f) \leq \det(g);$$

(7) *If f and g are morphisms of finite dimensional Hilbert $\mathcal{N}(G)$-modules, then*

$$\det(f \oplus g) = \det(f) \cdot \det(g).$$

Proof. (1) The first equation follows from partial integration for a continuously differentiable function g on $(0, \infty)$ for $\epsilon < a$ (see for instance [330, page 95])

$$\int_{\epsilon+}^{a} g(\lambda) \, dF = -\int_{\epsilon}^{a} g'(\lambda) \cdot F(\lambda) \cdot d\lambda + g(a) \cdot F(a) - g(\epsilon) \cdot F(\epsilon) \quad (3.16)$$

and the second and third from Levi's Theorem of monotone convergence. We conclude from the first three equations that

$$-\int_{0+}^{1} \frac{1}{\lambda} \cdot (F(\lambda) - F(0)) \, d\lambda \le \int_{0+}^{1} \ln(\lambda) \, dF,$$

and that it remains to show

$$\int_{0+}^{1} \ln(\lambda) \, dF > -\infty \Rightarrow \lim_{\lambda \to 0+} \ln(\lambda) \cdot (F(\lambda) - F(0)) = 0. \quad (3.17)$$

Suppose that $\lim_{\lambda \to 0+} \ln(\lambda) \cdot (F(\lambda) - F(0)) = 0$ is not true. Then we can find a number $C < 0$ and a monotone decreasing sequence $1 > \lambda_1 > \lambda_2 > \lambda_3 > \dots$ of positive real numbers converging to zero such that $\ln(\lambda_i) \cdot (F(\lambda_i) - F(0)) \le C$ holds for all $i \ge 0$. Since $\lim_{i \to \infty} F(\lambda_i) = F(0)$ holds we can assume without loss of generality $2 \cdot (F(\lambda_{i+1}) - F(0)) \le F(\lambda_i) - F(0)$, otherwise pass to an appropriate subsequence of $(\lambda_i)_i$. We have for each natural number n and each $\lambda \in (0, 1)$ that $\ln(\lambda) \le \sum_{i=1}^{n} \ln(\lambda_i) \cdot \chi_{(\lambda_{i+1}, \lambda_i]}(\lambda)$ holds, where $\chi_{(\lambda_{i+1}, \lambda_i]}(\lambda)$ is the characteristic function of $(\lambda_{i+1}, \lambda_i]$. This implies for all $n \ge 1$

$$\int_{0+}^{1} \ln(\lambda) \, dF \le \int_{0+}^{1} \sum_{i=1}^{n} \ln(\lambda_i) \cdot \chi_{(\lambda_{i+1}, \lambda_i]}(\lambda) \, dF$$

$$= \sum_{i=1}^{n} \ln(\lambda_i) \cdot (F(\lambda_i) - F(\lambda_{i+1}))$$

$$\le \sum_{i=1}^{n} \ln(\lambda_i) \cdot \frac{F(\lambda_i) - F(0)}{2}$$

$$\le n \cdot \frac{C}{2}.$$

We conclude $\int_{0+}^{1} \ln(\lambda) \, dF = -\infty$. Hence (3.17) and therefore assertion (1) are proved.

(2) We conclude from (1.65) or the more general fact that the trace is linear and ultra-weakly continuous

$$\operatorname{tr}(\ln(f^* f)) = \operatorname{tr}\left(\int_{0+}^{\infty} \ln(\lambda)\, dE_\lambda^{f^* f}\right)$$

$$= \int_{0+}^{\infty} \ln(\lambda)\, d\left(\operatorname{tr}(E_\lambda^{f^* f})\right)$$

$$= \int_{0+}^{\infty} \ln(\lambda^2)\, d\left(\operatorname{tr}(E_{\lambda^2}^{f^* f})\right)$$

$$= 2 \cdot \int_{0+}^{\infty} \ln(\lambda)\, dF(f).$$

(3) If $F^\perp$ is the spectral density function of $f^\perp$, then $F(\lambda) = F^\perp(\lambda) + F(0)$.

(4) This follows from assertion (1) and the conclusion from Lemma 2.4

$$F(f)(\lambda) - F(f)(0) = F(f^*)(\lambda) - F(f^*)(0)$$
$$= F(f^* f)(\lambda^2) - F(f^* f)(0) = F(f f^*)(\lambda^2) - F(f f^*)(0).$$

(5) We have $F(f + \epsilon \cdot \operatorname{id}_U)(\lambda) = F(f)(\lambda - \epsilon)$. Since $F(f)(0) = 0$, we get

$$\int_{0+}^{\infty} \ln(\lambda)\, dF(f + \epsilon \cdot \operatorname{id}_U) = \int_{(-\epsilon)+}^{\infty} \ln(\lambda + \epsilon)\, dF(f)$$

$$= \int_{(-\epsilon)+}^{0} \ln(\lambda + \epsilon)\, dF(f) + \int_{0+}^{\infty} \ln(\lambda + \epsilon)\, dF(f)$$

$$= \ln(\epsilon) \cdot F(f)(0) + \int_{0+}^{\infty} \ln(\lambda + \epsilon)\, dF(f)$$

$$= \int_{0+}^{\infty} \ln(\lambda + \epsilon)\, dF(f).$$

We conclude from Levi's Theorem of monotone convergence

$$\lim_{\epsilon \to 0+} \int_{0+}^{\infty} \ln(\lambda + \epsilon)\, dF(f) = \int_{0+}^{\infty} \ln(\lambda)\, dF(f).$$

This shows

$$\lim_{\epsilon \to 0+} \det(f + \epsilon \cdot \operatorname{id}_U) = \det(f).$$

(6) For $u \in U$ we get

$$\|f^{1/2}(u)\| = \langle f^{1/2}(u), f^{1/2}(u) \rangle^{1/2}$$
$$= \langle f(u), u \rangle^{1/2}$$
$$\le \langle g(u), u \rangle^{1/2}$$
$$= \langle g^{1/2}(u), g^{1/2}(u) \rangle^{1/2}$$
$$= \|g^{1/2}(u)\|.$$

This implies for the spectral density functions

$$F(g^{1/2})(\lambda) \leq F(f^{1/2})(\lambda).$$

Next we give the proof of the claim under the additional hypothesis that f and g are invertible. Then we can choose $\epsilon > 0$ such that $F(f)(\lambda) = F(g)(\lambda) = 0$ for $\lambda \leq \epsilon$. Fix $a \geq ||g|| \geq ||f||$. Since $F(f^{1/2})(\lambda) = F(g^{1/2})(\lambda) = \dim_{\mathcal{N}(G)}(U)$ for $\lambda \geq \sqrt{a}$, we conclude from assertion (1)

$$\int_0^\infty \ln(\lambda)\, dF(g^{1/2}) \geq \int_0^\infty \ln(\lambda)\, dF(f^{1/2}).$$

This implies $\det(f^{1/2}) \leq \det(g^{1/2})$. Now the claim for invertible f and g follows from assertion (4). The general case follows now from assertion (5) since $f + \epsilon \cdot \mathrm{id}_U$ and $g + \epsilon \cdot \mathrm{id}_U$ are invertible and $f + \epsilon \cdot \mathrm{id}_U \leq g + \epsilon \cdot \mathrm{id}_U$ holds for all $\epsilon > 0$.

(7) Obviously $F(f \oplus g) = F(f) + F(g)$. Now apply assertion (1). $\square$

For the proof of Theorem 3.14 we will need the next lemma where we will use holomorphic calculus [282, Theorem 3.3.5 on page 206].

Lemma 3.18. *Let $f\colon D \to \mathbb{C}$ be a holomorphic function defined on a domain D in $\mathbb{C}$ whose boundary is a smooth closed curve $\gamma\colon S^1 \to \mathbb{C}$. Let $X(t)$ for $0 \leq t \leq 1$ be a differentiable family of morphisms $X(t)\colon U \to U$ for a finite dimensional Hilbert $\mathcal{N}(G)$-module U, where differentiable is to be understood with respect to the operator norm. Suppose that the spectrum $\mathrm{spec}(X(t)) := \{z \in \mathbb{C} \mid z - X(t)$ is not invertible$\}$ of each $X(t)$ lies in the interior of D. We define (motivated by the Cauchy integral formula)*

$$f(X(t)) := \frac{1}{2\pi i} \int_\gamma f(z) \cdot (z - X(t))^{-1}\, dz.$$

Then $f(X(t))$ is differentiable with respect to t and

$$\mathrm{tr}\left(\frac{d}{dt} f(X(t))\right) = \mathrm{tr}(f'(X(t)) \circ X'(t)),$$

where $f'(z) = \frac{d}{dz} f(z)$ and $X'(t) = \frac{d}{dt} X(t)$.

Proof. From

$$\frac{f(X(t+h)) - f(X(t))}{h}$$

$$= \frac{1}{2\pi i} \int_\gamma f(z) \cdot \frac{(z - X(t+h))^{-1} - (z - X(t))^{-1}}{h}\, dz$$

$$= \frac{1}{2\pi i} \int_\gamma f(z) \cdot (z - X(t+h))^{-1} \circ \frac{X(t+h) - X(t)}{h} \circ (z - X(t))^{-1}\, dz$$

we conclude

$$\frac{d}{dt} f(X(t)) = \frac{1}{2\pi i} \int_\gamma f(z) \cdot (z - X(t))^{-1} \circ X'(t) \circ (z - X(t))^{-1} \, dz.$$

Since

$$\frac{d}{dz}(f(z) \cdot (z - X(t))^{-1}) = f'(z) \cdot (z - X(t))^{-1} - f(z) \cdot (z - X(t))^{-2}$$

and γ is closed, partial integration gives

$$f'(X(t)) = \frac{1}{2\pi i} \int_\gamma f(z) \cdot (z - X(t))^{-2} \, dz.$$

Since

$$\operatorname{tr}\left(f(z) \cdot (z - X(t))^{-1} \circ X'(t) \circ (z - X(t))^{-1}\right)$$
$$= \operatorname{tr}\left(f(z) \cdot (z - X(t))^{-2} \circ X'(t)\right)$$

holds and tr commutes with integration, Lemma 3.18 follows.   □

Now we are ready to give the proof of Theorem 3.14.

Proof. (1) Next we show for positive invertible morphisms $f, g \colon U \to U$

$$\operatorname{tr}(\ln(gf^2 g)) = 2 \cdot (\operatorname{tr}(\ln(g)) + \operatorname{tr}(\ln(f))). \tag{3.19}$$

Consider the families $g(t \cdot f^2 + (1 - t) \cdot \mathrm{id})g$ and $t \cdot f^2 + (1 - t) \cdot \mathrm{id}$. There are real numbers $0 < a < b$ such that the spectrum of each member of these families lies in $[a, b]$. Choose a domain D in the half plane of complex numbers with positive real part which is bounded by a smooth curve and contains $[a, b]$. Let ln be a holomorphic extension of the logarithm $(0, \infty) \to \mathbb{R}$ to a holomorphic function on $\mathbb{C}$ with the negative real numbers removed. Notice that D lies in the domain of ln. Since the definition of $\ln(h)$ of 1.64 and the one by holomorphic calculus in Lemma 3.18 for invertible h agree, Lemma 3.18 implies

$$\frac{d}{dt} \operatorname{tr}\left(\ln(g(t \cdot f^2 + (1 - t) \cdot \mathrm{id})g)\right)$$
$$= \operatorname{tr}\left(\frac{d}{dt} \ln(g(t \cdot f^2 + (1 - t) \cdot \mathrm{id})g)\right)$$
$$= \operatorname{tr}\left((g(t \cdot f^2 + (1 - t) \cdot \mathrm{id})g)^{-1} \circ (g(f^2 - \mathrm{id})g)\right)$$
$$= \operatorname{tr}\left((t \cdot f^2 + (1 - t) \cdot \mathrm{id})^{-1} \circ (f^2 - \mathrm{id})\right)$$
$$= \frac{d}{dt} \operatorname{tr}\left(\ln(t \cdot f^2 + (1 - t) \cdot \mathrm{id})\right).$$

Now (3.19) follows since $\ln(h^2) = 2 \cdot \ln(h)$ for invertible positive h and $\ln(\mathrm{id}) = 0$.

Next we show for injective positive morphisms $f, g \colon U \to U$

$$\det(gf^2g) = \det(f)^2 \cdot \det(g)^2. \tag{3.20}$$

Notice that (3.20) holds under the additional assumption that f and g are invertible because of Lemma 3.15 (2), the equation $\ln(h^2) = 2 \cdot \ln(h)$ for invertible positive h and (3.19). The general case is reduced to this special case as follows.

Choose a constant C such that for all $0 < \epsilon \le 1$

$$gf^2g \le g(f + \epsilon \cdot \mathrm{id}_U)^2 g \le gf^2g + C\epsilon \cdot \mathrm{id}_U$$

holds. We conclude from Lemma 3.15 (4), (5) and (6)

$$\det(gf^2g) = \lim_{\epsilon \to 0+} \det(g(f + \epsilon \cdot \mathrm{id}_U)^2 g);$$
$$\det(f) = \lim_{\epsilon \to 0+} \det(f + \epsilon \cdot \mathrm{id}_U);$$
$$\det(gf^2g) = \det(fg^2f).$$

Since $g + \epsilon$ and $f + \epsilon$ for $\epsilon > 0$ are invertible, we get

$$\begin{aligned}
\det(gf^2g) &= \lim_{\epsilon \to 0+} \det(g(f + \epsilon \cdot \mathrm{id}_U)^2 g) \\
&= \lim_{\epsilon \to 0+} \det((f + \epsilon \cdot \mathrm{id}_U)g^2(f + \epsilon \cdot \mathrm{id}_U)) \\
&= \lim_{\epsilon \to 0+} \lim_{\delta \to 0+} \det((f + \epsilon \cdot \mathrm{id}_U)(g + \delta \cdot \mathrm{id}_U)^2(f + \epsilon \cdot \mathrm{id}_U)) \\
&= \lim_{\epsilon \to 0+} \lim_{\delta \to 0+} \det(f + \epsilon \cdot \mathrm{id}_U)^2 \cdot \det(g + \delta \cdot \mathrm{id}_U)^2 \\
&= \det(f)^2 \cdot \det(g)^2.
\end{aligned}$$

Hence (3.20) holds for all injective positive morphisms f and g.

Given morphisms $f \colon U \to V$ and $g \colon V \to W$ of finite dimensional Hilbert $\mathcal{N}(G)$-modules such that f has dense image and g is injective, it remains to show

$$\det(g \circ f) = \det(f) \cdot \det(g).$$

We can assume without loss of generality that both f and g are injective. Otherwise replace f by the injective map $f^\perp \colon \ker(f)^\perp \to V$ induced by f and use the conclusion from Lemma 3.15 (7) that $\det(f^\perp) = \det(f)$ and $\det(g \circ f^\perp) = \det(g \circ f)$ holds.

If u is a unitary and v some morphism, then $u^{-1}vu$ and v have the same spectral density function and hence

$$\det(u^{-1}vu) = \det(v). \tag{3.21}$$

In order to prove assertion (1) we use the polar decomposition $f = au$ and $g = vb$ where u and v are unitary isomorphisms and a and b are positive. Notice that g is injective, f has dense image and hence f^* is injective. Therefore both a and b are injective. We compute using Lemma 3.15 (4), (3.20) and (3.21)

$$\det(gf) = \det(vbau) = \sqrt{\det((vbau)^*(vbau))} = \sqrt{\det(u^{-1}ab^2au)}$$

$$= \sqrt{\det(ab^2a)} = \sqrt{\det(a)^2 \cdot \det(b)^2} = \sqrt{\det(ff^*) \cdot \det(g^*g)} = \det(f) \cdot \det(g).$$

(2) The claim is already proved in Lemma 3.15 (7) if f_3 is trivial. Because of the equation

$$\begin{pmatrix} f_1 & f_3 \\ 0 & f_2 \end{pmatrix} = \begin{pmatrix} 1 & 0 \\ 0 & f_2 \end{pmatrix} \cdot \begin{pmatrix} 1 & f_3 \\ 0 & 1 \end{pmatrix} \cdot \begin{pmatrix} f_1 & 0 \\ 0 & 1 \end{pmatrix}$$

assertion (1) implies that it suffices to prove

$$\det \begin{pmatrix} 1 & f_3 & 0 \\ 0 & 1 & 0 \\ 0 & 0 & 1 \end{pmatrix} = 1.$$

Since this matrix can be written as a commutator,

$$\begin{pmatrix} 1 & f_3 & 0 \\ 0 & 1 & 0 \\ 0 & 0 & 1 \end{pmatrix} = \begin{pmatrix} 1 & 0 & 0 \\ 0 & 1 & 0 \\ 0 & -f_3 & 1 \end{pmatrix} \cdot \begin{pmatrix} 1 & 0 & 1 \\ 0 & 1 & 0 \\ 0 & 0 & 1 \end{pmatrix} \cdot \begin{pmatrix} 1 & 0 & 0 \\ 0 & 1 & 0 \\ 0 & -f_3 & 1 \end{pmatrix}^{-1} \cdot \begin{pmatrix} 1 & 0 & 1 \\ 0 & 1 & 0 \\ 0 & 0 & 1 \end{pmatrix}^{-1}$$

the claim follows from assertion (1).

(3) has already been proved in Lemma 3.15 (4).

(4) Because of Lemma 3.15 (1) it suffices to show:

$$\lim_{\epsilon \to 0+} \int_\epsilon^a \frac{1}{\lambda} \cdot (F(\lambda) - F(0)) \cdot d\lambda < \infty.$$

Since $\alpha(f)$ is assumed to be positive, there is $0 < \delta$ and $0 < \alpha < \alpha(f)$ such that

$$F(\lambda) - F(0) \le \lambda^\alpha$$

holds for $0 \le \lambda \le \delta$. Now assertion (4) follows from

$$\lim_{\epsilon \to 0+} \int_\epsilon^a \lambda^{\alpha-1} \, d\lambda = \lim_{\epsilon \to 0+} \frac{1}{\alpha} \cdot (a^\alpha - \epsilon^\alpha) = \frac{1}{\alpha} \cdot a^\alpha.$$

(5) This follows from Theorem 1.12 (6).

(6) This follows from (2.57). This finishes the proof of Theorem 3.14.   □

Example 3.22. Consider a non-trivial element $p \in \mathbb{C}[\mathbb{Z}]$. We want to compute the Fuglede-Kadison determinant of the morphism $R_p \colon l^2(\mathbb{Z}) \to l^2(\mathbb{Z})$ given by multiplication with $p \in \mathbb{C}[\mathbb{Z}]$. We can write

$$p(z) = C \cdot z^n \cdot \prod_{k=1}^l (z - a_k)$$

for complex numbers $C, a_0, a_1, \ldots, a_l$ and integers n, l with $C \ne 0$ and $l \ge 0$. We want to show

$$\ln(\det(R_p)) = \ln(|C|) + \sum_{1 \leq k \leq l, |a_k| > 1} \ln(|a_k|). \tag{3.23}$$

We get from Theorem 3.14 (1)

$$\det(R_p) = \det(R_C) \cdot \det(R_z)^n \cdot \prod_{k=1}^{l} \det(R_{(z-a_i)}).$$

Hence it remains to show

$$\det(R_{(z-a)}) = \begin{cases} |a| & \text{for } |a| \geq 1 \\ 1 & \text{for } |a| \leq 1 \end{cases} \tag{3.24}$$

Because of Example 3.13 it suffices to show

$$\int_{S^1} \ln((z-a)\overline{(z-a)}) \, d\text{vol} = \begin{cases} 2 \cdot \ln(|a|) & \text{for } |a| \geq 1 \\ 0 & \text{for } |a| \leq 1 \end{cases} \tag{3.25}$$

for $a \in \mathbb{C}$, where we equip S^1 with the obvious measure satisfying $\text{vol}(S^1) = 1$. Because $\int_{S^1} \ln((z-a)\overline{(z-a)}) \, d\text{vol} = \int_{S^1} \ln((z-|a|)(z^{-1}-|a|)) \, d\text{vol}$ we may suppose $a \in \mathbb{R}^{\geq 0}$ in the sequel.

We compute for $a \neq 1$ and the path $\gamma \colon [0,1] \to S^1$, $t \mapsto \exp(2\pi i t)$ using the Residue Theorem

$$\int_{S^1} \frac{d}{da} \ln((z-a)(z^{-1}-a)) \, d\text{vol}$$

$$= \int_{S^1} \frac{1}{a-z} + \frac{1}{a-z^{-1}} \, d\text{vol}$$

$$= 2 \cdot \int_{S^1} \frac{1}{a-z} \, d\text{vol}$$

$$= 2 \cdot \int_{S^1} \frac{1}{(a-z) \cdot 2\pi i z} \cdot 2\pi i z \cdot d\text{vol}$$

$$= \frac{2}{2\pi i} \cdot \int_{\gamma} \frac{1}{(a-z) \cdot z} \, dz$$

$$= \begin{cases} \frac{2}{a} & \text{for } a > 1 \\ 0 & \text{for } a < 1 \end{cases}.$$

This implies for $a \in \mathbb{R}^{\geq 0}, a \neq 1$

$$\frac{d}{da} \int_{S^1} \ln((z-a)(z^{-1}-a)) \, d\text{vol} = \begin{cases} \frac{2}{a} & \text{for } a > 1 \\ 0 & \text{for } a < 1 \end{cases}.$$

We conclude for an appropriate number C

$$\begin{aligned} \int_{S^1} \ln((z-a)(z^{-1}-a)) \, d\text{vol} &= 2 \cdot \ln(a) + C & \text{for } a > 1 \\ \int_{S^1} \ln((z-a)(z^{-1}-a)) \, d\text{vol} &= 0 & \text{for } a < 1 \end{aligned}.$$

We get from Levi's Theorem of monotone convergence

$$\int_{S^1} \ln((z-1)(z^{-1}-1))\, d\mathrm{vol} = C.$$

We get from Lebesgue's Theorem of majorized convergence

$$\int_{S^1} \ln((z-1)(z^{-1}-1))\, d\mathrm{vol} = 0.$$

This proves (3.25) and hence (3.24) and (3.23).

Example 3.26. The following examples show that the conditions appearing in Theorem 3.14 (1) and (2) are necessary. We will use the same notation as in Example 3.13. Let $u \in L^\infty(T^n)$ be a given function such that $\{z \in T^n \mid u(z) = 0\}$ has measure zero. Put

$$
\begin{aligned}
g:&\ L^2(T^n)^2 \to L^2(T^n)^2, & (a,b) &\mapsto (a,0);\\
f:&\ L^2(T^n)^2 \to L^2(T^n)^2, & (a,b) &\mapsto (u\cdot a + b, b);\\
f_1:&\ L^2(T^n) \to L^2(T^n), & a &\mapsto u\cdot a;\\
f_2:&\ L^2(T^n) \to L^2(T^n), & a &\mapsto 0;\\
f_3:&\ L^2(T^n) \to L^2(T^n), & a &\mapsto a.
\end{aligned}
$$

Then one easily checks using Lemma 3.15, Theorem 3.14 (2) and Example 3.13.

$$
\begin{aligned}
\det(g\circ f) &= \det\begin{pmatrix} u & 1 \\ 0 & 0 \end{pmatrix} = \left(\det\begin{pmatrix} u & 1 \\ 0 & 0 \end{pmatrix}\begin{pmatrix} u & 1 \\ 0 & 0 \end{pmatrix}^*\right)^{1/2}\\
&= \left(\det\begin{pmatrix} uu^*+1 & 0 \\ 0 & 0 \end{pmatrix}\right)^{1/2} = \exp\left(\frac{1}{2}\cdot\int_{T^n}\ln(1+|u(z)|^2)\, d\mathrm{vol}_z\right);
\end{aligned}
$$

$$
\det(f) = \det\begin{pmatrix} u & 1 \\ 0 & 1 \end{pmatrix} = \det(u) = \exp\left(\int_{T^n}\ln(|u(z)|)\, d\mathrm{vol}_z\right);
$$

$$
\det\begin{pmatrix} f_1 & f_3 \\ 0 & f_2 \end{pmatrix} = \det\begin{pmatrix} u & 1 \\ 0 & 0 \end{pmatrix} = \exp\left(\frac{1}{2}\cdot\int_{T^n}\ln(1+|u(z)|^2)\, d\mathrm{vol}_z\right);
$$

$$\det(g) = 1;$$

$$\det(f_1) = \exp\left(\int_{T^n}\ln(|u(z)|)\, d\mathrm{vol}_z\right);$$

$$\det(f_2) = 1.$$

If we put for instance $n = 1$ and $u(\exp(2\pi it)) = \exp(\frac{-1}{t})$ for $t \in (0,1]$, then we get $\int_{T^n}\ln(|u(z)|)\, d\mathrm{vol}_z = -\infty$ and $\int_{T^n}\ln(1+|u(z)|^2)\, d\mathrm{vol}_z \geq 0$ and hence

$$\det(g \circ f) \geq 1;$$
$$\det(f) = 0;$$
$$\det(g) = 1;$$
$$\det \begin{pmatrix} f_1 & f_3 \\ 0 & f_2 \end{pmatrix} \geq 1;$$
$$\det(f_1) = 0;$$
$$\det(f_2) = 1.$$

Notice that f has dense image and g is not injective and $\det(g \circ f) \neq \det(f) \cdot \det(g)$. Moreover, f^* is injective and g^* has not dense image and $\det(f^* \circ g^*) \neq \det(f^*) \cdot \det(g^*)$ because of Lemma 3.15 (4). A similar statement holds for $\begin{pmatrix} f_1 & f_3 \\ 0 & f_2 \end{pmatrix}$.

We have shown in Theorem 3.14 (4) for a morphisms $f: U \to V$ of finite dimensional Hilbert $\mathcal{N}(G)$-modules that $\alpha(f) > 0$ implies $\det(f) > 0$. The converse is not true in general as the following example shows.

Example 3.27. We give an example of a morphism of finitely generated Hilbert $\mathcal{N}(\mathbb{Z})$-modules such that f is of determinant class but its Novikov Shubin invariant is zero (cf Lemma 3.14 (4)). Fix $\epsilon > 0$. Define a monotone non-decreasing continuous function $F: [0, \infty) \to [0, \infty)$ by $F(0) = 0$, $F(\lambda) = |\ln(\lambda/(1 + \epsilon))|^{-1-\epsilon}$ for $0 < \lambda \leq 1$ and $F(\lambda) = |\ln(1/(1 + \epsilon))|^{-1-\epsilon}$ for $1 \leq \lambda$. Since $\lim_{\lambda \to 0+} \frac{\lambda^\alpha}{F(\lambda)} = 0$ holds for any $\alpha > 0$, we get $\alpha(F) = 0$. We conclude from Levi's Theorem of monotone convergence

$$\int_{0+}^{1} \frac{F(\lambda) - F(0)}{\lambda} \, d\lambda = \int_{0+}^{1} \frac{1}{\lambda \cdot |\ln(\lambda/(1 + \epsilon))|^{1+\epsilon}} \, d\lambda$$
$$= \lim_{\delta \to 0+} \int_{\delta}^{1} \frac{1}{\lambda \cdot |\ln(\lambda/(1 + \epsilon))|^{1+\epsilon}} \, d\lambda$$
$$= \lim_{\delta \to 0+} -\frac{1}{\epsilon \cdot |\ln(1/(1 + \epsilon))|} + \frac{1}{\epsilon \cdot \ln(\delta/(1 + \epsilon))}$$
$$= -\frac{1}{\epsilon \cdot \ln(1 + \epsilon)}$$
$$< \infty.$$

Define $f: S^1 \to \mathbb{R}$ by sending $\exp(2\pi i t)$ to $F(t)$ for $0 \leq t < 1$. Then F is the spectral density function of the morphism $M_f: L^2(S^1) \to L^2(S^1)$ of finitely generated Hilbert $\mathcal{N}(\mathbb{Z})$-modules (see Example 2.6). Lemma 3.15 (1) implies

$$\det(f) > 0;$$
$$\alpha(f) = 0.$$

Hence $\det(f) > 0$ does not imply $\alpha(f) > 0$. But we can get from $\det(f) > 0$ the following related conclusion for the asymptotic behaviour of the spectral density function of f at zero which of course is weaker than the condition $\alpha(f) > 0$

Theorem 3.28. *Let $f: U \to V$ be a morphism of finite dimensional Hilbert $\mathcal{N}(G)$-modules. Let F be its spectral density function. Put $K = \max\{1, \|f\|_\infty\}$. Suppose that $\det(f) > 0$. Then we get for all $\lambda \in (0, 1]$*

$$F(\lambda) - F(0) \leq (\ln(K) \cdot (\dim(U) - \dim(\ker(f))) - \ln(\det(f))) \cdot \frac{1}{-\ln(\lambda)}.$$

Proof. We conclude from Lemma 3.15 (1) that $\int_{0+}^{K} \frac{F(\xi) - F(0)}{\xi} d\xi < \infty$ and

$$\ln(\det(f)) = \ln(K) \cdot (\dim(U) - \dim(\ker(f))) - \int_{0+}^{K} \frac{F(\xi) - F(0)}{\xi} d\xi.$$

We estimate for $\lambda \in (0, 1]$

$$
\begin{aligned}
(F(\lambda) &- F(0)) \cdot (\ln(K) - \ln(\lambda)) \\
&= \int_{\lambda}^{K} \frac{F(\lambda) - F(0)}{\xi} d\xi \\
&\leq \int_{\lambda}^{K} \frac{F(\xi) - F(0)}{\xi} d\xi \\
&\leq \int_{0+}^{K} \frac{F(\xi) - F(0)}{\xi} d\xi \\
&= \ln(K) \cdot (\dim(U) - \dim(\ker(f))) - \ln(\det(f)).
\end{aligned}
$$

This implies

$$
\begin{aligned}
F(\lambda) - F(0) &\leq \frac{\ln(K) \cdot (\dim(U) - \dim(\ker(f))) - \ln(\det(f))}{\ln(K) - \ln(\lambda)} \\
&\leq (\ln(K) \cdot (\dim(U) - \dim(\ker(f))) - \ln(\det(f))) \cdot \frac{1}{-\ln(\lambda)}. \quad \square
\end{aligned}
$$

The obvious version of Theorem 3.28 for analytic spectral density functions is stated in Theorem 13.7.

3.3 L^2-Torsion of Hilbert Chain Complexes

In this section we introduce and study the L^2-torsion of finite Hilbert $\mathcal{N}(G)$-chain complexes.

3.3.1 Basic Definitions and Properties of L^2-Torsion

Definition 3.29 (L^2-torsion). *We call a Hilbert $\mathcal{N}(G)$-chain complex C_* dim-finite if $\dim(C_p) < \infty$ for all p and $C_p = 0$ for $|p| \geq N$ for some integer N. A dim-finite Hilbert $\mathcal{N}(G)$-chain complex C_* is of determinant class if the differential $c_p \colon C_p \to C_{p-1}$ is of determinant class in the sense of Definition 3.11 for each $p \in \mathbb{Z}$. A Hilbert $\mathcal{N}(G)$-chain complex C_* is called weakly acyclic or equivalently L^2-acyclic if its L^2-homology is trivial. It is called det-L^2-acyclic if C_* is of determinant class and weakly acyclic.*

If C_* is of determinant class, define its L^2-torsion by

$$\rho^{(2)}(C_*) := -\sum_{p \in \mathbb{Z}}(-1)^p \cdot \ln(\det(c_p)) \qquad \in \mathbb{R}.$$

Notice that we prefer to take the logarithm of the determinant in order to get later additive instead of multiplicative formulas and because this will fit better with the analytic version. The next lemma expresses the L^2-torsion in terms of the Laplace operator $\Delta_p \colon C_p \to C_p$ (see (1.17)) which will motivate the analytic definition later.

Lemma 3.30. *A dim-finite Hilbert $\mathcal{N}(G)$-chain complex C_* is of determinant class if and only if Δ_p is of determinant class for all $p \in \mathbb{Z}$. In this case*

$$\rho^{(2)}(C_*) = -\frac{1}{2} \cdot \sum_{p \in \mathbb{Z}}(-1)^p \cdot p \cdot \ln(\det(\Delta_p)).$$

Proof. From Lemma 1.18 we obtain an orthogonal decomposition

$$C_p = \ker(c_p)^\perp \oplus \mathrm{clos}(\mathrm{im}(c_{p+1})) \oplus \ker(\Delta_p);$$
$$\Delta_p = ((c_p^\perp)^* \circ c_p^\perp) \oplus (c_{p+1}^\perp \circ (c_{p+1}^\perp)^*) \oplus 0,$$

where $c_j^\perp \colon \ker(c_j)^\perp \to \mathrm{clos}(\mathrm{im}(c_j))$ is the weak isomorphism induced by c_j. Lemma 3.15 (4) and (7) imply that C_* is of determinant class if and only Δ_p is of determinant class for all $p \in \mathbb{Z}$ and that in this case

$$-\frac{1}{2} \cdot \sum_{p \in \mathbb{Z}}(-1)^p \cdot p \cdot \ln(\det(\Delta_p))$$

$$= -\frac{1}{2} \cdot \sum_{p \in \mathbb{Z}}(-1)^p \cdot p \cdot \ln\left(\det\left(((c_p^\perp)^* \circ c_p^\perp) \oplus (c_{p+1}^\perp \circ (c_{p+1}^\perp)^*) \oplus 0\right)\right)$$

$$= -\frac{1}{2} \cdot \sum_{p \in \mathbb{Z}}(-1)^p \cdot p \cdot \left(\ln\left(\det\left((c_p^\perp)^* \circ c_p^\perp\right)\right)\right.$$

$$\left. + \ln\left(\det\left(c_{p+1}^\perp \circ (c_{p+1}^\perp)^*\right)\right) + \ln(\det(0))\right)$$

$$= -\frac{1}{2} \cdot \sum_{p \in \mathbb{Z}}(-1)^p \cdot p \cdot \left(2 \cdot \ln\left(\det\left(c_p\right)\right) + 2 \cdot \ln\left(\det\left(c_{p+1}\right)\right)\right)$$

$$= -\sum_{p \in \mathbb{Z}}(-1)^p \cdot \ln\left(\det\left(c_p\right)\right). \qquad \square$$

Definition 3.31. *Let* $f_* \colon C_* \to D_*$ *be a chain map of dim-finite Hilbert* $\mathcal{N}(G)$-*chain complexes. We call it of determinant class if its mapping cone* $\mathrm{cone}_*(f_*)$ *is of determinant class. In this case we define the* L^2-*torsion of* f_* *by*

$$t^{(2)}(f_*) := \rho^{(2)}(\mathrm{cone}_*(f_*)).$$

Before we can state the main properties of these invariants $\rho^{(2)}(C_*)$ and $t^{(2)}(f_*)$, we need some preparations. Let $0 \to U \xrightarrow{i} V \xrightarrow{p} W \to 0$ be a weakly exact sequence of finite dimensional Hilbert $\mathcal{N}(G)$-modules. We call it *of determinant class* if the 2-dimensional weakly acyclic chain complex which it defines with W in dimension 0 is of determinant class, and we define in this case

$$\rho^{(2)}(U, V, W) \in \mathbb{R} \tag{3.32}$$

by the L^2-torsion of this chain complex in the sense of Definition 3.29. If the sequence is of determinant class we will later see in Lemma 3.41 or directly from Theorem 3.14 that for any choice of map $s \colon W \to V$ for which $p \circ s$ is a weak isomorphism of determinant class also $i \oplus s \colon U \oplus W \to V$ is a weak isomorphism of determinant class and

$$\begin{aligned}
\rho^{(2)}(U, V, W) &= -\ln(\det(i)) + \ln(\det(p)) \\
&= -\ln(\det(i \oplus s)) + \ln(\det(p \circ s)). \tag{3.33}
\end{aligned}$$

If $0 \to U \xrightarrow{i} V \xrightarrow{p} W \to 0$ is exact, then it is of determinant class. If i is isometric and p induces an isometric isomorphism $\ker(p)^\perp \to W$, then $\rho^{(2)}(U, V, W) = 0$. This applies to the canonical exact sequence $0 \to U \to U \oplus W \to W \to 0$.

If $0 \to C_* \to D_* \to E_* \to 0$ is a weakly exact sequence of dim-finite Hilbert $\mathcal{N}(G)$-chain complexes, we call it *of determinant class* if each exact sequence $0 \to C_p \to D_p \to E_p \to 0$ is of determinant class and define in this case

$$\rho^{(2)}(C_*, D_*, E_*) = \sum_{p \in \mathbb{Z}} (-1)^p \cdot \rho^{(2)}(C_p, D_p, E_p). \tag{3.34}$$

Let $LHS_*(C_*, D_*, E_*)$ be the weakly acyclic dim-finite Hilbert $\mathcal{N}(G)$-chain complex given by the weakly exact long homology sequence associated to an exact sequence $0 \to C_* \to D_* \to E_* \to 0$ of dim-finite Hilbert $\mathcal{N}(G)$-chain complexes (see Theorem 1.21), where we use the convention that $H_0^{(2)}(E_*)$ sits in dimension zero. A chain map $f_* \colon C_* \to D_*$ is called a *weak homology equivalence* if $H_p^{(2)}(f_*)$ is a weak isomorphism for all $p \in \mathbb{Z}$.

The next result reflects the main properties of the torsion invariants defined above. These properties are very similar to the one for the classical notions for finite based free acyclic chain complexes over a field. However, the proof, which will be given in Subsection 3.3.3 after some preliminaries in Subsection 3.3.2, is more complicated in our context. To understand the basic properties of L^2-torsion it suffices to study the next theorem, and the reader may skip its proof and pass directly to Section 3.4. The main ideas of its proof come from [345, Section 6]. A proof of Theorem 3.35 (1) can also be found in [82, Theorem 2.7 on page 40].

Theorem 3.35 (L^2-torsion of Hilbert chain complexes). *(1) Let $0 \to C_* \xrightarrow{i_*} D_* \xrightarrow{q_*} E_* \to 0$ be an exact sequence of dim-finite Hilbert $\mathcal{N}(G)$-chain complexes. Suppose that three of the Hilbert $\mathcal{N}(G)$-chain complexes C_*, D_*, E_* and LHS_* are of determinant class.*
Then all four are of determinant class and

$$\rho^{(2)}(C_*) - \rho^{(2)}(D_*) + \rho^{(2)}(E_*)$$
$$= \rho^{(2)}(C_*, D_*, E_*) - \rho^{(2)}(LHS_*(C_*, D_*, E_*));$$

(2) Let

$$
\begin{array}{ccccccccc}
0 & \longrightarrow & C_* & \xrightarrow{\ i_*\ } & D_* & \xrightarrow{\ p_*\ } & E_* & \longrightarrow & 0 \\
& & \Big\downarrow{\scriptstyle f_*} & & \Big\downarrow{\scriptstyle g_*} & & \Big\downarrow{\scriptstyle h_*} & & \\
0 & \longrightarrow & C'_* & \xrightarrow{\ i'_*\ } & D'_* & \xrightarrow{\ p'_*\ } & E'_* & \longrightarrow & 0
\end{array}
$$

be a commutative diagram of dim-finite Hilbert $\mathcal{N}(G)$-chain complexes which are of determinant class. Suppose that two of the chain maps f_*, g_* and h_* are weak homology equivalences of determinant class and that the rows are weakly exact and of determinant class.
Then all three chain maps f_*, g_* and h_* are weak homology equivalences of determinant class and

$$t^{(2)}(f_*) - t^{(2)}(g_*) + t^{(2)}(h_*) \ = \ \rho^{(2)}(C'_*, D'_*, E'_*) - \rho^{(2)}(C_*, D_*, E_*);$$

(3) Let $f_*, g_* \colon C_* \to D_*$ be weak homology equivalences of dim-finite $\mathcal{N}(G)$-chain complexes such that f_* or g_* is of determinant class. Suppose that f_* and g_* are homotopic. Then both are of determinant class and

$$t^{(2)}(f_*) \ = \ t^{(2)}(g_*);$$

(4) Let C_*, D_* and E_* be dim-finite Hilbert $\mathcal{N}(G)$-chain complexes and let $f_* \colon C_* \to D_*$ and $g_* \colon D_* \to E_*$ be chain maps. Suppose that two of the chain maps f_*, g_* and $g_* \circ f_*$ are weak homology equivalences of determinant class. Then all three are weak homology equivalences of determinant class and

$$t^{(2)}(g \circ f) \ = \ t^{(2)}(f) + t^{(2)}(g);$$

(5) Let C_* and D_* be dim-finite Hilbert $\mathcal{N}(G)$-chain complexes of determinant class and $f_* \colon C_* \to D_*$ be a weak homology equivalence. Then f_* is of determinant class if and only if $H_p^{(2)}(f_*)$ is of determinant class for all $p \in \mathbb{Z}$ and in this case

$$t^{(2)}(f_*) = \rho^{(2)}(D_*) - \rho^{(2)}(C_*) + \sum_{p \in \mathbb{Z}}(-1)^p \cdot \ln\left(\det\left(H_p^{(2)}(f_*)\right)\right);$$

(6) Let $f_*\colon C_* \to C_*'$ and $g_*\colon D_* \to D_*'$ be chain maps of dim-finite Hilbert $\mathcal{N}(G)$-chain complexes and $\mathcal{N}(H)$-chain complexes. Denote by $\chi^{(2)}(C_*) \in \mathbb{R}$ the L^2-Euler characteristic $\sum_{p \in \mathbb{Z}}(-1)^p \cdot b_p^{(2)}(C_*)$. Then

(a) If D_* is det-L^2-acyclic, then the dim-finite Hilbert $\mathcal{N}(G \times H)$-chain complex $C_* \otimes D_*$ is det-L^2-acyclic and

$$\rho^{(2)}(C_* \otimes D_*) = \chi^{(2)}(C_*) \cdot \rho^{(2)}(D_*);$$

(b) If C_* and D_* are of determinant class, then the dim-finite Hilbert $\mathcal{N}(G \times H)$-chain complex $C_* \otimes D_*$ is of determinant class and

$$\rho^{(2)}(C_* \otimes D_*) = \chi^{(2)}(C_*) \cdot \rho^{(2)}(D_*) + \chi^{(2)}(D_*) \cdot \rho^{(2)}(C_*);$$

(c) If f_* and g_* are weak homology equivalences of determinant class, then the chain map $f_* \otimes g_*$ of Hilbert $\mathcal{N}(G \times H)$-chain complexes is a weak homology equivalence of determinant class and

$$t^{(2)}(f_* \otimes g_*) = \chi^{(2)}(C_*) \cdot t^{(2)}(g_*) + \chi^{(2)}(D_*) \cdot t^{(2)}(f_*);$$

(7) Let $H \subset G$ be a subgroup of finite index $[G : H]$ and let C_* be a dim-finite Hilbert $\mathcal{N}(G)$-chain complex. Then C_* is det-L^2-acyclic if and only if the dim-finite Hilbert $\mathcal{N}(H)$-chain complex $\mathrm{res}\, C_*$ obtained from C_* by restriction is det-L^2-acyclic, and in this case

$$\rho^{(2)}(\mathrm{res}\, C_*) = [G : H] \cdot \rho^{(2)}(C_*);$$

(8) Let $i\colon H \to G$ be an inclusion of groups and let C_* be a dim-finite Hilbert $\mathcal{N}(H)$-chain complex. Then C_* is det-L^2-acyclic if and only if the dim-finite Hilbert $\mathcal{N}(G)$-chain complex i_*C_* obtained by induction with i (see Definition 1.23) is det-L^2-acyclic, and in this case

$$\rho^{(2)}(C_*) = \rho^{(2)}(i_*C_*).$$

Example 3.36. We give an example that in Theorem 3.35 (3) the condition that f_* and g_* are weak homology equivalences is necessary. The same is true for Theorem 3.35 (4) and (5). Let C_* and D_* respectively be the Hilbert $\mathcal{N}(G)$-chain complexes concentrated in dimensions 0 and 1 whose first differentials are $\mathrm{id}\colon \mathcal{N}(G) \to \mathcal{N}(G)$ and $0\colon \mathcal{N}(G) \to \mathcal{N}(G)$ respectively. For any morphism $\gamma\colon \mathcal{N}(G) \to \mathcal{N}(G)$ we obtain a chain map $f(\gamma)_*\colon C_* \to D_*$ by putting $f(\gamma)_1 = \gamma$ and $f(\gamma)_0 = 0$. The chain map $f(\gamma)_*$ is homotopic to $0\colon C_* \to D_*$, a chain homotopy is given by γ itself. However, one easily computes that $t^{(2)}(f(\gamma)_*)$ is not independent of γ, namely

$$t^{(2)}(f(\gamma)_*) = -\frac{1}{2} \cdot \ln\left(\det\left(\mathrm{id} + \gamma^*\gamma\right)\right).$$

3.3.2 L^2-Torsion and Chain Contractions

Before we give the proof of Theorem 3.35 in Subsection 3.3.3, we reformulate the definition of L^2-torsion in terms of weak chain contractions. This formulation will be useful for the proofs since it is more flexible. Moreover, it applies to more general situations and is closer to standard notions such as Whitehead torsion.

Lemma 3.37. *(1) Let $f : U \to V$ and $g : V \to W$ be morphisms of finite dimensional Hilbert $\mathcal{N}(G)$-modules. If two of the maps f, g and $g \circ f$ are weak isomorphisms (of determinant class), then also the third;*
(2) Let

$$0 \longrightarrow U_1 \overset{i}{\longrightarrow} U_0 \overset{p}{\longrightarrow} U_2 \longrightarrow 0$$
$$f_1 \downarrow \qquad f_0 \downarrow \qquad f_2 \downarrow$$
$$0 \longrightarrow V_1 \overset{j}{\longrightarrow} V_0 \overset{q}{\longrightarrow} V_2 \longrightarrow 0$$

be a commutative diagram of maps of finite dimensional Hilbert $\mathcal{N}(G)$-modules whose rows are weakly exact (and of determinant class). If two of the three maps f_1, f_0 and f_2 are weak isomorphisms (of determinant class), then also the third;
(3) Let $f_* : C_* \to D_*$ be a chain map of Hilbert $\mathcal{N}(G)$-chain complexes such that C_p and D_p have finite dimension for all $p \in \mathbb{Z}$. Then f_* is a weak homology equivalence if and only if $\mathrm{cone}(f_*)$ is weakly acyclic.

Proof. (1) This follows from Lemma 1.13 and Theorem 3.14 (1).

(2) The given diagram induces the following commutative diagram with exact rows

$$0 \longrightarrow \ker(p) \longrightarrow U_0 \longrightarrow \ker(p)^\perp \longrightarrow 0$$
$$f_1' \downarrow \qquad f_0 \downarrow \qquad f_2' \downarrow$$
$$0 \longrightarrow \ker(q) \longrightarrow V_0 \longrightarrow \ker(q)^\perp \longrightarrow 0 \,.$$

The induced maps $i : U_1 \to \ker(p)$, $p^\perp : \ker(p)^\perp \to U_2$, $j : V_1 \to \ker(q)$, and $q^\perp : \ker(q)^\perp \to V_2$ are weak isomorphisms (of determinant class) by assumption. Because of assertion (1) f_1' and f_2' respectively are weak isomorphisms (of determinant class) if and only if f_1 and f_2 respectively are weak isomorphisms (of determinant class.) We conclude from Theorem 1.21 and Theorem 3.14 (2) that already all three maps f_1', f_0 and f_2' are weak isomorphisms (of determinant class) if two of them are weak isomorphisms (of determinant class).

(3) follows from the long weakly exact homology equivalence (see Theorem 1.21) associated to the exact sequence $0 \to D_* \to \mathrm{cone}(f_*) \to \Sigma C_* \to 0$ since the boundary map is $H_*^{(2)}(f_*)$. $\qquad\square$

A chain map $f_* : C_* \to D_*$ of Hilbert $\mathcal{N}(G)$-chain complexes is called *weak chain isomorphism* if $f_p : C_p \to D_p$ is a weak isomorphism for all p.

Definition 3.38. *A weak chain contraction for a Hilbert $\mathcal{N}(G)$-chain complex C_* is a pair (γ_*, u_*) which consists of a weak chain isomorphism $u_*\colon C_* \to C_*$ and a chain homotopy $\gamma_*\colon u_* \simeq 0$ satisfying $\gamma_* \circ u_* = u_* \circ \gamma_*$.*

A chain contraction in the ordinary sense is just a weak chain contraction (γ_*, u_*) with $u_* = \mathrm{id}$.

Lemma 3.39. *The following statements are equivalent for a dim-finite Hilbert $\mathcal{N}(G)$-chain complex.*

(1) C_* is weakly acyclic (and of determinant class);
(2) $\Delta_p\colon C_p \to C_p$ is a weak isomorphism (of determinant class) for all $p \in \mathbb{Z}$;
(3) There is a weak chain contraction (γ_*, u_*) with $\gamma_* \circ \gamma_* = 0$ (such that u_* is of determinant class);
(4) There is a weak chain contraction (γ_*, u_*) (such that u_* is of determinant class).

Proof. (1) $\Rightarrow$ (2) This follows from Lemma 1.18 and Lemma 3.30 since a selfadjoint endomorphism of a Hilbert $\mathcal{N}(G)$-module (such as Δ_p) is a weak isomorphism if and only if it is injective.

(2) $\Rightarrow$ (3) Put $\gamma_p = c_p^*$. Then (γ_*, Δ_*) is the desired weak chain contraction.

(3) $\Rightarrow$ (4) is trivial.

(4) $\Rightarrow$ (1) Since γ_* is a chain homotopy between u_* and 0_*, we get $H_p^{(2)}(u_*) = 0$ for $p \in \mathbb{Z}$. Since u_* is a weak isomorphism, $H_p^{(2)}(u_*)$ is a weak isomorphism for all $p \in \mathbb{Z}$ by Lemma 3.44 which we will prove later. Hence C_* is weakly acyclic. We split

$$u_p = \begin{pmatrix} u_p^0 & u_p^1 \\ 0 & u_p^\perp \end{pmatrix} : C_p = \ker(c_p) \oplus \ker(c_p)^\perp \to C_p = \ker(c_p) \oplus \ker(c_p)^\perp;$$

$$c_p = \begin{pmatrix} 0 & c_p^\perp \\ 0 & 0 \end{pmatrix} : C_p = \ker(c_p) \oplus \ker(c_p)^\perp \to C_{p-1} = \ker(c_{p-1}) \oplus \ker(c_{p-1})^\perp;$$

$$\gamma_p = \begin{pmatrix} \gamma_p^0 & \gamma_p^1 \\ \gamma_p^2 & \gamma_p^3 \end{pmatrix} : C_p = \ker(c_p) \oplus \ker(c_p)^\perp \to C_{p+1} = \ker(c_{p+1}) \oplus \ker(c_{p+1})^\perp.$$

Since $u_p^\perp$ is an endomorphism of a finite dimensional Hilbert $\mathcal{N}(G)$-module with dense image, we conclude from Lemma 1.13 that $u_p^\perp$ is a weak isomorphism. From $c_{p+1} \circ \gamma_p + \gamma_{p-1} \circ c_p = u_p$ we conclude $\gamma_{p-1}^2 \circ c_p^\perp = u_p^\perp$. Since $c_p^\perp$ and $u_p^\perp$ are weak isomorphisms, γ_{p-1}^2 is a weak isomorphism by Lemma 3.37 (1). If u_p is of determinant class, c_p is of determinant class by Theorem 3.14 (1) and Lemma 3.15 (3). Hence C_* is of determinant class if u_* is of determinant class for all $p \in \mathbb{Z}$. $\square$

The proof of the next lemma is a direct calculation.

Lemma 3.40. *Let (γ_*, u_*) and (δ_*, v_*) be weak chain contractions for the dim-finite Hilbert $\mathcal{N}(G)$-chain complex C_*. Define $\Theta\colon C_{\mathrm{ev}} \to C_{\mathrm{ev}}$ by*

$$\Theta := (v_* \circ u_* + \delta_* \circ \gamma_*) = \begin{pmatrix} \ddots & \vdots & \vdots & \vdots & \iddots \\ \cdots & vu & 0 & 0 & \cdots \\ \cdots & \delta\gamma & vu & 0 & \cdots \\ \cdots & 0 & \delta\gamma & vu & \cdots \\ \iddots & \vdots & \vdots & \vdots & \ddots \end{pmatrix}.$$

Then the composition

$$\Theta' : C_{\mathrm{odd}} \xrightarrow{(uc+\gamma)_{\mathrm{odd}}} C_{\mathrm{ev}} \xrightarrow{\Theta} C_{\mathrm{ev}} \xrightarrow{(vc+\delta)_{\mathrm{ev}}} C_{\mathrm{odd}}$$

is given by the lower triangle matrix

$$\begin{pmatrix} \ddots & \vdots & \vdots & \vdots & \iddots \\ \cdots & (v^2u^2)_{2n-1} & 0 & 0 & \cdots \\ \cdots & * & (v^2u^2)_{2n+1} & 0 & \cdots \\ \cdots & * & * & (v^2u^2)_{2n+3} & \cdots \\ \iddots & \vdots & \vdots & \vdots & \ddots \end{pmatrix}.$$

Lemma 3.41. *Let C_* be a weakly acyclic dim-finite Hilbert $\mathcal{N}(G)$-chain complex of determinant class. Let (γ_*, u_*) and (δ_*, v_*) be weak chain contractions such that u_p and v_p are of determinant class for $p \in \mathbb{Z}$.*

Then the maps $(uc + \gamma)_{\mathrm{odd}} \colon C_{\mathrm{odd}} \to C_{\mathrm{ev}}$, $(uc + \gamma)_{\mathrm{ev}} \colon C_{\mathrm{ev}} \to C_{\mathrm{odd}}$, u_{odd} and u_{ev} are weak isomorphisms and of determinant class and we get

$$\ln\left(\det((uc+\gamma)_{\mathrm{odd}})\right) - \ln\left(\det(u_{\mathrm{odd}})\right) = -\ln\left(\det((vc+\delta)_{\mathrm{ev}})\right) + \ln\left(\det(v_{\mathrm{ev}})\right);$$
$$\ln\left(\det(u_{\mathrm{odd}})\right) = \ln\left(\det(u_{\mathrm{ev}})\right);$$
$$\rho^{(2)}(C_*) = \ln\left(\det((uc+\gamma)_{\mathrm{odd}})\right) - \ln\left(\det(u_{\mathrm{odd}})\right).$$

Proof. Since u_p and v_p are weak isomorphisms and of determinant class, we conclude from Lemma 1.13, Theorem 3.14 (1) and (2), Lemma 3.37 (1) and (2) and Lemma 3.40 that the maps $(uc + \gamma)_{\mathrm{odd}}$, $(vc + \delta)_{\mathrm{ev}}$, u_{odd}, v_{odd}, u_{ev} and v_{ev} are weak isomorphisms and of determinant class and that

$$2 \cdot \left(\ln\left(\det(v_{\mathrm{odd}})\right) + \ln\left(\det(u_{\mathrm{odd}})\right)\right) = \ln\left(\det((vc+\delta)_{\mathrm{ev}})\right) + \ln\left(\det(v_{\mathrm{ev}})\right)$$
$$+ \ln\left(\det(u_{\mathrm{ev}})\right) + \ln\left(\det((uc+\gamma)_{\mathrm{odd}})\right).$$

We conclude $\ln\left(\det(u_{\mathrm{odd}})\right) = \ln\left(\det(u_{\mathrm{ev}})\right)$ from Theorem 3.14 (1) applied to $(uc + \gamma)_{\mathrm{odd}} \circ u_{\mathrm{odd}} = u_{\mathrm{ev}} \circ (uc + \gamma)_{\mathrm{odd}}$ and analogously $\ln\left(\det(v_{\mathrm{odd}})\right) = \ln\left(\det(v_{\mathrm{ev}})\right)$. This proves the first two equations.

The first equation shows that $\ln\left(\det((uc+\gamma)_{\mathrm{odd}})\right) - \ln\left(\det(u_{\mathrm{odd}})\right)$ is independent of the choice of the weak chain contraction (γ_*, u_*) with u_p a weak isomorphism of determinant class for all $p \in \mathbb{Z}$. Hence it suffices to prove the third equation for the special choice $(\gamma_*, u_*) = ((c_*)^*, \Delta_*)$.

Let $f_p \colon C_p \to C_p$ be the p-fold composition $(\Delta_p)^p = \Delta_p \circ \ldots \circ \Delta_p$. Then the following square commutes

$$
\begin{array}{ccc}
C_{\mathrm{odd}} & \xrightarrow{(\Delta c + c^*)_{\mathrm{odd}}} & C_{\mathrm{ev}} \\
{\scriptstyle f_{\mathrm{odd}}}\Big\downarrow & & \Big\downarrow{\scriptstyle f_{\mathrm{ev}}} \\
C_{\mathrm{odd}} & \xrightarrow{(\Delta c^* + c)_{\mathrm{odd}}} & C_{\mathrm{ev}}
\end{array}
$$

We conclude from Theorem 3.14, Lemma 3.15 (4) and Lemma 3.30

$$
\begin{aligned}
2 \cdot \ln & \left(\det\left((\Delta c + c^*)_{\mathrm{odd}}\right)\right) \\
&= \ln\left(\det\left((\Delta c + c^*)^*_{\mathrm{odd}} \circ (\Delta c + c^*)_{\mathrm{odd}}\right)\right) \\
&= \ln\left(\det\left((\Delta c + c^*)^*_{\mathrm{odd}} \circ f_{\mathrm{ev}} \circ (\Delta c + c^*)_{\mathrm{odd}}\right)\right) - \ln\left(\det\left(f_{\mathrm{ev}}\right)\right) \\
&= \ln\left(\det\left((\Delta c + c^*)^*_{\mathrm{odd}} \circ (\Delta c^* + c)_{\mathrm{odd}} \circ f_{\mathrm{odd}}\right)\right) - \ln\left(\det\left(f_{\mathrm{ev}}\right)\right) \\
&= \ln\left(\det\left((\Delta c^* + c)_{\mathrm{ev}} \circ (\Delta c^* + c)_{\mathrm{odd}}\right)\right) + \ln\left(\det\left(f_{\mathrm{odd}}\right)\right) - \ln\left(\det\left(f_{\mathrm{ev}}\right)\right)
\end{aligned}
$$

and

$$
\begin{aligned}
2 \cdot & \left(\ln\left(\det\left((\Delta c + c^*)_{\mathrm{odd}}\right)\right) - \ln\left(\det\left(\Delta_{\mathrm{odd}}\right)\right)\right) \\
&= \ln\left(\det\left(f_{\mathrm{odd}}\right)\right) - \ln\left(\det\left(f_{\mathrm{ev}}\right)\right) + \ln\left(\det\left((\Delta c^* + c)_{\mathrm{ev}} \circ (\Delta c^* + c)_{\mathrm{odd}}\right)\right) \\
&\qquad - 2 \cdot \ln\left(\det\left(\Delta_{\mathrm{odd}}\right)\right) \\
&= -\sum_p (-1)^p \cdot p \cdot \ln\left(\det\left(\Delta_p\right)\right) + \ln\left(\det\left((\Delta c^* + c)_{\mathrm{ev}} \circ (\Delta c^* + c)_{\mathrm{odd}}\right)\right) \\
&\qquad - 2 \cdot \ln\left(\det\left(\Delta_{\mathrm{odd}}\right)\right) \\
&= 2 \cdot \rho^{(2)}(C_*) + \ln\left(\det\left((\Delta c^* + c)_{\mathrm{ev}}\right)\right) + \ln\left(\det\left((\Delta c^* + c)_{\mathrm{odd}}\right)\right) \\
&\qquad - 2 \cdot \ln\left(\det\left(\Delta_{\mathrm{odd}}\right)\right).
\end{aligned}
$$

Hence it remains to show

$$
\ln\left(\det\left((\Delta c^* + c)_{\mathrm{ev}}\right)\right) + \ln\left(\det\left((\Delta c^* + c)_{\mathrm{odd}}\right)\right) = 2 \cdot \ln\left(\det\left(\Delta_{\mathrm{odd}}\right)\right).
$$

The dual chain complex $(C_*)^*$ has the chain contraction (c_*, Δ_*). If we apply to it the first and second equation, which we have already proved, we obtain

$$
\ln\left(\det\left((\Delta c^* + c)_{\mathrm{odd}}\right)\right) - \ln\left(\det\left(\Delta_{\mathrm{odd}}\right)\right) = -\ln\left(\det\left((\Delta c^* + c)_{\mathrm{ev}}\right)\right)
$$
$$
+ \ln\left(\det\left(\Delta_{\mathrm{ev}}\right)\right);
$$
$$
\ln\left(\det\left(\Delta_{\mathrm{odd}}\right)\right) = \ln\left(\det\left(\Delta_{\mathrm{ev}}\right)\right).
$$

This finishes the proof of Lemma 3.41. □

3.3.3 Proofs of the Basic Properties of L^2-Torsion

This subsection is devoted to the proof of Theorem 3.35. We will need the following lemmas.

Lemma 3.42. *Let* $0 \to C_* \xrightarrow{i_*} D_* \xrightarrow{q_*} E_* \to 0$ *be an exact sequence of* dim-*finite* $\mathcal{N}(G)$-*chain complexes. Suppose that* (ϵ_*, w_*) *is a weak chain contraction for* E_* *(such that* w_p *is of determinant class for* $p \in \mathbb{Z}$*). Then there is a chain map* $s_*: E_* \to D_*$ *such that* $q_* \circ s_* = w_*: E_* \to E_*$ *and that* $i_p \oplus s_p: C_p \oplus E_p \to D_p$ *is a weak isomorphism (of determinant class) for* $p \in \mathbb{Z}$.

Proof. For each $p \in \mathbb{Z}$ choose a morphism $\sigma_p: E_p \to D_p$ with $q_p \circ \sigma_p = \mathrm{id}$. Now define

$$s_p := d_{p+1} \circ \sigma_{p+1} \circ \epsilon_p + \sigma_p \circ \epsilon_{p-1} \circ e_p: \ E_p \to D_p.$$

Then $s_*: D_* \to E_*$ is a chain map with $p_* \circ s_* = w_*$ and the following diagram commutes

$$
\begin{array}{ccccccccc}
0 & \longrightarrow & C_* & \longrightarrow & C_* \oplus E_* & \longrightarrow & E_* & \longrightarrow & 0 \\
& & \downarrow{\scriptstyle \mathrm{id}} & & \downarrow{\scriptstyle i_* \oplus s_*} & & \downarrow{\scriptstyle w_*} & & \\
0 & \longrightarrow & C_* & \xrightarrow{i_*} & D_* & \xrightarrow{q_*} & E_* & \longrightarrow & 0
\end{array}
$$

where the upper horizontal row is the canonical one. Now the claim follows from Lemma 3.37 (2). $\qquad\square$

Lemma 3.43. *Let* C_* *be a Hilbert* $\mathcal{N}(G)$-*chain complex with trivial differentials. Let* $u_*, v_*: C_* \to D_*$ *be chain maps to a* dim-*finite Hilbert* $\mathcal{N}(G)$-*chain complex* D_* *of determinant class with* $H_*^{(2)}(u_*) = H_*^{(2)}(v_*)$. *Then there is a weak chain isomorphism* $g_*: D_* \to D_*$ *such that each* $g_p: D_p \to D_p$ *is of determinant class,* $H_p^{(2)}(g_*) = \mathrm{id}$ *and there is a chain homotopy* $g_* \circ u_* \simeq g_* \circ v_*$.

Proof. Since $d_p^\perp: \ker(d_p)^\perp \to \mathrm{clos}(\mathrm{im}(d_p))$ is a weak isomorphism, we can choose an isomorphism $\psi_{p-1}: \mathrm{clos}(\mathrm{im}(d_p)) \xrightarrow{\cong} \ker(d_p)^\perp$ by the Polar Decomposition Theorem. Define $g_p: D_p \to D_p$ by the orthogonal sum of

$$d_{p+1}|_{\ker(d_{p+1})^\perp} \circ \psi_p: \mathrm{clos}(\mathrm{im}(d_{p+1})) \to \mathrm{clos}(\mathrm{im}(d_{p+1}));$$
$$\mathrm{id}: \ker(d_p) \cap \mathrm{clos}(\mathrm{im}(d_{p+1}))^\perp \to \ker(d_p) \cap \mathrm{clos}(\mathrm{im}(d_{p+1}))^\perp;$$
$$\psi_{p-1} \circ d_p^\perp: \ker(d_p)^\perp \to \ker(d_p)^\perp.$$

Since D_* is of determinant class, g_p is a weak isomorphism of determinant class by Lemma 3.15 (3) and (7) and Lemma 3.37 (1). Define $\gamma_p: C_p \to D_{p+1}$ by the composition of $u_p - v_p: C_p \to \mathrm{clos}(\mathrm{im}(d_{p+1}))$ and $\psi_p: \mathrm{clos}(\mathrm{im}(d_{p+1})) \to \ker(d_{p+1})^\perp \subset D_{p+1}$. Now one easily checks using Lemma 1.18 that we obtain a chain map $g_*: D_* \to D_*$ with $H_*^{(2)}(g_*) = \mathrm{id}$ and γ_* defines a chain homotopy between $g_* \circ u_*$ and $g_* \circ v_*$. $\qquad\square$

Lemma 3.44. *Let C_* and D_* be dim-finite Hilbert $\mathcal{N}(G)$-chain complexes. If $f_*\colon C_* \to D_*$ is a weak chain isomorphism (such that f_p is of determinant class for all $p \in \mathbb{Z}$), then $H_p^{(2)}(f_*)$ is a weak isomorphism (of determinant class) for all $p \in \mathbb{Z}$. If $f_*\colon C_* \to D_*$ is a weak chain isomorphism (such that f_p is of determinant class for all $p \in \mathbb{Z}$) and C_* or D_* is of determinant class, then both C_* and D_* are of determinant class and*

$$\rho^{(2)}(D_*) - \rho^{(2)}(C_*) = \sum_{p \in \mathbb{Z}} (-1)^p \cdot \ln\left(\det(f_p)\right) - \sum_{p \in \mathbb{Z}} (-1)^p \cdot \ln\left(\det\left(H_p^{(2)}(f_*)\right)\right).$$

Proof. Fix $n_0 \in \mathbb{Z}$ such that $C_p = D_p = 0$ for $p < n_0$. We use induction over n for which $C_p = D_p = 0$ for $p > n$. The induction beginning $n \le n_0$ is trivial since then C_* and D_* are concentrated in dimension n_0. The induction step from n to $n+1$ is done as follows.

Define C'_* as the subchain complex of C_* with $C'_p = C_p$ for $p \ge n+1$, $C_n = \mathrm{clos}(\mathrm{im}(c_{n+1}))$ and $C'_p = 0$ for $p \le n-1$. Define C''_* as the quotient chain complex of C_* with $C''_p = 0$ for $p \ge n+1$, $C''_n = \mathrm{im}(c_{n+1})^\perp$ and $C''_p = C_p$ for $p \le n-1$. There is an obvious exact sequence $0 \to C'_* \to C_* \to C''_* \to 0$. Lemma 3.15 (3) implies that both C'_* and C''_* are of determinant class if and only if C_* is of determinant class and in this case

$$\rho^{(2)}(C_*) = \rho^{(2)}(C'_*) + \rho^{(2)}(C''_*). \tag{3.45}$$

We obtain a commutative diagram with exact rows

$$
\begin{array}{ccccccccc}
0 & \longrightarrow & C'_n & \longrightarrow & C_n & \longrightarrow & C''_n & \longrightarrow & 0 \\
& & \downarrow{\scriptstyle f'_n} & & \downarrow{\scriptstyle f_n} & & \downarrow{\scriptstyle f''_n} & & \\
0 & \longrightarrow & D'_n & \longrightarrow & D_n & \longrightarrow & D''_n & \longrightarrow & 0
\end{array}
\tag{3.46}
$$

The map $f'_n\colon C'_n \to D'_n$ has dense image as its composition with $c'_{n+1}\colon C'_{n+1} \to C'_n$ is the composition of the maps which both have dense image $f_{n+1}\colon C_{n+1} \to D_{n+1}$ and $d_{n+1}\colon D_{n+1} \to \mathrm{clos}(\mathrm{im}(d_{n+1}))$. Since the middle vertical arrow in diagram (3.46) is a weak isomorphism and the left vertical arrow in diagram (3.46) has dense image and hence is a weak isomorphism, all three vertical arrows in diagram (3.46) are weak isomorphisms by Lemma 3.37 (2). Theorem 3.14 (2) applied to diagram (3.46) shows that for all $p \in \mathbb{Z}$ the maps f'_p and f''_p are weak isomorphisms of determinant class and

$$\ln(\det(f_p)) = \ln(\det(f'_p)) + \ln(\det(f''_p)), \tag{3.47}$$

provided that f_p is of determinant class for all $p \in \mathbb{Z}$.

Next we prove Lemma 3.44 for $f'_*\colon C'_* \to D'_*$. Suppose that $f'_*\colon C'_* \to D'_*$ is a weak chain isomorphism. Since $C'_p = 0$ for $p \ge n+2$, we obtain a commutative diagram with exact rows

$$
\begin{array}{ccccccc}
0 & \longrightarrow & H_{n+1}^{(2)}(C_*') & \longrightarrow & C_{n+1}' & \xrightarrow{\ \mathrm{pr}\ } & \ker(c_{n+1}')^{\perp} & \longrightarrow & 0 \\
& & {\scriptstyle H_{n+1}^{(2)}(f_*')}\downarrow & & {\scriptstyle f_{n+1}'}\downarrow & & {\scriptstyle f_{n+1}'^{\perp}}\downarrow & & \\
0 & \longrightarrow & H_{n+1}^{(2)}(D_*') & \longrightarrow & D_{n+1}' & \xrightarrow{\ \mathrm{pr}\ } & \ker(d_{n+1}')^{\perp} & \longrightarrow & 0
\end{array}
\qquad (3.48)
$$

and the commutative diagram

$$
\begin{array}{ccc}
\ker(c_{n+1}')^{\perp} & \xrightarrow{\ c_{n+1}'^{\perp}\ } & C_n' \\
{\scriptstyle f_{n+1}'^{\perp}}\downarrow & & {\scriptstyle f_n'}\downarrow \\
\ker(d_{n+1}')^{\perp} & \xrightarrow{\ d_{n+1}'^{\perp}\ } & D_n'
\end{array}
\qquad (3.49)
$$

Since the horizontal arrows and the right vertical arrow in diagram (3.49) are weak isomorphisms, $f_{n+1}'^{\perp}$ is a weak isomorphism by Lemma 3.37 (1). Lemma 3.37 (2) applied to diagram (3.48) shows that $H_{n+1}^{(2)}(f_*')$ is a weak isomorphism.

Now suppose that $f_p' \colon C_p' \to D_p'$ is a weak isomorphism of determinant class for all $p \in \mathbb{Z}$. Theorem 3.14 (2) applied to diagram (3.48) shows that $H_{n+1}^{(2)}(f_*')$ and $f_{n+1}'^{\perp}$ are of determinant class and satisfy

$$
\ln(\det(f_{n+1}')) = \ln\left(\det\left(H_{n+1}^{(2)}(f_*')\right)\right) + \ln\left(\det\left(f_{n+1}'^{\perp}\right)\right). \qquad (3.50)
$$

If C_*' or D_*' is of determinant class, we conclude from Lemma 3.15 (3) and Lemma 3.37 (1) applied to diagram (3.49) that both C_*' and D_*' are of determinant class. In this case Theorem 3.14 (1) and Lemma 3.15 (3) applied to (3.49) imply

$$
(-1)^n \cdot \ln(\det(f_n')) + \rho^{(2)}(C_*') = (-1)^n \cdot \ln(\det(f_{n+1}'^{\perp})) + \rho^{(2)}(D_*'). \qquad (3.51)
$$

We conclude from (3.50) and (3.51) that $H_{n+1}(f_*')$ is a weak isomorphism of determinant class and that both C_*' and D_*' are of determinant class and

$$
\rho^{(2)}(D_*') - \rho^{(2)}(C_*') = \sum_{p \in \mathbb{Z}}(-1)^p \cdot \ln\left(\det(f_p')\right)
$$
$$
- \sum_{p \in \mathbb{Z}}(-1)^{p+1} \cdot \ln\left(\det\left(H_{p+1}^{(2)}(f_*')\right)\right). \qquad (3.52)
$$

Notice that the induction hypothesis applies to $f_*'' \colon C_*'' \to D_*''$. We conclude that C_*'' and D_*'' are of determinant class and $H_p^{(2)}(f_*'')$ is a weak isomorphism of determinant class for all $p \in \mathbb{Z}$ and

$$
\rho^{(2)}(D_*'') - \rho^{(2)}(C_*'') = \sum_{p \in \mathbb{Z}}(-1)^p \cdot \ln\left(\det(f_p'')\right)
$$
$$
- \sum_{p \in \mathbb{Z}}(-1)^p \cdot \ln\left(\det\left(H_p^{(2)}(f_*'')\right)\right). \qquad (3.53)
$$

Now Lemma 3.44 follows from (3.45), (3.47), (3.52) and (3.53).   □
  The next lemma contains a kind of rotation principle.

Lemma 3.54. *Let* $0 \to C_* \xrightarrow{i_*} D_* \xrightarrow{q_*} E_* \to 0$ *be an exact sequence of dim-finite Hilbert* $\mathcal{N}(G)$-*chain complexes. Then there is an exact sequence of dim-finite Hilbert* $\mathcal{N}(G)$-*chain complexes* $0 \to D_* \to \widetilde{E}_* \to \widetilde{C}_* \to 0$ *with the following properties*

(1) C_*, E_* *and* $LHS_*(C_*, D_*, E_*)$ *(defined in Theorem 3.35 (1)) respectively are of determinant class if and only if* $\widetilde{C}_*$, $\widetilde{E}_*$ *and* $LHS_*(D_*, \widetilde{E}_*, \widetilde{C}_*)$ *respectively are of determinant class;*
(2) *There is a chain isomorphism*

$$LHS_*(C_*, D_*, E_*) \xrightarrow{\cong} \Sigma \left(LHS_*(D_*, \widetilde{E}_*, \widetilde{C}_*) \right);$$

(3) *If* C_*, D_*, E_* *and* $LHS_*(C_*, D_*, E_*)$ *are of determinant class, then* D_*, $\widetilde{E}_*$, $\widetilde{C}_*$ *and* $LHS_*(D_*, \widetilde{E}_*, \widetilde{C}_*)$ *are of determinant class and*

$$\rho^{(2)}(C_*) - \rho^{(2)}(D_*) + \rho^{(2)}(E_*)$$
$$-\rho^{(2)}(C_*, D_*, E_*) + \rho^{(2)}(LHS_*(C_*, D_*, E_*))$$
$$= -\rho^{(2)}(D_*) + \rho^{(2)}(\widetilde{E}_*) - \rho^{(2)}(\widetilde{C}_*)$$
$$+\rho^{(2)}(D_*, \widetilde{E}_*, \widetilde{C}_*) - \rho^{(2)}(LHS_*(D_*, \widetilde{E}_*, \widetilde{C}_*)).$$

Proof. The desired exact sequence is the canonical exact sequence $0 \to D_* \to \text{cyl}_*(q_*) \to \text{cone}_*(q_*) \to 0$. It remains to show the various claims.
  Notice for any dim-finite Hilbert $\mathcal{N}(G)$-chain complex F_*, that $\text{cone}_*(F_*) = \text{cone}_*(\text{id}: F_* \to F_*)$ is contractible and satisfies

$$\rho^{(2)}(\text{cone}_*(F_*)) = 0. \tag{3.55}$$

This follows from the fact that

$$\begin{pmatrix} 0 & 1 \\ 0 & 0 \end{pmatrix} : F_{p-1} \oplus F_p \to F_p \oplus F_{p+1}$$

is an explicit chain contraction for $\text{cone}_*(F_*)$. We have the canonical short exact sequences $0 \to \Sigma C_* \xrightarrow{j_*} \text{cone}_*(q_*) \to \text{cone}_*(E_*) \to 0$ and $0 \to E_* \xrightarrow{k_*} \text{cyl}_*(q_*) \to \text{cone}(D_*) \to 0$. From Lemma 3.42 we obtain chain isomorphisms $u_*: \Sigma C_* \oplus \text{cone}_*(E_*) \xrightarrow{\cong} \text{cone}_*(q_*)$ and $v_*: E_* \oplus \text{cone}(D_*) \xrightarrow{\cong} \text{cyl}_*(q_*)$ such that under the obvious identifications $H_p^{(2)}(\Sigma C_* \oplus \text{cone}_*(E_*)) = H_p(\Sigma C_*)$ and $H_p^{(2)}(E_* \oplus \text{cone}(D_*)) = H_p^{(2)}(E_*)$

$$H_p^{(2)}(u_*) = H_p^{(2)}(j_*); \tag{3.56}$$
$$H_p^{(2)}(v_*) = H_p^{(2)}(k_*). \tag{3.57}$$

Moreover, we conclude from (3.33)

$$\rho(\Sigma C_*, \mathrm{cone}_*(q_*), \mathrm{cone}_*(E_*)) = -\sum_{p\in\mathbb{Z}}(-1)^p \cdot \ln(\det(u_p)); \qquad (3.58)$$

$$\rho(E_*, \mathrm{cyl}_*(q_*), \mathrm{cone}(D_*)) = -\sum_{p\in\mathbb{Z}}(-1)^p \cdot \ln(\det(v_p)). \qquad (3.59)$$

One easily checks using Theorem 3.14 (2)

$$\rho(\Sigma C_*, \mathrm{cone}_*(q_*), \mathrm{cone}_*(E_*)) = -\rho(C_*, D_*, E_*); \qquad (3.60)$$
$$\rho(E_*, \mathrm{cyl}_*(q_*), \mathrm{cone}(D_*)) = 0; \qquad (3.61)$$
$$\rho^{(2)}(D_*, \mathrm{cyl}_*(q_*), \mathrm{cone}_*(q_*)) = 0. \qquad (3.62)$$

We conclude from (3.58), (3.59), (3.60) and (3.61)

$$\sum_{p\in\mathbb{Z}}(-1)^p \cdot \ln(\det(u_p)) = \rho(C_*, D_*, E_*); \qquad (3.63)$$

$$\sum_{p\in\mathbb{Z}}(-1)^p \cdot \ln(\det(v_p)) = 0. \qquad (3.64)$$

From Lemma 3.44 and equations (3.55), (3.56) and (3.57) we conclude that $\mathrm{cone}_*(q_*)$ and $\mathrm{cyl}_*(q_*)$ respectively are of determinant class if and only if C_* and E_* respectively are of determinant class and in this case

$$\rho^{(2)}(\mathrm{cone}_*(q_*)) - \rho^{(2)}(\Sigma C_*)$$
$$= \sum_{p\in\mathbb{Z}}(-1)^p \cdot \ln(\det(u_p)) - \sum_{p\in\mathbb{Z}}(-1)^p \cdot \ln\left(\det\left(H_p^{(2)}(j_*)\right)\right); \quad (3.65)$$

$$\rho^{(2)}(\mathrm{cyl}_*(q_*)) - \rho^{(2)}(E_*)$$
$$= \sum_{p\in\mathbb{Z}}(-1)^p \cdot \ln(\det(v_p)) - \sum_{p\in\mathbb{Z}}(-1)^p \cdot \ln\left(\det\left(H_p^{(2)}(k_*)\right)\right). \quad (3.66)$$

Recall that j_* and k_* induce isomorphisms $H_p^{(2)}(\Sigma C_*) \xrightarrow{\cong} H_p^{(2)}(\mathrm{cone}_*(q_*))$ and $H_p^{(2)}(E_*) \xrightarrow{\cong} H_p^{(2)}(\mathrm{cyl}_*(q_*))$ and we have the obvious isomorphisms $\mathrm{id}: H_p^{(2)}(D_*) \xrightarrow{\cong} H_p^{(2)}(D_*)$. They induce a chain isomorphism

$$LHS_*(C_*, D_*, E_*) \xrightarrow{\cong} \Sigma\left(LHS_*(D_*, \mathrm{cyl}_*(q_*), \mathrm{cone}_*(q_*))\right)$$

and Lemma 3.44 shows

$$-\rho^{(2)}(LHS_*(D_*, \mathrm{cyl}_*(q_*), \mathrm{cone}_*(q_*))) - \rho^{(2)}(LHS_*(C_*, D_*, E_*))$$
$$= -\sum_{p\in\mathbb{Z}}(-1)^p \cdot \ln\left(\det\left(H_p^{(2)}(j_*)\right)\right)$$
$$+ \sum_{p\in\mathbb{Z}}(-1)^p \cdot \ln\left(\det\left(H_p^{(2)}(k_*)\right)\right). \qquad (3.67)$$

Now Lemma 3.54 follows from (3.62), (3.63), (3.64), (3.65), (3.66) and (3.67).

□

The next lemma is the decisive step in the proof of Theorem 3.35 (1). It proves additivity and all other properties are consequences.

Lemma 3.68. *Let C_*, D_* and E_* be dim-finite Hilbert $\mathcal{N}(G)$-chain complexes. Let $0 \to C_* \xrightarrow{i_*} D_* \xrightarrow{q_*} E_* \to 0$ be a weakly exact sequence of determinant class. Suppose that two of the chain complexes C_*, D_* and E_* are weakly acyclic and of determinant class. Then all three are weakly acyclic and of determinant class and*

$$\rho^{(2)}(C_*) - \rho^{(2)}(D_*) + \rho^{(2)}(E_*) \;=\; \rho^{(2)}(C_*, D_*, E_*).$$

Proof. The given exact sequence induces weak isomorphisms $\bar{i}_* : C_* \to \ker(q_*)$ and $\bar{q}_* : \ker(q_*)^\perp \to E_*$ such that $\bar{i}_p$ and $\bar{q}_p$ are of determinant class for all $p \in \mathbb{Z}$ and

$$\rho^{(2)}(C_*, D_*, E_*) = \sum_{p \in \mathbb{Z}} (-1)^p \cdot \left(-\ln(\det(\bar{i}_p)) + \ln(\det(\bar{q}_p))\right). \quad (3.69)$$

From Lemma 3.44 we conclude that $\ker(q_*)$ and $\ker(q_*)^\perp$ respectively are weakly acyclic of determinant class if and only if C_* and E_* respectively are weakly acyclic of determinant class and in this case we get

$$\rho^{(2)}(\ker(q_*)) - \rho^{(2)}(C_*) = \sum_{p \in \mathbb{Z}} (-1)^p \cdot \ln(\det(\bar{i}_p)); \quad (3.70)$$

$$\rho^{(2)}(E_*) - \rho^{(2)}(\ker(q_*)^\perp) = \sum_{p \in \mathbb{Z}} (-1)^p \cdot \ln(\det(\bar{q}_p)) \quad (3.71)$$

respectively. Because of (3.69), (3.70) and (3.71) it remains to show the claim for $0 \to \ker(q_*) \to D_* \to \ker(q_*)^\perp \to 0$. Because of Lemma 3.54 it suffices to show under the assumption that $\ker(q_*)$ and $\ker(q_*)^\perp$ are weakly acyclic and of determinant class that D_* is weakly acyclic and of determinant class and

$$\rho^{(2)}(\ker(q_*)) - \rho^{(2)}(D_*) + \rho^{(2)}(\ker(q_*)^\perp) = 0 \quad (3.72)$$

holds.

We can write the differential d_p of D_* by

$$d_p = \begin{pmatrix} d'_p & \bar{d}_p \\ 0 & d''_p \end{pmatrix} : D_p = \ker(q_p) \oplus \ker(q_p)^\perp \to D_{p-1} = \ker(q_{p-1}) \oplus \ker(q_{p-1})^\perp,$$

where d'_p and d''_p are the differentials of $\ker(q_*)$ and $\ker(q_*)^\perp$. By Lemma 3.39 we can choose weak chain contractions (γ'_*, u'_*) and (γ''_*, u''_*) for $\ker(q_*)$ and $\ker(q_*)^\perp$ such that u'_p and u''_p are weak isomorphism of determinant class for $p \in \mathbb{Z}$. Define a chain homotopy $\gamma_* : u_* \simeq 0$ for D_* by

$$\gamma_p := \begin{pmatrix} \gamma_p' & 0 \\ 0 & \gamma_p'' \end{pmatrix} : D_p \to D_{p+1};$$

$$u_p := \begin{pmatrix} u_p' & \overline{d}_{p+1} \circ \gamma_p'' + \gamma_{p-1}' \circ \overline{d}_p \\ 0 & u_p'' \end{pmatrix} : D_p \to D_p$$

with respect to the orthogonal decomposition $D_p = \ker(q_p) \oplus \ker(q_p)^{\perp}$. We conclude from Theorem 3.14 (2), Lemma 3.37 (2) and Lemma 3.41 that (γ_*, u_*) is a weak chain contraction such that u_p is of determinant class for all $p \in \mathbb{Z}$ and that (3.72) holds. This finishes the proof of Lemma 3.68. $\square$

Now we are ready to give the proof of Theorem 3.35.

Proof. (1) Step 1: If E_* is weakly acyclic and of determinant class, then LHS_* is of determinant class and assertion (1) is true.

We get from Lemma 3.39 and Lemma 3.42 a chain map $s_* : E_* \to D_*$ such that $q_p \circ s_p : E_p \to E_p$ and $i_p \oplus s_p : C_p \oplus E_p \to D_p$ are weak isomorphisms of determinant class for $p \in \mathbb{Z}$. Because of Lemma 3.44 the induced map $H_p^{(2)}(i_* \oplus s_*)$ is a weak isomorphism of determinant class for $p \in \mathbb{Z}$. Hence the long weakly exact homology sequence LHS_* is of determinant class since $H_p^{(2)}(E_*) = 0$ for $p \in \mathbb{Z}$ by assumption and in particular $H_p^{(2)}(i_* \oplus s_*) = H_p^{(2)}(i_*)$. Provided that C_* or D_* is of determinant class, we conclude from Lemma 3.15 (7) and Lemma 3.44 that both C_* and D_* are of determinant class and

$$\rho^{(2)}(D_*) - \rho^{(2)}(C_*) - \rho^{(2)}(E_*) = \sum_{p \in \mathbb{Z}} (-1)^p \cdot \ln(\det(i_p \oplus s_p))$$
$$+ \rho^{(2)}(LHS_*(C_*, D_*, E_*)). \quad (3.73)$$

Since E_* is weakly acyclic and of determinant class by assumption and $q_* \circ s_* : E_* \to E_*$ is a weak chain isomorphism such that each $q_p \circ s_p$ is a weak isomorphism of determinant class, Lemma 3.44 applied to $q_* \circ s_*$ shows

$$\sum_{p \in \mathbb{Z}} (-1)^p \cdot \ln(\det(q_p \circ s_p)) = 0.$$

Hence (3.33) shows

$$\rho^{(2)}(C_*, D_*, E_*) = - \sum_{p \in \mathbb{Z}} (-1)^p \cdot \ln(\det(i_p \oplus s_p)). \quad (3.74)$$

Now Step 1 follows from (3.73) and (3.74).

Step 2: If one of the chain complexes C_*, D_* and E_* is weakly acyclic and of determinant class, then $LHS_*(C_*, D_*, E_*)$ is of determinant class and assertion (1) is true.

This follows from Lemma 3.54 and Step 1.

Step 3: Assertion (1) is true provided that the differentials of C_* and E_* are trivial.

Since i_* and q_* induce isomorphisms $C_* \to \ker(q_*)$ and $\ker(q_*) \to E_*$, (3.33) and Lemma 3.44 shows that we can assume without loss of generality that $D_p = C_p \oplus E_p$ and i_p and q_p are the obvious inclusions and projections for all $p \in \mathbb{Z}$. Obviously C_* and E_* are of determinant class and satisfy

$$\rho^{(2)}(C_*) = \rho^{(2)}(E_*) = 0. \tag{3.75}$$

If we write

$$d_p = \begin{pmatrix} 0 & x_p \\ 0 & 0 \end{pmatrix} : D_p = C_p \oplus E_p \to D_{p-1} = C_{p-1} \oplus E_{p-1},$$

then one easily checks that the associated long weakly exact homology sequence looks like

$$\cdots \xrightarrow{x_{p+1}} C_p \xrightarrow{j} C_p \oplus \ker(x_p) \xrightarrow{\text{pr}} E_p \xrightarrow{x_p} C_{p-1} \xrightarrow{j} \cdots,$$

where j is the canonical inclusion onto the first factor and pr is induced by the projection onto the second factor. Hence $LHS_*(C_*, D_*, E_*)$ is of determinant class if and only if x_p is of determinant class for all $p \in \mathbb{Z}$ and in this case

$$\rho^{(2)}(LHS_*(C_*, D_*, E_*)) = -\sum_{p \in \mathbb{Z}} (-1)^p \cdot \ln(\det(x_p)). \tag{3.76}$$

Lemma 3.15 (3) implies that D_* is of determinant class if and only if x_p is of determinant class for all $p \in \mathbb{Z}$ and in this case

$$\rho^{(2)}(D_*) = -\sum_{p \in \mathbb{Z}} (-1)^p \cdot \ln(\det(x_p)). \tag{3.77}$$

Now the claim follows from (3.75), (3.76) and (3.77).

Step 4: Assertion (1) is true provided that C_* is of determinant class and the differentials of E_* are trivial.

In the sequel we write $\overline{C}_p = \ker(\Delta_p)$ and $\overline{\overline{C}}_p = \ker(\Delta_p)^{\perp}$. Denote by $k_* : \overline{\overline{C}}_* \to C_*$ the canonical inclusion and by $\text{pr}_* : C_* \to \overline{C}_*$ the canonical projection. We conclude from Lemma 1.18 that we have an orthogonal decomposition of Hilbert $\mathcal{N}(G)$-chain complexes $C_* = \overline{C}_* \oplus \overline{\overline{C}}_*$, the differentials of $\overline{C}_*$ are all trivial and $\overline{\overline{C}}_*$ is weakly acyclic. From Lemma 3.15 (7) we conclude that $\overline{\overline{C}}_*$ is of determinant class and

$$\rho^{(2)}(\overline{\overline{C}}_*) = \rho^{(2)}(C_*). \tag{3.78}$$

Notice that the chain map $\text{pr}_* \oplus i_* : C_* \to \overline{C}_* \oplus D_*$ is injective and has closed image since this is true for i_*. Define a dim-finite Hilbert $\mathcal{N}(G)$-chain complex $\hat{D}_*$ by the orthogonal complement in $\overline{C}_* \oplus D_*$ of the image of $\text{pr}_* \oplus i_* : C_* \to \overline{C}_* \oplus D_*$. We obtain a commutative diagram of dim-finite Hilbert $\mathcal{N}(G)$-chain complexes with exact rows and columns

$$
\begin{array}{ccccccc}
& & 0 & & 0 & & 0 \\
& & \downarrow & & \downarrow & & \downarrow \\
0 & \longrightarrow & \overline{\overline{C}}_* & \xrightarrow{\text{id}} & \overline{\overline{C}}_* & \longrightarrow & 0 & \longrightarrow & 0 \\
& & \downarrow{\scriptstyle k_*} & & \downarrow{\scriptstyle i_* \circ k_*} & & \downarrow \\
0 & \longrightarrow & C_* & \xrightarrow{i_*} & D_* & \xrightarrow{q_*} & E_* & \longrightarrow & 0 \\
& & \downarrow{\scriptstyle \mathrm{pr}_*} & & \downarrow{\scriptstyle \widehat{\mathrm{pr}}_*} & & \downarrow{\scriptstyle \text{id}} \\
0 & \longrightarrow & \overline{C}_* & \xrightarrow{\widehat{i}_*} & \widehat{D}_* & \xrightarrow{\widehat{q}_*} & E_* & \longrightarrow & 0 \\
& & \downarrow & & \downarrow & & \downarrow \\
& & 0 & & 0 & & 0
\end{array}
$$

Since $\overline{\overline{C}}_*$ is weakly acyclic and of determinant class, Step 2 applied to the middle column shows that the induced map $H_p^{(2)}(\widehat{\mathrm{pr}}_*)\colon H_p^{(2)}(D_*) \to H_p^{(2)}(\widehat{D}_*)$ is a weak isomorphism of determinant class for all $p \in \mathbb{Z}$. Moreover, D_* is of determinant class if and only if $\widehat{D}_*$ is of determinant class, and in this case

$$
\rho^{(2)}(\overline{\overline{C}}_*) - \rho^{(2)}(D_*) + \rho^{(2)}(\widehat{D}_*)
$$
$$
= \rho^{(2)}(\overline{\overline{C}}_*, D_*, \widehat{D}_*) - \sum_{p \in \mathbb{Z}}(-1)^p \cdot \ln\left(\det\left(H_p^{(2)}(\widehat{\mathrm{pr}}_*)\right)\right). \quad (3.79)
$$

The map from the middle row to the lower row induces a weak chain isomorphism from $LHS_*(C_*, D_*, E_*)$ to $LHS_*(\overline{C}_*, \widehat{D}_*, E_*)$ which is in each dimension of determinant class. Lemma 3.44 implies that $LHS_*(C_*, D_*, E_*)$ is of determinant class if and only if $LHS_*(\overline{C}_*, \widehat{D}_*, E_*)$ is of determinant class, and in this case

$$
\rho^{(2)}(LHS_*(\overline{C}_*, \widehat{D}_*, E_*)) - \rho^{(2)}(LHS_*(C_*, D_*, E_*))
$$
$$
= -\sum_{p \in \mathbb{Z}}(-1)^p \cdot \ln\left(\det\left(H_p^{(2)}(\widehat{\mathrm{pr}}_*)\right)\right). \quad (3.80)
$$

Step 3 applied to the lower row shows that $\widehat{D}_*$ is of determinant class if and only if $LHS_*(\overline{C}_*, \widehat{D}_*, E_*)$ is of determinant class, and in this case

$$
-\rho^{(2)}(\widehat{D}_*) = \rho^{(2)}(\overline{C}_*, \widehat{D}_*, E_*) - \rho^{(2)}(LHS_*(\overline{C}_*, \widehat{D}_*, E_*)). \quad (3.81)
$$

One easily checks using Lemma 3.68

$$
\rho^{(2)}(\overline{C}_*, \widehat{D}_*, E_*) = \rho^{(2)}(C_*, D_*, E_*); \quad (3.82)
$$
$$
\rho^{(2)}(\overline{\overline{C}}_*, D_*, \widehat{D}_*) = 0. \quad (3.83)
$$

Now the claim follows from (3.78), (3.79), (3.80), (3.81), (3.82) and (3.83).

Step 5: Assertion (1) is true provided that C_* and E_* are of determinant class.

Define $\widehat{D}_*$ as the kernel of the chain map $q_* \oplus k_* \colon D_* \oplus \overline{E}_* \to E_*$. Then we obtain a commutative diagram with exact rows and columns

$$
\begin{array}{ccccccccc}
& & 0 & & 0 & & 0 & & \\
& & \downarrow & & \downarrow & & \downarrow & & \\
0 & \longrightarrow & C_* & \xrightarrow{\widehat{i}_* \circ \mathrm{pr}_*} & \widehat{D}_* & \longrightarrow & \overline{E}_* & \longrightarrow & 0 \\
& & \downarrow{\scriptstyle\mathrm{id}} & & \downarrow{\scriptstyle\widehat{k}_*} & & \downarrow{\scriptstyle k_*} & & \\
0 & \longrightarrow & C_* & \xrightarrow{i_*} & D_* & \xrightarrow{q_*} & E_* & \longrightarrow & 0 \\
& & \downarrow & & \downarrow{\scriptstyle\mathrm{pr}_* \circ q_*} & & \downarrow{\scriptstyle\mathrm{pr}_*} & & \\
0 & \longrightarrow & 0 & \longrightarrow & \overline{\overline{E}}_* & \xrightarrow{\mathrm{id}} & \overline{\overline{E}}_* & \longrightarrow & 0 \\
& & \downarrow & & \downarrow & & \downarrow & & \\
& & 0 & & 0 & & 0 & &
\end{array}
$$

Now we proceed analogously to Step 4 by applying Step 1 to the middle column and Step 4 to the upper row.

Step 6: Assertion (1) is true.

This follows from Lemma 3.54 and Step 5. This finishes the proof of assertion (1) of Theorem 3.35.

(2) follows from Lemma 3.37 (3) and Lemma 3.68 applied to the induced exact sequence $0 \to \mathrm{cone}_*(f_*) \to \mathrm{cone}_*(g_*) \to \mathrm{cone}_*(h_*) \to 0$.

(3) Let $\gamma_* \colon f_* \simeq g_*$ be a chain homotopy. Consider the isomorphism of dim-finite $\mathcal{N}(G)$-chain complexes $u_* \colon \mathrm{cone}_*(f_*) \to \mathrm{cone}_*(g_*)$ given by

$$
u_p = \begin{pmatrix} \mathrm{id} & 0 \\ \gamma_{p-1} & \mathrm{id} \end{pmatrix} \colon C_{p-1} \oplus D_p \to C_{p-1} \oplus D_p.
$$

We get from Lemma 3.37 (3) and Lemma 3.44 that both $\mathrm{cone}_*(f_*)$ and $\mathrm{cone}_*(g_*)$ are weakly acyclic and of determinant class and

$$
t^{(2)}(g_*) - t^{(2)}(f_*) = \rho^{(2)}(\mathrm{cone}_*(g_*)) - \rho^{(2)}(\mathrm{cone}_*(f_*))
$$
$$
= \sum_{p \in \mathbb{Z}} (-1)^p \cdot \ln(\det(u_p))
$$
$$
= 0.
$$

(4) Consider the chain map $h_* \colon \Sigma^{-1} \mathrm{cone}_*(g_*) \to \mathrm{cone}_*(f_*)$ given by

$$
h_p = \begin{pmatrix} 0 & 0 \\ -\mathrm{id} & 0 \end{pmatrix} \colon D_p \oplus E_{p+1} \to C_{p-1} \oplus D_p.
$$

There are obvious exact sequences $0 \to \mathrm{cone}_*(f_*) \to \mathrm{cone}_*(h_*) \to \mathrm{cone}_*(g_*) \to$ 0 and $0 \to \mathrm{cone}_*(g_* \circ f_*) \xrightarrow{i_*} \mathrm{cone}_*(h_*) \to \mathrm{cone}_*(D_*) \to 0$, where i_* is given by

$$i_p := \begin{pmatrix} f_{p-1} & 0 \\ 0 & 1 \\ 1 & 0 \\ 0 & 0 \end{pmatrix} : C_{p-1} \oplus D_p \to D_{p-1} \oplus E_p \oplus C_{p-1} \oplus D_p$$

and the other maps are the canonical ones. Because of Lemma 3.68 f_*, h_*, g_* and $g_* \circ f_*$ are weak homology equivalences of determinant class. One easily checks

$$\rho^{(2)}(\mathrm{cone}_*(f_*), \mathrm{cone}_*(h_*), \mathrm{cone}_*(g_*)) = 0;$$
$$\rho^{(2)}(\mathrm{cone}_*(g_* \circ f_*), \mathrm{cone}_*(h_*), \mathrm{cone}_*(D_*)) = 0.$$

Hence we get from Lemma 3.68

$$\rho^{(2)}(\mathrm{cone}_*(f_*)) - \rho^{(2)}(\mathrm{cone}_*(h_*)) + \rho^{(2)}(\mathrm{cone}_*(g_*)) = 0; \quad (3.84)$$
$$\rho^{(2)}(\mathrm{cone}_*(g_* \circ f_*)) - \rho^{(2)}(\mathrm{cone}_*(h_*)) + \rho^{(2)}(\mathrm{cone}_*(D_*)) = 0. \quad (3.85)$$

Now assertion (4) follows from (3.55), (3.84) and (3.85).

(5) Consider $H_*^{(2)}(C_*)$ and $H_*^{(2)}(D_*)$ as chain complexes with the trivial differential. We get from Lemma 1.18 chain maps $i_* : H_*^{(2)}(C_*) \to C_*$ and $j_* : H_*^{(2)}(D_*) \to D_*$ such that i_p and j_p are isometric inclusions for $p \in \mathbb{Z}$ and in particular both i_* and j_* are of determinant class, $H_*^{(2)}(i_*) = \mathrm{id}$ and $H_*^{(2)}(j_*) = \mathrm{id}$. Moreover, we get from the definitions and Lemma 3.15 (3) and (7)

$$\rho^{(2)}(C_*) = t^{(2)}(i_*); \quad (3.86)$$
$$\rho^{(2)}(D_*) = t^{(2)}(j_*). \quad (3.87)$$

We obtain from Lemma 3.43 applied to $u_* = f_* \circ i_*$ and $v_* = j_* \circ H_*^{(2)}(f_*)$ a chain map $g_* : D_* \to D_*$ such that $H_p^{(2)}(g_*) = \mathrm{id}$, each g_p is a weak isomorphism of determinant class and $g_* \circ f_* \circ i_*$ and $g_* \circ j_* \circ H_*^{(2)}(f_*)$ are chain homotopic. Because of assertion (3) and (4) f_* is of determinant class if and only if $H_*^{(2)}(f_*)$ is of determinant class and in this case

$$t^{(2)}(f_*) + t^{(2)}(i_*) = t^{(2)}(H_*^{(2)}(f_*)) + t^{(2)}(j_*). \quad (3.88)$$

Since $t^{(2)}(H_*^{(2)}(f_*)) = \sum_{p \in \mathbb{Z}} (-1)^p \cdot \ln\left(\det\left(H_p^{(2)}(f_*)\right)\right)$ holds by assertion (2), assertion (5) follows from (3.86), (3.87) and (3.88).

(6a) Fix n_0 with $C_p = 0$ for $p < n_0$. We use induction over n with $C_p = 0$ for $p > n$. For the induction beginning $n = n_0$ we have to show for a finite dimensional Hilbert $\mathcal{N}(G)$-module U and a weak isomorphism of determinant class $f : V \to W$ for finite dimensional $\mathcal{N}(H)$-modules V and W that $\mathrm{id} \otimes f : U \otimes V \to U \otimes W$ is a weak isomorphism of determinant class and

$$\ln(\det_{\mathcal{N}(G \times H)}(\mathrm{id} \otimes f)) = \dim(U) \cdot \ln(\det_{\mathcal{N}(H)}(f)).$$

This follows from the equation of spectral density functions

$$F_{\mathcal{N}(G \times H)}(\mathrm{id} \otimes f) = \dim(U) \cdot F_{\mathcal{N}(H)}(f),$$

which is a consequence of the equality $\mathrm{id} \otimes E_\lambda^{f^* f} = E_\lambda^{(\mathrm{id} \otimes f)^* (\mathrm{id} \otimes f)}$. Next we explain the induction step from n to $n + 1$. Notice that Theorem 1.12 (2) implies

$$\chi^{(2)}(C_*) = \sum_{p \in \mathbb{Z}} (-1)^p \cdot \dim(C_p). \tag{3.89}$$

Let $C_* | n$ be obtained from C_* by truncating in dimensions $> n$ and let $(n+1)[C_*]$ be the chain complex whose $(n+1)$-th chain module is C_{n+1} and whose other chain modules are all trivial. Then the induction step follows from Lemma 3.68 applied to the short exact sequence $0 \to C_* | n \otimes D_* \to C_* \otimes D_* \to (n+1)[C_*] \otimes D_* \to 0$ and the induction beginning applied to $(n+1)[C_*] \otimes D_*$ and the induction hypothesis applied to $C_* | n \otimes D_*$. This finishes the proof of assertion (6a).

(6b) Recall that we have introduced $\overline{C}_*$ and $\overline{\overline{C}}_*$ before (see (3.78)). We conclude from (3.78) and assertion (6a)

$$\rho^{(2)}(C_* \otimes D_*) = \rho^{(2)}(\overline{C}_* \otimes \overline{D}_*) + \rho^{(2)}(\overline{C}_* \otimes \overline{\overline{D}}_*) + \rho^{(2)}(\overline{\overline{C}}_* \otimes \overline{D}_*)$$
$$+ \rho^{(2)}(\overline{\overline{C}}_* \otimes \overline{\overline{D}}_*)$$
$$= 0 + \chi^{(2)}(C_*) \cdot \rho^{(2)}(D_*) + \chi^{(2)}(D_*) \cdot \rho^{(2)}(C_*) + 0.$$

(6c) Since $f_* \otimes g_* = f_* \otimes \mathrm{id} \circ \mathrm{id} \otimes g_*$ holds, it suffices because of assertion (4) to show

$$t^{(2)}(\mathrm{id} \otimes g_*) = \chi^{(2)}(C_*) \cdot t^{(2)}(g_*).$$

This is done as above by induction over the dimension of C_*.

(7) This follows from Theorem 3.14 (5).

(8) This follows from Theorem 3.14 (6). This finishes the proof of Theorem 3.35. □

Remark 3.90. Theorem 3.35 remains true if one replaces "of determinant class" everywhere by "with positive Novikov-Shubin invariant". In view of Theorem 3.14 (4) it remains to check that the property "with positive Novikov-Shubin invariant" is inherited as claimed in all the lemmas and theorems. Notice that the proof of inheritance of the property "of determinant class" is a formal consequence of the following two facts: i.) given two weak isomorphisms $f\colon U \to V$ and $g\colon V \to W$ of finite dimensional Hilbert $\mathcal{N}(G)$-modules, both f and g are of determinant class if and only if $g \circ f$ is of determinant class (see Theorem 3.14 (1)) and ii.) given a commutative diagram of maps of finite dimensional Hilbert $\mathcal{N}(G)$-modules whose rows are weakly exact and of determinant class and whose vertical arrows are weak isomorphisms

$$
\begin{array}{ccccccccc}
0 & \longrightarrow & U_1 & \xrightarrow{\;i\;} & U_0 & \xrightarrow{\;p\;} & U_2 & \longrightarrow & 0 \\
 & & \Big\downarrow{\scriptstyle f_1} & & \Big\downarrow{\scriptstyle f_0} & & \Big\downarrow{\scriptstyle f_2} & & \\
0 & \longrightarrow & V_1 & \xrightarrow{\;j\;} & V_0 & \xrightarrow{\;q\;} & V_2 & \longrightarrow & 0
\end{array}
$$

then f_0 is of determinant class if and only if both f_1 and f_2 are of determinant class (see Theorem 3.14 (2)). The corresponding statements i.) and ii.) remain true if one replaces "of determinant class" everywhere by "with positive Novikov-Shubin invariant" because of Lemma 2.14 and Lemma 2.15.

3.4 Cellular L^2-Torsion

In this section we introduce and study cellular L^2-torsion. Essentially we apply the material of Section 3.3 to the cellular L^2-chain complex of a finite free G-CW-complex. There are two interesting cases, the case where the L^2-homology vanishes and the case where the underlying space is a cocompact free proper G-manifold with a G-invariant Riemannian metric. Cellular L^2-torsion has been introduced in [93] and [345]. The definition of determinant class is taken from [84, Definition 4.1 on page 800].

3.4.1 Cellular L^2-Torsion in the Weakly-Acyclic Case

Definition 3.91 (L^2-torsion). *Let X be a finite free G-CW-complex. We call it* det-L^2-acyclic *if its cellular L^2-chain complex $C_*^{(2)}(X)$ is det-L^2-acyclic (see Definition 3.29). In this case we define its cellular L^2-torsion*

$$
\rho^{(2)}(X; \mathcal{N}(G)) \; := \; \rho^{(2)}(C_*^{(2)}(X))
$$

by the L^2-torsion of $C_*^{(2)}(X)$ (see Definition 3.29). Often we omit $\mathcal{N}(G)$ from the notation. □

Since for two equivariant smooth triangulations $f\colon K \to M$ and $g\colon L \to M$ of a cocompact free proper G-manifold M the Whitehead torsion of $g^{-1} \circ f\colon K \to L$ in $\mathrm{Wh}(G)$ is trivial, K is det-L^2-acyclic if and only if L is det-L^2-acyclic and in this case $\rho^{(2)}(K) = \rho^{(2)}(L)$ by Theorem 3.93 (1). Hence we can define M to be det-L^2-acyclic if K is det-L^2-acyclic for some (and hence each) equivariant smooth triangulation and put in this case $\rho^{(2)}(M) = \rho^{(2)}(K)$. The notions above extend in the obvious way to pairs.

We obtain a homomorphism

$$\Phi = \Phi^G \colon \mathrm{Wh}(G) \to \mathbb{R} \tag{3.92}$$

by sending the class of an invertible (n,n) matrix A over $\mathbb{Z}G$ to the logarithm of the Fuglede-Kadison determinant $\det(R_A)$ (see Definition 3.11) of the morphism $R_A \colon l^2(G)^n \to l^2(G)^n$ induced by A. This is well-defined by Theorem 3.14. The next theorem presents the basic properties of L^2-torsion.

Theorem 3.93 (Cellular L^2-torsion). *(1) Homotopy invariance*

Let $f \colon X \to Y$ be a G-homotopy equivalence of finite free G-CW-complexes. Let $\tau(f) \in \mathrm{Wh}(G)$ be its Whitehead torsion (see (3.2)). Suppose that X or Y is det-L^2-acyclic. Then both X and Y are det-L^2-acyclic and

$$\rho^{(2)}(Y) - \rho^{(2)}(X) \;=\; \Phi^G(\tau(f));$$

(2) Sum formula

Consider the G-pushout of finite free G-CW-complexes such that j_1 is an inclusion of G-CW-complexes, j_2 is cellular and X inherits its G-CW-complex structure from X_0, X_1 and X_2

$$
\begin{array}{ccc}
X_0 & \xrightarrow{\;j_1\;} & X_1 \\[2pt]
{\scriptstyle j_2}\Big\downarrow & & \Big\downarrow{\scriptstyle i_1} \\[2pt]
X_2 & \xrightarrow{\;i_2\;} & X
\end{array}
$$

Assume that three of the G-CW-complexes X_0, X_1, X_2 and X are det-L^2-acyclic. Then all four G-CW-complexes X_0, X_1, X_2 and X are det-L^2-acyclic and

$$\rho^{(2)}(X) \;=\; \rho^{(2)}(X_1) + \rho^{(2)}(X_2) - \rho^{(2)}(X_0);$$

(3) Poincaré duality

Let M be a cocompact free proper G-manifold without boundary of even dimension which is orientable and det-L^2-acyclic. Then

$$\rho^{(2)}(M) \;=\; 0;$$

(4) Product formula

Let X be a finite free G-CW-complex and let Y be a finite free H-CW-complex. Suppose that X is det-L^2-acyclic. Then the finite free $G \times H$-CW-complex $X \times Y$ is det-L^2-acyclic and

$$\rho^{(2)}(X \times Y, \mathcal{N}(G \times H)) \;=\; \chi(H \backslash Y) \cdot \rho^{(2)}(X, \mathcal{N}(G));$$

(5) Restriction

Let X be a finite free G-CW-complex and let $H \subset G$ be a subgroup of finite index $[G : H]$. Let $\mathrm{res}^H_G X$ be the finite H-CW-complex obtained from X by restricting the G-action to an H-action. Then X is det-L^2-acyclic if and only if $\mathrm{res}^H_G X$ is det-L^2-acyclic, and in this case

$$\rho^{(2)}(X; \mathcal{N}(G)) = [G : H] \cdot \rho^{(2)}(\mathrm{res}^H_G X; \mathcal{N}(H));$$

(6) Induction

Let H be a subgroup of G and let X be a finite free H-CW-complex. Then the finite free G-CW-complex $G \times_H X$ is det-L^2-acyclic if and only if X is det-L^2-acyclic, and in this case

$$\rho^{(2)}(G \times_H X; \mathcal{N}(G)) = \rho^{(2)}(X; \mathcal{N}(H));$$

(7) Positive Novikov-Shubin invariants and determinant class

If X is a finite free G-CW-complex with $b_p^{(2)}(X) = 0$ and $\alpha_p(X) > 0$ for all $p \geq 0$, then X is det-L^2-acyclic.

Proof. (1) This follows from Theorem 3.35 (5) and Lemma 3.41.

(2) We obtain an exact sequence of Hilbert $\mathcal{N}(G)$-chain complexes $0 \to C_*^{(2)}(X_0) \to C_*^{(2)}(X_1) \oplus C_*^{(2)}(X_2) \to C_*^{(2)}(X) \to 0$. Now apply Theorem 3.35 (1).

(3) There is a subgroup $G_0 \subset G$ of index 1 or 2 which acts orientation preserving on M. Since $\rho^{(2)}(M; \mathcal{N}(G_0)) = [G : G_0] \cdot \rho^{(2)}(M; \mathcal{N}(G_0))$ by assertion (5) we can assume without loss of generality that $G = G_0$, i.e. $G \backslash M$ is orientable. Fix an equivariant smooth triangulation $f\colon K \to M$ of M. Put $\pi = \pi_1(K)$ and $n = \dim(M)$. Let $[G \backslash K]$ be the image of the fundamental class of $[G \backslash M]$ under the isomorphism $H_n(G \backslash M) \to H_n(G \backslash K)$ induced by $G \backslash f^{-1}$. The Poincaré $\mathbb{Z}G$-chain homotopy equivalence

$$\cap [G \backslash K]\colon C^{n-*}(K) \to C_*(K)$$

has trivial Whitehead torsion with respect to the cellular basis [510, Theorem 2.1 on page 23]. It induces a chain homotopy equivalence of finite Hilbert $\mathcal{N}(G)$-chain complexes $f_*\colon l^2(G) \otimes_{\mathbb{Z}G} C^{n-*}(K) \to C_*^{(2)}(K)$ with $t^{(2)}(f_*) = 0$. We get from Theorem 3.35 (5) and Lemma 3.41 that $l^2(G) \otimes_{\mathbb{Z}G} C^{n-*}(K)$ is det-L^2-acyclic and $\rho^{(2)}(l^2(G) \otimes_{\mathbb{Z}G} C^{n-*}(K)) = \rho^{(2)}(C_*^{(2)}(K))$. We conclude $\rho^{(2)}(M) = (-1)^{n+1} \cdot \rho^{(2)}(M)$ from Theorem 3.35 (5). Since n is even by assumption, assertion (3) follows.

(4) This follows from Theorem 3.35 (6a).

(5) This follows from Theorem 3.35 (7).

(6) This follows from Theorem 3.35 (8).

(7) This follows from Theorem 3.14 (4). This finishes the proof of Theorem 3.93. □

Part (1) and (2) of the following conjecture is taken from [330, Conjecture 1.5]. We will later prove it for a large class of groups (see Lemma 13.6 and Theorem 13.3).

Conjecture 3.94 (Homotopy invariance of L^2-torsion). *We have for any group G:*

(1) The homomorphism

$$\Phi = \Phi^G \colon \mathrm{Wh}(G) \to \mathbb{R}$$

sending the class $[A]$ of an invertible matrix $A \in GL_n(\mathbb{Z}G)$ to $\ln(\det(r_A^{(2)}))$ (which we have already defined in (3.92)) is trivial;

(2) If X and Y are det-L^2-acyclic finite G-CW-complexes, which are G-homotopy equivalent, then their L^2-torsion agree

$$\rho^{(2)}(X; \mathcal{N}(G)) = \rho^{(2)}(Y; \mathcal{N}(G));$$

(3) Let $A \in M_n(\mathbb{Z}G)$ be a (n, n)-matrix over $\mathbb{Z}G$. Then $r_A^{(2)} \colon l^2(G)^n \to l^2(G)^n$ is of determinant class (see Definition 3.11).

Conjecture 3.94 (1) is obviously true if $\mathrm{Wh}(G)$ vanishes. There is the conjecture that the Whitehead group $\mathrm{Wh}(G)$ vanishes if G is torsionfree.

In most applications X will occur as the universal covering of a finite CW-complex. Therefore we will discuss this special case here. Since we also want to deal with non-connected CW-complexes, we introduce

Notation 3.95. *Let X be a (not necessarily connected) finite CW-complex. We say that $\widetilde{X}$ is det-L^2-acyclic, if the universal covering $\widetilde{C}$ of each connected component C of X is det-L^2-acyclic in the sense of Definition 3.91. In this case we write*

$$\rho^{(2)}(\widetilde{X}) := \sum_{C \in \pi_0(X)} \rho^{(2)}(\widetilde{C}),$$

where $\rho^{(2)}(\widetilde{C})$ is the L^2-torsion of the finite free $\pi_1(C)$-CW-complex $\widetilde{C}$ of Definition 3.91.

The next theorem presents the basic properties of $\rho^{(2)}(\widetilde{X})$. It is a consequence of Theorem 3.93. Notice the formal analogy between the behaviour of $\rho^{(2)}(\widetilde{X})$ and the ordinary Euler characteristic $\chi(X)$.

Theorem 3.96. (Cellular L^2-torsion for universal coverings).

(1) Homotopy invariance

Let $f \colon X \to Y$ be a homotopy equivalence of finite CW-complexes. Let $\tau(f) \in \mathrm{Wh}(\pi_1(Y))$ be its Whitehead torsion. Suppose that $\widetilde{X}$ or $\widetilde{Y}$ is det-L^2-acyclic. Then both $\widetilde{X}$ and $\widetilde{Y}$ are det-L^2-acyclic and

$$\rho^{(2)}(\widetilde{Y}) - \rho^{(2)}(\widetilde{X}) = \Phi^{\pi_1(Y)}(\tau(f)),$$

where $\Phi^{\pi_1(Y)} \colon \mathrm{Wh}(\pi_1(Y)) = \bigoplus_{C \in \pi_0(Y)} \mathrm{Wh}(\pi_1(C)) \to \mathbb{R}$ is the sum of the maps $\Phi^{\pi_1(C)}$ of (3.92);

(2) Sum formula

Consider the pushout of finite CW-complexes such that j_1 is an inclusion of CW-complexes, j_2 is cellular and X inherits its CW-complex structure from X_0, X_1 and X_2

$$
\begin{array}{ccc}
X_0 & \xrightarrow{\ j_1\ } & X_1 \\
{\scriptstyle j_2}\big\downarrow & & \big\downarrow{\scriptstyle i_1} \\
X_2 & \xrightarrow{\ i_2\ } & X
\end{array}
$$

Assume $\widetilde{X_0}$, $\widetilde{X_1}$, and $\widetilde{X_2}$ are det-L^2-acyclic and that for $k = 0, 1, 2$ the map $\pi_1(i_k) \colon \pi_1(X_k) \to \pi_1(X)$ induced by the obvious map $i_k \colon X_k \to X$ is injective for all base points in X_k.
Then $\widetilde{X}$ is det-L^2-acyclic and we get

$$\rho^{(2)}(\widetilde{X}) = \rho^{(2)}(\widetilde{X_1}) + \rho^{(2)}(\widetilde{X_2}) - \rho^{(2)}(\widetilde{X_0});$$

(3) Poincaré duality

Let M be a closed manifold of even dimension such that $\widetilde{M}$ is det-L^2-acyclic. Then

$$\rho^{(2)}(\widetilde{M}) = 0;$$

(4) Product formula

Let X and Y be finite CW-complexes. Suppose that $\widetilde{X}$ is det-L^2-acyclic. Then $\widetilde{X \times Y}$ is det-L^2-acyclic and

$$\rho^{(2)}(\widetilde{X \times Y}) = \chi(Y) \cdot \rho^{(2)}(\widetilde{X});$$

(5) Multiplicativity

Let $X \to Y$ be a finite covering of finite CW-complexes with d sheets. Then $\widetilde{X}$ is det-L^2-acyclic if and only if $\widetilde{Y}$ is det-L^2-acyclic and in this case

$$\rho^{(2)}(\widetilde{X}) = d \cdot \rho^{(2)}(\widetilde{Y});$$

(6) Positive Novikov-Shubin invariants and determinant class

If X is a finite CW-complex with $b_p^{(2)}(\widetilde{X}) = 0$ and $\alpha_p(\widetilde{X}) > 0$ for all $p \geq 0$, then $\widetilde{X}$ is det-L^2-acyclic.

Next we want to deal with the behaviour of the cellular L^2-torsion under a fibration $p \colon E \to B$ with fiber F. We begin with introducing simple structures. A *simple structure* $\xi = [(X, f)]$ on a topological space Z is an equivalence class of pairs (X, f) consisting of a finite CW-complex X and a homotopy equivalence $f\colon X \to Z$, where we call two such pairs (X, f) and (Y, g) equivalent if the Whitehead torsion $\tau(g^{-1} \circ f)$ vanishes, i.e. $g^{-1} \circ f$ is a simple homotopy equivalence. If $g\colon Z_1 \to Z_2$ is a homotopy equivalence of topological spaces and we have specified simple structures $\xi_i = [(X_i, f_i)]$ on Z_i, we can still define the Whitehead torsion

$$\tau(g) \ \in \ \mathrm{Wh}(\pi_1(Z_2))$$

by the image of $\tau(f_2^{-1} \circ g \circ f_1)$ under the isomorphism $f_{2*}\colon \mathrm{Wh}(\pi_1(X_2)) \to \mathrm{Wh}(\pi_1(Z_2))$. If for some representative (X_1, f_1) of ξ_1 and a given G-covering $\overline{Z_1} \to Z_1$ the total space of the pullback $\overline{X_1} \to X_1$ with f_1 is det-L^2-acyclic with respect to the action of the group of deck transformations, then this is true for all representatives by Theorem 3.93 (1), and we say that $(\overline{Z_1}, \xi_1)$ is det-L^2-acyclic. In this case we can still define the cellular L^2-torsion of $\overline{Z_1}$ with respect to ξ_1 by

$$\rho^{(2)}(\overline{Z_1}, \xi_1) := \rho^{(2)}(\overline{X_1}). \tag{3.97}$$

This is independent of the choice of the representative (X_1, f_1) by Theorem 3.93 (1).

Let $E \xrightarrow{p} B$ be a fibration such that the fiber F has the homotopy type of a finite CW-complex and B is a connected finite CW-complex. Recall that F is only determined up to homotopy. We can associate to p an element

$$\theta(p) \in H^1(B; \mathrm{Wh}(\pi_1(E))) \tag{3.98}$$

by specifying a homomorphism $\pi_1(B, b) \to \mathrm{Wh}(\pi_1(E))$ for a fixed base point $b \in B$ as follows. The fiber transport [488, 15.12 on page 343] of an element $w \in \pi_1(B, b)$ determines a homotopy class of selfhomotopy equivalences $t_w\colon F_b \to F_b$ of $F_b := p^{-1}(b)$. If we choose a simple structure $\xi(F_b)$ on F_b, we can take the Whitehead torsion $\tau(t_w)$ in $\mathrm{Wh}(\pi_1(F_b))$ and push it forward to $\mathrm{Wh}(\pi_1(E))$ using the map induced by the inclusion $k_b\colon F_b \to E$. It turns out that the class of $\theta(p)$ is independent of the base point $b \in B$ and the simple structure on F_b. We call p *simple* if $\theta(p) = 0$.

Suppose that p is simple. If we fix a base point $b \in B$ and a simple structure $\xi(F_b)$ on F_b, there is a preferred simple structure $\xi_{b,\xi(F_b)}(E)$ on E. If we choose another base point b' and simple structure $\xi(F_{b'})$ on $F_{b'}$, we obtain another simple structure on $\xi_{b',\xi(F_{b'})}(E)$. If w is any path in B from b' to b, the fiber transport yields a homotopy class of homotopy equivalences $t_w\colon F_b \to F_{b'}$ and we get

$$\tau\left(\mathrm{id}\colon (E, \xi_{b,\xi(F_b)}(E)) \to (E, \xi_{b',\xi(F_{b'})}(E))\right)$$
$$= \chi(B) \cdot k_{b'*}\left(\tau\left(t_w\colon (F_b, \xi(F_b)) \to (F_{b'}, \xi(F_{b'}))\right)\right). \tag{3.99}$$

Notice that the right side is independent of the choice of w because of the assumption $\theta(p) = 0$.

Details of the construction of $\xi_{b,\xi(F_b)}(E)$ and the claims above can be found in [325, section 2] or in the more general context of equivariant CW-complexes in [326, section 15]. It is based on the observation that over a cell e in B there is an up to fiber homotopy unique (strong) fiber homotopy equivalence from the restriction of E to the cell e to the trivial fibration $e \times F_b \to e$, provided we have fixed a homotopy class of paths from b to some point in e. The point is that after a choice of $b \in B$ and simple structure $\xi(F_b)$ on F_b we obtain uniquely a simple structure $\xi_{b,\xi(F_b)}(E)$ on E and $\rho^{(2)}(\widetilde{E}, \xi_{b,\xi(F_b)}(E))$ is defined, and we can ask for the relation of $\rho^{(2)}(\widetilde{E}, \xi_{b,\xi(F_b)}(E))$ and $\rho^{(2)}(\widetilde{F_b}, \xi(F_b))$.

Theorem 3.100 (L^2-torsion and fibrations). *Let $p\colon E \to B$ be a fibration with $\theta(p) = 0$ such that B is a connected finite CW-complex and the fiber has the homotopy type of a finite CW-complex. Suppose that the inclusion of F_b into E induces an injection on the fundamental groups for all base points in F_b. Fix $b \in B$ and a simple structure $\xi(F_b)$ on F_b. Suppose that $\widetilde{F_b}$ is det-L^2-acyclic.*

Then E is det-L^2-acyclic and

$$\rho^{(2)}(\widetilde{E}, \xi_{b,\xi(F_b)}(E)) \;=\; \chi(B) \cdot \rho^{(2)}(\widetilde{F_b}, \xi(F_b)).$$

Before we give the proof of Theorem 3.100, we discuss some interesting special cases.

Example 3.101. If we make the additional assumption that $\chi(B) = 0$, then E has a preferred simple structure $\xi(E)$ independent of the choices of $b \in B$ and $\xi(F_b)$ because of (3.99) and Theorem 3.100 says $\rho(\widetilde{E}, \xi(E)) = 0$.

Remark 3.102. Suppose for one (and hence) all base points $b \in B$ that the composition $\mathrm{Wh}(\pi_1(F_b)) \xrightarrow{k_b} \mathrm{Wh}(\pi_1(E)) \xrightarrow{\Phi^{\pi_1(E)}} \mathbb{R}$ is trivial (cf. Conjecture 3.94 (1)). Then $\rho^{(2)}(\widetilde{E}, \xi_{b,\xi(F_b)}(E))$ and $\chi(B) \cdot \rho^{(2)}(\widetilde{F_b}, \xi(F_b))$ are independent of the choice of $b \in B$ and $\xi(F_b)$ by Theorem 3.96 (1).

Let $F \to E \xrightarrow{p} B$ be a (locally trivial) fiber bundle of finite CW-complexes with connected base space B. Then $\theta(p) = 0$ and $\rho^{(2)}(\widetilde{F})$ and $\rho^{(2)}(\widetilde{E})$ are independent of the choice of a finite CW-structure on F and E since the Whitehead torsion of a homeomorphism of finite CW-complexes is always trivial [99], [100] and Theorem 3.96 (1) holds. It turns out that $\xi_{b,\xi(F_b)}(E)$ is the simple structure on E given by any finite CW-structure if $\xi(F_b)$ is the simple structure on F_b given by any finite CW-structure. Hence Theorem 3.100 yields

Corollary 3.103. *Suppose that $F \to E \xrightarrow{p} B$ is a (locally trivial) fiber bundle of finite CW-complexes with connected B. Suppose that for one (and*

hence all) $b \in B$ the inclusion of F_b into E induces an injection on the fundamental groups for all base points in F_b and $\widetilde{F_b}$ is det-L^2-acyclic. Then $\widetilde{E}$ is det-L^2-acyclic and

$$\rho^{(2)}(\widetilde{E}) \;=\; \chi(B) \cdot \rho^{(2)}(\widetilde{F}).$$

Next we give the proof of Theorem 3.100. Theorem 3.93 (6) implies that $\widetilde{E}|_{F_b}$ is det-L^2-acyclic if and only if $\widetilde{F_b}$ is det-L^2-acyclic and in this case $\rho^{(2)}(\widetilde{E}|_{F_b}) = \rho(\widetilde{F_b})$ since by assumption the inclusion of F into E induces an injection on the fundamental groups for all base points in F_b. Hence Theorem 3.100 is the special case $\overline{E} = \widetilde{E}$ of the following slightly stronger statement.

Lemma 3.104. *Let* $p \colon E \to B$ *be a fibration with* $\theta(p) = 0$ *such that* B *is a connected finite CW-complex and the fiber has the homotopy type of a finite CW-complex. Let* $q \colon \overline{E} \to E$ *be a G-covering and let* $\overline{F_b} \to F_b$ *be its restriction to* F_b *for some fixed* $b \in B$. *Suppose that* $\overline{F_b}$ *is det-L^2-acyclic. Then* $\overline{E}$ *is det-L^2-acyclic and*

$$\rho^{(2)}(\overline{E}, \xi_{b,\xi(F_b)}(E)) \;=\; \chi(B) \cdot \rho^{(2)}(\overline{F_b}, \xi(F_b)).$$

Proof. We use induction over the number of cells of B. The induction beginning $B = \emptyset$ is trivial. We have to deal with the situation that B is obtained from B_0 by attaching a cell, i.e. there is a pushout

$$\begin{array}{ccc}
S^{n-1} & \xrightarrow{\;q\;} & B_0 \\[2pt]
\scriptstyle i \downarrow & & \scriptstyle j \downarrow \\[2pt]
D^n & \xrightarrow{\;Q\;} & B
\end{array}$$

The pullback construction applied to $p \circ q \colon \overline{E} \to B$ yields a G-pushout

$$\begin{array}{ccc}
\overline{E_S} & \xrightarrow{\;\overline{q}\;} & \overline{E_0} \\[2pt]
\scriptstyle \overline{i} \downarrow & & \scriptstyle \overline{j} \downarrow \\[2pt]
\overline{E_D} & \xrightarrow{\;\overline{Q}\;} & \overline{E}
\end{array}$$

Let Y be a finite CW-complex and $g \colon Y \to F_b$ be a homotopy equivalence representing the given simple structure $\xi(F_b)$ on F_b. The pullback of $q \colon \overline{E} \to E$ with g yields a G-covering $\overline{Y} \to Y$. By inspecting the construction of the simple structure $\xi_{b,\xi(F_b)}(E)$ on E and applying the induction hypothesis to $\overline{E_0}|_C \to C$ for each component C of B_0 taking (3.99) into account, one checks that there is a commutative square of finite G-CW-complexes

$$\begin{array}{ccc}
\overline{Y} \times S^{n-1} & \xrightarrow{\;j_1\;} & \overline{X_0} \\[2pt]
\scriptstyle j_2 \downarrow & & \scriptstyle i_1 \downarrow \\[2pt]
\overline{Y} \times D^n & \xrightarrow{\;i_2\;} & \overline{X}
\end{array}$$

such that j_1 is an inclusion of G-CW-complexes, j_2 is cellular and $\overline{X}$ inherits its G-CW-complex structure from $\overline{Y} \times S^{n-1}$, $\overline{Y} \times D^n$ and $\overline{X_0}$, the finite G-CW-complexes $\overline{Y} \times S^{n-1}$, $\overline{Y} \times D^n$ and $\overline{X_0}$ are det-L^2-acyclic and

$$\rho^{(2)}(\overline{X_0}) = \chi(B_0) \cdot \rho^{(2)}(\overline{F_b}, \xi(F_b));$$
$$\rho^{(2)}(\overline{E}, \xi_{b,\xi(F_b)}(E)) = \rho^{(2)}(\overline{X}).$$

Theorem 3.93 (2) and (4) imply that $\overline{X}$ is det-L^2-acyclic and

$$\begin{aligned}
\rho^{(2)}(\overline{E}, \xi_{b,\xi(F_b)}(E)) &= \rho^{(2)}(\overline{X}) \\
&= \rho^{(2)}(\overline{X_0}) + \rho^{(2)}(\overline{Y} \times D^n) - \rho^{(2)}(\overline{Y} \times S^{n-1}) \\
&= \rho^{(2)}(\overline{X_0}) + \chi(D^n) \cdot \rho^{(2)}(\overline{Y}) - \chi(S^{n-1}) \cdot \rho^{(2)}(\overline{Y}) \\
&= \chi(B_0) \cdot \rho^{(2)}(\overline{F_b}, \xi(F_b)) + \chi(D^n) \cdot \rho^{(2)}(\overline{F_b}, \xi(F_b)) \\
&\qquad\qquad\qquad\qquad - \chi(S^{n-1}) \cdot \rho^{(2)}(\overline{F_b}, \xi(F_b)) \\
&= \chi(B) \cdot \rho^{(2)}(\overline{F_b}, \xi(F_b)). \qquad \square
\end{aligned}$$

Theorem 3.105 (L^2-torsion and S^1-actions). *Let X be a connected S^1-CW-complex of finite type. Suppose that for one orbit S^1/H (and hence for all orbits) the inclusion into X induces a map on π_1 with infinite image. (In particular the S^1-action has no fixed points.) Let $\widetilde{X}$ be the universal covering of X with the canonical $\pi_1(X)$-action. Then $\widetilde{X}$ is det-L^2-acyclic and*

$$\alpha_p(\widetilde{X}) \geq 1 \qquad \text{for all } p;$$
$$\rho^{(2)}(\widetilde{X}) = 0.$$

Proof. Theorem 2.61 shows that $b_p^{(2)}(\widetilde{X}) = 0$ for $p \geq 0$ and $\alpha_p(\widetilde{X}) \geq 1$ for $p \geq 1$. In particular $\widetilde{X}$ is det-L^2-acyclic by Theorem 3.96 (6). The proof of $\rho^{(2)}(\widetilde{X}) = 0$ is analogous to the one of Theorem 1.40 using Theorem 3.93 (2) and (4) and the conclusion $\rho^{(2)}(\widetilde{S^1}) = 0$ from (3.24) appearing in Example 3.22. $\qquad \square$

Next we deal with the mapping torus T_f of a selfmap $f\colon X \to X$ of a connected finite CW-complex. If $p\colon T_f \to S^1$ is the canonical projection, let $\pi_1(T_f) \xrightarrow{\phi} G \xrightarrow{\psi} \mathbb{Z}$ be a factorization of the epimorphism $\pi_1(T_f) \xrightarrow{\pi_1(p)} \pi_1(S^1) = \mathbb{Z}$ into epimorphisms ϕ and ψ. If $i\colon X \to T_f$ is the obvious inclusion, let $L \subset G$ be the image of the composition $\pi_1(X) \xrightarrow{\pi_1(i)} \pi_1(T_f) \xrightarrow{\phi} G$. Let $\overline{T_f}$ be the covering of T_f associated to ϕ which is a free G-CW-complex. Denote by $\overline{X} \to X$ the L-covering of X associated to the epimorphism $\iota := \phi \circ \pi_1(i)\colon \pi_1(X) \to L$. There is an automorphism $\mu\colon L \to L$ uniquely determined by the property that $\iota \circ \pi_1(f) = \mu \circ \iota$. Then G is the semidirect product $L \rtimes_\mu \mathbb{Z}$. Let $\overline{f}\colon \overline{X} \to \overline{X}$ be a $(\mu : L \to L)$-equivariant lift of f. Then $\overline{T_f}$ is the mapping telescope of $\overline{f}$ infinite to both sides, i.e., the identification space

$$\overline{T_f} = \coprod_{n \in \mathbb{Z}} \overline{X} \times [n, n+1]/\sim,$$

where the identification $\sim$ is given by $(x, n+1) \sim (\overline{f}(x), n)$. The group of deck transformations G is the semidirect product $L \rtimes_\mu \mathbb{Z}$ and acts in the obvious way.

Let $j \colon L \to G$ be the inclusion. We obtain a $\mathbb{Z}G$-chain map, a Hilbert $\mathcal{N}(G)$-chain map and a morphism of Hilbert $\mathcal{N}(G)$-modules

$$\overline{C_*(\overline{f})_*} \colon \mathbb{Z}G \otimes_{\mathbb{Z}L} C_*(\overline{X}) \to \mathbb{Z}G \otimes_{\mathbb{Z}L} C_*(\overline{X});$$

$$\overline{C_*(\overline{f})}_*^{(2)} \colon l^2(G) \otimes_{\mathbb{Z}L} C_*(\overline{X}) \to l^2(G) \otimes_{\mathbb{Z}L} C_*(\overline{X});$$

$$\overline{H_p^{(2)}(\overline{f})} \colon j_* H_p^{(2)}(\overline{X}) = \mathbb{C}G \otimes_{\mathbb{C}L} H_p^{(2)}(\overline{X}) \to j_* H_p^{(2)}(\overline{X}) = \mathbb{C}G \otimes_{\mathbb{C}L} H_p^{(2)}(\overline{X})$$

by sending $g \otimes u$ to $g \otimes u - gt \otimes C_*(\overline{f})(u)$ or $g \otimes u - gt \otimes H_p^{(2)}(\overline{f})(u)$ respectively. Then the cellular $\mathbb{Z}G$-chain complex $C_*(\widetilde{T_f})$ is the mapping cone $\mathrm{cone}_*(\overline{C_*(\overline{f})_*})$. Under the obvious identification of $j_* H_p^{(2)}(\overline{X})$ with $H_p^{(2)}(l^2(G) \otimes_{\mathbb{Z}L} C_*(\overline{X}))$ the map $\overline{H_p^{(2)}(\overline{f})}$ becomes the endomorphism of $H_p^{(2)}(l^2(G) \otimes_{\mathbb{Z}L} C_*(\overline{X}))$ induced by $\overline{C_*(\overline{f})}_*^{(2)}$.

Theorem 3.106 (L^2-torsion of mapping tori). *Let $f \colon X \to X$ and ϕ, ψ be given as above. Suppose that the G-CW-complex $\overline{T_f}$ is of determinant class. Then $\overline{T_f}$ is det-L^2-acyclic, for any $p \geq 0$ the endomorphism of finitely generated Hilbert $\mathcal{N}(\pi_1(T_f))$-modules $\overline{H_p^{(2)}(\overline{f})} \colon j_* H_p^{(2)}(\overline{X}) \to j_* H_p^{(2)}(\overline{X})$ is a weak isomorphism of determinant class and*

$$\rho^{(2)}(\widetilde{T_f}) = \sum_{p \geq 0} (-1)^p \cdot \ln\left(\det\left(\overline{H_p^{(2)}(\overline{f})}\right)\right).$$

Proof. We conclude from Theorem 1.39 that $H_p^{(2)}(\widetilde{T_f}) = H_p^{(2)}(\overline{C_*(\overline{f})}_*^{(2)})$ vanishes for all $p \geq 0$. Hence the long weakly exact homology sequences associated to $0 \to l^2(G) \otimes_{\mathbb{Z}L} C_*(\overline{X}) \to \mathrm{cone}_*(\overline{f}_*^{(2)}) \to \Sigma \, l^2(G) \otimes_{\mathbb{Z}L} C_*(\overline{X}) \to 0$ looks like

$$\ldots \to 0 \to H_p^{(2)}(l^2(G) \otimes_{\mathbb{Z}L} C_*(\overline{X}))$$

$$\xrightarrow{H_p^{(2)}(\overline{C_*(\overline{f})}_*^{(2)})} H_p^{(2)}(l^2(G) \otimes_{\mathbb{Z}L} C_*(\overline{X})) \to 0 \to \ldots$$

Now Theorem 3.35 (1) implies that $H_p^{(2)}(\overline{C_*(\overline{f})}_*^{(2)})$ is a weak isomorphism of determinant class and

$$\rho^{(2)}(\widetilde{T_f}) = \sum_{p \geq 0} (-1)^p \cdot \ln\left(\det\left(H_p^{(2)}(\overline{C_*(\overline{f})}_*^{(2)})\right)\right).$$

Since $H_p^{(2)}(\overline{C_*(\overline{f})}_*^{(2)})$ and $\overline{H_p^{(2)}(\overline{f})}$ are conjugated by an isomorphism, the claim follows from Theorem 3.14 (1). $\qquad\square$

The assumption in Theorem 3.106 that $\overline{T_f}$ is of determinant class is for instance satisfied if $\phi\colon \pi_1(T_f) \to G$ is bijective and $\pi_1(X)$ belongs to the class $\mathcal{G}$ (see Definition 13.9, Lemma 13.6, Lemma 13.11 (4) and Theorem 13.3) because $\pi_1(T_f)$ is the mapping torus group of the endomorphism of $\pi_1(X)$ induced by f appearing in Lemma 13.11 (4).

Example 3.107. Let $f\colon X \to X$ be a selfmap of a connected finite CW-complex. Let $\overline{T_f} \to T_f$ be the canonical infinite cyclic covering of the mapping torus. It is the covering associated to the canonical epimorphism $\pi_1(T_f) \to \mathbb{Z}$ or, equivalently, the pullback of the universal covering of S^1 with respect to the canonical map $T_f \to S^1$. Let $H_p(f;\mathbb{C})\colon H_p(X;\mathbb{C}) \to H_p(X;\mathbb{C})$ be the endomorphism of a finite dimensional complex vector space induced by f. By the Jordan Normal Form Theorem it is conjugated by an isomorphism $H_p(X;\mathbb{C}) \to \mathbb{C}^n$ to an automorphism of $\mathbb{C}^n$ which is a direct sum of automorphisms of the form

$$B(\lambda_{i_p}, n_{i_p}) = \begin{pmatrix} \lambda_{i_p} & 1 & 0 & \dots & 0 \\ 0 & \lambda_{i_p} & 1 & \dots & 0 \\ 0 & 0 & \lambda_{i_p} & \dots & 0 \\ \vdots & \vdots & \vdots & \ddots & \vdots \\ 0 & 0 & 0 & \dots & \lambda_{i_p} \end{pmatrix}$$

where the size of the block $B(\lambda_{i_p}, n_{i_p})$ is n_{i_p}, $\lambda_{i_p} \in \mathbb{C}$ and $i_p = 1, 2 \dots, r_p$. One easily checks that then $H_p(\overline{T_f};\mathbb{C}) = \bigoplus_{i_p=1}^{r_p} \mathbb{C}[\mathbb{Z}]/((t - \lambda_{i_p})^{n_{i_p}})$. We conclude from Lemma 2.58 that $\overline{T_f}$ is of determinant class and

$$b_p^{(2)}(\overline{T_f};\mathcal{N}(G)) = 0; \tag{3.108}$$

$$\alpha_{p+1}(\overline{T_f};\mathcal{N}(G)) = \max\left\{ \frac{1}{n_{i_p}} \ \middle| \ i_p \in \{1, 2 \dots, r_p\}, |\lambda_{i_p}| = 1 \right\}, \tag{3.109}$$

where the maximum over the empty set is defined to be ∞^+. In the notation of Theorem 3.106 we see that $H_p^{(2)}(\overline{f})$ is conjugate to a direct sum of automorphisms of the shape

$$\begin{pmatrix} 1 - \lambda_{i_p} t & -t & 0 & \dots & 0 \\ 0 & 1 - \lambda_{i_p} t & -t & \dots & 0 \\ 0 & 0 & 1 - \lambda_{i_p} t & \dots & 0 \\ \vdots & \vdots & \vdots & \ddots & \vdots \\ 0 & 0 & 0 & \dots & 1 - \lambda_{i_p} t \end{pmatrix} \colon l^2(\mathbb{Z})^{n_{i_p}} \to l^2(\mathbb{Z})^{n_{i_p}}.$$

We conclude from Theorem 3.14 (2), (3.24) appearing in Example 3.22 and Theorem 3.106

$$\rho^{(2)}(\overline{T_f};\mathcal{N}(G)) = \sum_{p\geq 0}(-1)^p \cdot \left(\sum_{1\leq i_p \leq r_p, |\lambda_{i_p}|>1} n_{i_p} \cdot \ln(|\lambda_{i_p}|) \right).$$

Example 3.110. We construct a 4-dimensional non-orientable cocompact free proper $\mathbb{Z}$-manifold M without boundary such that M is det-L^2-acyclic, $\alpha_2(M; \mathcal{N}(\mathbb{Z})) \neq \alpha_3(M; \mathcal{N}(\mathbb{Z}))$ and $\rho^{(2)}(M; \mathcal{N}(\mathbb{Z})) \neq 0$. This shows that the condition that M is orientable is necessary in the statements about Poincaré duality in Theorem 2.55 (2) and Theorem 3.93 (3).

Equip T^3 with the orientation reversing free $\mathbb{Z}/2$-action which sends $(z_1, z_2, z_3) \in S^1 \times S^1 \times S^1$ to $(\overline{z_1}, -z_2, -z_3)$, where we think of $S^1 \subset \mathbb{C}$ and $\overline{z_1}$ is the complex conjugate. The map $f \colon T^2 \xrightarrow{\cong} T^2$, which sends (z_2, z_3) to $(z_2^4 z_3^7, z_2 z_3^2)$, is an automorphism, since the integral $(2, 2)$-matrix $A = \begin{pmatrix} 4 & 7 \\ 1 & 2 \end{pmatrix}$ has determinant 1. The automorphism $\mathrm{id} \times f \colon T^3 \to T^3$ is $\mathbb{Z}/2$-equivariant and hence induces an automorphism $g \colon (\mathbb{Z}/2)\backslash T^3 \xrightarrow{\cong} (\mathbb{Z}/2)\backslash T^3$ of the non-orientable closed 3-manifold $(\mathbb{Z}/2)\backslash T^3$. Let M be $\overline{T_g}$ for the infinite cyclic covering $\overline{T_g} \to T_g$ associated to the canonical epimorphism $\pi_1(T_f) \to \mathbb{Z}$. Obviously M is a 4-dimensional non-orientable cocompact free proper $\mathbb{Z}$-manifold without boundary. Let $h \colon T^2 \to M$ be the composition of the map $T^2 \to T^3$ sending (z_2, z_3) to $(1, z_2, z_3)$ and the projection $T^3 \to (\mathbb{Z}/2)\backslash T^3$. Then the following diagram commutes and has isomorphisms as vertical maps

$$
\begin{array}{ccc}
H_p(T^2; \mathbb{C}) & \xrightarrow{\;H_p(f;\mathbb{C})\;} & H_p(T^2; \mathbb{C}) \\
\Big\downarrow{\scriptstyle H_p(h;\mathbb{C})} & & \Big\downarrow{\scriptstyle H_p(h;\mathbb{C})} \\
H_p((\mathbb{Z}/2)\backslash T^3; \mathbb{C}) & \xrightarrow{\;H_p(g;\mathbb{C})\;} & H_p((\mathbb{Z}/2)\backslash T^3; \mathbb{C})
\end{array}
$$

The vertical maps are isomorphisms by the following argument. The trivial $\mathbb{C}[\mathbb{Z}/2]$-module $\mathbb{C}$ is $\mathbb{C}[\mathbb{Z}/2]$-projective and hence $H_p((\mathbb{Z}/2)\backslash T^3; \mathbb{C}) = H_p(T^3; \mathbb{C}) \otimes_{\mathbb{C}[\mathbb{Z}/2]} \mathbb{C}$ holds. The map S^1 to S^1 sending z to $\overline{z}$ or $-z$ respectively induces $-\mathrm{id}$ or id respectively on $H_1(S^1; \mathbb{C})$ and id on $H_0(S^1; \mathbb{C})$. Now apply the Künneth formula. The endomorphism $H_p(f; \mathbb{C})$ is the identity on $\mathbb{C}$ for $p = 0, 2$ and is the automorphism of $\mathbb{C}^2$ given by the matrix A for $p = 1$. Since the complex eigenvalues of the matrix A are $3 - \sqrt{8}$ and $3 + \sqrt{8}$, we conclude from Example 3.107

$$
\begin{aligned}
\alpha_p(M; \mathcal{N}(\mathbb{Z})) &= 1 && \text{for } p = 1, 3; \\
\alpha_p(M; \mathcal{N}(\mathbb{Z})) &= \infty^+ && \text{otherwise}; \\
\rho^{(2)}(M; \mathcal{N}(\mathbb{Z})) &= -\ln(3 + \sqrt{8}).
\end{aligned}
$$

3.4.2 Cellular L^2-Torsion in the Weakly-Acyclic and Aspherical Case

Theorem 3.111. *Let M be an aspherical closed manifold with non-trivial S^1-action. Then the action has no fixed points and the inclusion of any orbit* into M induces an injection on the fundamental groups. Moreover, $\widetilde{M}$ is det-L^2-acyclic and

$$\alpha_p(\widetilde{M}) \geq 1 \qquad \text{for all } p;$$
$$\rho^{(2)}(\widetilde{M}) = 0.$$

Proof. This follows from Corollary 1.43 and Theorem 3.105. $\square$

Definition 3.112 (det $\geq$ 1-class). *A group G is of* det $\geq$ 1-class *if for any $A \in M(m, n, \mathbb{Z}G)$ the Fuglede-Kadison determinant (see Definition 3.11) of the induced morphism $r_A^{(2)} \colon l^2(G)^m \to l^2(G)^n$ satisfies*

$$\det(r_A^{(2)}) \geq 1.$$

Schick [462] uses the phrase "has semi-integral determinant" instead of the phrase "of det $\geq$ 1-class" which we prefer. There is no group G known which is not of det $\geq$ 1-class. We will later present in Subsection 13.1.3 a class of groups for which it is known that they are of det $\geq$ 1-class. It includes amenable groups and countable residually finite groups. The assertion for the L^2-torsion in the theorem below is the main result of [515] (see also [516]). Its proof is interesting as it preshadows a more ring theoretic approach to L^2-Betti numbers, which we will present in Chapter 6, and localization techniques for non-commutative rings, which will play a role in Chapter 8 and Chapter 10.

Theorem 3.113 (L^2-torsion and aspherical CW-complexes). *Let X be an aspherical finite CW-complex. Suppose that its fundamental group $\pi_1(X)$ contains an elementary amenable infinite normal subgroup H and $\pi_1(X)$ is of* det $\geq$ 1-class. *Then*

$$b_p^{(2)}(\widetilde{X}) = 0 \quad \text{for } p \geq 0;$$
$$\alpha_p(\widetilde{X}) \geq 1 \quad \text{for } p \geq 1;$$
$$\rho^{(2)}(\widetilde{X}) = 0.$$

The claims for the L^2-Betti numbers and Novikov-Shubin invariants have already been proved in Theorem 2.63. We conclude from Theorem 3.93 (7) that $\widetilde{X}$ is det-L^2-acyclic. The proof for the claim about the L^2-torsion needs some preparation.

Lemma 3.114. *Let G be a group which is of* det $\geq$ 1-class. *Let Y be a finite free G-CW-complex which is det-L^2-acyclic. Let $S \subset \mathbb{Z}G$ be a multiplicatively closed subset such that for any $s \in S$ the induced morphism $r_s^{(2)} \colon l^2(G) \to l^2(G)$ sending u to us is a weak isomorphism and its Fuglede-Kadison determinant satisfies $\det(r_s^{(2)}) = 1$. Assume that $(\mathbb{Z}G, S)$ satisfies the Ore condition (see Definition 8.14). Suppose that the Ore localization (see Definition 8.14) $S^{-1}H_p(X)$ vanishes for all $p \geq 0$. Then*

$$\rho^{(2)}(Y; \mathcal{N}(G)) = 0.$$

Proof. Let $f: \mathcal{N}(G)^m \to \mathcal{N}(G)^n$ be an $\mathcal{N}(G)$-homomorphism of (left) $\mathcal{N}(G)$-modules. Choose a (m, n)-matrix $A \in M(m, n, \mathcal{N}(G))$ such that f sends x to xA. Define

$$\nu(f): l^2(G)^m \to l^2(G)^n, \quad y \mapsto \left(A^* y^t\right)^t$$

where y^t is obtained from y by transposing and applying elementwise the involution $l^2(G) \to l^2(G)$ which sends $\sum_{g \in G} \lambda_g \cdot g$ to $\sum_{g \in G} \overline{\lambda_g} \cdot g$, the matrix A^* is obtained from A by transposing and applying the involution $*: \mathcal{N}(G) \to \mathcal{N}(G)$ to each entry, and Ay^t is understood in the sense of matrices and plugging y_j into an element $a: l^2(G) \to l^2(G)$ in $\mathcal{N}(G)$. Notice that $\nu(g \circ f) = \nu(g) \circ \nu(f)$ and that ν is compatible with direct sums and, more generally, with block decompositions of matrices. Moreover $\nu(\lambda \cdot f) = \lambda \cdot \nu(f)$ for $\lambda \in \mathbb{C}$ and $\nu(f + g) = \nu(f) + \nu(g)$. (This construction will be analysed further and be extended to finitely generated projective $\mathcal{N}(G)$-modules in Section 6.2).

Given a $\mathbb{Z}G$-homomorphism $f: \mathbb{Z}G^m \to \mathbb{Z}G^n$, define $\det_{\mathbb{Z}G}(f) \in [0, \infty)$ by the Fuglede-Kadison determinant of $\nu(\mathrm{id}_{\mathcal{N}(G)} \otimes_{\mathbb{Z}G} f): l^2(G)^m \to l^2(G)^n$. We call f a weak isomorphism or of determinant class if $\nu(\mathrm{id}_{\mathcal{N}(G)} \otimes_{\mathbb{Z}G} f)$ has this property. A based free finite $\mathbb{Z}G$-chain complex C_* is called det-L^2-acyclic, if $\nu(\mathcal{N}(G) \otimes_{\mathbb{Z}G} C_*)$ is det-L^2-acyclic, and in this case we define $\rho^{(2)}(C_*)$ by $\rho^{(2)}(\nu(\mathcal{N}(G) \otimes_{\mathbb{Z}G} C_*))$. The point is that these notions can be extended to $S^{-1}\mathbb{Z}G$-modules and $S^{-1}\mathbb{Z}G$-chain complexes as follows.

Let $f: S^{-1}\mathbb{Z}G^m \to S^{-1}\mathbb{Z}G^n$ be a $S^{-1}\mathbb{Z}G$-homomorphism. Then there exist elements $s_1, s_2 \in S$ such that $r_{s_1} \circ f \circ r_{s_2}$ maps $\mathbb{Z}G^m \subset S^{-1}\mathbb{Z}G^m$ to $\mathbb{Z}G^n \subset S^{-1}\mathbb{Z}G^n$, where r_{s_1} and r_{s_2} are given by right multiplication with s_1 and s_2. (It is possible to choose $s_1 = 1$ or $s_2 = 1$.) This follows from the fact that any element in $S^{-1}\mathbb{Z}G$ can be written as us^{-1} or $t^{-1}v$ for $u, v \in \mathbb{Z}G$ and $s, t \in S$. Fix such elements s_1 and s_2. We say that f is a weak isomorphism resp. is of determinant class if the $\mathbb{Z}G$-homomorphism $r_{s_1} \circ f \circ r_{s_2}: \mathbb{Z}G^m \to \mathbb{Z}G^n$ has this property. If f is a weak isomorphism, we define

$$\det_{S^{-1}\mathbb{Z}G}(f) := \det_{\mathbb{Z}G}(r_{s_1} \circ f \circ r_{s_2}) \qquad \in [0, \infty). \qquad (3.115)$$

We conclude from Theorem 3.14 (1) that this is independent of the choice of s_1 and s_2 and assertions (1) and (2) of Theorem 3.14 for the Fuglede-Kadison determinant carry over to both $\det_{\mathbb{Z}G}$ and $\det_{S^{-1}\mathbb{Z}G}$. Moreover, we get

$$\det_{S^{-1}\mathbb{Z}G}(f) \geq 1 \qquad (3.116)$$

from the following argument. Choose $s \in S$ such that $r_s \circ f = S^{-1}g$ holds for an appropriate $\mathbb{Z}G$-homomorphisms $g: \mathbb{Z}G^m \to \mathbb{Z}G^n$. Since G is of det ≥ 1-class by assumption, we conclude from the definition (3.115)

$$\det_{S^{-1}\mathbb{Z}G}(f) = \det_{\mathbb{Z}G}(g) \geq 1.$$

We call a based free finite $S^{-1}\mathbb{Z}G$-chain complex D_* det-L^2-acyclic if there is a weak chain contraction (δ_*, v_*) in the sense that δ_* is a $S^{-1}\mathbb{Z}G$-chain homotopy $\delta_*: v_* \simeq 0$ satisfying $\delta_* \circ v_* = v_* \circ \delta_*$ and $v_p: D_p \to D_p$ is a weak $S^{-1}\mathbb{Z}G$-isomorphism of determinant class for all $p \in \mathbb{Z}$ (cf. Definition 3.38). In this case we define the L^2-torsion of D_*

$$\rho^{(2)}(D_*) := \ln\left(\det\nolimits_{S^{-1}\mathbb{Z}G}\left((vd + \delta)_{\mathrm{odd}}\right)\right) - \ln\left(\det\nolimits_{S^{-1}\mathbb{Z}G}\left(v_{\mathrm{odd}}\right)\right) \in \mathbb{R}.$$

This is independent of the choice of (δ_*, v_*) by the same argument as in Subsection 3.3.2 (cf. Definition 3.38, Lemma 3.40 and Lemma 3.41). If E_* is a det-L^2-acyclic based free finite $\mathbb{Z}G$-chain complex, then $S^{-1}E_*$ is a det-L^2-acyclic based free finite $S^{-1}\mathbb{Z}G$-chain complex and $\rho^{(2)}(E_*) = \rho^{(2)}(S^{-1}E_*)$.

Now suppose that C_* is a based free finite $\mathbb{Z}G$-chain complex such that $S^{-1}H_p(C_*) = 0$ for $p \in \mathbb{Z}$ and C_* is det-L^2-acyclic. Lemma 8.15 (3) implies that $H_p(S^{-1}C_*) = 0$ for $p \in \mathbb{Z}$. Hence we can find a chain contraction $\gamma_*: S^{-1}C_* \to S^{-1}C_{*+1}$. Both compositions of the maps $(S^{-1}c + \gamma)_{\mathrm{odd}}: S^{-1}C_{\mathrm{odd}} \to S^{-1}C_{\mathrm{ev}}$ and $(S^{-1}c + \gamma)_{\mathrm{ev}}: S^{-1}C_{\mathrm{ev}} \to S^{-1}C_{\mathrm{odd}}$ are given by upper triangular automorphisms of $S^{-1}C_{\mathrm{odd}}$ or $S^{-1}C_{\mathrm{ev}}$ with identity maps on the diagonal (cf. Lemma 3.40). Hence both maps are isomorphisms and satisfy

$$\det\nolimits_{S^{-1}\mathbb{Z}G}\left((S^{-1}c + \gamma)_{\mathrm{odd}}\right) \cdot \det\nolimits_{S^{-1}\mathbb{Z}G}\left((S^{-1}c + \gamma)_{\mathrm{ev}}\right) = 1. \quad (3.117)$$

Since $\det_{S^{-1}\mathbb{Z}G}\left((S^{-1}c + \gamma)_{\mathrm{odd}}\right) \geq 1$ and $\det_{S^{-1}\mathbb{Z}G}\left((S^{-1}c + \gamma)_{\mathrm{ev}}\right) \geq 1$ holds by (3.116), we get $\det_{S^{-1}\mathbb{Z}G}\left((S^{-1}c + \gamma)_{\mathrm{odd}}\right) = 1$. Since γ_* is a chain contraction (and not only a weak chain contraction), we conclude

$$\rho^{(2)}(C_*) = \ln\left(\det\nolimits_{S^{-1}\mathbb{Z}G}\left((S^{-1}c + \gamma)_{\mathrm{odd}}\right)\right) = 0.$$

Now apply this to $C_* = C_*(Y)$ with a cellular $\mathbb{Z}G$-basis and Lemma 3.114 follows. $\square$

Remark 3.118. The assumption in Lemma 3.114 that Y is det-L^2-acyclic is not necessary. With a little effort and some additional knowledge it follows from the other assumptions by the following argument. Obviously it suffices to show that $b_p^{(2)}(Y) = 0$ for all $p \in \mathbb{Z}$ since G is of det ≥ 1-class by assumption. We will introduce the algebra $\mathcal{U}(G)$ of affiliated operators in Chapter 8. We only have to know that $\mathcal{U}(G)$ is the Ore localization of $\mathcal{N}(G)$ with respect to the set of non-zero-divisors of $\mathcal{N}(G)$ (see Theorem 8.22 (1)). Hence we get an embedding $S^{-1}\mathbb{Z}G \subset \mathcal{U}(G)$. Since $S^{-1}C_*(Y)$ is contractible, $\mathcal{U}(G) \otimes_{\mathbb{Z}G} C_*(Y)$ is contractible and hence $H_p(\mathcal{U}(G) \otimes_{\mathbb{Z}G} C_*(Y)) = 0$ for $p \geq 0$. We will later show for an appropriate notion of dimension that $b_p^{(2)}(Y) = \dim_{\mathcal{U}(G)}\left(H_p\left(\mathcal{U}(G) \otimes_{\mathbb{Z}G} C_*(Y)\right)\right)$ (see Theorem 8.31). This implies $b_p^{(2)}(Y) = 0$.

Lemma 3.119. *Let G be a group and let $A \subset G$ be a normal abelian infinite torsionfree subgroup. Put $S = \{x \in \mathbb{Z}A \mid x \neq 0, \det(r_x^{(2)}) = 1\}$, where $r_x^{(2)} \colon l^2(G) \to l^2(G)$ is given by right multiplication with x.*

Then for each element $x \in S$ the morphism $r_x^{(2)} \colon l^2(G) \to l^2(G)$ is a weak isomorphism. The set S is multiplicatively closed. The pair $(\mathbb{Z}G, S)$ satisfies the Ore condition (see Definition 8.14). The trivial $\mathbb{Z}G$-module $\mathbb{Z}$ satisfies $S^{-1}\mathbb{Z} = 0$.

Proof. Consider $x \in \mathbb{Z}A$. There is a finitely generated subgroup $B \subset A$ with $x \in \mathbb{Z}B$. Since B is a finitely generated torsionfree abelian group, it is isomorphic to $\mathbb{Z}^n$ for some $n \in \mathbb{Z}$. We conclude from Lemma 1.34 (1) that $\dim_{\mathcal{N}(B)}\left(\ker\left(r_x^{(2)} \colon l^2(B) \to l^2(B)\right)\right) \in \{0, 1\}$. Since $\ker\left(r_x^{(2)}\right)$ is a proper closed subspace of $l^2(B)$ because of $x \neq 0$, its von Neumann dimension cannot be $\dim_{\mathcal{N}(B)}(l^2(B)) = 1$. This implies that its dimension is zero. We conclude from Lemma 1.24 (2) and (3) that $\dim_{\mathcal{N}(G)}\left(\ker\left(r_x^{(2)} \colon l^2(G) \to l^2(G)\right)\right) = 0$. Hence $r_x^{(2)} \colon l^2(G) \to l^2(G)$ is a weak isomorphism by Lemma 1.13.

We conclude from Theorem 3.14 (1) that S is multiplicatively closed.

Next we show that $(\mathbb{Z}G, S)$ satisfies the right Ore condition. (Notice $\mathbb{Z}G$ is a ring with involution which leaves S invariant so that then also the left Ore condition is satisfied.) Since S contains no non-trivial zero-divisors, it suffices to show for $(r, s) \in \mathbb{Z}G \times S$ that there exists $(r', s') \in \mathbb{Z}G \times S$ satisfying $rs' = sr'$. Let $\{g_i \mid i \in I\}$ be a complete system of representatives for the cosets $Ag \in A \backslash G$. We can write $r = \sum_{i \in I} f_i g_i$ for $f_i \in \mathbb{Z}A$, where almost all f_i are zero. Since A is normal in G, we get $g^{-1}sg \in \mathbb{Z}A$ for each $g \in G$. Obviously $g^{-1}sg \neq 0$. We have

$$\det\left(r_{g^{-1}sg}^{(2)}\right) = \det\left(r_g^{(2)} \circ r_s^{(2)} \circ r_{g^{-1}}^{(2)}\right)$$
$$= \det\left(r_g^{(2)}\right) \cdot \det\left(r_s^{(2)}\right) \cdot \det\left(r_{g^{-1}}^{(2)}\right)$$
$$= \det(r_s^{(2)}) = 1.$$

This shows $g^{-1}sg \in S$ for all $g \in G$. Since S is abelian, we can define

$$s' = \prod_{\{j \in I \mid f_j \neq 0\}} g_j^{-1}sg_j;$$

$$s_i' = \prod_{\{j \in I \mid f_j \neq 0, j \neq i\}} g_i g_j^{-1} s g_j g_i^{-1};$$

$$r' = \sum_{i \in I} s_i' f_i g_i.$$

Since S is multiplicatively closed, we have $s', s_i' \in S$. Since $\mathbb{Z}A$ is abelian, we conclude

$$rs' = \sum_{i\in I} f_i g_i s' = \sum_{\{i\in I | f_i \neq 0\}} f_i g_i s' g_i^{-1} g_i = \sum_{\{i\in I | f_i \neq 0\}} g_i s' g_i^{-1} f_i g_i$$

$$= \sum_{\{i\in I | f_i \neq 0\}} s s_i' f_i g_i = s r'.$$

Consider a subgroup $\mathbb{Z} \subset A$. Fix a generator $t \in \mathbb{Z}$. We conclude from Lemma 3.14 (6) and Example 3.13

$$\det\left(r_{1-t}^{(2)} \colon l^2(G) \to l^2(G)\right) = \det\left(r_{1-t}^{(2)} \colon l^2(\mathbb{Z}) \to l^2(\mathbb{Z})\right) = 1.$$

Hence $1 - t \in S$. If $\mathbb{Z}$ is the trivial $\mathbb{Z}G$-module, then $(1 - t)$ acts by multiplication with zero on $\mathbb{Z}$. Hence $S^{-1}\mathbb{Z} = 0$. This finishes the proof of Lemma 3.119. □

Now we are ready to finish the proof of Theorem 3.113. By Lemma 2.62 we can assume without loss of generality that $\pi_1(X)$ contains a normal abelian infinite torsionfree subgroup $A \subset \pi_1(X)$. Notice that $H_p(\widetilde{X})$ is zero for $p \neq 0$ and is the trivial $\mathbb{Z}[\pi_1(X)]$-module $\mathbb{Z}$ for $p = 0$. Now Theorem 3.113 follows from Lemma 3.114 and Lemma 3.119. □

We refer to Conjecture 11.3, which deals with the L^2-torsion of universal coverings of aspherical closed manifolds.

3.4.3 Topological L^2-Torsion for Riemannian Manifolds

In this subsection we introduce another variant of L^2-torsion where we assume that the underlying G-space is a cocompact free proper G-manifold without boundary and with G-invariant Riemannian metric. The G-invariant Riemannian metric together with the L^2-Hodge-de Rham Theorem 1.59 will allow us to drop the condition that all L^2-Betti numbers vanish.

Definition 3.120 (Topological L^2-torsion). *Let M be a cocompact free proper G-manifold without boundary and with G-invariant Riemannian metric. Let $f \colon K \to M$ be an equivariant smooth triangulation. Assume that M is of determinant class. Hence $\rho^{(2)}(K) := \rho^{(2)}(C_*^{(2)}(K)) \in \mathbb{R}$ is defined (see Definition 3.29). Let*

$$A_K^p \colon \mathcal{H}_{(2)}^p(M) \xrightarrow{\cong} H_{(2)}^p(K)$$

be the L^2-Hodge-de Rham isomorphism of Theorem 1.59, where we use on $\mathcal{H}_{(2)}^p(M)$ the Hilbert $\mathcal{N}(G)$-structure coming from the Riemannian metric. Define the topological L^2-torsion

$$\rho_{\mathrm{top}}^{(2)}(M) = \rho^{(2)}(K) - \sum_{p\geq 0}(-1)^p \cdot \ln\left(\det\left(A_K^p \colon \mathcal{H}_{(2)}^p(M) \xrightarrow{\cong} H_{(2)}^p(K)\right)\right).$$

We have to check that this is independent of the choice of equivariant smooth triangulation. Let $g\colon L \to M$ be another choice. Then the composition of $H^p_{(2)}(g^{-1} \circ f)\colon H^p_{(2)}(L) \to H^p_{(2)}(K)$ with A^p_L is just A^p_K. Theorem 3.14 (1) implies

$$\ln(\det(A^p_K)) = \ln(\det(A^p_L)) + \ln\left(\det\left(H^p_{(2)}(g^{-1} \circ f)\right)\right). \quad (3.121)$$

We conclude from Theorem 3.35 (5)

$$\rho^{(2)}(L) - \rho^{(2)}(K) = -\sum_{p \geq 0}(-1)^p \cdot \ln\left(\det\left(H^{(2)}_p(g^{-1} \circ f)\right)\right). \quad (3.122)$$

We obtain from Lemma 1.18 a commutative square of finitely generated Hilbert $\mathcal{N}(G)$-modules with isometric isomorphisms as vertical arrows

$$
\begin{array}{ccc}
H^p_{(2)}(L) & \xrightarrow{\;H^p_{(2)}(g^{-1}\circ f)\;} & H^p_{(2)}(K) \\[4pt]
\cong \downarrow & & \cong \downarrow \\[4pt]
H^{(2)}_p(L) & \xrightarrow{\;\left(H^{(2)}_p(g^{-1}\circ f)\right)^*\;} & H^{(2)}_p(K)
\end{array}
$$

We conclude from Theorem 3.14 (1) and Lemma 3.15 (4)

$$\ln\left(\det\left(H^{(2)}_p(g^{-1} \circ f)\right)\right) = \ln\left(\det\left(H^p_{(2)}(g^{-1} \circ f)\right)\right). \quad (3.123)$$

We get from (3.121), (3.122) and (3.123)

$$\rho^{(2)}(K) - \sum_{p \geq 0} \det\left(A^p_K \colon \mathcal{H}^p_{(2)}(M) \xrightarrow{\cong} H^p_{(2)}(K)\right)$$
$$= \rho^{(2)}(L) - \sum_{p \geq 0} \det\left(A^p_L \colon \mathcal{H}^p_{(2)}(M) \xrightarrow{\cong} H^p_{(2)}(L)\right). \quad (3.124)$$

Hence Definition 3.120 makes sense since $\rho^{(2)}_{\mathrm{top}}(M)$ is independent of the choice of equivariant smooth triangulation by (3.124). Notice that $\rho^{(2)}(M)$ does depend on the choice of Riemannian metric. If M happens to be det-L^2-acyclic, then $\rho^{(2)}(M)$ of Definition 3.91 and $\rho^{(2)}_{\mathrm{top}}(M)$ of Definition 3.120 agree.

There is an obvious analog of Theorem 3.93 for $\rho^{(2)}_{\mathrm{top}}(M)$. If $f\colon M \to N$ is a G-homotopy equivalence of cocompact free proper G manifolds without boundary and with G-invariant Riemannian metrics, then we get

$$\rho^{(2)}_{\mathrm{top}}(N) - \rho^{(2)}_{\mathrm{top}}(M) = \phi^G(\tau(f)) - \sum_{p \geq 0}(-1)^p \cdot$$
$$\ln\left(\det\left(\mathcal{H}^p_{(2)}(f)\colon \mathcal{H}^p_{(2)}(N) \to \mathcal{H}^p_{(2)}(M)\right)\right), \quad (3.125)$$

where $\mathcal{H}^p_{(2)}(f)$ is obtained from $H^p_{(2)}(f)$ by conjugating with the L^2-Hodge-de Rham isomorphism. We still have Poincaré duality, i.e. $\rho^{(2)}_{\mathrm{top}}(M) = 0$ if M is orientable and has even dimension. There is a product formula

$$\rho^{(2)}_{\mathrm{top}}(M \times N) = \chi(G\backslash M) \cdot \rho^{(2)}_{\mathrm{top}}(N) + \chi(H\backslash N) \cdot \rho^{(2)}_{\mathrm{top}}(M), \quad (3.126)$$

where M and N respectively are Riemannian manifolds without boundary and with cocompact free proper actions by isometries of the group G and H respectively. Restriction and induction also carry over in the obvious way. Given a closed Riemannian manifold M, $\rho^{(2)}_{\mathrm{top}}(\widetilde{M})$ has the obvious meaning and properties.

3.5 Analytic L^2-Torsion

In this section we introduce the analytic version of L^2-torsion and compute it for universal coverings of closed hyperbolic manifolds.

3.5.1 Definition of Analytic L^2-Torsion

The next definition of analytic L^2-torsion is taken from Lott [316] and Mathai [358]. The notion of analytic determinant class is due to Burghelea-Friedlander-Kappeler-McDonald [84, Definition 4.1 on page 800]. Recall the definition of the Γ-function as a holomorphic function

$$\Gamma(s) = \int_0^\infty t^{s-1} e^{-t}\, dt \quad (3.127)$$

for $\Re(s) > 0$, where $\Re(s)$ denotes the real part of the complex number s. It has a meromorphic extension to $\mathbb{C}$ whose set of poles is $\{n | n \in \mathbb{Z}, n \leq 0\}$ and which satisfies $\Gamma(s+1) = s \cdot \Gamma(s)$. All poles have order one. We have $\Gamma(z) \neq 0$ for all $z \in \mathbb{C} - \{0, -1, -2, \ldots\}$ and $\Gamma(n+1) = n!$ for $n \in \mathbb{Z}, n \geq 0$, where we use the standard convention $0! = 1$.

Definition 3.128 (Analytic L^2-torsion). *Let M be a cocompact free proper G-manifold without boundary and with G-invariant Riemannian metric. Define*

$$\theta_p(M)(t) := \int_{\mathcal{F}} \mathrm{tr}_{\mathbb{C}}(e^{-t\Delta_p}(x, x))\, d\mathrm{vol}, \quad (3.129)$$

where $e^{-t\Delta_p}(x, y)$ denotes the heat kernel associated to the Laplacian acting on p-forms and $\mathcal{F}$ is a fundamental domain for the G-action. We call M of analytic determinant class if for any $0 \leq p \leq \dim(M)$ and for some (and hence all) $\epsilon > 0$

$$\int_\epsilon^\infty t^{-1} \cdot \left(\theta_p(M)(t) - b_p^{(2)}(M) \right) \, dt < \infty.$$

In this case we define the analytic L^2-torsion *of M for any choice of $\epsilon > 0$ by*

$$\rho_{an}^{(2)}(M) := \frac{1}{2} \cdot \sum_{p \geq 0} (-1)^p \cdot p \cdot \left(\frac{d}{ds} \frac{1}{\Gamma(s)} \int_0^\epsilon t^{s-1} \cdot \left(\theta_p(M)(t) - b_p^{(2)}(M) \right) dt \Big|_{s=0} \right.$$
$$\left. + \int_\epsilon^\infty t^{-1} \cdot \left(\theta_p(M)(t) - b_p^{(2)}(M) \right) \, dt \right).$$

Some comments are necessary in order to show that this definition makes sense and how it is motivated.

The expression $\frac{d}{ds} \frac{1}{\Gamma(s)} \int_0^\epsilon t^{s-1} \cdot \left(\theta_p(M)(t) - b_p^{(2)}(M) \right) dt \Big|_{s=0}$ is to be understood in the sense that $\frac{1}{\Gamma(s)} \int_0^\epsilon t^{s-1} \cdot \left(\theta_p(M)(t) - b_p^{(2)}(M) \right) dt$ is holomorphic for $\Re(s) > \dim(M)/2$ and has a meromorphic extension to $\mathbb{C}$ with no pole in 0 [316, Section 3]. This fact is shown by comparing the heat kernel on M with the heat kernel on $G \backslash M$ [316, Lemma 4] and using the corresponding statement for $G \backslash M$.

Definition 3.128 is independent of the choice of ϵ by the following calculation. In the sequel we abbreviate

$$\theta_p^\perp(t) := \theta_p(M)(t) - b_p^{(2)}(M). \tag{3.130}$$

We compute for $0 < \epsilon \leq \delta$

$$\frac{d}{ds} \frac{1}{\Gamma(s)} \int_\epsilon^\delta t^{s-1} \cdot \theta_p^\perp(t) \, dt \Big|_{s=0}$$
$$= \frac{d}{ds} s \cdot \frac{1}{\Gamma(s+1)} \cdot \int_\epsilon^\delta t^{s-1} \cdot \theta_p^\perp(t) \, dt \Big|_{s=0}$$
$$= \frac{d}{ds} s \Big|_{s=0} \cdot \frac{1}{\Gamma(s+1)} \cdot \int_\epsilon^\delta t^{s-1} \cdot \theta_p^\perp(t) \, dt \Big|_{s=0}$$
$$+ 0 \cdot \frac{d}{ds} \frac{1}{\Gamma(s+1)} \cdot \int_\epsilon^\delta t^{s-1} \cdot \theta_p^\perp(t) \, dt \Big|_{s=0}$$
$$= \int_\epsilon^\delta t^{-1} \cdot \theta_p^\perp(t) \, dt. \tag{3.131}$$

We first explain the relation of the definition above to the classical Ray-Singer torsion (3.9) for a closed manifold M. We have for $s \in \mathbb{C}$ with $\Re(s) > 0$ and real number $\lambda > 0$

$$\frac{1}{\Gamma(s)} \cdot \int_0^\infty t^{s-1} e^{-\lambda t} \, dt = \frac{1}{\Gamma(s)} \cdot \int_0^\infty (\lambda^{-1} u)^{s-1} e^{-\lambda \cdot (\lambda^{-1} u)} \lambda^{-1} \, du$$

$$= \frac{1}{\Gamma(s)} \cdot \lambda^{-s} \cdot \int_0^\infty u^{s-1} e^{-u} \, du$$

$$= \lambda^{-s}. \tag{3.132}$$

From (3.132) we get in the setting of Ray-Singer torsion the following equation, where λ runs over the eigenvalues of the Laplace operator in dimension p listed with multiplicity.

$$\sum_{\lambda > 0} \lambda^{-s} = \sum_{\lambda > 0} \frac{1}{\Gamma(s)} \cdot \int_0^\infty t^{s-1} e^{-t\lambda} \, dt$$

$$= \frac{1}{\Gamma(s)} \cdot \int_0^\infty t^{s-1} \cdot \sum_{\lambda > 0} e^{-\lambda t} \, dt$$

$$= \frac{1}{\Gamma(s)} \cdot \int_0^\infty t^{s-1} \cdot \left(\mathrm{tr}_{\mathbb{C}} \left(e^{-t\Delta_p} \right) - \dim_{\mathbb{C}}(H_p(M; V)) \right) \, dt. \tag{3.133}$$

Hence we can rewrite the analytic Ray-Singer torsion (3.9), where $\theta_p^\perp$ is $\mathrm{tr}_{\mathbb{C}} \left(e^{-t\Delta_p} \right) - \dim_{\mathbb{C}}(H_p(M; V))$

$$\rho_{\mathrm{an}}(M; V) := \frac{1}{2} \cdot \sum_{p \geq 0} (-1)^p \cdot p \cdot \frac{d}{ds} \frac{1}{\Gamma(s)} \cdot \int_0^\infty t^{s-1} \cdot \theta_p^\perp \, dt \Big|_{s=0}. \tag{3.134}$$

Notice that in the setting of Ray-Singer torsion one has only to deal with convergence problems connected with the asymptotics of the large eigenvalues of the Laplacian, whereas in the L^2-setting there is an additional convergence problem connected with small eigenvalues and which causes us to require the condition to be of determinant class. Therefore one cannot define the L^2-torsion directly using expression (3.134) because there is no guarantee that $\theta_p^\perp$ decays fast enough in the L^2-setting. In the Ray-Singer situation or, more generally, under the assumption that the Laplacian Δ_p has a gap in the spectrum at zero, $\theta_p^\perp$ decays exponentially as we will see in Lemma 3.139 (5). If $\theta_p^\perp$ decays exponentially, then (3.131) is also true for a real number $\epsilon > 0$ and $\delta = \infty$. Hence we can rewrite in the setting of Ray-Singer torsion

$$\rho_{\mathrm{an}}(M; V) := \frac{1}{2} \cdot \sum_{p \geq 0} (-1)^p \cdot p \cdot \left(\frac{d}{ds} \frac{1}{\Gamma(s)} \int_0^\epsilon t^{s-1} \cdot \theta_p^\perp(t) \, dt \Big|_{s=0} \right.$$

$$\left. + \int_\epsilon^\infty t^{-1} \cdot \theta_p(t)^\perp \, dt \right). \tag{3.135}$$

This expression (3.135) for the analytic Ray-Singer torsion is the one which can be extended to the L^2-setting.

Next we want to show

Theorem 3.136. *Let M be a cocompact free proper G-manifold without boundary with invariant Riemannian metric. Then*

(1) We have
$$b_p^{(2)}(M) = F(0) = \lim_{t \to \infty} \theta_p(M)(t);$$

(2) M is of analytic determinant class in the sense of Definition 3.128 if and only if it is of determinant class in the sense of Definition 3.120;

(3) The following statements are equivalent:
   (a) The Laplace operator Δ_p has a gap in its spectrum at zero;
   (b) $\alpha_p^\Delta(M) = \infty^+$;
   (c) There exists $\epsilon > 0$ such that $F_p^\Delta(M)(\lambda) = F_p^\Delta(M)(0)$ holds for $0 < \lambda < \epsilon$;
   (d) There exists $\epsilon > 0$ and a constant $C(\epsilon)$ such that $\theta_p(M)(t) \leq C(\epsilon) \cdot e^{-\epsilon t}$ holds for $t > 0$;

(4) Suppose that $\alpha_p^\Delta(M) \neq \infty^+$. Then
$$\alpha_p^\Delta(M) = \liminf_{t \to \infty} \frac{-\ln(\theta_p(M)(t) - b_p^{(2)}(M))}{\ln(t)}.$$

We will prove Theorem 3.136 in Subsection 3.5.2 after we have dealt with the Laplace transform of a density function. Theorem 3.136 (2) has already been proved in [84, Proposition 5.6 on page 815].

3.5.2 The Laplace Transform of a Density Function

Let $F : [0, \infty) \to [0, \infty)$ be a density function (see Definition 2.7). Its *Laplace transform* is defined by

$$\theta_F(t) = \int_0^\infty e^{-t\lambda} \, dF(\lambda). \tag{3.137}$$

This definition is motivated by

Lemma 3.138. *Let M be a cocompact free proper G-manifold without boundary with invariant Riemannian metric. Then the Laplace transform in the sense of (3.137) of the spectral density function $F_p^\Delta(M)$ (see Definition 2.64) is $\theta_p(M)$ defined in (3.129).*

Proof. We get from (1.65) and [9, Proposition 4.16 on page 63]

$$\theta_{F_p^\Delta}(t) = \int_0^\infty e^{-t\lambda} \, dF_p^\Delta(\lambda)$$
$$= \operatorname{tr}_{\mathcal{N}(G)}\left(e^{-t(\Delta_p)_{\min}}\right)$$
$$= \theta_p(M). \quad \square$$

The asymptotic behaviour of the Laplace transform $\theta_F(t)$ for $t \to \infty$ can be read of from the asymptotic behaviour of $F(\lambda)$ for $\lambda \to 0$ and vice versa as explained in the next lemma (cf. [240, Appendix])

Lemma 3.139. *Let* $F: [0, \infty) \to [0, \infty)$ *be a density function with Laplace transform* $\theta_F(t)$. *Then*

(1) *We have for* $\lambda > 0$

$$F(\lambda) \leq \theta_F(t) \cdot e^{t\lambda};$$

(2) *Suppose that for all* $t > 0$ *there is a constant* $C(t)$ *such that* $F(\lambda) \leq C(t) \cdot e^{t\lambda}$ *holds for all* $\lambda \geq 0$. *Then we get for* $t > 0$

$$\theta_F(t) = t \cdot \int_0^\infty e^{-t\lambda} \cdot F(\lambda) \, d\lambda;$$

(3) *The Laplace transform* $\theta_F(t)$ *is finite for all* $t > 0$ *if and only if for all* $t > 0$ *there is a constant* $C(t)$ *such that* $F(\lambda) \leq C(t) \cdot e^{t\lambda}$ *holds for all* $\lambda \geq 0$;

(4) *Suppose that* $\theta_F(t) < \infty$ *for all* $t > 0$. *Then*

$$F(0) = \lim_{t \to \infty} \theta_F(t).$$

(5) *Suppose that* $\theta_F(t) < \infty$ *for all* $t > 0$. *Let* $\epsilon > 0$ *be a real number. Then* $F(\lambda) = F(0)$ *for all* $\lambda < \epsilon$ *if and only if there is a constant* $C(\epsilon)$ *such that* $\theta_F(t) - F(0) \leq C(\epsilon) \cdot e^{-\epsilon t}$ *holds for* $t > 0$;

(6) *Suppose that* $F(\lambda) > F(0)$ *for all* $\lambda > 0$ *and that* $\theta_F(t) < \infty$ *for all* $t > 0$. *Then*

$$\liminf_{\lambda \to 0} \frac{F(\lambda) - F(0)}{\ln(\lambda)} = \liminf_{t \to \infty} \frac{-\ln(\theta_F(t) - F(0))}{\ln(t)};$$

(7) *Suppose that* F *and* G *are dilationally equivalent density functions. Then* $\int_{0+}^1 \ln(\lambda) \, dF(\lambda) > -\infty$ *if and only if* $\int_{0+}^1 \ln(\lambda) \, dG(\lambda) > -\infty$;

(8) *Suppose that* $\theta_F(t) < \infty$ *for all* $t > 0$. *Then* $\int_{0+}^1 \ln(\lambda) \, dF(\lambda) > -\infty$ *if and only if* $\int_1^\infty t^{-1} \cdot (\theta_F(t) - F(0)) \, dt < \infty$;

Proof. (1) Integration by part (see (3.16)) yields for $0 < \epsilon < K < \infty$

$$\int_{\epsilon+}^K e^{-t\lambda} \, dF(\lambda) = e^{-tK} \cdot F(K) - e^{-t\epsilon} \cdot F(\epsilon) + t \cdot \int_\epsilon^K e^{-t\lambda} \cdot F(\lambda) \, d\lambda.$$

Since F is right continuous, Levi's Theorem of Monotone Convergence implies

$$\int_{0+}^K e^{-t\lambda} \, dF(\lambda) = e^{-tK} \cdot F(K) - F(0) + t \cdot \int_0^K e^{-t\lambda} \cdot F(\lambda) \, d\lambda, \quad (3.140)$$

and hence

$$\int_0^K e^{-t\lambda} \, dF(\lambda) = e^{-tK} \cdot F(K) + t \cdot \int_0^K e^{-t\lambda} \cdot F(\lambda) \, d\lambda. \quad (3.141)$$

We conclude $F(\lambda) \le \theta_F(t) \cdot e^{t\lambda}$ for all $t, \lambda > 0$ from (3.141).

(2) By assumption we have $F(\lambda) \le C(t/2) \cdot e^{-t/2 \cdot \lambda}$. This implies for $t > 0$

$$\lim_{\lambda \to \infty} e^{-t\lambda} \cdot F(\lambda) = 0. \tag{3.142}$$

Now assertion (2) follows from (3.141), (3.142) and Levi's Theorem of Monotone Convergence.

(3) Suppose that $F(\lambda) \le C(t/2) \cdot e^{-t/2 \cdot \lambda}$. Then $\theta_F(t) < \infty$ follows from the following estimate based on assertion (2).

$$\theta_F(t) = t \cdot \int_0^\infty e^{-t\lambda} \cdot F(\lambda) \, d\lambda$$

$$\le t \cdot \int_0^\infty e^{-t\lambda} \cdot C(t/2) \cdot e^{t/2 \cdot \lambda} \, d\lambda$$

$$= 2 \cdot C(t/2) \cdot \int_0^\infty t/2 \cdot e^{-t/2 \cdot \lambda} \, d\lambda$$

$$= 2 \cdot C(t/2).$$

The other implication follows from assertion (1).

(4) This follows from Lebesgue's Theorem of Majorized Convergence.

(5) Suppose that $F(\epsilon) = F(0)$. Consider $\lambda < \epsilon$. We conclude from assertions (1) and (2) for $t > 2$

$$\theta_F(t) - F(0) = t \cdot \int_0^\infty e^{-t\lambda} \cdot (F(\lambda) - F(0)) \, d\lambda$$

$$= t \cdot \int_\epsilon^\infty e^{-t\lambda} \cdot (F(\lambda) - F(0)) \, d\lambda$$

$$\le t \cdot \int_\epsilon^\infty e^{-t\lambda} \cdot F(\lambda) \, d\lambda$$

$$\le t \cdot \int_\epsilon^\infty e^{-t\lambda} \cdot \theta_F(1) \cdot e^\lambda \, d\lambda$$

$$= \theta_F(1) \cdot t \cdot \int_\epsilon^\infty e^{(-t+1)\lambda} \, d\lambda$$

$$= \theta_F(1) \cdot \frac{t}{t-1} \cdot e^{(-t+1)\epsilon}$$

$$\le \theta_F(1) \cdot 2 \cdot e^\epsilon \cdot e^{-t\epsilon}.$$

This implies the existence of a constant $C(\epsilon) > 0$ such that $\theta_F(t) - F(0) \le C(\epsilon) \cdot e^{-t\epsilon}$ holds for all $t > 0$.

Now suppose that $\theta_F(t) - F(0) \le C(\epsilon) \cdot e^{-t\epsilon}$ holds for $t > 0$. Fix $\mu \in (0, 1)$. Then we get for $t > 0$

$$(F(\mu \cdot \epsilon) - F(0)) \cdot \left(e^{(1-\mu)\epsilon t} - 1\right) \cdot e^{-\epsilon t} = (F(\mu \cdot \epsilon) - F(0)) \cdot \left(e^{-\mu \epsilon t} - e^{-\epsilon t}\right)$$

$$= t \cdot \int_{\mu\epsilon}^{\epsilon} (F(\mu \cdot \epsilon) - F(0)) \cdot e^{-\lambda t} \, d\lambda$$

$$\leq t \cdot \int_{\mu\epsilon}^{\epsilon} (F(\lambda) - F(0)) \cdot e^{-\lambda t} \, d\lambda$$

$$\leq t \cdot \int_{0}^{\infty} (F(\lambda) - F(0)) \cdot e^{-\lambda t} \, d\lambda$$

$$= \left(t \cdot \int_{0}^{\infty} F(\lambda) \cdot e^{-\lambda t} \, d\lambda \right) - F(0).$$

We conclude from assertion (2)

$$(F(\mu \cdot \epsilon) - F(0)) \cdot \left(e^{(1-\mu)\epsilon t} - 1\right) \cdot e^{-\epsilon t} \leq \theta_F(t) - F(0)$$

$$\leq C(\epsilon) \cdot e^{-t\epsilon}.$$

This implies for all $t > 0$

$$(F(\mu \cdot \epsilon) - F(0)) \cdot \left(e^{(1-\mu)\epsilon t} - 1\right) \leq C(\epsilon).$$

Hence $F(\mu \cdot \epsilon) - F(0) = 0$ for all $\mu \in (0,1)$ and assertion (5) is proved.

(6) We can assume without loss of generality that $F(0) = 0$, otherwise consider the new density function $F(\lambda) - F(0)$. We first show

$$\liminf_{\lambda \to 0} \frac{\ln(F(\lambda))}{\ln(\lambda)} \leq \liminf_{t \to \infty} \frac{-\ln(\theta_F(t))}{\ln(t)}. \tag{3.143}$$

Obviously it suffices to treat the case, where the left-hand side is different from zero. Consider $0 < \alpha$ such that $\alpha < \liminf_{\lambda \to 0} \frac{\ln(F(\lambda))}{\ln(\lambda)}$. Then there is $\epsilon > 0$ such that $F(\lambda) \leq \lambda^{\alpha}$ for all $\lambda \in (0, \epsilon)$. We get from assertions (1) and (2) and (3.132) for $t > 1$

$$\theta_F(t) = t \cdot \int_{0}^{\epsilon} e^{-\lambda t} \cdot F(\lambda) \, d\lambda + t \cdot \int_{\epsilon}^{\infty} e^{-\lambda t} \cdot F(\lambda) \, d\lambda$$

$$\leq t \cdot \int_{0}^{\epsilon} e^{-\lambda t} \cdot \lambda^{\alpha} \, d\lambda + t \cdot \int_{\epsilon}^{\infty} e^{-\lambda t} \cdot \theta_F(1) \cdot e^{\lambda} \, d\lambda$$

$$\leq t \cdot \int_{0}^{\infty} e^{-\lambda t} \cdot \lambda^{\alpha} \, d\lambda + \theta_F(1) \cdot t \cdot \int_{\epsilon}^{\infty} e^{\lambda \cdot (-t+1)} \, d\lambda$$

$$= \Gamma(\alpha + 1) \cdot t^{-\alpha} + \theta_F(1) \cdot \frac{t}{t-1} \cdot e^{\epsilon \cdot (-t+1)}.$$

We conclude $\alpha \leq \liminf_{t \to \infty} \frac{-\ln(\theta_F(t))}{\ln(t)}$. This proves (3.143). It remains to show

$$\liminf_{\lambda \to 0} \frac{\ln(F(\lambda))}{\ln(\lambda)} \geq \liminf_{t \to \infty} \frac{-\ln(\theta_F(t))}{\ln(t)}. \tag{3.144}$$

Obviously it suffices to treat the case, where the right-hand side is greater than zero. Fix $0 < \alpha$ satisfying $\alpha < \liminf_{t \to \infty} \frac{-\ln(\theta_F(t))}{\ln(t)}$. Then we can find $K > 0$ such that $\theta_F(t) \leq t^{-\alpha}$ for $t > K$ holds. We conclude from assertion (1) for $t > K$

$$F(\lambda) \leq e^{t\lambda} \cdot \theta_F(t) \leq e^{t\lambda} \cdot t^{-\alpha}.$$

If we take $t = \lambda^{-1}$, we get $F(\lambda) \leq e \cdot \lambda^{\alpha}$. This implies

$$\alpha \leq \liminf_{\lambda \to 0} \frac{\ln(F(\lambda))}{\ln(\lambda)}.$$

This finishes the proof of (3.144) and hence of assertion (6).

(7) We conclude from Lemma 3.15 (1) that $\int_{0+}^{1} \ln(\lambda) \, dF(\lambda) > -\infty$ is equivalent to $\int_{0+}^{1} \frac{F(\lambda) - F(0)}{\lambda} \, d\lambda < \infty$ and analogous for G.

(8) Obviously the claim is true for $F - F(0)$ if and only if it is true for F. Hence we can assume without loss of generality that $F(0) = 0$ in the sequel. Assertion (2) implies

$$\int_{1}^{\infty} t^{-1} \cdot \theta_F(t) \, dt = \int_{1}^{\infty} \left(\int_{0}^{\infty} e^{-t\lambda} \cdot F(\lambda) \, d\lambda \right) dt. \tag{3.145}$$

We get

$$\int_{1}^{\infty} \left(\int_{1}^{\infty} e^{-t\lambda} \cdot F(\lambda) \, d\lambda \right) dt < \infty \tag{3.146}$$

from the inequality $F(\lambda) \leq C(1) \cdot e^{\lambda}$ of assertion (1). Using the inequality $F(\lambda) \leq C(t) \cdot e^{t \cdot \lambda}$ of assertion (1) and the standard results about commuting differentiation and integration we conclude

$$\frac{d}{dt} \int_{0}^{1} \frac{e^{-t\lambda}}{-\lambda} \cdot F(\lambda) \, d\lambda = \int_{0}^{1} e^{-t\lambda} \cdot F(\lambda) \, d\lambda.$$

This implies using Lebesgue's Theorem of Majorized Convergence and Levi's Theorem of Monotone Convergence

$$\int_{1}^{\infty} \left(\int_{0}^{1} e^{-t\lambda} \cdot F(\lambda) \, d\lambda \right) dt$$

$$= \lim_{K \to \infty} \int_{1}^{K} \left(\int_{0}^{1} e^{-t\lambda} \cdot F(\lambda) \, d\lambda \right) dt$$

$$= \lim_{K \to \infty} \int_{0}^{1} \frac{e^{-K\lambda}}{-\lambda} \cdot F(\lambda) \, d\lambda - \int_{0}^{1} \frac{e^{-1\lambda}}{-\lambda} \cdot F(\lambda) \, d\lambda$$

$$= \int_{0}^{1} \lim_{K \to \infty} \frac{e^{-K\lambda}}{-\lambda} \cdot F(\lambda) \, d\lambda - \int_{0}^{1} \frac{e^{-\lambda}}{-\lambda} \cdot F(\lambda) \, d\lambda$$

$$= \int_{0}^{1} \frac{e^{-\lambda}}{\lambda} \cdot F(\lambda) \, d\lambda. \tag{3.147}$$

We have

$$e^{-1} \cdot \int_0^1 \frac{F(\lambda)}{\lambda} \, d\lambda \ \leq \ \int_0^1 \frac{e^{-\lambda}}{\lambda} \cdot F(\lambda) \, d\lambda \ \leq \ \int_0^1 \frac{F(\lambda)}{\lambda} \, d\lambda. \qquad (3.148)$$

We conclude from Lemma 3.15 (1), (3.145), (3.146), (3.147) and (3.148) that $\int_1^\infty t^{-1} \cdot \theta_F(t) \, dt < \infty$ is equivalent to $\int_{0+}^1 \ln(\lambda) \, dF(\lambda) > -\infty$. This finishes the proof of Lemma 3.139.   □

Now Theorem 3.136 follows from Theorem 2.68, Lemma 3.138 and Lemma 3.139.

3.5.3 Comparison of Topological and Analytic L^2-Torsion

Next we cite the deep result of Burghelea, Friedlander, Kappeler and McDonald [84] that the topological and analytic L^2-torsion $\rho_{\mathrm{top}}^{(2)}(M)$ (see Definition 3.120) and $\rho_{\mathrm{an}}^{(2)}(M)$ (see Definition 3.128) agree. The main idea is to perform the Witten deformation of the Laplacian with a suitable Morse function and investigate the splitting of the de Rham complex according to small and large eigenvalues. We do not give the long and complicated proof here. The main technical tools are asymptotic expansions and a Mayer-Vietoris type formula for determinants (see [79], [80]). A survey is given in [81]. See also [83].

Theorem 3.149 (Analytic and topological L^2-torsion). *Let M be a cocompact free proper G-manifold without boundary and with G-invariant Riemannian metric. Suppose that M is of analytic determinant class in the sense of Definition 3.128, or equivalently, of determinant class in the sense of Definition 3.120 (see Theorem 3.136 (2)). Then*

$$\rho_{\mathrm{top}}^{(2)}(M) = \rho_{\mathrm{an}}^{(2)}(M).$$

3.5.4 Analytic L^2-Torsion for Hyperbolic Manifolds

In general it is easier to work with topological L^2-torsion because it has nice properties as stated for instance in Theorem 3.96, Theorem 3.100 and Theorem 3.106 and it can be computed combinatorially without investigating the spectral density functions of certain operators (see Theorem 3.172). However, there are some special cases where it is easier to deal with analytic L^2-torsion because the Riemannian metrics have special properties and there is a proportionality principle (see Theorem 3.183). Examples are hyperbolic manifolds which are treated in the next theorem due to Hess and Schick [254], the special case of dimension 3 is proved in [316, Proposition 16] and [358, Corollary 6.7]. Recall that the L^2-torsion of the universal covering of a closed even-dimensional Riemannian manifold is always trivial so that only the odd-dimensional case is interesting.

Consider the polynomial with integer coefficients for $j \in \{0, 1, 2, \ldots, n-1\}$

$$P_j^n(\nu) := \frac{\prod_{i=0}^n (\nu^2 + i^2)}{\nu^2 + (n-j)^2} = \sum_{k \geq 0}^{2n} K_{k,j}^n \cdot \nu^{2k}. \tag{3.150}$$

Define

$$C_d := \sum_{j=0}^{n-1} (-1)^{n+j+1} \frac{n!}{(2n)! \cdot \pi^n} \cdot \binom{2n}{j}$$

$$\cdot \sum_{k=0}^n K_{k,j}^n \cdot \frac{(-1)^{k+1}}{2k+1} \cdot (n-j)^{2k+1}. \tag{3.151}$$

The first values of C_d are computed in [254, Theorem 2]

$$C_3 = \tfrac{1}{6\pi} \approx 0.05305;$$

$$C_5 = \tfrac{31}{45\pi^2} \approx 0.06980;$$

$$C_7 = \tfrac{221}{70\pi^3} \approx 0.10182;$$

$$C_{39} \approx 2.4026 \cdot 10^7,$$

and the constants C_d are positive, strictly increasing and grow very fast, namely they satisfy [254, Proposition 6]

$$C_{2n+1} \geq \frac{n}{2\pi} \cdot C_{2n-1};$$

$$C_{2n+1} \geq \frac{2n!}{3(2\pi)^n}.$$

Theorem 3.152 (Analytic L^2-torsion of hyperbolic manifolds).
Let M be a closed hyperbolic d-dimensional manifold for odd $d = 2n + 1$. Let $C_d > 0$ be the constant introduced in (3.151). It can be written as $C_d = \pi^{-n} \cdot r_d$ for a rational number $r_d > 0$. Then

$$\rho^{(2)}(\widetilde{M}) = (-1)^n \cdot C_d \cdot \mathrm{vol}(M).$$

Proof. In the sequel we abbreviate

$$D := \frac{(4\pi)^{-n-1/2}}{\Gamma(n+1/2)} = \frac{n!}{2 \cdot \pi^{n+1} \cdot (2n)!}.$$

Notice that $\widetilde{M}$ can be identified with the hyperbolic space $\mathbb{H}^d$ since M is by assumption hyperbolic and that there are no harmonic L^2-integrable p-forms on $\mathbb{H}^d$ for $p \geq 0$ by Theorem 1.62 Let $\Delta_j \colon d^{j-1}(\Omega_c^{j-1}(\mathbb{H}^d))^\perp \to d^{j-1}(\Omega_c^{j-1}(\mathbb{H}^d))^\perp$ be the operator induced by the Laplacian $\Delta_j \colon L^2\Omega^j(\mathbb{H}^d) \to L^2\Omega^j(\mathbb{H}^d)$ (see Theorem 1.57) and $e^{t\Delta_j}(x,x)$ be the kernel of the operator $e^{t\Delta_j}$. Recall that $\mathbb{H}^d$ is homogeneous, i.e. for two points x and y in $\mathbb{H}^d$ there is an isometry $\mathbb{H}^d \to \mathbb{H}^d$ mapping x to y. Hence $\operatorname{tr}_{\mathbb{C}}(e^{t\Delta_j}|(x,x))$ is independent of $x \in \mathbb{H}^d$. Put

$$
L_j = \frac{d}{ds} \frac{1}{\Gamma(s)} \int_0^1 t^{s-1} \cdot \operatorname{tr}_{\mathbb{C}}(e^{t\Delta_j}|(x,x)) \, dt \bigg|_{s=0}
$$
$$
+ \int_1^\infty t^{-1} \cdot \operatorname{tr}_{\mathbb{C}}(e^{t\Delta_j}|(x,x)) \, dt,
$$

where the first summand is to be understood as before, namely it is a holomorphic function for large $\Re(s)$ and has a meromorphic extension to $\mathbb{C}$ without pole in 0. Now a calculation analogous to the proof of Lemma 3.30 and using the Hodge-de Rham decomposition (see Theorem 1.57) and the fact that the Laplacian is compatible with the Hodge star-operator (1.48) shows

$$
\frac{\rho_{\mathrm{an}}^{(2)}(M)}{\mathrm{vol}(M)} = \sum_{j=0}^n (-1)^{j+1} L_j. \tag{3.153}
$$

The following equality is taken from [316, Proposition 15] where it is derived as a special case of [371]

$$
\operatorname{tr}_{\mathbb{C}}(e^{t\Delta_j}|(x,x)) = D \cdot \binom{2n}{j} \cdot \int_{-\infty}^\infty e^{-t(\nu^2 + (n-j)^2)} \cdot P_j^n(\nu) \, d\nu,
$$

where the polynomial $P_j^n(\nu)$ and its coefficients $K_{k,j}^n$ are defined in (3.150). We compute

$$
\operatorname{tr}_{\mathbb{C}}(e^{t\Delta_j}|(x,x)) = D \cdot \binom{2n}{j} \cdot \sum_{k=0}^n K_{k,j}^n e^{-t(n-j)^2} \cdot t^{-k-1/2} \cdot \Gamma(k+1/2).
$$

Hence we can write

$$
L_j = D \cdot \binom{2n}{j} \cdot \sum_{k=0}^n K_{k,j}^n \cdot \Gamma(k+1/2) \cdot
$$
$$
\left(\frac{d}{ds} \frac{1}{\Gamma(s)} \int_0^1 e^{-t(n-j)^2} \cdot t^{s-k-3/2} \, dt \bigg|_{s=0} \right.
$$
$$
\left. + \int_1^\infty e^{-t(n-j)^2} \cdot t^{-k-3/2} \, dt \right). \tag{3.154}
$$

A computation analogous to the one in (3.131) shows for $(n-j) > 0$

$$
\frac{d}{ds} \frac{1}{\Gamma(s)} \int_1^\infty e^{-t(n-j)^2} \cdot t^{s-k-3/2} \, dt \bigg|_{s=0}
$$
$$
= \int_1^\infty e^{-t(n-j)^2} \cdot t^{-k-3/2} \, dt. \tag{3.155}
$$

One easily checks for $j < n$

$$\frac{d}{ds}\frac{1}{\Gamma(s)}\int_0^\infty e^{-t(n-j)^2}\cdot t^{s-k-3/2}\ dt\bigg|_{s=0}$$

$$= \frac{d}{ds}\frac{s}{\Gamma(s+1)}\cdot(n-j)^{-2s+2k+1}\cdot\Gamma(s-k-1/2)\bigg|_{s=0}$$

$$= (n-j)^{2k+1}\cdot\Gamma(-k-1/2). \tag{3.156}$$

We conclude from (3.154), (3.155) and (3.156) for $j \neq n$

$$L_j = D\cdot\binom{2n}{j}\cdot\sum_{k=0}^n K_{k,j}^n\cdot\Gamma(k+1/2)\cdot\Gamma(-k-1/2)\cdot(n-j)^{2k+1}$$

$$= D\cdot\binom{2n}{j}\cdot\sum_{k=0}^n K_{k,j}^n\cdot(-1)^{k+1}\cdot\frac{2\pi}{2k+1}\cdot(n-j)^{2k+1}. \tag{3.157}$$

We get

$$L_n = 0 \tag{3.158}$$

from the following calculation for $\alpha > 0$

$$\frac{d}{ds}\frac{1}{\Gamma(s)}\int_0^1 t^{s-1-\alpha}\ dt\bigg|_{s=0} + \int_1^\infty t^{-1-\alpha}\ dt$$

$$= \frac{d}{ds}\frac{s}{\Gamma(s+1)}\cdot\frac{1}{s-\alpha}\bigg|_{s=0} + \frac{1}{\alpha}$$

$$= \frac{1}{-\alpha}+\frac{1}{\alpha}$$

$$= 0.$$

We get from (3.153), (3.157) and (3.158)

$$\rho_{an}^{(2)}(M) = (-1)^n\cdot C_d\cdot\text{vol}(M). \tag{3.159}$$

This finishes the proof of Theorem 3.152.   □

Theorem 3.152 will be extended to closed Riemannian manifolds whose universal coverings are symmetric spaces of non-compact type in Theorem 5.12.

3.6 L^2-Torsion of Manifolds with Boundary

In this section we discuss what happens for a cocompact free proper G-manifold M with G-invariant Riemannian metric whose boundary ∂M is the disjoint union of the (possibly empty) G-spaces $\partial_0 M$ and $\partial_1 M$. Choose an equivariant smooth triangulation $(K; K_0, K_1) \to (M; \partial_0 M, \partial_1 M)$. We have introduced the L^2-Hodge-de Rham isomorphism $A_K^p \colon \mathcal{H}_{(2)}^p(M, \partial_0 M) \xrightarrow{\cong} H_{(2)}^p(K, K_0)$ in Theorem 1.89. We define analogously to the case where the boundary is empty the notions of determinant class and of *topological L^2-torsion*

$$\rho_{\text{top}}^{(2)}(M, \partial_0 M) = \rho^{(2)}(K, K_0) - \sum_{p \geq 0} (-1)^p \cdot$$
$$\ln \left(\det \left(A_K^p \colon \mathcal{H}_{(2)}^p(M, \partial_0 M) \xrightarrow{\cong} H_{(2)}^p(K, K_0) \right) \right). \quad (3.160)$$

We define the notions of analytic determinant class and of *analytic L^2-torsion* $\rho_{\text{an}}^{(2)}(M, \partial_0 M)$ as in Definition 3.128 but now imposing Dirichlet boundary conditions on $\partial_0 M$ and Neumann boundary conditions on $\partial_1 M$, i.e. we use the Laplacian $\Delta_p \colon L^2 \Omega^p(M) \to L^2 \Omega^p(M)$ which is the closure of the operator $\Delta_p \colon \Omega_2^p(M) \to \Omega_c^p(M)$ where $\Omega_2^p(M) \subset \Omega_c^p(M)$ was introduced in 1.83. Analogously we obtain the notions of *topological Reidemeister torsion* $\rho_{\text{top}}(M, \partial_0 M; V)$ and of *analytic Reidemeister torsion or Ray-Singer torsion* $\rho_{\text{an}}(M, \partial_0 M; V)$. We say that the G-invariant Riemannian metric is a *product near the boundary*, if there are $\epsilon > 0$ and a G-invariant neighborhood U of ∂M in M together with an isometric G-diffeomorphism $\partial M \times [0, \epsilon) \to U$ inducing the identity on ∂M, where ∂M is equipped with the G-invariant Riemannian metric induced from M, $[0, \epsilon)$ with the standard Riemannian metric and $\partial M \times [0, \epsilon)$ with the product Riemannian metric.

Theorem 3.161 (L^2-torsion of manifolds with boundary). *Let M be a cocompact free proper G-manifold M with G-invariant Riemannian metric whose boundary ∂M is the disjoint union of the (possibly empty) G-spaces $\partial_0 M$ and $\partial_1 M$. Suppose that the G-invariant Riemannian metric is a product near the boundary. Suppose that $(M, \partial_0 M)$, $(M, \partial M)$ and ∂M are of determinant class. Then*

$$\rho_{\text{an}}^{(2)}(M, \partial_0 M) = \rho_{\text{top}}^{(2)}(M, \partial_0 M) + \frac{\ln(2)}{4} \cdot \chi(G \backslash \partial M);$$
$$\rho_{\text{an}}(G \backslash M, G \backslash \partial_0 M; V) = \rho_{\text{top}}(G \backslash M, G \backslash \partial_0 M; V)$$
$$+ \frac{\ln(2)}{4} \cdot \chi(G \backslash \partial M) \cdot \dim_{\mathbb{R}}(V).$$

Proof. The first equation is proved by Burghelea, Friedlander and Kappeler [82, Theorem 4.1 on page 34] and the second independently by Lück [327, Theorem 4.5 on page 266] and Vishik [505]. □

In the case of the analytic Ray-Singer-torsion the general boundary correction term (without the condition that the metric is a product near the boundary) is given in [126]. One would expect to get the same correction term in the L^2-case.

Remark 3.162. We have seen in the context of L^2-Betti numbers (see Theorem 1.89) and of Novikov-Shubin invariants (see (2.84)) that the combinatorial and analytic versions also agree if M has a boundary. In the context of L^2-torsion there are two main differences, there is a correction term involving the Euler characteristic of the boundary and the assumption that the Riemannian metric is a product near the boundary. The additional Euler characteristic term is related to the observation that the boundary ∂M is a zero set in M and hence on the analytical side certain integration processes on the double $M \cup_{\partial M} M$ cannot feel the boundary ∂M. On the other hand equivariant cells sitting in the boundary ∂M do contribute to the cellular chain complex of $M \cup_{\partial M} M$ and affect therefore the topological side. The condition that the metric is a product near the boundary cannot be dropped for both Reidemeister torsion and L^2-torsion because otherwise examples of Lück and Schick [346, Appendix A] based on [60] show that the formulas become wrong. The condition that the metric is a product near the boundary ensures that the double $M \cup_{\partial M} M$ inherits a G-invariant Riemannian metric and that one can rediscover the information about $(M, \partial M)$ by inspecting the manifold without boundary $M \cup_{\partial M} M$ but now with the obvious $G \times \mathbb{Z}/2$-action induced by the flip map (see also Section 2.6).

The necessity of the condition that the Riemannian metric is a product near the boundary also shows that the next result is rather delicate since one has to chop the manifold into compact pieces with boundary and the relevant comparison formulas and glueing formulas are not a priori true any more since this chopping cannot be done such that metrics are product metrics near the boundaries.

The next result is proved in [346, Theorem 0.5].

Theorem 3.163. L^2-torsion of hyperbolic manifolds with boundary.
Let $\overline{M}$ be a compact connected manifold with boundary of odd dimension $d = 2n + 1$ such that the interior M comes with a complete hyperbolic metric of finite volume. Then $\widetilde{\overline{M}}$ is of determinant class and

$$\rho_{\text{top}}^{(2)}(\widetilde{\overline{M}}) \;=\; \rho_{\text{an}}^{(2)}(\widetilde{M}) \;=\; (-1)^n \cdot C_d \cdot \text{vol}(M),$$

where $\rho_{\text{an}}^{(2)}(\widetilde{M})$ is defined as in the cocompact case (which makes sense since $\widetilde{M}$ is homogeneous and has some cocompact free proper action of some discrete group by isometries and M has finite volume) and $C_d > 0$ is the dimension constant of 3.151.

Poincaré duality still holds for manifolds with boundary provided that M is orientable (cf. Theorem 3.93 (3)) i.e.

$$\rho_{\text{an}}^{(2)}(M, \partial_0 M) = (-1)^{\dim(M)+1} \cdot \rho_{\text{an}}^{(2)}(M, \partial_1 M); \tag{3.164}$$

$$\rho_{\text{top}}^{(2)}(M, \partial_0 M) = (-1)^{\dim(M)+1} \cdot \rho_{\text{top}}^{(2)}(M, \partial_1 M), \tag{3.165}$$

and analogously for Reidemeister torsion. We mention the following glueing formula. Define $\rho^{(2)}(C^*)$ of a finite Hilbert $\mathcal{N}(G)$-cochain complex C^* of determinant class by

$$\rho^{(2)}(C^*) := -\frac{1}{2} \cdot \sum_{p \in \mathbb{Z}} (-1)^p \cdot p \cdot \ln(\det(\Delta_p))$$

$$= \sum_{p \in \mathbb{Z}} (-1)^p \cdot \ln(\det(c^p)). \tag{3.166}$$

Theorem 3.167 (Glueing formula for L^2-torsion). *Let M and N be cocompact free proper G-manifolds with G-invariant Riemannian metrics which are products near the boundary. Their boundaries come with decompositions $\partial M = \partial_0 M \coprod \partial_1 M \coprod \partial_2 M$ and $\partial N = \partial_0 N \coprod \partial_1 N \coprod \partial_2 N$. Let $f \colon \partial_2 M \to \partial_2 N$ be an isometric G-diffeomorphism. Let $M \cup_f N$ be the cocompact free proper G-manifold with G-invariant Riemannian metric obtained by glueing M and N together along f. Suppose that $(M, \partial_0 M)$, $(M, \partial M)$, $(N, \partial_0 N)$, $(N, \partial N)$, ∂M and ∂N are of determinant class. Then $(M \cup_f N, \partial_0 M \coprod \partial_0 N)$ is of determinant class. We obtain a long weakly exact sequence of finitely generated Hilbert $\mathcal{N}(G)$-modules*

$$\ldots \to \mathcal{H}^{p-1}_{(2)}(\partial_2 M) \to \mathcal{H}^p_{(2)}(M \cup_f N, \partial_0 M \coprod \partial_0 N)$$

$$\to \mathcal{H}^p_{(2)}(M, \partial_0 M) \oplus \mathcal{H}^p_{(2)}(N, \partial_0 N) \to \mathcal{H}^p_{(2)}(\partial_2 M) \to \ldots$$

where we use the Hilbert structures coming from the Riemannian metrics and the maps are given by comparing the corresponding weakly exact L^2-cohomology sequence associated to equivariant smooth triangulations with the various L^2-Hodge-de Rham isomorphisms. We view it as a weakly acyclic Hilbert $\mathcal{N}(G)$-cochain complex LHS^* with $\mathcal{H}^0_{(2)}(M \cup_f N, \partial_0 M \coprod \partial_0 N)$ in dimension zero. Then LHS^* is weakly $\det$-L^2-acyclic and

$$\rho^{(2)}_{\mathrm{top}}(M \cup_f N, \partial_0 M \coprod \partial_0 N) = \rho^{(2)}_{\mathrm{top}}(M, \partial_0 M) + \rho^{(2)}_{\mathrm{top}}(N, \partial_0 N) - \rho^{(2)}_{\mathrm{top}}(\partial_2 M)$$

$$- \rho^{(2)}(LHS^*);$$

$$\rho^{(2)}_{\mathrm{an}}(M \cup_f N, \partial_0 M \coprod \partial_0 N) = \rho^{(2)}_{\mathrm{an}}(M, \partial_0 M) + \rho^{(2)}_{\mathrm{an}}(N, \partial_0 N) - \rho^{(2)}_{\mathrm{an}}(\partial_2 M)$$

$$- \rho^{(2)}(LHS^*) - \frac{\ln(2)}{2} \cdot \chi(G \backslash \partial_2 M).$$

The analogous result is true for the topological Reidemeister torsion and analytic Reidemeister torsion.

Proof. The claim about the analytic L^2-torsion is proved in [82, Theorem 4.4 on page 67]. The claim for the topological L^2-torsion can also be derived from Theorem 3.35 (1) applied to the exact sequence of Hilbert $\mathcal{N}(G)$-chain complexes $0 \to C_*(\partial_2 M) \to C_*(M, \partial_0 M) \oplus C_*(N, \partial_0 N) \to$

$C_*(M \cup_f N, \partial_0 M \coprod \partial_0 N) \to 0$ and the fact that the L^2-torsion of the weakly exact long homology sequence is the negative of the torsion of the weakly exact long cohomology sequence because $\rho^{(2)}(C_*) = \rho^{(2)}((C_*)^*)$ holds for a finite Hilbert $\mathcal{N}(G)$-chain complex C_*. The claim for the analytic L^2-torsion is then a consequence of Theorem 3.161.

The claim for the topological and analytical Reidemeister torsion is proved in [327, Proposition 3.11 on page 290 and Proposition 5.9 on page 313]. $\square$

3.7 Combinatorial Computations of L^2-Invariants

In this section we want to give a more combinatorial approach to the L^2-invariants such as L^2-Betti numbers, Novikov-Shubin invariants and L^2-torsion. The point is that it is in general very hard to compute the spectral density function of some morphism of finitely generated Hilbert $\mathcal{N}(G)$-modules. However in the geometric situation these morphisms are induced by matrices over the integral group ring $\mathbb{Z}G$. We want to exploit this information to get an algorithm which produces a sequence of rational numbers which converges to the L^2-Betti number or the L^2-torsion and whose members are computable in an algorithmic way.

Let $A \in M(n, m, \mathbb{C}G)$ be a (n, m)-matrix over $\mathbb{C}G$. It induces by right multiplication a $\mathbb{C}G$-homomorphism of left $\mathbb{C}G$-modules

$$R_A \colon \bigoplus_{i=1}^{n} \mathbb{C}G \to \bigoplus_{i=1}^{m} \mathbb{C}G, \qquad x \mapsto xA$$

and by completion a bounded G-equivariant operator

$$R_A^{(2)} \colon \bigoplus_{i=1}^{n} l^2(G) \to \bigoplus_{i=1}^{m} l^2(G).$$

Notice for the sequel that $R_{AB} = R_B \circ R_A$ holds. We define an involution of rings on $\mathbb{C}G$ by

$$\overline{\sum_{g \in G} \lambda_g \cdot g} := \sum_{g \in G} \overline{\lambda_g} \cdot g^{-1}, \tag{3.168}$$

where $\overline{\lambda_g}$ is the complex conjugate of λ_g. Denote by A^* the (m, n)-matrix obtained from A by transposing and applying the involution above to each entry. As the notation suggests, the bounded G-equivariant operator $R_{A^*}^{(2)}$ is the adjoint $(R_A^{(2)})^*$ of the bounded G-equivariant operator $R_A^{(2)}$. Define the $\mathbb{C}G$-trace of an element $u = \sum_{g \in G} \lambda_g \cdot g \in \mathbb{C}G$ by

$$\mathrm{tr}_{\mathbb{C}G}(u) := \lambda_e \in \mathbb{C} \tag{3.169}$$

for e the unit element in G. This extends to a square (n, n)-matrix A over $\mathbb{C}G$ by

$$\mathrm{tr}_{\mathbb{C}G}(A) := \sum_{i=1}^{n} \mathrm{tr}_{\mathbb{C}G}(a_{i,i}). \tag{3.170}$$

We get directly from the definitions that the $\mathbb{C}G$-trace $\mathrm{tr}_{\mathbb{C}G}(A)$ agrees with the von Neumann trace $\mathrm{tr}_{\mathcal{N}(G)}(R_A^{(2)})$ introduced in Definition 1.2.

Let $A \in M(n, m, \mathbb{C}G)$ be a (n, m)-matrix over $\mathbb{C}G$. In the sequel let K be any positive real number satisfying

$$K \geq ||R_A^{(2)}||_\infty,$$

where $||R_A^{(2)}||_\infty$ is the operator norm of the bounded G-equivariant operator $R_A^{(2)}$. For $u = \sum_{g \in G} \lambda_g \cdot g \in \mathbb{C}G$ define $||u||_1$ by $\sum_{g \in G} |\lambda_g|$. Then a possible choice for K is given by:

$$K = \sqrt{(2n-1)m} \cdot \max \{||a_{i,j}||_1 \mid 1 \leq i \leq n, 1 \leq j \leq m\}.$$

The bounded G-equivariant operator $1 - K^{-2} \cdot (R_A^{(2)})^* R_A^{(2)} : \bigoplus_{i=1}^{n} l^2(G) \to \bigoplus_{i=1}^{n} l^2(G)$ is positive. Let $\left(1 - K^{-2} \cdot AA^*\right)^p$ be the p-fold product of matrices and let $\left(1 - K^{-2} \cdot (R_A^{(2)})^* R_A^{(2)}\right)^p$ be the p-fold composition of operators.

Definition 3.171. *The* characteristic sequence *of a matrix $A \in M(n, m, \mathbb{C}G)$ and a non-negative real number K satisfying $K \geq ||R_A^{(2)}||_\infty$ is the sequence of real numbers*

$$c(A, K)_p := \mathrm{tr}_{\mathbb{C}G}\left(\left(1 - K^{-2} \cdot AA^*\right)^p\right) = \mathrm{tr}_{\mathcal{N}(G)}\left(\left(1 - K^{-2} \cdot (R_A^{(2)})^* R_A^{(2)}\right)^p\right).$$

We have defined $b^{(2)}(R_A^{(2)})$ in Definition 2.1 and $\det(R_A^{(2)})$ in Definition 3.11.

Theorem 3.172. (Combinatorial computation of L^2-invariants).
Let $A \in M(n, m, \mathbb{C}G)$ be a (n, m)-matrix over $\mathbb{C}G$. Denote by F the spectral density function of $R_A^{(2)}$. Let K be a positive real number satisfying $K \geq ||R_A^{(2)}||_\infty$. Then

(1) The characteristic sequence $c(A, K)_p$ is a monotone decreasing sequence of non-negative real numbers;
(2) We have
$$b^{(2)}(R_A^{(2)}) = \lim_{p \to \infty} c(A, K)_p;$$

(3) Define $\beta(A) \in [0, \infty]$ by

$$\beta(A) := \sup\left\{ \beta \in [0, \infty) \mid \lim_{p \to \infty} p^\beta \cdot \left(c(A, K)_p - b^{(2)}(R_A^{(2)}) \right) = 0 \right\}.$$

If $\alpha(R_A^{(2)}) < \infty$, then $\alpha(R_A^{(2)}) \leq \beta(A)$ and if $\alpha(R_A^{(2)}) \in \{\infty, \infty^+\}$, then $\beta(A) = \infty$;

(4) The sum of positive real numbers

$$\sum_{p=1}^{\infty} \frac{1}{p} \cdot \left(c(A, K)_p - b^{(2)}(R_A^{(2)}) \right)$$

converges if and only if $R_A^{(2)}$ is of determinant class and in this case

$$2 \cdot \ln(\det(R_A^{(2)})) = 2 \cdot (n - b^{(2)}(R_A^{(2)})) \cdot \ln(K) - \sum_{p=1}^{\infty} \frac{1}{p} \cdot \left(c(A, K)_p - b^{(2)}(R_A^{(2)}) \right);$$

(5) Suppose $\alpha(R_A^{(2)}) > 0$. Then $R_A^{(2)}$ is of determinant class. Given a real number α satisfying $0 < \alpha < \alpha(R_A^{(2)})$, there is a real number C such that we have for all $L \geq 1$

$$0 \leq c(A, K)_L - b^{(2)}(R_A^{(2)}) \leq \frac{C}{L^\alpha}$$

and

$$0 \leq -2 \cdot \ln(\det(R_A^{(2)})) + 2 \cdot (n - b^{(2)}(R_A^{(2)})) \cdot \ln(K)$$

$$- \sum_{p=1}^{L} \frac{1}{p} \cdot \left(c(A, K)_p - b^{(2)}(R_A^{(2)}) \right) \leq \frac{C}{L^\alpha}.$$

Remark 3.173. Before we give the proof of Theorem 3.172, we discuss its meaning. Let X be a finite free G-CW-complex. Describe the p-th differential $c_p \colon C_p(X) \to C_{p-1}(X)$ of its cellular $\mathbb{Z}G$-chain complex with respect to a cellular basis by the matrix $A_p \in M(n_p, n_{p-1}, \mathbb{Z}G)$. Then $R_{A_p}^{(2)}$ is just the p-th differential $c_p^{(2)}$ of the cellular Hilbert $\mathcal{N}(G)$-chain complex $C_*^{(2)}(X)$ and for the p-th Laplace operator $\Delta_p \colon C_p^{(2)}(X) \to C_p^{(2)}(X)$ of (1.17) we get $R_{A_p A_p^* + A_{p+1}^* A_{p+1}}^{(2)}$. This implies

$$b_p^{(2)}(X) = b^{(2)}(R_{A_p A_p^* + A_{p+1}^* A_{p+1}}^{(2)}); \tag{3.174}$$

$$\alpha_p(X) = \alpha(R_{A_p}^{(2)}); \tag{3.175}$$

$$\alpha_p^{\Delta}(X) = \alpha(R_{A_p A_p^* + A_{p+1}^* A_{p+1}}^{(2)}); \tag{3.176}$$

$$\rho^{(2)}(C_*^{(2)}(X)) = -\sum_{p \geq 0}(-1)^p \cdot \det(R_{A_p}) \tag{3.177}$$

$$= -\frac{1}{2} \cdot \sum_{p \geq 0}(-1)^p \cdot p \cdot \det(R_{A_p A_p^* + A_{p+1}^* A_{p+1}}^{(2)}), \quad (3.178)$$

where $\rho^{(2)}(C_*^{(2)}(X))$ is introduced in Definition 3.29 and agrees with $\rho^{(2)}(X)$ if X is det-L^2-acyclic (see Definition 3.91) and we have used Lemma 1.18 and Lemma 3.30. Notice that in all cases the relevant L^2-invariant is given by the corresponding L^2-invariant of a morphism of the shape $R_B^{(2)}$ for some matrix over $\mathbb{Z}G$. Hence Theorem 3.172 applies to the geometric situation.

Each term of the characteristic sequences $c(A, K)_p$ can be computed by an algorithm as long as the word problem for G has a solution. Because of Theorem 3.172 (1) and (2) one can use the following strategy to show the vanishing of $b_p^{(2)}(X)$ provided one knows that there is an integer such that $n \cdot b_p^{(2)}(X) \in \mathbb{Z}$ (see the Strong Atiyah Conjecture 10.2). One computes $c(A, K)_p$ for $p = 1, 2, 3, \ldots$ and hopes that for some p its value becomes smaller than $1/n$. This would imply $b_p^{(2)}(X) = 0$.

Notice that the knowledge of the Novikov-Shubin invariants gives information about the speed of convergence of the relevant sequences or sums converging to the L^2-Betti number or L^2-torsion because of Theorem 3.172 (5).

We need the following lemma for the proof of Theorem 3.172.

Lemma 3.179. *If $F(\lambda)$ is the spectral density function of $(R_A^{(2)})^* R_A^{(2)}$ for $A \in M(n, m, \mathbb{C}G)$ and K satisfies $K \geq ||R_A^{(2)}||_\infty = \sqrt{||(R_A^{(2)})^* R_A^{(2)}||_\infty}$, then we get for all $\lambda \in [0, 1]$*

$$(1 - \lambda)^p \cdot (F(K^2 \cdot \lambda) - F(0)) \leq c(A, K)_p - b^{(2)}(R_A^{(2)})$$
$$\leq F(K^2 \cdot \lambda) - F(0) + (1 - \lambda)^p \cdot (n - F(0)).$$

Proof. We have for $\mu \in [0, ||R_A^{(2)}||_\infty^2]$

$$(1 - \lambda)^p \cdot \chi_{[0,\lambda]}(K^{-2} \cdot \mu) \leq (1 - K^{-2} \cdot \mu)^p \leq \chi_{[0,\lambda]}(K^{-2} \cdot \mu) + (1 - \lambda)^p.$$

Hence we get by integrating over μ

$$\int_{0+}^{||R_A^{(2)}||_\infty^2} (1-\lambda)^p \cdot \chi_{[0,\lambda]}(K^{-2} \cdot \mu) \, dF$$

$$\leq \int_{0+}^{||R_A^{(2)}||_\infty^2} (1 - K^{-2} \cdot \mu)^p \, dF$$

$$\leq \int_{0+}^{||R_A^{(2)}||_\infty^2} \chi_{[0,\lambda]}(K^{-2} \cdot \mu) + (1-\lambda)^p \, dF.$$

We have

$$\int_{0+}^{||R_A^{(2)}||_\infty^2} (1-\lambda)^p \cdot \chi_{[0,\lambda]}(K^{-2} \cdot \mu) \, dF = (1-\lambda)^p \cdot \left(F(K^2 \cdot \lambda) - F(0) \right);$$

$$\int_{0+}^{||R_A^{(2)}||_\infty^2} (1 - K^{-2} \cdot \mu)^p \, dF = \mathrm{tr}_{\mathcal{N}(G)} \left(\left(1 - K^{-2} \cdot (R_A^{(2)})^* R_A^{(2)} \right)^p \right)$$
$$- \dim(\ker((R_A^{(2)})^* R_A^{(2)}));$$

$$\int_{0+}^{||R_A^{(2)}||_\infty^2} \chi_{[0,\lambda]}(K^{-2} \cdot \mu) + (1-\lambda)^p \, dF = F(K^2 \cdot \lambda) - F(0)$$
$$+ (1-\lambda)^p \cdot (F(||R_A^{(2)}||_\infty^2) - F(0)).$$

This finishes the proof of Lemma 3.179.   □

Now we can give the proof of Theorem 3.172.

Proof. (1) The bounded G-equivariant operator

$$1 - K^{-2} \cdot (R_A^{(2)})^* R_A^{(2)} : \bigoplus_{i=1}^{n} l^2(G) \to \bigoplus_{i=1}^{n} l^2(G)$$

is positive and satisfies

$$0 \leq 1 - K^{-2} \cdot (R_A^{(2)})^* R_A^{(2)} \leq 1.$$

This implies for $0 \leq p \leq q$

$$0 \leq \left(1 - K^{-2} \cdot (R_A^{(2)})^* R_A^{(2)}\right)^q \leq \left(1 - K^{-2} \cdot (R_A^{(2)})^* R_A^{(2)}\right)^p \leq 1$$

and the first assertion follows as the trace is monotone.

(2) If we apply Lemma 3.179 to the value $\lambda = 1 - \sqrt[p]{\frac{1}{p}}$ we obtain for all positive integers p

$$0 \leq c(A,K)_p - b^{(2)}(R_A^{(2)}) \leq F\left(K^2 \cdot \left(1 - \sqrt[p]{\frac{1}{p}}\right)\right) - F(0) + \frac{n - F(0)}{p}.$$

We get $\lim_{x \to 0+} x \cdot \ln(x) = 0$ from l'Hospital's rule. This implies

$$\lim_{p\to\infty} 1 - \sqrt[p]{\frac{1}{p}} = 0.$$

Since the spectral density function is right continuous, assertion (2) of Theorem 3.172 follows.

(3) Let β and α be any real numbers satisfying

$$0 < \beta < \alpha < \alpha(R_A^{(2)}).$$

Choose a real number γ satisfying

$$\frac{\beta}{\alpha} < \gamma < 1.$$

We conclude from Lemma 3.179 for $\lambda = p^{-\gamma}$

$$0 \le c(A, K)_p - b^{(2)}(R_A^{(2)}) \le F(K^2 \cdot p^{-\gamma}) - F(0) + (1 - p^{-\gamma})^p \cdot (n - F(0)).$$

By the definition of $\alpha(R_A^{(2)})$ there is $\delta > 0$ such that we have for $0 < \lambda < \delta$

$$F(\lambda) - F(0) \le \lambda^\alpha.$$

The last two inequalities imply for p satisfying $K^2 p^{-\gamma} < \delta$

$$\begin{aligned}
0 &\le p^\beta \cdot (c(A, K)_p - b^{(2)}(R_A^{(2)})) \\
&\le p^\beta \cdot ((K^2 \cdot p^{-\gamma})^\alpha + (1 - p^{-\gamma})^p \cdot (n - F(0))).
\end{aligned} \qquad (3.180)$$

We get using l'Hospital's rule

$$\lim_{x\to\infty} x \cdot \ln(1 - x^{-\gamma}) = -\infty;$$

$$\lim_{x\to\infty} \frac{\ln(x)}{x \ln(1 - x^{-\gamma})} = 0;$$

$$\lim_{x\to\infty} \left(\frac{\beta \ln(x)}{x \ln(1 - x^{-\gamma})} + 1 \right) = 1;$$

$$\lim_{x\to\infty} \beta \ln(x) + x \ln(1 - x^{-\gamma}) = -\infty;$$

$$\lim_{x\to\infty} x^\beta (1 - x^{-\gamma})^x = 0.$$

Since $\beta - \gamma\alpha < 0$ holds we have $\lim_{x\to\infty} x^\beta (K^2 \cdot x^{-\gamma})^\alpha = 0$. Hence we get

$$\lim_{p\to\infty} p^\beta \cdot ((K^2 \cdot p^{-\gamma})^\alpha + (1 - p^{-\gamma})^p \cdot (n - F(0))) = 0.$$

This implies using the inequality (3.180)

$$\lim_{p\to\infty} p^\beta \cdot \left(c(A, K)_p - b^{(2)}(R_A^{(2)}) \right) = 0.$$

We have shown $\beta \leq \beta(A)$. Since β was an arbitrary number satisfying $0 < \beta < \alpha(R_A^{(2)})$, assertion (3) follows.

(4) We get the chain of equations where a sum or integral is put to be $-\infty$ if it does not converge, and $\{E_\lambda \mid \lambda\}$ is the spectral family of $(R_A^{(2)})^* R_A^{(2)}$

$$2 \cdot (n - b^{(2)}(R_A^{(2)})) \cdot \ln(K) - \sum_{p=1}^{\infty} \frac{1}{p} \cdot \left(c(A, K)_p - b^{(2)}(R_A^{(2)}) \right)$$

$$= 2 \cdot (n - b^{(2)}(R_A^{(2)})) \cdot \ln(K) -$$
$$\sum_{p=1}^{\infty} \frac{1}{p} \cdot \left(\mathrm{tr}_{\mathcal{N}(G)} \left(\left(1 - K^{-2} \cdot (R_A^{(2)})^* R_A^{(2)} \right)^p \right) - b^{(2)}(R_A^{(2)}) \right)$$

$$= 2 \cdot (n - b^{(2)}(R_A^{(2)})) \cdot \ln(K)$$
$$- \sum_{p=1}^{\infty} \frac{1}{p} \cdot \mathrm{tr}_{\mathcal{N}(G)} \left(\int_{0+}^{||R_A^{(2)}||_\infty^2} (1 - K^{-2} \cdot \lambda)^p \; dE_\lambda \right).$$

The trace is linear, monotone and ultra-weakly continuous.

$$= 2 \cdot (n - b^{(2)}(R_A^{(2)})) \cdot \ln(K) - \sum_{p=1}^{\infty} \frac{1}{p} \cdot \int_{0+}^{||R_A^{(2)}||_\infty^2} (1 - K^{-2} \cdot \lambda)^p \; d\mathrm{tr}_{\mathcal{N}(G)}(E_\lambda)$$

$$= 2 \cdot (n - b^{(2)}(R_A^{(2)})) \cdot \ln(K) - \sum_{p=1}^{\infty} \frac{1}{p} \cdot \int_{0+}^{||R_A^{(2)}||_\infty^2} (1 - K^{-2} \cdot \lambda)^p \; dF$$

We can put the sum under the integral sign because of Levi's Theorem of Monotone Convergence since $(1 - K^{-2} \cdot \lambda)^p$ is non-negative for $0 < \lambda \leq ||R_A^{(2)}||_\infty^2 \leq K^2$.

$$= 2 \cdot (n - b^{(2)}(R_A^{(2)})) \cdot \ln(K) - \int_{0+}^{||R_A^{(2)}||_\infty^2} \sum_{p=1}^{\infty} \frac{1}{p} \cdot (1 - K^{-2} \cdot \lambda)^p \; dF$$

The Taylor series $-\sum_{p=1}^{\infty} \frac{1}{p} \cdot (1 - \mu)^p$ of $\ln(\mu)$ about 1 converges for $|1 - \mu| < 1$.

$$= 2 \cdot (n - b^{(2)}(R_A^{(2)})) \cdot \ln(K) + \int_{0+}^{||R_A^{(2)}||_\infty^2} \ln(K^{-2} \cdot \lambda) \; dF$$

$$= 2 \cdot (n - b^{(2)}(R_A^{(2)})) \cdot \ln(K) + \int_{0+}^{||R_A^{(2)}||_\infty^2} \ln(\lambda) \; dF - \int_{0+}^{||R_A^{(2)}||_\infty^2} \ln(K^2) \; dF$$

$$= 2 \cdot (n - b^{(2)}(R_A^{(2)})) \cdot \ln(K) + \int_{0+}^{||R_A^{(2)}||_\infty^2} \ln(\lambda) \; dF - \ln(K^2) \cdot (n - b^{(2)}(R_A^{(2)}))$$

$$= \int_{0+}^{||R_A^{(2)}||_\infty^2} \ln(\lambda) \; dF \; = \; \int_{0+}^{\infty} \ln(\lambda) \; dF$$

$$= \ln(\det(R_A^{(2)})^* R_A^{(2)}) \; = \; 2 \cdot \ln(\det(R_A^{(2)})),$$

where the last equation follows from Lemma 3.15 (4). This finishes the proof of assertion (4).

(5) Let α be any number satisfying $\alpha < \alpha(R_A^{(2)})$. Then we conclude from assertion (3) $\lim_{p\to\infty} p^\alpha \left(c(A, K)_p - b^{(2)}(R_A^{(2)}) \right) = 0$. Let C be any positive number such that $p^\alpha \left(c(A, K)_p - b^{(2)}(R_A^{(2)}) \right) \leq C$ holds for all p. We conclude

$$0 \leq c(A, K)_p - b^{(2)}(R_A^{(2)}) \leq \frac{C}{p^\alpha}.$$

Next we estimate using assertion (4)

$$0 \leq -2 \cdot \ln(\det(R_A^{(2)})) + 2(n - b^{(2)}(R_A^{(2)})) \cdot \ln(K)$$
$$- \sum_{p=1}^{L} \frac{1}{p} \cdot \left(c(A, K)_p - b^{(2)}(R_A^{(2)}) \right)$$
$$= \sum_{p=L+1}^{\infty} \frac{1}{p} \cdot \left(c(A, K)_p - b^{(2)}(R_A^{(2)}) \right)$$
$$= \sum_{p=L+1}^{\infty} p^{-1-\alpha} \cdot \left(p^\alpha \cdot \left(c(A, K)_p - b^{(2)}(R_A^{(2)}) \right) \right)$$
$$\leq C \cdot \sum_{p=L+1}^{\infty} p^{-1-\alpha}$$
$$\leq C \cdot \int_{L}^{\infty} x^{-1-\alpha} dx$$
$$= \frac{C}{\alpha} \cdot L^{-\alpha}.$$

This finishes the proof of Theorem 3.172.   □

Remark 3.181. We conjecture that the inequality $\alpha(R_A^{(2)}) \leq \beta(A)$ of Theorem 3.172.3 is an equality where we do not distinguish between ∞ and ∞^+ as value for $\alpha(R_A^{(2)})$. If the spectral density function F of $R_A^{(2)}$ has the limit property (see Definition 2.41), then this is true.

We will give more explicit calculations for 3-manifolds later which depend only on a presentation of the fundamental group (see Theorem 4.9).

3.8 Miscellaneous

The following result is taken from [106, Proposition 6.4 on page 149].

Lemma 3.182. *Let M be a simply connected Riemannian manifold and $f: M \to \mathbb{R}$ be a function which is invariant under the isometries of M. Then there is a constant $C(f)$ with the property that for any cocompact free proper action of a discrete group G by isometries and any fundamental domain $\mathcal{F}$*

$$\int_{\mathcal{F}} f \, dvol_M = C(f) \cdot vol(G \backslash M)$$

holds.

Lemma 3.182 is obvious if M is homogeneous since then G acts transitively on M and hence any function on M which is invariant under the isometries of M is constant. Lemma 3.182 implies the following proportionality principle.

Theorem 3.183 (Proportionality Principle for L^2-invariants). *Let M be a simply connected Riemannian manifold. Then there are constants $B_p^{(2)}(M)$ for $p \geq 0$, $A_p^{(2)}(M)$ for $p \geq 1$ and $T^{(2)}(M)$ depending only on the Riemannian manifold M with the following property: For any discrete group G with a cocomapct free proper action on M by isometries the following holds*

$$b_p^{(2)}(M; \mathcal{N}(G)) = B_p^{(2)}(M) \cdot vol(G \backslash M);$$
$$\alpha_p^{(2)}(M; \mathcal{N}(G)) = A_p^{(2)}(M);$$
$$\rho^{(2)}(M; \mathcal{N}(G)) = T^{(2)}(M) \cdot vol(G \backslash M),$$

where for the third equality we assume that M with this G-action (and hence for all cocompact free proper actions) is of determinant class.

Proof. Obviously the following function

$$M \to \mathbb{R}, \qquad x \mapsto tr_{\mathbb{R}}(E_\lambda^{(\delta^{p+1} d^p)_{\min}^\perp}(x, x))$$

is invariant under isometries since d^p and δ^p are compatible with isometries and hence the endomorphisms $E_\lambda^{(\delta^{p+1} d^p)_{\min}^\perp}(x, x)$ and $E_\lambda^{(\delta^{p+1} d^p)_{\min}^\perp}(\phi(x), \phi(x))$ are conjugate for any isometry $\phi: M \to M$ and $x \in X$. Let $C_p(M)(\lambda)$ be the constants associated to this function for each $\lambda \geq 0$ in Lemma 3.182. We conclude from Lemma 3.182 and Lemma 2.66

$$F_p(M; \mathcal{N}(G))(\lambda) = C_p(M)(\lambda) \cdot vol(G \backslash M),$$

where $F(M; \mathcal{N}(G))$ is the analytic spectral density function associated to the cocompact free proper G-manifold M with G-invariant Riemannian metric. Notice that $C_p(M)(\lambda)$ is independent of G. Now the claim for the L^2-Betti numbers and the Novikov-Shubin invariants follow. Analogously we conclude that there is also a function $\Theta_p(M)(t)$ depending only on M such that

$$\int_{\mathcal{F}} tr_{\mathbb{R}}(e^{-t\Delta_p}(x, x)) \, dvol_x = \Theta_p(M)(t) \cdot vol(G \backslash M)$$

holds. Now the claim for the L^2-torsion follows. $\qquad\qquad\square$

Another proportionality principle for L^2-Betti numbers will be given in Theorem 7.34 and Corollary 7.37.

Remark 3.184. We call a finite free G-CW-complex X NS-L^2-*acyclic* if $b_p^{(2)}(X) = 0$ and $\alpha_p(X) > 0$ for all $p \geq 0$. Notice that NS-L^2-acyclic implies det-L^2-acyclic because of Theorem 3.14 (4). We conclude from Remark 3.90 that all the results of Section 3.4 remain true if one replaces det-L^2-acyclic by NS-L^2-acyclic. The point is that the property NS-L^2-acyclic is enherited like the property det-L^2-acyclic.

The next result is well-known.

Lemma 3.185. *A closed hyperbolic manifold does not carry a non-trivial S^1-action.*

Proof. Suppose that the closed hyperbolic manifold M carries a non-trivial S^1-action. Since the universal covering of M is the hyperbolic space and hence contractible, $\widetilde{M}$ is det-L^2-acyclic and $\rho^{(2)}(\widetilde{M}) = 0$ by Corollary 3.111. This contradicts Theorem 1.62 if $\dim(M)$ is even and Theorem 3.152 if n is odd. □

One may ask whether Theorem 3.111 extends from S^1-actions to S^1-foliations, namely

Question 3.186. (S^1-foliations and L^2-torsion for closed aspherical manifolds).
Let M be an aspherical closed manifold which admits an S^1-foliation. Is then $\widetilde{M}$ of determinant class and

$$b_p^{(2)}(\widetilde{M} = 0 \qquad \text{for all } p;$$
$$\alpha_p(\widetilde{M}) \geq 1 \qquad \text{for all } p;$$
$$\rho^{(2)}(\widetilde{M}) = 0 \ ?$$

If the answer is positive, one would get a negative answer to the (to the author's knowledge) open problem whether there is a S^1-foliation on a closed hyperbolic manifold.

The notion of spectral density function makes also sense for proper G-CW-complexes of finite type because in this setting the cellular L^2-chain complex is still defined as a Hilbert $\mathcal{N}(G)$-chain complex as explained in Section 1.6. Hence the notion of det-L^2-acyclic and of L^2-torsion $\rho^{(2)}(X)$ for a proper finite G-CW-complex in the sense of Definition 3.91 and the notion of topological L^2-torsion $\rho_{\text{top}}^{(2)}(M)$ for a cocompact proper G-manifold M with G-invariant Riemannian metric in the sense of Definition 3.120 are defined. Theorem 3.93 (1) remains true if one replaces $\text{Wh}(G)$ by $K_1(\mathbb{Q}G)/T(G)$, where $T(G)$ is the subgroup generated by elements represented by automorphisms of the shape $\pm\mathbb{Q}[\phi]\colon \mathbb{Q}[G/H] \to \mathbb{Q}[G/H]$ for finite subgroups $H \subset G$

and G-maps $\phi\colon G/H \to G/H$, and the Whitehead torsion $\tau(f)$ by the corresponding element in $K_1(\mathbb{Q}G)/T(G)$ given by the $\mathbb{Q}G$-chain homotopy equivalence $\mathbb{Q}\otimes_{\mathbb{Z}}C_*(f)\colon \mathbb{Q}\otimes_{\mathbb{Z}}C_*(X) \to \mathbb{Q}\otimes_{\mathbb{Z}}C_*(Y)$. Notice that the map induced by the logarithm of the Fuglede-Kadison determinant $\Phi^G\colon K_1(\mathbb{Q}G)/T(G) \to \mathbb{R}$ is definitely non-trivial. But it may still be true that $\Phi^G(\tau(f))$ is always trivial for a G-homotopy equivalence f and hence that $\rho^{(2)}(X)$ is a G-homotopy invariant for det-L^2-acyclic finite proper G-CW-complexes. This can be proved, provided that G is of det $\geq$ 1-class or if $\mathcal{G}$ belongs to the class $\mathcal{G}$ (see Definition 13.9). Theorem 3.93 (2), (3), (4), (5) and (6) carry over word by word for a det-L^2-acyclic finite proper G-CW-complex. The proof of Poincaré duality stated in Theorem 3.93 (3) is non-trivial because the Poincaré $\mathbb{Q}G$-chain homotopy equivalence $\mathbb{Q}\otimes_{\mathbb{Z}} C^{n-*}(M) \to \mathbb{Q}\otimes_{\mathbb{Z}} C_*(M)$ has in general non-trivial Whitehead torsion in $K_1(\mathbb{Q}G)/T(G)$ [327, Definition 3.19 and Example 3.25], but one can show that its image under the homomorphism $K_1(\mathbb{Q}G)/T(G) \to \mathbb{R}$ given by the logarithm of the Fuglede-Kadison determinant is trivial. The notion of analytic L^2-torsion $\rho_{\mathrm{an}}^{(2)}(M)$ for a cocompact proper G-manifold with G-invariant Riemannian metric introduced in Definition 3.128 still makes sense. Without having checked the details we claim that Theorem 3.149 of Burghela, Friedlander, Kappeler and McDonald [84] is still true if one drops the assumption free.

One can try to get an improved L^2-torsion working in K-theory, instead of applying the Fuglede-Kadison determinant from the very beginning. Fix a set $\mathcal{W}$ of morphisms of Hilbert $\mathcal{N}(G)$-modules $f\colon l^2(G)^n \to l^2(G)^n$ with the following properties: (i) Any isomorphism $l^2(G)^n \to l^2(G)^n$ belongs to $\mathcal{W}$, (ii) any element in $\mathcal{W}$ is a weak isomorphism. (iii) If two of the morphisms f, g and $g \circ f$ belong to $\mathcal{W}$, then also the third. (iv) If both $f \circ g$ and $g \circ f$ belong to $\mathcal{W}$, then f and g belong to $\mathcal{W}$. (v) If two of the morphisms $f_1\colon l^2(G)^m \to l^2(G)^m$, $f_2\colon l^2(G)^n \to l^2(G)^n$ and $\begin{pmatrix} f_1 & g \\ 0 & f_2 \end{pmatrix}$ belong to $\mathcal{W}$, then all three. Define $K_1^{\mathcal{W}}(\mathcal{N}(G))$ to be the abelian group whose generators are classes $[f]$ of endomorphisms $f\colon l^2(G)^n \to l^2(G)^n$ belonging to $\mathcal{W}$ such that the following relations are satisfied: (i) $[\mathrm{id}\colon l^2(G) \to l^2(G)] = 0$, (ii) $[g \circ f] = [f] + [g]$ for $f, g\colon l^2(G)^n \to l^2(G)^n$ in $\mathcal{W}$ and (iii) $\left[\begin{pmatrix} f_1 & g \\ 0 & f_2 \end{pmatrix}\right] = [f_1] + [f_2]$ for $f_1, f_2 \in \mathcal{W}$. Let $\widetilde{K}_1^{\mathcal{W}}(\mathcal{N}(G))$ be the quotient of $K_1^{\mathcal{W}}(\mathcal{N}(G))$ by the subgroup generated by $[-\mathrm{id}\colon l^2(G) \to l^2(G)]$.

We call a finite Hilbert $\mathcal{N}(G)$-chain complex C_* $\mathcal{W}$-acyclic if there is a weak chain contraction (γ_*, u_*) in the sense of Definition 3.38, where we now require that $u_p\colon C_p \to C_p$ belongs to $\mathcal{W}$. Then we can associate to a finite $\mathcal{W}$-acyclic Hilbert $\mathcal{N}(G)$-chain complex C_* analogously to the formulas appearing in Lemma 3.41 an element

$$\rho^{\mathcal{W}}(C_*) := [(uc + \gamma)_{\mathrm{odd}}] - [u_{\mathrm{odd}}] \qquad \in \widetilde{K}_1^{\mathcal{W}}(\mathcal{N}(G)). \qquad (3.187)$$

By the same arguments we used for the L^2-torsion one shows that $\rho^{\mathcal{W}}(C_*)$ is well-defined, i.e. independent of the choice of (γ_*, u_*) and that a lot of the good properties remain true. For instance, if $0 \to C_* \to D_* \to E_* \to 0$ is an exact sequence of finite Hilbert $\mathcal{N}(G)$-chain complexes and two of them are $\mathcal{W}$-acyclic, then all three are $\mathcal{W}$-acyclic and we get analogously to Lemma 3.68

$$\rho^{\mathcal{W}}(C_*) - \rho^{\mathcal{W}}(D_*) + \rho^{\mathcal{W}}(E_*) = \rho^{\mathcal{W}}(C_*, D_*, E_*). \qquad (3.188)$$

The problem is to find out how much information $K_1^{\mathcal{W}}(\mathcal{N}(G))$ contains depending of the various choices of $\mathcal{W}$.

If we take for instance $\mathcal{W}$ to be the class of isomorphisms, $\widetilde{K}_1^{\mathcal{W}}(\mathcal{N}(G))$ is just the ordinary $\widetilde{K}_1(\mathcal{N}(G))$ of the ring $\mathcal{N}(G)$ which has been computed in [344] (see also Subsection 9.2.2). However, $\mathcal{W}$-acyclic means in this situation that C_* is contractible and this condition is extremely restrictive as we will see in Chapter 12 on the Zero-in-the-Spectrum-conjecture. If we take $\mathcal{W}$ to be the class of weak isomorphisms of determinant class, then the logarithm of the Fuglede-Kadison determinant gives a homomorphism $\det: \widetilde{K}_1^{\mathcal{W}}(\mathcal{N}(G)) \to \mathbb{R}$ which maps $\rho^{\mathcal{W}}(C_*)$ to $\rho^{(2)}(C_*)$. This homomorphism is split surjective but we do not know whether it is bijective. Finally we discuss the case where $\mathcal{W}$ consists of all weak isomorphisms. Then $K_1^{\mathcal{W}}(\mathcal{N}(G))$ can be identified with $K_1^{\mathrm{inj}}(\mathcal{N}(G))$ (see Definition 9.16). If G contains $\mathbb{Z}^n$ as subgroup of finite index, $K_1^{\mathcal{W}}(\mathcal{N}(G))$ has been computed in [345] (see also Section 9.3) and there it is shown that $\rho^{\mathcal{W}}(C_*)$ contains essentially the same information as the classical Alexander polynomial. If G is finitely generated and does not contain $\mathbb{Z}^n$ as subgroup of finite index, then $K_1^{\mathcal{W}}(\mathcal{N}(G)) = 0$ [344] (see also Section 9.3) and hence $\rho^{\mathcal{W}}(C_*)$ carries no information. This shows why a condition such as of determinant class has to appear to get a meaningful invariant.

Lott [321] defines analytically delocalized L^2-torsion for the universal covering $\widetilde{M}$ of a closed Riemannian manifold M, which gives a number for each conjugacy class of the fundamental group $\pi_1(M)$. The value of the conjugacy class of the unit element is the L^2-torsion. This invariant is presently only defined under certain technical convergence assumptions. At least for universal coverings of closed hyperbolic manifolds of odd dimension the delocalized L^2-torsion is defined and the marked length spectrum can be recovered from it.

Formulas for the variation of the L^2-torsion under varying the Riemannian metric are given for instance in [316, page 480] and [346, section 7].

We will give further computations of L^2-torsion for universal coverings of compact 3-manifolds in Theorem 4.3, of knot complements in Section 4.3 and of closed locally symmetric spaces in Theorem 5.12 and in (5.13).

We will discuss the behaviour of L^2-Betti numbers, Novikov-Shubin invariants and L^2-torsion of groups under quasi-isometry and measure equivalence in Section 7.5.

In Chapter 11 we will deal with the Conjecture 11.3 which makes a prediction about the parity of the L^2-torsion of the universal covering of a closed manifold M, provided M is aspherical or carries a Riemannian metric with negative sectional curvature.

We will relate the L^2-torsion of the universal covering of an aspherical closed orientable manifold to its simplicial volume in Chapter 14.

Further references about L^2-torsion are [92], [93], [137], [138], [139], [360], [362], [363].

Exercises

3.1. Let C_* be a finite free $\mathbb{Z}$-chain complex. Choose a $\mathbb{Z}$-basis for C_p and for $H_p(C_*)/\operatorname{tors}(H_p(C_*))$ for each $p \in \mathbb{Z}$. They induce $\mathbb{R}$-bases and thus Hilbert space structures on $C_p \otimes_{\mathbb{Z}} \mathbb{R}$ and on $H_p(C_* \otimes_{\mathbb{Z}} \mathbb{R}) \cong H_p(C_*) \otimes_{\mathbb{Z}} \mathbb{R}$ for all $p \in \mathbb{Z}$. Let $\rho^{\mathbb{Z}}(C_*)$ be the real number $\rho(C_* \otimes_{\mathbb{Z}} \mathbb{R})$ whose definition is the obvious variation of (3.5). Prove

$$\rho^{\mathbb{Z}}(C_*) \;=\; \sum_{p \in \mathbb{Z}} (-1)^p \cdot \ln\left(|\operatorname{tors}(H_p(C_*))|\right).$$

3.2. Let X and Y be finite CW-complexes. Let $j_i \colon X_i \to X$ and $k_i \colon Y_i \to Y$ be inclusions of CW-subcomplexes for $i = 0, 1, 2$ such that $X = X_1 \cup X_2$, $X_0 = X_1 \cap X_2$, $Y = Y_1 \cup Y_2$ and $Y_0 = Y_1 \cap Y_2$ holds. Let $f \colon X \to Y$ be a map which induces homotopy equivalences $f_i \colon X_i \to Y_i$ for $i = 0, 1, 2$. Show that f is a homotopy equivalence and

$$\tau(\widetilde{f}) = k_{1*}(\tau(\widetilde{f_1})) + k_{2*}(\tau(\widetilde{f_2})) - k_{0*}(\tau(\widetilde{f_0})).$$

3.3.   Let $f \colon X \to Y$ and $g \colon Y \to Z$ be homotopy equivalences of finite CW-complexes. Prove

$$\tau(\widetilde{g \circ f}) = \tau(\widetilde{g}) + g_*(\tau(\widetilde{f})).$$

3.4. Let $f \colon X' \to X$ and $g \colon Y' \to Y$ be homotopy equivalences of connected finite CW-complexes. Denote by $k_X \colon X \to X \times Y$ and $k_Y \colon Y \to X \times Y$ the canonical inclusions for some choice of base points in X and Y. Prove

$$\tau(\widetilde{f \times g}) = \chi(X) \cdot k_{Y*}(\tau(\widetilde{g})) + \chi(Y) \cdot k_{X*}(\tau(\widetilde{f})).$$

3.5. Show for S^n equipped with some Riemannian metric

$$\rho_{\text{top}}(S^n; \mathbb{R}) = \frac{(1 + (-1)^{n+1})}{2} \cdot \ln(\text{vol}(S^n)).$$

3.6. Equip $\mathbb{R}$ with the standard Riemannian metric and $\mathbb{Z}$-operation. Denote by V the trivial orthogonal 1-dimensional $\mathbb{Z}$-representation whose underlying vector space is $\mathbb{R}$ with the standard Hilbert space structure. Let $\zeta_R(s) = \sum_{n \geq 1} n^{-s}$ be the Riemannian Zeta-function. Show for the Zeta-function $\zeta_1(\mathbb{R}; V)(s)$ defined in (3.8)

$$\zeta_1(\mathbb{R}; V)(s) = 2 \cdot (2\pi)^{-2s} \cdot \zeta_R(2s).$$

3.7. Define a function $F \colon \mathbb{R} \to [0, \infty)$ by $F(\lambda) = \frac{1}{-\ln(-\ln(\lambda)) \cdot \ln(\lambda)}$ for $0 < \lambda \leq e^{-e}$, $F(\lambda) = e^{-1}$ for $\lambda \geq e^{-e}$ and $F(\lambda) = 0$ for $\lambda \leq 0$. Show that F is a density function with

$$\lim_{\lambda \to 0+} \ln(\lambda) \cdot F(\lambda) = 0;$$

$$\int_{0+}^{1} \ln(\lambda) \, dF = -\infty.$$

3.8. Let A be a (k, k)-matrix over $\mathbb{C}[\mathbb{Z}^n]$. Denote its determinant over $\mathbb{C}[\mathbb{Z}^n]$ by $\det_{\mathbb{C}[\mathbb{Z}^n]}(A) \in \mathbb{C}[\mathbb{Z}^n]$. Let $R_A \colon l^2(\mathbb{Z}^n)^k \to l^2(\mathbb{Z}^n)^k$ and $R_{\det_{\mathbb{C}[\mathbb{Z}^n]}(A)} \colon l^2(\mathbb{Z}^n) \to l^2(\mathbb{Z}^n)$ be the morphism given by right multiplication with A and $\det_{\mathbb{C}[\mathbb{Z}]}(A)$. Show that the following statements are equivalent: i.) R_A is a weak isomorphism, ii.) $R_{\det_{\mathbb{C}[\mathbb{Z}^n]}(A)}$ is a weak isomorphism and iii.) $\det_{\mathbb{C}[\mathbb{Z}^n]}(A) \neq 0$. Show that in this case $\det(R_A) = \det(R_{\det_{\mathbb{C}[\mathbb{Z}^n]}(A)})$.

3.9. Let X be a finite free $\mathbb{Z}$-CW-complex. Since $\mathbb{C}[\mathbb{Z}]$ is a principal ideal domain [15, Proposition V.5.8 on page 151 and Corollary V.8.7 on page 162], we can write

$$H_p(X; \mathbb{C}) = \mathbb{C}[\mathbb{Z}]^{n_p} \oplus \left(\bigoplus_{i_p=1}^{s_p} \mathbb{C}[\mathbb{Z}]/((z - a_{p,i_p})^{r_{p,i_p}}) \right)$$

for $a_{p,i_p} \in \mathbb{C}$ and $n_p, s_p, r_{p,i_p} \in \mathbb{Z}$ with $n_p, s_p \geq 0$ and $r_{p,i_p} \geq 1$ where $z \in \mathbb{Z}$ is a fixed generator.

Prove for $p \geq 0$

$$b_p^{(2)}(X) = n_p;$$

$$\alpha_{p+1}(X) = \begin{cases} \min\{\frac{1}{r_{p,i_p}} \mid i_p = 1, 2 \ldots, s_p, |a_{p,i_p}| = 1\} \\ \quad \text{if } s_p \geq 1 \text{ and } \{i_p = 1, 2 \ldots, s_p, |a_{p,i_p}| = 1\} \neq \emptyset \\ \infty^+ \\ \quad \quad \quad \quad \quad \text{otherwise} \end{cases} .$$

Show that X is det-L^2-acyclic if and only if $n_p = 0$ for all $p \geq 0$.

Prove that $\rho^{(2)}(X)$ cannot be read off from the $\mathbb{C}[\mathbb{Z}]$-modules $H_p(X; \mathbb{C})$ for all $p \in \mathbb{Z}$, but that $\rho^{(2)}(X)$ is determined by the $\mathbb{Z}[\mathbb{Z}]$-modules $H_p(X; \mathbb{C})$ for $p \in \mathbb{Z}$ and can be written as $\ln(|a|) - \ln(|b|)$ for algebraic integers $a, b \in \mathbb{C}$ with $|a|, |b| \geq 1$.

3.10.   Give an example of finite Hilbert $\mathcal{N}(G)$-chain complexes C_* and D_* which are both of determinant class and a weak homology equivalence $f_* \colon C_* \to D_*$ such that f_p is of determinant class for all $p \in \mathbb{Z}$, but neither f_* is of determinant class nor $H_p^{(2)}(f_*)$ is of determinant class for all $p \in \mathbb{Z}$. (cf. Theorem 3.35 (5) and Lemma 3.44).

3.11.   Let $(C_{*,*}, d_{*,*})$ be a bicomplex of finitely generated Hilbert $\mathcal{N}(G)$-chain modules such that $C_{p,q} = 0$ for $|p|, |q| \geq N$ holds for some number N. Suppose for $p \in \mathbb{Z}$ that the chain complex $C_{p,*}$ given by the p-th column and the chain complex $C_{*,q}$ given by the q-th row are det-L^2-acyclic. Let T_* be the associated total chain complex with $T_n = \bigoplus_{p+q=n} C_{p,q}$. Show that T_* is det-L^2-acyclic and

$$\sum_{p \in \mathbb{Z}} (-1)^p \cdot \rho^{(2)}(C_{p,*}) \;=\; \rho^{(2)}(T_*) \;=\; \sum_{q \in \mathbb{Z}} (-1)^q \cdot \rho^{(2)}(C_{*,q}).$$

3.12.   Show that the following statements are equivalent:

(1) The map $\Phi^G \colon \mathrm{Wh}(G) \to \mathbb{R}$ of (3.92) is trivial for all groups G;
(2) For all finitely generated groups G and G-homotopy equivalences $f \colon X \to Y$ of det-L^2-acyclic free finite G-CW-complexes $\rho^{(2)}(X) = \rho^{(2)}(Y)$ holds;
(3) For all finitely generated groups G and G-homotopy equivalences $f \colon M \to N$ of det-L^2-acyclic cocompact free proper G-manifolds with G-invariant Riemannian metric and without boundary $\rho_{\mathrm{an}}^{(2)}(M) = \rho_{\mathrm{an}}^{(2)}(N)$ holds.

3.13.   Show that the composition of the obvious map given by induction $\bigoplus_{H \subset G, |H| < \infty} \mathrm{Wh}(H) \to \mathrm{Wh}(G)$ with $\phi^G \colon \mathrm{Wh}(G) \to \mathbb{R}$ is trivial.

3.14.   Let G be a countable group. Show that the following sets are countable

$$\{b_p^{(2)}(X; \mathcal{N}(G)) \mid X \text{ connected free } G\text{-}CW\text{-complex of finite type}, p \geq 0\};$$

$$\{a_p^{(2)}(X; \mathcal{N}(G)) \mid X \text{ connected free } G\text{-}CW\text{-complex of finite type}, p \geq 1\};$$

$$\{\rho^{(2)}(X; \mathcal{N}(G)) \mid X \text{ det-}L^2\text{-acyclic connected finite free } G\text{-}CW\text{-complex}\}.$$

Show that the first set is closed under addition in $\mathbb{R}^{\geq 0}$, and that the third set is an additive subgroup of $\mathbb{R}$ provided that the third set is non-empty.

3.15.   Show that the following sets are countable

$$\{b_p^{(2)}(\widetilde{X}) \mid X \text{ connected } CW\text{-complex of finite type}, p \geq 0\};$$

$$\{a_p(\widetilde{X}) \mid X \text{ connected } CW\text{-complex of finite type}, p \geq 1\};$$

$$\{\rho^{(2)}(\widetilde{X}) \mid X \text{ det-}L^2\text{-acyclic connected finite } CW\text{-complex}\}.$$

3.16. Give an example of a fibration of connected finite CW-complexes which is not simple, i.e. $\theta(p) \neq 0$.

3.17. Let $f \colon X \to X$ be a selfmap of a connected finite CW-complex and let T_f be its mapping torus. Let L be the colimit of the system

$$\cdots \xrightarrow{\pi_1(f)} \pi_1(X) \xrightarrow{\pi_1(f)} \pi_1(X) \xrightarrow{\pi_1(f)} \cdots$$

There is a canonical epimorphism $\iota \colon \pi_1(X) \to L$ and an automorphism $\mu \colon L \to L$ satisfying $\mu \circ \iota = \iota \circ \pi_1(f)$. Show that $\pi_1(T_f)$ is isomorphic to the semidirect product $L \rtimes_\mu \mathbb{Z}$. Prove that ι is bijective and $\pi_1(T_f) \cong \pi_1(X) \rtimes_{\pi_1(f)} \mathbb{Z}$, provided that $\pi_1(f)$ is bijective.

3.18. Let $p \colon E \to S^1$ be a fibration with fiber $S^n \vee S^n$ for $n \geq 2$. Suppose that $\pi_1(S^1) = \mathbb{Z}$ acts on $H_n(S^n \vee S^n; \mathbb{Z})$ by an automorphism $H_1(f)$ with determinant 1. Let $\mathrm{tr} \in \mathbb{Z}$ be the trace of $H_1(f)$. Show that all L^2-Betti numbers of $\widetilde{T_f}$ vanish, $\alpha_1(\widetilde{T_f}) = 1$, $\alpha_p(\widetilde{T_f}) = \infty^+$ for $p \neq 1, n+1$ and that precisely one of the following cases occurs

(1) $H_1(f)$ is periodic. Then $\mathrm{tr} \in \{-1, 0, +1\}$ or $H_1(f) = \pm \mathrm{id}$ and we have $\alpha_{n+1}(\widetilde{T_f}) = 1$ and $\rho^{(2)}(\widetilde{T_f}) = 0$;

(2) $H_1(f)$ is parabolic, i.e. $H_1(f)$ is not periodic and all complex eigenvalues have norm 1. Then we have $\mathrm{tr} \in \{-2, +2\}$, $\alpha_{n+1}(\widetilde{T_f}) = 1/2$ and $\rho^{(2)}(\widetilde{T_f}) = 0$;

(3) $H_1(f)$ is hyperbolic, i.e. there is one (and hence two) complex eigenvalue whose norm is not 1. Then $|\mathrm{tr}| > 2$, $\alpha_{n+1}(\widetilde{T_f}) = \infty^+$ and $\rho^{(2)}(\widetilde{T_f}) = \ln\left(\frac{\mathrm{tr}}{2} + \sqrt{\frac{\mathrm{tr}^2}{4} - 1}\right)$.

3.19. Give examples of two fibrations $F \to E_0 \to B$ and $F \to E_1 \to B$ of connected finite CW-complexes with the same fiber and base space such that the fiber is simply connected, $\widetilde{E_0}$ and $\widetilde{E_1}$ are of determinant class and have trivial L^2-Betti numbers, but $\widetilde{E_0}$ and $\widetilde{E_1}$ have both different Novikov-Shubin invariants and different L^2-torsion.

3.20. Let $F \colon [0, \infty) \to [0, \infty)$ be a density function with $F(0) = 0$ for which there is $K > 0$ with $F(\lambda) = F(K)$. Let θ_F be its Laplace transform. Fix $\epsilon > 0$. Show:

(1) We have

$$\theta_F(t) = \sum_{n \geq 0} a_n t^n$$

for a power series $\sum_{n \geq 0} a_n t^n$ which converges for all $t \in \mathbb{R}$ and satisfies $a_0 = F(K)$;

(2) The function

$$\frac{1}{\Gamma(s)} \cdot \int_0^\epsilon t^{s-1} \theta_F(t) \, dt$$

is holomorphic for $\Re(s) > 0$ and has a meromorphic extension to $\mathbb{C}$ which has poles at $s = -1, -2, -3, \ldots$, all of order 1;

(3) Suppose $0 < \alpha(F)$. The function

$$\frac{1}{\Gamma(s)} \cdot \int_\epsilon^\infty t^{s-1} \cdot \theta_F(t) \, dt$$

is holomorphic for $\Re(s) < \alpha(F)$. We have

$$\frac{d}{ds} \frac{1}{\Gamma(s)} \cdot \int_\epsilon^\infty t^{s-1} \theta_F(t) \, dt \bigg|_{s=0} = \int_\epsilon^\infty t^{-1} \cdot \theta_F(t) \, dt < \infty;$$

(4) Suppose $0 < \alpha(F)$. Then the function $\int_0^\infty \lambda^{-s} \, dF(\lambda)$ is holomorphic for $\Re(s) < \alpha(F)$ and we get

$$\frac{d}{ds} \int_0^\infty \lambda^{-s} \, dF(\lambda) \bigg|_{s=0} = \int_0^\infty \ln(\lambda) \, dF(\lambda) < \infty;$$

(5) We have for $0 < \Re(s) < \alpha(F)$

$$\frac{1}{\Gamma(s)} \cdot \int_0^\infty t^{s-1} \cdot \theta_F(t) \, dt = \int_0^\infty \lambda^{-s} \, dF(\lambda);$$

(6) If $0 < \alpha(F)$, we get

$$\frac{d}{Dr} \frac{1}{\Gamma(s)} \int_0^\epsilon t^{s-1} \cdot \theta_F(t) \, du \bigg|_{s=0} + \int_\epsilon^\infty t^{-1} \cdot \theta_F(t) \, dt$$

$$= \frac{d}{ds} \frac{1}{\Gamma(s)} \int_0^\infty t^{s-1} \cdot \theta_F(t) \, dt \bigg|_{s=0}$$

$$= \int_0^\infty \ln(\lambda) \, dF(\lambda) < \infty.$$

3.21. Show directly using Definition 3.128 that $\rho_{\mathrm{an}}^{(2)}(\widetilde{S^1}) = 0$.

3.22. Compute $\rho_{\mathrm{an}}([0,1]) = \frac{\ln(2)}{2}$ and $\rho_{\mathrm{top}}([0,1]) = 0$ for $[0,1]$ equipped with the standard Riemannian metric directly from the definitions without using Theorem 3.161 but using the facts for the Riemannian Zeta-function $\xi_R(s)$ that $\xi_R(0) = -1/2$ and $\frac{d}{ds}\xi_R(0) = -\ln(2)/2$.

3.23. Let M be a cocompact proper free G-manifold of even dimension which is orientable. Suppose that M is det-L^2-acyclic. Show that ∂M is det-L^2-acyclic and

$$\rho^{(2)}(M) = \frac{\rho^{(2)}(\partial M)}{2}.$$

3.24. Compute the characteristic sequence $c(A, K)_p$ of the $(1, 1)$-matrix $(z - \lambda)$ over $\mathbb{C}[\mathbb{Z}]$ for $z \in \mathbb{Z}$ a fixed generator and real numbers $\lambda \geq 0$ and $K = \lambda + 1$ and conclude

$$\sum_{p=1}^{\infty} \frac{1}{p} \cdot \left(\frac{(2 \cdot \lambda)^p}{(1+\lambda)^{2p}} \cdot \sum_{k=0}^{[p/2]} 4^{-k} \cdot \frac{p!}{(p-2k)! \cdot k! \cdot k!} \right)$$
$$= \begin{cases} 2 \cdot \ln(\lambda + 1) - 2 \cdot \ln(|\lambda|) & \text{if } \lambda \geq 1 \\ 2 \cdot \ln(\lambda + 1) & \text{if } \lambda \leq 1 \end{cases} .$$

4. L^2-Invariants of 3-Manifolds

Introduction

In this section we compute all L^2-Betti numbers and the L^2-torsion and give some values and estimates for the Novikov-Shubin invariants for the universal covering of a compact connected orientable 3-manifold. This will use both the general properties of these L^2-invariants which we have developed in the preceding chapters and the geometry of 3-manifolds. In particular our computations will be based on Thurston's Geometrization Conjecture. The necessary input of the theory of 3-manifolds will be given in Section 4.1 and the actual computations and sketches of their proofs in Section 4.2. In our opinion they combine analytic, geometric and topological methods in a beautiful way. Moreover, these computations will give evidence for various general conjectures about L^2-invariants such as Conjecture 2.82 about the positivity and rationality of Novikov-Shubin invariants, the Strong Atiyah Conjecture 10.2, the Singer Conjecture 11.1, Conjecture 11.3 about the parity of the L^2-torsion of the universal covering of an aspherical closed manifold and the zero-in-the-spectrum Conjecture 12.1.

4.1 Survey on 3-Manifolds

In this section we give a brief survey about connected compact orientable 3-manifolds. For more information we refer for instance to [252], [466], [491], [492].

In the sequel 3-manifold means connected compact orientable 3-manifold possibly with boundary. A 3-manifold M is *prime* if for any decomposition of M as a connected sum $M_1 \# M_2$, M_1 or M_2 is homeomorphic to S^3. It is *irreducible* if every embedded 2-sphere bounds an embedded 3-disk. Any prime 3-manifold is either irreducible or is homeomorphic to $S^1 \times S^2$ [252, Lemma 3.13]. A 3-manifold M has a prime decomposition, i.e. one can write M as a connected sum

$$M = M_1 \# M_2 \# \ldots \# M_r,$$

where each M_j is prime, and this prime decomposition is unique up to renumbering and orientation preserving homeomorphism [252, Theorems 3.15, 3.21].

By the Sphere Theorem [252, Theorem 4.3], an irreducible 3-manifold is aspherical if and only if it is a 3-disk or has infinite fundamental group.

Given a 3-manifold M, a compact connnected orientable surface F which is properly embedded in M, i.e. $\partial M \cap F = \partial F$, or embedded in ∂M, is called *incompressible* if it is not a 2-sphere and the inclusion $F \to M$ induces an injection on the fundamental groups. One says that ∂M is *incompressible in* M if and only if ∂M is empty or any component C of ∂M is incompressible in the sense above. An irreducible 3-manifold is *Haken* if it contains an embedded orientable incompressible surface. If $H_1(M)$ is infinite, which is implied if ∂M contains a surface other than S^2, and M is irreducible, then M is Haken [252, Lemma 6.6 and 6.7].

The fundamental group plays a dominant role in the theory of 3-manifolds, as explained by the next results. Let M be a 3-manifold with incompressible boundary whose fundamental group admits a splitting $\alpha\colon \pi_1(M) \to \Gamma_1 * \Gamma_2$. Kneser's Conjecture, whose proof can be found in [252, chapter 7], says that there are manifolds M_1 and M_2 with Γ_1 and Γ_2 as fundamental groups and a homeomorphism $M \to M_1 \# M_2$ inducing α on the fundamental groups. Kneser's conjecture fails even in the closed case in dimensions ≥ 5 by results of Cappell [87], [88] and remains true in dimension 4 stably but not unstably [298], [299].

Let $(f, \partial f)\colon (M, \partial M) \to (N, \partial N)$ is a map of (compact connected orientable) Haken 3-manifolds such that $\pi_1(f, x)\colon \pi_1(M, x) \to \pi_1(N, f(x))$ and $\pi_1(\partial f, y)\colon \pi_1(\partial M, y) \to \pi_1(\partial N, f(y))$ are isomorphisms for any choice of base points $x \in M$ and $y \in \partial M$, then f is homotopic to a homeomorphism. This is a result of Waldhausen [252, Corollary 13.7 on page 148], [507]. One can read off from the fundamental group of a 3-manifold M whether M is the total space of a fiber bundle [252, Chapter 11]. One knows which finite groups or abelian groups occur as fundamental groups of 3-manifolds [252, Chapter 9]. Notice that for $n \geq 4$ any finitely presented group is the fundamental group of a closed connected orientable n-dimensional manifold but not any finitely presented group occurs as the fundamental group of a compact connected 3-manifold. For instance the fundamental group of a 3-manifold whose prime factors are all non-exceptional is residually finite [253]. (The notion of exceptional 3-manifold will be introduced in Section 4.2.)

Recall that a manifold (possible with boundary) is called *hyperbolic* if its interior admits a complete Riemannian metric whose sectional curvature is constant -1. We use the definition of *Seifert fibered 3-manifold* or briefly *Seifert manifold* given in [466], which we recommend as a reference on Seifert manifolds. If a 3-manifold M has infinite fundamental group and empty or incompressible boundary, then it is Seifert if and only if it admits a finite covering $\overline{M}$ which is the total space of a S^1-principal bundle over a compact orientable surface [466, page 436]. The work of Casson and Gabai shows that an irreducible 3-manifold with infinite fundamental group π is Seifert if and only if π contains a normal infinite cyclic subgroup [213, Corollary 2 on page 395].

A *geometry* on a 3-manifold M is a complete locally homogeneous Riemannian metric on its interior. The universal cover of the interior has a complete homogeneous Riemannian metric, meaning that the isometry group acts transitively [476]. Thurston has shown that there are precisely eight simply connected 3-dimensional geometries having compact quotients, namely S^3, $\mathbb{R}^3$, $S^2 \times \mathbb{R}$, $H^2 \times \mathbb{R}$, Nil, $\widetilde{SL_2(\mathbb{R})}$, Sol and H^3. If a closed 3-manifold admits a geometric structure modelled on one of these eight geometries then the geometry involved is unique. In terms of the Euler class e of the Seifert bundle and the Euler characteristic χ of the base orbifold, the geometric structure of a closed Seifert manifold M is determined as follows [466, Theorem 5.3]

$$
\begin{array}{c|ccc}
 & \chi > 0 & \chi = 0 & \chi < 0 \\
\hline
e = 0 & S^2 \times \mathbb{R} & \mathbb{R}^3 & H^2 \times \mathbb{R} \\
e \neq 0 & S^3 & Nil & \widetilde{SL_2(\mathbb{R})}
\end{array}
$$

If M has a S^3-structure then $\pi_1(M)$ is finite. In all other cases M is finitely covered by the total space $\overline{M}$ of an S^1-principal bundle over an orientable closed surface F. Moreover, $e(M) = 0$ if and only if $e(\overline{M}) = 0$, and the Euler characteristic χ of the base orbifold of M is negative, zero or positive according to the same condition for $\chi(\overline{M}/S^1)$ [466, page 426, 427 and 436].

Next we summarize what is known about Thurston's Geometrization Conjecture for irreducible 3-manifolds with infinite fundamental groups. (Again, our 3-manifolds are understood to be compact, connected and orientable.) Johannson [278] and Jaco and Shalen [276] have shown that given an irreducible 3-manifold M with incompressible boundary, there is a finite family of disjoint, pairwise-nonisotopic incompressible tori in M which are not isotopic to boundary components and which split M into pieces that are Seifert manifolds or are *geometrically atoroidal*, meaning that they admit no embedded incompressible torus (except possibly parallel to the boundary). A minimal family of such tori is unique up to isotopy, and we will say that it gives a *toral splitting* of M. We will say that the toral splitting is a *geometric toral splitting* if the geometrically atoroidal pieces which do not admit a Seifert structure are hyperbolic. *Thurston's Geometrization Conjecture* for irreducible 3-manifolds with infinite fundamental groups states that such manifolds have geometric toral splittings.

For completeness we mention that Thurston's Geometrization Conjecture says for a closed 3-manifold with finite fundamental group that its universal covering is homeomorphic to S^3, the fundamental group of M is a subgroup of $SO(4)$ and the action of it on the universal covering is conjugated by a homeomorphism to the restriction of the obvious $SO(4)$-action on S^3. This implies, in particular, the *Poincaré Conjecture* that any homotopy 3-sphere is homeomorphic to S^3.

Suppose that M is Haken. The pieces in its toral splitting are certainly Haken. Let N be a geometrically atoroidal piece. The Torus Theorem says that N is a special Seifert manifold or is *homotopically atoroidal*, i.e. any subgroup of $\pi_1(N)$ which is isomorphic to $\mathbb{Z} \times \mathbb{Z}$ is conjugate to the fundamental group of a boundary component. McMullen following Thurston has shown that a homotopically atoroidal Haken manifold is a twisted I-bundle over the Klein bottle (which is Seifert), or is hyperbolic [368]. Thus the only case in which Thurston's Geometrization Conjecture for an irreducible 3-manifold M with infinite fundamental group is still open is when M is a closed non-Haken irreducible 3-manifold with infinite fundamental group which is not Seifert. The conjecture states that such a manifold is hyperbolic.

4.2 L^2-Invariants of 3-Manifolds

In this section we state the values of the various L^2-invariants for universal coverings of compact connected orientable 3-manifolds. Notice that the assumption orientable is not a serious restriction, since any non-orientable 3-manifold has a connected two-sheeted covering which is orientable and we know how the L^2-invariants behave under finite coverings (see Theorem 1.35 (9), Theorem 2.55 (6) and Theorem 3.96 (5)). Recall that we have already computed the L^2-invariants for the universal covering of a compact connected orientable surface F (see Example 1.36, Example 2.70 and Theorem 3.105). Notice in the context of L^2-torsion that the universal covering of a compact orientable surface F is L^2-acyclic if and only if F is T^2 or $S^1 \times D^1$.

Let us say that a prime 3-manifold is *exceptional* if it is closed and no finite covering of it is homotopy equivalent to a Haken, Seifert or hyperbolic 3-manifold. No exceptional prime 3-manifolds are known, and Thurston's Geometrization Conjecture and *Waldhausen's Conjecture* that any 3-manifold is finitely covered by a Haken manifold imply that there are none. Notice that any exceptional manifold has infinite fundamental group.

Theorem 4.1 (L^2-Betti numbers of 3-manifolds). *Let M be the connected sum $M_1 \# \ldots \# M_r$ of (compact connected orientable) prime 3-manifolds M_j which are non-exceptional. Assume that $\pi_1(M)$ is infinite. Then the L^2-Betti numbers of the universal covering $\widetilde{M}$ are given by*

$$b_0^{(2)}(\widetilde{M}) = 0;$$

$$b_1^{(2)}(\widetilde{M}) = (r-1) - \sum_{j=1}^{r} \frac{1}{\mid \pi_1(M_j) \mid} + \left| \{C \in \pi_0(\partial M) \mid C \cong S^2\} \right| - \chi(M);$$

$$b_2^{(2)}(\widetilde{M}) = (r-1) - \sum_{j=1}^{r} \frac{1}{\mid \pi_1(M_j) \mid} + \left| \{C \in \pi_0(\partial M) \mid C \cong S^2\} \right|;$$

$$b_3^{(2)}(\widetilde{M}) = 0.$$

In particular, $\widetilde{M}$ has trivial L^2-cohomology if and only if M is homotopy equivalent to $RP^3 \# RP^3$ or a prime 3-manifold with infinite fundamental group whose boundary is empty or a union of tori.

Proof. We give a sketch of the strategy of proof. Details can be found in [322, Sections 5 and 6]. Since the fundamental group is infinite, we get $b_0^{(2)}(\widetilde{M}) = 0$ from Theorem 1.35 (8). If M is closed, we get $b_3^{(2)}(\widetilde{M}) = 0$ because of Poincaré duality (see Theorem 1.35 (3)). If M has boundary, it is homotopy equivalent to a 2-dimensional CW-complex and hence $b_3^{(2)}(\widetilde{M}) = 0$. It remains to compute the second L^2-Betti number, because the first one is then determined by the Euler-Poincaré formula of Theorem 1.35 (2).

Using the formula for connected sums of Theorem 1.35 (6) we reduce the claim to prime 3-manifolds. Since a prime 3-manifold is either irreducible or $S^1 \times S^2$, it remains to treat the irreducible case. If the boundary is compressible, we use the Loop Theorem [252, Theorem 4.2 on page 39] to reduce the claim to the incompressible case. By doubling M we can reduce the claim further to the case of an irreducible 3-manifold with infinite fundamental group and incompressible torus boundary. Because of the toral splitting and the assumptions about Thurston's Geometrization Conjecture it suffices to show that the L^2-Betti numbers vanish if M is Seifert with infinite fundamental group or is hyperbolic with incompressible torus boundary. All these steps use the weakly exact Mayer-Vietoris sequence for L^2-(co)homology (see Theorem 1.21). In the Seifert case we can assume by the multiplicative property (see Theorem 1.35 (9)) that M is a S^1-principal bundle over a 2-dimensional manifold. Then we can apply Theorem 1.40.

The hyperbolic case follows directly from Theorem 1.62 provided that the manifold has no boundary. One of the hard parts in the proof is to reduce the case of a hyperbolic 3-manifold with incompressible torus boundary to the closed case by a careful analysis of the manifold near its boundary using explicit models and the fact that the volume is finite. $\square$

Let $\chi_{\mathrm{virt}}(\pi_1(M))$ be the $\mathbb{Q}$-valued virtual group Euler characteristic of the group $\pi_1(M)$ in the sense of [69, IX.7], [509]. (This will be the same as the L^2-Euler characteristic of $\pi_1(M)$ as explained in Remark 6.81). Then the conclusion in Theorem 4.1 is equivalent to

$$b_1^{(2)}(\widetilde{M}) = -\chi_{\mathrm{virt}}(\pi_1(M));$$
$$b_2^{(2)}(\widetilde{M}) = \chi(M) - \chi_{\mathrm{virt}}(\pi_1(M)).$$

This is proved in [322, page 53 - 54].

Next we state what is known about the values of the Novikov-Shubin invariants of 3-manifolds (see [322, Theorem 0.1]).

Theorem 4.2 (Novikov-Shubin invariants of 3-manifolds). *Let M be the connected sum $M_1 \# \ldots \# M_r$ of (compact connected orientable) prime 3-manifolds M_j which are non-exceptional. Assume that $\pi_1(M)$ is infinite. Then*

(1) We have $\alpha_p(\widetilde{M}) > 0$ for $p \geq 1$;

(2) Let the Poincaré associate $P(M)$ be the connected sum of the M_j's which are not 3-disks or homotopy 3-spheres. Then $\alpha_p(\widetilde{P(M)}) = \alpha_p(\widetilde{M})$ for $p \leq 2$. We have $\alpha_1(\widetilde{M}) = \infty^+$ except for the following cases:

   (a) $\alpha_1(\widetilde{M}) = 1$ if $P(M)$ is $S^1 \times D^2$, $S^1 \times S^2$ or homotopy equivalent to $RP^3 \# RP^3$;

   (b) $\alpha_1(\widetilde{M}) = 2$ if $P(M)$ is $T^2 \times I$ or a twisted I-bundle over the Klein bottle K;

   (c) $\alpha_1(\widetilde{M}) = 3$ if $P(M)$ is a closed $\mathbb{R}^3$-manifold;

   (d) $\alpha_1(\widetilde{M}) = 4$ if $P(M)$ is a closed Nil-manifold;

   (e) $\alpha_1(\widetilde{M}) = \infty$ if $P(M)$ is a closed Sol-manifold;

(3) If M is a closed hyperbolic 3-manifold then $\alpha_2(M) = 1$. If M is a closed Seifert 3-manifold then $\alpha_2(M)$ is given in terms of the Euler class e of the bundle and the Euler characteristic χ of the base orbifold by

	$\chi > 0$	$\chi = 0$	$\chi < 0$
$e = 0$	∞^+	3	1
$e \neq 0$	∞^+	2	1

   If M is a Seifert 3-manifold with boundary then $\alpha_2(M) = \infty^+$ if $M = S^1 \times D^2$, $\alpha_2(M) = 2$ if M is $T^2 \times I$ or a twisted I-bundle over K, and $\alpha_2(M) = 1$ otherwise. If M is a closed Sol-manifold then $\alpha_2(M) \geq 1$.

(4) If ∂M contains an incompressible torus then $\alpha_2(\widetilde{M}) \leq 2$. If one of the M_j's is closed with infinite fundamental group and does not admit an $\mathbb{R}^3$, $S^2 \times \mathbb{R}$ or Sol-structure, then $\alpha_2(\widetilde{M}) \leq 2$.

(5) If M is closed then $\alpha_3(\widetilde{M}) = \alpha_1(\widetilde{M})$. If M is not closed then $\alpha_3(\widetilde{M}) = \infty^+$.

Proof. The strategy of the proof is similar to the one of Theorem 4.1 using now Theorem 2.20, Theorem 2.55, Theorem 2.61, Theorem 2.68 and Theorem 3.183 together with explicit computations of heat kernels on the various spaces occuring in Thurston's list of eight geometries with compact quotients in dimension 3. Details can be found in [322, Sections 5 and 6]. □

   Finally we state the values for the L^2-torsion (see [346, Theorem 0.6]).

Theorem 4.3 (L^2-torsion of 3-manifolds). *Let M be a compact connected orientable prime 3-manifold with infinite fundamental group such that the boundary of M is empty or a disjoint union of incompressible tori. Suppose that M satisfies Thurston's Geometrization Conjecture, i.e. there is a geometric toral splitting along disjoint incompressible 2-sided tori in M whose pieces are Seifert manifolds or hyperbolic manifolds. Let M_1, M_2, ..., M_r be the hyperbolic pieces. They all have finite volume [385, Theorem B on page 52]. Then $\widetilde{M}$ is det-L^2-acyclic and*

$$\rho^{(2)}(\widetilde{M}) \;=\; -\frac{1}{6\pi} \cdot \sum_{i=1}^{r} \mathrm{vol}(M_i).$$

In particular, $\rho^{(2)}(\widetilde{M})$ is 0 if and and only if there are no hyperbolic pieces.

Proof. The strategy of the proof is similar to the one of Theorem 4.1 using now Theorem 3.96, Theorem 3.105, Theorem 3.161 and Theorem 3.163.   □

4.3 L^2-Invariants of Knot Complements

Let $K \subset S^3$ be a knot, i.e. a smooth embedding of S^1 into S^3. Let $N(K)$ be a closed tubular neighboorhood. Notice that $N(K)$ is diffeomorphic to $S^1 \times D^2$. Define the *knot complement* $M(K)$ to be the 3-manifold $S^3 - \mathrm{int}(N(K))$. The complement of the trivial knot is $S^1 \times D^2$.

Lemma 4.4. *The knot complement $M(K)$ of a non-trivial knot is an irreducible compact connected oriented 3-manifold whose boundary is an incompressible torus T^2.*

Proof. Everything is obvious except for the fact that the boundary is incompressible and $M(K)$ is irreducible. Incompressibility is for instance proved in [75, Proposition 3.17 on page 39]. Next we show irreducibility. Let $S^2 \subset M(K)$ be an embedded 2-sphere. In particular we can think of S^2 as embedded in S^3. By the Alexander-Schönflies Theorem [75, Theorem 1.8 on page 5], [381] there are embbeded balls D_1^3 and D_2^3 in S^3 such that $S^3 = D_1^3 \cup D_2^3$ and $\partial D_1^3 = \partial D_2^3 = D_1^3 \cap D_2^3 = S^2$. Since the knot is connected and does not meet the embedded S^2, it is contained in one of the balls, let us say D_2^3. Then D_1^3 is a ball embedded in $M(K)$ whose boundary is the given S^2.   □

In particular $M(K)$ is Haken and Theorem 4.3 applies to $M(K)$ for a nontrivial knot. If we choose a different tubular neighborhood, then the corresponding knot complements are diffeomorphic. This implies that $\rho^{(2)}(\widetilde{M(K)})$ is defined for all knots and depends only on the ambient isotopy class of the knot.

Definition 4.5 (L^2-torsion of a knot). *Define the L^2-torsion of a knot $K \subset S^3$ to be the real number*

$$\rho^{(2)}(K) \;:=\; \rho^{(2)}(\widetilde{M(K)}).$$

Notice that $\rho^{(2)}(\widetilde{M}) = 0$ for the trivial knot by Theorem 3.96 (5). We get from Theorem 4.3.

Theorem 4.6. *Let K be a non-trivial knot. Then the boundary of $M(K)$ is incompressible and there is a geometric toral splitting of $M(K)$ along disjoint* incompressible 2-sided tori in M whose pieces are Seifert manifolds or hyperbolic manifolds. Let M_1, M_2, ..., M_r be the hyperbolic pieces. They all have finite volume. We have

$$\rho^{(2)}(K) = -\frac{1}{6\pi} \cdot \sum_{i=1}^{r} \text{vol}(M_i);$$

The next result follows from [225, Corollary 4.2 on page 696], [395, Lemma 5.5 and Lemma 5.6 on page 102]. For the notions of connected sum (sometimes also called product) of knots and cabling of knots were refer for instance to [75, 2.7 on page 19 and 2.9 on page 20]. The Alexander polynomial $\Delta(K)$ of a knot is explained for instance in [75, Definition 8.10 on page 109], [374], [376], [497].

Theorem 4.7. *(1) Let K be a knot. Then $\rho^{(2)}(K) = 0$ if and only if K is obtained from the trivial knot by applying a finite number of times the operation "connected sum" and "cabling";*
(2) A knot is trivial if and only if both its L^2-torsion $\rho^{(2)}(\widetilde{M})$ and its Alexander polynomial $\Delta(K)$ are trivial.

This shows that invariants of Reidemeister torsion type, namely the L^2-torsion and the Alexander polynomial, detect whether a knot is trivial.

There is the following conjecture due to Kashaev [286] and H. and J. Murakami [395, Conjecture 5.1 on page 102].

Conjecture 4.8 (Volume Conjecture). *Let K be a knot and denote by $J_N(K)$ the normalized colored Jones polynomial at the primitive N-th root of unity as defined in (see [395]). Then*

$$\rho^{(2)}(K) = \frac{-1}{3} \cdot \lim_{N \to \infty} \frac{\ln(|J_N(K)|)}{N}.$$

Kashaev states his conjecture in terms of the sum of the volumes of the hyperbolic pieces in the geometric toral splitting and H. and J. Murakami in terms of the simplicial volume. These lead to equivalent conjectures by Theorem 4.6 and Theorem 14.18 (3). Maybe there is a link between Conjecture 4.8 above and Question 13.73.

The Volume Conjecture 4.8 implies by Theorem 4.7 (2) the version of Vassiliev's conjecture that a knot K is trivial if and only if every Vassiliev finite type invariant of K agrees with the one of the trivial knot. The point is that the colored Jones polynomials and the Alexander polynomial are determined by the Vassiliev finite type invariants.

4.4 Miscellaneous

The combinatorial computations of Theorem 3.172 and Remark 3.173 enables us to compute the L^2-torsion of the universal covering of a 3-manifold (and hence in view of Theorem 4.3 the sum of the volumes of its hyperbolic pieces in its geometric toral splitting) directly from a presentation of its fundamental group. Namely, we have (see [330, Theorem 2.4 on page 84])

Theorem 4.9. *Let M be a compact connected orientable irreducible 3-manifold with infinite fundamental group G. Let*

$$G = \langle s_1, s_2, \ldots s_g \mid R_1, R_2, \ldots R_r \rangle$$

be a presentation of G. Let the (r, g)-matrix

$$F = \begin{pmatrix} \frac{\partial R_1}{\partial s_1} & \cdots & \frac{\partial R_1}{\partial s_g} \\ \vdots & \ddots & \vdots \\ \frac{\partial R_r}{\partial s_1} & \cdots & \frac{\partial R_r}{\partial s_g} \end{pmatrix}$$

be the Fox matrix of the presentation (see [75, 9B on page 123], [200], [330, page 84]). Now there are two cases:

(1) Suppose ∂M is non-empty. We make the assumption that ∂M is a union of incompressible tori and that $g = r + 1$. Then M is $\det$-L^2-acyclic. Define A to be the $(g-1, g-1)$-matrix with entries in $\mathbb{Z}G$ obtained from the Fox matrix F by deleting one of the columns. Let α be any real number satisfying $0 < \alpha < \frac{2 \cdot \alpha_2(\widetilde{M})}{\alpha_2(\widetilde{M})+2}$;

(2) Suppose ∂M is empty. We make the assumption that a finite covering of M is homotopy equivalent to a hyperbolic, Seifert or Haken 3-manifold and that the given presentation comes from a Heegaard decomposition. Then M is $\det$-L^2-acyclic and $g = r$. Define A to be the $(g - 1, g - 1)$-matrix with entries in $\mathbb{Z}G$ obtained from the Fox matrix F by deleting one of the columns and one of the rows. Let α be any real number satisfying $0 < \alpha < \frac{2 \cdot \alpha_2(\widetilde{M})}{\alpha_2(\widetilde{M})+1}$.

Let K be any positive real number satisfying $K \geq \|R_A^{(2)}\|$. A possible choice for K is the product of $(g - 1)^2$ and the maximum over the word length of those relations R_i whose Fox derivatives appear in A.

Then the sum of non-negative rational numbers

$$\sum_{p=1}^{L} \frac{1}{p} \cdot \mathrm{tr}_{\mathbb{Z}G}\left(\left(1 - K^{-2} \cdot AA^*\right)^p \right)$$

converges for $L \to \infty$ to the real number $2 \cdot \rho^{(2)}(\widetilde{M}) + 2(g - 1) \cdot \ln(K)$. More precisely, there is a constant $C > 0$ such that we get for all $L \geq 1$

$$0 \leq 2 \cdot \rho^{(2)}(\widetilde{M}) + 2(g - 1) \cdot \ln(K) - \sum_{p=1}^{L} \frac{1}{p} \cdot \mathrm{tr}_{\mathbb{Z}G}\left(\left(1 - K^{-2} \cdot AA^*\right)^p \right) \leq \frac{C}{L^\alpha}.\;\square$$

The case of an automorphism of a compact connected orientable surface, which yields a 3-manifold by taking its mapping torus is discussed in Subsection 7.4.2

Exercises

4.1.   Let M be a compact connected orientable 3-manifold. Show that it is aspherical if and only its prime decompositions has precisely one factor which is not a homotopy sphere and this factor is either D^3 or an irreducible 3-manifold with infinite fundamental group.

4.2.   Show that two connected closed 3-manifolds possessing the same geometry have the same L^2-Betti numbers and Novikov-Shubin invariants, provided their fundamental groups are infinite. Which of the eight geometries can be distinguished from one another by the knowledge of all L^2-Betti numbers and Novikov-Shubin invariants of the universal coverings?

4.3.   Let M be a closed connected orientable 3-manifold which possesses a geometry. Show that the relevant geometry can be read off from the fundamental group π as follows:

(1) $\mathbb{H}^3$: π is not virtually cyclic and contains no subgroup isomorphic to $\mathbb{Z} \oplus \mathbb{Z}$;
(2) S^3: π is finite;
(3) $S^2 \times \mathbb{R}$: π is virtually cyclic and infinite;
(4) $\mathbb{R}^3$: π is contains $\mathbb{Z}^3$ as subgroup of finite index;
(5) Nil: π contains a subgroup of finite index G which can be written as an extension $1 \to \mathbb{Z} \to G \to \mathbb{Z}^2 \to 1$ but π does not contain $\mathbb{Z}^3$ as subgroup of finite index;
(6) $\mathbb{H}^2 \times \mathbb{R}$: π contains a subgroup of finite index which is isomorphic to $\mathbb{Z} \times G$ for some group G and π is not solvable;
(7) $\widetilde{Sl_2(\mathbb{R})}$: π is not solvable, contains $\mathbb{Z} \oplus \mathbb{Z}$ as subgroup and contains no subgroup of finite index which is isomorphic to $\mathbb{Z} \times G$ for some group G;
(8) Sol: π is not virtually abelian and contains a subgroup G of finite index which is an extension $0 \to \mathbb{Z}^2 \to G \to \mathbb{Z} \to 0$.

4.4.   Compute the L^2-Betti numbers and the Novikov-Shubin invariants of the universal covering of a compact connected 3-manifold whose fundamental group is finite.

4.5.   Let M be the connected sum $M_1 \# \ldots \# M_r$ of compact connected orientable 3-manifolds M_j which are non-exceptional and prime. Show that then for some $p \geq 0$ the Laplace operator $(\Delta_p)_{\min}$ of Definition 2.64 acting on smooth p-forms on the universal covering $\widetilde{M}$ has zero in its spectrum.

4.6. Give a proof of the conclusion which is stated in the last sentence of Theorem 4.1.

4.7. Let $f \colon F \to F$ be a selfhomeomorphism of a closed orientable surface of genus ≤ 1. Show that $\widetilde{T}_f$ is det-L^2-acyclic and $\rho^{(2)}(\widetilde{T}_f) = 0$.

5. L^2-Invariants of Symmetric Spaces

Introduction

In this chapter we state the values of the L^2-Betti numbers, the Novikov-Shubin invariants and the L^2-torsion for universal coverings of closed locally symmetric spaces. We give a brief survey about locally symmetric and symmetric spaces in Section 5.1 and state the values in Section 5.2 and 5.3. These computations will give evidence for various general conjectures about L^2-invariants such as Conjecture 2.82 about the positivity and rationality of Novikov-Shubin invariants, the Strong Atiyah Conjecture 10.2, the Singer Conjecture 11.1, Conjecture 11.3 about the parity of the L^2-torsion of the universal covering of an aspherical closed manifold and the zero-in-the-spectrum Conjecture 12.1.

5.1 Survey on Symmetric Spaces

In this section we collect some basic facts about symmetric spaces so that the reader will be able to understand the results on the computations of L^2-invariants of Section 5.2 and 5.3. A reader who is familiar with symmetric spaces should pass directly to Section 5.2.

Let M be a complete Riemannian manifold and let $x \in M$ be a point. A *normal neighborhood* of M at x is an open neighborhood V of x in M such that there is an open neighborhood U of 0 in the tangent space $T_x M$ with the properties that for any $u \in U$ and $t \in [-1, 1]$ also tu belongs to U and the exponential map $\exp_x : T_x M \to M$ induces a diffeomorphism $U \to V$. The *geodesic symmetry* of a normal neighborhood V at x is the diffeomorphism $s_p : V \to V$ which sends $\exp_x(u)$ to $\exp_x(-u)$ for $u \in U$.

Definition 5.1 (Locally symmetric space). *A complete Riemannian manifold M is called a* locally symmetry space *if for any $x \in M$ there exists a normal neighborhood V of x such that the geodesic symmetry is an isometry.*

A complete Riemannian manifold is a locally symmetric space if and only if the sectional curvature is invariant under parallel transports with respect to the Levi-Civita connection [251, Theorem 1.3 in IV.1 on page 201].

Definition 5.2 (Symmetric space). *A complete Riemannian manifold is called* (globally) *symmetric space if for each $x \in M$ there is an isometric diffeomorphism $t_x \colon M \to M$ which is an involution, i.e. $t_x \circ t_x = \mathrm{id}$, and has x as isolated fixed point, i.e. $t_x(x) = x$ and there is a neighborhood W of x in M such that $y \in W, t_x(y) = y$ implies $x = y$.*

Examples of symmetric spaces are S^n, $\mathbb{RP}^n$, $\mathbb{CP}^n$, $\mathbb{R}^n$, $\mathbb{H}^n$.

A symmetric space is always locally symmetric [251, Lemma 3.1 in IV.3 on page 205]. On the other hand any simply connected locally symmetric space is a symmetric space [251, Theorem 5.6. in IV.5 on page 222]. In particular the universal covering of a locally symmetric space is a symmetric space.

Let M be a symmetric space. Denote by $\mathrm{Isom}(M)$ the *group of isometries* $M \to M$. This group inherits the structure of a topological group by the compact-open topology coming from the topology of M. Denote for a Lie group L its *identity component* by L^0. We get from [251, Lemma 3.2 in IV.3 on page 205, Theorem 3.3 in IV.3 on page 208]

Theorem 5.3. *The group $\mathrm{Isom}(M)$ has the unique structure of an (analytic) Lie group. Given a point $x \in M$, let $\mathrm{Isom}(M)_x^0$ be the stabilizer of $\mathrm{Isom}(M)^0$ at x, i.e. the subgroup of elements $f \in \mathrm{Isom}(M)^0$ with $f(x) = x$. Then $\mathrm{Isom}(M)^0$ acts transitively on M, $\mathrm{Isom}(M)_x^0$ is compact and we get a (analytic) diffeomorphism*

$$\overline{\mathrm{ev}}_x \colon \mathrm{Isom}(M)^0 / \mathrm{Isom}(M)_x^0 \xrightarrow{\cong} M, \quad f \cdot \mathrm{Isom}(M)_x^0 \mapsto f(x).$$

The *Killing form* of a Lie algebra $\mathfrak{g}$ is defined by

$$B \colon \mathfrak{g} \times \mathfrak{g} \to \mathbb{R}, \quad (a, b) \mapsto \mathrm{tr}_{\mathbb{R}}(\mathrm{ad}(a)\,\mathrm{ad}(b)),$$

where $\mathrm{ad}(x) \colon \mathfrak{g} \to \mathfrak{g}$ denotes the *adjoint representation* sending z to $[x, z]$ and $\mathrm{tr}_{\mathbb{R}}$ is the trace of an endomorphism of a finite dimensional real vector space. The Lie algebra $\mathfrak{g}$ is *semisimple* if the Killing form is non-degenerate. This is equivalent to the condition that $\mathfrak{g}$ contains no non-trivial solvable ideals, where solvable means that the commutators series ends at $\{0\}$ [293, page 668]. A Lie algebra $\mathfrak{g}$ is *simple* if it is non-abelian and contains no proper ideals.

The Killing form on a semisimple Lie algebra $\mathfrak{g}$ is strictly negative-definite if and only if there is a connected compact Lie group G whose Lie algebra is $\mathfrak{g}$ [251, Corollary 6.7 in II.6 on page 133].

An involution of Lie algebras $\theta \colon \mathfrak{g} \to \mathfrak{g}$ is a *Cartan involution* if for the associated decomposition $\mathfrak{g} = \mathfrak{k} \oplus \mathfrak{p}$ with $\mathfrak{k} = \ker(\theta - \mathrm{id})$ and $\mathfrak{p} = \ker(\theta + \mathrm{id})$ the Killing form B is strictly negative-definite on $\mathfrak{k}$ and strictly positive-definite on $\mathfrak{p}$. In this context $\mathfrak{g} = \mathfrak{k} \oplus \mathfrak{p}$ is called *Cartan decomposition*. Any semisimple Lie algebra $\mathfrak{g}$ has a Cartan decomposition and two Cartan decompositions are conjugate [251, III.7]. Notice that $\mathfrak{k}$ is a subalgebra and $\mathfrak{p}$ a vector subspace.

Example 5.4. A *connected linear reductive Lie group* G is a closed connected subgroup of $GL(n, \mathbb{R})$ or $GL(n, \mathbb{C})$, which is stable under taking transpose conjugate matrices. Examples are $SL(n, \mathbb{C}) \subset GL(n, \mathbb{C})$ and $SO(n) \subset GL(n, \mathbb{R})$. Such a group G is a Lie group. It is semisimple if and only if its center is finite. Taking inverse conjugate transpose matrices induces an involution of Lie groups $\Theta \colon G \to G$. The group $K = \{g \in G \mid \Theta(g) = g\}$ turns out to be a maximal compact subgroup. Denote by $\mathfrak{g}$ the Lie algebra of G and by $\theta \colon \mathfrak{g} \to \mathfrak{g}$ the differential of Θ at the identity matrix. Notice that $\mathfrak{gl}_n(\mathbb{R})$ is $M_n(\mathbb{R})$ and $\mathfrak{gl}_n(\mathbb{C})$ is $M_n(\mathbb{C})$ and that $\mathfrak{g}$ is a subalgebra. The involution of Lie algebras θ sends a matrix to its negative conjugate transpose. The Lie bracket of $\mathfrak{g}$ is given by taking commutators. The involution θ is a Cartan involution. In the Cartan decomposition $\mathfrak{g} = \mathfrak{k} \oplus \mathfrak{p}$ the subalgebra $\mathfrak{k}$ is the Lie algebra of $K \subset G$.

Let M be a symmetric space. In the sequel we abbreviate $G = \mathrm{Isom}(M)^0$ and $K = \mathrm{Isom}(M)^0_x$. Let $\sigma \colon G \to G$ be given by conjugation with the geodesic symmetry $s_x \in K$. Let $\mathfrak{g}$ be the Lie algebra of G. Put

$$\mathfrak{k} = \{a \in \mathfrak{g} \mid T_1\sigma(a) = a\};$$
$$\mathfrak{p} = \{a \in \mathfrak{g} \mid T_1\sigma(a) = -a\}.$$

We get a decomposition of Lie algebras $\mathfrak{g} = \mathfrak{k} \oplus \mathfrak{p}$. Notice that $\mathfrak{k}$ is the kernel of the differential at $1 \in G$ of the evaluation map $\mathrm{ev}_x \colon G \to M$ which sends f to $f(x)$.

Definition 5.5 (Type of a symmetric space). *The symmetric space $M = G/K$ is of* compact type *if $\mathfrak{g}$ is semisimple and has strictly negative-definite Killing form. It is of* non-compact type *if $\mathfrak{g}$ is semisimple and σ is a Cartan involution. It is called of* Euclidean type *if $\mathfrak{p}$ is an abelian ideal in $\mathfrak{g}$.*

A symmetric space M is of compact type if and only $\mathrm{Isom}(M)$ is compact and semisimple. A symmetric space M of compact type is a compact manifold.

If $M = G/K$ for $G = \mathrm{Isom}(M)^0$ and $K = \mathrm{Isom}(M)^0_x$ is a symmetric space of non-compact type, then K is connected and is a maximal compact subgroup in G and M is diffeomorphic to $\mathbb{R}^n$ [251, Theorem 1. in Chaper VI on page 252]. A symmetric space M is of non-compact type if and only if the Lie algebra of $\mathrm{Isom}(M)$ is semisimple and has no compact ideal [157, Proposition 2.1.1 on page 69], [251, page 250]. On the other hand, given a connected semisimple Lie group G with finite center such that its Lie algebra has no compact ideal, the homogeneous space G/K for a maximal compact subgroup $K \subset G$ equipped with a G-invariant Riemannian metric is a symmetric space of non-compact type with $G = \mathrm{Isom}(M)^0$ and $K = \mathrm{Isom}(M)^0_x$ [157, Section 2.2 on page 70].

A *θ-stable Cartan subalgebra* $\mathfrak{h} \subset \mathfrak{g}$ of a semisimple Lie algebra $\mathfrak{g}$ is a maximal abelian θ-stable abelian subalgebra. All θ-stable Cartan subalgebras have the same dimension. Hence we can define the *complex rank* $\mathrm{rk}_{\mathbb{C}}(\mathfrak{g})$ by the dimension of a θ-stable Cartan subalgebra of $\mathfrak{g}$. [293, page 128f]. (This should not be confused with the real rank which is the dimension of the term $\mathfrak{a}$ in the Iwasawa decomposition $\mathfrak{g} = \mathfrak{k} \oplus \mathfrak{a} \oplus \mathfrak{n}$.) Recall that a Lie group G is *semisimple* if its Lie algebra is semisimple. The *complex rank* of a semisimple Lie group $\mathrm{rk}_{\mathbb{C}}(G)$ is defined to be the complex rank of its Lie algebra $\mathfrak{g}$. The complex rank of a connected compact Lie group is the same as the dimension of a maximal torus.

Definition 5.6. *Let* $M = G/K$ *be a symmetric space such that* $G = \mathrm{Isom}(M)^0$ *is semisimple. Define its* fundamental rank

$$\text{f-rk}(M) := \mathrm{rk}_{\mathbb{C}}(G) - \mathrm{rk}_{\mathbb{C}}(K).$$

This is the notion of rank which will be relevant for our considerations. It should not be confused with the following different notion.

A Riemannian submanifold $N \subset M$ of a complete Riemannian manifold M is *totally geodesic* if for any geodesic $\gamma \colon \mathbb{R} \to M$, for which $\gamma(0) \in N$ and $\gamma'(0)$ lies in the tangent space $T_{\gamma(0)}N$, the image of γ lies in N. Recall that a Riemannian manifold M is *flat* if the sectional curvature is identically zero.

Definition 5.7. *Let* M *be a symmetric space. Its* rank *is the maximal dimension of a flat totally geodesic complete submanifold of* M.

The rank of M in the sense of Definition 5.7 is always greater or equal to the fundamental rank. This follows from [56, Formula (3) in III.4 on page 99].

The next two results are taken from [251, Proposition 4.2 in V.4 on page 244, Theorem 3.1 in V.3 on page 241].

Theorem 5.8. *Let* M *be a simply connected symmetric space. Then it can be written as a product*

$$M = M_{\mathrm{cp}} \times M_{\mathrm{Eucl}} \times M_{\mathrm{ncp}},$$

where M_{cp} *is of compact type,* M_{Eucl} *of Euclidean type and* M_{ncp} *of noncompact type.*

Theorem 5.9. *Let* M *be a symmetric space. Then*

(1) If M *is of compact type, then the sectional curvature of* M *is nonnegative:* $\sec(M) \geq 0$;

(2) If M *is of Euclidean type, then* M *is flat:* $\sec(M) = 0$;

(3) If M *is of non-compact type, then the sectional curvature of* M *is nonpositive:* $\sec(M) \leq 0$;

Lemma 5.10. *A simply connected symmetric space* M *is contractible if and only if in the decomposition* $M = M_{\mathrm{cp}} \times M_{\mathrm{Eucl}} \times M_{\mathrm{ncp}}$ *of Theorem 5.8 the factor* M_{cp} *of compact type is trivial. In this case* M *is diffeomorphic to* $\mathbb{R}^n$.

Proof. The factor M_{cp} is a closed manifold and hence contractible if and only if M_{cp} is a point. Since M_{Eucl} and M_{ncp} carry Riemannian metrics of non-positive sectional curvature, they are diffeomorphic to $\mathbb{R}^n$ for appropriate n by Hadamard's Theorem. □

There is an important duality between symmetric spaces of non-compact type and symmetric spaces of compact type [251, V.2]. Let $M = G/K$ be a symmetric space of non-compact type. Let $\mathfrak{g}$ and $\mathfrak{k}$ be the Lie algebras of G and K and let $\mathfrak{g} = \mathfrak{k} \oplus \mathfrak{p}$ be the Cartan decomposition. The complexification $G_{\mathbb{C}}$ of G is the simply connected Lie group with the complexification $\mathbb{C} \otimes_{\mathbb{R}} \mathfrak{g}$ of $\mathfrak{g}$ as Lie algebra. Obviously $\mathfrak{g}^d = \mathfrak{k} \oplus i \cdot \mathfrak{p}$ is a real subalgebra of $\mathbb{C} \otimes_{\mathbb{R}} \mathfrak{g}$. Let G^d be the corresponding analytic subgroup of the complexification $G_{\mathbb{C}}$ of G. Then G^d is a compact group. Let $K' \subset G^d$ be the subgroup corresponding to $\mathfrak{k} \subset \mathfrak{g}^d$. The *dual symmetric space* is defined to be $M^d = G^d/K'$ with respect to the G^d-invariant Riemannian metric for which multiplication with i induces an isometry $T_{1K}G/K \to T_{1K}G^d/K'$. M. Olbricht pointed out to us that one can assume without loss of generality that G is linear, i.e. $G \subset GL(n, \mathbb{R})$. Put G^d to be the analytic subgroup in $GL(n, \mathbb{C})$ corresponding to $\mathfrak{g}^d$. Then K is also a subgroup of G^d and M^d agrees with G^d/K. The symmetric space M^d is of compact type. Analogously one can associate to a symmetric space of compact type M a symmetric space of non-compact type M^d. In both cases $(M^d)^d = M$. The following example is taken from [251, Example 1 in V.2 on page 238].

Example 5.11. Denote by by $SO(p,q)$ the group of real $(p+q)$-$(p+q)$-matrices of determinant 1 which leave the quadratic form $-x_1^2 - \ldots - x_p^2 + x_{p+1}^2 + \ldots x_{p+q}^2$ invariant. Denote by $SO(p,q)^0$ the identity component of $SO(p,q)$. Let $SO(n)$ be the Lie group $\{A \in GL(n, \mathbb{R}) \mid AA^t = I, \det(A) = 1\}$. This agrees with $SO(0,n)$. There are obvious embeddings of $SO(p) \times SO(q)$ into both $SO(p,q)^0$ and $SO(p+q)$. Equip $SO(p,q)^0/SO(p) \times SO(q)$ and $SO(p+q)/SO(p) \times SO(q)$ with a $SO(p,q)^0$-invariant and $SO(p+q)$-invariant Riemannian metric. These are uniquely determined up to scaling with a constant. Then $SO(p,q)^0/SO(p) \times SO(q)$ is a symmetric space of non-compact type and $SO(p+q)/SO(p) \times SO(q)$ is a symmetric space of compact type. They are dual to one another possibly after scaling the Riemannian metric with a constant.

5.2 L^2-Invariants of Symmetric Spaces of Non-Compact Type

In this section we state the values of the L^2-Betti numbers, the Novikov-Shubin invariants and the L^2-torsion of the universal covering $\widetilde{M}$ of a closed Riemannian manifold M provided that $\widetilde{M}$ is a symmetric space of non-compact type. Notice that a symmetric space comes with a preferred Riemannian metric so that its L^2-torsion is defined without the assumption that all L^2-Betti numbers of $\widetilde{M}$ vanish. Recall that it does not matter whether we work with the topological version (see Definition 3.120) or with the analytic version (see Definition 3.128) of L^2-torsion because of Theorem 3.149.

Theorem 5.12 (L^2-invariants of symmetric spaces). *Let M be a closed Riemannian manifold whose universal covering $\widetilde{M}$ is a symmetric space of non-compact type. Let $\text{f-rk}(\widetilde{M}) := \text{rk}_{\mathbb{C}}(G) - \text{rk}_{\mathbb{C}}(K)$ be the fundamental rank of the universal covering $\widetilde{M}$ for $G = \text{Isom}(\widetilde{M})^0$ and $K \subset G$ a maximal compact subgroup. Then*

(1) We have $b_p^{(2)}(\widetilde{M}) \neq 0$ if and only if $\text{f-rk}(\widetilde{M}) = 0$ and $2p = \dim(M)$. Let $\widetilde{M}^d$ be the to $\widetilde{M}$ dual symmetric space. If $\text{f-rk}(\widetilde{M}) = 0$, then $\dim(M)$ is even and for $2p = \dim(M)$ we get

$$0 < b_p^{(2)}(\widetilde{M}) = (-1)^p \cdot \chi(M) = \frac{\text{vol}(M)}{\text{vol}(\widetilde{M}^d)} \cdot \chi(\widetilde{M}^d);$$

(2) We have $\alpha_p(\widetilde{M}) \neq \infty^+$ if and only if $\text{f-rk}(\widetilde{M}) > 0$ and p belongs to $[\frac{\dim(M)-\text{f-rk}(\widetilde{M})}{2} + 1, \frac{\dim(M)+\text{f-rk}(\widetilde{M})}{2}]$. If $\alpha_p(\widetilde{M}) \neq \infty^+$, then $\alpha_p(\widetilde{M}) = \text{f-rk}(\widetilde{M})$.

The number $\dim(M) - \text{f-rk}(\widetilde{M})$ is even and positive if M is not the one-point-space $\{*\}$;

(3) We have $\rho^{(2)}(\widetilde{M}) \neq 0$ if and only if $\text{f-rk}(\widetilde{M}) = 1$;

(4) Suppose that $\text{f-rk}(\widetilde{M}) = 1$. Then $\widetilde{M} = X_0 \times X_1$, where X_0 is a symmetric space of non-compact type with $\text{f-rk}(X_0) = 0$ and $X_1 = X_{p,q} := SO(p,q)^0/SO(p) \times SO(q)$ for p, q odd or $X_1 = SL(3, \mathbb{R})/SO(3)$. We have

$$\rho^{(2)}(\widetilde{M}) = \text{vol}(M) \cdot T^{(2)}(\widetilde{M}),$$

where the number $T^{(2)}(\widetilde{M})$ is given below.

(a) If X_0^d is the symmetric space dual to X_0, then $\dim(X_0)$ is even, $\chi(X_0^d) > 0$ and we have

$$T^{(2)}(\widetilde{M}) := (-1)^{\dim(X_0)/2} \cdot \frac{\chi(X_0^d)}{\text{vol}(X_0^d)} \cdot T^{(2)}(X_1);$$

(b) Let C_n be the positive constant introduced in (3.151). Denote by $(\mathbb{H}^n)^d$ the symmetric space dual to the hyperbolic space $\mathbb{H}^n$. (Then $(\mathbb{H}^n)^d$ is the sphere S^n with the Riemannian metric whose sectional curvature is constant 1). We have

$$T^{(2)}(X_{p,q}) = (-1)^{\frac{pq-1}{2}} \cdot \chi\left((X_{p-1,q-1})^d\right) \cdot \frac{\text{vol}\left((\mathbb{H}^{p+q-1})^d\right)}{\text{vol}\left((X_{p,q})^d\right)} \cdot C_{p+q-1}$$

and

$$\chi\left((X_{p-1,q-1})^d\right) = \begin{cases} 2\cdot\begin{pmatrix} \frac{p+q-2}{2} \\ \frac{p-1}{2} \end{pmatrix} & p,q > 1 \\ 1 & p = 1, q > 1 \end{cases} ;$$

(c) We have

$$T^{(2)}(SL(3,\mathbb{R})/SO(3)) = \frac{2\pi}{3\,\mathrm{vol}\left((SL(3,\mathbb{R})/SO(3))^d\right)}.$$

Here are some explanations. Of course $\widetilde{M}$ inherits its Riemannian metric from the one on M. The symmetric spaces $X_{p,q} = SO(p,q)^0/SO(p) \times SO(q)$ and $SL(3,\mathbb{R})/SO(3)$ are equipped with the Riemannian metrics coming from $\widetilde{M} = X_0 \times X_1$. These Riemannian metrics are $SO(p,q)^0$-invariant and $SL(3,\mathbb{R})$-invariant. Two such invariant Riemannian metrics on $X_{p,q} = SO(p,q)^0/SO(p) \times SO(q)$ and $SL(3,\mathbb{R})/SO(3)$ differ only by scaling with a constant. Notice that $X_{1,q}$ is isometric to $\mathbb{H}^q$ after possibly scaling the metric with a constant. If we scale the Riemannian metric on M by a constant C, then $\mathrm{vol}(M)$ is scaled by $C^{\dim(M)}$ and $T^{(2)}(\widetilde{M})$ by $C^{-\dim(M)}$. Hence $\rho^{(2)}(\widetilde{M})$ is unchanged.

The result for L^2-Betti numbers has been proved in [54]. The computations for the Novikov-Shubin invariants have been carried out in [314, Section 11], [404, Theorem 1.1], partial results have already been obtained in [316, VII.B]. (The reader should be aware of the fact that range of the finiteness of $\alpha_p(\widetilde{M})$ is misprinted in [314, Section 11] and correct in [316, VII.B] and [404, Theorem 1.1].) Notice that there is a shift by one in the range where $\alpha_p(\widetilde{M}) \neq \infty^+$ in Theorem 5.12 (2) in comparison with [404, Theorem 1.1], since the definition of $\alpha_p(\widetilde{M})$ here and in [404, Theorem 1.1] differ by 1 concerning the index p. The result about L^2-torsion is proved in [404, Theorem 1.1, Proposition 1.3 and Proposition 1.4]. The basic input is the Harish-Chandra Plancherel Theorem (see [248], [293, Theorem 3.11 in Chapter XIII on page 511]) and $(\mathfrak{g}, K)$-cohomology.

5.3 L^2-Invariants of Symmetric Spaces

Let M be a simply connected symmetric space which is not necessarily of noncompact type. We conclude from Theorem 3.183 that there are constants $B_p^{(2)}(M)$ for $p \geq 0$, $A_p^{(2)}(M)$ for $p \geq 1$ and $T^{(2)}(M)$ depending only on the Riemannian structure on M such that for any cocompact free proper action of a group G by isometries

$$b_p^{(2)}(M;\mathcal{N}(G)) = B_p^{(2)}(M) \cdot \mathrm{vol}(G\backslash M);$$
$$\alpha_p^{(2)}(M;\mathcal{N}(G)) = A_p^{(2)}(M);$$
$$\rho^{(2)}(M;\mathcal{N}(G)) = T^{(2)}(M) \cdot \mathrm{vol}(G\backslash M).$$

We want to compute these numbers explicitly. Recall that M splits as a product $M_{\mathrm{cp}} \times M_{\mathrm{Eucl}} \times M_{\mathrm{ncp}}$, where M_{cp}, M_{Eucl} and M_{ncp} respectively are symmetric spaces of compact type, Euclidean type and non-compact type respectively (see Theorem 5.8). Notice that these numbers have already been computed for M_{ncp} in Theorem 5.12 and that M_{Eucl} is isometric to $\mathbb{R}^n$ with the standard Euclidean Riemannian metric [218, Theorem 3.82 on page 131]. We want to extend these computations to M. Suppose that G_0 and G_1 respectively are groups with a cocompact free proper action by isometries on M_{Eucl} and M_{ncp} respectively. We conclude from the product formula for L^2-torsion (see Theorem 3.93 (4) and (3.126)) that $\rho^{(2)}(M_{\mathrm{cp}} \times M_{\mathrm{Eucl}} \times M_{\mathrm{ncp}}; \mathcal{N}(\{1\} \times G_0 \times G_1))$ is given by

$$\chi(M_{\mathrm{cp}}) \cdot \rho^{(2)}(M_{\mathrm{ncp}}; \mathcal{N}(G_1)) + \chi(G_1 \backslash M_{\mathrm{ncp}}) \cdot \rho^{(2)}(M_{\mathrm{cp}}; \mathcal{N}(\{1\})),$$

if $M_{\mathrm{Eucl}} = *$ and by 0 if $M_{\mathrm{Eucl}} \neq *$. Since by Hirzebruch's proportionality principle [262]. $\frac{\chi(G_1 \backslash M_{\mathrm{ncp}})}{\mathrm{vol}(G_1 \backslash M_{\mathrm{ncp}})} = \frac{\chi(M_{\mathrm{ncp}}^d)}{\mathrm{vol}(M_{\mathrm{ncp}}^d)}$, we conclude

$$T^{(2)}(M) = \begin{cases} \frac{\chi(M_{\mathrm{ncp}}^d)}{\mathrm{vol}(M_{\mathrm{ncp}}^d)} \cdot T^{(2)}(M_{\mathrm{cp}}) & \text{if } M_{\mathrm{Eucl}} = \{*\}, \text{f-rk}(M_{\mathrm{ncp}}) = 0; \\ \frac{\chi(M_{\mathrm{cp}})}{\mathrm{vol}(M_{\mathrm{cp}})} \cdot T^{(2)}(M_{\mathrm{ncp}}) & \text{if } M_{\mathrm{Eucl}} = \{*\}, \text{f-rk}(M_{\mathrm{ncp}}) = 1; \\ 0 & \text{otherwise.} \end{cases} \tag{5.13}$$

Notice that we have already given the value of $T^{(2)}(M_{\mathrm{ncp}})$ in Theorem 5.12 (4). Similar one gets from the product formula for the Novikov-Shubin invariants (see Theorem 2.55 (3)), where the necessary assumption about the limit property follows from the computations in [404]

$$A_p(M) = \begin{cases} a & \text{if } b_{p-i}(M_{\mathrm{cp}}) \neq 0 \text{ for some integer } i \text{ satisfying} \\ & \frac{\dim(M_{\mathrm{ncp}}) - \text{f-rk}(M_{\mathrm{ncp}})}{2} + 1 \leq i \text{ and} \\ & i \leq \frac{\dim(M_{\mathrm{ncp}}) + \text{f-rk}(M_{\mathrm{ncp}})}{2} + \dim(M_{\mathrm{Eucl}}); \\ \infty^+ & \text{otherwise.} \end{cases} \tag{5.14}$$

where $a = \text{f-rk}(M_{\mathrm{ncp}}) + \dim(M_{\mathrm{Eucl}})$. We get from the product formula for L^2-Betti numbers (see Theorem 1.35 (4))

$$B_p^{(2)}(M) = \begin{cases} b & \text{if f-rk}(M_{\mathrm{ncp}}) = 0 \text{ and } M_{\mathrm{Eucl}} = \{*\} \\ 0 & \text{otherwise} \end{cases} ; \tag{5.15}$$

where $b = \frac{b_{(p-\dim(M_{\mathrm{ncp}})/2)}(M_{\mathrm{cp}})}{\mathrm{vol}(M_{\mathrm{cp}})} \cdot (-1)^{\dim(M_{\mathrm{ncp}})/2} \cdot \frac{\chi(M_{\mathrm{ncp}}^d)}{\mathrm{vol}(M_{\mathrm{ncp}}^d)}$.

Suppose that the closed locally symmetric space M has negative sectional curvature. Then its universal covering $\widetilde{M}$ must be of non-compact type with fundamental rank f-rk$(\widetilde{M}) \leq 1$ since its rank in the sense of Definition 5.7 is always greater or equal to the fundamental rank [56, Formula (3) in III.4 on page 99] and is equal to 1 for a Riemannian metric with negative sectional curvature. We conclude from Lemma 5.10, Theorem 5.12 and equations (5.13), (5.14) and (5.15)

Corollary 5.16. *Let M be an aspherical closed Riemannian manifold whose universal covering $\widetilde{M}$ is a symmetric space. Then*

(1) We have $b_p^{(2)}(\widetilde{M}) = 0$ if $2p \neq \dim(M)$. If $\widetilde{M}$ has negative sectional curvature and has even dimension, then $b_{\dim(M)/2}^{(2)}(\widetilde{M}) > 0$, f-rk$(\widetilde{M}) = 0$;

(2) One of the L^2-Betti numbers $b_p^{(2)}(\widetilde{M})$ is different from zero or one of the Novikov Shubin invariants $\alpha_p(\widetilde{M})$ is different from ∞^+;

(3) If $\rho^{(2)}(\widetilde{M}) \neq 0$, then $\widetilde{M}$ is of non-compact type, $\dim(M)$ is odd and we have

$$(-1)^{\frac{\dim(M)-1}{2}} \cdot \rho^{(2)}(M) > 0.$$

If M carries a metric of negative sectional curvature and has odd dimension, then

$$(-1)^{\frac{\dim(M)-1}{2}} \cdot \rho^{(2)}(M) > 0.$$

If M carries a metric of negative sectional curvature and has even dimension, then

$$\alpha_p(\widetilde{M}) = \infty^+ \qquad for \ p \geq 1.$$

Corollary 5.16 implies that various conjectures for aspherical closed manifolds (see Section 11.1.3 and Subsection 12.2.2) turn out to be true in the case that the universal covering is a symmetric space.

5.4 Miscellaneous

Consider a closed Riemannian manifold M with non-positive sectional curvature such that $\widetilde{M}$ is det-L^2-acyclic. Suppose that N is a closed manifold which is homotopy equivalent to M. The Whitehead group of the fundamental group of a closed Riemannian manifold with non-positive sectional curvature is known to be zero [192, page 61]. (Actually Farrell and Jones prove the stronger statement that for any closed aspherical topological manifold N with $\dim(N) \neq 3, 4$ and $\pi_1(M) \cong \pi_1(N)$ any homotopy equivalence $M \to N$ is homotopic to a homeomorphism [195, Theorem 0.1].) Hence also $\widetilde{N}$ is det-L^2-acyclic and $\rho^{(2)}(\widetilde{M}) = \rho^{(2)}(\widetilde{N})$ by Theorem 3.96 (1).

Notice that this applies by Lemma 5.9, Lemma 5.10 and Theorem 5.12 (2) to the case, where M is an aspherical closed manifold which admits the structure of a locally symmetric space and satisfies $b_p^{(2)}(\widetilde{M}) = 0$ for $p \geq 0$. In particular we conclude from Lemma 5.10, from Theorem 5.12 (1) and (3) and from (5.13) that for two aspherical closed locally symmetric Riemannian manifolds M and N with isomorphic fundamental groups $\rho^{(2)}(\widetilde{M}) = \rho^{(2)}(\widetilde{N})$ holds.

Let M be a closed topological manifold with $\dim(M) \neq 3, 4$. Then M carries the structure of a locally symmetric Riemannian manifold whose universal covering is a symmetric space of non-compact type if and only if M is aspherical and $\pi_1(M)$ is isomorphic to a cocompact discrete subgroup of a linear semisimple Lie group with finitely many path components [195, Theorem 0.2].

The famous rigidity result of Mostow [388] says that two closed locally symmetric spaces M and N with non-positive sectional curvature are isometrically diffeomorphic if and only if $\pi_1(M) \cong \pi_1(N)$ and $\mathrm{vol}(M) = \mathrm{vol}(N)$ hold, provided that $\widetilde{N}$ is irreducible, i.e. not a product of two Riemannian manifolds of positive dimension, and $\dim(N) \geq 3$. The following rigidity result is proved in [20, Theorem 1 on page i]. Let M be a closed locally symmetric space such that its rank (in the sense of Definition 5.7) is greater or equal to 2 and its universal covering is irreducible. Let N be a closed Riemannian manifold with non-positive sectional curvature. Suppose that $\pi_1(M) \cong \pi_1(N)$ and $\mathrm{vol}(M) = \mathrm{vol}(N)$. Then M and N are isometrically diffeomorphic.

A classification of symmetric spaces is given in [251, Chapter X]. More information about Lie groups, Lie algebras and symmetric spaces can be found for instance in [17], [19], [20], [157], [158], [251], [268] and [293].

Exercises

5.1. Let M be a symmetric space and let $x \in M$. Let $t_x \colon M \to M$ be an isometric involution which has x as isolated fixed point. Show that there is a normal neighborhood U of x such that t induces the geodesic symmetry on U.

5.2. Show that the following groups are connected linear reductive Lie groups:

$$GL(n, \mathbb{C}) = \{A \in M_n(\mathbb{C}) \mid A \text{ invertible}\};$$
$$SL(n, \mathbb{C}) = \{A \in GL(n, \mathbb{C}) \mid \det(A) = 1\};$$
$$U(n) = \{A \in GL(n, \mathbb{C}) \mid A\overline{A}^t = 1\};$$
$$SU(n) = \{A \in GL(n, \mathbb{C}) \mid A\overline{A}^t = 1, \det(A) = 1\};$$
$$SO(n, \mathbb{C}) = \{A \in GL(n, \mathbb{C}) \mid AA^t = 1, \det(A) = 1\};$$
$$SO(n) = \{A \in GL(n, \mathbb{R}) \mid AA^t = 1\}.$$

Show that their Lie algebras are given by

$$\mathfrak{gl}(n, \mathbb{C}) = M_n(\mathbb{C});$$
$$\mathfrak{sl}(n, \mathbb{C}) = \{A \in M_n(\mathbb{C}) \mid \mathrm{tr}(A) = 0\};$$
$$\mathfrak{u}(n) = \{A \in M_n(\mathbb{C}) \mid A + \overline{A}^t = 0\};$$
$$\mathfrak{su}(n) = \{A \in M_n(\mathbb{C}) \mid A + \overline{A}^t = 0 \text{ and } \mathrm{tr}(A) = 0\};$$
$$\mathfrak{so}(n, \mathbb{C}) = \{A \in M_n(\mathbb{C}) \mid A + A^t = 0\};$$
$$\mathfrak{so}(n) = \{A \in M_n(\mathbb{R}) \mid A + A^t = 0\}.$$

Show that $GL(n, \mathbb{C})$ for $n \geq 1$ and $SO(2)$ are not semisimple. Show that $SL(n, \mathbb{C})$, $SO(n, \mathbb{C})$ and $SU(n)$ for $n \geq 2$, and $SO(n)$ for $n \geq 3$ are semisimple. Prove that $U(n)$, $SU(n)$ and $SO(n)$ is the group $K = \{g \in G \mid \Theta(g) = g\}$ for $G = GL(n, \mathbb{C})$, $SL(n, \mathbb{C})$ and $SO(n, \mathbb{C})$ and is in particular a maximal compact subgroup.

5.3. Let G be a connected linear reductive Lie group. Show that $s \colon \mathfrak{g} \times \mathfrak{g} \to \mathbb{R}$ sending (A, B) to the real part $\Re(\mathrm{tr}(AB))$ of $\mathrm{tr}(AB)$ is an inner product on the real vector space $\mathfrak{g}$. Show for any $a \in \mathfrak{g}$ that the adjoint $\mathrm{ad}(A)^*$ of $\mathrm{ad}(A) \colon \mathfrak{g} \to \mathfrak{g}$ with respect to this inner product is $- \mathrm{ad}(A)$. Conclude for the Killing form B that $-B(A, \theta(A)) = \mathrm{tr}(\mathrm{ad}(A) \mathrm{ad}(A)^*)$ holds and that hence the symmetric bilinear form $B(A, \theta(A))$ on $\mathfrak{g}$ is strictly negative-definite, where θ is defined in Example 5.4. Conclude that θ is a Cartan involution.

5.4. Show that a connected linear reductive Lie group is semisimple if and only if its center is finite.

5.5. Let G and H be connected linear reductive Lie groups. Show that $G \times H$ is again a connected linear reductive Lie group. Prove that $G \times H$ is semisimple if and only if both G and H are semisimple.

5.6. Show that S^n, $\mathbb{R}\mathrm{P}^n$ and $\mathbb{C}\mathrm{P}^n$ are symmetric spaces of compact type, $\mathbb{R}^n$ is a symmetric space of Euclidean type and $\mathbb{H}^n$ is a symmetric space of non-compact type.

5.7. Let $G_p(\mathbb{R}^{p+q})$ be the topological space of oriented p-dimensional linear subspaces of $\mathbb{R}^{p+q}$. Show that it carries the structure of a symmetric space of compact type. Determine its dual symmetric space.

5.8. Let M and N be closed Riemannian manifolds whose universal coverings $\widetilde{M}$ and $\widetilde{N}$ are symmetric spaces of non-compact type. Suppose that $\pi_1(M) \cong \pi_1(N)$. Show that then $\mathrm{f\text{-}rk}(\widetilde{M}) = \mathrm{f\text{-}rk}(\widetilde{N})$.

5.9. Let M be a closed locally symmetric Riemannian manifold. Show that $\pi_1(M)$ is amenable if and only if $\widetilde{M}_{\mathrm{ncp}} = \{*\}$.

5.10. Construct two closed Riemannian manifolds M and N whose universal coverings are symmetric spaces of non-compact type such that $\pi_1(M)$ and $\pi_1(N)$ are not isomorphic but all L^2-Betti numbers, all Novikov-Shubin invariants and the L^2-torsion of $\widetilde{M}$ and $\widetilde{N}$ with respect to the $\pi_1(M)$- and $\pi_1(N)$-action agree.

5.11. Let M be an aspherical closed locally symmetric Riemannian manifold of even dimension $\dim(M) = 2p$. Show that $(-1)^p \cdot \chi(M) \geq 0$. Prove that $(-1)^p \cdot \chi(M) > 0$ if and only if $\widetilde{M}_{\mathrm{Eucl}} = \{*\}$ and f-rk$(\widetilde{M}_{\mathrm{ncp}}) = 0$.

5.12. Let M be an aspherical closed locally symmetric Riemannian manifold. Show that scaling the Riemannian metric by a constant does not change $\rho^{(2)}(\widetilde{M})$ if and only if $\rho^{(2)}(\widetilde{M}) = \rho^{(2)}(\widetilde{N})$ holds for any closed Riemannian manifold N which is homotopy equivalent to M.

6. L^2-Invariants for General Spaces with Group Action

Introduction

In this chapter we will extend the definition of L^2-Betti numbers for free G-CW-complexes of finite type to arbitrary G-spaces. Of course then the value may be infinite, but we will see that in surprisingly many interesting situations the value will be finite or even zero. This will be applied to problems in geometry, topology, group theory and K-theory.

The first elementary observation is that the $\mathbb{C}$-category of finitely generated Hilbert $\mathcal{N}(G)$-modules is isomorphic to the $\mathbb{C}$-category of finitely generated projective $\mathcal{N}(G)$-modules. Notice that the second category does not involve any functional analytic structure of $\mathcal{N}(G)$, only the ring structure comes in. The second observation is that the ring $\mathcal{N}(G)$ is semihereditary, i.e. any finitely generated $\mathcal{N}(G)$-submodule of a projective module is projective again. This implies that the $\mathbb{C}$-category of finitely presented $\mathcal{N}(G)$-modules is abelian. Thus a finitely generated Hilbert $\mathcal{N}(G)$-chain complex C_* defines a finitely generated projective $\mathcal{N}(G)$-chain complex denoted by $\nu^{-1}(C_*)$ and the homology of $\nu^{-1}(C_*)$ consists of finitely presented $\mathcal{N}(G)$-modules. Any finitely presented $\mathcal{N}(G)$-module M splits as $\mathbf{T}M \oplus \mathbf{P}M$, where $\mathbf{P}M$ is finitely generated projective. Then the L^2-homology of C_* corresponds to $\mathbf{P}H_*(\nu^{-1}(C_*)) = \nu^{-1}(H^{(2)}(C_*))$. These facts will be explained and proved in Section 6.2 and we will compare them with the approach of Farber [182] in Section 6.8.

Now the main technical result of Section 6.1 is that the von Neumann dimension, which is a priori defined for finitely generated projective $\mathcal{N}(G)$-modules, has an extension to arbitrary $\mathcal{N}(G)$-modules which takes values in $[0, \infty]$ and is uniquely determined by three important and desired properties, namely, Additivity, Cofinality and Continuity. This will allow us to define in Section 6.5 L^2-Betti numbers for an arbitrary G-space X by the extended von Neumann dimension of the $\mathcal{N}(G)$-module $H_p^G(X; \mathcal{N}(G))$, which is the homology of the $\mathcal{N}(G)$-chain complex $\mathcal{N}(G) \otimes_{\mathbb{Z}G} C_*^{\mathrm{sing}}(X)$. This is the L^2-Betti number defined in Section 1.2 if X happens to be a free G-CW-complex of finite type. Thus we can define the L^2-Betti number of an arbitrary group G by applying this construction to the classifying space EG. Notice that after we have established Assumption 6.2 for the von Neumann algebra and the von Neumann dimension, no more functional analysis will appear, the rest is pure homological algebra and ring theory. We mention the slogan that the von Neumann algebra $\mathcal{N}(G)$ is very similar to the ring $\mathbb{Z}$ of integers with two exceptions, namely, $\mathcal{N}(G)$ is not Noetherian, unless G is finite, and $\mathcal{N}(G)$ has non-trivial zero-divisors, unless G is finite. We recommend the reader to test statements and results about modules and their dimensions over $\mathcal{N}(G)$ by the corresponding ones for modules over $\mathbb{Z}$.

This algebraic approach is very convenient because it is very flexible and constructions like taking kernels, cokernels, quotients and homology are of course available. One may also try to extend the notion of L^2-Betti numbers within the category of Hilbert $\mathcal{N}(G)$-modules, the von Neumann dimension is defined for any Hilbert $\mathcal{N}(G)$-module. The problem is that one would have to take the Hilbert completion of the cellular chain complex. To get Hilbert $\mathcal{N}(G)$-chain modules one would have to restrict to proper G-CW-complexes. More serious problems occur with the differentials. Only under very restrictive conditions the differentials become bounded operators in the case of a proper G-CW-complex (not necessarily of finite type), which for instance are not satisfied for the bar-model of EG. If the differentials are not bounded one could take their minimal closures, but then it becomes very difficult or impossible to do certain constructions and establish certain proofs. The same problems arise with the chain maps induced by G-maps of G-spaces.

We show in Section 6.3 for an injective group homomorphism $i\colon H \to G$ that induction with the induced ring homomorphism $i\colon \mathcal{N}(H) \to \mathcal{N}(G)$ is faithfully flat and compatible with von Neumann dimension. This will be used all over the place when one wants to pass from the universal covering of a space X to a regular covering associated to an injective group homomorphism $\pi_1(X) \to G$. We will also prove that $\mathcal{N}(G) \otimes_{\mathbb{C}G} \mathbb{C}$ is non-trivial if and only if G is amenable. This may be viewed as an extension of the result of Brooks [68] that the Laplacian on functions on the universal covering of a closed Riemannian manifold M has zero in its spectrum if and only if $\pi_1(M)$ is amenable.

In Section 6.4 we investigate the von Neumann dimension for amenable G. A survey of amenable groups is presented in Subsection 6.4.1. The von Neumann algebra $\mathcal{N}(G)$ is known to be flat over $\mathbb{C}G$ only for virtually cyclic groups (and conjecturally these are the only ones), but from a dimension point of view it looks like a flat module over $\mathbb{C}G$, more precisely, the von Neumann dimension of $\mathrm{Tor}_p^{\mathbb{C}G}(\mathcal{N}(G), M)$ vanishes for all $\mathbb{C}G$-modules and $p \geq 1$, provided that G is amenable. This implies that the p-th L^2-Betti number of a G-space X for amenable G can be read off from the $\mathbb{C}G$-module $H_p(X; \mathbb{C})$, namely, it is the von Neumann dimension of $\mathcal{N}(G) \otimes_{\mathbb{C}G} H_p(X; \mathbb{C})$. Since $H_p(EG; \mathbb{C})$ vanishes for $p \geq 1$, one obtains the result of Cheeger and Gromov [107] that all the L^2-Betti numbers of an amenable group G vanish. The dimension-flatness of the von Neumann algebra $\mathcal{N}(G)$ over $\mathbb{C}G$ for amenable G will play also an important role in applications to G-theory in Subsection 9.5.3.

In Section 6.5 we will prove the basic properties of the L^2-Betti numbers for arbitrary G-spaces such as homotopy or more generally homology invariance, Künneth formula and behaviour under induction and restriction. We will investigate its behaviour under fibrations and S^1-actions. In Section 6.6 we will introduce the L^2-Euler characteristic and prove its basic properties like Euler-Poincaré formula, homology invariance, Künneth formula, sum formula and behaviour under induction and restriction. We will compare the various L^2-Euler characteristics of the fixed point sets of a finite proper G-CW-complex with the equivariant Euler characteristic which takes value in the so called Burnside group. This will be applied in particular to the classifying space $E(G, \mathcal{FIN})$. The definition of the Burnside group $A(G)$ of a group G, which should not be confused with the Burnside group $B(m, n)$ appearing in group theory, is analogous to the definition of the Burnside ring of a finite group, but the Burnside group $A(G)$ for infinite G does not inherit an internal multiplication.

To get a quick overview one should read through Theorem 6.7, Theorem 6.24, Theorem 6.37 and then start immediately with Section 6.5. To understand the basics of this Chapter 6 only some knowledge about Section 1.1 from the preceding chapters is necessary.

We briefly will mention in Section 6.8. that a similar approach can be used to extend the notion of Novikov-Shubin invariants to arbitray G-spaces.

This chapter is based on the papers [333] and [334] which have been motivated by the paper of Cheeger and Gromov [107].

6.1 Dimension Theory for Arbitrary Modules

In this section we show that the von Neumann dimension for finitely generated projective $\mathcal{N}(G)$-modules has a unique extension to all $\mathcal{N}(G)$-modules which has nice properties like Additivity, Cofinality and Continuity. Ring will always mean associative ring with unit and R-module will mean left R-module unless explicitly stated differently.

Recall that the dual M^* of a left or right respectively R-module M is the right or left respectively R-module $\hom_R(M, R)$ where the R-multiplication is given by $(fr)(x) = f(x)r$ or $(rf)(x) = rf(x)$ respectively for $f \in M^*$, $x \in M$ and $r \in R$.

Definition 6.1 (Closure of a submodule). *Let M be an R-submodule of N. Define the* closure *of M in N to be the R-submodule of N*

$$\overline{M} = \{x \in N \mid f(x) = 0 \text{ for all } f \in N^* \text{ with } M \subset \ker(f)\}.$$

For an R-module M define the R-submodule $\mathbf{T}M$ and the quotient R-module $\mathbf{P}M$ by

$$\mathbf{T}M := \{x \in M \mid f(x) = 0 \text{ for all } f \in M^*\};$$
$$\mathbf{P}M := M/\mathbf{T}M.$$

We call a sequence of R-modules $L \xrightarrow{i} M \xrightarrow{q} N$ *weakly exact if* $\overline{\text{im}(i)} = \ker(q)$.

Notice that $\mathbf{T}M$ is the closure of the trivial submodule in M. It can also be described as the kernel of the canonical map $i(M) \colon M \to (M^*)^*$ which sends $x \in M$ to the map $M^* \to R$, $f \mapsto f(x)$. Notice that $\mathbf{T}\mathbf{P}M = 0$, $\mathbf{P}\mathbf{P}M = \mathbf{P}M$, $M^* = (\mathbf{P}M)^*$ and that $\mathbf{P}M = 0$ is equivalent to $M^* = 0$.

Assumption 6.2. *We assume that there is a dimension function* dim *which assigns to any finitely generated projective R-module P a non-negative real number*

$$\dim(P) \in [0, \infty)$$

with the following properties:

(1) If P and Q are finitely generated projective R-modules, then

$$P \cong_R Q \;\Rightarrow\; \dim(P) = \dim(Q);$$
$$\dim(P \oplus Q) \;=\; \dim(P) + \dim(Q);$$

(2) Let $K \subset Q$ be a submodule of the finitely generated projective R-module Q. Then its closure $\overline{K}$ (see Definition 6.1) is a direct summand in Q and

$$\dim(\overline{K}) = \sup\{\dim(P) \mid P \subset K \text{ finitely generated projective submodule}\}.$$

Let $\mathcal{N}(G)$ be the group von Neumann algebra of the discrete group G (see Definition 1.1). Let P be any finitely generated projective $\mathcal{N}(G)$-module. (Here we view $\mathcal{N}(G)$ just as a ring, there is no Hilbert structure involved in P). Choose any (n, n)-matrix $A \in M_n(\mathcal{N}(G))$ such that $A^2 = A$ and the image of the $\mathcal{N}(G)$-linear map induced by right multiplication with A

$$r_A \colon \mathcal{N}(G)^n \to \mathcal{N}(G)^n, \qquad x \mapsto xA$$

is isomorphic as an $\mathcal{N}(G)$-module to P. (It is not necessary but possible to require $A = A^*$.) Define the *von Neumann dimension*

$$\dim_{\mathcal{N}(G)}(P) \;:=\; \text{tr}_{\mathcal{N}(G)}(A), \tag{6.3}$$

where $\text{tr}_{\mathcal{N}(G)} \colon M_n(\mathcal{N}(G)) \to \mathbb{C}$ sends A to the sum of the von Neumann traces (see Definition 1.2) of its diagonal entries. This is independent of the choice of A by the following standard argument.

Suppose $B \in M_p(\mathcal{N}(G))$ is a second square matrix with $B^2 = B$ and $\text{im}(r_B) \cong P$. By possibly taking the direct sum with a zero square-matrix we can achieve without changing $\text{tr}_{\mathcal{N}(G)}(A)$ and $\text{tr}_{\mathcal{N}(G)}(B)$ and the isomorphism class of $\text{im}(r_A)$ and $\text{im}(r_B)$ that $n = p$ and that $\text{im}(r_{1-A})$ and $\text{im}(r_{1-B})$ are isomorphic. Let $C \in M_n(\mathcal{N}(G))$ be an invertible matrix such that r_C maps $\mathrm{im}(r_A)$ to $\mathrm{im}(r_B)$ and $\mathrm{im}(r_{1-A})$ to $\mathrm{im}(r_{1-B})$. Then $r_B \circ r_C = r_B \circ r_C \circ r_A = r_C \circ r_A$ and hence $CBC^{-1} = A$. This implies

$$\mathrm{tr}_{\mathcal{N}(G)}(B) = \mathrm{tr}_{\mathcal{N}(G)}(C^{-1}CB) = \mathrm{tr}_{\mathcal{N}(G)}(CBC^{-1}) = \mathrm{tr}_{\mathcal{N}(G)}(A). \quad (6.4)$$

We will explain in Theorem 6.24 that this notion is essentially the same as the notion of the von Neumann dimension of a finitely generated Hilbert $\mathcal{N}(G)$-module of Definition 1.10.

The proof of the following Theorem 6.5 will be given in Section 6.2. Notice that in the extension of the von Neumann dimension from finitely generated projective $\mathcal{N}(G)$-modules to arbitrary $\mathcal{N}(G)$-modules the functional analytic aspects only enter in the proof of Theorem 6.5 the rest is purely algebraic ring and module theory.

Theorem 6.5. *The pair $(\mathcal{N}(G), \dim_{\mathcal{N}(G)})$ satisfies Assumption 6.2.*

Definition 6.6 (Extended dimension). *If $(R, \dim)$ satisfies Assumption 6.2, we define for an R-module M its* extended dimension

$$\dim'(M) := \sup\{\dim(P) \mid P \subset M \text{ finitely generated projective submodule}\}$$
$$\in [0, \infty].$$

We will later drop the prime in $\dim'$ (see Notation 6.11).

The next result is one of the basic results of this chapter.

Theorem 6.7. (Dimension function for arbitrary $\mathcal{N}(G)$-modules). *Suppose that $(R, \dim)$ satisfies Assumption 6.2. Then*

(1) R is semihereditary, i.e. any finitely generated submodule of a projective module is projective;

(2) If $K \subset M$ is a submodule of the finitely generated R-module M, then $M/\overline{K}$ is finitely generated projective and $\overline{K}$ is a direct summand in M;

(3) If M is a finitely generated R-module, then $\mathbf{P}M$ is finitely generated projective and

$$M \cong \mathbf{P}M \oplus \mathbf{T}M;$$

(4) The dimension $\dim'$ has the following properties:

    (a) Extension Property

       If M is a finitely generated projective R-module, then

$$\dim'(M) = \dim(M);$$

    (b) Additivity

       If $0 \to M_0 \xrightarrow{i} M_1 \xrightarrow{p} M_2 \to 0$ is an exact sequence of R-modules, then

$$\dim'(M_1) = \dim'(M_0) + \dim'(M_2),$$

       where for $r, s \in [0, \infty]$ we define $r + s$ by the ordinary sum of two real numbers if both r and s are not ∞, and by ∞ otherwise;

(c) *Cofinality*

Let $\{M_i \mid i \in I\}$ be a cofinal system of submodules of M, i.e. $M = \bigcup_{i \in I} M_i$ and for two indices i and j there is an index k in I satisfying $M_i, M_j \subset M_k$. Then

$$\dim'(M) = \sup\{\dim'(M_i) \mid i \in I\};$$

(d) *Continuity*

If $K \subset M$ is a submodule of the finitely generated R-module M, then

$$\dim'(K) = \dim'(\overline{K});$$

(e) If M is a finitely generated R-module, then

$$\dim'(M) = \dim(\mathbf{P}M);$$
$$\dim'(\mathbf{T}M) = 0;$$

(f) The dimension $\dim'$ is uniquely determined by the Extension Property, Additivity, Cofinality and Continuity.

Proof. (1) Let $M \subset P$ be a finitely generated R-submodule of the projective R-module P. Choose a homomorphism $q \colon R^n \to P$ with $\operatorname{im}(q) = M$. Then $\ker(q) = \overline{\ker(q)}$. Hence $\ker(q)$ is by Assumption 6.2 a direct summand. This shows that M is projective.

(2) Choose an epimorphism $q \colon R^n \to M$. One easily checks that $q^{-1}(K) = \overline{q^{-1}(K)}$ and that $R^n/q^{-1}(K)$ and M/K are R-isomorphic. By Assumption 6.2 $R^n/\overline{q^{-1}(K)}$ and hence $M/\overline{K}$ are finitely generated projective.

(3) This follows from (2) by taking $K = 0$.

(4a) If $P \subset M$ is a finitely generated projective R-submodule of the finitely generated projective R-module M, we conclude

$$\dim(P) \leq \dim(M) \tag{6.8}$$

from the following calculation based on Assumption 6.2

$$\dim(P) \leq \dim(\overline{P}) = \dim(M) - \dim(M/\overline{P}) \leq \dim(M).$$

This implies $\dim(M) = \dim'(M)$.

(4b) Let $P \subset M_2$ be a finitely generated projective submodule. We obtain an exact sequence $0 \to M_0 \to p^{-1}(P) \to P \to 0$. Since $p^{-1}(P) \cong M_0 \oplus P$, we conclude

$$\dim'(M_0) + \dim(P) \leq \dim'(p^{-1}(P)) \leq \dim'(M_1).$$

Since this holds for all finitely generated projective submodules $P \subset M_2$, we get

$$\dim'(M_0) + \dim'(M_2) \leq \dim'(M_1). \tag{6.9}$$

Let $Q \subset M_1$ be finitely generated projective. Let $\overline{i(M_0) \cap Q}$ be the closure of $i(M_0) \cap Q$ in Q. We obtain exact sequences

$$0 \to i(M_0) \cap Q \to Q \to \quad p(Q) \quad \to 0;$$
$$0 \to \overline{i(M_0) \cap Q} \to Q \to Q/\overline{i(M_0) \cap Q} \to 0.$$

By Assumption 6.2 $\overline{i(M_0) \cap Q}$ is a direct summand in Q. We conclude

$$\dim(Q) = \dim(\overline{i(M_0) \cap Q}) + \dim(Q/\overline{i(M_0) \cap Q}).$$

From Assumption 6.2 and the fact that there is an epimorphism from $p(Q)$ onto the finitely generated projective R-module $Q/\overline{i(M_0) \cap Q}$, we conclude

$$\dim(\overline{i(M_0) \cap Q}) = \dim'(i(M_0) \cap Q);$$
$$\dim(Q/\overline{i(M_0) \cap Q}) \le \dim'(p(Q)).$$

Since obviously $\dim'(M) \le \dim'(N)$ holds for R-modules M and N with $M \subset N$, we get

$$\begin{aligned} \dim(Q) &= \dim(\overline{i(M_0) \cap Q}) + \dim(Q/\overline{i(M_0) \cap Q}) \\ &\le \dim'(i(M_0) \cap Q) + \dim'(p(Q)) \\ &\le \dim'(M_0) + \dim'(M_2). \end{aligned}$$

Since this holds for all finitely generated projective submodules $Q \subset M_1$, we get

$$\dim'(M_1) \le \dim'(M_0) + \dim'(M_2). \tag{6.10}$$

Now assertion (4b) follows from (6.9) and (6.10).

(4c) If $P \subset M$ is a finitely generated projective submodule, then there is an index $i \in I$ with $P \subset M_i$ by cofinality.

(4d) Choose an epimorphism $q \colon R^n \to M$. Since $q^{-1}(\overline{K}) = \overline{q^{-1}(K)}$, we get from Assumption 6.2 and assertion (4a)

$$\dim'(q^{-1}(K)) = \dim'(\overline{q^{-1}(K)}) = \dim'(q^{-1}(\overline{K})).$$

If L is the kernel of q, we conclude from assertions (4a) and (4b)

$$\begin{aligned} \dim'(q^{-1}(\overline{K})) &= \dim'(L) + \dim'(\overline{K}); \\ \dim'(q^{-1}(K)) &= \dim'(L) + \dim'(K); \\ \dim'(q^{-1}(K)) &\le \dim(R^n) < \infty. \end{aligned}$$

This proves assertion (4d).

(4e) This follows from (4d) applied to the special case $K = 0$ and assertions (3), (4a) and (4b).

(4f) Let $\dim''$ be another function satisfying Extension Property, Additivity, Cofinality and Continuity. We want to show for an R-module M

$$\dim''(M) = \dim'(M).$$

Since assertion (4e) is a consequence of assertions (3), (4a), (4b) and (4d), it holds also for $\dim''$. Hence we get $\dim'(M) = \dim''(M)$ for any finitely generated R-module. Since the system of finitely generated submodules of a module is cofinal, the claim follows from Cofinality. This finishes the proof of Theorem 6.7.   □

Notation 6.11. *In view of Theorem 6.7 we will not distinguish between* $\dim'$ *and* $\dim$ *in the sequel.*   □

Example 6.12. Let R be a principal ideal domain. Then any finitely generated projective R-module P is isomorphic to R^n for a unique $n \geq 0$ and we consider the dimension $\dim(P) := n$. Notice that for a submodule $M \subset R^n$ we have

$$\overline{M} = \{x \in R^n \mid r \cdot x \in M \text{ for appropriate } r \in R, r \neq 0\}.$$

One easily checks that Assumption 6.2 is satisfied. Let F be the quotient field of R. Then we get for any R-module M for the dimension defined in Definition 6.6

$$\dim(M) = \dim_F(F \otimes_R M),$$

where $\dim_F(F \otimes_R M)$ is the dimension of the F-vector space $F \otimes_R M$.

Notice that $\dim(P)$ is finite for a projective R-module P if and only if P is finitely generated. This is the crucial difference to the case, where we consider the von Neumann algebra $\mathcal{N}(G)$ and the extension $\dim_{\mathcal{N}(G)}$ of Definition 6.6 of the von Neumann dimension $\dim_{\mathcal{N}(G)}$ of (6.3). If for instance $H_1, H_2, \ldots$ is a sequence of finite subgroups of G, then $\bigoplus_{i=1}^{\infty} \mathcal{N}(G) \otimes_{\mathbb{C}H_i} \mathbb{C}$ is a projective not finitely generated $\mathcal{N}(G)$-module and

$$\dim_{\mathcal{N}(G)} \left(\bigoplus_{i=1}^{\infty} \mathcal{N}(G) \otimes_{\mathbb{C}H_i} \mathbb{C} \right) = \sum_{i=1}^{\infty} \frac{1}{|H_i|},$$

and this infinite sum may converge to a finite real number.

Recall that a directed set I is a set with a partial ordering $\leq$ such that for two elements i_0 and i_1 there exists an element i with $i_0 \leq i$ and $i_1 \leq i$. We can consider I as a category with I as set of objects, where the set of morphisms from i_0 to i_1 consists of precisely one element, if $i_0 \leq i_1$, and is empty otherwise. A *directed system* or an *inverse system* respectively $\{M_i \mid i \in I\}$ of R-modules indexed by I is a functor from I into the category of R-modules which is covariant or contravariant respectively. We denote by $\operatorname{colim}_{i \in I} M_i$ the *colimit* of the directed and by $\lim_{i \in I} M_i$ the *limit* of the inverse system $\{M_i \mid i \in I\}$ which is again an R-module. We mention that colimit is sometimes also called inductive limit or direct limit and that limit is sometimes also called inverse limit or projective limit in the literature. The colimit is characterized by the following property. For any $i \in I$ there is an R-homomorphism $\psi_i \colon M_i \to \operatorname{colim}_{i \in I} M_i$ such that for any two elements i_0 and i_1 in I with $i_0 \leq i_1$ we have $\psi_{i_1} \circ \phi_{i_0,i_1} = \psi_{i_0}$, where $\phi_{i_0,i_1} \colon M_{i_0} \to M_{i_1}$ comes from functoriality. For any R-module N together with R-homomorphisms $f_i \colon M_i \to N$ such that for any two elements i_0 and i_1 in I with $i_0 \leq i_1$ we have $f_{i_1} \circ \phi_{i_0,i_1} = f_{i_0}$, there is precisely one R-homomorphism $f \colon \operatorname{colim}_{i \in I} M_i \to N$ satisfying $f \circ \psi_i = f_i$ for all $i \in I$. The limit of an inverse system has an analogous characterization, one has to reverse all arrows.

Next we investigate the behaviour of dimension under colimits indexed by a directed set.

Theorem 6.13 (Dimension and colimits). *Let $\{M_i \mid i \in I\}$ be a directed system of R-modules over the directed set I. For $i \leq j$ let $\phi_{i,j} \colon M_i \to M_j$ be the associated morphism of R-modules. For $i \in I$ let $\psi_i \colon M_i \to \operatorname{colim}_{i \in I} M_i$ be the canonical morphism of R-modules. Then*

(1) We get for the dimension of the R-module given by the colimit $\operatorname{colim}_{i \in I} M_i$

$$\dim\left(\operatorname{colim}_{i \in I} M_i\right) = \sup\left\{\dim(\operatorname{im}(\psi_i)) \mid i \in I\right\};$$

(2) Suppose for each $i \in I$ that there is $i_0 \in I$ with $i \leq i_0$ such that $\dim(\operatorname{im}(\phi_{i,i_0})) < \infty$ holds. Then

$$\dim\left(\operatorname{colim}_{i \in I} M_i\right)$$
$$= \sup\left\{\inf\left\{\dim(\operatorname{im}(\phi_{i,j} \colon M_i \to M_j)) \mid j \in I, i \leq j\right\} \mid i \in I\right\}.$$

Proof. (1) Recall that the colimit $\operatorname{colim}_{i \in I} M_i$ is $\coprod_{i \in I} M_i / \sim$ for the equivalence relation for which $M_i \ni x \sim y \in M_j$ holds precisely if there is $k \in I$ with $i \leq k$ and $j \leq k$ with the property $\phi_{i,k}(x) = \phi_{j,k}(y)$. With this description one easily checks

$$\operatorname{colim}_{i \in I} M_i = \bigcup_{i \in I} \operatorname{im}(\psi_i \colon M_i \to \operatorname{colim}_{j \in I} M_j).$$

Now apply Cofinality (see Theorem 6.7 (4c)).

(2) It remains to show for $i \in I$

$$\dim(\operatorname{im}(\psi_i)) = \inf\left\{\dim(\operatorname{im}(\phi_{i,j} \colon M_i \to M_j)) \mid j \in I, i \leq j\right\}. \quad (6.14)$$

By assumption there is $i_0 \in I$ with $i \leq i_0$ such that $\dim(\operatorname{im}(\phi_{i,i_0}))$ is finite. Let $K_{i_0,j}$ be the kernel of the map $\operatorname{im}(\phi_{i,i_0}) \to \operatorname{im}(\phi_{i,j})$ induced by $\phi_{i_0,j}$ for $i_0 \leq j$ and K_{i_0} be the kernel of the map $\operatorname{im}(\phi_{i,i_0}) \to \operatorname{im}(\psi_i)$ induced by ψ_{i_0}. Then $K_{i_0} = \bigcup_{j \in I, i_0 \leq j} K_{i_0,j}$ and hence by Cofinality (see Theorem 6.7 (4c))

$$\dim(K_{i_0}) = \sup\{\dim(K_{i_0,j}) \mid j \in I, i_0 \leq j\}.$$

Since $\dim(\operatorname{im}(\phi_{i,i_0}))$ is finite, we get from Additivity (see Theorem 6.7 (4b))

$$
\begin{aligned}
&\dim(\operatorname{im}(\psi_i)) \\
&= \dim\left(\operatorname{im}\left(\psi_{i_0}|_{\operatorname{im}(\phi_{i,i_0})}\colon \operatorname{im}(\phi_{i,i_0}) \to \operatorname{colim}_{i\in I} M_i\right)\right) \\
&= \dim(\operatorname{im}(\phi_{i,i_0})) - \dim(K_{i_0}) \\
&= \dim(\operatorname{im}(\phi_{i,i_0})) - \sup\{\dim(K_{i_0,j}) \mid j \in I, i_0 \le j\} \\
&= \inf\left\{\dim(\operatorname{im}(\phi_{i,i_0})) - \dim(K_{i_0,j}) \mid j \in I, i_0 \le j\right\} \\
&= \inf\left\{\dim\left(\operatorname{im}(\phi_{i_0,j}|_{\operatorname{im}(\phi_{i,i_0})}\colon \operatorname{im}(\phi_{i,i_0}) \to \operatorname{im}(\phi_{i,j}))\right) \mid j \in I, i_0 \le j\right\} \\
&= \inf\left\{\dim\left(\operatorname{im}(\phi_{i,j})\right) \mid j \in I, i_0 \le j\right\}. \qquad (6.15)
\end{aligned}
$$

Given $j_0 \in J$ with $i \le j_0$, there is $j \in I$ with $i_0 \le j$ and $j_0 \le j$. We conclude $\dim(\operatorname{im}(\phi_{i,j_0})) \ge \dim(\operatorname{im}(\phi_{i,j}))$ from Additivity (see Theorem 6.7 (4b)). This implies

$$
\begin{aligned}
\inf\{\dim(\operatorname{im}(\phi_{i,j})) \mid j \in J, i \le j\} \\
= \inf\{\dim(\operatorname{im}(\phi_{i,j})) \mid j \in J, i_0 \le j\}. \quad (6.16)
\end{aligned}
$$

Now (6.14) follows from (6.15) and (6.16). This finishes the proof of Theorem 6.13. $\qquad\square$

Example 6.17. The condition in Theorem 6.13 (2) above that for each $i \in I$ there is $i_0 \in I$ with $i \le i_0$ and $\dim(\operatorname{im}(\phi_{i,i_0})) < \infty$ is necessary as the following example shows. Take $I = \mathbb{N}$. Define $M_j = \bigoplus_{n=j}^{\infty} R$ and $\phi_{j,k}\colon \bigoplus_{m=j}^{\infty} R \to \bigoplus_{m=k}^{\infty} R$ to be the obvious projection for $j \le k$. Then $\dim(\operatorname{im}(\phi_{j,k})) = \infty$ for all $j \le k$, but $\operatorname{colim}_{i\in I} M_i$ is trivial and hence has dimension zero.

Next we state the version of Theorem 6.13 about dimension and colimits for limits over inverse systems. Since we do not need it elsewhere, we do not give its proof which is, however, much harder than the one of Theorem 6.13.

Theorem 6.18 (Dimension and limits). *Let $\{M_i \mid i \in I\}$ be an inverse system of $\mathcal{N}(G)$-modules over the directed set I. For $i \le j$ let $\phi_{i,j}\colon M_j \to M_i$ be the associated morphism of $\mathcal{N}(G)$-modules. For $i \in I$ let $\psi_i\colon \lim_{i\in I} M_i \to M_i$ be the canonical map. Suppose that there is a countable sequence $i_1 \le i_2 \le \dots$ such that for each $j \in I$ there is $n \ge 0$ with $j \le i_n$. Let $\dim_{\mathcal{N}(G)}$ be the extension of Definition 6.6 of the von Neumann dimension of (6.3). Then*

(1)

$$
\dim_{\mathcal{N}(G)}\left(\lim_{i\in I} M_i\right) = \sup\left\{\dim_{\mathcal{N}(G)}(\operatorname{im}(\psi_i\colon \lim_{i\in I} M_i \to M_i)) \,\Big|\, i \in I\right\};
$$

(2) *Suppose that for each index $i \in I$ there is an index $i_0 \in I$ with $i \leq i_0$ and $\dim_{\mathcal{N}(G)}(\mathrm{im}(\phi_{i,i_0})) < \infty$. Then*

$$\dim_{\mathcal{N}(G)}\left(\lim_{i \in I} M_i\right) = \sup\left\{\inf\{\dim_{\mathcal{N}(G)}(\mathrm{im}(\phi_{i,j})) \mid j \in I, i \leq j\} \mid i \in I\right\}.$$

Example 6.19. In this example we want to show that the condition that there is a countable sequence $i_1 \leq i_2 \leq \ldots$ such that for each $j \in I$ there is $n \geq 0$ with $j \leq i_n$ appearing in Theorem 6.18 is necessary. (Compare this with the claim in [107, page 210] that equation (A1) in [107, page 210] holds for arbitrary (not necessarily countable) intersections of Γ-weakly closed submodules.) Let $G = \mathbb{Z}$. Then $\mathcal{N}(\mathbb{Z}) = L^\infty(S^1)$ by Example 1.4. For $u \in S^1$ let $(z-u)$ be the $\mathcal{N}(\mathbb{Z})$-ideal which is generated by the function $S^1 \to \mathbb{C}$ sending $z \in S^1 \subset \mathbb{C}$ to $z - u$. Next we prove

$$\bigcap_{u \in S^1}(z-u) = 0.$$

Consider $f \in \bigcap_{u \in S^1}(z-u)$. Define $s \in [0,1]$ to be the supremum over all $r \in [0,1]$ such that $|f(\exp(2\pi it))| \leq 1$ holds for almost all $t \in [0,r]$. For almost all means for all elements with the exception of the elements of a subset of Lebesgue measure zero. Notice that the definition of s makes sense, its value does not change if we add to f a measurable function which vanishes outside a set of measure zero. We want to show $s = 1$ by contradiction. Suppose $s < 1$. Put $u = \exp(2\pi is)$. Since f belongs to $(z-u)$ there is $g \in L^\infty(S^1)$ with $f(z) = (z-u) \cdot g(z)$. Choose $\epsilon > 0$ with $||g||_\infty \leq \epsilon^{-1}$ and $s + \epsilon \leq 1$. Then we have $|f(z)| \leq 1$ for almost all elements z in $\{\exp(2\pi it) \mid s - \epsilon \leq t \leq s + \epsilon\}$. This implies that $|f(z)| \leq 1$ holds for almost all elements z in $\{\exp(2\pi it) \mid 0 \leq t \leq s + \epsilon\}$. We conclude $s \geq s + \epsilon$, a contradiction. We get $s = 1$. Hence we have $||f||_\infty \leq 1$ for all $f \in \bigcap_{u \in S^1}(z-u)$. This implies that $\bigcap_{u \in S^1}(z-u)$ is zero.

This shows that the obvious $\mathcal{N}(\mathbb{Z})$-map $L^\infty(S^1) \to \prod_{u \in S^1} L^\infty(S^1)/(z-u)$ is injective. From Additivity (see Theorem 6.7 (4b)) we conclude

$$\dim_{\mathcal{N}(\mathbb{Z})}\left(\prod_{u \in S^1} L^\infty(S^1)/(z-u)\right) \geq 1.$$

Notice that $\prod_{u \in S^1} L^\infty(S^1)/(z-u)$ is the limit over the obvious inverse system $\{\prod_{u \in J} L^\infty(S^1)/(z-u)) \mid J \in I\}$, where I is the set of finite subsets of S^1 ordered by inclusion, and that $\dim_{\mathcal{N}(\mathbb{Z})}\left(\prod_{u \in J} L^\infty(S^1)/(z-u)\right) = 0$ holds for all $J \in I$.

Also the condition that for each index $i \in I$ there is an index $j \in I$ with $\dim_{\mathcal{N}(\mathbb{Z})}(\mathrm{im}(\phi_{i,j})) < \infty$ appearing in Theorem 6.18 (2) is necessary as the following example shows. Take $I = \mathbb{N}$. Define $M_j = \prod_{n=j}^\infty \mathcal{N}(G)$ and $\phi_{j,k} \colon \prod_{m=k}^\infty \mathcal{N}(G) \to \prod_{m=j}^\infty \mathcal{N}(G)$ to be the canonical inclusion for $j \leq k$.

Then for all $j \leq k$ we get $\dim_{\mathcal{N}(\mathbb{Z})}(\mathrm{im}(\phi_{j,k})) = \infty$, but $\lim_{i \in I} M_i$ is trivial and hence has dimension zero.

Theorem 6.18 (2) has only been stated for $R = \mathcal{N}(G)$ and the extended von Neumann dimension $\dim_{\mathcal{N}(G)}$, in contrast to Theorem 6.13 about dimension and colimits. Indeed Theorem 6.18 fails for $R = \mathbb{Z}$ with the ordinary rank of abelian groups as dimension. For instance the limit of the inverse system

$$\mathbb{Z} \supset 2 \cdot \mathbb{Z} \supset 2^2 \cdot \mathbb{Z} \supset 2^3 \cdot \mathbb{Z} \supset \dots$$

is zero and hence has dimension 0, but the image of any structure map $2^j \cdot \mathbb{Z} \to 2^i \cdot \mathbb{Z}$ has dimension 1.

Definition 6.20 (Extended von Neumann dimension). *Let M be an $\mathcal{N}(G)$-module. Define its* extended von Neumann dimension

$$\dim_{\mathcal{N}(G)}(M) \in [0, \infty]$$

by the extension of Definition 6.6 of the von Neumann dimension of (6.3).

We will extend in Theorem 8.29 the dimension function for $\mathcal{N}(G)$-modules to $\mathcal{U}(G)$-modules, where $\mathcal{U}(G)$ is the algebra of affiliated operators, or, equivalently, the Ore localization of $\mathcal{N}(G)$ with respect to all non-zero-divisors in $\mathcal{N}(G)$.

6.2 Comparison of Modules and Hilbert Modules

In this section we want to show that the category of finitely generated projective $\mathcal{N}(G)$-modules (with inner product) and the category of finitely generated Hilbert $\mathcal{N}(G)$-modules are equivalent as $\mathbb{C}$-categories (with involution) and that this equivalence preserves weakly exact and exact sequences and dimension (see Theorem 6.24). This will be the key step to come from the operator theoretic approach to a purely algebraic approach to L^2-Betti numbers. Recall that we use the convention that groups and rings act from the left unless stated explicitly differently.

We need some notations to formulate the main result of this section. A $\mathbb{C}$-*category* $\mathcal{C}$ is a category such that for any two objects the set of morphisms between them carries the structure of a complex vector space for which composition of morphisms is bilinear and $\mathcal{C}$ has a (strict) sum which is compatible with the complex vector space structures above. A *(strict) involution* on a $\mathbb{C}$-category $\mathcal{C}$ is an assignment which associates to each morphism $f \colon x \to y$ a morphism $f^* \colon y \to x$ and has the following properties

$$(f^*)^* = f;$$
$$(\lambda \cdot f + \mu \cdot g)^* = \overline{\lambda} \cdot f^* + \overline{\mu} \cdot g^*;$$
$$(f \circ g)^* = g^* \circ f^*;$$
$$(f \oplus g)^* = f^* \oplus g^*,$$

where f, g are morphisms, λ and μ complex numbers. There is a canonical structure of a $\mathbb{C}$-category with involution on the category of finitely generated Hilbert $\mathcal{N}(G)$-modules {fin. gen. Hilb. $\mathcal{N}(G)$-mod.} (see Definition 1.5), where the involution is given by taking adjoint operators. We call an endomorphism f in $\mathcal{C}$ *selfadjoint* if $f = f^*$. We call an isomorphism f in $\mathcal{C}$ *unitary* if $f^* = f^{-1}$. A *functor of $\mathbb{C}$-categories (with involution)* is a functor compatible with the complex vector space structures on the set of morphisms between two objects and the sums (and the involutions). A natural equivalence T of functors of $\mathbb{C}$-categories with involution is called *unitary* if the evaluation of T at each object is a unitary isomorphism. An *equivalence of $\mathbb{C}$-categories (with involution)* is a functor of such categories such that there is a functor of such categories in the other direction with the property that both compositions are (unitarily) naturally equivalent to the identity. If a functor F of $\mathbb{C}$-categories (with involution) induces a bijection on the sets of (unitary) isomorphism classes of objects and for any two objects x and y it induces a bijection between the set of morphisms from x to y to the set of morphisms from $F(x)$ to $F(y)$, then it is an equivalence of $\mathbb{C}$-categories (with involution) and vice versa. (cf [351, Theorem 1 in IV.9 on page 91]).

Given a finitely generated projective (left) $\mathcal{N}(G)$-module P, an *inner product* on P is a map $\mu \colon P \times P \to \mathcal{N}(G)$ satisfying (cf. [514, Definition 15.1.1 on page 232])

(1) μ is $\mathcal{N}(G)$-linear in the first variable;
(2) μ is symmetric in the sense $\mu(x,y) = \mu(y,x)^*$;
(3) μ is positive-definite in the sense that $\mu(p,p)$ is a positive element in $\mathcal{N}(G)$, i.e. of the form a^*a for some $a \in \mathcal{N}(G)$, and $\mu(p,p) = 0 \Leftrightarrow p = 0$;
(4) The induced map $\overline{\mu} \colon P \to P^* = \hom_{\mathcal{N}(G)}(P, \mathcal{N}(G))$, defined by $\overline{\mu}(y)(x) = \mu(x,y)$, is bijective.

Notice that we have already introduced an $\mathcal{N}(G)$-right module structure on P^* given by $(fr)(x) = f(x)r$ for $r \in \mathcal{N}(G)$. Using the involution on $\mathcal{N}(G)$, we can define also a left $\mathcal{N}(G)$-module structure by $rf(x) = f(x)r^*$. Then $\overline{\mu}$ is an isomorphism of left $\mathcal{N}(G)$-modules. Moreover, $\overline{\mu}$ agrees with the composition $P \xrightarrow{i(P)} (P^*)^* \xrightarrow{\overline{\mu}^*} P^*$, where $i(P) \colon P \to (P^*)^*$ is the bijection which sends $x \in P$ to the map $P^* \to \mathcal{N}(G)$, $f \mapsto f(x)^*$. Let {fin. gen. proj. $\mathcal{N}(G)$-mod. with $\langle\ \rangle$} be the $\mathbb{C}$-category with involution, whose objects are finitely generated projective $\mathcal{N}(G)$-modules with inner product (P, μ) and whose morphisms are $\mathcal{N}(G)$-linear maps. We get an involution on it if we specify $f^* \colon (P_1, \mu_1) \to (P_0, \mu_0)$ for $f \colon (P_0, \mu_0) \to (P_1, \mu_1)$ by requiring $\mu_1(f(x),y) = \mu_0(x, f^*(y))$ for all $x \in P_0$ and $y \in P_1$. In other words, we define $f^* := \overline{\mu_0}^{-1} \circ f^* \circ \overline{\mu_1}$ where the second f^* refers to the $\mathcal{N}(G)$-map $f^* = \hom_{\mathcal{N}(G)}(f, \mathrm{id}) \colon P_1^* \to P_0^*$. In the sequel we will use the symbol f^* for both $f^* \colon P_1 \to P_0$ and $f^* \colon P_1^* \to P_0^*$.

Let {$\mathcal{N}(G)^n$} $\subset$ {fin. gen. proj. $\mathcal{N}(G)$-mod. with $\langle\ \rangle$} be the full subcategory whose objects are $(\mathcal{N}(G)^n, \mu_{st})$, where the *standard inner product* is given by

$$\mu_{\mathrm{st}} \colon \mathcal{N}(G)^n \times \mathcal{N}(G)^n \to \mathcal{N}(G), \qquad (x,y) \mapsto \sum_{i=1}^n x_i y_i^*. \qquad (6.21)$$

Denote by $\{l^2(G)^n\} \subset \{\text{fin. gen. Hilb. } \mathcal{N}(G)\text{-mod.}\}$ the full subcategory whose objects are $l^2(G)^n$. We define an isomorphism of $\mathbb{C}$-categories with involution

$$\nu \colon \{\mathcal{N}(G)^n\} \to \{l^2(G)^n\} \qquad (6.22)$$

as follows. It sends an object $\mathcal{N}(G)^n$ to the object $l^2(G)^n$. Let $f \colon \mathcal{N}(G)^m \to \mathcal{N}(G)^n$ be an $\mathcal{N}(G)$-homomorphism of (left) $\mathcal{N}(G)$-modules. Choose an (m,n)-matrix $A \in M(m,n,\mathcal{N}(G))$ such that f sends x to xA. Define

$$\nu(f) \colon l^2(G)^m \to l^2(G)^n, \qquad y \mapsto \left(A^* y^t\right)^t$$

where y^t is obtained from y by transposing and applying elementwise the involution $l^2(G) \to l^2(G)$ which sends $\sum_{g \in G} \lambda_g \cdot g$ to $\sum_{g \in G} \overline{\lambda_g} \cdot g$, the matrix A^* is obtained from A by transposing and applying the involution $* \colon \mathcal{N}(G) \to \mathcal{N}(G)$ to each entry, and Ay^t is understood in the sense of matrices and plugging y_j into an element $a \colon l^2(G) \to l^2(G)$ in $\mathcal{N}(G)$. The involutions appearing in the definition above come from the fact that we have decided to consider always left group actions and left module structures and have defined $\mathcal{N}(G)$ to be $\mathcal{B}(l^2(G))^G$. We will extend ν to finitely generated $\mathcal{N}(G)$-modules with inner product by a completion process below.

Lemma 6.23. *For any finitely generated projective $\mathcal{N}(G)$-module P there is an $\mathcal{N}(G)$-map $p \colon \mathcal{N}(G)^n \to \mathcal{N}(G)^n$ with $p \circ p = p$ and $p^* = p$ such that $\mathrm{im}(p)$ is $\mathcal{N}(G)$-isomorphic to P. Any finitely generated projective $\mathcal{N}(G)$-module P has an inner product. Two finitely generated projective $\mathcal{N}(G)$-modules with inner product are unitarily $\mathcal{N}(G)$-isomorphic if and only if the underlying $\mathcal{N}(G)$-modules are $\mathcal{N}(G)$-isomorphic.*

Proof. If P is finitely generated projective, we can find an $\mathcal{N}(G)$-map $q \colon \mathcal{N}(G)^n \to \mathcal{N}(G)^n$ with $q \circ q = q$ and an $\mathcal{N}(G)$-isomorphism $f \colon P \xrightarrow{\cong} \mathrm{im}(q)$. Let $p \colon \mathcal{N}(G)^n \to \mathcal{N}(G)^n$ be the $\mathcal{N}(G)$-map for which $\nu(p)$ is the orthogonal projection onto the image of $\nu(q)$. Then p satisfies $p \circ p = p$, $p^* = p$ and $\mathrm{im}(p) = \mathrm{im}(q)$. The standard inner product μ_{st} on $\mathcal{N}(G)^n$ defined in (6.21) restricts to an inner product on $\mathrm{im}(p)$, also denoted by μ_{st}, and this restriction can be pulled back to an inner product on P by the isomorphism $f \colon P \to \mathrm{im}(p)$. It remains to show for an $\mathcal{N}(G)$-map $p \colon \mathcal{N}(G)^n \to \mathcal{N}(G)^n$ with $p \circ p = p$ and $p^* = p$ that for any inner product μ on $\mathrm{im}(p)$ there is a unitary isomorphism $g \colon (\mathrm{im}(p), \mu) \to (\mathrm{im}(p), \mu_{\mathrm{st}})$.

Let $f \colon P \to P$ be the $\mathcal{N}(G)$-isomorphism $(\overline{\mu_{\mathrm{st}}})^{-1} \circ \overline{\mu}$. It satisfies $\mu(x,y) = \mu_{\mathrm{st}}(x, f(y))$ for all $x, y \in P$. Since $\mu_{\mathrm{st}}(x, f(x)) = \mu(x,x) \geq 0$ holds, f is positive with respect to μ_{st}. Consider the composition $i \circ f \circ p \colon \mathcal{N}(G)^n \to \mathcal{N}(G)^n$

where $i \colon P \to \mathcal{N}(G)^n$ is the inclusion, which is the adjoint of $p \colon \mathcal{N}(G)^n \to P$ with respect to μ_{st}. This composition is positive with respect to μ_{st} on $\mathcal{N}(G)^n$. Recall that we can consider $\mathbb{C}G$ as a dense subset of $l^2(G)$ in the obvious way and as a subset of $\mathcal{N}(G)$ if we identify $g \in G$ with the element $R_{g^{-1}} \colon l^2(G) \to l^2(G)$ in $\mathcal{N}(G)$ which is given by right multiplication with g^{-1}. Under this identification one easily checks for $u, v \in \mathbb{C}G^n$

$$\mathrm{tr}_{\mathcal{N}(G)}(\mu_{st}(u, i \circ f \circ p(v))) \ = \ \langle u, \nu(i \circ f \circ p)(v) \rangle_{l^2(G)^n}.$$

Hence $\langle u, \nu(i \circ f \circ p)(u) \rangle_{l^2(G)^n} \geq 0$ for all $u \in \mathbb{C}G^n$ and therefore for all $u \in l^2(G)^n$. So $\nu(i \circ f \circ p)$ is positive. Let $g' \colon \mathcal{N}(G)^n \to \mathcal{N}(G)^n$ be defined by the property that $\nu(g') \colon l^2(G)^n \to l^2(G)^n$ is positive and $\nu(g') \circ \nu(g') = \nu(i \circ f \circ p)$. Define $g \colon P \to P$ by $p \circ g' \circ i$. Then g is invertible, is selfadjoint with respect to μ_{st} and $g^2 = f$. Now the claim follows from

$$\mu_{st}(g(x), g(y)) \ = \ \mu_{st}(x, g^2(y)) \ = \ \mu_{st}(x, f(y)) \ = \ \mu(x, y). \square$$

Given a finitely generated projective $\mathcal{N}(G)$-module (P, μ) with inner product μ, we obtain a pre-Hilbert structure on P by $\mathrm{tr}_{\mathcal{N}(G)} \circ \mu \colon P \times P \to \mathbb{C}$ for $\mathrm{tr}_{\mathcal{N}(G)}$ the standard trace (see Definition 1.2). Let $\nu(P, \mu)$ be the associated Hilbert space. The group G acts from the left by unitary operators on P and hence on $\nu(P, \mu)$ by putting $g \cdot x := R_{g^{-1}} \cdot x$ for $x \in P$ and $R_{g^{-1}} \colon l^2(G) \to l^2(G) \in \mathcal{N}(G)$ given by right multiplication with g^{-1}. This is a finitely generated Hilbert $\mathcal{N}(G)$-module since one can find another finitely generated projective $\mathcal{N}(G)$-module with inner product (P_0, μ_0) and a unitary isomorphism $(P, \mu) \oplus (P_0, \mu_0) \to (\mathcal{N}(G)^n, \mu_{st})$ (see Lemma 6.23) which induces an isometric G-isomorphism $\nu(P, \mu) \oplus \nu(P_0, \mu_0) \to \nu(\mathcal{N}(G)^n, \mu_{st}) = l^2(G)^n$. Let (P, μ) and (P', μ') be finitely generated projective $\mathcal{N}(G)$-modules with inner product and let $f \colon P \to P'$ be an $\mathcal{N}(G)$-homomorphism. Then f extends to a morphism of finitely generated Hilbert $\mathcal{N}(G)$-modules $\nu(f) \colon \nu(P, \mu) \to \nu(P', \mu')$.

Theorem 6.24 (Modules over $\mathcal{N}(G)$ and Hilbert $\mathcal{N}(G)$-modules).

(1) We obtain an equivalence of $\mathbb{C}$-categories with involution

$$\nu \colon \{\textit{fin. gen. proj. } \mathcal{N}(G)\textit{-mod. with } \langle \, \rangle\} \to \{\textit{fin. gen. Hilb. } \mathcal{N}(G)\textit{-mod.}\};$$

(2) The forgetful functor

$$F \colon \{\textit{fin. gen. proj. } \mathcal{N}(G)\textit{-mod. with } \langle \, \rangle\} \to \{\textit{fin. gen. proj. } \mathcal{N}(G)\textit{-mod.}\}$$

is an equivalence of $\mathbb{C}$-categories. We obtain functors of $\mathbb{C}$-categories

$$\nu \circ F^{-1} \colon \{\textit{fin. gen. proj. } \mathcal{N}(G)\textit{-mod.}\} \to \{\textit{fin. gen. Hilb. } \mathcal{N}(G)\textit{-mod.}\}$$
$$F \circ \nu^{-1} \colon \{\textit{fin. gen. Hilb. } \mathcal{N}(G)\textit{-mod.}\} \to \{\textit{fin. gen. proj. } \mathcal{N}(G)\textit{-mod.}\}$$

which are unique up to natural equivalence of functors of $\mathbb{C}$-categories and inverse to one another up to natural equivalence. We will denote $\nu \circ F^{-1}$ and $F \circ \nu^{-1}$ by ν and ν^{-1} again;

(3) The functors ν and ν^{-1} preserve exact sequences and weakly exact sequences;

(4) If P is a finitely generated projective $\mathcal{N}(G)$-module, then

$$\dim_{\mathcal{N}(G)}(P) = \dim_{\mathcal{N}(G)}(\nu(P))$$

for the two notions of $\dim_{\mathcal{N}(G)}$ defined in Definition 1.10 and in (6.3).

Proof. (1) The *idempotent completion* $\mathrm{Idem}(\mathcal{C})$ of a $\mathbb{C}$-category $\mathcal{C}$ with involution has as objects (V, p) selfadjoint idempotents $p \colon V \to V$. A morphism from (V_0, p_0) to (V_1, p_1) is a morphism $f \colon V_0 \to V_1$ satisfying $p_1 \circ f \circ p_0 = f$. The identity on (V, p) is given by $p \colon (V, p) \to (V, p)$. The idempotent completion $\mathrm{Idem}(\mathcal{C})$ inherits from $\mathcal{C}$ the structure of a $\mathbb{C}$-category with involution in the obvious way. There are unitary equivalences of $\mathbb{C}$-categories with involutions

$$\mathrm{IM} \colon \mathrm{Idem}(\{\mathcal{N}(G)^n\}) \to \{\text{fin. gen. proj. } \mathcal{N}(G)\text{-mod. with } \langle\ \rangle\};$$
$$\mathrm{IM} \colon \mathrm{Idem}(\{l^2(G)^n\}) \to \{\text{fin. gen. Hilb. } \mathcal{N}(G)\text{-mod.}\},$$

which sends $(\mathcal{N}(G)^n, p)$ or $(l^2(G)^n, p)$ to the image of p where the inner product μ_{st} on $\mathrm{im}(p)$ is given by restricting the standard inner product μ_{st} on $\mathcal{N}(G)^n$. The functor ν defined in (6.22) induces an isomorphism

$$\mathrm{Idem}(\nu) \colon \mathrm{Idem}(\{\mathcal{N}(G)^n\}) \to \mathrm{Idem}(\{l^2(G)^n\}).$$

The following diagram commutes

$$
\begin{array}{ccc}
\mathrm{Idem}(\{\mathcal{N}(G)^n\}) & \xrightarrow{\ \mathrm{Idem}(\nu)\ } & \mathrm{Idem}(\{l^2(G)^n\} \\
\Big\downarrow{\scriptstyle \mathrm{IM}} & & \Big\downarrow{\scriptstyle \mathrm{IM}} \\
\{\text{fin. gen. proj. } \mathcal{N}(G)\text{-mod. with } \langle\ \rangle\} & \xrightarrow{\ \nu\ } & \{\text{fin. gen. Hilb. } \mathcal{N}(G)\text{-mod.}\}
\end{array}
$$

up to unitary natural equivalence which is induced from the inclusion $\mathcal{N}(G)^n \to l^2(G)^n$ sending $(a_1, \ldots, a_n)$ to $(\overline{a_1^*(e)}, \ldots, \overline{a_n^*(e)})$, where for $u = \sum_{g \in G} \lambda_g \cdot g \in l^2(G)$ we put $\overline{u} := \sum_{g \in G} \overline{\lambda_g} \cdot g$. Now (1) follows.

(2) follows from Lemma 6.23 and assertion (1).

(3) A sequence $U \xrightarrow{f} V \xrightarrow{g} W$ of finitely generated Hilbert $\mathcal{N}(G)$-modules is weakly exact at V if and only if the following holds: $g \circ f = 0$ and for any finitely generated Hilbert $\mathcal{N}(G)$-modules P and Q and morphisms $u \colon V \to P$ and $v \colon Q \to V$ with $u \circ f = 0$ and $g \circ v = 0$ we get $u \circ v = 0$. It is exact at V if and only if the following holds: $g \circ f = 0$ and for any finitely generated Hilbert $\mathcal{N}(G)$-module P and morphism $v \colon P \to V$ with $g \circ v = 0$ there is a morphism $u \colon P \to U$ satisfying $f \circ u = v$. The analogous statements are true if one considers finitely generated projective $\mathcal{N}(G)$-modules instead of finitely generated Hilbert $\mathcal{N}(G)$-modules. Now ν and ν^{-1} obviously preserve

these criterions for weak exactness and exactness and the claim follows.

(4) This follows from the definitions.   □

Next we can give the proof of Theorem 6.5.

Proof. Part (1) of Assumption 6.2 is obvious, it remains to prove Part (2). In the sequel we use Theorem 6.24 and the functors ν and ν^{-1} appearing there.

First we show that any finitely generated submodule $M \subset P$ of a finitely generated projective $\mathcal{N}(G)$-module is finitely generated projective. Namely, choose an $\mathcal{N}(G)$-map $f\colon \mathcal{N}(G)^n \to P$ with $\mathrm{im}(f) = M$. Let $p\colon \mathcal{N}(G)^n \to \mathcal{N}(G)^n$ be the $\mathcal{N}(G)$-map for which $\nu(p)\colon l^2(G)^n \to l^2(G)^n$ is the orthogonal projection onto the kernel of $\nu(f)\colon \nu(\mathcal{N}(G)^n) = l^2(G)^n \to \nu(P)$. Then p is an idempotent with $\mathrm{im}(p) = \ker(f)$ and hence M is projective.

Let $\mathcal{P} = \{P_i \mid i \in I\}$ be the directed system of finitely generated projective $\mathcal{N}(G)$-submodules of K. Notice that $\mathcal{P}$ is indeed directed by inclusion since the submodule of P generated by two finitely generated projective submodules is again finitely generated and hence finitely generated projective. Let $j_i\colon P_i \to Q$ be the inclusion. Equip Q and each P_i with a fixed inner product (Lemma 6.23). Let $\mathrm{pr}_i\colon \nu(Q) \to \nu(Q)$ be the orthogonal projection satisfying $\mathrm{im}(\mathrm{pr}_i) = \overline{\mathrm{im}(\nu(j_i))}$ and $\mathrm{pr}\colon \nu(Q) \to \nu(Q)$ be the orthogonal projection satisfying $\mathrm{im}(\mathrm{pr}) = \overline{\bigcup_{i \in I} \mathrm{im}(\mathrm{pr}_i)}$. Next we show

$$\mathrm{im}(\nu^{-1}(\mathrm{pr})) = \overline{K}. \tag{6.25}$$

Let $f\colon Q \to \mathcal{N}(G)$ be an $\mathcal{N}(G)$-map with $K \subset \ker(f)$. Then $f \circ j_i = 0$ and therefore $\nu(f) \circ \nu(j_i) = 0$ for all $i \in I$. We get $\mathrm{im}(\mathrm{pr}_i) \subset \ker(\nu(f))$ for all $i \in I$. Since the kernel of $\nu(f)$ is closed we conclude $\mathrm{im}(\mathrm{pr}) \subset \ker(\nu(f))$. This shows $\mathrm{im}(\nu^{-1}(\mathrm{pr})) \subset \ker(f)$ and hence $\mathrm{im}(\nu^{-1}(\mathrm{pr})) \subset \overline{K}$. As $K \subset \ker(\mathrm{id} - \nu^{-1}(\mathrm{pr})) = \mathrm{im}(\nu^{-1}(\mathrm{pr}))$, we conclude $\overline{K} \subset \mathrm{im}(\nu^{-1}(\mathrm{pr}))$. This finishes the proof of (6.25). In particular $\overline{K}$ is a direct summand in Q.

Next we prove

$$\mathrm{dim}'_{\mathcal{N}(G)}(K) = \mathrm{dim}_{\mathcal{N}(G)}(\overline{K}). \tag{6.26}$$

The inclusion j_i induces a weak isomorphism $\nu(P_i) \to \mathrm{im}(\mathrm{pr}_i)$ of finitely generated Hilbert $\mathcal{N}(G)$-modules. If we apply the Polar Decomposition Theorem to it we obtain a unitary $\mathcal{N}(G)$-isomorphism from $\nu(P_i)$ to $\mathrm{im}(\mathrm{pr}_i)$. This implies $\mathrm{dim}_{\mathcal{N}(G)}(P_i) = \mathrm{tr}_{\mathcal{N}(G)}(\mathrm{pr}_i)$. Therefore it remains to prove

$$\mathrm{tr}_{\mathcal{N}(G)}(\mathrm{pr}) = \sup\{\mathrm{tr}_{\mathcal{N}(G)}(\mathrm{pr}_i) \mid i \in I\}. \tag{6.27}$$

Let $\epsilon > 0$ and $x \in \nu(Q)$ be given. Choose $i(\epsilon) \in I$ and $x_{i(\epsilon)} \in \mathrm{im}(\mathrm{pr}_{i(\epsilon)})$ with $\|\mathrm{pr}(x) - x_{i(\epsilon)}\| \leq \epsilon/2$. Since $\mathrm{im}(\mathrm{pr}_{i(\epsilon)}) \subset \mathrm{im}(\mathrm{pr}_i) \subset \mathrm{im}(\mathrm{pr})$ we get for all $i \geq i(\epsilon)$

$$\|\operatorname{pr}(x) - \operatorname{pr}_i(x)\| \leq \|\operatorname{pr}(x) - \operatorname{pr}_{i(\epsilon)}(x)\|$$
$$\leq \|\operatorname{pr}(x) - \operatorname{pr}_{i(\epsilon)}(x_{i(\epsilon)})\| + \|\operatorname{pr}_{i(\epsilon)}(x_{i(\epsilon)}) - \operatorname{pr}_{i(\epsilon)}(x)\|$$
$$= \|\operatorname{pr}(x) - x_{i(\epsilon)}\| + \|\operatorname{pr}_{i(\epsilon)}(x_{i(\epsilon)} - \operatorname{pr}(x))\|$$
$$\leq \|\operatorname{pr}(x) - x_{i(\epsilon)}\| + \|\operatorname{pr}_{i(\epsilon)}\| \cdot \|x_{i(\epsilon)} - \operatorname{pr}(x)\|$$
$$\leq 2 \cdot \|\operatorname{pr}(x) - x_{i(\epsilon)}\|$$
$$\leq \epsilon.$$

Now (6.27) and hence (6.26) follow from Theorem 1.9 (2). This finishes the proof of Theorem 6.5.   □

We will sometimes use

Lemma 6.28. *(1) Let $f: P \to Q$ be an $\mathcal{N}(G)$-map of finitely generated projective $\mathcal{N}(G)$-modules. If f is a weak isomorphism, then P and Q are $\mathcal{N}(G)$-isomorphic;*

(2) Let $f: P \to Q$ be an $\mathcal{N}(G)$-map of finitely generated projective $\mathcal{N}(G)$-modules with $\dim_{\mathcal{N}(G)}(P) = \dim_{\mathcal{N}(G)}(Q)$. Then the following assertions are equivalent:

(a) f is injective;

(b) f has dense image;

(c) f is a weak isomorphism;

(3) A projective $\mathcal{N}(G)$-module P is trivial if and only if $\dim_{\mathcal{N}(G)}(P) = 0$;

(4) Let M be a finitely presented $\mathcal{N}(G)$-module. Then $\dim_{\mathcal{N}(G)}(M) = 0$ if and only if there is an exact sequence $0 \to \mathcal{N}(G)^n \xrightarrow{f} \mathcal{N}(G)^n \to M \to 0$ for some $\mathcal{N}(G)$-map f which is positive, i.e. $f = h^*h$ for some $\mathcal{N}(G)$-map $h: \mathcal{N}(G)^n \to \mathcal{N}(G)^n$.

Proof. (1) and (2) follow from Lemma 1.13 and Theorem 6.24.

(3) Since P is the colimit of its finitely generated submodules, it suffices to prove the claim for a finitely generated projective $\mathcal{N}(G)$-module by Theorem 6.7 (1) and (4c). Now apply Theorem 1.12 (1) and Theorem 6.24.

(4) Suppose that $\dim_{\mathcal{N}(G)}(M) = 0$. Since $\mathcal{N}(G)$ is semihereditary (see Theorem 6.7 (1)) and M is finitely presented, there is an exact sequence $0 \to P \xrightarrow{g} \mathcal{N}(G)^n \xrightarrow{q} M \to 0$ of $\mathcal{N}(G)$-modules such that P is finitely generated projective. We get from Additivity (see Theorem 6.7 (4b)) that $\dim_{\mathcal{N}(G)}(P) = \dim_{\mathcal{N}(G)}(\mathcal{N}(G)^n)$. Since by assertion (2) g is a weak isomorphism, $\nu(g)$ is a weak isomorphism by Theorem 6.24 (3). Let $f: \mathcal{N}(G)^n \to \mathcal{N}(G)^n$ be the $\mathcal{N}(G)$-homomorphism for which $\nu(f)$ is the positive part in the polar decomposition of $\nu(g)$. Now $0 \to \mathcal{N}(G)^n \xrightarrow{f} \mathcal{N}(G)^n \to M \to 0$ is exact by Theorem 6.24 (3). The other implication follows from Additivity (see Theorem 6.7 (4b)).   □

6.3 Induction and the Extended von Neumann Dimension

In this section we show that induction for the ring homomorphism $\mathcal{N}(H) \to \mathcal{N}(G)$ induced by an inclusion of groups $i \colon H \to G$ is faithfully flat and compatible with the extended von Neumann dimension. This will be important when we will compare L^2-invariants of a regular covering with the ones of the universal covering of a given space.

Recall from Theorem 6.5 and Theorem 6.7 that $\dim_{\mathcal{N}(G)}$ introduced in Definition 6.20 satisfies Cofinality, Additivity and Continuity and that for a finitely generated $\mathcal{N}(G)$-module M we get a finitely generated projective $\mathcal{N}(G)$-module $\mathbf{P}M$ such that $M = \mathbf{P}M \oplus \mathbf{T}M$ and $\dim_{\mathcal{N}(G)}(M) = \dim_{\mathcal{N}(G)}(\mathbf{P}M)$.

We have associated to an injective group homomorphism $i \colon H \to G$ a ring homomorphism $i \colon \mathcal{N}(H) \to \mathcal{N}(G)$ (see Definition 1.23). Given an $\mathcal{N}(H)$-module M, the *induction* with i is the $\mathcal{N}(G)$-module $i_* M = \mathcal{N}(G) \otimes_{\mathcal{N}(H)} M$. Obviously i_* is a covariant functor from the category of $\mathcal{N}(H)$-modules to the category of $\mathcal{N}(G)$-modules, preserves direct sums and the properties "finitely generated" and "projective" and sends $\mathcal{N}(H)$ to $\mathcal{N}(G)$.

Theorem 6.29. *Let $i \colon H \to G$ be an injective group homomorphism. Then*

(1) Induction with i is a faithfully flat functor from the category of $\mathcal{N}(H)$-modules to the category of $\mathcal{N}(G)$-modules, i.e. a sequence of $\mathcal{N}(H)$-modules $M_0 \to M_1 \to M_2$ is exact at M_1 if and only if the induced sequence of $\mathcal{N}(G)$-modules $i_* M_0 \to i_* M_1 \to i_* M_2$ is exact at $i_* M_1$;
(2) For any $\mathcal{N}(H)$-module M we have:

$$\dim_{\mathcal{N}(H)}(M) = \dim_{\mathcal{N}(G)}(i_* M).$$

Proof. It is enough to show for any $\mathcal{N}(H)$-module M

$$\dim_{\mathcal{N}(H)}(M) = \dim_{\mathcal{N}(G)}(i_* M); \tag{6.30}$$

$$\operatorname{Tor}_p^{\mathcal{N}(H)}(\mathcal{N}(G), M) = 0 \qquad \text{for } p \geq 1; \tag{6.31}$$

$$i_* M = 0 \Rightarrow M = 0. \tag{6.32}$$

We begin with the case where M is finitely generated projective. Let $A \in M_n(\mathcal{N}(H))$ be a matrix such that $A = A^*$, $A^2 = A$ and the image of the $\mathcal{N}(H)$-linear map $R_A \colon \mathcal{N}(H)^n \to \mathcal{N}(H)^n$ induced by right multiplication with A is $\mathcal{N}(H)$-isomorphic to M. Let $i_* A$ be the matrix in $M_n(\mathcal{N}(G))$ obtained from A by applying i to each entry. Then $i_* A = (i_* A)^*$, $(i_* A)^2 = i_* A$ and the image of the $\mathcal{N}(G)$-linear map $R_{i_* A} \colon \mathcal{N}(G)^n \to \mathcal{N}(G)^n$ induced by right multiplication with $i_* A$ is $\mathcal{N}(G)$-isomorphic to $i_* M$. Hence we get from the definition (6.3)

$$\dim_{\mathcal{N}(H)}(M) = \operatorname{tr}_{\mathcal{N}(H)}(A);$$
$$\dim_{\mathcal{N}(G)}(i_* M) = \operatorname{tr}_{\mathcal{N}(G)}(i_* A).$$

Since $\operatorname{tr}_{\mathcal{N}(G)}(i(a)) = \operatorname{tr}_{\mathcal{N}(H)}(a)$ holds for $a \in \mathcal{N}(H)$ (see Lemma 1.24 (1)), we get (6.30). Since for a finitely generated projective $\mathcal{N}(H)$-module we have $M = 0 \Leftrightarrow \dim_{\mathcal{N}(H)}(M) = 0$ and $i_* M$ is a finitely generated projective $\mathcal{N}(G)$-module, (6.32) follows from (6.30). If M is projective, (6.31) is obviously true.

Next we treat the case where M is finitely presented. Then M splits as $M = \mathbf{T}M \oplus \mathbf{P}M$ where $\mathbf{P}M$ is finitely generated projective and $\mathbf{T}M$ is finitely presented (see Theorem 6.7 (3)). By Lemma 6.28 (4) we obtain an exact sequence $0 \to \mathcal{N}(H)^n \xrightarrow{f} \mathcal{N}(H)^n \xrightarrow{q} \mathbf{T}M \to 0$ with $f = f^*$. If we apply the right exact functor given by induction with i to it, we get an exact sequence $\mathcal{N}(G)^n \xrightarrow{i_* f} \mathcal{N}(G)^n \to i_* \mathbf{T}M \to 0$ with $(i_* f)^* = i_* f$. Since we know (6.30) and (6.31) already for finitely generated projective $\mathcal{N}(H)$-modules, we conclude from Theorem 6.7 (4b) and (4e) and the definition of Tor that (6.30) and (6.31) hold for the finitely presented $\mathcal{N}(H)$-module M provided that we can show that $i_* f$ is injective.

We have $i_*(\nu(f)) = \nu(i_* f)$, where $i_*(\nu(f))$ was introduced in Definition 1.23. Because ν respects weak exactness (see Theorem 6.24 (3)), $\nu(f)$ has dense image since $\mathcal{N}(H)^n \xrightarrow{f} \mathcal{N}(H)^n \to 0$ is weakly exact. Then one easily checks that $\nu(i_* f) = i_*(\nu(f))$ has dense image since $\mathbb{C}G \otimes_{\mathbb{C}H} l^2(H)$ is a dense subspace of $l^2(G)$. Since the kernel of a bounded operator of Hilbert spaces is the orthogonal complement of the image of its adjoint and $\nu(i_* f)$ is selfadjoint, $\nu(i_* f)$ is injective. Since ν^{-1} respects exactness (see Theorem 6.24 (3)) $i_* f$ is injective.

Next we show (6.32) for finitely presented M. It suffices to show for an $\mathcal{N}(H)$-map $\mathcal{N}(H)^m \to \mathcal{N}(H)^n$ that g is surjective if $i_* g$ is surjective. Since the functors ν^{-1} and ν of Theorem 6.24 are exact we have to show for an H-equivariant bounded operator $h \colon l^2(H)^m \to l^2(H)^n$ that h is surjective if $i_* h \colon l^2(G)^m \to l^2(G)^n$ is surjective. Let $\{E_\lambda \mid \lambda \geq 0\}$ be the spectral family of the positive operator $h \circ h^*$. Then $\{i_* E_\lambda \mid \lambda \geq 0\}$ is the spectral family of the positive operator $i_* h \circ (i_* h)^*$. Notice that h or $i_* h$ respectively is surjective if and only if $E_\lambda = 0$ or $i_* E_\lambda = 0$ respectively for some $\lambda > 0$. Because $E_\lambda = 0$ or $i_* E_\lambda = 0$ respectively is equivalent to $\operatorname{tr}_{\mathcal{N}(H)}(E_\lambda) = 0$ or $\operatorname{tr}_{\mathcal{N}(G)}(i_* E_\lambda) = 0$ respectively and $\operatorname{tr}_{\mathcal{N}(H)}(E_\lambda) = \operatorname{tr}_{\mathcal{N}(G)}(i_* E_\lambda)$ holds by Lemma 1.24 (1), the claim follows. This finishes the proof of (6.30), (6.31) and (6.32) in the case where M is finitely presented.

Next we explain how we can derive the case of a finitely generated $\mathcal{N}(H)$-module M from the case of a finitely presented $\mathcal{N}(H)$-module. Notice that from now on the argument is purely algebraic, no more functional analysis will enter. If $f \colon \mathcal{N}(H)^n \to M$ is an epimorphism with kernel K, then M is the colimit over the directed system of $\mathcal{N}(H)$-modules $\mathcal{N}(H)^n / K_j$ indexed by the set $\{K_j \mid j \in J\}$ of finitely generated submodules of K. Now (6.30) follows from Additivity and Cofinality (see Theorem 4 (4b) and (4c)) or directly from Theorem 6.13 about dimension and colimits since i_* commutes with colimits. The functor Tor commutes in both variables with colimits over directed systems [94, Proposition VI.1.3. on page 107] and hence (6.31) follows.

Next we prove (6.32) for finitely generated M. As above we write $M = \operatorname{colim}_{j \in J} \mathcal{N}(H)^n / K_j$. Since each structure map $\mathcal{N}(H)^n / K_{j_0} \to \mathcal{N}(H)^n / K_{j_1}$ is an epimorphism of finitely generated $\mathcal{N}(H)$-modules, we have $M = 0$ if and only if for each $j_0 \in J$ there is $j_1 \in J$ such that $\mathcal{N}(H)^n / K_{j_1}$ vanishes. Recall that i_* commutes with colimits and is right exact. Hence $i_* M = \operatorname{colim}_{j \in J} i_*(\mathcal{N}(H)^n / K_j)$ and $i_* M = 0$ if and only if for each $j_0 \in J$ there is $j_1 \in J$ such that $i_*(\mathcal{N}(H)^n / K_{j_1})$ vanishes. Since $\mathcal{N}(H)^n / K_{j_1}$ is finitely presented, we know already that $\mathcal{N}(H)^n / K_{j_1} = 0$ if and only if $i_*(\mathcal{N}(H)^n / K_{j_1}) = 0$. Hence $i_* M = 0$ if and only if $M = 0$.

The argument that (6.30), (6.31) and (6.32) are true for all $\mathcal{N}(H)$-modules if they are true for all finitely generated $\mathcal{N}(H)$-modules is analogous to the proof above that they are true for all finitely generated $\mathcal{N}(H)$-modules if they are true for all finitely presented $\mathcal{N}(H)$-modules. Namely, repeat the argument for the directed system of finitely generated submodules of a given $\mathcal{N}(H)$-module. This finishes the proof of Theorem 6.29. $\square$

The proof of Theorem 6.29 would be obvious if we knew that $\mathcal{N}(G)$ viewed as an $\mathcal{N}(H)$-module were projective. Notice that this is a stronger statement than the one proved in Theorem 6.29. One would have to show that the higher Ext-groups instead of the Tor-groups vanish to get this stronger statement. However, the proof for the Tor-groups does not go through directly since the Ext-groups are not compatible with colimits.

The rather easy proof of the next lemma can be found in [333, Lemma 3.4 on page 149].

Lemma 6.33. *Let $H \subset G$ be a subgroup. Then*

(1) $\dim_{\mathcal{N}(G)} (\mathcal{N}(G) \otimes_{\mathbb{C}G} \mathbb{C}[G/H]) = |H|^{-1}$, *where $|H|^{-1}$ is defined to be zero if H is infinite;*

(2) *If G is infinite and V is a $\mathbb{C}G$-module which is finite dimensional over $\mathbb{C}$, then*

$$\dim_{\mathcal{N}(G)}(\mathcal{N}(G) \otimes_{\mathbb{C}G} V) = 0.$$

6.4 The Extended Dimension Function and Amenable Groups

In this section we deal with the special case of an amenable group G. The main result (Theorem 6.37) will be that $\mathcal{N}(G)$ is flat over $\mathbb{C}G$ from the point of view of the extended dimension function, although $\mathcal{N}(G)$ is flat over $\mathbb{C}G$ in the strict sense only for very few groups, conjecturally exactly for virtually cyclic groups (see Conjecture 6.49).

6.4.1 Survey on Amenable Groups

In this subsection we give a brief survey about amenable and elementary amenable groups. Let $l^\infty(G, \mathbb{R})$ be the space of bounded functions from G to $\mathbb{R}$ with the supremum norm. Denote by 1 the constant function with value 1.

Definition 6.34 (Amenable group). *A group G is called amenable if there is a (left) G-invariant linear operator $\mu \colon l^\infty(G, \mathbb{R}) \to \mathbb{R}$ with $\mu(1) = 1$ which satisfies for all $f \in l^\infty(G, \mathbb{R})$*

$$\inf\{f(g) \mid g \in G\} \le \mu(f) \le \sup\{f(g) \mid g \in G\}.$$

The latter condition is equivalent to the condition that μ is bounded and $\mu(f) \ge 0$ if $f(g) \ge 0$ for all $g \in G$. Let $\mathcal{AM}$ be the class of amenable groups.

The class $\mathcal{EAM}$ of elementary amenable groups is defined as the smallest class of groups which has the following properties:

(1) It contains all finite and all abelian groups;
(2) It is closed under taking subgroups;
(3) It is closed under taking quotient groups;
(4) It is closed under extensions, i.e. if $1 \to H \to G \to K \to 1$ is an exact sequence of groups and H and K belong to $\mathcal{EAM}$, then also $G \in \mathcal{EAM}$;
(5) It is closed under directed unions, i.e. if $\{G_i \mid i \in I\}$ is a directed system of subgroups such that $G = \bigcup_{i \in I} G_i$ and each G_i belongs to $\mathcal{EAM}$, then $G \in \mathcal{EAM}$. (Directed means that for two indices i and j there is a third index k with $G_i, G_j \subset G_k$.)

We give an overview of some basic properties of these notions. A group G is amenable if and only if each finitely generated subgroup is amenable [419, Proposition 0.16 on page 14]. The class of amenable groups satisfies the conditions appearing in the Definition 6.34 of elementary amenable groups, namely, it contains all finite and all abelian groups, and is closed under taking subgroups, forming factor groups, group extensions, and directed unions [419, Proposition 0.15 and 0.16 on page 14]. Hence any elementary amenable group is amenable. Grigorchuk [228] has constructed a finitely presented group which is amenable but not elementary amenable. A group which contains the free group on two letters $\mathbb{Z} * \mathbb{Z}$ as a subgroup is not amenable [419, Example 0.6 on page 6]. There exist examples of finitely presented non-amenable groups which do not contain $\mathbb{Z} * \mathbb{Z}$ as a subgroup [407]. Any countable amenable group is *a-T-menable*, i.e. it admits an affine, isometric action on a real inner product space which is metrically proper in the sense that $\lim_{g \to \infty} \|gv\| = \infty$ holds for every $v \in V$ [33]. For a-T-menable groups the Baum-Connes-Conjecture has been proved by Higson and Kasparov [257]. An infinite a-T-menable group and in particular an infinite amenable group does not have Kazhdan's property T.

A useful geometric characterization of amenable groups is given by the Følner condition [35, Theorem F.6.8 on page 308].

Lemma 6.35. *A group G is amenable if and only if it satisfies the Følner condition, i.e. for any finite set $S \subset G$ with $s \in S \Rightarrow s^{-1} \in S$ and $\epsilon > 0$ there exists a finite non-empty subset $A \subset G$ such that for its S-boundary $\partial_S A = \{a \in A \mid \text{ there is } s \in S \text{ with } as \notin A\}$ we have*

$$|\partial_S A| \leq \epsilon \cdot |A|.$$

Another version of the Følner criterion is that a finitely presented group G is amenable if and only if for any positive integer n, any connected closed n-dimensional Riemannian manifold M with fundamental group $\pi_1(M) = G$ and any $\epsilon > 0$ there is a domain $\Omega \subset \widetilde{M}$ with $(n-1)$-measurable boundary such that the $(n-1)$-measure of $\partial \Omega$ does not exceed ϵ times the measure of Ω [35, Theorem F.6.8 on page 308]. Such a domain can be constructed by an appropriate finite union of translations of a fundamental domain if G is amenable.

The fundamental group of a closed connected manifold is not amenable if M admits a Riemannian metric of non-positive curvature which is not constant zero [16]. Any finitely generated group which is not amenable has exponential growth [35, Proposition F.6.24 on page 318]. A group G is called *good* in the sense of Freedmann if the so called π_1-null disk lemma holds for G which implies the topological s-Cobordism Theorem for 4-manifolds with G as fundamental group. Any group with subexponential growth is good [204, Theorem 0.1 on page 511] and amenable. No amenable group is known which is not good and it may be true that the classes of good groups and amenable groups coincide. The following conditions on a group G are equivalent:

(1) G is amenable;
(2) The canonical map from the full C^*-algebra of G to the reduced C^*-algebra of G is an isomorphism [420, Theorem 7.3.9 on page 243];
(3) The reduced C^*-algebra of G is nuclear [303], provided that G is countable;
(4) G is finite or for any connected free G-CW-complex X the first Novikov-Shubin invariant $\alpha_1(X)$ is not ∞^+ (see Theorem 2.55 (5b));
(5) For a closed connected Riemannian manifold M with $G = \pi_1(M)$ the Laplacian $\Delta_0 \colon L^2\Omega^0(\widetilde{M}) \to L^2\Omega^0(\widetilde{M})$ acting on functions on the universal covering has zero in its spectrum, provided that G is finitely presented (see [68] or Section 2.7);
(6) $H^1(G, l^2(G))$ is not Hausdorff [242, Corollary III.2.4 on page 188];
(7) $\mathcal{N}(G) \otimes_{\mathbb{C}G} \mathbb{C}$ is not trivial (see Lemma 6.36).

Conjecture 6.48 implies another characterization of amenability in terms of $G_0(\mathbb{C}G)$, namely $[\mathbb{C}G] = 0$ would hold in $G_0(\mathbb{C}G)$ if and only if G is not amenable (see Conjecture 9.67).

It is sometimes easier to deal with elementary amenable groups than with amenable groups since elementary amenable groups have a useful description in terms of transfinite induction (see Lemma 10.40).

For more information about amenable groups we refer to [419].

6.4.2 Amenability and the Coinvariants of the Group von Neumann Algebra

Lemma 6.36. $\mathcal{N}(G) \otimes_{\mathbb{C}G} \mathbb{C}[G/H]$ *is trivial if and only if H is non-amenable*

Proof. Since $\mathcal{N}(G) \otimes_{\mathbb{C}G} \mathbb{C}[G/H]$ and $\mathcal{N}(G) \otimes_{\mathcal{N}(H)} \mathcal{N}(H) \otimes_{\mathbb{C}H} \mathbb{C}$ are $\mathcal{N}(G)$-isomorphic, we conclude from Theorem 6.29 (1) that its suffices to prove the claim in the special case $G = H$.

Let S be a set of generators of G. Then $\bigoplus_{s \in S} \mathbb{C}G \xrightarrow{\bigoplus_{s \in S} r_{s-1}} \mathbb{C}G \xrightarrow{\epsilon} \mathbb{C} \to 0$ is exact where $\epsilon(\sum_{g \in G} \lambda_g \cdot g) = \sum_{g \in G} \lambda_g$ and r_u denotes right multiplication with $u \in \mathbb{C}G$. We obtain an exact sequence $\bigoplus_{s \in S} \mathcal{N}(G) \xrightarrow{\bigoplus_{s \in S} r_{s-1}}$ $\mathcal{N}(G) \xrightarrow{\epsilon} \mathcal{N}(G) \otimes_{\mathbb{C}G} \mathbb{C} \to 0$ since the tensor product is right exact. Hence $\mathcal{N}(G) \otimes_{\mathbb{C}G} \mathbb{C}$ is trivial if and only if $\bigoplus_{s \in S} \mathcal{N}(G) \xrightarrow{\bigoplus_{s \in S} r_{s-1}} \mathcal{N}(G)$ is surjective. This is equivalent to the existence of a finite subset $T \subset S$ such that $\bigoplus_{s \in T} \mathcal{N}(G) \xrightarrow{\bigoplus_{s \in T} r_{s-1}} \mathcal{N}(G)$ is surjective. Let $G_0 \subset G$ be the subgroup generated by T. Then the map above is induction with the inclusion of $G_0 \subset G$ applied to $\bigoplus_{t \in T} \mathcal{N}(G_0) \xrightarrow{\bigoplus_{t \in T} r_{t-1}} \mathcal{N}(G_0)$. Hence we conclude from Theorem 6.29 (1) that $\mathcal{N}(G) \otimes_{\mathbb{C}G} \mathbb{C}$ is trivial if and only if $\mathcal{N}(G_0) \otimes_{\mathbb{C}G_0} \mathbb{C}$ is trivial for some finitely generated subgroup $G_0 \subset G$. The group G is amenable if and only if each of its finitely generated subgroups is amenable [419, Proposition 0.16 on page 14]. Hence we can assume without loss of generality that G is finitely generated and S is finite. Moreover, we can also assume that S is symmetric, i.e. $s \in S$ implies $s^{-1} \in S$.

Because the functor ν of Theorem 6.24 (3) is exact, $\mathcal{N}(G) \otimes_{\mathbb{C}G} \mathbb{C}$ is trivial if and only if the operator $f \colon \bigoplus_{s \in S} l^2(G) \xrightarrow{\bigoplus_{s \in S} r_{s-1}} l^2(G)$ is surjective. This is equivalent to the bijectivity of the operator

$$\frac{1}{2 \cdot |S|} f \circ f^* \colon l^2(G) \xrightarrow{\operatorname{id} - \sum_{s \in S} \frac{1}{|S|} \cdot r_s} l^2(G).$$

It is bijective if and only if the spectral radius of the operator $l^2(G) \xrightarrow{\sum_{s \in S} \frac{1}{|S|} \cdot r_s}$ $l^2(G)$ is different from 1. Since this operator is convolution with a probability distribution whose support contains S, namely

$$P \colon G \to [0,1], \qquad g \mapsto \begin{cases} |S|^{-1}, \ g \in S \\ 0, \qquad g \notin S \end{cases}$$

the spectral radius is 1 precisely if G is amenable, by a result of Kesten [290]. $\square$

6.4.3 Amenability and Flatness Properties of the Group von Neumann Algebra over the Group Ring

In this subsection we investigate the flatness properties of the group von Neumann algebra $\mathcal{N}(G)$ over the complex group ring $\mathbb{C}G$. The next statement says roughly that for an amenable group the group von Neumann algebra is flat over the group ring from the point of view of dimension theory.

Theorem 6.37. (Dimension-flatness of $\mathcal{N}(G)$ over $\mathbb{C}G$ for amenable G). *Let G be amenable and M be a $\mathbb{C}G$-module. Then*

$$\dim_{\mathcal{N}(G)}\left(\operatorname{Tor}_p^{\mathbb{C}G}(\mathcal{N}(G), M)\right) = 0 \qquad \text{for } p \geq 1,$$

where we consider $\mathcal{N}(G)$ as an $\mathcal{N}(G)$-$\mathbb{C}G$-bimodule.

Proof. In the first step we show for a finitely presented $\mathcal{N}(G)$-module M

$$\dim_{\mathcal{N}(G)}\left(\operatorname{Tor}_1^{\mathbb{C}G}(\mathcal{N}(G), M)\right) = 0. \tag{6.38}$$

Choose a finite presentation $\mathbb{C}G^m \xrightarrow{f} \mathbb{C}G^n \xrightarrow{p} M \to 0$. For an element $u = \sum_g \lambda_g \cdot g$ in $l^2(G)$ define its *support* by $\operatorname{supp}(u) := \{g \in G \mid \lambda_g \neq 0\} \subset G$. Let $B = (b_{i,j}) \in M(m,n,\mathbb{C}G)$ be the matrix describing f, i.e. f is given by right multiplication with B. Define the finite subset S of G by

$$S := \{g \in G \mid g \text{ or } g^{-1} \in \bigcup_{i,j} \operatorname{supp}(b_{i,j})\}.$$

Let $f^{(2)} \colon l^2(G)^m \to l^2(G)^n$ be the bounded G-equivariant operator induced by f. Denote by K the G-invariant linear subspace of $l^2(G)^m$ which is the image of the kernel of f under the canonical inclusion $k \colon \mathbb{C}G^m \to l^2(G)^m$. Next we show for the closure $\overline{K}$ of K

$$\overline{K} = \ker(f^{(2)}). \tag{6.39}$$

Obviously $\overline{K} \subset \ker(f^{(2)})$. Let $\operatorname{pr} \colon l^2(G)^m \to l^2(G)^m$ be the orthogonal projection onto the closed G-invariant subspace $\overline{K}^\perp \cap \ker(f^{(2)})$. The von Neumann dimension of $\operatorname{im}(\operatorname{pr})$ is zero if and only if pr itself is zero (see Theorem 1.9 (3)). Hence (6.39) will follow if we can prove

$$\operatorname{tr}_{\mathcal{N}(G)}(\operatorname{pr}) = 0. \tag{6.40}$$

Let $\epsilon > 0$ be given. We conclude from Lemma 6.35 that there is a finite non-empty subset $A \subset G$ satisfying

$$\frac{m \cdot (|S| + 1) \cdot |\partial_S A|}{|A|} \leq \epsilon, \tag{6.41}$$

where $\partial_S A$ is defined by $\{a \in A \mid \text{there is } s \in S \text{ with } as \notin A\}$. Define

$$\Delta := \{g \in G \mid g \in \partial_S A \text{ or } gt \in \partial_S A \text{ for some } t \in S\}$$

$$= \partial_S A \bigcup \left(\bigcup_{t \in S} \partial_S A \cdot t\right).$$

Let $\mathrm{pr}_A \colon l^2(G) \to l^2(G)$ be the projection which sends $\sum_{g \in G} \lambda_g \cdot g$ to $\sum_{g \in A} \lambda_g \cdot g$. Define $\mathrm{pr}_\Delta \colon l^2(G) \to l^2(G)$ analogously. Next we show for $s \in S$ and $u \in l^2(G)$

$$\mathrm{pr}_A \circ r_s(u) = r_s \circ \mathrm{pr}_A(u) \qquad \text{if } \mathrm{pr}_\Delta(u) = 0, \tag{6.42}$$

where $r_s \colon l^2(G) \to l^2(G)$ is right multiplication with s. Since $t \in S$ implies $t^{-1} \in S$, we get the following equality of subsets of G

$$\{g \in G \mid gs \in A, g \notin \Delta\} = \{g \in A \mid g \notin \Delta\}.$$

Now (6.42) follows from the following calculation for $u = \sum_{g \in G, g \notin \Delta} \lambda_g \cdot g \in l^2(G)$

$$\mathrm{pr}_A \circ r_s(u) = \mathrm{pr}_A \left(\sum_{g \in G, g \notin \Delta} \lambda_g \cdot gs \right) = \sum_{g \in G, gs \in A, g \notin \Delta} \lambda_g \cdot gs$$

$$= \sum_{g \in A, g \notin \Delta} \lambda_g \cdot gs = \left(\sum_{g \in A, g \notin \Delta} \lambda_g \cdot g \right) \cdot s = r_s \circ \mathrm{pr}_A(u).$$

We have defined S such that each entry in the matrix B describing f is a linear combination of elements in S. Hence (6.42) implies

$$\left(\bigoplus_{j=1}^n \mathrm{pr}_A \right) \circ f^{(2)}(u) = f^{(2)} \circ \left(\bigoplus_{i=1}^m \mathrm{pr}_A \right)(u)$$

$$\text{if } \mathrm{pr}_\Delta(u_i) = 0 \text{ for } i = 1, 2 \ldots, m.$$

Notice that the image of $\bigoplus_{i=1}^m \mathrm{pr}_A$ lies in $\mathbb{C}G^m$. We conclude

$$\bigoplus_{i=1}^m \mathrm{pr}_A(u) \in K \qquad \text{if } u \in \ker(f^{(2)}), \mathrm{pr}_\Delta(u_i) = 0 \text{ for } i = 1, 2 \ldots, m.$$

This shows

$$\left(\mathrm{pr} \circ \bigoplus_{i=1}^m \mathrm{pr}_A \right) \left(\ker(f^{(2)}) \cap \bigoplus_{i=1}^m \ker(\mathrm{pr}_\Delta) \right) = 0.$$

Since $\ker(\mathrm{pr}_\Delta)$ has complex codimension $|\Delta|$ in $l^2(G)$ and $|\Delta| \leq (|S| + 1) \cdot |\partial_S A|$, we conclude for the complex dimension $\dim_\mathbb{C}$ of complex vector spaces

$$\dim_\mathbb{C} \left(\left(\mathrm{pr} \circ \bigoplus_{i=1}^m \mathrm{pr}_A \right) \left(\ker(f^{(2)}) \right) \right) \leq m \cdot (|S| + 1) \cdot |\partial_S A|. \tag{6.43}$$

Since $\mathrm{pr} \circ \mathrm{pr}_A$ is an endomorphism of Hilbert spaces with finite dimensional image, it is trace-class and its trace $\mathrm{tr}_\mathbb{C}(\mathrm{pr} \circ \mathrm{pr}_A)$ is defined. We get

$$\mathrm{tr}_{\mathcal{N}(G)}(\mathrm{pr}) = \frac{\mathrm{tr}_{\mathbb{C}}\left(\mathrm{pr} \circ \bigoplus_{i=1}^{m} \mathrm{pr}_A\right)}{|A|} \tag{6.44}$$

from the following computation for $e \in G \subset l^2(G)$ the unit element

$$\mathrm{tr}_{\mathcal{N}(G)}(\mathrm{pr}) = \sum_{i=1}^{m} \langle \mathrm{pr}_{i,i}(e), e \rangle$$

$$= \frac{1}{|A|} \sum_{i=1}^{m} |A| \cdot \langle \mathrm{pr}_{i,i}(e), e \rangle$$

$$= \frac{1}{|A|} \sum_{i=1}^{m} \sum_{g \in A} \langle \mathrm{pr}_{i,i}(g), g \rangle$$

$$= \frac{1}{|A|} \sum_{i=1}^{m} \sum_{g \in A} \langle \mathrm{pr}_{i,i} \circ \mathrm{pr}_A(g), g \rangle$$

$$= \frac{1}{|A|} \sum_{i=1}^{m} \sum_{g \in G} \langle \mathrm{pr}_{i,i} \circ \mathrm{pr}_A(g), g \rangle$$

$$= \frac{1}{|A|} \sum_{i=1}^{m} \mathrm{tr}_{\mathbb{C}}(\mathrm{pr}_{i,i} \circ \mathrm{pr}_A)$$

$$= \frac{1}{|A|} \mathrm{tr}_{\mathbb{C}}\left(\mathrm{pr} \circ \left(\bigoplus_{i=1}^{m} \mathrm{pr}_A\right)\right).$$

If H is a Hilbert space and $f \colon H \to H$ is a bounded operator with finite dimensional image, then $\mathrm{tr}_{\mathbb{C}}(f) \leq \|f\| \cdot \dim_{\mathbb{C}}(f(\mathrm{im}(f)))$. Since the image of pr is contained in $\ker(f^{(2)})$ and pr and pr_A have operator norm 1, we conclude

$$\mathrm{tr}_{\mathbb{C}}\left(\mathrm{pr} \circ \bigoplus_{i=1}^{m} \mathrm{pr}_A\right) \leq \dim_{\mathbb{C}}\left(\left(\mathrm{pr} \circ \bigoplus_{i=1}^{m} \mathrm{pr}_A\right)\left(\ker(f^{(2)})\right)\right). \tag{6.45}$$

Equations (6.41), (6.43), (6.44) and (6.45) imply

$$\mathrm{tr}_{\mathcal{N}(G)}(\mathrm{pr}) \leq \epsilon.$$

Since this holds for all $\epsilon > 0$, we get (6.40) and hence (6.39) is true.

Let $\mathrm{pr}_{\overline{K}} \colon l^2(G)^m \to l^2(G)^m$ be the projection onto $\overline{K}$. Let $i \colon \ker(f) \to \mathbb{C}G^m$ be the inclusion. It induces a map

$$\mathrm{id}_{\mathcal{N}(G)} \otimes_{\mathbb{C}G} i \colon \mathcal{N}(G) \otimes_{\mathbb{C}G} \ker(f) \to \mathcal{N}(G)^m,$$

where here and in the sequel we will identify $\mathcal{N}(G) \otimes_{\mathbb{C}G} \mathbb{C}G^m = \mathcal{N}(G)^m$ by the obvious isomorphism. Next we want to show

$$\mathrm{im}(\nu^{-1}(\mathrm{pr}_{\overline{K}})) = \overline{\mathrm{im}(\mathrm{id}_{\mathcal{N}(G)} \otimes_{\mathbb{C}G} i)}. \tag{6.46}$$

Let $x \in \ker(f)$. Then

$$l \circ (\text{id} - \nu^{-1}(\text{pr}_{\overline{K}})) \circ (\text{id}_{\mathcal{N}(G)} \otimes_{\mathbb{C}G} i)(1 \otimes x) = (\text{id} - \text{pr}_{\overline{K}}) \circ k \circ i(x), \quad (6.47)$$

where $k \colon \mathbb{C}G^m \to l^2(G)^m$ and $l \colon \mathcal{N}(G)^m \to l^2(G)^m$ are the obvious inclusions. Since $(\text{id} - \text{pr}_{\overline{K}})$ is trivial on K we get $(\text{id} - \text{pr}_{\overline{K}}) \circ k \circ i = 0$. Now we conclude from (6.47) that $\text{im}(\text{id}_{\mathcal{N}(G)} \otimes_{\mathbb{C}G} i) \subset \ker(\text{id} - \nu^{-1}(\text{pr}_{\overline{K}}))$ and hence $\text{im}(\text{id}_{\mathcal{N}(G)} \otimes_{\mathbb{C}G} i) \subset \text{im}(\nu^{-1}(\text{pr}_{\overline{K}}))$ holds. This shows $\overline{\text{im}(\text{id}_{\mathcal{N}(G)} \otimes_{\mathbb{C}G} i)} \subset \text{im}(\nu^{-1}(\text{pr}_{\overline{K}}))$. It remains to prove for any $\mathcal{N}(G)$-map $g \colon \mathcal{N}(G)^m \to \mathcal{N}(G)$ with $\text{im}(\text{id}_{\mathcal{N}(G)} \otimes_{\mathbb{C}G} i) \subset \ker(g)$ that $g \circ \nu^{-1}(\text{pr}_{\overline{K}})$ is trivial. Obviously $K \subset \ker(\nu(g))$. Since $\ker(\nu(g))$ is a closed subspace, we get $\overline{K} \subset \ker(\nu(g))$. We conclude $\nu(g) \circ \text{pr}_{\overline{K}} = 0$ and hence $g \circ \nu^{-1}(\text{pr}_{\overline{K}}) = 0$. This finishes the proof of (6.46).

Since ν^{-1} preserves exactness by Theorem 6.24 (3) and $\text{id}_{\mathcal{N}(G)} \otimes_{\mathbb{C}G} f = \nu^{-1}(f^{(2)})$, we conclude from (6.39) and (6.46) that the sequence

$$\mathcal{N}(G) \otimes_{\mathbb{C}G} \ker(f) \xrightarrow{\text{id}_{\mathcal{N}(G)} \otimes_{\mathbb{C}G} i} \mathcal{N}(G)^m \xrightarrow{\text{id}_{\mathcal{N}(G)} \otimes_{\mathbb{C}G} f} \mathcal{N}(G)^m$$

is weakly exact. Continuity of the dimension function (see Theorem 6.7 (4d)) implies

$$\dim_{\mathcal{N}(G)} \left(\ker(\text{id}_{\mathcal{N}(G)} \otimes_{\mathbb{C}G} f) / \text{im}(\text{id}_{\mathcal{N}(G)} \otimes_{\mathbb{C}G} i) \right) = 0.$$

Since $\text{Tor}_1^{\mathbb{C}G}(\mathcal{N}(G), M) = \ker(\text{id}_{\mathcal{N}(G)} \otimes_{\mathbb{C}G} f) / \text{im}(\text{id}_{\mathcal{N}(G)} \otimes_{\mathbb{C}G} i)$ holds, we have proved (6.38) provided that M is finitely presented.

Next we prove (6.38) for arbitrary $\mathbb{C}G$-modules M. Obviously M is the union of its finitely generated submodules. Any finitely generated module N is a colimit over a directed system of finitely presented modules, namely, choose an epimorphism from a finitely generated free module F to N with kernel K. Since K is the union of its finitely generated submodules, N is the colimit of the directed system F/L where L runs over the finitely generated submodules of K. The functor Tor commutes in both variables with colimits over directed systems [94, Proposition VI.1.3. on page 107]. Now (6.38) follows from Additivity and Cofinality (see Theorem 6.7 (4c) and 4b)) or directly from Theorem 6.13 about dimension and colimits.

Next we show for any $\mathbb{C}G$-module, that $\dim_{\mathcal{N}(G)} \left(\text{Tor}_p^{\mathbb{C}G}(\mathcal{N}(G), M) \right) = 0$ holds for all $p \geq 1$ by induction over p. The induction step $p = 1$ has been proved above. Choose an exact sequence $0 \to N \to F \to M \to 0$ of $\mathcal{N}(G)$-modules such that F is free. Then we obtain form the associated long exact sequence of Tor-groups an isomorphism $\text{Tor}_p^{\mathbb{C}G}(\mathcal{N}(G), M) \cong \text{Tor}_{p-1}^{\mathbb{C}G}(\mathcal{N}(G), N)$ and the induction step follows. This finishes the proof of Theorem 6.37. $\square$

Conjecture 6.48. (Amenability and dimension-flatness of $\mathcal{N}(G)$ over $\mathbb{C}G$). *A group G is amenable if and only if for each $\mathbb{C}G$-module M*

$$\dim_{\mathcal{N}(G)} \left(\text{Tor}_p^{\mathbb{C}G}(\mathcal{N}(G), M) \right) = 0 \qquad \text{for } p \geq 1$$

holds.

Theorem 6.37 gives the "only if"-statement. Evidence for the "if"-statement comes from the following observation. Suppose that G contains a free group $\mathbb{Z} * \mathbb{Z}$ of rank 2 as subgroup. Notice that $S^1 \vee S^1$ is a model for $B(\mathbb{Z} * \mathbb{Z})$. Its cellular $\mathbb{C}[\mathbb{Z} * \mathbb{Z}]$-chain complex yields an exact sequence $0 \to \mathbb{C}[\mathbb{Z} * \mathbb{Z}]^2 \to \mathbb{C}[\mathbb{Z} * \mathbb{Z}] \to \mathbb{C} \to 0$, where $\mathbb{C}$ is equipped with the trivial $\mathbb{Z} * \mathbb{Z}$-action. We conclude from Additivity (see Theorem 6.7 (4b)), Theorem 6.29 (1) and Lemma 6.36

$$\dim_{\mathcal{N}(\mathbb{Z}*\mathbb{Z})}\left(\operatorname{Tor}_1^{\mathbb{C}[\mathbb{Z}*\mathbb{Z}]}(\mathcal{N}(\mathbb{Z}*\mathbb{Z}),\mathbb{C})\right) = 1;$$

$$\mathcal{N}(G) \otimes_{\mathcal{N}(\mathbb{Z}*\mathbb{Z})} \operatorname{Tor}_1^{\mathbb{C}[\mathbb{Z}*\mathbb{Z}]}(\mathcal{N}(\mathbb{Z}*\mathbb{Z}),\mathbb{C})) = \operatorname{Tor}_1^{\mathbb{C}G}(\mathcal{N}(G),\mathbb{C}G \otimes_{\mathbb{C}[\mathbb{Z}*\mathbb{Z}]} \mathbb{C}).$$

We conclude from Theorem 6.29 (2)

$$\dim_{\mathcal{N}(G)}\left(\operatorname{Tor}_1^{\mathbb{C}G}(\mathcal{N}(G),\mathbb{C}G \otimes_{\mathbb{C}[\mathbb{Z}*\mathbb{Z}]} \mathbb{C})\right) \neq 0.$$

One may ask for which groups the von Neumann algebra $\mathcal{N}(G)$ is flat as a $\mathbb{C}G$-module. This is true if G is virtually cyclic. There is some evidence for the conjecture

Conjecture 6.49 (Flatness of $\mathcal{N}(G)$ over $\mathbb{C}G$). *The group von Neumann algebra $\mathcal{N}(G)$ is flat over $\mathbb{C}G$ if and only if G is virtually cyclic, i.e. G is finite or contains $\mathbb{Z}$ as normal subgroup of finite index.*

6.5 L^2-Betti Numbers for General Spaces with Group Actions

In this section we extend the notion of L^2-Betti numbers for free G-CW-complexes of finite type introduced in Definition 1.30 to arbitrary G-spaces, where we will use the extended dimension function $\dim_{\mathcal{N}(G)}$ of Definition 6.20. We prove the main properties of these notions.

Definition 6.50 (L^2-Betti numbers). *Let X be a (left) G-space. Equip $\mathcal{N}(G)$ with the obvious $\mathcal{N}(G)$-$\mathbb{Z}G$-bimodule structure. The singular homology $H_p^G(X;\mathcal{N}(G))$ of X with coefficients in $\mathcal{N}(G)$ is the homology of the $\mathcal{N}(G)$-chain complex $\mathcal{N}(G) \otimes_{\mathbb{Z}G} C_*^{\mathrm{sing}}(X)$, where $C_*^{\mathrm{sing}}(X)$ is the singular chain complex of X with the induced $\mathbb{Z}G$-structure. Define the p-th L^2-Betti number of X by*

$$b_p^{(2)}(X;\mathcal{N}(G)) := \dim_{\mathcal{N}(G)}(H_p^G(X;\mathcal{N}(G))) \qquad \in [0,\infty],$$

where $\dim_{\mathcal{N}(G)}$ is the extended dimension function of Definition 6.20.

If G and its action on X are clear from the context, we often omit $\mathcal{N}(G)$ in the notation above.

Define for any (discrete) group G its p-th L^2-Betti number by

$$b_p^{(2)}(G) := b_p^{(2)}(EG,\mathcal{N}(G)).$$

We first show that this new definition and the old one agree when both apply. For that purpose we will need the following two lemmas. The proof of the first one can be found for instance in [333, Lemma 4.2 on page 152].

Lemma 6.51. *Let X be a G-CW-complex and $C_*(X)$ be its cellular $\mathbb{Z}G$-chain complex. Then there exists a $\mathbb{Z}G$-chain homotopy equivalence*

$$f_*(X)\colon C_*(X) \to C_*^{\mathrm{sing}}(X),$$

which is, up to $\mathbb{Z}G$-homotopy, uniquely defined and natural in X. In particular we get a natural isomorphism of $\mathcal{N}(G)$-modules

$$H_p(\mathcal{N}(G) \otimes_{\mathbb{Z}G} C_*(X)) \xrightarrow{\cong} H_p^G(X;\mathcal{N}(G)) := H_p(\mathcal{N}(G) \otimes_{\mathbb{Z}G} C_*^{\mathrm{sing}}(X)).$$

Lemma 6.52. *Let C_* be a finitely generated Hilbert $\mathcal{N}(G)$-chain complex. Then there is an isomorphism, natural in C_*,*

$$h_p(C_*)\colon \nu^{-1}(H_p^{(2)}(C_*)) \xrightarrow{\cong} \mathbf{P}H_p(\nu^{-1}(C_*)),$$

where ν^{-1} is the functor appearing in Theorem 6.24.

Proof. We define $h_p(C_*)$ by the following commutative diagram whose columns are exact and whose middle and lower vertical arrows are isomorphisms by Theorem 6.7 (2) and Theorem 6.24 (3)

$$
\begin{array}{ccc}
0 & & 0 \\
\uparrow & & \uparrow \\
\nu^{-1}(H_p^{(2)}(C_*)) & \xrightarrow{h_p(C_*)} & \mathbf{P}H_p(\nu^{-1}(C_*)) \\
{\scriptstyle \nu^{-1}(q)}\uparrow & & {\scriptstyle r}\uparrow \\
\nu^{-1}(\ker(c_p)) & \xrightarrow{\nu^{-1}(i)} & \ker(\nu^{-1}(c_p)) \\
{\scriptstyle \nu^{-1}(j)}\uparrow & & {\scriptstyle l}\uparrow \\
\nu^{-1}(\overline{\mathrm{im}(c_{p+1})}) & \xrightarrow{\nu^{-1}(k)} & \overline{\mathrm{im}(\nu^{-1}(c_{p+1}))} \\
\uparrow & & \uparrow \\
0 & & 0
\end{array}
$$

where $i\colon \ker(c_p) \to C_p$, $j\colon \overline{\mathrm{im}(c_{p+1})} \to \ker(c_p)$, $k\colon \overline{\mathrm{im}(c_{p+1})} \to C_p$ and l are the obvious inclusions and q and r the obvious projections. Then $h_p(C_*)$ is an isomorphism by the five-lemma. □

Lemma 6.53. *Let X be a G-CW-complex of finite type. Then Definition 1.30 and Definition 6.50 of L^2-Betti numbers $b_p^{(2)}(X;\mathcal{N}(G))$ agree.*

Proof. We can identify $\mathcal{N}(G) \otimes_{\mathbb{Z}G} C_*(X)$ with $\nu^{-1}(C_*^{(2)}(X))$. Now apply Theorem 6.7 (4e), Theorem 6.24 (4), Lemma 6.51 and Lemma 6.52. $\square$

Next we collect some basic properties of L^2-Betti numbers.

Theorem 6.54. L^2-Betti numbers for arbitrary spaces).

(1) Homology invariance

We have for a G-map $f\colon X \to Y$

(a) Suppose for $n \geq 1$ that for each subgroup $H \subset G$ the induced map $f^H\colon X^H \longrightarrow Y^H$ is $\mathbb{C}$-homologically n-connected, i.e. the map $H_p^{\mathrm{sing}}(f^H;\mathbb{C})\colon H_p^{\mathrm{sing}}(X^H;\mathbb{C}) \to H_p^{\mathrm{sing}}(Y^H;\mathbb{C})$ induced by f^H on singular homology with complex coefficients is bijective for $p < n$ and surjective for $p = n$. Then the induced map $f_*\colon H_p^G(X;\mathcal{N}(G)) \longrightarrow H_p^G(Y;\mathcal{N}(G))$ is bijective for $p < n$ and surjective for $p = n$ and we get

$$b_p^{(2)}(X) = b_p^{(2)}(Y) \qquad \text{for } p < n;$$
$$b_p^{(2)}(X) \geq b_p^{(2)}(Y) \qquad \text{for } p = n;$$

(b) Suppose that for each subgroup $H \subset G$ the induced map $f^H\colon X^H \to Y^H$ is a $\mathbb{C}$-homology equivalence, i.e. $H_p^{\mathrm{sing}}(f^H;\mathbb{C})$ is bijective for $p \geq 0$. Then for all $p \geq 0$ the map $f_*\colon H_p^G(X;\mathcal{N}(G)) \to H_p^G(Y;\mathcal{N}(G))$ induced by f is bijective and we get

$$b_p^{(2)}(X) = b_p^{(2)}(Y) \qquad \text{for } p \geq 0;$$

(2) Comparison with the Borel construction

Let X be a G-CW-complex. Suppose that for all $x \in X$ the isotropy group G_x is finite or satisfies $b_p^{(2)}(G_x) = 0$ for all $p \geq 0$. Then

$$b_p^{(2)}(X;\mathcal{N}(G)) = b_p^{(2)}(EG \times X;\mathcal{N}(G)) \qquad \text{for } p \geq 0;$$

(3) Invariance under non-equivariant $\mathbb{C}$-homology equivalences

Suppose that $f\colon X \to Y$ is a G-equivariant map of G-CW-complexes such that the induced map $H_p^{\mathrm{sing}}(f;\mathbb{C})$ on singular homology with complex coefficients is bijective. Suppose that for all $x \in X$ the isotropy group G_x is finite or satisfies $b_p^{(2)}(G_x) = 0$ for all $p \geq 0$, and analogously for all $y \in Y$. Then we have for all $p \geq 0$

$$b_p^{(2)}(X;\mathcal{N}(G)) = b_p^{(2)}(Y;\mathcal{N}(G));$$

(4) Independence of equivariant cells with infinite isotropy

Let X be a G-CW-complex. Let $X[\infty]$ be the G-CW-subcomplex consisting of those points whose isotropy subgroups are infinite. Then we get for all $p \geq 0$

$$b_p^{(2)}(X;\mathcal{N}(G)) = b_p^{(2)}(X,X[\infty];\mathcal{N}(G));$$

(5) *Künneth formula*

Let X be a G-space and Y be an H-space. Then $X \times Y$ is a $G \times H$-space and we get for all $n \geq 0$

$$b_n^{(2)}(X \times Y) = \sum_{p+q=n} b_p^{(2)}(X) \cdot b_q^{(2)}(Y),$$

where we use the convention that $0 \cdot \infty = 0$, $r \cdot \infty = \infty$ for $r \in (0, \infty]$ and $r + \infty = \infty$ for $r \in [0, \infty]$;

(6) *Restriction*

Let $H \subset G$ be a subgroup of finite index $[G : H]$. Then

(a) Let M be an $\mathcal{N}(G)$-module and $\mathrm{res}_G^H M$ be the $\mathcal{N}(H)$-module obtained from M by restriction. Then

$$\dim_{\mathcal{N}(H)}(\mathrm{res}_G^H M) = [G : H] \cdot \dim_{\mathcal{N}(G)}(M),$$

where $[G : H] \cdot \infty$ is understood to be ∞;

(b) Let X be a G-space and let $\mathrm{res}_G^H X$ be the H-space obtained from X by restriction. Then

$$b_p^{(2)}(\mathrm{res}_G^H X; \mathcal{N}(H)) = [G : H] \cdot b_p^{(2)}(X; \mathcal{N}(G));$$

(7) *Induction*

Let $i \colon H \to G$ be an inclusion of groups and let X be an H-space. Let $i \colon \mathcal{N}(H) \to \mathcal{N}(G)$ be the induced ring homomorphism (see Definition 1.23). Then

$$H_p^G(G \times_H X; \mathcal{N}(G)) = i_* H_p^H(X; \mathcal{N}(H));$$
$$b_p^{(2)}(G \times_H X; \mathcal{N}(G)) = b_p^{(2)}(X; \mathcal{N}(H));$$

(8) *Zero-th homology and L^2-Betti number*

Let X be a path-connected G-space. Then

(a) There is an $\mathcal{N}(G)$-isomorphism $H_0^G(X; \mathcal{N}(G)) \xrightarrow{\cong} \mathcal{N}(G) \otimes_{\mathbb{C}G} \mathbb{C}$;

(b) $b_0^{(2)}(X; \mathcal{N}(G)) = |G|^{-1}$, where $|G|^{-1}$ is defined to be zero if the order $|G|$ of G is infinite;

(c) $H_0^G(X; \mathcal{N}(G))$ is trivial if and only if G is non-amenable;

Proof. (1) The proof of assertion (1) which generalizes Theorem 1.35 (1) can be found in [333, Lemma 4.8 on page 153].

(2) Because of Additivity (see Theorem 6.7 (4b)) it suffices to prove that the dimension of the kernel and the cokernel of the map induced by the projection

$$\mathrm{pr}_* \colon H_p^G(EG \times X; \mathcal{N}(G)) \to H_p^G(X; \mathcal{N}(G))$$

are trivial. Notice that X is the directed colimit of its finite G-CW-sub-complexes. Since $H_p^G(-, \mathcal{N}(G))$ is compatible with directed colimits and directed colimits preserve exact sequences, we can assume by Theorem 6.13

about dimension and colimits that X itself is finite. By induction over the number of equivariant cells, the long exact homology sequence and Additivity (see Theorem 6.7 (4b)) the claim reduces to the case where X is of the shape G/H. Because of assertion (7) it suffices to prove for the map $\mathrm{pr}_*\colon H_p^H(EH; \mathcal{N}(H)) \to H_p^H(\{*\}; \mathcal{N}(H))$ that its kernel and cokernel have trivial dimension, provided that H is finite or $b_p^{(2)}(H; \mathcal{N}(H)) = 0$ for $p \geq 0$. This follows from assertion (8a).

(3) Since $EG \times X$ and $EG \times Y$ are free G-CW-complexes and $\mathrm{id} \times f\colon EG \times X \to EG \times Y$ induces an isomorphism on homology with $\mathbb{C}$-coefficients, the claim follows from assertions (1b) and (2).

(4) We get an exact sequence of cellular $\mathbb{Z}G$-chain complexes $0 \to C_*(X[\infty]) \to C_*(X) \to C_*(X, X[\infty]) \to 0$ which is a split exact sequence of $\mathbb{Z}G$-modules in each dimension. Hence it stays exact after applying $\mathcal{N}(G) \otimes_{\mathbb{Z}G} -$, and we get a long exact sequence of $\mathcal{N}(G)$-homology modules

$$\ldots \to H_p^G(X[\infty]; \mathcal{N}(G)) \to H_p^G(X; \mathcal{N}(G)) \to H_p^G(X, X[\infty]; \mathcal{N}(G))$$

$$\to H_{p-1}^G(X[\infty]; \mathcal{N}(G)) \to \ldots.$$

Because of Additivity (see Theorem 6.7 (4b)) it suffices to prove that $\dim_{\mathcal{N}(G)}(\mathcal{N}(G) \otimes C_p(X[\infty])) = 0$ for all $p \geq 0$. This follows from Additivity (see Theorem 6.7 (4b)), Cofinality (see Theorem 6.7 (4c)) and Lemma 6.33 (1), since $C_p(X[\infty])$ is a direct sum of $\mathbb{Z}G$-modules of the shape $\mathbb{Z}[G/H]$ for infinite subgroups $H \subset G$.

(5) This assertion is a generalization of Theorem 1.35 (4).

For any G-space Z there is a G-CW-complex Z' together with a G-map $f\colon Z \to Z'$ such that f^H is a weak homotopy equivalence and hence induces an isomorphism on singular homology with complex coefficients for all $H \subset G$ [326, Proposition 2.3 iii) on page 35]. Hence we can assume without loss of generality that X is a G-CW-complex and Y is a H-CW-complex by assertion (1b).

The G-CW-complex X is the colimit of the directed system $\{X_i \mid i \in I\}$ of its finite G-CW-subcomplexes. Analogously Y is the colimit of the directed system $\{Y_j \mid j \in J\}$ of its finite H-CW-subcomplexes. Since we are working in the category of compactly generated spaces, the $G \times H$-CW-complex $X \times Y$ is the colimit of the directed system $\{X_i \times Y_j \mid (i,j) \in I \times J\}$. Since homology and tensor products are compatible with colimits over directed systems, we conclude from Theorem 6.13 (2) about dimension and colimits

$b_p^{(2)}(X; \mathcal{N}(G))$

$$= \sup\left\{\inf\left\{\dim_{\mathcal{N}(G)}\left(\operatorname{im}(\phi_{i_0,i_1} : H_p^G(X_{i_0}; \mathcal{N}(G)) \to H_p^G(X_{i_1}; \mathcal{N}(G)))\right) \\ \mid i_1 \in I, i_0 \le i_1\right\} \mid i_0 \in I\right\};$$

(6.55)

$b_p^{(2)}(Y; \mathcal{N}(G))$

$$= \sup\left\{\inf\left\{\dim_{\mathcal{N}(H)}\left(\operatorname{im}(\psi_{j_0,j_1} : H_p^H(Y_{j_0}; \mathcal{N}(H)) \to H_p^H(Y_{j_1}; \mathcal{N}(H)))\right) \\ \mid j_1 \in J, j_0 \le j_1\right\} \mid j_0 \in J\right\};$$

(6.56)

$b_p^{(2)}(X \times Y; \mathcal{N}(G \times H))$

$$= \sup\left\{\inf\left\{\dim_{\mathcal{N}(G\times H)}\left(\operatorname{im}(\mu_{(i_0,j_0),(i_1,j_1)} : \right.\right.\right.$$
$$H_p^{G\times H}(X_{i_0} \times Y_{j_0}; \mathcal{N}(G \times H)) \to H_p^{G\times H}(X_{i_1} \times Y_{j_1}; \mathcal{N}(G \times H))))$$
$$\left.\left.\mid (i_1, j_1) \in I \times J, i_0 \le i_1, j_0 \le j_1\right\} \mid (i_0, j_0) \in I \times J\right\},$$

(6.57)

where here and in the sequel ϕ_{i_0,i_1}, ψ_{j_0,j_1} and $\mu_{(i_0,j_0),(i_1,j_1)}$ are the maps induced by the obvious inclusions.

Next we show for $n \ge 0$ and $i_0, i_1 \in I$ with $i_0 \le i_1$ and $j_0, j_1 \in J$ with $j_0 \le j_1$

$$\dim_{\mathcal{N}(G\times H)}\left(\operatorname{im}\left(\mu_{(i_0,j_0),(i_1,j_1)} : H_n^{G\times H}(X_{i_0} \times Y_{j_0}; \mathcal{N}(G \times H))\right.\right.$$
$$\to H_n(X_{i_1} \times X_{j_1}; \mathcal{N}(G \times H))))$$
$$= \sum_{p+q=n} \dim_{\mathcal{N}(G)}\left(\operatorname{im}(\phi_{i_0,i_1} : H_p^G(X_{i_0}; \mathcal{N}(G)) \to H_p^G(X_{i_1}; \mathcal{N}(G)))\right)$$
$$\cdot \dim_{\mathcal{N}(H)}\left(\operatorname{im}(\psi_{j_0,j_1} : H_q^H(Y_{j_0}; \mathcal{N}(G)) \to H_q^H(Y_{j_1}; \mathcal{N}(G)))\right). \quad (6.58)$$

For $k = 0, 1$ let $C_*[k]$ be the cellular $\mathbb{Z}G$-chain complex of the G-CW-pair $(X_{i_k}, X_{i_k}[\infty])$, where $X_{i_k}[\infty]$ has been introduced in Theorem 6.54 (4). For $k = 0, 1$ let $D_*[k]$ be the cellular $\mathbb{Z}H$-chain complex of the H-CW-pair $(Y_{j_k}, Y_{j_k}[\infty])$. Then $C_*[k] \otimes_{\mathbb{Z}} D_*[k]$ is the cellular $\mathbb{Z}[G \times H]$-chain complex of the $G \times H$-CW-pair $(X \times Y, (X \times Y)[\infty])$. Notice that $C_p[k]$ is a direct sum of $\mathbb{Z}G$-modules of the shape $\mathbb{Z}[G/H]$ for finite subgroups $H \subset G$. Hence the G-CW-complex structure on $(X_{i_k}, X_{i_k}[\infty])$ induces the structure of a Hilbert $\mathcal{N}(G)$-chain complex on $C_*^{(2)}[k] := l^2(G) \otimes_{\mathbb{Z}G} C_*[k]$. The analogous statements are true for $D_*^{(2)}[k] := l^2(H) \otimes_{\mathbb{Z}H} D_*[k]$ and $l^2(G \times H) \otimes_{\mathbb{Z}[G\times H]}(C_*[k] \otimes_{\mathbb{Z}} D_*[k])$. Notice that the last Hilbert $\mathcal{N}(G \times H)$-chain complex is isomorphic to the Hilbert tensor product $C_*^{(2)}[k] \otimes D_*^{(2)}[k]$. We conclude from Theorem 6.7 (2) and (4d), Theorem 6.24 and Theorem 6.54 (4)

$$\dim_{\mathcal{N}(G)} \left(\mathrm{im}(\phi_{i_0,i_1} : H_p^G(X_{i_0}; \mathcal{N}(G)) \to H_p^G(X_{i_1}; \mathcal{N}(G))) \right)$$

$$= \dim_{\mathcal{N}(G)} \left(\mathrm{clos}(\mathrm{im}(\phi_{i_0,i_1} : H_p^{(2)}(C_*^{(2)}[0]) \to H_p^{(2)}(C_*^{(2)}[1]))) \right); \quad (6.59)$$

$$\dim_{\mathcal{N}(H)} \left(\mathrm{im}(\psi_{j_0,j_1} : H_p^H(Y_{j_0}; \mathcal{N}(H)) \to H_p^H(Y_{j_1}; \mathcal{N}(H))) \right)$$

$$= \dim_{\mathcal{N}(H)} \left(\mathrm{clos}(\mathrm{im}(\psi_{j_0,j_1} : H_p^{(2)}(D_*^{(2)}[0]) \to H_p^{(2)}(D_*^{(2)}[1]))) \right); \quad (6.60)$$

$$\dim_{\mathcal{N}(G \times H)} \left(\mathrm{im}(\mu_{(i_0,j_0),(i_1,j_1)} : H_p^{G \times H}(X_{i_0} \times Y_{j_0}; \mathcal{N}(G \times H)) \right.$$

$$\left. \to H_p^{G \times H}(X_{i_1} \times Y_{j_1}; \mathcal{N}(G \times H))) \right)$$

$$= \dim_{\mathcal{N}(G \times H)} \left(\mathrm{clos}(\mathrm{im}(\mu_{(i_0,j_0),(i_1,j_1)} : H_p^{(2)}(C_*^{(2)}[0] \otimes D_*^{(2)}[0]) \right.$$

$$\left. \to H_p^{(2)}(C_*^{(2)}[1] \otimes D_*^{(2)}[1]))) \right). \quad (6.61)$$

There is a commutative diagram of Hilbert $\mathcal{N}(G \times H)$-modules

$$
\begin{array}{ccc}
\bigoplus_{p+q=n} H_p^{(2)}(C_*^{(2)}[0]) \otimes H_q((D_*^{(2)}[0]) & \xrightarrow{\cong} & H_n^{(2)}(C_*^{(2)}[0] \otimes D_*^{(2)}[0]) \\
\bigoplus_{p+q=n} \phi_{i_0,i_1} \otimes \psi_{j_0,j_1} \downarrow & & \mu_{(i_0,j_0),(i_1,j_1)} \downarrow \\
\bigoplus_{p+q=n} H_p^{(2)}(C_*^{(2)}[1]) \otimes H_q^{(2)}((D_*^{(2)}[1]) & \xrightarrow{\cong} & H_n^{(2)}(C_*^{(2)}[1] \otimes D_*^{(2)}[1])
\end{array}
$$

where the horizontal isomorphisms are the ones appearing in Lemma 1.22. Thus we obtain an isomorphism of closure of the images of the two vertical maps. The closure of the image of a Hilbert tensor product of two morphisms is the Hilbert tensor product of the closures of the individual images. Now (6.58) follows from Theorem 1.12 (5), (6.59), (6.60) and (6.61). Finally assertion (5) follows from (6.55), (6.56), (6.57) and (6.58).

(6) generalizes Theorem 1.35 (9).

We begin with the proof of assertion (6a) Notice for the sequel that $\mathrm{res}_G^H \mathcal{N}(G)$ is $\mathcal{N}(H)$-isomorphic to $\mathcal{N}(H)^{[G:H]}$. Since M is the colimit of the directed system of its finitely generated $\mathcal{N}(G)$-submodules, it suffices to consider the case where M is finitely generated because of Cofinality (see Theorem 6.7 (4c)). Since for a finitely generated $\mathcal{N}(G)$-module we have $\mathbf{T}(\mathrm{res}_G^H M) = \mathrm{res}_G^H(\mathbf{T}M)$ and hence $\mathbf{P}(\mathrm{res}_G^H M) = \mathrm{res}_G^H(\mathbf{P}M)$, we conclude from Theorem 6.7 (3) and (4e) that it suffices to treat the case of a finitely generated projective $\mathcal{N}(G)$-module M. The functor ν appearing in Theorem 6.24 is compatible with restriction for $\mathcal{N}(G)$-modules and restriction from G to H for Hilbert modules. Using Theorem 6.24 (4) we reduce the claim to the assertion that for any finitely generated Hilbert $\mathcal{N}(G)$-module V we have

$$\dim_{\mathcal{N}(H)}(\mathrm{res}_G^H V) = [G : H] \cdot \dim_{\mathcal{N}(G)}(V).$$

This has already been proved in Theorem 1.12 (6).

Next we prove assertion (6b). We obtain an isomorphism of $\mathcal{N}(H)$-$\mathbb{C}G$-bimodules $\mathcal{N}(H) \otimes_{\mathbb{C}H} \mathbb{C}G \xrightarrow{\cong} \mathcal{N}(G)$ by sending $u \otimes g$ to $i_*(u) \circ R_{g^{-1}}$. It induces an $\mathcal{N}(H)$-isomorphism

$$\mathcal{N}(H) \otimes_{\mathbb{C}H} C_*(\mathrm{res}_G^H X) \xrightarrow{\cong} \mathrm{res}_G^H(\mathcal{N}(G) \otimes_{\mathbb{C}G} C_*(X)).$$

Now assertion (6b) follows from assertion (6a).

(7) is a generalization of Theorem 1.35 (10) and follows from Theorem 6.29.

(8) This follows from Lemma 6.33 and Lemma 6.36 and generalizes Theorem 1.35 (8). This finishes the proof of Theorem 6.54.   □

We will investigate the class $\mathcal{B}_\infty$ of groups G, for which $b_p^{(2)}(G) = 0$ holds for all $p \geq 0$, in Section 7.1.

Remark 6.62. Let M be a closed Riemannian manifold. We have already mentioned in Section 2.7 that the analytic Laplace operator $\Delta_p \colon L^2\Omega^p(\widetilde{M}) \to L^2\Omega^p(\widetilde{M})$ on the universal covering does not have zero in its spectrum if and only if we get for its analytic p-th L^2-Betti number $b_p^{(2)}(\widetilde{M}) = 0$ and for its analytic p-th Novikov-Shubin invariant $\alpha_p^\Delta(\widetilde{M}) = \infty^+$. In the case $p = 0$ this is equivalent to the condition that $H_0^G(X; \mathcal{N}(G)) = 0$ by Theorem 1.59, Lemma 2.11 (8) Theorem 2.68 and Theorem 6.24 (3). The latter condition is by Theorem 6.54 (8c) equivalent to the condition that $\pi_1(M)$ is amenable. Thus we rediscover (and in some sense generalize to arbitrary G-spaces) the result of Brooks [68] that for a closed Riemannian manifold M the Laplacian $\Delta_0 \colon L^2\Omega^0(\widetilde{M}) \to L^2\Omega^0(\widetilde{M})$ acting on functions on the universal covering has zero in its spectrum if and only if $\pi_1(M)$ is amenable. Notice that both Brook's and our proof are based on Kesten [290].

The next result is a generalization of Theorem 1.39 and was conjectured for a closed aspherical manifold fibering over the circle in [237, page 229].

Theorem 6.63 (Vanishing of L^2-Betti numbers of mapping tori).
Let $f \colon X \to X$ be a cellular selfmap of a connected CW-complex X and let $\pi_1(T_f) \xrightarrow{\phi} G \xrightarrow{\psi} \mathbb{Z}$ be a factorization of the canonical epimorphism into epimorphisms ϕ and ψ. Suppose for given $p \geq 0$ that $b_p^{(2)}(G \times_{\phi \circ i} \widetilde{X}; \mathcal{N}(G)) < \infty$ and $b_{p-1}^{(2)}(G \times_{\phi \circ i} \widetilde{X}; \mathcal{N}(G)) < \infty$ holds, where $i \colon \pi_1(X) \to \pi_1(T_f)$ is the map induced by the obvious inclusion of X into T_f. Let $\overline{T_f}$ be the covering of T_f associated to ϕ, which is a free G-CW-complex. Then we get

$$b_p^{(2)}(\overline{T_f}; \mathcal{N}(G)) = 0.$$

Proof. The proof is analogous to that of Theorem 1.39 except the following changes. Instead of Theorem 1.35 (1) and Theorem 1.35 (9) one has to refer to Theorem 6.54 (1b) and Theorem 6.54 (6b) and one has to give a different argument for the existence of a constant C which is independent of n and satisfies

$$b_p^{(2)}(\overline{T_{f^n}}; \mathcal{N}(G_n)) \leq C. \tag{6.64}$$

In the proof of Theorem 1.39 we have taken $C = \beta_p + \beta_{p-1}$, where β_p is the number of p-cells in X. Next we explain why we can take for (6.64) the constant

$$C = b_p^{(2)}(G \times_{\phi \circ i} \widetilde{X}; \mathcal{N}(G)) + b_{p-1}^{(2)}(G \times_{\phi \circ i} \widetilde{X}; \mathcal{N}(G)).$$

First one checks that the image of $\phi \circ i \colon \pi_1(X) \to G$ is contained in $\psi^{-1}(0)$ and hence in G_n. One constructs an exact sequence of $\mathbb{Z}G_n$-chain complexes

$$0 \to C_*(G_n \times_{\phi \circ i} \widetilde{X}) \to C_*(\overline{T_{f^n}}) \to \Sigma C_*(G_n \times_{\phi \circ i} \widetilde{X}) \to 0.$$

Tensoring with $\mathcal{N}(G_n)$ and taking the long homology sequence yields the exact sequence of $\mathcal{N}(G_n)$-modules

$$H_p(\mathcal{N}(G_n) \otimes_{\mathbb{Z}G_n} C_*(G_n \times_{\phi \circ i} \widetilde{X})) \to H_p(\mathcal{N}(G_n) \otimes_{\mathbb{Z}G_n} C_*(\overline{T_{f^n}}))$$
$$\to H_{p-1}(\mathcal{N}(G_n) \otimes_{\mathbb{Z}G_n} C_*(G_n \times_{\phi \circ i} \widetilde{X})).$$

Additivity (see Theorem 6.7 (4b)) and Theorem 6.54 (7) imply

$$b_p^{(2)}(\overline{T_{f^n}}; \mathcal{N}(G_n)) \leq b_p^{(2)}(G_n \times_{\phi \circ i} \widetilde{X}; \mathcal{N}(G_n))$$
$$+ b_{p-1}^{(2)}(G_n \times_{\phi \circ i} \widetilde{X}; \mathcal{N}(G_n));$$
$$b_p^{(2)}(G_n \times_{\phi \circ i} \widetilde{X}; \mathcal{N}(G_n)) = b_p^{(2)}(G \times_{\phi \circ i} \widetilde{X}; \mathcal{N}(G));$$
$$b_{p-1}^{(2)}(G_n \times_{\phi \circ i} \widetilde{X}; \mathcal{N}(G_n)) = b_{p-1}^{(2)}(G \times_{\phi \circ i} \widetilde{X}; \mathcal{N}(G)).$$

This finishes the proof of (6.64) and thus of Theorem 6.63.   □

The next result is a generalization of Theorem 1.40. Its proof is analogous to that one of Theorem 1.40.

Theorem 6.65. (L^2-Betti numbers and S^1-actions). *Let X be a connected S^1-CW-complex. Suppose that for one orbit S^1/H (and hence for all orbits) the inclusion into X induces a map on π_1 with infinite image. (In particular the S^1-action has no fixed points.) Let $\widetilde{X}$ be the universal covering of X with the canonical $\pi_1(X)$-action. Then we get for all $p \geq 0$*

$$b_p^{(2)}(\widetilde{X}) = 0.$$

The next Lemma 6.66 is a generalization of Lemma 1.41. Although it is more general — we have dropped the finiteness conditions on the CW-complexes F, E and B — its proof is now simpler because our algebraic approach to dimension theory and L^2-Betti numbers is so flexible that we can use standard methods from algebra and topology such as the Serre spectral sequence for fibrations and local coefficients.

Lemma 6.66. *Let $F \xrightarrow{i} E \to B$ be a fibration of connected CW-complexes. Let $\phi \colon \pi_1(E) \to G$ be a group homomorphisms and let $i_* \colon \pi_1(F) \to \pi_1(E)$ be the homomorphism induced by the inclusion i. Suppose that for a given integer $d \geq 0$ the L^2-Betti number $b_p^{(2)}(G \times_{\phi \circ i_*} \widetilde{F}; \mathcal{N}(G))$ vanishes for all $p \leq d$. Then the L^2-Betti number $b_p^{(2)}(G \times_\phi \widetilde{E}; \mathcal{N}(G))$ vanishes for all $p \leq d$.*

Proof. There is a spectral homology sequence of $\mathcal{N}(G)$-modules which converges to $H^G_{p+q}(G \times_\phi \widetilde{E}; \mathcal{N}(G))$ and whose E^1-term is

$$E^1_{p,q} = H^G_q(G \times_{\phi o i_*} \widetilde{F}; \mathcal{N}(G)) \otimes_{\mathbb{Z}\pi_1(B)} C_p(\widetilde{B}),$$

where the right $\pi_1(B)$-action on $H^G_q(G \times_{\phi o i_*} \widetilde{F}; \mathcal{N}(G))$ is induced by the fiber transport. Since

$$b^{(2)}_q(G \times_{\phi o i_*} \widetilde{F}; \mathcal{N}(G)) = \dim_{\mathcal{N}(G)} \left(H^G_q(G \times_{\phi o i_*} \widetilde{F}; \mathcal{N}(G)) \right) = 0$$

holds for $q \leq d$ by assumption and $C_p(\widetilde{B})$ is a free $\mathbb{Z}\pi_1(B)$-module, we conclude from Additivity (see Theorem 6.7 (4b)) and Cofinality (see Theorem 6.7 (4c)) that $\dim_{\mathcal{N}(G)}(E^1_{p,q}) = 0$ holds for all $p \geq 0$ and $q \leq d$. Hence $b^{(2)}_p(G \times_\phi \widetilde{E}; \mathcal{N}(G)) = 0$ holds for all $p \leq d$. □

Theorem 6.67 (L^2-Betti numbers and fibrations). *Let $F \xrightarrow{i} E \xrightarrow{p} B$ be a fibration of connected CW-complexes. Let $p_* \colon \pi_1(E) \xrightarrow{\phi} G \xrightarrow{\psi} \pi_1(B)$ be a factorization of the map induced by p into epimorphisms ϕ and ψ. Let $i_* \colon \pi_1(F) \to \pi_1(E)$ be the homomorphism induced by the inclusion i. Suppose for a given integer $d \geq 1$ that $b^{(2)}_p(G \times_{\phi o i_*} \widetilde{F}; \mathcal{N}(G)) = 0$ for $p \leq d-1$ and $b^{(2)}_d(G \times_{\phi o i_*} \widetilde{F}; \mathcal{N}(G)) < \infty$ holds. Suppose that $\pi_1(B)$ contains an element of infinite order or finite subgroups of arbitrarily large order. Then $b^{(2)}_p(G \times_\phi \widetilde{E}; \mathcal{N}(G)) = 0$ for $p \leq d$.*

Proof. Since $b^{(2)}_q(G \times_{\phi o i_*} \widetilde{F}; \mathcal{N}(G)) = 0$ holds for $q \leq d-1$ by assumption and $C_p(\widetilde{B})$ is a free $\mathbb{Z}\pi_1(B)$-module, we conclude by the same spectral sequence argument as in the proof of Lemma 6.66

$$b^{(2)}_p(G \times_\phi \widetilde{E}; \mathcal{N}(G)) = 0 \qquad 0 \leq p \leq d-1$$

and

$$\begin{aligned}
b^{(2)}_d&(G \times_\phi \widetilde{E}; \mathcal{N}(G)) \\
&= \dim_{\mathcal{N}(G)} \left(E^2_{0,d} \right) \\
&= \dim_{\mathcal{N}(G)} \left(H^{\pi_1(B)}_0(\widetilde{B}; H^G_d(G \times_{\phi o i_*} \widetilde{F}; \mathcal{N}(G))) \right) \\
&= \dim_{\mathcal{N}(G)} \left(H^G_d(G \times_{\phi o i_*} \widetilde{F}; \mathcal{N}(G)) \otimes_{\mathbb{Z}\pi_1(B)} \mathbb{Z} \right), \qquad (6.68)
\end{aligned}$$

where the right $\pi_1(B)$-action on $H^G_d(G \times_{\phi o i_*} \widetilde{F}; \mathcal{N}(G))$ is induced by the fiber transport. It remains to prove

$$b^{(2)}_d(G \times_\phi \widetilde{E}; \mathcal{N}(G)) = 0. \qquad (6.69)$$

Let $\Delta \subset \pi_1(B)$ be a subgroup which is either $\mathbb{Z}$ or finite. Let X be S^1 if $\Delta = \mathbb{Z}$ and a connected finite 2-dimensional CW-complex X with $\pi_1(X) = \Delta$ if Δ is finite. In any case we can choose a map $f \colon X \to B$ which induces on fundamental groups the inclusion of Δ into $\pi_1(B)$. Let $F \xrightarrow{i_0} E_0 \xrightarrow{p_0} X$ be the fibration obtained from $F \xrightarrow{i} E \xrightarrow{p} B$ by applying the pullback construction with respect to f. Let $\overline{f} \colon E_0 \to E$ be the canonical map. Since there is an obvious $\mathcal{N}(G)$-epimorphism from $H_d^G(G \times_{\phi o i_*} \widetilde{F}; \mathcal{N}(G)) \otimes_{\mathbb{Z}\Delta} \mathbb{Z}$ to $H_d^G(G \times_{\phi o i_*} \widetilde{F}; \mathcal{N}(G)) \otimes_{\mathbb{Z}\pi_1(B)} \mathbb{Z}$, we get from Additivity (see Theorem 6.7 (4b))

$$\dim_{\mathcal{N}(G)} \left(H_d^G(G \times_{\phi o i_*} \widetilde{F}; \mathcal{N}(G)) \otimes_{\mathbb{Z}\Delta} \mathbb{Z} \right)$$
$$\geq \dim_{\mathcal{N}(G)} \left(H_d^G(G \times_{\phi o i_*} \widetilde{F}; \mathcal{N}(G)) \otimes_{\mathbb{Z}\pi_1(B)} \mathbb{Z} \right). \quad (6.70)$$

Define $G_0 = \psi^{-1}(\Delta)$. Then ϕ and ψ induce epimorphisms ϕ_0 and ψ_0 which make the following diagram commute

$$
\begin{array}{ccccc}
\pi_1(E_0) & \xrightarrow{\phi_0} & G_0 & \xrightarrow{\psi_0} & \pi_1(X) \\
{\scriptstyle \overline{f}_*}\downarrow & & \downarrow & & \downarrow{\scriptstyle f_*} \\
\pi_1(E) & \xrightarrow{\phi} & G & \xrightarrow{\psi} & \pi_1(B)
\end{array}
$$

Notice that the fiber transport is compatible with pullbacks. From Theorem 6.29, from (6.68) applied to p and ϕ and applied to p_0 and ϕ_0 and from (6.70) we conclude

$$b_d^{(2)}(G \times_\phi \widetilde{E}; \mathcal{N}(G)) \leq b_d^{(2)}(G_0 \times_{\phi_0} \widetilde{E_0}; \mathcal{N}(G_0)). \quad (6.71)$$

We firstly consider the case where $\Delta = \mathbb{Z}$ and $X = S^1$. Then E_0 is up to homotopy the mapping torus of the selfhomotopy equivalence $F \to F$ given by the fiber transport of p_0 along a generator of $\pi_1(S^1)$ and hence $b_d^{(2)}(G_0 \times_\phi \widetilde{E_0}; \mathcal{N}(G_0)) = 0$ by Theorem 6.63. Now (6.69) follows from (6.71). Next we consider the case where Δ is finite. Let $F \xrightarrow{i_1} E_1 \xrightarrow{p_1} \widetilde{X}$ be the fibration obtained from $p_0 \colon E_0 \to X$ by the pullback construction with respect to the universal covering $\widetilde{X} \to X$. Put $G_1 = \psi^{-1}(1)$. This is a subgroup of G_0 of index $|\Delta|$. Let $\phi_1 \colon \pi_1(E_1) \to G_1$ be the epimorphism induced by ϕ_0. Notice that the canonical map $E_1 \to E_0$ is a $|\Delta|$-sheeted covering. We conclude from Theorem 6.54 (7) and (6b) and from applying (6.68) to $\overline{p_1}$ and $\phi_1 \colon \pi_1(E_1) \to G_1$

$$|\Delta| \cdot b_d^{(2)}(G_0 \times_\phi \widetilde{E_0}; \mathcal{N}(G_0)) = b_d^{(2)}\left(\mathrm{res}_{G_0}^{G_1}(G_0 \times_{\phi_0} \widetilde{E_0}); \mathcal{N}(G_1)\right)$$
$$= b_d^{(2)}(G_1 \times_{\phi_1} \widetilde{E_1}; \mathcal{N}(G_1))$$
$$= \dim_{\mathcal{N}(G_1)}\left(H_d^{G_1}(G_1 \times_{\phi_1 \circ (i_1)_*} \widetilde{F}; \mathcal{N}(G_1))\right)$$
$$= b_d^{(2)}(G_1 \times_{\phi_1 \circ (i_1)_*} \widetilde{F}; \mathcal{N}(G));$$
$$= b_d^{(2)}(G \times_{\phi \circ i_*} \widetilde{F}; \mathcal{N}(G)).$$

This implies together with (6.71)

$$b_d^{(2)}(G \times_\phi \widetilde{E}; \mathcal{N}(G)) \leq \frac{b_d^{(2)}(G \times_{\phi \circ i_*} \widetilde{F}; \mathcal{N}(G))}{|\Delta|}.$$

Since by assumption $\pi_1(B)$ either contains $\mathbb{Z}$ as a subgroup or finite subgroups Δ of arbitrarily large order, (6.69) follows. This finishes the proof of Theorem 6.67.   □

As far as the assumption about $\pi_1(B)$ in Theorem 6.67 is concerned, we mention that there is for any prime number $p > 10^{75}$ an infinite finitely generated group all of whose proper subgroups are finite of order p [405]. To the author's knowledge it is not known whether there is an infinite finitely presented group with finite exponent. For more information about this so called Burnside problem we refer to [469]. Using Theorem 6.29 one obtains the following special case of Theorem 6.67.

Corollary 6.72. *Let $F \xrightarrow{i} E \xrightarrow{p} B$ be a fibration of connected CW-complexes such that $i_* \colon \pi_1(F) \to \pi_1(E)$ is injective, $\pi_1(F)$ is infinite and $b_1^{(2)}(\widetilde{F}) < \infty$. Suppose that $\pi_1(B)$ contains an element of infinite order or finite subgroups of arbitrarily large order. Then*

$$b_0(\widetilde{E}) = b_1(\widetilde{E}) = 0.$$

Corollary 6.72 has been conjectured in [237, page 235] for fibrations $F \to E \to B$ of connected aspherical CW-complexes of finite type with non-trivial fundamental groups. It will play a role in applications of L^2-Betti numbers to deficiencies of finitely presented groups in Section 7.3.

The next Theorem shows that for amenable G the p-th L^2-Betti number of a G-space X depends only on the $\mathbb{C}G$-modules given by the singular homology of X with complex coefficients. The key ingredient will be Theorem 6.37 which shows that the von Neumann algebra of an amenable group is not flat over the complex group ring in the strict sense but is "dimension-flat".

Theorem 6.73. *Let G be an amenable group and X be a G-space. Then*

$$b_p^{(2)}(X; \mathcal{N}(G)) = \dim_{\mathcal{N}(G)}\left(\mathcal{N}(G) \otimes_{\mathbb{C}G} H_p^{\mathrm{sing}}(X; \mathbb{C})\right),$$

where $H_p^{\mathrm{sing}}(X; \mathbb{C})$ is the $\mathbb{C}G$-module given by the singular homology of X with complex coefficients.

Proof. We will show for an arbitrary $\mathbb{C}G$-chain complex C_* with $C_p = 0$ for $p < 0$

$$\dim_{\mathcal{N}(G)} \left(H_p(\mathcal{N}(G) \otimes_{\mathbb{C}G} C_*) \right) = \dim_{\mathcal{N}(G)} \left(\mathcal{N}(G) \otimes_{\mathbb{C}G} H_p(C_*) \right). \quad (6.74)$$

We begin with the case where C_* is projective. Then there is a universal co-efficient spectral sequence converging to $H_{p+q}(\mathcal{N}(G) \otimes_{\mathbb{C}G} C_*)$ [518, Theorem 5.6.4 on page 143] whose E^2-term is $E_{p,q}^2 = \mathrm{Tor}_p^{\mathbb{C}G}(\mathcal{N}(G), H_q(C_*))$. Now Additivity (see Theorem 6.7 (4b)) together with Theorem 6.37 imply (6.74) if C_* is projective.

Next we prove (6.74) in the case where C_* is acyclic. If C_* is 2-dimensional, we conclude $\dim_{\mathcal{N}(G)} \left(H_p(\mathcal{N}(G) \otimes_{\mathbb{C}G} C_*) \right) = 0$ for $p \geq 0$ using the long exact Tor-sequences of the exact sequence $0 \to C_2 \to C_1 \to C_0 \to 0$, Additivity (see Theorem 6.7 (4b)) and Theorem 6.37. Now the claim for any acyclic $\mathbb{C}G$-chain complexes C_* with $C_p = 0$ for $p < 0$ follows from by inspecting the various short exact sequences $0 \to \mathrm{im}(c_{p+1}) \to C_p \xrightarrow{c_p} \mathrm{im}(c_p) \to 0$.

In the general case one chooses a projective $\mathbb{C}G$-chain complex P_* together with a $\mathbb{C}G$-chain map $f_* : P_* \to C_*$ which induces an isomorphism on homology. Notice that the mapping cylinder $\mathrm{cyl}(f_*)$ is $\mathbb{C}G$-chain homotopy equivalent to C_* and the mapping cone $\mathrm{cone}(f_*)$ of f_* is acyclic. Hence (6.74) is true for P_* and $\mathrm{cone}(f_*)$. Thus we get (6.74) for C_* from Additivity (see Theorem 6.7 (4b)) and the long exact homology sequences of the short exact sequence of $\mathcal{N}(G)$-chain complexes which we obtain by applying $\mathcal{N}(G) \otimes_{\mathbb{C}G} -$ to the short exact (in each dimension split exact) $\mathbb{C}G$-sequence $0 \to P_* \to \mathrm{cyl}(f_*) \to \mathrm{cone}(f_*) \to 0$. This finishes the proof of Theorem 6.73. $\square$

We conclude from Theorem 6.54 (8b) and Theorem 6.73

Corollary 6.75. *If G is an infinite amenable group, then $b_p^{(2)}(G) = 0$ for all $p \geq 0$.*

We will investigate the class $\mathcal{B}_\infty$ of groups G, for which $b_p^{(2)}(G) = 0$ holds for all $p \geq 0$, in Section 7.1.

Remark 6.76. Next we compare our approach with the one of Cheeger and Gromov [107, section 2], where Corollary 6.75 has already been proved in [107, Theorem 0.2 on page 191]. We begin with the case of a countable simplicial complex X with free simplicial G-action. Then for any exhaustion $X_0 \subset X_1 \subset X_2 \subset \ldots \subset X$ by G-equivariant simplicial subcomplexes for which $G \backslash X$ is compact, the p-th L^2-Betti number in the sense and notation of [107, 2.8 on page 198] is given by

$$b_p^{(2)}(X : G) = \lim_{j \to \infty} \lim_{k \to \infty} \dim_{\mathcal{N}(G)} \left(\mathrm{im} \left(\overline{H}_{(2)}^p(X_k : G) \xrightarrow{i_{j,k}^*} \overline{H}_{(2)}^p(X_j : G) \right) \right),$$

where $i_{j,k} : X_j \to X_k$ is the inclusion for $j \leq k$. We get an identification $\overline{H}_{(2)}^p(X_j : G) = H_{(2)}^p(X_j; \mathcal{N}(G))$ from Lemma 1.76. Notice that for a G-map

$f: Y \to Z$ of G-CW-complexes of finite type $H_p^{(2)}(Y; \mathcal{N}(G))$ can be identified with $H_{(2)}^p(Y; \mathcal{N}(G))$ and analogously for Z and that under these identifications $H_{(2)}^p(f) = (H_{(2)}^p(f))^*$ (see Lemma 1.18). Moreover, we conclude from Additivity (see Theorem 6.7 (4b)) that $\dim_{\mathcal{N}(G)} \left(\mathrm{im} \left(H_{(2)}^p(f) \right) \right) = \dim_{\mathcal{N}(G)} \left(\mathrm{im} \left(H_p^{(2)}(f) \right) \right)$. Now we get from Lemma 6.52, Lemma 6.53 and Theorem 6.7

$$\dim_{\mathcal{N}(G)} \left(\mathrm{im} \left(\overline{H}_{(2)}^p(X_k : G) \xrightarrow{i_{j,k}^*} \overline{H}_{(2)}^p(X_j : G) \right) \right)$$
$$= \dim_{\mathcal{N}(G)} \left(\mathrm{im} \left(H_p^G(X_j; \mathcal{N}(G)) \xrightarrow{(i_{j,k})_*} H_p^G(X_k; \mathcal{N}(G)) \right) \right).$$

Hence we conclude from Theorem 6.13 about dimension and colimits that the definitions in [107, 2.8 on page 198] and in Definition 6.50 for a countable free simplicial complex X with free simplicial G-action agree

$$b_p^{(2)}(X : G) = b_p^{(2)}(X; \mathcal{N}(G)). \tag{6.77}$$

If G is countable and X is a countable simplicial complex with simplicial (not necessarily free) G-action, then by [107, Proposition 2.2 on page 198] and by (6.77)

$$b_p^{(2)}(X : G) = b_p^{(2)}(EG \times X : G) = b_p^{(2)}(EG \times X; \mathcal{N}(G)). \tag{6.78}$$

Cheeger and Gromov [107, Section 2] define L^2-cohomology and L^2-Betti numbers of a G-space X by considering the category whose objects are G-maps $f: Y \longrightarrow X$ for a simplicial complex Y with cocompact free simplicial G-action and then using inverse limits to extend the classical notions for finite free G-CW-complexes such as Y to X. Their approach is technically more complicated for instance because they work with cohomology instead of homology and therefore have to deal with inverse limits instead of directed limits (see Theorem 6.13 and Theorem 6.18). Our approach is closer to standard notions, the only non-standard part is the verification of the properties of the extended dimension function (Theorem 6.7). Moreover, the notion of Cheeger and Gromov [107, Section 2] does only give $b_p^{(2)}(EG \times X; \mathcal{N}(G))$, where we can and do also consider $b_p^{(2)}(X; \mathcal{N}(G))$.

6.6 L^2-Euler Characteristic

In this section we introduce the notion of L^2-Euler characteristic and investigate its relation to the equivariant Euler characteristic of a cocompact proper G-CW-complex and the Burnside group.

6.6.1 Definition and Basic Properties of L^2-Euler Characteristic

If X is a G-CW-complex, denote by $I(X)$ the set of its equivariant cells. For a cell $c \in I(X)$ let (G_c) be the conjugacy class of subgroups of G given by its orbit type and let $\dim(c)$ be its dimension. Denote by $|G_c|^{-1}$ the inverse of the order of any representative of (G_c), where $|G_c|^{-1}$ is to be understood to be zero if the order is infinite.

Definition 6.79 (L^2-Euler characteristic). *Let G be a group and let X be a G-space. Define*

$$h^{(2)}(X; \mathcal{N}(G)) := \sum_{p \geq 0} b_p^{(2)}(X; \mathcal{N}(G)) \in [0, \infty];$$

$$\chi^{(2)}(X; \mathcal{N}(G)) := \sum_{p \geq 0} (-1)^p \cdot b_p^{(2)}(X; \mathcal{N}(G)) \in \mathbb{R} \quad \text{if } h^{(2)}(X; \mathcal{N}(G)) < \infty;$$

$$m(X; G) := \sum_{c \in I(X)} |G_c|^{-1} \in [0, \infty] \qquad \text{if } X \text{ is a } G\text{-}CW\text{-complex;}$$

$$h^{(2)}(G) := h^{(2)}(EG; \mathcal{N}(G)) \in [0, \infty];$$
$$\chi^{(2)}(G) := \chi^{(2)}(EG; \mathcal{N}(G)) \in \mathbb{R} \qquad \text{if } h^{(2)}(G) < \infty.$$

We call $\chi^{(2)}(X; \mathcal{N}(G))$ and $\chi^{(2)}(G)$ the L^2-Euler characteristic of X and G.

The condition $h^{(2)}(X; \mathcal{N}(G)) < \infty$ ensures that the sum which appears in the definition of $\chi^{(2)}(X; \mathcal{N}(G))$ converges absolutely and that the following results are true. The reader should compare the next theorem with [107, Theorem 0.3 on page 191]. We will investigate the class $\mathcal{B}_\infty$ of groups whose L^2-Betti number all vanish in Section 7.1.

Theorem 6.80 (L^2-Euler characteristic).

(1) Generalized Euler-Poincaré formula

Let X be a G-CW-complex with $m(X; G) < \infty$. Then

$$h^{(2)}(X; \mathcal{N}(G)) < \infty;$$
$$\sum_{c \in I(X)} (-1)^{\dim(c)} \cdot |G_c|^{-1} = \chi^{(2)}(X; \mathcal{N}(G));$$

(2) Sum formula

Consider the following G-pushout

$$
\begin{array}{ccc}
X_0 & \xrightarrow{\ i_1\ } & X_1 \\
{\scriptstyle i_2}\downarrow & & \downarrow{\scriptstyle j_1} \\
X_2 & \xrightarrow{\ j_2\ } & X
\end{array}
$$

such that i_1 is a G-cofibration. Suppose that $h^{(2)}(X_i; \mathcal{N}(G)) < \infty$ for $i = 0, 1, 2$. Then

$$h^{(2)}(X; \mathcal{N}(G)) < \infty;$$
$$\chi^{(2)}(X; \mathcal{N}(G)) = \chi^{(2)}(X_1; \mathcal{N}(G)) + \chi^{(2)}(X_2; \mathcal{N}(G)) - \chi^{(2)}(X_0; \mathcal{N}(G));$$

(3) *Amenable groups*

Suppose that G is amenable and that X is a G-CW-complex with $m(X; G) < \infty$. Then

$$\sum_{c \in I(X)} (-1)^{\dim(c)} \cdot |G_c|^{-1} = \sum_{p \geq 0} (-1)^p \cdot \dim_{\mathcal{N}(G)} \left(\mathcal{N}(G) \otimes_{\mathbb{C}G} H_p(X; \mathbb{C}) \right),$$

where $H_p(X; \mathbb{C})$ is the $\mathbb{C}G$-module given by the cellular or the singular homology of X with complex coefficients. In particular $\sum_{c \in I(X)} (-1)^{\dim(c)} \cdot |G_c|^{-1}$ depends only the isomorphism class of the $\mathbb{C}G$-modules $H_n(X; \mathbb{C})$ for all $n \geq 0$;

(4) *Comparison with the Borel construction*

Let X be a G-CW-complex. If for all $c \in I(X)$ the group G_c is finite or $b_p^{(2)}(G_c) = 0$ for all $p \geq 0$, then

$$b_p^{(2)}(X; \mathcal{N}(G)) = b_p^{(2)}(EG \times X; \mathcal{N}(G)) \qquad \text{for } p \geq 0;$$
$$h^{(2)}(X; \mathcal{N}(G)) = h^{(2)}(EG \times X; \mathcal{N}(G));$$
$$\chi^{(2)}(X; \mathcal{N}(G)) = \chi^{(2)}(EG \times X; \mathcal{N}(G)), \text{ if } h^{(2)}(X; \mathcal{N}(G)) < \infty;$$
$$\sum_{c \in I(X)} (-1)^{\dim(c)} \cdot |G_c|^{-1} = \chi^{(2)}(EG \times X; \mathcal{N}(G)), \text{ if } m(X; G) < \infty;$$

(5) *Invariance under non-equivariant $\mathbb{C}$-homology equivalences*

Suppose that $f \colon X \to Y$ is a G-equivariant map of G-CW-complexes with $m(X; G) < \infty$ and $m(Y; G) < \infty$, such that the induced map $H_p(f; \mathbb{C})$ on homology with complex coefficients is bijective. Suppose that for all $c \in I(X)$ the group G_c is finite or $b_p^{(2)}(G_c) = 0$ for all $p \geq 0$, and analogously for all $d \in I(Y)$. Then

$$\chi^{(2)}(X; \mathcal{N}(G)) = \sum_{c \in I(X)} (-1)^{\dim(c)} \cdot |G_c|^{-1}$$
$$= \sum_{d \in I(Y)} (-1)^{\dim(d)} \cdot |G_d|^{-1}$$
$$= \chi^{(2)}(Y; \mathcal{N}(G));$$

(6) *Künneth formula*

Let X be a G-CW-complex and Y be an H-CW-complex. Then we get for the $G \times H$-CW-complex $X \times Y$

$$m(X \times Y; \mathcal{N}(G \times H)) = m(X; G) \cdot m(Y; H);$$
$$h^{(2)}(X \times Y; \mathcal{N}(G \times H)) = h^{(2)}(X; \mathcal{N}(G)) \cdot h^{(2)}(Y; \mathcal{N}(H));$$
$$\chi^{(2)}(X \times Y; \mathcal{N}(G \times H)) = \chi^{(2)}(X; \mathcal{N}(G)) \cdot \chi^{(2)}(Y; \mathcal{N}(H)),$$
$$\text{if } h^{(2)}(X; \mathcal{N}(G)), h^{(2)}(Y; \mathcal{N}(H)) < \infty,$$

where we use the convention that $0 \cdot \infty = 0$ and $r \cdot \infty = \infty$ for $r \in (0, \infty]$;

(7) *Restriction*

Let $H \subset G$ be a subgroup of finite index $[G : H]$. Let X be a G-space and let $\mathrm{res}_G^H X$ be the H-space obtained from X by restriction. Then

$$m(\mathrm{res}_G^H X; H) = [G : H] \cdot m(X; G);$$
$$h^{(2)}(\mathrm{res}_G^H X; \mathcal{N}(H)) = [G : H] \cdot h^{(2)}(X; \mathcal{N}(G));$$
$$\chi^{(2)}(\mathrm{res}_G^H X; \mathcal{N}(H)) = [G : H] \cdot \chi^{(2)}(X; \mathcal{N}(G)), \quad \text{if } h^{(2)}(X; \mathcal{N}(G)) < \infty,$$

where $[G : H] \cdot \infty$ is understood to be ∞;

(8) *Induction*

Let $H \subset G$ be a subgroup and let X be an H-space. Then

$$m(G \times_H X; \mathcal{N}(G)) = m(X; H);$$
$$h^{(2)}(G \times_H X; \mathcal{N}(G)) = h^{(2)}(X; \mathcal{N}(H));$$
$$\chi^{(2)}(G \times_H X; \mathcal{N}(G)) = \chi^{(2)}(X; \mathcal{N}(H)), \quad \text{if } h^{(2)}(X; \mathcal{N}(H)) < \infty.$$

Proof. (1) Additivity (see Theorem 6.7 (4b)) and Lemma 6.33 (1) imply

$$b_p^{(2)}(X; \mathcal{N}(G)) = \dim_{\mathcal{N}(G)} \left(H_p \left(\mathcal{N}(G) \otimes_{\mathbb{Z}G} C_p(X) \right) \right)$$
$$\leq \dim_{\mathcal{N}(G)} \left(\mathcal{N}(G) \otimes_{\mathbb{Z}G} C_p(X) \right)$$
$$= \sum_{c \in I(X), \dim(c) = p} |G_c|^{-1}.$$

This shows $h^{(2)}(X; \mathcal{N}(G)) \leq m(X; G)$. Additivity (see Theorem 6.7 (4b)) implies

$$\chi^{(2)}(X; \mathcal{N}(G)) = \sum_{p=0}^{\infty} (-1)^p \cdot \dim_{\mathcal{N}(G)} \left(\mathcal{N}(G) \otimes_{\mathbb{Z}G} C_p(X; \mathcal{N}(G)) \right),$$

and thus assertion (1) follows.

(2) Next we prove that the given square induces a long exact Mayer-Vietoris sequence of $\mathcal{N}(G)$-modules

$$\ldots \to H_{p+1}^G(X; \mathcal{N}(G)) \to H_p^G(X_0; \mathcal{N}(G)) \to H_p^G(X_1; \mathcal{N}(G)) \oplus H_p^G(X_2; \mathcal{N}(G))$$

$$\to H_p^G(X; \mathcal{N}(G)) \to \ldots.$$

This is clear if the given pushout consists of G-CW-complexes with i_1 an inclusion of G-CW-complexes, i_2 a cellular G-map and X equipped with the

G-CW-complex structure induced by the ones on X_0, X_1 and X_2, and we use the cellular $\mathbb{Z}G$-chain complexes. Namely, in this situation we obtain a short exact sequence of $\mathbb{Z}G$-chain complexes $0 \to C_*(X_0) \to C_*(X_1) \oplus C_*(X_2) \to C_*(X) \to 0$ which stays exact after applying $\mathcal{N}(G) \otimes_{\mathbb{Z}G} -$ and we can take the associated long homology sequence. The general case follows from Lemma 6.51, Theorem 6.54 (1b) and the fact that one can construct such a G-pushout of G-CW-complexes together with G-maps from the cellular G-pushout to the given G-pushout such that the maps induce on each H-fixed point set a weak homotopy equivalence (see [326, Proposition 2.3 on page 35 and Lemma 2.13 on page 38]).

Notice that the alternating sums defining the L^2-Euler characteristics converge absolutely so that one can reorder the summands without changing the limit. Now assertion (2) follows from Additivity (see Theorem 6.7 (4b)).

(3) This follows from assertion (1) and Theorem 6.73.

(4) This follows from assertion (1) and Theorem 6.54 (2).

(5) This follows from assertion (1) and Theorem 6.54 (3).

(6) This follows from Theorem 6.54 (5).

(7) This follows from Theorem 6.54 (6).

(8) This follows from Theorem 6.54 (7). This finishes the proof of Theorem 6.80.   □

Remark 6.81. Let X be a CW-complex which is *virtually homotopy finite*, i.e. there is a d-sheeted covering $p \colon \overline{X} \to X$ for some positive integer d such that $\overline{X}$ is homotopy equivalent to a finite CW-complex. Define the *virtual Euler characteristic* following Wall [509]

$$\chi_{\mathrm{virt}}(X) := \frac{\chi(\overline{X})}{d}.$$

One easily checks that this is independent of the choice of $p \colon \overline{X} \to X$ since the ordinary Euler characteristic is multiplicative under finite coverings. Moreover, we conclude from Theorem 6.80 (1) and (7) that for virtually homotopy finite X

$$m(\widetilde{X}; \mathcal{N}(\pi_1(X))) < \infty;$$
$$\chi^{(2)}(\widetilde{X}; \mathcal{N}(\pi_1(X))) = \chi_{\mathrm{virt}}(X).$$

6.6.2 L^2-Euler Characteristic, Equivariant Euler Characteristic and the Burnside Group

In this subsection we introduce the Burnside group and the equivariant Euler characteristic and relate these notions to the L^2-Euler characteristic. The elementary proof of the following lemma is left to the reader.

Lemma 6.82. *Let H and K be subgroups of G. Let NK be the normalizer of K in G and WK be its Weyl group NK/K. Then*

(1) $G/H^K = \{gH \mid g^{-1}Kg \subset H\}$;
(2) The map

$$\phi: G/H^K \to \mathrm{consub}(H), \qquad gH \mapsto g^{-1}Kg$$

induces an injection

$$WK\backslash(G/H^K) \to \mathrm{consub}(H),$$

where $\mathrm{consub}(H)$ is the set of conjugacy classes in H of subgroups of H;
(3) The WK-isotropy group of $gH \in G/H^K$ is $(gHg^{-1}\cap NK)/K \subset NK/K = WK$;
(4) If H is finite, then G/H^K is a finite union of WK-orbits of the shape WK/L for finite subgroups $L \subset WK$. □

Definition 6.83. *Define the* Burnside group *$A(G)$ of a group G to be the Grothendieck group which is associated to the abelian monoid under disjoint union of G-isomorphism classes of proper cocompact G-sets S, i.e. G-sets S for which the isotropy group of each element in S and the quotient $G\backslash S$ are finite.* □

Notice that $A(G)$ is the free abelian group generated by G-isomorphism classes of orbits G/H for finite subgroups $H \subset G$ and that G/H and G/K are G-isomorphic if and only if H and K are conjugate in G. If G is a finite group, $A(G)$ is the classical *Burnside ring* [494, section 5], [495, chapter IV]. If G is infinite, then the cartesian product of two proper cocompact G-sets with the diagonal action is not necessarily cocompact anymore so that the cartesian product does not induce a ring structure on $A(G)$. At least there is a bilinear map induced by the cartesian product $A(G_1) \otimes A(G_2) \to A(G_1 \times G_2)$.

Definition 6.84. *Let X be a finite proper G-CW-complex. Define its* equivariant Euler characteristic

$$\chi^G(X) := \sum_{c \in I(X)} (-1)^{\dim(c)} \cdot [G/G_c] \qquad \in A(G).$$

An *additive invariant* (A, a) for finite proper G-CW-complexes consists of an abelian group A and a function a which assigns to any finite proper G-CW-complex X an element $a(X) \in A$ such that the following three conditions hold, i.) if X and Y are G-homotopy equivalent, then $a(X) = a(Y)$, ii.) if X_0, X_1 and X_2 are G-CW-subcomplexes of X with $X = X_1 \cup X_2$ and $X_0 = X_1 \cap X_2$, then $a(X) = a(X_1) + a(X_2) - a(X_0)$, and iii.) $a(\emptyset) = 0$. We call an additive invariant (U, u) *universal*, if for any additive invariant (A, a) there is precisely one homomorphism $\psi: U \to A$ such that $\psi(u(X)) = a(X)$ holds for all finite proper G-CW-complexes. One easily checks using induction over the number of equivariant cells

Lemma 6.85. $(A(G), \chi^G)$ *is the universal additive invariant for finite proper G-CW-complexes and we get for a finite proper G-CW-complex X*

$$\chi^G(X) = \sum_{(H),|H|<\infty} \chi(WH\backslash(X^H, X^{>H})) \cdot [G/H],$$

where $\chi(WH\backslash(X^H, X^{>H}))$ is the ordinary Euler characteristic of the pair of finite CW-complexes $WH\backslash(X^H, X^{>H})$.

Definition 6.86. *Define for a finite subgroup $K \subset G$ the L^2-character map*

$$\mathrm{ch}_K^G \colon A(G) \to \mathbb{Q}, \qquad [S] \mapsto \sum_{i=1}^{r} |L_i|^{-1}$$

if $WK/L_1, WK/L_2, \ldots, WK/L_r$ are the WK-orbits of S^K. Define the global L^2-character map by

$$\mathrm{ch}^G := \prod_{(K)} \mathrm{ch}_K^G \colon A(G) \to \prod_{(K)} \mathbb{Q}$$

where (K) runs over the conjugacy classes of finite subgroups of G.   □

Lemma 6.87. *Let X be a finite proper G-CW-complex and $K \subset G$ be a finite subgroup. Then X^K is a finite proper WK-CW-complex and*

$$\chi^{(2)}(X^K; \mathcal{N}(WK)) = \mathrm{ch}_K^G(\chi^G(X)).$$

Proof. The WK-space X^K is a finite proper WK-CW-complex because for finite $H \subset G$ the WK-set G/H^K is proper and cocompact by Lemma 6.82 (4). Since the assignment which associates to a finite proper G-CW-complex X the element $\chi^{(2)}(X^K; \mathcal{N}(WK))$ in $\mathbb{Q}$ is an additive invariant by Theorem 6.80 (2) and (5), it suffices by Lemma 6.85 to check the claim for $X = G/H$ for finite $H \subset G$. This follows from the conclusion of Lemma 6.33 (1) that $\chi^{(2)}(WK/L; \mathcal{N}(WK)) = |L|^{-1}$ holds for finite $L \subset WK$.   □

Notice that one gets from Lemma 6.82 the following explicit formula for the value of $\mathrm{ch}_K^G(G/H)$. Namely, define

$$\mathcal{L}_K(H) := \{(L) \in \mathrm{consub}(H) \mid L \text{ conjugate to } K \text{ in } G\}.$$

For $(L) \in \mathcal{L}_K(H)$ choose $L \in (L)$ and $g \in G$ with $g^{-1}Kg = L$. Then

$$g(H \cap NL)g^{-1} = gHg^{-1} \cap NK;$$

$$|(gHg^{-1} \cap NK)/K|^{-1} = \frac{|K|}{|H \cap NL|}.$$

This implies

$$\mathrm{ch}_K^G(G/H) = \sum_{(L)\in\mathcal{L}_K(H)} \frac{|K|}{|H \cap NL|}. \tag{6.88}$$

Lemma 6.89. *The global L^2-character map of Definition 6.86 induces a map denoted by*

$$\mathrm{ch}^G \otimes_{\mathbb{Z}} \mathbb{Q} : \ A(G) \otimes_{\mathbb{Z}} \mathbb{Q} \to \prod_{(K)} \mathbb{Q}.$$

It is injective. If G has only finitely many conjugacy classes of finite subgroups, then it is bijective.

Proof. Consider an element $\sum_{i=1}^n r_i \cdot [G/H_i]$ in the kernel of $\mathrm{ch}^G \otimes_{\mathbb{Z}} \mathbb{Q}$. We show by induction on n that the element must be trivial. The induction beginning $n = 0$ is trivial, the induction step is done as follows. We can choose the numeration such that H_i subconjugated to H_j implies $i \geq j$. We get from (6.88)

$$\mathrm{ch}_K^G(G/H) = 1 \qquad \text{if } H = K;$$
$$\mathrm{ch}_K^G(G/H) = 0 \qquad \text{if } K \text{ is not subconjugated to } H \text{ in } G.$$

This implies

$$\mathrm{ch}_{H_1}^G \left(\sum_{i=1}^n r_i \cdot [G/H_i] \right) = r_1$$

and hence $r_1 = 0$. Hence the global L^2-character map is injective. If G has only finitely many conjugacy classes of finite subgroups, then the source and target of $\mathrm{ch}^G \otimes_{\mathbb{Z}} \mathbb{Q}$ are rational vector spaces of the same finite dimension and hence $\mathrm{ch}^G \otimes_{\mathbb{Z}} \mathbb{Q}$ must be bijective. □

Remark 6.90. Suppose that there are only finitely many conjugacy classes $(H_1), (H_2), \ldots, (H_r)$ of finite subgroups in G. Without loss of generality we can assume that H_i subconjugated to H_j implies $i \geq j$. With respect to the obvious ordered basis for the source and target, the map $\mathrm{ch}^G \otimes_{\mathbb{Z}} \mathbb{Q}$ is described by an upper triangular matrix A with ones on the diagonal. One can get an explicit inverse A^{-1} which again has ones on the diagonal. This leads to a characterization of the image of $A(G)$ under the global L^2-character map χ^G. Namely, an element in $\eta \in \prod_{i=1}^r \mathbb{Q}$ lies in $\mathrm{ch}^G(A(G))$ if and only if the following *Burnside integrality conditions* are satisfied

$$A^{-1}\eta \in \prod_{i=1}^r \mathbb{Z}. \tag{6.91}$$

Now suppose that G is finite. Then the global L^2-character map is related to the classical character map by the factor $|WK|^{-1}$, i.e. we have for each subgroup K of G and any finite G-set S

$$\mathrm{ch}_K^G(S) = |WK|^{-1} \cdot |S^K|. \tag{6.92}$$

One easily checks that under the identification (6.92) the integrality conditions (6.91) correspond to the classical Burnside ring congruences for finite groups [494, section 5.8], [495, section IV.5]. □

Let $E(G, \mathcal{FIN})$ be the classifying G-space for the family $\mathcal{FIN}$ of finite subgroups (see Definition 1.28).

Lemma 6.93. *Suppose that there is a model for $E(G, \mathcal{FIN})$ which is a finite G-CW-complex. Then there are only finitely many conjugacy classes of finite subgroups and for a finite subgroup $K \subset G$*

$$\mathrm{ch}_K^G(\chi^G(E(G, \mathcal{FIN}))) = \chi^{(2)}(WK).$$

If G is amenable, then we get for a finite subgroup $K \subset G$

$$\mathrm{ch}_K^G(\chi^G(E(G, \mathcal{FIN}))) = |WK|^{-1},$$

where $|WK|^{-1}$ is to be understood as 0 for infinite WK.

Proof. We get from Theorem 6.80 (4) and Lemma 6.87 since $E(G, \mathcal{FIN})^K$ is a model for $E(WK, \mathcal{FIN})$ and $E(WK) \times E(WK, \mathcal{FIN})$ is a model for $E(WK)$

$$\mathrm{ch}_K^G(\chi^G(E(G, \mathcal{FIN}))) = \chi^{(2)}(E(WK, \mathcal{FIN}); \mathcal{N}(WK)) = \chi^{(2)}(WK).$$

In the case where G is amenable apply Theorem 6.54 (8b) and Corollary 6.75.   □

Example 6.94. Let $1 \to \mathbb{Z}^n \to G \to \mathbb{Z}/p \to 1$ be an extension of groups for $n \geq 1$ and a prime number p. The conjugation action of G on the normal subgroup $\mathbb{Z}^n$ factorizes through the projection $G \to \mathbb{Z}/p$ to an operation ρ of $\mathbb{Z}/p$ onto $\mathbb{Z}^n$. There exists a finite G-CW-complex model for $E(G, \mathcal{FIN})$ by the following argument. If G contains a finite subgroup, then G is a semidirect product of $\mathbb{Z}^n$ and $\mathbb{Z}/p$ and one can construct a finite G-CW-complex as model for $E(G, \mathcal{FIN})$ with $\mathbb{R}^n$ as underlying space. Suppose that G contains no finite subgroup. Then $H_1(G; \mathbb{Q}) \cong_{\mathbb{Q}} (\mathbb{Z}^n)^{\mathbb{Z}/p} \otimes_{\mathbb{Z}} \mathbb{Q} \neq \{0\}$ and hence G admits an epimorphism onto $\mathbb{Z}$. Now one can show inductively over n that there is a finite model for BG.

We want to compute $\chi^G(E(G, \mathcal{FIN}))$. If this operation has a non-trivial fixed point, then WH is infinite for any finite subgroup H of G and we conclude from Lemma 6.89 and Theorem 6.93 that

$$\chi^G(E(G, \mathcal{FIN})) = 0.$$

Now suppose that this operation ρ has no non-trivial fixed points. Let H_0 be the trivial subgroup and $H_1, H_2, \ldots, H_r$ be a complete set of representatives of the conjugacy classes of finite subgroups. Each H_i is isomorphic to $\mathbb{Z}/p$. One easily checks that there is a bijection

$$H^1(\mathbb{Z}/p; \mathbb{Z}_\rho^n) \to \{(H) \mid H \subset G, 1 < |H| < \infty\}$$

and in particular $r \geq 1$, where $\mathbb{Z}_\rho^n$ denotes the $\mathbb{Z}[\mathbb{Z}/p]$-module given by $\mathbb{Z}^n$ and ρ. We compute using (6.88)

$$\mathrm{ch}^G_{H_0}(G/H_0) = 1;$$
$$\mathrm{ch}^G_{H_0}(G/H_j) = \frac{1}{p} \quad j = 1, 2, \dots, r;$$
$$\mathrm{ch}^G_{H_i}(G/H_j) = 1 \quad i = j, \ i, j = 1, 2, \dots, r;$$
$$\mathrm{ch}^G_{H_i}(G/H_j) = 0 \quad i \neq j, \ i, j = 1, 2, \dots, r.$$

We conclude

$$\chi^G(E(G, \mathcal{FIN})) = -\frac{r}{p} \cdot [G/H_0] + \sum_{i=1}^{r} [G/H_i].$$

The integrality conditions of (6.91) become in this case

$$\eta_0 - \frac{1}{p} \cdot \sum_{i=1}^{r} \eta_i \in \mathbb{Z};$$
$$\eta_i \in \mathbb{Z} \qquad i = 1, 2, \dots, r.$$

6.7 Finitely Presented Torsion Modules and Novikov-Shubin Invariants

In this section we explain how the Novikov-Shubin invariants can be read off from $H_p(X; \mathcal{N}(G))$ for a free G-CW-complex X of finite type.

Let M be a finitely presented $\mathcal{N}(G)$-module. Then we can choose an exact sequence $0 \to \mathcal{N}(G)^n \xrightarrow{f} \mathcal{N}(G)^n \to \mathbf{T}M \to 0$ for some positive morphism f by Lemma 6.28. Now define the Novikov-Shubin invariant

$$\alpha(M) := \alpha(\nu(f)) \in [0, \infty] \coprod \{\infty^+\} \tag{6.95}$$

by the Novikov-Shubin invariant of the morphism $\nu(f) \colon l^2(G)^n \to l^2(G)^n$ (see Definition 2.8), where ν has been introduced in (6.22). This is independent of the choice of f because of homotopy invariance (see Theorem 2.19). Moreover, for any finitely generated projective $\mathcal{N}(G)$-resolution C_* of M we have $\alpha_1(\nu(C_*)) = \alpha(\nu(c_1)) = \alpha(M)$. Obviously $\alpha(M) = \alpha(\mathbf{T}M)$. Recall that the Novikov-Shubin invariant of a morphism $f \colon U \to V$ of finitely generated Hilbert $\mathcal{N}(G)$-modules introduced in Definition 2.8 measures the deviation of the image of f to be closed. Analogously, the Novikov-Shubin invariant of the cokernel of an $\mathcal{N}(G)$-map $g \colon P \to Q$ of finitely generated projective $\mathcal{N}(G)$-modules measures the difference between $\mathrm{im}(g)$ and $\overline{\mathrm{im}(g)}$.

Let $0 \to M_0 \to M_1 \to M_2 \to 0$ be an exact sequence of $\mathcal{N}(G)$-modules such that two of them are finitely presented and have von Neumann dimension zero. Then M_i is finitely presented and $\mathbf{T}M_i = M_i$ for $i = 0, 1, 2$ by Theorem 6.5 and Theorem 6.7 and we conclude from Theorem 2.20

$$\frac{1}{\alpha(M_1)} \leq \frac{1}{\alpha(M_0)} + \frac{1}{\alpha(M_2)}. \tag{6.96}$$

In particular we get $\frac{1}{\alpha(M)} \le \frac{1}{\alpha(N)}$, if M and N are finitely presented $\mathcal{N}(G)$-modules, $\dim_{\mathcal{N}(G)}(N) = 0$ and $M \subset N$. Moreover, Lemma (2.11) (8) and Theorem 6.24 (3) imply that $\alpha(M) = \infty^+ \Leftrightarrow M = 0$ holds for a finitely presented $\mathcal{N}(G)$-module M with $\dim_{\mathcal{N}(G)}(M) = 0$. Hence we may say that $\frac{1}{\alpha(M)}$ measures the size of finitely presented $\mathcal{N}(G)$-modules M with $\dim_{\mathcal{N}(G)}(M) = 0$.

Lemma 6.97. *Let X be a free G-CW-complex of finite type. Then its $(p-1)$-th L^2-Betti number $b_{p-1}^{(2)}(X; \mathcal{N}(G))$ of Definition 1.30 and its p-th Novikov-Shubin invariant $\alpha_p(X; \mathcal{N}(G))$ of Definition 2.54 can be read off from $H_{p-1}^G(X; \mathcal{N}(G))$ by*

$$b_{p-1}^{(2)}(X; \mathcal{N}(G)) = \dim_{\mathcal{N}(G)}(H_{p-1}^G(X; \mathcal{N}(G)));$$
$$\alpha_p(X; \mathcal{N}(G)) = \alpha(H_{p-1}^G(X; \mathcal{N}(G))).$$

Proof. The statement about the L^2-Betti numbers has already been proved in Lemma 6.53. We conclude from Lemma 2.11 (9), Theorem 6.7 (3) and Theorem 6.24

$$\alpha_p(X; \mathcal{N}(G)) = \alpha\left(c_p^{(2)} \colon C_p^{(2)}(X) \to C_{p-1}^{(2)}(X)\right)$$
$$= \alpha\left(c_p^{(2)}|_{\ker(c_p^{(2)})^\perp} \colon \ker(c_p^{(2)})^\perp \to \mathrm{clos}(\mathrm{im}(c_p^{(2)}))\right)$$
$$= \alpha\left(\mathrm{coker}\left(\nu^{-1}\left(c_p^{(2)}|_{\ker(c_p^{(2)})^\perp} \colon \ker(c_p^{(2)})^\perp \to \mathrm{clos}(\mathrm{im}(c_p^{(2)}))\right)\right)\right)$$
$$= \alpha\left(\overline{\mathrm{im}(\nu^{-1}(c_p^{(2)}))}/\mathrm{im}(\nu^{-1}(c_p^{(2)}))\right)$$
$$= \alpha\left(\overline{\mathrm{im}(\mathcal{N}(G) \otimes_{\mathbb{Z}G} c_p)}/\mathrm{im}(\mathcal{N}(G) \otimes_{\mathbb{Z}G} c_p)\right)$$
$$= \alpha(\mathbf{T}H_{p-1}^G(X; \mathcal{N}(G)))$$
$$= \alpha(H_{p-1}^G(X; \mathcal{N}(G))),$$

where coker denotes the cokernel. $\qquad\square$

Notice that Lemma 6.97 gives a good explanation for Theorem 2.55 (1).

The next lemma will be interesting in connection with the zero-in-the-spectrum Conjecture 12.1 (see Lemma 12.3).

Lemma 6.98. *Let X be a free G-CW-complex of finite type and let $p \ge 0$ be an integer. Then $H_p^G(X; \mathcal{N}(G)) = 0$ if and only if $b_p^{(2)}(X; \mathcal{N}(G)) = 0$ and $\alpha_{p+1}(X; \mathcal{N}(G)) = \infty^+$.*

Proof. Lemma 6.97 implies that $b_p^{(2)}(X; \mathcal{N}(G)) = 0$ and $\alpha_{p+1}(X; \mathcal{N}(G)) = \infty^+$ if and only if $\dim_{\mathcal{N}(G)}(H_p(X; \mathcal{N}(G))) = 0$ and $\alpha(H_p(X; \mathcal{N}(G))) = \infty^+$. Hence $H_p(X; \mathcal{N}(G)) = 0$ implies $b_p^{(2)}(X; \mathcal{N}(G)) = 0$ and $\alpha_{p+1}(X; \mathcal{N}(G)) = \infty^+$. Now suppose that $b_p^{(2)}(X; \mathcal{N}(G)) = 0$ and $\alpha_{p+1}(X; \mathcal{N}(G)) = \infty^+$. Because of Theorem 6.7 (3) and (4e) we conclude $\dim_{\mathcal{N}(G)}(\mathbf{P}H_p(X; \mathcal{N}(G))) = 0$

and $\alpha(\mathbf{T}H_p(X;\mathcal{N}(G))) = \infty^+$ and therefore $\mathbf{P}H_p(X;\mathcal{N}(G)) = 0$ and $\mathbf{T}H_p(X;\mathcal{N}(G)) = 0$. This shows $H_p(X;\mathcal{N}(G)) = 0$. $\square$

6.8 Miscellaneous

We will deal with the Atiyah Conjecture for a group G in Chapter 10. It says for a finite free G-CW-complex that $b_p^{(2)}(X;\mathcal{N}(G)) \in \mathbb{Q}$. If furthermore there is an integer d such that the order of any finite subgroup $H \subset G$ divides d, then the strong Atiyah Conjecture 10.2 predicts $d \cdot b_p^{(2)}(X;\mathcal{N}(G)) \in \mathbb{Z}$ for a finite free G-CW-complex X. The following result is proved in [334, Theorem 5.2 on page 233]

Theorem 6.99. *(1) Suppose that there is no bound on the order of finite subgroups of G. Then for any sequence β_3, β_4, ... of elements in $[0,\infty]$, there is a free simply connected G-CW-complex X satisfying*

$$b_p^{(2)}(X;\mathcal{N}(G)) = \beta_p \qquad \text{for } p \geq 3.$$

If G is countable, one can arrange that X has countably many G-equivariant cells;

(2) Suppose that there is an integer d such that the order of any finite subgroup of G divides d and that the strong Atiyah Conjecture 10.2 holds for G. Then we get for any G-space X and $p \geq 0$

$$d \cdot b_p^{(2)}(X;\mathcal{N}(G)) \in \mathbb{Z} \cup \{\infty\}.$$

In [343] the notion of Novikov-Shubin invariants for free G-CW-complexes of finite type is extended to arbitrary G-spaces. In particular one can talk of the Novikov-Shubin invariants $\alpha_p(G) := \alpha_p(EG;\mathcal{N}(G))$ of a group G. Recall that a group is *locally finite*, if any finitely generated subgroup is finite. In [343, Theorem 3.9 on page 174] it is proved that $\alpha_p(G) \geq 1$ for $p \geq 1$ if G contains $\mathbb{Z}^n$ as normal subgroup for some $n \geq 1$, and that $\alpha_p(G) \geq 1$ for $p = 1, 2$ if G contains a normal subgroup, which is infinite elementary-amenable and which does not contain an infinite locally finite subgroup. This implies in particular that $\alpha_p(X;\mathcal{N}(G)) \geq 1$ for $p = 1, 2$ holds for a G-CW-complex X of finite type if G contains a normal subgroup, which is infinite elementary-amenable and which does not contain an infinite locally finite subgroup.

Farber [182] constructs a category $\mathcal{E}(\mathcal{N}(G))$ which contains the category {fin. gen. Hilb. $\mathcal{N}(G)$-mod.} as a subcategory. The point is that $\mathcal{E}(\mathcal{N}(G))$ is an abelian category, it is an abelian extension of {fin. gen. Hilb. $\mathcal{N}(G)$-mod.} in the sense of [205]. An object in $\mathcal{E}(\mathcal{N}(G))$ is a map of finitely generated Hilbert $\mathcal{N}(G)$-modules $(\alpha\colon A' \to A)$. A morphism in $\mathcal{E}(\mathcal{N}(G))$ from $(\alpha\colon A' \to A)$ to $(\beta\colon B' \to B)$ is an equivalence class of maps $f\colon A \to B$ such that there exists a map $g\colon A' \to B'$ with $f \circ \alpha = \beta \circ g$. Here f and f' are called equivalent if and only if $f - f' = \beta \circ h$ for some morphism $h\colon A \to B'$. Let {fin. pres. $\mathcal{N}(G)$-mod.} be the category of finitely presented $\mathcal{N}(G)$-modules. The two approaches are unified by the fact that there is an equivalence of abelian categories

$$\overline{\nu^{-1}}\colon \mathcal{E}(\mathcal{N}(G)) \to \{\text{fin. pres. } \mathcal{N}(G)\text{-mod.}\}$$

which induces the equivalence appearing in Theorem 6.24

$$\nu^{-1}\colon \{\text{fin. gen. Hilb. } \mathcal{N}(G)\text{-mod.}\} \to \{\text{fin. gen. proj. } \mathcal{N}(G)\text{-mod.}\}.$$

The equivalence $\overline{\nu^{-1}}$ sends an object $(\alpha\colon A' \to A)$ to the cokernel of $\nu^{-1}(\alpha)$.

Any von Neumann algebra is semihereditary (cf. Theorem 6.7 (1)). This follows from the facts that any von Neumann algebra is a Baer $*$-ring and hence in particular a Rickart C^*-algebra [36, Definition 1, Definition 2 and Proposition 9 in Chapter 1.4] and that a C^*-algebra is semihereditary if and only if it is Rickart [8, Corollary 3.7 on page 270].

The material of Subsection 6.6.2 can be extended from finite proper G-CW-complexes to G-endomorphisms $f\colon X \to X$ of finite proper G-CW-complexes by replacing Euler characteristics by Lefschetz invariants. Define the *equivariant Lefschetz invariant*

$$L^G(f) := \sum_{(H),|H|<\infty} \sum_{p\geq 0} (-1)^p \cdot \mathrm{tr}_{\mathbb{Z}WH}\left(C_p(f^H, f^{>H})\colon C_p(X^H, X^{>H})\right.$$
$$\left. \to C_p(X^H, X^{>H})\right) \cdot [G/H] \quad \in A(G), \tag{6.100}$$

where $\mathrm{tr}_{\mathbb{Z}WH}$ if the trace defined for endomorphisms of finitely generated projective $\mathbb{Z}WH$-modules which is essentially determined by $\mathrm{tr}_{\mathbb{Z}WK}(R_u\colon \mathbb{Z}WH \to \mathbb{Z}WH) = \lambda_e$ for $u = \sum_{w\in WH} \lambda_w \cdot w \in \mathbb{Z}WH$. One can also define an L^2-Lefschetz invariant

$$L^{(2)}(f^K; \mathcal{N}(WK))$$
$$:= \sum_{p\geq 0} (-1)^p \cdot \mathrm{tr}_{\mathcal{N}(WK)}\left(\mathrm{id} \otimes_{\mathbb{Z}WK} C_p(f^K)\colon \mathcal{N}(WK) \otimes_{\mathbb{Z}WK} C_p(X^K)\right.$$
$$\left. \to \mathcal{N}(WK) \otimes_{\mathbb{Z}WK} C_p(X^K)\right) \quad \in \mathbb{C}. \tag{6.101}$$

Notice that $L^G(\mathrm{id}\colon X \to X) = \chi^G(X)$ and $L^{(2)}(\mathrm{id}\colon X^K \to X^K; \mathcal{N}(WK)) = \chi^{(2)}(X^K; \mathcal{N}(WK))$. We can extend Theorem 6.80 (1) and Lemma 6.87 to

$$\mathrm{ch}_K^G(L^G(f)) = L^{(2)}(f^K; \mathcal{N}(WK)) \tag{6.102}$$
$$= \sum_{p\geq 0} (-1)^p \cdot \mathrm{tr}_{\mathcal{N}(WK)}\left(\mathbf{P}H_p(f^K; \mathcal{N}(WK))\colon\right.$$
$$\mathbf{P}H_p^{WK}(X^K; \mathcal{N}(WK)) \to \mathbf{P}H_p^{WK}(X^K; \mathcal{N}(WK))\right). \tag{6.103}$$

For more sophisticated Lefschetz type invariants see for instance [301] and [335].

Dimension functions and the role of the torsion part $\mathbf{T}M$ for an $\mathcal{N}(G)$-module will be further explained in Chapter 8 when we extend the group von Neumann algebra $\mathcal{N}(G)$ to the algebra $\mathcal{U}(G)$ of affiliated operators. This corresponds for a principal ideal domain R to the passage to its quotient field.

The dimension-flatness of the von Neumann algebra $\mathcal{N}(G)$ over $\mathbb{C}G$ for amenable G in the sense of Theorem 6.37 will play an important role in computations about $G_0(\mathbb{C}G)$ in Subsection 9.5.3.

We state without proof the next result which yields another proof of the Künneth formula (see Theorem 6.54 (5)).

Theorem 6.104. *Let G and H be two groups.*

(1) Let $0 \to M_0 \to M_1 \to M_2 \to 0$ be an exact sequence of $\mathcal{N}(G)$-modules and let N be an $\mathcal{N}(H)$-module. Then

$$\dim_{\mathcal{N}(G \times H)} \left(\ker \left(\mathcal{N}(G \times H) \otimes_{\mathcal{N}(G) \otimes_{\mathbb{C}} \mathcal{N}(H)} M_0 \otimes_{\mathbb{C}} N \to \right. \right.$$
$$\left. \left. \mathcal{N}(G \times H) \otimes_{\mathcal{N}(G) \otimes_{\mathbb{C}} \mathcal{N}(H)} M_1 \otimes_{\mathbb{C}} N \right) \right) = 0;$$

(2) Let M be an $\mathcal{N}(G)$ and N be an $\mathcal{N}(H)$-module. Then

$$\dim_{\mathcal{N}(G \times H)} \left(\mathcal{N}(G \times H) \otimes_{\mathcal{N}(G) \otimes_{\mathbb{C}} \mathcal{N}(H)} M \otimes_{\mathbb{C}} N \right)$$
$$= \dim_{\mathcal{N}(G)}(M) \cdot \dim_{\mathcal{N}(H)}(N)$$

with the convention $0 \cdot \infty = \infty \cdot 0 = 0$.

Elek [173] defines a dimension for finitely generated KG-modules for a finitely generated amenable group G and a field K which satisfies additivity and sends KG to 1.

Exercises

6.1.   Let M be a submodule of a projective $\mathcal{N}(G)$-module. Prove that $\dim_{\mathcal{N}(G)}(M) = 0$ holds if and only if $M = 0$.

6.2.   Let M be a finitely generated $\mathcal{N}(G)$-module and $\epsilon > 0$. Then there is a finitely generated projective $\mathcal{N}(G)$-module Q together with an epimorphism $p \colon Q \to M$ such that $\dim_{\mathcal{N}(G)}(\ker(p)) \le \epsilon$.

6.3.   Let M be a submodule of the finitely generated projective $\mathcal{N}(G)$-module P. Given $\epsilon > 0$, there is a submodule $P' \subset M$ which is a direct summand in P and satisfies $\dim_{\mathcal{N}(G)}(M) \le \dim_{\mathcal{N}(G)}(P') + \epsilon$.

6.4.   Let M be a countably generated $\mathcal{N}(G)$-module and M^* be its dual $\mathcal{N}(G)$-module $\hom_{\mathcal{N}(G)}(M, \mathcal{N}(G))$. Prove $\dim_{\mathcal{N}(G)}(M) = \dim_{\mathcal{N}(G)}(M^*)$.

6.5.   Let I be the set of finite subsets of S^1 directed by inclusion. Define the directed system $\{N_J \mid J \in I\}$ by $N_J = L^\infty(S^1)$ with the associated maps

$\phi_{J,K} \colon L^\infty(S^1) \to L^\infty(S^1)$ for $J \subset K$ which are given by multiplication with $\prod_{u \in K-J} \overline{z-u}$ for $J \subset K$, where $\overline{z-u} \in \mathcal{N}(\mathbb{Z})$ is given by the function $S^1 \to \mathbb{C}$ sending $z \in S^1$ to the complex conjugate of $z-u$. Define the $\mathcal{N}(\mathbb{Z})$-module $M = \operatorname{colim}_{j \in I} N_j$. Let M^* be the dual $\mathcal{N}(\mathbb{Z})$-module $\hom_{\mathcal{N}(\mathbb{Z})}(M, \mathcal{N}(\mathbb{Z}))$. Show

$$\dim_{\mathcal{N}(\mathbb{Z})}(M) = 1;$$
$$M^* = 0;$$
$$\dim_{\mathcal{N}(\mathbb{Z})}(M^*) = 0.$$

6.6. Show that $\mathcal{N}(\mathbb{Z})$ is not Noetherian. Deduce that $\mathcal{N}(G)$ is not Noetherian if G contains an element of infinite order.

6.7. Let G be a group and let A be a ring with $\mathbb{Z} \subset A \subset \mathbb{C}$. Let P be a projective AG-module such that for some finitely generated AG-submodule $M \subset P$ we have $\dim_{\mathcal{N}(G)}(\mathcal{N}(G) \otimes_{AG} P/M) = 0$. Show that P is finitely generated.

6.8. Let M be an $\mathcal{N}(G)$-module. Define $\mathbf{T}_{\dim} M$ to be the union of all $\mathcal{N}(G)$-submodules $N \subset M$ with $\dim_{\mathcal{N}(G)}(N) = 0$. Show that $\mathbf{T}_{\dim} M$ is the largest $\mathcal{N}(G)$-submodule of M with vanishing von Neumann dimension and that this definition coincides with the Definition 6.1 of $\mathbf{T}M$ provided that M is finitely generated, but not for arbitrary $\mathcal{N}(G)$-modules M.

6.9. Let G be a group with the property that for any $\mathbb{C}G$-module M $\dim_{\mathcal{N}(G)}\left(\operatorname{Tor}_p^{\mathbb{C}G}(\mathcal{N}(G), M)\right) = 0$ holds for $p \geq 1$. Show that then any subgroup $H \subset G$ inherits this property, i.e. $\dim_{\mathcal{N}(H)}\left(\operatorname{Tor}_p^{\mathbb{C}H}(\mathcal{N}(H), N)\right) = 0$ holds for any $\mathbb{C}H$-module N and any $p \geq 1$.

6.10. Show for a virtually cyclic group G that $\mathcal{N}(G)$ is flat over $\mathbb{C}G$.

6.11. Let G be a group for which $\mathcal{N}(G)$ is flat over $\mathbb{C}G$. Prove

(1) For any subgroup $H \subset G$ the von Neumann algebra $\mathcal{N}(H)$ is flat over $\mathbb{C}H$;
(2) If the group K contains G as a subgroup of finite index, then $\mathcal{N}(K)$ is flat over $\mathbb{C}K$;
(3) $H_p^G(EG; \mathcal{N}(G)) = 0$ for $p \geq 1$;
(4) $b_p^{(2)}(G) = 0$ for $p \geq 1$;
(5) If there is a CW-model for BG of finite type, then $\alpha_p(EG; \mathcal{N}(G)) = \infty^+$ for $p \geq 1$;
(6) G does not contain a subgroup which is $\mathbb{Z}^n$ or $*_{i=1}^n \mathbb{Z}$ for some $n \geq 2$;
(7) G does not contain the fundamental group of an aspherical closed manifold whose universal covering is a symmetric space;
(8) G does not contain the fundamental group of a connected sum $M_1 \# \ldots \# M_r$ of (compact connected orientable) non-exceptional prime 3-manifolds M_j.

6.12.   Let G be a group such that there is no bound on the order of its finite subgroups. Construct for $\beta \in [0, \infty]$ a countably generated projective $\mathbb{Z}G$-module P satisfying $\dim_{\mathcal{N}(G)}(\mathcal{N}(G) \otimes_{\mathbb{Z}G} P) = \beta$. Moreover, construct for a sequence $\beta_3, \beta_4, \beta_5, \dots$ of elements in $[0, \infty]$ a free simply connected G-CW-complex X with $b_p^{(2)}(X; \mathcal{N}(G)) = \beta_p$ for $p \geq 3$. If G happens to be countable, find X with only countably many equivariant cells.

6.13. Given a sequence $\beta_1, \beta_2, \beta_3, \dots$ of elements in $[0, \infty]$, construct a group G with $b_p^{(2)}(G) = \beta_p$ for $p \geq 1$.

6.14.   Let X be a G-CW-complex with $m(X; G) < \infty$. Show that then X has at most countably many equivariant cells with finite isotropy groups.

6.15.   Let $F \to E \to B$ be a fibration of connected CW-complexes and $\phi \colon \pi_1(E) \to G$ be a group homomorphism. Let $\overline{F} \to F$ and $\overline{E} \to E$ be the coverings with G as deck transformation group associated to the homomorphisms $\pi_1(F) \xrightarrow{i_*} \pi_1(E) \xrightarrow{\phi} G$ and $\phi \colon \pi_1(E) \to G$. Suppose that B is a finite CW-complex and $h^{(2)}(\overline{F}; \mathcal{N}(G)) < \infty$. Show

$$h^{(2)}(\overline{E}, \mathcal{N}(G)) < \infty;$$
$$\chi^{(2)}(\overline{E}, \mathcal{N}(G)) = \chi^{(2)}(\overline{F}, \mathcal{N}(G)) \cdot \chi(B),$$

where $\chi(B)$ is the ordinary Euler characteristic of the finite CW-complex B.

6.16. Show by constructing a counterexample that the condition $b_p^{(2)}(G \times_{\phi \circ i} \widetilde{X}; \mathcal{N}(G)) < \infty$ and $b_{p-1}^{(2)}(G \times_{\phi \circ i} \widetilde{X}; \mathcal{N}(G)) < \infty$ in Theorem 6.63 and the condition $b_1^{(2)}(\widetilde{F}) < \infty$ in Corollary 6.72 are necessary.

6.17.   Let M be the connected sum $M_1 \# \dots \# M_r$ of (compact connected orientable) non-exceptional prime 3-manifolds M_j. Assume that $\pi_1(M)$ is infinite. Show

$$b_1^{(2)}(\widetilde{M}) = -\chi^{(2)}(\pi_1(M));$$
$$b_2^{(2)}(\widetilde{M}) = \chi(M) - \chi^{(2)}(\pi_1(M));$$
$$b_p^{(2)}(\widetilde{M}) = 0 \qquad \text{for } p \neq 1, 2.$$

6.18. Let G be a group for which there is a d-dimensional G-CW-model for $E(G, \mathcal{FIN})$. Show $b_p^{(2)}(G) = 0$ for $p > d$.

6.19.   Let G be a group for which there is a finite G-CW-model for $E(G, \mathcal{FIN})$. Show that then the $\mathbb{C}G$-module $\mathbb{C}$, which is given by $\mathbb{C}$ with the trivial G-operation, has a finite projective $\mathbb{C}G$-resolution P_*. The class $\sum_{n \geq 0}(-1)^p \cdot [P_n] \in K_0(\mathbb{C}G)$ is independent of the choice of P_* and denoted by $[\mathbb{C}]$. Prove that it is the image of $\chi^G(E(G; \mathcal{FIN}))$ under the homomorphism $A(G) \to K_0(\mathbb{C}G)$, which sends a finite proper G-set S to the class of the finitely generated projective $\mathbb{C}G$-module $\mathbb{C}[S]$ given by the complex vector space with S as basis.

6.20.   Let X be a G-CW-complex which is contractible (after forgetting the group action). Suppose that each isotropy group G_x is finite or satisfies $b_p^{(2)}(G_x) = 0$ for $p \geq 0$. Prove $b_p^{(2)}(X; \mathcal{N}(G)) = b_p^{(2)}(G)$ for $p \geq 0$.

6.21.   Equip $\mathbb{Z}[\mathbb{Z}/5]/(N) \cong \mathbb{Z}[\exp(2\pi i/5)]$ with the obvious $\mathbb{Z}/5$-action, where $N = \sum_{i=0}^{4} t^i$ for $t \in \mathbb{Z}/5$ a fixed generator and (N) is the $\mathbb{Z}[\mathbb{Z}/5]$-ideal generated by N. Let G be the associated semidirect product. Compute $A(G)$ and $\chi^G(E(G, \mathcal{FIN}))$ explicitly.

6.22.   Let F_g be the free group of rank $g \geq 1$. Using the fact that $\mathbb{C}[F_g]$ is a so called fir (= free ideal ring), i.e. any submodule of a free module is free again (see [117, Corollary 3 on page 68]), show that for a F_g-CW-complex X of a finite type each $\mathbb{C}[F_g]$-module $H_p^{\text{sing}}(X; \mathbb{C})$ has a 1-dimensional finite free resolution. Then prove that $b_p^{(2)}(X; \mathcal{N}(F_g))$ and $\alpha_{p+1}(X; \mathcal{N}(F_g))$ depend only on the $\mathbb{C}[F_g]$-isomorphism class of $H_p^{\text{sing}}(X; \mathbb{C})$.

6.23. Show for the group $G = \prod_{i=1}^{\infty} \mathbb{Z} * \mathbb{Z}$ that $H_p^G(EG; \mathcal{N}(G)) = 0$ for $p \geq 0$.

6.24.   Give a counterexample to the following statement by inspecting the special case $G = H = \mathbb{Z}$ and $X = Y = \widetilde{S^1}$: If X is a G-CW-complex of finite type and Y an H-CW-complex of finite type, then there is an isomorphism

$$H_n^{G \times H}(X \times Y; \mathcal{N}(G \times H))$$
$$\cong \bigoplus_{p+q=n} \mathcal{N}(G \times H) \otimes_{\mathcal{N}(G) \otimes_{\mathbb{C}} \mathcal{N}(H)} \left(H_p^G(X; \mathcal{N}(G)) \otimes_{\mathbb{C}} H_q^H(Y; \mathcal{N}(H)) \right).$$

6.25. Prove Theorem 6.104.

6.26. Show that Theorem 6.54 (5) follows from Theorem 6.104.

7. Applications to Groups

Introduction

In this chapter we apply the results of Chapter 6 to questions about group theory, mainly about deficiency and Euler characteristic.

In Section 7.1 we investigate the class $\mathcal{B}_1$ and $\mathcal{B}_\infty$ respectively of groups for which $b_p^{(2)}(G)$ vanishes for $p = 0, 1$ and for $p \geq 0$ respectively. The classes $\mathcal{B}_1$ and $\mathcal{B}_\infty$ turn out to be surprisingly large. For instance we prove that a group containing a normal infinite amenable subgroup belongs to $\mathcal{B}_\infty$. We also show that a group G belongs to $\mathcal{B}_1$ if it is an extension $1 \to H \to G \to K \to 1$ of an infinite finitely generated group H and a group K which is infinite elementary amenable or which contains an element of infinite order or which contains finite subgroups of arbitrary large order.

The motivation to investigate the class $\mathcal{B}_1$ is that a finitely presented group which belongs to $\mathcal{B}_1$ has the following two properties. Its deficieny satisfies $\operatorname{def}(G) \leq 1$, and for any closed oriented smooth 4-manifold M with $\pi_1(M) = G$ we have $|\operatorname{sign}(M)| \leq \chi(M)$. This will be explained in Section 7.3. A survey on deficiency is presented in Subsection 7.3.1.

In Section 7.2 we discuss the L^2-Euler characteristic of a group. It vanishes if G belongs to $\mathcal{B}_\infty$ and satisfies all the properties which are known for the classical Euler characteristic $\chi(BG)$ of a group for which BG has a finite CW-model. The classical Euler characteristic and the virtual Euler characteristic due to Wall are special cases of the L^2-Euler characteristic.

In Section 7.4 we define for a group automorphism $f \colon G \to G$ of a group with finite model for BG a real number $\rho^{(2)}(f)$ using L^2-torsion. It behaves like the Euler characteristic of BG. Applied to $\pi_1(f)$ for a pseudo-Anosov self-homeomorphism $f \colon S \to S$ of a closed hyperbolic oriented surface S it detects the volume of the hyperbolic 3-manifold given by the mapping torus of f.

7.1 Groups with Vanishing L^2-Betti Numbers

In this section we investigate the following classes of groups. Recall that we have introduced the L^2-Betti numbers of a group G in Definition 6.50.

Definition 7.1. *Let d be a non-negative integer or $d = \infty$. Define the class of groups*

$$\mathcal{B}_d := \{G \mid b_p^{(2)}(G) = 0 \text{ for } 0 \le p \le d\}.$$

Notice that $\mathcal{B}_0$ is the class of infinite groups by Theorem 6.54 (8b).

7.1.1 General Criterions for the Vanishing of the L^2-Betti Numbers of a Group

Theorem 7.2. *Let d be a non-negative integer or $d = \infty$. Then*

(1) The class $\mathcal{B}_\infty$ contains all infinite amenable groups;

(2) If G contains a normal subgroup H with $H \in \mathcal{B}_d$, then $G \in \mathcal{B}_d$;

(3) If G is the union of a directed system of subgroups $\{G_i \mid i \in I\}$ such that each G_i belongs to $\mathcal{B}_d$, then $G \in \mathcal{B}_d$;

(4) Suppose that there are groups G_1 and G_2 and group homomorphisms $\phi_i : G_0 \to G_i$ for $i = 1, 2$ such that ϕ_1 and ϕ_2 are injective, G_0 belongs to $\mathcal{B}_{d-1}$, G_1 and G_2 belong to $\mathcal{B}_d$ and G is the amalgamated product $G_1 *_{G_0} G_2$ with respect to ϕ_1 and ϕ_2. Then G belongs to $\mathcal{B}_d$;

(5) Let $1 \to H \xrightarrow{i} G \xrightarrow{p} K \to 1$ be an exact sequence of groups such that $b_p^{(2)}(H)$ is finite for all $p \le d$. Suppose that K is infinite amenable or suppose that BK has finite d-skeleton and there is an injective endomorphism $j : K \to K$ whose image has finite index, but is not equal to K. Then $G \in \mathcal{B}_d$;

(6) Let $1 \to H \xrightarrow{i} G \xrightarrow{p} K \to 1$ be an exact sequence of groups such that $H \in \mathcal{B}_{d-1}$, $b_d^{(2)}(H) < \infty$ and K contains an element of infinite order or finite subgroups of arbitrary large order. Then $G \in \mathcal{B}_d$;

(7) Let $1 \to H \xrightarrow{i} G \xrightarrow{p} K \to 1$ be an exact sequence of infinite countable groups such that $b_1^{(2)}(H) < \infty$. Then $G \in \mathcal{B}_1$.

Proof. (1) This has already been proved in Corollary 6.75.

(2) This follows from Theorem 6.54 (7) and Lemma 6.66 applied to the fibration $BH \to BG \to B(G/H)$ and the obvious isomorphism $\phi : \pi_1(BG) \to G$.

(3) Inspecting for instance the bar-resolution or the infinite join model for EG, one sees that EG is the colimit of a directed system of G-CW-subcomplexes of the form $G \times_{G_i} EG_i$ directed by I. Hence

$$H_p^G(EG; \mathcal{N}(G)) = \operatorname{colim}_{i \in I} H_p^G(G \times_{G_i} EG_i; \mathcal{N}(G)).$$

Now the claim follows from Theorem 6.13 about dimension and colimits and Theorem 6.54 (7) about compatibility of L^2-Betti numbers with induction.

(4) Using the Seifert-van Kampen Theorem one easily checks that there is a G-pushout

$$G \times_{G_0} EG_0 \longrightarrow G \times_{G_1} EG_1$$

$$\downarrow \qquad\qquad\qquad \downarrow$$

$$G \times_{G_2} EG_2 \longrightarrow \qquad EG$$

Now apply Theorem 6.54 (7) and Additivity (see Theorem 6.7 (4b)) to the associated long exact homology sequence for $H_*^G(-, \mathcal{N}(G))$.

(5) This will be proved in Theorem 7.4 (5) and (7).

(6) This follows from Theorem 6.54 (7) and Theorem 6.67 applied to the fibration $BH \to BG \to BK$ and the obvious isomorphism $\phi \colon \pi_1(BG) \to G$.

(7) This is proved by Gaboriau [214, Theorem 6.8].   □

Next we prove a fibered version of Theorem 7.2.

Definition 7.3. *Let d be a non-negative integer or $d = \infty$. Define $\mathcal{BQ}_d$ to be the class of groups G such that for any extension of groups $1 \to H \to K \to G \to 1$ with $b_p^{(2)}(H) < \infty$ for $p \le d$ the group K belongs to $\mathcal{B}_d$. Define $\mathcal{BF}_d$ to be the class of groups G with the property that for any fibration $F \to E \to BG$ for which F is path-connected and $b_p^{(2)}(\widetilde{F}) < \infty$ holds for $p \le d$, we have $b_p^{(2)}(\widetilde{E}) = 0$ for $p \le d$.*

Theorem 7.4. *Let d be a non-negative integer or $d = \infty$. Then*

(1) $\mathcal{BF}_d \subset \mathcal{BQ}_d \subset \mathcal{B}_d$;

(2) If G contains a normal subgroup H which belongs to $\mathcal{BQ}_d$ or $\mathcal{BF}_d$ respectively , then G belongs to $\mathcal{BQ}_d$ or $\mathcal{BF}_d$ respectively;

(3) If G is the union of a directed system of subgroups $\{G_i \mid i \in I\}$ such that each G_i belongs to $\mathcal{BQ}_d$ or $\mathcal{BF}_d$ respectively, then G belongs $\mathcal{BQ}_d$ or $\mathcal{BF}_d$ respectively;

(4) Suppose that there are groups G_1 and G_2 and group homomorphisms $\phi_i \colon G_0 \to G_i$ for $i = 1, 2$ such that ϕ_1 and ϕ_2 are injective, G_0 belongs to $\mathcal{BQ}_{d-1}$ or $\mathcal{BF}_{d-1}$ respectively, G_1 and G_2 belong to $\mathcal{BQ}_d$ or $\mathcal{BF}_d$ respectively and G is the amalgamated product $G_1 *_{G_0} G_2$ with respect to ϕ_1 and ϕ_2. Then G belongs to $\mathcal{BQ}_d$ or $\mathcal{BF}_d$ respectively;

(5) Suppose that BG has finite d-skeleton and that there is an injective endomorphism $j \colon G \to G$ whose image has finite index, but is not equal to G. Then G belongs to $\mathcal{BF}_d$;

(6) The class $\mathcal{BF}_d$ contains all infinite elementary amenable groups.

(7) The class $\mathcal{BQ}_d$ contains all infinite amenable groups.

Proof. (1) This is obvious.

(2) This follows for $\mathcal{BQ}_d$ from Theorem 7.2 (2) applied to $p^{-1}(H) \subset L$ for a given extension $1 \to K \to L \xrightarrow{p} G \to 1$ with $b_p^{(2)}(K) < \infty$ for $p \le d$. For $\mathcal{BF}_d$, the pullback construction for $BH \to BG$ yields a fibration $F \to E_0 \to BH$ and by hypothesis $b_p^{(2)}(\widetilde{E_0}) = 0$ for $p \le d$. Since F is path-connected, both E_0 and E are path-connected. We obtain an exact sequence $1 \to \pi_1(E_0) \xrightarrow{\pi_1(i)}$

$\pi_1(E) \to G/H \to 1$ for $i: E_0 \to E$ the inclusion. The $\pi_1(E_0)$-space $\widetilde{E_0}$ is the restriction of the $\pi_1(E)$ space $\widetilde{E}$ with $\pi_1(i) : \pi_1(E_0) \to \pi_1(E)$. Now we get $b_p^{(2)}(\widetilde{E}) = 0$ for $p \leq d$ from Theorem 6.54 (6b).

(3) This follows for $\mathcal{BQ}_d$ from Theorem 7.2 (3). For $\mathcal{BF}_d$, we can arrange that BG is the directed union of CW-subcomplexes BG_i. Let $F \to E_i \to BG_i$ be the restriction of a given fibration $F \to E \to BG$ with $b_p^{(2)}(\widetilde{F}) < \infty$ for $p \leq d$. By inspecting the maps between the long homotopy sequences associated to these fibrations one shows that $\pi_1(E_i) \to \pi_1(E)$ is injective for all $i \in I$. Obviously E is the union of the E_i. Since any compact subset of BG is contained in a finite CW-subcomplex and hence in one of the BG_i, any compact subset of E is contained in one of the E_i. This implies

$$H_p^{\pi_1(E)}(\widetilde{E}; \mathcal{N}(\pi_1(E)))$$
$$= \mathrm{colim}_{i \in I} \, H_p^{\pi_1(E)}(\pi_1(E) \times_{\pi_1(E_i)} \widetilde{E_i}; \mathcal{N}(\pi_1(E))). \tag{7.5}$$

Because of Theorem 6.13 (1) about dimension and colimits and Theorem 6.54 (7) it suffices to show $b_p^{(2)}(\widetilde{E_i}) = 0$ for $p \leq d$ and $i \in I$. But this follows from the assumption $G_i \in \mathcal{BF}_d$.

(4) This follows for $\mathcal{BQ}_d$ from Theorem 7.2 (4) since for an epimorphism $p: K \to G$ we can write K as the amalgamated product $p^{-1}(G_1) *_{p^{-1}(G_0)} p^{-1}(G_2)$. The proof for $\mathcal{BF}_d$ is analogous (using [326, Lemma 1.26 on page 19]).

(5) Fix an integer $n \geq 1$. Put $G_n = \mathrm{im}(j^n)$. If k is the index of $\mathrm{im}(j)$ in G, then k^n is the index of G_n in G. Let $F \to E \to BG$ be a fibration with $b_p^{(2)}(\widetilde{F}) < \infty$ for $p \leq d$. We get a k^n-sheeted covering $BG_n \to BG$. The pullback construction yields a fibration $F \to E_n \to BG_n$ together with a k^n-sheeted covering $E_n \to E$. We conclude from Theorem 6.54 (6b)

$$b_p^{(2)}(\widetilde{E}) = \frac{b_p^{(2)}(\widetilde{E_n})}{k^n}. \tag{7.6}$$

Let i_p be the number of p-cells in BG. Since G_n is isomorphic to G, we get from the Leray-Serre spectral sequence applied to $E_n \to BG_n$ and Additivity (see Theorem 6.7 (4b))

$$b_p^{(2)}(\widetilde{E_n}) \leq \sum_{q=0}^{p} b_q^{(2)}(\widetilde{F}) \cdot i_{p-q}. \tag{7.7}$$

Equations (7.6) and (7.7) imply

$$b_p^{(2)}(\widetilde{E}) = \frac{\sum_{q=0}^{p} b_q^{(2)}(\widetilde{F}) \cdot i_{p-q}}{k^n}. \tag{7.8}$$

Since $k > 1$ and (7.8) holds for all $n \geq 1$ and $\sum_{q=0}^{p} b_q^{(2)}(\widetilde{F}) \cdot i_{p-q}$ is finite for $p \leq d$ by assumption, assertion (5) follows.

(6) We first show $G \in \mathcal{BF}_\infty$ provided that G is infinite and locally finite. Let $F \to E \to BG$ be a fibration with $b_p^{(2)}(\widetilde{F}) < \infty$ for $p \leq d$. We still have (7.5), if we take $\{G_i \mid i \in I\}$ as the system of finite subgroups of G. From Theorem 6.54 (7) and (6b) we conclude

$$b_p^{(2)}(\pi_1(E) \times_{\pi_1(E_i)} \widetilde{E_i}; \mathcal{N}(\pi_1(E))) = b_p^{(2)}(\widetilde{E_i}) = \frac{b_p^{(2)}(\widetilde{F})}{|G_i|}.$$

Since G is infinite and locally finite, $|G_i|$ becomes arbitrary large for appropriate $i \in I$. Additivity (see Theorem 6.7 (4b)) and Theorem 6.13 (2) about dimension and colimits implies together with (7.5) that $b_p^{(2)}(\widetilde{E}) = 0$ for $p \leq d$.

We want to show that the class of elementary amenable groups is contained in $\mathcal{FIN} \cup \mathcal{BF}_d$, where $\mathcal{FIN}$ is the class of finite groups. By Lemma 10.40 it suffices to show the following two claims. i.) If all finitely generated subgroups of G belong to $\mathcal{FIN} \cup \mathcal{BF}_d$, then $G \in \mathcal{FIN} \cup \mathcal{BF}_d$, and ii.) for any extension $1 \to H \to G \to K \to 1$, for which $H \in \mathcal{FIN} \cup \mathcal{BF}_d$ and K contains $\mathbb{Z}^n$ as normal subgroup of finite index for some $n \geq 0$, we have $G \in \mathcal{FIN} \cup \mathcal{BF}_d$. We begin with i.). We have already shown for locally finite G that G belongs to $\mathcal{FIN} \cup \mathcal{BF}_d$. It remains to treat the case where G is not locally finite. Then G can be written as the union of the directed system of its infinite finitely generated subgroups G_i. By induction hypothesis each G_i belongs to $\mathcal{BF}_d$. Then $G \in \mathcal{BF}_d$ by assertion (3). Finally we prove ii.) If H is finite, G contains $\mathbb{Z}^n$ for some $n \geq 0$ as normal subgroup of finite index and hence belongs to $\mathcal{FIN} \cup \mathcal{BF}_d$ by assertions (2) and (5). It remains to treat the case, where H is infinite and hence by induction hypothesis belongs to $\mathcal{BF}_d$. If F_0 is the fiber of the composition $E \to BG \to BK$, then we obtain a fibration $F \to F_0 \to BH$. Hence $b_p^{(2)}(\widetilde{F_0}) = 0$ for $p \leq d$. From Lemma 6.66 we conclude $b_p^{(2)}(\widetilde{E}) = 0$ for $p \leq d$.

(7) Because of assertion (3) it suffices to treat the case of a finitely generated (and hence countable) amenable group. This case follows from [214, Theorem 6.6]. This finishes the proof of Theorem 7.4. $\qquad\square$

The next question is related to Question 1.95 and arises from Theorem 7.4.

Question 7.9. (Vanishing of L^2-Betti numbers of groups and epimorphism of groups).
Is $\mathcal{B}_d = \mathcal{BQ}_d = \mathcal{BF}_d$?

7.1.2 The Vanishing of the L^2-Betti Numbers of Thompson's Group

Finally we explain the following observation about *Thompson's group* F. It is the group of orientation preserving dyadic PL-automorphisms of $[0, 1]$ where dyadic means that all slopes are integral powers of 2 and the break points are contained in $\mathbb{Z}[1/2]$. It has the presentation

$$F = \langle x_0, x_1, x_2, \ldots \mid x_i^{-1} x_n x_i = x_{n+1} \text{ for } i < n \rangle.$$

This group has some very interesting properties. Its classifying space BF is of finite type [70] but is not homotopy equivalent to a finite dimensional CW-complex since F contains $\mathbb{Z}^n$ as subgroup for all $n \geq 0$ [70, Proposition 1.8]. It is not elementary amenable and does not contain a subgroup which is free on two generators [66], [85]. Hence it is a very interesting question whether F is amenable or not. We conclude from Theorem 7.2 (1) that a necessary condition for F to be amenable is that $b_p^{(2)}(F)$ vanishes for all $p \geq 0$. This motivates the following result.

Theorem 7.10. (L^2-Betti numbers of Thompson's group).
All L^2-Betti numbers $b_p^{(2)}(F)$ of Thompson's group F vanish.   □

Proof. There is a subgroup $F_1 \subset F$ together with a monomorphism $\Phi \colon F_1 \to F_1$ such that F_1 is isomorphic to F and F is the HNN-extension of F_1 with respect to Φ with one stable letter [70, Proposition 1.7 on page 370]. From the topological description of HNN-extensions [350, page 180] we conclude that F is the fundamental group of the mapping torus $T_{B\Phi}$ of the map $B\Phi \colon BF_1 \to BF_1$ induced by Φ. The inclusion $BF_1 \to T_{B\Phi}$ induces on the fundamental groups the inclusion of F_1 in F. One easily checks that the cellular $\mathbb{Z}F$-chain complex of the universal covering $\widetilde{T_{B\Phi}}$ of $T_{B\Phi}$ is the mapping cone of a certain $\mathbb{Z}F$-chain map from $\mathbb{Z}F \otimes_{\mathbb{Z}F_1} C_*(EF_1)$ to itself. Since $\mathbb{Z}F$ is free over $\mathbb{Z}F_1$, we conclude for $p \geq 1$

$$H_p(\mathbb{Z}F \otimes_{\mathbb{Z}F_1} C_*(EF_1)) = \mathbb{Z}F \otimes_{\mathbb{Z}F_1} H_p(C_*(EF_1)) = 0.$$

This implies $H_p(\widetilde{T_{B\Phi}}; \mathbb{Z}) = 0$ for $p \geq 2$. Hence $T_{B\Phi}$ is a model for BF. Now the claim follows from Theorem 1.39.   □

7.2 Euler Characteristics of Groups

We have introduced the L^2-Euler characteristic $\chi^{(2)}(G)$ of a group already in Definition 6.79. It encompasses the rational valued virtual Euler characteristic of Wall (see Remark 6.81). We have related it to the equivariant Euler characteristic of $E(G, \mathcal{FIN})$ provided that there is a finite G-CW-model for $E(G, \mathcal{FIN})$. Namely, Lemma 6.93 implies

$$\chi^{(2)}(G) = \mathrm{ch}_{\{1\}}^G(\chi^G(E(G, \mathcal{FIN}))), \tag{7.11}$$

where $\mathrm{ch}_{\{1\}}^G \colon A(G) \to \mathbb{Q}$ sends $[G/H]$ to $|H|^{-1}$.

Lemma 7.12. *If G belongs to $\mathcal{B}_\infty$, then $\chi^{(2)}(G) = 0$.*

Recall that we have given criterions for $G \in \mathcal{B}_\infty$ in Theorem 7.2. Now it becomes clear why it is worth while to extend the classical notion of the Euler characteristic $\chi(G) := \chi(BG)$ for groups G with finite BG to arbitrary groups. For instance it may very well happen for a group G with finite BG that G contains a normal group H which is not even finitely generated and has in particular no finite model for BH and which belongs to $\mathcal{B}_\infty$ (for instance, H is amenable). Then the classical Euler characteristic is not defined any more for H, but we can still conclude that the classical Euler characteristic of G vanishes.

The standard product and amalgamation formulas for the classical Euler characteristic carry over to the L^2-Euler characteristic. Namely, let G_0, G_1 and G_2 be groups with $h^{(2)}(G_i) < \infty$ for $i = 0, 1, 2$ and $\phi_i : G_0 \to G_1$ be injective group homomorphisms for $i = 1, 2$. Then the direct product $G_1 \times G_2$ and the amalgamated product $G_1 *_{G_0} G_2$ with respect to the homomorphisms ϕ_1 and ϕ_2 satisfy

$$h^{(2)}(G_1 *_{G_0} G_2) < \infty;$$
$$\chi^{(2)}(G_1 *_{G_0} G_2) = \chi^{(2)}(G_1) + \chi^{(2)}(G_2) - \chi^{(2)}(G_0); \qquad (7.13)$$
$$h^{(2)}(G_1 \times G_2) < \infty;$$
$$\chi^{(2)}(G_1 \times G_2) = \chi^{(2)}(G_1) \cdot \chi^{(2)}(G_1). \qquad (7.14)$$

This follows from Theorem 6.80 (2) and (6). More information about the classical Euler characteristic and the virtual Euler characteristic of a group can be found in [69, Chapter IX].

7.3 Deficiency of Groups

7.3.1 Survey on Deficiency of Groups

Definition 7.15 (Deficiency). *Let G be a finitely presented group. Define its deficiency $\mathrm{def}(G)$ to be the maximum $g(P) - r(P)$, where P runs over all presentations P of G and $g(P)$ is the number of generators and $r(P)$ is the number of relations of a presentation P.*

Next we reprove the well-known fact that the maximum appearing in Definition 7.15 does exist.

Lemma 7.16. *Let G be a group with finite presentation*

$$P = \langle s_1, s_2, \ldots, s_g \mid R_1, R_2, \ldots, R_r \rangle$$

Let $\phi : G \to K$ be any group homomorphism. Then

$$g(P) - r(P) \leq 1 - b_0^{(2)}(K \times_\phi EG; \mathcal{N}(K)) + b_1^{(2)}(K \times_\phi EG; \mathcal{N}(K))$$
$$- b_2^{(2)}(K \times_\phi EG; \mathcal{N}(K)).$$

Proof. Given a presentation P with g generators and r relations, let X be the associated finite 2-dimensional CW-complex. It has one 0-cell, g 1-cells, one for each generator, and r 2-cells, one for each relation. The attaching map of the 2-cell associated to a relation is a map from S^1 to the 1-skeleton, which is a wedge of g 1-dimensional spheres, and given by the word defining the relation. There is an obvious isomorphism from $\pi_1(X)$ to G so that we can choose a map $f \colon X \to BG$ which induces an isomorphism on the fundamental groups. It induces a 2-connected K-equivariant map $\overline{f} \colon K \times_\phi \widetilde{X} \to K \times_\phi \widetilde{EG}$. Theorem 6.54 (1a) implies

$$b_p^{(2)}(K \times_\phi \widetilde{X}; \mathcal{N}(K)) = b_p^{(2)}(K \times_\phi EG; \mathcal{N}(K)) \qquad \text{for } p = 0, 1; \quad (7.17)$$

$$b_2^{(2)}(K \times_\phi \widetilde{X}; \mathcal{N}(K)) \geq b_2^{(2)}(K \times_\phi EG; \mathcal{N}(K)). \qquad (7.18)$$

We conclude from the L^2-Euler-Poincaré formula (see Theorem 6.80 (1)) and from (7.17) and (7.18)

$$
\begin{aligned}
g - r &= 1 - \chi^{(2)}(K \times_\phi \widetilde{X}; \mathcal{N}(K)) \\
&= 1 - b_0^{(2)}(K \times_\phi \widetilde{X}; \mathcal{N}(K)) + b_1^{(2)}(K \times_\phi \widetilde{X}; \mathcal{N}(K)) \\
&\qquad - b_2^{(2)}(K \times_\phi \widetilde{X}; \mathcal{N}(K)) \\
&\leq 1 - b_0^{(2)}(K \times_\phi EG; \mathcal{N}(K)) + b_1^{(2)}(K \times_\phi EG; \mathcal{N}(K)) \\
&\qquad - b_2^{(2)}(K \times_\phi EG; \mathcal{N}(K)). \qquad \square
\end{aligned}
$$

Example 7.19. We give some examples of groups, where the deficiency is realized by the "obvious" presentation.

The free group F_g of rank g has the obvious presentation $\langle s_1, s_2, \ldots, s_g \mid \emptyset \rangle$ and its deficiency is realized by this presentation, namely $\operatorname{def}(F_g) = g$. This follows from Lemma 7.16 because of $b_0^{(2)}(F_g) = 0$ and $b_1^{(2)}(F_g) = g - 1$. One also can apply the analog of Lemma 7.16 for the classical Betti numbers instead of the L^2-Betti numbers since $b_0(F_g) = 1$ and $b_1(F_g) = g$.

If G is a finite group, $\operatorname{def}(G) \leq 0$ by Lemma 7.16 because we get for the classical Betti numbers $b_0(G) = 1$ and $b_1(G) = 0$ or because we get for the L^2-Betti numbers $b_0^{(2)}(G) = |G|^{-1}$ and $b_1^{(2)}(G) = 0$. The deficiency of a cyclic group $\mathbb{Z}/n$ is 0, the obvious presentation $\langle s \mid s^n \rangle$ realizes the deficiency. It is not hard to check using homology with coefficients in the finite field $\mathbb{F}_p$ of prime order p that the deficiency of $\mathbb{Z}/n \times \mathbb{Z}/n$ is -1. The obvious presentation $\langle s, t \mid s^n, t^n, [s, t] \rangle$ realizes the deficiency, where $[s, t]$ denotes the commutator $s t s^{-1} t^{-1}$.

The inequality in Lemma 7.16 is actually an equality and in particular $\operatorname{def}(G) = 1 - \chi(BG)$ if BG is a finite 2-dimensional CW-complex. If G is a torsion-free one-relator group, the 2-dimensional CW-complex associated with any presentation with one relation and g generators is aspherical and hence BG is a finite 2-dimensional CW-complex [350, chapter III §§9 -11] and G has deficiency $g - 1$.

We conjecture that for a torsion-free group having a presentation with $g \geq 2$ generators and one non-trivial relation

$$b_2^{(2)}(G) = 0;$$
$$b_1^{(2)}(G) = \operatorname{def}(G) - 1 = g - 2$$

holds (compare [237, page 235]). This would follow from the strong Atiyah Conjecture 10.2, which says that the L^2-Betti numbers of the universal covering of a finite CW-complex with torsion-free fundamental group are integers, by the following argument. Namely, the kernel of the second differential of the L^2-chain complex of BG is a proper Hilbert $\mathcal{N}(G)$-submodule of $l^2(G)$ so that its dimension $b_2^{(2)}(G)$ is less than one and hence by the Atiyah Conjecture zero. Since G must be infinite and hence $b_0^{(2)}(G) = 0$ (see Theorem 6.54 (8b)), the Euler-Poincaré formula (see Theorem 6.80 (1)) implies $b_1^{(2)}(G) = g - 2$.

The following result is a direct consequence of [177, Theorem 2.5]. It is proved using homology with coefficients in $\mathbb{Z}/2$.

Theorem 7.20. *Let M be a connected compact orientable 3-manifold with fundamental group π and prime decomposition*

$$M = M_1 \# M_2 \# \ldots \# M_r.$$

Let $s(M)$ be the number of prime factors M_i with non-empty boundary and $t(M)$ be the number of prime factors which are S^2-bundles over S^1. Denote by $\chi(M)$ the Euler characteristic. Then

$$\operatorname{def}(\pi_1(M)) = \dim_{\mathbb{Z}/2}(H_1(\pi; \mathbb{Z}/2)) - \dim_{\mathbb{Z}/2}(H_2(\pi; \mathbb{Z}/2))$$

$$= s(M) + t(M) - \chi(M).\square$$

Example 7.21. One may expect that the deficiency is additive under free products. This is not true as the following example, which is taken from [264, Theorem 3 on page 162], shows. It plays an important role in the construction of a counterexample up to homotopy of the Kneser Conjecture in dimension four [298]. There, a closed connected orientable smooth 4-manifold M is constructed whose fundamental group is the free product of two non-trivial groups such that M is not homotopy equivalent to $M_0 \# M_1$ unless M_0 or M_1 is homeomorphic to S^4. We mention that a stable version of the Kneser Conjecture remains true in dimension four [299], where stable means that one has to allow connected sums with copies of $S^2 \times S^2$.

Suppose that m_i, r_i, n_i and q_i for $i = 0, 1$ are integers satisfying

$$r_i > 1, \ r_i^{m_i} - 1 = n_i q_i, \ r_i \equiv 1 \bmod n_i, \ (m_i, n_i) \neq 1, \ (q_0, q_1) = 1.$$

Then the group

$$G = (\mathbb{Z}/m_0 \times \mathbb{Z}/n_0) * (\mathbb{Z}/m_1 \times \mathbb{Z}/n_1)$$

has the "obvious" presentation

$$G = \langle a_0, b_0, a_1, b_1 \mid a_0^{m_0}, b_0^{n_0}, [a_0, b_0], a_1^{m_1}, b_1^{n_1}, [a_1, b_1] \rangle.$$

But its deficiency is not realized by this presentation. Namely its deficiency is -1 and is realized by the following presentation

$$G = \langle a_0, b_0, a_1, b_1 \mid$$
$$a_0^{m_0} = 1, [a_0, b_0] = b_0^{r_0 - 1}, a_1^{m_1} = 1, [a_1, b_1] = b_1^{r_1 - 1}, b_0^{n_0} = b_1^{n_1} \rangle.$$

To show that this is indeed a presentation of G, it suffices to show that the relation $b_0^{n_0} = 1$ follows from the other relations. We start by proving inductively for $k = 1, 2, \ldots$ the relation $a_i^k b_i a_i^{-k} = b_i^{-r_i^k}$ for $i = 0, 1$. The induction step follows from the calculation

$$a_i^{k+1} b_i a_i^{-(k+1)} = a_i a_i^k b_i a_i^{-k} a_i^{-1} = a_i b_i^{r_i^k} a^{-1} = \left(a_i b_i a_i^{-1}\right)^{r_i^k} = \left(b_i^{r_i}\right)^{r_i^k} = b_i^{r_i^{k+1}}.$$

This implies for $k = m_i$ and $i = 0, 1$

$$(b_i^{n_i})^{q_i} = b_i^{r_i^{m_i} - 1} = 1.$$

Since $b_0^{n_0} = b_1^{n_1}$ holds, we conclude

$$(b_0^{n_0})^{q_0} = (b_0^{n_0})^{q_1} = 1.$$

Since q_0 and q_1 are prime, we get $b_0^{n_0} = 1$.

Notice that groups appearing in the Example 7.21 above contain torsion. It may still be true that the deficiency is additive under free products of torsionfree groups.

Finally we mention the following result [30, Theorem 2] which is in a certain sense complementary to our results (see Lemma 7.22 and Theorem 7.25). If G is a finitely presented group with $\mathrm{def}(G) \geq 2$, then G can be written as an amalgamated product $G = A *_C B$, where A, B and C are finitely generated, C is proper subgroup of both A and B and has index greater than two in A or B. In particular G contains a free subgroup of rank 2 and is not amenable. This implies that an amenable finitely presented group has deficiency less or equal to one (see also [28], [160, Corollary 2.5']). This also follows from Theorem 7.25. We mention that not every group of this particular shape $A *_C B$ has deficiency ≥ 2, take for example $\mathbb{Z} *_{3 \cdot \mathbb{Z}} \mathbb{Z} = \langle s, t \mid t^3 = s^3 \rangle$ which has deficiency 1. This follows from Lemma 7.22 (1) and Theorem 7.2 (4).

Another test for bounds on deficiencies is given in [349] using Fox ideals.

7.3.2 Applications of L^2-Betti Numbers to Deficiency and to Signatures of 4-Manifolds

Lemma 7.22. *Let G be a finitely presented group and $\phi\colon G \to K$ a homomorphism such that $b_1^{(2)}(K \times_\phi EG; \mathcal{N}(G)) = 0$. Then*

(1) $\operatorname{def}(G) \leq 1$;
(2) Let M be a closed oriented 4-manifold with G as fundamental group. Then
$$|\operatorname{sign}(M)| \leq \chi(M);$$

Proof. (1) Follows directly from Lemma 7.16.

(2) By the L^2-Signature Theorem (see [9]) applied to the regular covering $\overline{M} \to M$ associated to ϕ, the signature $\sigma(M)$ is the difference of the von Neumann dimensions of two complementary subspaces of the space of L^2-integrable harmonic smooth 2-forms $\mathcal{H}^p_{(2)}(\overline{M})$ and hence
$$|\operatorname{sign}(M)| \leq \dim_{\mathcal{N}(K)}(\mathcal{H}^p_{(2)}(\overline{M})).$$

We conclude from the L^2-Hodge-de Rham Theorem 1.59 and Lemma 6.53
$$|\operatorname{sign}(M)| \leq b_2^{(2)}(\overline{M}, \mathcal{N}(K)). \tag{7.23}$$

We get from the assumption and Poincaré duality (see Theorem 1.35 (3) together with Lemma 6.53) that $b_p^{(2)}(K \times_\phi EG; \mathcal{N}(G)) = 0$ for $p = 1, 3$. The Euler-Poincaré formula (see Theorem 1.35 (2)) implies
$$\chi(M) = \sum_{j=0}^{2} b_{2j}^{(2)}(\overline{M}, \mathcal{N}(K)). \tag{7.24}$$

Now assertion (2) follows from (7.23) and (7.24). $\square$

Theorem 7.25. *Let $1 \to H \xrightarrow{i} G \xrightarrow{q} K \to 1$ be an exact sequence of infinite groups. Suppose that G is finitely presented and one of the following conditions is satisfied.*

(1) $b_1^{(2)}(H) < \infty$;
(2) The ordinary first Betti number of H satisfies $b_1(H) < \infty$ and K belongs to $\mathcal{B}_1$;

Then

(i) $\operatorname{def}(G) \leq 1$;
(ii) Let M be a closed oriented 4-manifold with G as fundamental group. Then
$$|\operatorname{sign}(M)| \leq \chi(M).$$

Proof. If condition (1) is satisfied, then $b_p^{(2)}(G) = 0$ for $p = 0, 1$ by Theorem 7.2 (7), and the claim follows from Lemma 7.22.

Suppose that condition (2) is satisfied. There is a spectral sequence converging to $H_{p+q}^K(K \times_q EG; \mathcal{N}(K))$ with E^2-term

$$E_{p,q}^2 = \operatorname{Tor}_p^{\mathbb{C}K}(H_q(BH; \mathbb{C}), \mathcal{N}(K))$$

[518, Theorem 5.6.4 on page 143]. Since $H_q(BH; \mathbb{C})$ is $\mathbb{C}$ with the trivial K-action for $q = 0$ and finite dimensional as complex vector space by assumption for $q = 1$, we conclude $\dim_{\mathcal{N}(K)}(E_{p,q}^2) = 0$ for $p + q = 1$ from the assumption $b_1^{(2)}(K) = 0$ and Lemma 6.33 (2). This implies $b_1^{(2)}(K \times_q EG; \mathcal{N}(K)) = 0$ and the claim follows from Lemma 7.22. $\qquad\square$

Theorem 7.25 generalizes results in [162], [279], where also some other information is given. See also [250], [297]. We mention the result of Hitchin [263] that a connected closed oriented smooth 4-manifold which admits an Einstein metric satisfies the stronger inequality $|\operatorname{sign}(M)| \leq \frac{2}{3} \cdot \chi(M)$.

7.4 Group Automorphisms and L^2-Torsion

In this section we explain that for a group automorphism $f \colon G \to G$ the L^2-torsion applied to the $(G \times_f \mathbb{Z})$-CW-complex $E(G \times_f \mathbb{Z})$ gives an interesting new invariant, provided that G is of det ≥ 1-class and satisfies certain finiteness assumptions, for instance, that there is a finite G-CW-model for EG or more generally for $E(G, \mathcal{FIN})$. We will investigate the basic properties of this invariant.

7.4.1 Automorphisms of Groups G with Finite Models for BG

Let G be a group. We assume that there is a finite CW-model for its classifying space BG and that G is of det ≥ 1-class (see Definition 3.112). By Theorem 13.3 (2) G is of det ≥ 1-class if G belongs to the class $\mathcal{G}$ which will be dealt with in Subsection 13.1.3. It contains all residually amenable groups and in particular all residually finite groups.

Suppose that $f \colon G \to G$ is an automorphism. Let $G_f := G \times_f \mathbb{Z}$ be the semidirect product of G and $\mathbb{Z}$ with respect to the automorphism f. By assumption BG has a finite CW-model. We pick one. Let $Bf \colon BG \to BG$ be the map induced by f which is up to homotopy uniquely determined by the property that $\pi_1(f)$ is conjugate to f under the identification $G \cong \pi_1(BG)$. The mapping torus T_{Bf} is a finite CW-model for $B(G \times_f \mathbb{Z})$, since there is a fibration $BG \to T_{Bf} \to S^1$. We conclude from Theorem 1.39 that $b_p^{(2)}(B\widetilde{(G \times_f \mathbb{Z})}) = 0$ for $p \geq 0$. Since G is of det ≥ 1-class by assumption, $G \times_f \mathbb{Z}$ is of det ≥ 1-class by Theorem 13.3 (3). Notice that the construction of $B(G \times_f \mathbb{Z})$ is unique up to homotopy. We conclude from Lemma

13.6 that $B(\widetilde{G \rtimes_f \mathbb{Z}})$ is a det-L^2-acyclic finite free $(G \rtimes_f \mathbb{Z})$-CW-complex and the L^2-torsion $\rho^{(2)}(B(\widetilde{G \rtimes_f \mathbb{Z}}))$ is well-defined and depends only on the automorphism $f \colon G \to G$.

Definition 7.26 (L^2-torsion of group automorphisms). *Let $f \colon G \to G$ be a group automorphism. Suppose that there is a finite CW-model for BG and G is of det ≥ 1-class. Define the L^2-torsion of f by*

$$\rho^{(2)}(f \colon G \to G) := \rho^{(2)}(B(\widetilde{G \rtimes_f \mathbb{Z}})) \quad \in \mathbb{R}.$$

Next we present the basic properties of this invariant. Notice that its behaviour is similar to the Euler characteristic $\chi(G) := \chi(BG)$.

Theorem 7.27. *Suppose that all groups appearing below have finite CW-models for their classifying spaces and are of det ≥ 1-class.*

(1) Suppose that G is the amalgamated product $G_1 *_{G_0} G_2$ for subgroups $G_i \subset G$ and the automorphism $f \colon G \to G$ is the amalgamated product $f_1 *_{f_0} f_2$ for automorphisms $f_i \colon G_i \to G_i$. Then

$$\rho^{(2)}(f) = \rho^{(2)}(f_1) + \rho^{(2)}(f_2) - \rho^{(2)}(f_0);$$

(2) Let $f \colon G \to H$ and $g \colon H \to G$ be isomorphisms of groups. Then

$$\rho^{(2)}(f \circ g) = \rho^{(2)}(g \circ f).$$

In particular $\rho^{(2)}(f)$ is invariant under conjugation with automorphisms;

(3) Suppose that the following diagram of groups

$$
\begin{array}{ccccccccc}
1 & \longrightarrow & G_1 & \longrightarrow & G_2 & \longrightarrow & G_3 & \longrightarrow & 1 \\
& & \Big\downarrow{f_1} & & \Big\downarrow{f_2} & & \Big\downarrow{\mathrm{id}} & & \\
1 & \longrightarrow & G_1 & \longrightarrow & G_2 & \longrightarrow & G_3 & \longrightarrow & 1
\end{array}
$$

commutes, has exact rows and its vertical arrows are automorphisms. Then

$$\rho^{(2)}(f_2) = \chi(BG_3) \cdot \rho^{(2)}(f_1);$$

(4) Let $f \colon G \to G$ be an automorphism of a group. Then for all integers $n \geq 1$

$$\rho^{(2)}(f^n) = n \cdot \rho^{(2)}(f);$$

(5) Suppose that G contains a subgroup G_0 of finite index $[G : G_0]$. Let $f \colon G \to G$ be an automorphism with $f(G_0) = G_0$. Then

$$\rho^{(2)}(f) = \frac{1}{[G : G_0]} \cdot \rho^{(2)}(f|_{G_0});$$

(6) *Let $f\colon G \to G$ be an automorphism of a group G. Then $\rho^{(2)}(f)$ depends only on the map $H_p^{(2)}(\widetilde{Bf})\colon H_p^{(2)}(\widetilde{BG}) \to H_p^{(2)}(\widetilde{BG})$ induced by f on the L^2-homology of the universal covering of BG. More precisely,*

$$\rho^{(2)}(f) = \sum_{p \geq 0}(-1)^p \cdot \ln\left(\det{}_{G \rtimes_f \mathbb{Z}}\left(\overline{H_p^{(2)}(\widetilde{Bf})}\right)\right),$$

where $j_* H_p^{(2)}(\widetilde{BG})$ is the finitely generated Hilbert $\mathcal{N}(G \rtimes_f \mathbb{Z})$-Hilbert module, which is obtained from the $\mathcal{N}(G)$-Hilbert module $H_p^{(2)}(\widetilde{BG})$ by induction with the canonical inclusion $j\colon G \to G \rtimes_f \mathbb{Z}$, and the morphism $\overline{H_p^{(2)}(\widetilde{Bf})}\colon j_* H_p^{(2)}(\widetilde{BG}) \to j_* H_p^{(2)}(\widetilde{BG})$ is the completion of the map $\mathbb{C}[G \rtimes_f \mathbb{Z}] \otimes_{\mathbb{C}G} H_p^{(2)}(\widetilde{BG}) \to \mathbb{C}[G \rtimes_f \mathbb{Z}] \otimes_{\mathbb{C}G} H_p^{(2)}(\widetilde{BG})$ which sends $\gamma \otimes u$ to $\gamma \otimes u - \gamma t \otimes H_p^{(2)}(\widetilde{Bf})(u)$ for $t \in \mathbb{Z}$ a fixed generator, $\gamma \in \mathbb{C}[G \rtimes_f \mathbb{Z}]$ and $u \in H_p^{(2)}(\widetilde{BG})$;

(7) *We have $\rho^{(2)}(f) = 0$ if G satisfies one of the following conditions:*
   (a) All L^2-Betti numbers of the universal covering of BG vanish;
   (b) G contains an amenable infinite normal subgroup.

Proof. (1) One constructs finite CW-models BG_i for $i = 0, 1, 2$ and BG such that $BG_i \subset BG$ is a CW-subcomplex for $i = 0, 1, 2$ and $BG = BG_1 \cup BG_2$ and $BG_0 = BG_1 \cap BG_2$ and the inclusion $BG_i \to BG$ induces on the fundamental groups the inclusions $G_i \to G$ for $i = 0, 1, 2$. Then one constructs self-homotopy equivalences $Bf_i\colon BG_i \to BG_i$ for $i = 0, 1, 2$ and $Bf\colon BG \to BG$ such that Bf restricts to Bf_i on BG_i and Bf and Bf_i induce on the fundamental groups f and f_i for $i = 0, 1, 2$. Then the mapping torus T_{Bf} contains T_{Bf_i} for $i = 0, 1, 2$ as subcomplex, $T_{Bf} = T_{Bf_1} \cup T_{Bf_2}$ and $T_{Bf_0} = T_{Bf_1} \cap T_{Bf_2}$ and the inclusion $T_{Bf_i} \to T_{Bf}$ induces an injection on the fundamental groups for $i = 0, 1, 2$. The sum formula (see Theorem 3.96) (2) implies

$$\rho^{(2)}(\widetilde{T_{Bf}}) = \rho^{(2)}(\widetilde{T_{Bf_1}}) + \rho^{(2)}(\widetilde{T_{Bf_2}}) - \rho^{(2)}(\widetilde{T_{Bf_0}}).$$

(2) This follows from the fact that for maps $u\colon X \to Y$ and $v\colon Y \to X$ of CW-complexes the mapping tori $T_{u \circ v}$ and $T_{v \circ u}$ are homotopy equivalent [326, (7.31) on page 129].

(3) There is an induced fibration $B(G_1 \rtimes_{f_1} \mathbb{Z}) \to B(G_2 \rtimes_{f_2} \mathbb{Z}) \to BG_3$ such that the inclusion of the fiber into the total space induces the obvious injection on the fundamental groups. Now apply the fibration formula (see Theorem 3.100 and Remark 3.102).

(4) Since there is a n-sheeted covering $T_{(Bf)^n} \to T_{Bf}$, the claim follows from multiplicativity of L^2-torsion under finite coverings (see Theorem 3.96 (5)).

(5) There is a finite covering with $[G : G_0]$-sheets $T_{Bf_0} \to T_{Bf}$ since there is

a bijection from G/G_0 to $(G \rtimes_f \mathbb{Z})/(G_0 \rtimes_{f|_{G_0}} \mathbb{Z})$. Now the assertion follows from multiplicativity of L^2-torsion under finite coverings (see Theorem 3.96 (5)).

(6) This follows from Theorem 3.106.

(7) This follows from assertion (6) and Theorem 7.2 (1) and (2). This finishes the proof of Theorem 7.27. $\square$

7.4.2 Automorphisms of Surfaces

Let S be a compact connected orientable 2-dimensional manifold, possibly with boundary. Let $f \colon S \to S$ be an orientation preserving homeomorphism. The mapping torus T_f is a compact connected orientable 3-manifold whose boundary is empty or a disjoint union of 2-dimensional tori. Then T_f is an irreducible Haken manifold with infinite fundamental group and incompressible boundary if S is different from S^2 and D^2. In this case T_f satisfies Thurston's Geometrization Conjecture, i.e. there is a maximal family of embedded incompressible tori, which are pairwise not isotopic and not boundary parallel, such that it decomposes T_f into pieces, which are Seifert or hyperbolic. Let M_1, M_2, ..., M_r be the hyperbolic pieces. They all have finite volume $\mathrm{vol}(M_i)$. (We have explained the notions and facts above in Section 4.1). Then Theorem 4.3 shows

Theorem 7.28. *If S is S^2, D^2, or T^2, then $\rho^{(2)}(f) = 0$. Otherwise we get*

$$\rho^{(2)}(\pi_1(f) \colon \pi_1(S) \to \pi_1(S)) = \frac{-1}{6\pi} \cdot \sum_{i=1}^{r} \mathrm{vol}(M_i).$$

Let S be a closed orientable hyperbolic surface and $f \colon S \to S$ be a self-homeomorphism. It is called *irreducible* if f is not homotopic to an automorphism which leaves some essential closed 1-dimensional submanifold invariant. Essential means that none of the components is nullhomotopic in S. It is called *periodic* if it is homotopic (or, equivalently, isotopic) to a map $g \colon S \to S$ for which there is a positive integer n with $g^n = \mathrm{id}$. The notion *pseudo-Anosov* in terms of transverse singular foliations can be found for instance in [96, page 95]. It is important to know that the following statements for an irreducible selfhomeomorphism $f \colon S \to S$ are equivalent: i.) f is pseudo-Anosov, ii.) f is not periodic and iii.) the mapping torus T_f is hyperbolic [96, Theorem 6.3], [369, Theorem 3.6 on page 47, Theorem 3.9 on page 50]. We know from Theorem 4.3 that $\rho^{(2)}(\pi_1(f))$ is $-\frac{1}{6\pi} \cdot \mathrm{vol}(T_f)$ and hence different from zero, provided that T_f is hyperbolic. If T_f is not hyperbolic, f must be, up to homotopy, periodic and hence $\rho^{(2)}(\widetilde{T_f}) = 0$ by Theorem 7.27 (4). Hence f is pseudo-Anosov if and only if $\rho^{(2)}(f) < 0$, and f is periodic if and only if $\rho^{(2)}(f) = 0$.

Recall that we can read off $\rho^{(2)}(f) := \rho^{(2)}(\widetilde{T_f})$ from $H_1^{(2)}(\overline{f})$ by Theorem 3.106 and thus, because of Theorem 4.9, from the map induced by f

on the fundamental group $\pi_1(F)$. So we have in principle a procedure to decide whether f is pseudo-Anosov by inspecting the map induced on the fundamental group.

7.4.3 A Combinatorial Approach for the L^2-Torsion of an Automorphism of a Finitely Generated Free Group

In this Subsection we discuss how the combinatorial approach to L^2-torsion of Subsection 3.7 specializes in the case of an automorphism $f \colon F \to F$ of the free group F on r letters $s_1, s_2, \ldots, s_r$. Write $G = F \rtimes_f \mathbb{Z}$ for the semidirect product associated to f. Let $t \in \mathbb{Z}$ be a generator and denote the corresponding element in G also by t. Define a (r, r)-matrix A over $\mathbb{Z}F$ by

$$A = \left(\frac{\partial}{\partial s_j} f(s_i) \right)_{1 \le i, j \le r},$$

where $\frac{\partial}{\partial s_j}$ denotes the Fox derivative. Choose a real number $K > 0$ which is greater than or equal to the operator norm of the morphism

$$r_{I-tA} \colon \bigoplus_{i=1}^{r} l^2(G) \to \bigoplus_{i=1}^{r} l^2(G)$$

given by right multiplication with $(I - tA)$. A possible choice for K is

$$K = \sqrt{(2r-1)r} \cdot \max\{\|1 - tA_{i,j}\|_1 \mid 1 \le i, j \le r\},$$

where $\| \sum_{g \in F \rtimes_f \mathbb{Z}} \lambda_g \cdot g \|_1$ is defined by $\sum_{g \in G} |\lambda_g|$. Denote by A^* the matrix obtained from A by transposing and applying the standard involution $\mathbb{Z}F \to \mathbb{Z}F$, which sends $\sum_{u \in F} \lambda_u \cdot u$ to $\sum_{u \in F} \lambda_u \cdot u^{-1}$, to each entry. Denote by

$$\operatorname{tr}_{\mathbb{Z}G} \colon \mathbb{Z}G \to \mathbb{Z}, \qquad \sum_{g \in G} \lambda_g \cdot g \mapsto \lambda_e$$

the standard trace on $\mathbb{Z}G$, where λ_e is the coefficient of the unit element e in G. It extends to square matrices over $\mathbb{Z}G$ by taking the sum of the traces of the diagonal entries. Define

$$c(A, K)_p = \operatorname{tr}_{\mathbb{Z}G}\left(\left(1 - K^{-2} \cdot (1 - tA)(1 - A^* t^{-1})\right)^p \right).$$

Theorem 7.29. *In the setting above the sequence $c(A, K)_p$ is a monotone decreasing sequence of non-negative real numbers, and the L^2-torsion of f satisfies*

$$\rho^{(2)}(f) = -r \cdot \ln(K) + \frac{1}{2} \cdot \sum_{p=1}^{\infty} \frac{1}{p} \cdot c(A, K)_p \le 0.$$

Proof. Choose $\bigvee_{i=1}^r S^1$ with the obvious base point x as model for BF. Realize f as an endomorphism $Bf \colon BF \to BF$ respecting the base point. Fix a base point $\widetilde{x} \in \widetilde{BF}$ which is sent under the projection $\widetilde{BF} \to BF$ to x. Choose a lift $\widetilde{Bf} \colon \widetilde{BF} \to \widetilde{BF}$ uniquely determined by the property that it sends $\widetilde{x}$ to itself. In the sequel we identify F, $\pi_1(BF)$ and the group of deck transformations of $\widetilde{BF}$ with respect to the given base points x and $\widetilde{x}$. Then $\widetilde{Bf}$ is $(f \colon F \to F)$-equivariant. The cellular $\mathbb{Z}F$-chain complex of $\widetilde{BF}$ looks like

$$\bigoplus_{i=1}^r \mathbb{Z}F \xrightarrow{\sum_{i=1}^r r_{s_i-1}} \mathbb{Z}F$$

where $r_{s_i-1} \colon \mathbb{Z}F \to \mathbb{Z}F$ sends u to $u(s_1 - 1)$. The matrix A defines a $(\mathbb{Z}f \colon \mathbb{Z}F \to \mathbb{Z}F)$-equivariant homomorphism

$$\widehat{r}_A \colon \bigoplus_{i=1}^r \mathbb{Z}F \;\to\; \bigoplus_{i=1}^r \mathbb{Z}F,$$

$$\left(\sum_{u \in F} \lambda_{u,j} \cdot u\right)_{1 \le j \le r} \;\mapsto\; \left(\sum_{u \in F} \sum_j \lambda_{u,j} \cdot f(u) A_{j,k}\right)_{1 \le k \le r}$$

and analogously for the unit (r, r)-matrix I. The $(\mathbb{Z}f \colon \mathbb{Z}F \to \mathbb{Z}F)$-equivariant chain map $C_*(\widetilde{Bf}) \colon C_*(\widetilde{BF}) \to C_*(\widetilde{BF})$ looks like

$$
\begin{array}{ccc}
\bigoplus_{i=1}^r \mathbb{Z}F & \xrightarrow{\sum_{i=1}^r r_{s_i-1}} & \mathbb{Z}F \\[4pt]
\widehat{r}_A \downarrow & & \widehat{r}_I \downarrow \\[4pt]
\bigoplus_{i=1}^r \mathbb{Z}F & \xrightarrow{\sum_{i=1}^r r_{s_i-1}} & \mathbb{Z}F
\end{array}
$$

Then the cellular $\mathbb{Z}G$-chain complex of $\widetilde{T_{Bf}} = \widetilde{BG}$ is the mapping cone of the following chain endomorphism of $\mathbb{Z}G \otimes_{\mathbb{Z}F} C_*(\widetilde{BF})$

$$
\begin{array}{ccc}
\bigoplus_{i=1}^r \mathbb{Z}G & \xrightarrow{\sum_{i=1}^r r_{s_i-1}} & \mathbb{Z}G \\[4pt]
r_{I-tA} \downarrow & & r_{1-t} \downarrow \\[4pt]
\bigoplus_{i=1}^r \mathbb{Z}G & \xrightarrow{\sum_{i=1}^r r_{s_i-1}} & \mathbb{Z}G
\end{array}
$$

where r_{I-tA} and r_{1-t} are given by right multiplication with the square matrices $I - tA$ and $(1 - t)$ over $\mathbb{Z}G$. It is det-L^2-acyclic after tensoring with $l^2(G)$ since G belongs to the class $\mathcal{G}$ (see Definition 13.9) by Lemma 13.11, and Theorem 1.39 and Theorem 13.3 (2) hold. The 1-dimensional $\mathbb{Z}G$-chain complex $\mathbb{Z}F \xrightarrow{r_{1-t}} \mathbb{Z}F$ is det-L^2-acyclic and has trivial L^2-torsion after tensoring with $l^2(G)$ (see (3.24) and Theorem 3.14 (6)). Moreover, it is a subcomplex of the mapping cone of the chain endomorphism above. The weakly exact long

L^2-homology sequence (see Theorem 1.21) and the sum formula (see Theorem 3.35 (1)) imply that the L^2-torsion of f, which is the L^2-torsion of this mapping cone after tensoring with $l^2(G)$, is the L^2-torsion of the quotient $\mathbb{Z}G$-chain complex concentrated in dimension 2 and 1 after tensoring with $l^2(G)$

$$\cdots \to 0 \to \bigoplus_{i=1}^{r} \mathbb{Z}G \xrightarrow{r_{I-tA}} \bigoplus_{i=1}^{r} \mathbb{Z}G \to 0.$$

This implies

$$\rho^{(2)}(f) = -\ln\left(\det\nolimits_{\mathcal{N}(G)}(\mathrm{id}_{l^2(G)} \otimes_{\mathbb{Z}G} r_{I-tA})\right).$$

Since G belongs to the class $\mathcal{G}$ (see Definition 13.9) by Lemma 13.11, Theorem 13.3 (2) implies that $\rho^{(2)}(f)$ is non-positive. The other claims follow from Theorem 3.172 (1) and (4). $\qquad\square$

Suppose that the second Novikov-Shubin invariant $\alpha_2(B(\widetilde{F \rtimes_f \mathbb{Z}}))$ is positive, what is conjectured to be true for any group (see Conjecture 2.82) and proved if f is induced by a surface homeomorphism (see Theorem 4.2 (1)). If α is a number with $0 < \alpha < \alpha_2(B(F \rtimes_f \mathbb{Z}))$, then there is a real number $C > 0$ such that for all p

$$0 \leq \rho^{(2)}(f) + 2r \cdot \ln(K) - \sum_{p=1}^{L} \frac{1}{p} \cdot c(A, K)_p \leq \frac{C}{\alpha^L}$$

holds (see Theorem 3.172 (5)). In other words, the speed of convergence is exponential.

7.4.4 Generalizations

So far we have assumed that there is a finite model for BG. This implies that G is torsionfree. Similar to the notion of the virtual Euler characteristic (see Remark 6.81) of a group G, which possesses a subgroup $G'_0 \subset G$ of finite index with a finite CW-model for BG_0, one can extend the definition of $\rho^{(2)}(f)$ to automorphisms $f\colon G \to G$ of such a group G. Namely, choose a subgroup $G_0 \subset G$ of finite index such that there is a finite CW-model for BG_0 and $f(G_0) = G_0$. Then define

$$\rho^{(2)}_{\mathrm{virt}}(f\colon G \to G) = \frac{\rho^{(2)}(f|_{G_0}\colon G_0 \to G_0)}{[G : G_0]}. \qquad (7.30)$$

Given a subgroup G'_0 of finite index in G with a finite CW-model for BG'_0, one obtains the desired subgroup G_0 by $\bigcap_{n \in \mathbb{Z}} f^n(G'_0)$. We have to show that this is independent of the choice of G_0. Let G_1 be another such choice. Then we obtain from Theorem 7.27 (5) if we put $G_2 = G_0 \cap G_1$

$$\frac{\rho^{(2)}(f|_{G_0})}{[G:G_0]} = \frac{\rho^{(2)}(f|_{G_2})}{[G:G_0] \cdot [G_0:G_2]} = \frac{\rho^{(2)}(f|_{G_2})}{[G:G_1] \cdot [G_1:G_2]} = \frac{\rho^{(2)}(f|_{G_1})}{[G:G_1]}.$$

Another more general possibility is to use the classifying space $E(G \rtimes_f \mathbb{Z}, \mathcal{FIN})$ of $G \rtimes_f \mathbb{Z}$ for the family $\mathcal{FIN}$ of finite subgroups. Then one can define under the assumption that $E(G, \mathcal{FIN})$ has a finite G-CW-model and G is of det $\geq$ 1-class

$$\overline{\rho}^{(2)}(f: G \to G) = \rho^{(2)}(E(G \rtimes_f \mathbb{Z}; \mathcal{FIN}); \mathcal{N}(G \rtimes_f \mathbb{Z})). \qquad (7.31)$$

Here we use the fact that there is a finite $G \rtimes_f \mathbb{Z}$-model for $E(G \rtimes_f \mathbb{Z}; \mathcal{FIN})$, provided that there is a finite G-CW-model for $E(G, \mathcal{FIN})$. Namely, the to both ends infinite mapping telescope of the $f: G \to G$-equivariant homotopy equivalence $E(G, \mathcal{FIN}) \to E(G, \mathcal{FIN})$ induced by f is a model for $E(G \rtimes_f \mathbb{Z}; \mathcal{FIN})$. Moreover, one has to use the fact that the definition of L^2-torsion carries over to finite proper G-CW-complexes and depends only on the G-homotopy type provided that G is of det $\geq$ 1-class (see Section 3.8).

Any discrete cocompact subgroup of a connected Lie group and any word-hyperbolic group has a finite G-CW-model for $E(G, \mathcal{FIN})$ (see [1, Corollary 4.14], [370]). For more information about $E(G, \mathcal{FIN})$ we refer for instance to [336].

If G is torsionfree, then these three notions $\rho^{(2)}(f)$, $\rho^{(2)}_{\mathrm{virt}}(f)$ and $\overline{\rho}^{(2)}(f)$ are all defined if and only if one of them is defined, and in this case they coincide. Suppose that G possesses a torsionfree subgroup G_0 of finite index. Then BG_0 has a finite CW-model if there is a finite G-CW-model for $E(G, \mathcal{FIN})$. If both $\rho^{(2)}_{\mathrm{virt}}(f)$ and $\overline{\rho}^{(2)}(f)$ are defined, they coincide.

We mention without giving the proof that Theorem 7.27 carries directly over to $\overline{\rho}^{(2)}(f: G \to G)$, provided we assume in assertion (3) that G_3 is torsionfree.

7.5 Miscellaneous

If one takes the third Novikov-Shubin invariant into account, one can improve Lemma 7.22 (1). Namely, if BG has finite 3-skeleton, $b_1^{(2)}(K \times_\phi EG; \mathcal{N}(G)) = 0$ and $\alpha_3(K \times_\phi EG; \mathcal{N}(G)) \neq \infty^+$, then

$$\mathrm{def}(G) \leq 0. \qquad (7.32)$$

This can be seen as follows. It suffices to improve the inequality in Lemma 7.16 to a strict inequality

$$g - r < 1 - b_0^{(2)}(K \times_\phi EG; \mathcal{N}(K)) + b_1^{(2)}(K \times_\phi EG; \mathcal{N}(K)) - b_2^{(2)}(K \times_\phi EG; \mathcal{N}(K))$$

by the following modification of the proof of Lemma 7.16. Namely, since $\alpha_3(K \times_\phi EG; \mathcal{N}(G)) \neq \infty^+$, we have $\mathbf{T}H_2^K(K \times_\phi EG; \mathcal{N}(K)) \neq 0$. Notice that $H_2^K(K \times_\phi \widetilde{X}; \mathcal{N}(K)) \to H_2^K(K \times_\phi EG; \mathcal{N}(K))$ is surjective and

$H_2^K(K \times_\phi \tilde{X}; \mathcal{N}(K))$ is a finitely generated projective $\mathcal{N}(K)$-module since X is 2-dimensional and $\mathcal{N}(K)$ semihereditary by Theorem 6.7 (1). We conclude from Theorem 6.7 and Lemma 6.28 (3) that (7.18) becomes a strict inequality

$$b_2^{(2)}(K \times_\phi \tilde{X}; \mathcal{N}(K)) > b_2^{(2)}(K \times_\phi EG; \mathcal{N}(K))$$

and (7.32) follows.

We mention the following result of Lott [320, Theorem 2] which generalizes a result of Lubotzky [324]. The statement we present here is a slight improvement of Lott's result due to Hillman [259].

Theorem 7.33. *Let L be a connected Lie group. Let G be a lattice in L. If* $\operatorname{def}(G) > 0$, *then one of the following assertions holds*

(1) G is a lattice in $PSL_2(\mathbb{C})$;
(2) $\operatorname{def}(G) = 1$. Moreover, either G is isomorphic to a torsionfree non-uniform lattice in $\mathbb{R} \times PSL_2(\mathbb{R})$ or $PSL_2(\mathbb{C})$, or G is $\mathbb{Z}$ or $\mathbb{Z}^2$.

More information about the class $\mathcal{B}_1$ can be found in [34]. For instance the first L^2-Betti number of a group having Kazhdan's property (T) vanishes [34, Corollary 6].

The following result is due to Gaboriau [214, Theorem 6.3]. (An alternative proof using the dimension theory of Chapter 6 can be found in the Ph. D. thesis of Roman Sauer [456].) Two countable groups G_0 and G_1 are called *measure equivalent* if there exist commuting measure-preserving free actions of G_0 and G_1 on some infinite Lebesgue measure space (Ω, m) such that the actions of both G_0 and G_1 admit finite measure fundamental domains (see [237, 0.5E], [211] and [212]).

Theorem 7.34 (Measure equivalence and L^2-Betti numbers). *Let G_0 and G_1 be two countable groups which are measure equivalent. Then there is a constant $C > 0$ such that for all $p \geq 0$*

$$b_p^{(2)}(G_0) = C \cdot b_p^{(2)}(G_1).$$

Since any infinite amenable group is measure equivalent to $\mathbb{Z}$ [408], this gives another proof of the fact that the L^2-Betti numbers of infinite amenable countable groups vanish. Notice that $\mathbb{Z}^m$ and $\mathbb{Z}^n$ have different Novikov-Shubin invariants for $m \neq n$ (see Example 2.59) so that Novikov-Shubin invariants are not invariant under measure equivalence.

The notion of measure equivalence can be viewed as the measure theoretic analog of the metric notion of quasi-isometric groups. Namely, two finitely generated groups G_0 and G_1 are quasi-isometric if and only if there exist commuting proper (continuous) actions of G_0 and G_1 on some locally compact space such that each action has a compact fundamental domain [237, 0.2 C_2' on page 6].

If the finitely generated groups G_0 and G_1 are quasi-isometric and there exist finite models for BG_0 and BG_1 then $b_p^{(2)}(G_0) = 0 \Leftrightarrow b_p^{(2)}(G_1) = 0$ holds (see [237, page 224], [410]). But in general it is not true that there is a constant $C > 0$ such that $b_p^{(2)}(G_0) = C \cdot b_p^{(2)}(G_1)$ holds for all $p \geq 0$ (cf. [215, page 7], [237, page 233], [523]). If F_g denotes the free group on g generators, then define $G_n := (F_3 \times F_3) * F_n$ for $n \geq 2$. The groups G_m and G_n are quasi-isometric for $m, n \geq 2$ (see [132, page 105 in IV-B.46], [523, Theorem 1.5]) and have finite models for their classifying spaces. One easily checks that $b_1^{(2)}(G_n) = n$ and $b_2^{(2)}(G_n) = 4$ (see Example 1.38).

Gaboriau's Theorem 7.34 implies that G_n and G_m are measure equivalent if and only if $m = n$ holds. Hence there are finitely presented groups which are quasi-isometric but not measure equivalent. Another example is pointed out in [215, page 7]. There exist quasi-isometric finitely generated groups G_0 and G_1 such that G_0 has Kazhdan's property (T) and G_1 does not, and Kazhdan's property (T) is an invariant under measure equivalence [211, Theorem 8.2].

The converse is also true. The groups $\mathbb{Z}^n$ and $\mathbb{Z}^m$ are infinite amenable and hence measure equivalent. But they are not quasi-isometric for different m and n since n is the growth rate of $\mathbb{Z}^n$ and the growth rate is a quasi-isometry invariant.

Notice that Theorem 7.34 implies that the sign of the Euler characteristic of a group G is an invariant under measure equivalence, what is not true for quasi-isometry by the example above.

We mention the following not yet published result of Monod and Shalom that the non-vanishing of the second bounded cohomology $H_b^2(G; l^2(G))$ of a countable group G with coefficients in $l^2(G)$ is an invariant of the measure equivalence class of G.

The following questions arose in discussions with R. Sauer in view of Theorem 3.113 and [343].

Question 7.35. (L^2-torsion of groups and quasi-isometry and measure equivalence).
Let G_i for $i = 0, 1$ be a group such that there is a finite CW-model for BG_i and EG_i is det-L^2-acyclic. Suppose that G_0 and G_1 are measure equivalent or that G_0 and G_1 are quasi-isometric. Does then $\rho^{(2)}(EG_0; \mathcal{N}(G_0)) = 0 \Leftrightarrow \rho^{(2)}(EG_1; \mathcal{N}(G_1)) = 0$ hold?

Question 7.36. (Novikov-Shubin invariants and quasi-isometry).
Let G_i for $i = 0, 1$ be a finitely generated group. Suppose that G_0 and G_1 are quasi-isometric. Does then $\alpha_p(EG_0; \mathcal{N}(G_0)) = \alpha_p(EG_1; \mathcal{N}(G_1))$ hold for $p \geq 1$?

A *lattice* G in a locally compact second countable topological group T is a discrete subgroup such that the measure on T/G induced by a right invariant Haar measure on T has finite volume. (This implies that T is unimodular.) Since two lattices in the same locally compact second countable topological group T are measure equivalent [211, Example 1.2 on page 1059], Theorem 7.34 implies (see also Theorem 3.183)

Corollary 7.37. *Let G_1 and G_2 be two lattices in the same locally compact second countable topological group T. Then for all $p \geq 0$*

$$\frac{b_p^{(2)}(G_1)}{\text{vol}(T/G_1)} = \frac{b_p^{(2)}(G_2)}{\text{vol}(T/G_2)}.$$

A survey article about orbit equivalent measure preserving actions and L^2-Betti numbers is written by Gaboriau [215].

For each Artin group A, Davis and Leary [129] compute the L^2-cohomology of the universal covering of its so called Salvetti complex. This is a finite CW-complex which is conjectured to be a model for the classifying space BA. In the many cases when this conjecture is known to hold their calculation describes the reduced L^2-cohomology of EA, or, equivalently, the L^2-Betti numbers $b_p^{(2)}(A)$.

Exercises

7.1. Let d be a non-negative integer. Suppose that the group G has a descending series of infinite subgroups $G = G_0 \supset G_1 \supset G_2 \supset \ldots \supset G_d$ such that BG_{n+1} has finite $(d-n)$-skeleton for each $n = 0, 1, 2, \ldots, d-1$, G_{n+1} is normal in G_n and the quotient G_n/G_{n+1} contains an element of infinite order or contains finite subgroups of arbitrary large order for $n = 0, 1, \ldots, d-1$. Prove $G \in \mathcal{B}_d$.

7.2. Suppose that the group G has a descending series of infinite subgroups $G = G_0 \supset G_1 \supset G_2 \supset \ldots$ such that BG_{n+1} is of finite type, G_{n+1} is normal in G_n and the quotient G_n/G_{n+1} contains an element of infinite order or contains finite subgroups of arbitrary large order for $n = 0, 1, \ldots$. Prove $G \in \mathcal{B}_\infty$.

7.3. Let G_n for $n = 0, 1, 2, \ldots$ be a sequence of non-trivial groups. Prove

$$b_p^{(2)}(\ast_{n=0}^\infty G_n) = \sum_{n=0}^\infty b_p^{(2)}(G_n).$$

for $p \geq 2$ and

$$b_1^{(2)}(\ast_{n=0}^\infty G_n) = \infty.$$

7.4. Let $G_1, G_2, \ldots, G_n$ be finitely many groups. Prove

$$b_p^{(2)}\left(\prod_{k=1}^{n} G_k\right) = \sum_{\substack{0 \le j_1, j_2, \ldots, j_n, \\ \sum_{k=1}^{n} j_k = p}} \prod_{k=1}^{n} b_{j_k}^{(2)}(G_k),$$

where we use the convention that $0 \cdot \infty = 0$, $r \cdot \infty = \infty$ for $r \in (0, \infty]$ and $r + \infty = \infty$ for $r \in [0, \infty]$.

7.5. Let $\{G_i \mid i \in I\}$ be a family of non-trivial groups for an infinite index set I. Put $G = \prod_{i \in I} G_i$. Prove $G \in \mathcal{B}_\infty$. Show by giving an example that it is not necessarily true that $H_p^{(2)}(EG; \mathcal{N}(G)) = 0$ for all $p \ge 0$.

7.6. Let G be an infinite locally finite group. Show that $H_p^G(EG; \mathcal{N}(G))$ vanishes for $p \ge 1$ and does not vanish for $p = 0$ and that $b_p^{(2)}(G) = 0$ holds for all $p \ge 0$.

7.7. Let $1 \to H \to G \to K \to 1$ be an extension of groups such that H or K respectively is finite. Show that K or H respectively belongs to $\mathcal{B}_d$, $\mathcal{BQ}_d$ or $\mathcal{BF}_d$ if and only if G does.

7.8. Let d be a non-negative integer or $d = \infty$. Let $F \xrightarrow{i} E \xrightarrow{p} B$ be a fibration of connected CW-complexes such that $b_p^{(2)}(\pi_1(E) \times_{\pi_1(i)} \widetilde{F}; \mathcal{N}(\pi_1(E))) < \infty$ for $p \le d$. Suppose that $H_p(\widetilde{B}; \mathbb{C})$ is a finite dimensional complex vector space for $p \le d$ and that $\pi_1(B) \in \mathcal{BF}_d$. Show that then $b_p^{(2)}(\widetilde{E}) = 0$ for $p \le d$.

7.9. Let G and H be non-trivial groups. Show that $b_1^{(2)}(G * H) = 0$ if and only if $G = H = \mathbb{Z}/2$.

7.10. Suppose there is a finite G-CW-model for $E(G; \mathcal{FIN})$. Let d be the least common multiple of the order of finite subgroups of G. Show

$$d \cdot \chi^{(2)}(G) \in \mathbb{Z}.$$

7.11. Prove $\chi^{(2)}(SL(2, \mathbb{Z})) = -\frac{1}{12}$.

7.12. Let G be a finitely presented group. Show

$$\mathrm{def}(G) \le b_1(G; \mathbb{Z}/p) - b_2(G; \mathbb{Z}/p),$$

where $b_p(G; \mathbb{Z}/p)$ is the p-th Betti number of EG with respect to singular homology with coefficients in $\mathbb{Z}/p$.

7.13. Let $f : G \to G$ be a (not necessarily bijective) endomorphism of a group G which possesses a finite CW-model for BG and is of det ≥ 1-class. Let K be the colimit of the directed system $\ldots \xrightarrow{f} G \xrightarrow{f} G \xrightarrow{f} \ldots$ indexed by $\mathbb{Z}$. Let G_f be the semidirect product $K \rtimes \mathbb{Z}$ with respect to the shift automorphism of K. Show

(1) There is a finite CW-model for BG_f, namely the mapping torus of $Bf\colon BG \to BG$;

(2) BG_f is det-L^2-acyclic and the L^2-torsion $\rho^{(2)}(\widetilde{BG_f})$ depends only on f. Define

$$\rho^{(2)}(f) := \rho^{(2)}(\widetilde{BG_f});$$

(3) Show that this invariant reduces to the invariant $\rho^{(2)}(f)$ of Definition 7.26 in the case where f is an automorphism;

(4) Show that Theorem 7.27 does hold also for group endomorphisms if one makes the following modifications or additional assumptions. Assertion (1), (2), (3) and (4) remain true. For assertion (5) one needs the condition that f induces a bijection $G/G_0 \xrightarrow{\cong} G/G_0$. In the formulation of assertion (6) one must replace the universal covering of BG by the covering associated to the canonical projection $G \to K$. In assertion (7) the conditions must be required for K instead of G.

8. The Algebra of Affiliated Operators

Introduction

In this chapter we introduce and study the algebra $\mathcal{U}(G)$ of operators affiliated to $\mathcal{N}(G)$ for a group G. A *G-operator* $f\colon \operatorname{dom}(f) \subset V \to W$ of Hilbert $\mathcal{N}(G)$-modules is an operator whose domain $\operatorname{dom}(f)$ is a linear G-invariant subspace and which satisfies $f(gx) = gf(x)$ for all $x \in \operatorname{dom}(f)$ and $g \in G$. The algebra $\mathcal{U}(G)$ consists of densely defined closed G-operators $a\colon \operatorname{dom}(a) \subset l^2(G) \to l^2(G)$ and contains $\mathcal{N}(G)$ as a subalgebra. It is constructed in such a way that an element $f\colon l^2(G) \to l^2(G)$ in $\mathcal{N}(G)$ is a weak isomorphism if and only it is invertible in $\mathcal{U}(G)$. This is reflected algebraically by the fact that $\mathcal{U}(G)$ is the Ore localization of $\mathcal{N}(G)$ with respect to the set of non-zero divisors (see Theorem 8.22 (1)). It does not come with a natural topology anymore but has nice ring theoretic properties. Namely, $\mathcal{U}(G)$ is von Neumann regular, i.e. for any $r \in \mathcal{U}(G)$ there is $s \in \mathcal{U}(G)$ with $rsr = r$, or, equivalently, any finitely generated submodule of a projective $\mathcal{U}(G)$-module is a direct summand (see Theorem 8.22 (3)). We have already mentioned in Example 6.12 that $\mathcal{N}(G)$ behaves in several ways like a principal ideal domain except that $\mathcal{N}(G)$ is not Noetherian and has zero-divisors if G is infinite. Any principal ideal domain R has a quotient field F, and in this analogy $\mathcal{U}(G)$ should be thought of as F. Recall that the (extended) dimension of an arbitrary R-module M is the same as the F-dimension of the F-vector space $F \otimes_R M$ (see Example 6.12). In terms of this analogy it is not surprising that the extended dimension function $\dim_{\mathcal{N}(G)}$ for arbitrary $\mathcal{N}(G)$-modules comes from an extended $\mathcal{U}(G)$-dimension function $\dim_{\mathcal{U}(G)}$ over $\mathcal{U}(G)$ in the sense that $\dim_{\mathcal{N}(G)}(M) = \dim_{\mathcal{U}(G)}(\mathcal{U}(G) \otimes_{\mathcal{N}(G)} M)$ holds for any $\mathcal{N}(G)$-module M (see Theorem 8.29). The algebra $\mathcal{U}(G)$ will play a role in the proof of the Atiyah conjecture in Chapter 10. It will be the largest algebra attached to a given group which we will consider in this book. All other algebras attached to a given group will be contained in it.

8.1 The Algebra of Affiliated Operators

In this section we introduce the algebra $\mathcal{U}(G)$ of operators affiliated to the group von Neumann algebra $\mathcal{N}(G)$. It was originally introduced and studied in [396, Chapter XVI].

We have already studied the $\mathbb{C}$-category {fin. gen. Hilb. $\mathcal{N}(G)$-mod.} with involution of finitely generated Hilbert $\mathcal{N}(G)$-modules with bounded G-operators as morphisms (see Section 6.2). We have already seen that weak isomorphisms play an important role. Recall that an operator is a weak isomorphism if its kernel is trivial and its image is dense and that a weak isomorphism is in general not invertible. The basic idea for the following construction is to enlarge the set of morphisms in {fin. gen. Hilb. $\mathcal{N}(G)$-mod.} to get a new $\mathbb{C}$-category with involution {fin. gen. Hilb. $\mathcal{N}(G)$-mod.}$_{\mathcal{U}}$ with the same objects such that weak isomorphisms become invertible. To motivate it, let us consider a bounded G-operator $f \colon V \to W$ of finitely generated Hilbert $\mathcal{N}(G)$-modules, which is a weak isomorphism, and check what is needed to find an "inverse". The polar decomposition $f = us$ consists of a unitary invertible G-operator u and a positive G-operator s which is a weak isomorphism. So it suffices to "invert" s. Since s is a weak isomorphism, its kernel is trivial and we obtain an unbounded densely defined closed G-operator $\int \lambda^{-1} dE_\lambda$, if $\{E_\lambda \mid \lambda \in [0, \infty)\}$ is the spectral family of s (see Subsection 1.4.1). This should become the inverse of s. Hence we must enlarge the morphisms to include unbounded densely defined G-operators like $\int \lambda^{-1} dE_\lambda$. The difficulty will be that the composition and sum of unbounded densely defined closed operators (see Notation 1.69) is in general not again densely defined and closed. The main problem is that the intersection of two dense subspaces is not necessarily dense again. It will turn out that in the specific situation, which we are interested in, this problem can be solved. We need some preparation to handle this.

Definition 8.1 (Affiliated operator). *An unbounded operator $f \colon \mathrm{dom}(f) \subset V \to W$ of finitely generated Hilbert $\mathcal{N}(G)$-modules is called affiliated (to $\mathcal{N}(G)$) if f is densely defined with domain $\mathrm{dom}(f) \subset V$, is closed and is a G-operator, i.e. $\mathrm{dom}(f)$ is a linear G-invariant subspace and $f(gx) = gf(x)$ for all $x \in \mathrm{dom}(f)$ and $g \in G$.*

Definition 8.2. *Let V be a finitely generated Hilbert $\mathcal{N}(G)$-module. A G-invariant linear subspace $L \subset V$ is called essentially dense if for any $\epsilon > 0$ there is a finitely generated Hilbert $\mathcal{N}(G)$-submodule $P \subset V$ with $P \subset L$ and $\dim_{\mathcal{N}(G)}(V) - \dim_{\mathcal{N}(G)}(P) \leq \epsilon$.*

Notice that essentially dense implies dense.

Lemma 8.3. *Let $f, f' \colon V \to W$ be affiliated operators of finitely generated Hilbert $\mathcal{N}(G)$-modules. Then*

(1) *A countable intersection of essentially dense linear G-subspaces of V is again essentially dense;*

(2) *If $L \subset W$ is essentially dense, then $f^{-1}(L) \subset \mathrm{dom}(f)$ is an essentially dense subspace of V. In particular $\mathrm{dom}(f)$ is essentially dense;*

(3) *If $f' \subset f$, then $f' = f$. In particular $f = f'$, if there is a G-invariant subset S of $\mathrm{dom}(f) \cap \mathrm{dom}(f')$ which is dense in V and on which f and f' agree;*

(4) *If f is bounded on its domain, i.e. there is $C < 0$ with $|f(x)| \leq C|x|$ for all $x \in \mathrm{dom}(f)$, then $\mathrm{dom}(f) = V$.*

Proof. (1) Let $\{L_n \mid n \geq 0\}$ be a countable set of essentially dense linear G-subspaces of V. Given $\epsilon > 0$, choose a Hilbert $\mathcal{N}(G)$-submodule $P_n \subset V$ with $P_n \subset L_n$ and $\dim_{\mathcal{N}(G)}(V) - \dim_{\mathcal{N}(G)}(P_n) \leq 2^{-n-1}\epsilon$. Put $P = \bigcap_{n \geq 0} P_n$. Then P is a Hilbert $\mathcal{N}(G)$-submodule of V with $P \subset \bigcap_{n \geq 0} L_n$ and we conclude from Theorem 1.12 (2) and (4)

$$
\dim_{\mathcal{N}(G)}(V) - \dim_{\mathcal{N}(G)}(P)
$$

$$
= \dim_{\mathcal{N}(G)}(V) - \lim_{m \to \infty} \dim_{\mathcal{N}(G)}\left(\bigcap_{n=0}^{m} P_n\right)
$$

$$
= \lim_{m \to \infty}\left(\dim_{\mathcal{N}(G)}(V) - \dim_{\mathcal{N}(G)}\left(\bigcap_{n=0}^{m} P_n\right)\right)
$$

$$
\leq \lim_{m \to \infty} \sum_{n=0}^{m}\left(\dim_{\mathcal{N}(G)}(V) - \dim_{\mathcal{N}(G)}(P_n)\right)
$$

$$
\leq \lim_{m \to \infty}\left(\sum_{n=0}^{m} 2^{-n-1}\epsilon\right)
$$

$$
= \epsilon.
$$

(2) We can write f using polar decomposition as the composition

$$
f \colon V \xrightarrow{\mathrm{pr}} \ker(f)^{\perp} \xrightarrow{u} \overline{\mathrm{im}(f)} \xrightarrow{s} \overline{\mathrm{im}(f)} \xrightarrow{i} W,
$$

where pr and i are the canonical projection and inclusion, u is a unitary isomorphism and s is a densely defined unbounded closed G-operator which is positive and a weak isomorphism. Let $\{E_\lambda \mid \lambda \in [0, \infty)\}$ be the spectral family of s. Since E_λ converges for $\lambda \to \infty$ strongly to id, we get $\lim_{\lambda \to \infty} \dim_{\mathcal{N}(G)}(\mathrm{im}(E_\lambda)) = \dim_{\mathcal{N}(G)}(\overline{\mathrm{im}(f)})$. Fix $\epsilon > 0$. Choose λ_0 with $\dim_{\mathcal{N}(G)}(\overline{\mathrm{im}(f)}) - \dim_{\mathcal{N}(G)}(\mathrm{im}(E_{\lambda_0})) \leq \epsilon/2$. Since L is essentially dense in W, we can find a Hilbert $\mathcal{N}(G)$-submodule $P \subset W$ with $P \subset L$ and $\dim_{\mathcal{N}(G)}(W) - \dim_{\mathcal{N}(G)}(P) \leq \epsilon/2$. We conclude from Theorem 1.12 (2)

$$
\dim_{\mathcal{N}(G)}(\overline{\mathrm{im}(f)}) - \dim_{\mathcal{N}(G)}(P \cap \mathrm{im}(E_{\lambda_0})) \leq \epsilon. \tag{8.4}
$$

Notice that $\operatorname{im}(E_{\lambda_0}) \subset \operatorname{dom}(s)$ follows from (1.66) since each $y \in \operatorname{im}(E_{\lambda_0})$ lies in the domain of $\int_{-\infty}^{\infty} \lambda dE_\lambda$. Moreover, s induces a bounded G-operator $s_{\lambda_0} : \operatorname{im}(E_{\lambda_0}) \to \operatorname{im}(E_{\lambda_0})$ which is a weak isomorphism, and $s^{-1}(P \cap \operatorname{im}(E_{\lambda_0})) = s^{-1}(P \cap \operatorname{im}(E_{\lambda_0})) \cap \operatorname{im}(E_{\lambda_0}) = (s_{\lambda_0})^{-1}(P \cap \operatorname{im}(E_{\lambda_0}))$ is a closed G-invariant subspace of $\operatorname{im}(f)$. We conclude from Lemma 2.12

$$\dim_{\mathcal{N}(G)}(s^{-1}(P \cap \operatorname{im}(E_{\lambda_0}))) = \dim_{\mathcal{N}(G)}(P \cap \operatorname{im}(E_{\lambda_0})). \qquad (8.5)$$

We conclude from Additivity (see Theorem 1.12 (2))

$$\dim_{\mathcal{N}(G)}\left(s^{-1}(P \cap \operatorname{im}(E_{\lambda_0}))\right) + \dim_{\mathcal{N}(G)}(\ker(f))$$
$$= \dim_{\mathcal{N}(G)}\left((s \circ u \circ \operatorname{pr})^{-1}(P \cap \operatorname{im}(E_{\lambda_0}))\right); \qquad (8.6)$$
$$\dim_{\mathcal{N}(G)}(V) = \dim_{\mathcal{N}(G)}(\ker(f)) + \dim_{\mathcal{N}(G)}(\overline{\operatorname{im}(f)}). \qquad (8.7)$$

Hence $(s \circ u \circ \operatorname{pr})^{-1}(P \cap \operatorname{im}(E_{\lambda_0}))$ is a Hilbert $\mathcal{N}(G)$-submodule of V, is contained in $f^{-1}(L)$ and because of (8.4), (8.5), (8.6) and (8.7)

$$\dim_{\mathcal{N}(G)}(V) - \dim_{\mathcal{N}(G)}\left((s \circ u \circ \operatorname{pr})^{-1}(P \cap \operatorname{im}(E_{\lambda_0}))\right) \le \epsilon.$$

Hence $f^{-1}(L)$ is essentially dense.

(3) Let V be a finitely generated Hilbert $\mathcal{N}(G)$-module. We first show that an affiliated operator $h \colon V \to V$ which is symmetric is already selfadjoint. We do this using the *Cayley transform* $\kappa(s)$ of a symmetric densely defined operator $s \colon \operatorname{dom}(s) \subset H \to H$, which is the operator $(s - i)(s + i)^{-1}$ with domain $\operatorname{dom}(\kappa(s)) = (s + i)(\operatorname{dom}(s))$. Let us summarize the basic properties of the Cayley transform (see [421, Section 5.2]). It is an isometry on its domain $\operatorname{dom}(\kappa(s))$. The operator $\kappa(s) - 1$ with the same domain as $\kappa(s)$ is injective and has range $\operatorname{im}(\kappa(s) - 1) = \operatorname{dom}(s)$. If in addition s is closed, then the subspaces $(s \pm i)(\operatorname{dom}(s))$ are closed. The operator s is selfadjoint if and only if $(s + i)(\operatorname{dom}(s)) = (s - i)(\operatorname{dom}(s)) = H$.

We want to use the latter criterion to show that h is selfadjoint. Notice that $\kappa(h)$ is a G-operator as h is a G-operator, and $\operatorname{dom}(\kappa(h)) = (h + i)(\operatorname{dom}(h))$ is closed as h is closed. Let p denote the projection onto the G-invariant closed subspace $\operatorname{dom}(\kappa(h))$. Then $\kappa(h) \circ p - p$ is everywhere defined and $\operatorname{im}(\kappa(h) - 1)) = \operatorname{im}(\kappa(h) \circ p - p)$. Since

$$(\kappa(h) \circ p - p)^* = ((\kappa(h) \circ p - p) \circ p)^* = p^* \circ (\kappa(h) \circ p - p)^* = p \circ (\kappa(h) \circ p - p)^*,$$

we conclude using Additivity (see Theorem 1.12 (2))

$$\dim_{\mathcal{N}(G)}(\operatorname{dom}(h)) = \dim_{\mathcal{N}(G)}(\operatorname{im}(\kappa(h) - 1))$$
$$= \dim_{\mathcal{N}(G)}(\operatorname{im}(\kappa(h) \circ p - p))$$
$$= \dim_{\mathcal{N}(G)}(\operatorname{im}((\kappa(h) \circ p - p)^*))$$
$$\le \dim_{\mathcal{N}(G)}(\operatorname{im}(p))$$
$$= \dim_{\mathcal{N}(G)}((h + i)(\operatorname{dom}(h))).$$

This and assertion (2) imply

$$\dim_{\mathcal{N}(G)}((h+i)(\operatorname{dom}(h))) = \dim_{\mathcal{N}(G)}(V).$$

Since $(h + i)(\operatorname{dom}(h)) \subset V$ is closed, we conclude $(h + i)(\operatorname{dom}(h)) = \operatorname{dom}(\kappa(h)) = V$ from Theorem 1.12 (1) and (2). Since $\kappa(h)$ is an isometry on its domain and a G-operator, we get $\dim_{\mathcal{N}(G)}(\operatorname{im}(\kappa(h))) = \dim_{\mathcal{N}(G)}(V)$. As $(h - i)(\operatorname{dom}(h)) = \operatorname{im}(\kappa(h))$ is closed, we conclude $(h - i)(\operatorname{dom}(h)) = V$. This finishes the proof that h is selfadjoint.

Now we can give the proof of (3). Let $f = uh$ be the polar decomposition of f. Put $h' = u^* f'$. Since u^* is a bounded G-operator and f' is affiliated, h' is affiliated. We have $h' = u^* f' \subset u^* f = h$. We conclude $h' \subset h = h^* \subset (h')^*$, i.e. h' is symmetric. Hence h' is selfadjoint by the argument above. This implies $h' = h = h^* = (h')^*$ and hence $\operatorname{dom}(f') = \operatorname{dom}(h') = \operatorname{dom}(h) = \operatorname{dom}(f)$, i.e. $f = f'$.

(4) Any closed densely defined operator which is bounded on its domain is automatically everywhere defined.   □

Now we are ready to define the desired $\mathbb{C}$-category with involution {fin. gen. Hilb. $\mathcal{N}(G)$-mod.}$_{\mathcal{U}}$ as follows. Objects are finitely generated Hilbert $\mathcal{N}(G)$-modules. Given two finitely generated Hilbert $\mathcal{N}(G)$-modules V and W, the set of morphisms from V to W consists of all affiliated operators $f\colon \operatorname{dom}(f) \subset V \to W$. The identity element of V is given by the identity operator $\operatorname{id}\colon V \to V$. Given an affiliated operator $f\colon \operatorname{dom}(f) \subset V \to W$, its adjoint $f^*\colon \operatorname{dom}(f^*) \subset W \to V$ is affiliated, since the adjoint of a closed densely defined operator is always closed and densely defined and the adjoint of a G-operator is always a G-operator. Thus we can define the involution by taking the adjoint.

Given morphisms $f\colon \operatorname{dom}(f) \subset U \to V$ and $g\colon \operatorname{dom}(g) \subset V \to W$, define their composition $g \circ f$ in {fin. gen. Hilb. $\mathcal{N}(G)$-mod.}$_{\mathcal{U}}$ by the minimal closure of the unbounded operator $g \circ f\colon U \to W$ with domain $f^{-1}(\operatorname{dom}(g))$ as defined in Notation 1.69. We have to check that this is well-defined. From Lemma 8.3 (2) we conclude that $f^{-1}(\operatorname{dom}(g))$ is essentially dense and in particular dense in U. We know already that the adjoints g^* and f^* are affiliated again and hence $f^* \circ g^*$ with domain $(g^*)^{-1}(\operatorname{dom}(f^*))$ is densely defined. Since the domain of the adjoint of $g \circ f$ contains $(g^*)^{-1}(\operatorname{dom}(f^*))$, $(g \circ f)^*$ is densely defined. This implies that $g \circ f$ is closable, namely, $((g \circ f)^*)^*$ is its minimal closure. Obviously $g \circ f$ is a G-operator. Hence its minimal closure is indeed affiliated. Analogously one defines the complex vector space structure on the set of morphisms from V to W by taking the minimal closure of the addition and scalar multiplication of unbounded operators defined in Notation 1.69. We leave it to the reader to check that the various axioms like associativity of the composition and so on are immediate consequences of Lemma 8.3 (3). Obviously {fin. gen. Hilb. $\mathcal{N}(G)$-mod.} is a $\mathbb{C}$-subcategory with involution of {fin. gen. Hilb. $\mathcal{N}(G)$-mod.}$_{\mathcal{U}}$. Next we show that {fin. gen. Hilb. $\mathcal{N}(G)$-mod.}$_{\mathcal{U}}$ has the desired properties.

Lemma 8.8. *(1) An affiliated operator $f\colon \operatorname{dom}(f) \subset V \to W$ represents an isomorphism in $\{\text{fin. gen. Hilb. } \mathcal{N}(G)\text{-mod.}\}_{\mathcal{U}}$ if and only if f is a weak isomorphism, i.e. f has trivial kernel and dense image $f(\operatorname{dom}(f))$;*
(2) Let $f\colon \operatorname{dom}(f) \subset V \to W$ be an affiliated operator. Then there are bounded (everywhere defined) G-operators $a\colon V \to W$ and $b\colon V \to V$ with the properties that b is a weak isomorphism and $f = ab^{-1}$ holds in $\{\text{fin. gen. Hilb. } \mathcal{N}(G)\text{-mod.}\}_{\mathcal{U}}$.

Proof. (1) Suppose that $f\colon \operatorname{dom}(f) \subset V \to W$ is an affiliated operator which is a weak isomorphism. Let $f = us$ be its polar decomposition. Then $u\colon V \to W$ is a unitary bijective G-operator and $s\colon \operatorname{dom}(s) \subset V \to V$ is an affiliated operator which is positive and a weak isomorphism. Let $h(s)\colon V \to V$ be the operator given by the functional calculus applied to s for the Borel function $h\colon \mathbb{R} \to \mathbb{R}$ which sends $\lambda \neq 0$ to λ^{-1} and 0 to, let us say, 0. Notice that $h(s)$ is a densely defined selfadjoint G-operator and in particular affiliated. If $\{E_\lambda \mid \lambda \in [0, \infty)\}$ is the spectral family of s, then $\operatorname{im}(E_n) \cap \operatorname{im}(E_{1/n}^\perp)$ is contained in $\operatorname{dom}(h(s))$ for all integers $n \geq 1$. Since s has trivial kernel, $\lim_{n \to \infty} \dim_{\mathcal{N}(G)}(\operatorname{im}(E_n) \cap \operatorname{im}(E_{1/n}^\perp)) = \dim_{\mathcal{N}(G)}(V)$. Since both s and $h(s)$ map $\operatorname{im}(E_n) \cap \operatorname{im}(E_{1/n})^\perp$ to itself and define inverse operators there, Lemma 8.3 (3) implies that s and $h(s)$ define inverse morphisms in $\{\text{fin. gen. Hilb. } \mathcal{N}(G)\text{-mod.}\}_{\mathcal{U}}$. Since u obviously defines an isomorphism in $\{\text{fin. gen. Hilb. } \mathcal{N}(G)\text{-mod.}\}_{\mathcal{U}}$, f is an isomorphism in $\{\text{fin. gen. Hilb. } \mathcal{N}(G)\text{-mod.}\}_{\mathcal{U}}$.

Now suppose that the affiliated operator $f\colon \operatorname{dom}(f) \subset V \to W$ defines an isomorphism in $\{\text{fin. gen. Hilb. } \mathcal{N}(G)\text{-mod.}\}_{\mathcal{U}}$. Let $h\colon \operatorname{dom}(h) \subset W \to V$ be an inverse. For $x \in \ker(f)$ we have $x \in f^{-1}(\operatorname{dom}(h))$ and hence $x = \operatorname{id}(x) = h \circ f(x) = 0$. Hence f has trivial kernel. Since $f \circ h = \operatorname{id}$ in $\mathcal{U}$, the image of f contains the subspace $h^{-1}(\operatorname{dom}(f))$ which is dense by Lemma 8.3 (2). Hence the image of f is dense. This shows that f is a weak isomorphism.

(2) Let $f = us$ be the polar decomposition of f. Let $p = 1 - u^*u$. This is the projection onto $\ker(u^*u) = \ker(us) = \ker(s)$. We have $f = u(s+p)$ and $(s+p)$ is an affiliated positive operator $\operatorname{dom}(s + p) \subset V \to V$ with trivial kernel. Define Borel functions $h_1, h_2\colon \mathbb{R} \to \mathbb{R}$ by $h_1(\lambda) = \lambda$ and $h_2(\lambda) = 1$ for $\lambda \leq 1$ and by $h_1(\lambda) = 1$ and $h_2(\lambda) = 1/\lambda$ for $\lambda \geq 1$. Then $u \circ h_1(s+p)$ and $h_2(s+p)$ are bounded G-operators, $h_2(s + p)$ represents an invertible morphism in $\{\text{fin. gen. Hilb. } \mathcal{N}(G)\text{-mod.}\}_{\mathcal{U}}$ and $f = (u \circ h_1(s + p)) \circ (h_2(s + p))^{-1}$ in $\{\text{fin. gen. Hilb. } \mathcal{N}(G)\text{-mod.}\}_{\mathcal{U}}$. $\quad\square$

Definition 8.9 (Algebra of affiliated operators). *Define $\mathcal{U}(G)$ to be the ring with involution given by the endomorphisms of $l^2(G)$ in the $\mathbb{C}$-category with involution $\{\text{fin. gen. Hilb. } \mathcal{N}(G)\text{-mod.}\}_{\mathcal{U}}$, i.e. the ring of affiliated operators $f\colon \operatorname{dom}(f) \subset l^2(G) \to l^2(G)$.*

Let $\{l^2(G)^n\}_{\mathcal{U}} \subset \{\text{fin. gen. Hilb. } \mathcal{N}(G)\text{-mod.}\}_{\mathcal{U}}$ be the full subcategory, whose objects are $l^2(G)^n$ for $n \geq 0$. Let $\{\mathcal{U}(G)^n\}$ be the category whose objects are the (left) $\mathcal{U}(G)$-modules $\mathcal{U}(G)^n$ for $n \geq 0$ and whose morphisms are $\mathcal{U}(G)$-homomorphisms. Then we obtain an isomorphism of $\mathbb{C}$-categories with involution (compare with 6.22)

$$\nu_{\mathcal{U}} \colon \{\mathcal{U}(G)^n\} \to \{l^2(G)^n\}_{\mathcal{U}}. \tag{8.10}$$

Example 8.11. Let $G = \mathbb{Z}^n$. Using the identifications of Example 1.4 we get an identification of $\mathcal{U}(\mathbb{Z}^n)$ with the complex algebra $L(T^n)$ of equivalence classes of measurable functions $f \colon T^n \to \mathbb{C}$. Recall that two such functions are equivalent if they differ on a subset of measure zero only.

Let $i \colon H \to G$ be an injective group homomorphism. We have associated to it a covariant functor of $\mathbb{C}$-categories with involution

$$i_* \colon \{\text{fin. gen. Hilb. } \mathcal{N}(H)\text{-mod.}\} \to \{\text{fin. gen. Hilb. } \mathcal{N}(G)\text{-mod.}\}$$

in Definition 1.23. The same construction induces a covariant functor of $\mathbb{C}$-categories with involution

$$i_* \colon \{\text{fin. gen. Hilb. } \mathcal{N}(H)\text{-mod.}\}_{\mathcal{U}} \to \{\text{fin. gen. Hilb. } \mathcal{N}(G)\text{-mod.}\}_{\mathcal{U}}$$

because for an affiliated operator $f \colon U \to V$ of finitely generated Hilbert $\mathcal{N}(H)$-modules the densely defined G-operator $\text{id} \otimes_{\mathbb{C}H} f \colon \text{dom}(\text{id} \otimes_{\mathbb{C}H} f) = \mathbb{C}G \otimes_{\mathbb{C}H} \text{dom}(f) \subset i_* U \to i_* V$ has a densely defined adjoint and therefore its minimal closure exists and is affiliated. Functoriality follows from Lemma 8.3 (3). In particular we get a ring homomorphism

$$i_* \colon \mathcal{U}(H) \to \mathcal{U}(G). \tag{8.12}$$

Notice that $\mathcal{N}(G)$ is a $*$-subring of $\mathcal{U}(G)$ and that $\mathcal{U}(G)$ does not carry a natural topology anymore. However, $\mathcal{U}(G)$ has nice properties as a ring what we will explain below.

8.2 Basic Properties of the Algebra of Affiliated Operators

In this section we explain and prove various ring-theoretic properties of $\mathcal{U}(G)$.

8.2.1 Survey on Ore Localization

Definition 8.13. *Let R be a ring. Given a set $S \subset R$, a ring homomorphism $f \colon R \to R'$ is called S-inverting if $f(s)$ is invertible in R' for all $s \in S$. An S-inverting ring homomorphism $f \colon R \to R_S$ is called* universal S-inverting *if for any S-inverting ring homomorphism $g \colon R \to R'$ there is precisely one ring homomorphism $\overline{g} \colon R_S \to R'$ satisfying $\overline{g} \circ f = g$.*

The universal property implies as usual that the universal S-inverting homomorphism $f\colon R \to R_S$ is unique up to unique isomorphism. One can construct such a universal S-inverting ring homomorphism by writing down a suitable ring with generators and relations. If $\overline{S}$ is the set of elements $s \in R$ for which $f(s) \in R_S$ is a unit, then $\overline{S}$ is *multiplicatively closed*, i.e $s, t \in \overline{S}$ implies $st \in \overline{S}$ and $1 \in \overline{S}$, and the ring homomorphism $\phi\colon R_S \to R_{\overline{S}}$, which is uniquely determined by $f_{\overline{S}} \circ \phi = f_S$, is an isomorphism. Hence we can assume in the sequel without loss of generality that S is multiplicatively closed. If R is commutative and $S \subset R$ multiplicatively closed, one can describe R_S in terms of fractions rs^{-1} with the obvious rules for addition and multiplications. Under a certain condition this nice approach works also for non-commutative rings.

Definition 8.14 (Ore localization). *Let $S \subset R$ be a multiplicatively closed subset of the ring R. The pair (R, S) satisfies the (right) Ore condition if i.) for $(r, s) \in R \times S$ there exists $(r', s') \in R \times S$ satisfying $rs' = sr'$ and ii.) if for $r \in R$ and $s \in S$ with $sr = 0$ there is $t \in S$ with $rt = 0$.*

If (R, S) satisfies the Ore condition, define the (right) Ore localization to be the following ring RS^{-1}. Elements are represented by pairs $(r, s) \in R \times S$, where two such pairs (r, s) and (r', s') are called equivalent if there are $u, u' \in R$ such that $ru = r'u'$ and $su = s'u'$ hold and $su = s'u'$ belongs to S. The addition and multiplication is given on representatives by $(r, s) + (r', s') = (rc + r'd, t)$, where $t = sc = s'd \in S$ and by $(r, s) \cdot (r', s') = (rc, s't)$, where $sc = r't$ for $t \in S$. The unit element under addition is represented by $(0, 1)$ and under multiplication by $(1, 1)$. Let $f\colon R \to RS^{-1}$ be the ring homomorphism which sends $r \in R$ to the class of $(r, 1)$.

One may try to remember part i.) of the Ore condition by saying that for any left (= wrong way) fraction $s^{-1}r$ there is a right fraction $r'(s')^{-1}$ with $s^{-1}r = r'(s')^{-1}$. Notice that part ii.) of the Ore condition is automatically satisfied if S contains no zero-divisor.

Lemma 8.15. *Suppose that (R, S) satisfies the Ore condition. Then*

(1) The map $f\colon R \to RS^{-1}$ is the universal S-inverting ring homomorphism;
(2) The kernel of $f\colon R \to RS^{-1}$ is $\{r \in R \mid rs = 0 \text{ for some } s \in S\}$;
(3) The functor $RS^{-1} \otimes_R -$ is exact;
(4) The pair $(M_n(R), S \cdot I_n)$ satisfies the Ore condition, where $M_n(R)$ is the ring of (n, n)-matrices over R and I_n is the unit matrix. The canonical ring homomorphism induced by the universal property $M_n(R)(S \cdot I_n)^{-1} \xrightarrow{\cong} M_n(RS^{-1})$ is an isomorphism.

Proof. [484, Proposition II.1.4 on page 51].   □

The next example is based on [489].

Example 8.16. Let G be a torsionfree amenable group. Suppose that the Kaplansky Conjecture 10.14 holds for G with coefficients in the field F, i.e.

FG has no non-trivial zero-divisors. We want to show that the set S of non-zero-divisor satisfies the left Ore condition. Since FG is a ring with involution the right Ore condition follows as well. Notice that then the Ore localization $(FG)S^{-1}$ is a skewfield, in which FG embeds.

Consider elements β, γ in FG such that γ is a non-zero-divisor. Choose a finite subset $S \subset G$ such that $s \in S \Rightarrow s^{-1} \in S$ and β and γ can be written as $\beta = \sum_{s \in S} b_s \cdot s$ and $\gamma = \sum_{s \in S} c_s \cdot s$. Since G satisfies the Følner condition (see Lemma 6.35), we can find a non-empty subset $A \subset G$ satisfying $|\partial_S A| \cdot |S| < |A|$. We want to find elements $\delta = \sum_{a \in A} d_a \cdot a$ and $\epsilon = \sum_{a \in A} e_a \cdot a$ such that ϵ is a non-zero-divisor and $\epsilon\beta = \delta\gamma$. Let $AS \subset G$ be the subset $\{a \cdot s \mid a \in A, s \in S\}$. Then the latter equation is equivalent to the following set of equations indexed by elements $u \in AS$

$$\sum_{a \in A, s \in S, s \cdot a = u} e_a b_s - d_a c_s = 0.$$

This is a system of $|AS|$ homogeneous equations over F in $2|A|$ variables d_a and e_a. Suppose that $|AS| < 2|A|$. Then we can find a solution different from zero, in other words, we can find δ and ϵ such that $\epsilon\beta = \delta\gamma$ and at least one of the elements δ and ϵ is different from zero. Since γ is a non-zero-divisor, $\epsilon = 0$ implies $\delta = 0$. Therefore $\epsilon \neq 0$. By assumption ϵ must be a non-zero-divisor. It remains to prove $|AS| < 2|A|$. From

$$AS \subset A \cup \bigcup_{s \in S} \{a \in A \mid a \cdot s^{-1} \notin A\} \cdot s \subset A \cup \bigcup_{s \in S} \partial_S A \cdot s.$$

we conclude

$$|AS| \leq |A| + |S| \cdot |\partial_S A| < 2|A|.$$

and the claim follows.

Let F_g be the free group with $g \geq 2$ generators. Let S be the multiplicatively closed subset of $\mathbb{C}F_g$ consisting of all non-zero-divisors. We mention that $\mathbb{C}F_g$ has no non-trivial zero-divisors and that $(\mathbb{C}F_g, S)$ does not satisfy the Ore condition.

The lamplighter group L is amenable but the set S of non-zero-divisors does not satisfies the left Ore condition [311].

8.2.2 Survey on von Neumann Regular Rings

Definition 8.17 (Von Neumann regular). *A ring R is called* von Neumann regular *if for any $r \in R$ there is an $s \in R$ with $rsr = r$.*

This notion should not be confused with the notion of a *regular* ring, i.e. a Noetherian ring for which every R-module has a projective resolution of finite dimension. For instance a principal ideal domain R is always regular since submodules of free R-modules are free (see [15, Corollary 1.2 in chapter 10 on page 353]), but it is von Neumann regular if and only if it is a field. The definition above is appropriate to check whether a ring is von Neumann regular. However, for structural questions the following equivalent characterizations are more useful.

Lemma 8.18. *The following assertions are equivalent for a ring R.*

(1) R is von Neumann regular;
(2) Every principal (left or right) ideal in R is generated by an idempotent;
(3) Every finitely generated (left or right) ideal in R is generated by an idempotent;
(4) Every finitely generated submodule of a finitely generated projective (left or right) R-module is a direct summand;
(5) Any finitely presented (left or right) R-module is projective;
(6) Every (left or right) R-module is R-flat.

Proof. see [444, Lemma 4.15, Theorem 4.16 and Theorem 9.15], [518, Theorem 4.2.9 on page 98]. □

Several processes preserve the property of a ring to be von Neumann regular.

Lemma 8.19. *(1) If R is von Neumann regular, then the matrix ring $M_n(R)$ is von Neumann regular;*
(2) The center of a von Neumann regular ring is von Neumann regular;
(3) A directed union of von Neumann regular rings is von Neumann regular.

Proof. Assertions (1) and (2) follow from [224, Theorem 1.7 and Theorem 1.14], whereas assertion (3) follows directly from the Definition 8.17 of von Neumann regular. □

Recall that an R-module is *semisimple* if any submodule of a module is a direct summand. A ring R is called *semisimple* if any R-module is semisimple. A ring R is semisimple if and only if R considered as an R-module is semisimple.

Lemma 8.20. *Let R be von Neumann regular. Then*

(1) Any element in R is either a zero-divisor or a unit;
(2) If a von Neumann regular ring is Noetherian or Artinian, then it is already semisimple.

Proof. (1) This follows from Lemma 8.18 (2). Assertion (2) is proved in [224, page 21]. □

In our situation the rings are *rings with involution* $*: R \to R$, sometimes also called *$*$-rings*. The involution is required to satisfy i.) $* \circ * = $ id; ii.) $*(1) = 1$, iii.) $*(r + s) = *(r) + *(s)$ and iv.) $*(r \cdot s) = *(s) \cdot *(r)$ for $r, s \in R$. We often write $*(r) = r^*$.

Example 8.21. The complex group ring $\mathbb{C}G$ is von Neumann regular if and only if G is locally finite. If G is locally finite this follows from Lemma 8.19 (3) since G is the directed union of its finitely generated subgroups and the complex group ring of a finite group is semisimple and hence in particular von Neumann regular.

Suppose that $\mathbb{C}G$ is von Neumann regular. Let $H \subset G$ be a finitely generated subgroup. Let $s_1, s_2, \ldots, s_r$ be a set of generators of H. There is an exact sequence of $\mathbb{C}H$-modules $\mathbb{C}H^n \xrightarrow{i} \mathbb{C}H \xrightarrow{\epsilon} \mathbb{C} \to 0$, where $\mathbb{C}$ comes with the trivial H-action, i is given by the $(r, 1)$ matrix $(s_1 - 1, \ldots, s_r - 1)$ and ϵ sends $\sum_{h \in H} \lambda_h h$ to $\sum_{h \in H} \lambda_h$. This sequence stays exact after applying $\mathbb{C}G \otimes_{\mathbb{C}H} -$. From Lemma 8.18 (5) we conclude that $\mathbb{C}[G/H] = \mathbb{C}G \otimes_{\mathbb{C}H} \mathbb{C}$ is a projective $\mathbb{C}G$-module. This implies that H is finite. This shows that G is locally finite.

More information about von Neumann regular rings can be found for instance in [224].

8.2.3 Basic Properties of the Algebra of Affiliated Operators

The next result summarizes the main properties of the algebra of affiliated operators.

Theorem 8.22 (The algebra of affiliated operators).

(1) *The set of non-zero-divisors S in $\mathcal{N}(G)$ satisfies the right Ore condition and the Ore localization $\mathcal{N}(G)S^{-1}$ is canonically isomorphic to $\mathcal{U}(G)$;*
(2) *$\mathcal{U}(G)$ is flat as $\mathcal{N}(G)$-module;*
(3) *$\mathcal{U}(G)$ is von Neumann regular;*
(4) *Let M be a finitely presented $\mathcal{N}(G)$-module. Then $\mathcal{U}(G) \otimes_{\mathcal{N}(G)} M$ is a finitely generated projective $\mathcal{U}(G)$-module. If additionally $\dim_{\mathcal{N}(G)}(M) = 0$, then $\mathcal{U}(G) \otimes_{\mathcal{N}(G)} M = 0$;*
(5) *A sequence $P_0 \to P_1 \to P_2$ of finitely generated projective $\mathcal{N}(G)$-modules is weakly exact at P_1 if and only if the induced sequence $\mathcal{U}(G) \otimes_{\mathcal{N}(G)} P_0 \to \mathcal{U}(G) \otimes_{\mathcal{N}(G)} P_1 \to \mathcal{U}(G) \otimes_{\mathcal{N}(G)} P_2$ is exact at $\mathcal{U}(G) \otimes_{\mathcal{N}(G)} P_1$;*
(6) *If $q \in M_n(\mathcal{U}(G))$ is a projection, then $q \in M_n(\mathcal{N}(G))$. If $e \in M_n(\mathcal{U}(G))$ is an idempotent, then there is a projection $p \in M_n(\mathcal{N}(G))$ such that $pe = e$ and $ep = p$ holds in $M_n(\mathcal{U}(G))$;*
(7) *Given a finitely generated projective $\mathcal{U}(G)$-module Q, there is a finitely generated projective $\mathcal{N}(G)$-module P such that $\mathcal{U}(G) \otimes_{\mathcal{N}(G)} P$ and Q are $\mathcal{U}(G)$-isomorphic;*
(8) *If P_0 and P_1 are two finitely generated projective $\mathcal{N}(G)$-modules, then $P_0 \cong_{\mathcal{N}(G)} P_1 \Leftrightarrow \mathcal{U}(G) \otimes_{\mathcal{N}(G)} P_0 \cong_{\mathcal{U}(G)} \mathcal{U}(G) \otimes_{\mathcal{N}(G)} P_1$.*

Proof. (1) An element $f \in \mathcal{N}(G) = \mathcal{B}(l^2(G))^G$ is not a zero-divisor if and only if f is a weak isomorphism (see Lemma 1.13). Hence the right Ore condition for the set S of non-zero-divisors of $\mathcal{N}(G)$ is satisfied by Lemma 8.8 (2). From Lemma 8.8 (2) and Lemma 8.15 (1) we obtain a canonical homomorphism of rings with involution $\mathcal{N}(G)S^{-1} \to \mathcal{U}(G)$ which is bijective.

(2) This follows from assertion (1) and Lemma 8.15 (3).

(3) Consider $a \in \mathcal{U}(G)$. Let $a = us$ be its polar decomposition into a G-equivariant partial isometry u and a positive affiliated operator s. Let $f \colon \mathbb{R} \to \mathbb{R}$ be the function which sends λ to λ^{-1} if $\lambda > 0$ and to zero if $\lambda \leq 0$. Then $f(s) \in \mathcal{U}$ and $sf(s)s = s$ holds in $\mathcal{U}(G)$. If we put $b = f(s)u^*$, we conclude

$$aba = usf(s)u^*us = usf(s)s = us = a.$$

(4) Because of Theorem 6.5 and Theorem 6.7 (3) and (4e) it suffices to prove the claim that $\dim_{\mathcal{N}(G)}(M) = 0$ implies $\mathcal{U}(G) \otimes_{\mathcal{N}(G)} M = 0$. From Lemma 6.28 we get an exact sequence $0 \to \mathcal{N}(G)^n \xrightarrow{s} \mathcal{N}(G)^n \to M \to 0$ with a weak isomorphism s. By Theorem 6.24 (3) $\nu(s)$ is a weak isomorphism. We get from Lemma 8.8 (1) that $\mathcal{U}(G) \otimes_{\mathcal{N}(G)} s$ is an isomorphism. Hence $\mathcal{U}(G) \otimes_{\mathcal{N}(G)} M = 0$.

(5) We firstly show for a finitely generated projective $\mathcal{N}(G)$-module P

$$P = 0 \Leftrightarrow \mathcal{U}(G) \otimes_{\mathcal{N}(G)} P = 0. \tag{8.23}$$

Choose an idempotent $e \colon \mathcal{N}(G)^n \to \mathcal{N}(G)^n$ whose image is $\mathcal{N}(G)$-isomorphic to P. Then $\mathcal{U}(G) \otimes_{\mathcal{N}(G)} e$ is an idempotent whose image is $\mathcal{U}(G)$-isomorphic to $\mathcal{U}(G) \otimes_{\mathcal{N}(G)} P$. The matrix in $M_n(\mathcal{U}(G))$ describing $\mathcal{U}(G) \otimes_{\mathcal{N}(G)} e$ is the image of the matrix in $M_n(\mathcal{N}(G))$ describing e under the injection $M_n(\mathcal{N}(G)) \to M_n(\mathcal{U}(G))$ induced by the inclusion of rings $\mathcal{N}(G) \subset \mathcal{U}(G)$. Hence e is zero if and only if $\mathcal{U}(G) \otimes_{\mathcal{N}(G)} e = 0$. Therefore (8.23) is true.

Consider a sequence of finitely generated projective $\mathcal{N}(G)$-modules $P_0 \xrightarrow{f_0} P_1 \xrightarrow{f_1} P_2$. Since $\mathcal{N}(G)$ is semihereditary (see Theorem 6.5 and Theorem 6.7 (1)), the image of $f_1 \circ f_0$ is a finitely generated projective $\mathcal{N}(G)$-module. We conclude from assertion (2) and (8.23) that $f_1 \circ f_0$ is zero if and only if $\mathcal{U}(G) \otimes_{\mathcal{N}(G)} f_1 \circ \mathcal{U}(G) \otimes_{\mathcal{N}(G)} f_0$ is zero. Hence we can assume without loss of generality that $\mathrm{im}(f_0) \subset \ker(f_1)$. Recall that the given sequence $P_0 \xrightarrow{f_0} P_1 \xrightarrow{f_1} P_2$ is weakly exact if and only if the finitely generated projective $\mathcal{N}(G)$-module $\mathbf{P}(\ker(f_1)/\mathrm{im}(f_0))$ is trivial and that $\ker(f_1)/\mathrm{im}(f_0) = \mathbf{P}(\ker(f_1)/\mathrm{im}(f_0)) \oplus \mathbf{T}(\ker(f_1)/\mathrm{im}(f_0))$ (see Definition 6.1, Theorem 6.5 and Theorem 6.7 (3)). Now apply Theorem 6.7 (4e), assertion (2), assertion (4) and (8.23).

(6) Let $q \in M_n(\mathcal{U}(G))$ be a projection. Then the induced affiliated operator $l^2(G)^n \to l^2(G)^n$ is bounded on its domain, namely by 1, and hence everywhere defined by Lemma 8.3 (4). Therefore $q \in M_n(\mathcal{N}(G))$.

Let $e \in M_n(\mathcal{U}(G))$ be an idempotent. Put $z = 1 - (e^* - e)^2 = 1 + (e^* - e)(e^* - e)^*$. This element z is invertible in $M_n(\mathcal{U}(G))$ by Lemma 1.13 and Lemma 8.8 (1) because the induced affiliated operator $l^2(G) \to l^2(G)$ is obviously injective. We have $z = z^*$, $ze = ez = ee^*e$, $ze^* = e^*ee^* = e^*z$. We also get $z^{-1}e = ez^{-1}$ and $z^{-1}e^* = e^*z^{-1}$. Put $p = ee^*z^{-1}$. Then one easily checks $p^* = p$, $p^2 = p$, $pe = e$ and $ep = p$.

(7) Let Q be a finitely generated projective $\mathcal{U}(G)$-module. Let $e \in M_n(\mathcal{U}(G))$ be an idempotent whose image is $\mathcal{U}(G)$-isomorphic to Q. By assertion (6) there is a projection $p \in M_n(\mathcal{N}(G))$ such that $pe = p$ and $ep = e$ holds in $M_n(\mathcal{U}(G))$. Put $P = \mathrm{im}(p)$. Then P is a finitely generated projective $\mathcal{N}(G)$-module with $\mathcal{U}(G) \otimes_{\mathcal{N}(G)} P \cong_{\mathcal{U}(G)} Q$.

(8) Let P_0 and P_1 be two finitely generated projective $\mathcal{N}(G)$-modules with $\mathcal{U}(G) \otimes_{\mathcal{N}(G)} P_0 \cong_{\mathcal{U}(G)} \mathcal{U}(G) \otimes_{\mathcal{N}(G)} P_0$. Because of assertion (1) there is an $\mathcal{N}(G)$-map $f \colon P_0 \to P_1$ such that $\mathcal{U}(G) \otimes_{\mathcal{N}(G)} f$ is an isomorphism. By assertion (5) $f \colon P_0 \to P_1$ is a weak isomorphism. Hence P_0 and P_1 are $\mathcal{N}(G)$-isomorphic by Lemma 6.28 (1). $\qquad\square$

8.3 Dimension Theory and L^2-Betti Numbers over the Algebra of Affiliated Operators

In this section we construct a dimension function $\dim_{\mathcal{U}(G)}$ for arbitrary $\mathcal{U}(G)$-modules such that Additivity, Cofinality and Continuity hold and for any $\mathcal{N}(G)$-module M we recapture $\dim_{\mathcal{N}(G)}(M)$ by $\dim_{\mathcal{U}(G)}(\mathcal{U}(G) \otimes_{\mathcal{N}(G)} M)$. This allows us to define L^2-Betti numbers over $\mathcal{U}(G)$.

Notation 8.24. *Denote by $L^{\mathrm{dr}}(R^n)$ the set of direct summands in R^n with the partial ordering $\leq$ induced by inclusion. Denote by $L^{\mathrm{dr}}(l^2(G)^n)$ the set of Hilbert $\mathcal{N}(G)$-submodules of $l^2(G)^n$ with the partial ordering $\leq$ induced by inclusion.*

Define for a finitely generated projective $\mathcal{U}(G)$-module Q

$$\dim_{\mathcal{U}(G)}(Q) := \dim_{\mathcal{N}(G)}(P) \qquad \in [0, \infty), \tag{8.25}$$

where P is any finitely generated projective $\mathcal{N}(G)$-module P with $\mathcal{U}(G) \otimes_{\mathcal{N}(G)} P \cong_{\mathcal{U}(G)} Q$. This definition makes sense because of Lemma 8.22 (7) and (8)

Lemma 8.26. *(1) Let $j \colon L^{\mathrm{dr}}(\mathcal{N}(G)^n) \to L^{\mathrm{dr}}(\mathcal{U}(G)^n)$ be the map which sends $P \subset \mathcal{N}(G)^n$ to the image of the composition*

$$\mathcal{U}(G) \otimes_{\mathcal{N}(G)} P \xrightarrow{\mathcal{U}(G) \otimes_{\mathcal{N}(G)} i} \mathcal{U}(G) \otimes_{\mathcal{N}(G)} \mathcal{N}(G)^n \xrightarrow{l} \mathcal{U}(G)^n,$$

where $i \colon P \to \mathcal{N}(G)^n$ is the inclusion and l the obvious isomorphism. Then j preserves the partial ordering and is bijective;

(2) The map $k \colon L^{\mathrm{dr}}(\mathcal{N}(G)^n) \to L^{\mathrm{dr}}(l^2(G)^n)$ induced by the functor ν (see (6.22)) is bijective and preserves the ordering;

(3) For a subset $S \subset L^{\mathrm{dr}}(\mathcal{U}(G)^n)$ there is a least upper bound $\sup(S) \in L^{\mathrm{dr}}(\mathcal{U}(G)^n)$. If S is directed under inclusion, then

$$\dim_{\mathcal{U}(G)}(\sup(S)) = \sup\{\dim_{\mathcal{U}(G)}(P) \mid P \in S\}.$$

The analogous statement holds for $\mathcal{N}(G)$ instead of $\mathcal{U}(G)$.

Proof. (1) Obviously j preserves the partial ordering. Bijectivity follows from Lemma 8.22 (6).

(2) This is obvious.

(3) Because of assertions (1) and (2) it suffices to show for any subset $S \subset L^{\mathrm{dr}}(l^2(G)^n))$ that a least upper bound $\sup(S)$ exists in $L^{\mathrm{dr}}(l^2(G)^n)$ and that $\dim_{\mathcal{N}(G)}(\sup(S)) = \sup\{\dim_{\mathcal{N}(G)}(V) \mid V \in S\}$ holds if S is directed under $\leq$. We construct $\sup(S)$ as the intersection of all Hilbert $\mathcal{N}(G)$-submodules V of $l^2(G)^n$ for which each element of S is contained in V. Suppose that S is directed under inclusion. Then $\sup(S)$ is the same as the closure of the union of all elements in S. Now the second claim follows from Lemma 1.12 (3). This finishes the proof of Lemma 8.26. $\qquad\square$

Lemma 8.27. *The pair* $(\mathcal{U}(G), \dim_{\mathcal{U}(G)})$ *satisfies Assumption 6.2.*

Proof. We already know that $(\mathcal{N}(G), \dim_{\mathcal{N}(G)})$ satisfies Assumption 6.2 (see Theorem 6.5). Hence $(\mathcal{U}(G), \dim_{\mathcal{U}(G)})$ satisfies Assumption 6.2 (1). It remains to prove Assumption 6.2 (2). One easily checks that it suffices to do this for a submodule $K \subset \mathcal{U}(G)^n$ for some positive integer n. Let $S = \{P \mid P \subset K, P \text{ fin. gen. } \mathcal{U}(G)\text{-module}\}$. Notice that each $P \in S$ belongs to $L^{\mathrm{dr}}(\mathcal{U}(G)^n)$ since $\mathcal{U}(G)$ is von Neumann regular (see Lemma 8.18 (4) and Lemma 8.22 (3)). By assertion (3) of the previous Lemma 8.26 the supremum $\sup(S) \in L^{\mathrm{dr}}(\mathcal{U}(G)^n)$ exists and satisfies

$$\dim_{\mathcal{U}(G)}(\sup(S)) = \sup\{\dim_{\mathcal{U}(G)}(P) \mid P \in S\}.$$

Hence it remains to prove $\overline{K} = \sup(S)$. Consider $f \colon \mathcal{U}(G)^n \to \mathcal{U}(G)$ with $K \subset \ker(f)$. Then $\ker(f) \in L^{\mathrm{dr}}(\mathcal{U}(G)^n)$ and $P \subset \ker(f)$ for $P \in S$. This implies $\sup(S) \subset \ker(f)$. Since any finitely generated submodule of K belongs to S and hence is contained in $\sup(S)$, we get $K \subset \sup(S)$ and hence $\overline{K} \subset \sup(S)$. $\qquad\square$

Definition 8.28. *Let M be a $\mathcal{U}(G)$-module. Define its* extended von Neumann dimension

$$\dim_{\mathcal{U}(G)}(M) \in [0, \infty]$$

by the extension constructed in Definition 6.6 of $\dim_{\mathcal{U}(G)}(Q)$ which we have introduced for any finitely generated projective $\mathcal{U}(G)$-module Q in (8.25).

Specializing Theorem 6.7 about dimension functions for arbitrary modules to this case and using Lemma 6.28 (3) and Theorem 8.22 (see also [435, Chapter 3]) we obtain

Theorem 8.29. [**Dimension function for arbitrary $\mathcal{U}(G)$-modules**].
The dimension $\dim_{\mathcal{U}(G)}$ satisfies Additivity, Cofinality and Continuity. Given any $\mathcal{N}(G)$-module M, we get

$$\dim_{\mathcal{N}(G)}(M) = \dim_{\mathcal{U}(G)}(\mathcal{U}(G) \otimes_{\mathcal{N}(G)} M).$$

If P is a projective $\mathcal{U}(G)$-module, then $\dim_{\mathcal{U}(G)}(P) = 0 \Leftrightarrow P = 0$.

Now we can define L^2-Betti numbers working with $\mathcal{U}(G)$ instead of $\mathcal{N}(G)$. This corresponds in the classical setting for Betti numbers of a CW-complex Y to define the p-th Betti number as the dimension of the rational vector space $H_p(Y;\mathbb{Q})$ instead of the rank of the abelian group $H_p(Y;\mathbb{Z})$.

Definition 8.30. *Let X be a (left) G-space. Define the singular homology $H_p^G(X;\mathcal{U}(G))$ of X with coefficients in $\mathcal{U}(G)$ to be the homology of the $\mathcal{U}(G)$-chain complex $\mathcal{U}(G) \otimes_{\mathbb{Z}G} C_*^{\mathrm{sing}}(X)$, where $C_*^{\mathrm{sing}}(X)$ is the singular chain complex of X with the induced $\mathbb{Z}G$-structure. Define the p-th L^2-Betti number of X by*

$$b_p^{(2)}(X;\mathcal{U}(G)) \;:=\; \dim_{\mathcal{U}(G)}(H_p^G(X;\mathcal{U}(G))) \qquad \in [0,\infty],$$

where $\dim_{\mathcal{U}(G)}$ is the extended dimension function of Definition 8.28.

We conclude from Lemma 8.22 (2) and the above Theorem 8.29

Theorem 8.31. *Let X be a G-space with G-action. Then we get*

$$b_p^{(2)}(X;\mathcal{N}(G)) \;:=\; b_p^{(2)}(X;\mathcal{U}(G)).$$

8.4 Various Notions of Torsion Modules over a Group von Neumann Algebra

In this section we consider three different notions of torsion modules in the category of $\mathcal{N}(G)$-modules and analyse their relationship. We have already introduced for an $\mathcal{N}(G)$-module M its submodule $\mathbf{T}M$ in Definition 6.1 as the closure of $\{0\}$ in M, or, equivalently, as the kernel of the canonical map $i(M): M \to (M^*)^*$. Next we define

Definition 8.32. *Let M be an $\mathcal{N}(G)$-module. Define $\mathcal{N}(G)$-submodules*

$$\mathbf{T}_{\dim}M := \bigcup\{N \mid N \text{ is an } \mathcal{N}(G)\text{-submodule of } M \text{ with } \dim_{\mathcal{N}(G)}(N) = 0\};$$
$$\mathbf{T}_{\mathcal{U}}M := \ker(j) \quad \text{for } j: M \to \mathcal{U}(G) \otimes_{\mathcal{N}(G)} M, \quad m \mapsto 1 \otimes m.$$

Notice that $\mathbf{T}_{\dim}(M) \subset M$ is indeed an $\mathcal{N}(G)$-submodule by Additivity (see Theorem 6.7 (4b)) and can be characterized to be the largest $\mathcal{N}(G)$-submodule of M of dimension zero. Because of Lemma 8.15 (2) or [484, Corollary II.3.3 on page 57] one can identify $\mathbf{T}_{\mathcal{U}}M$ with the set of elements $m \in M$ for which there is a non-zero-divisor $r \in \mathcal{N}(G)$ with $rm = 0$. Notice that

$$\begin{aligned}
M = \mathbf{T}M &\Leftrightarrow M^* = \hom_{\mathcal{N}(G)}(M,\mathcal{N}(G)) = 0; \\
M = \mathbf{T}_{\dim}M &\Leftrightarrow \dim_{\mathcal{N}(G)}(M) = 0; \\
M = \mathbf{T}_{\mathcal{U}}M &\Leftrightarrow \mathcal{U}(G) \otimes_{\mathcal{N}(G)} M = 0.
\end{aligned}$$

Lemma 8.33. *(1) If M is an $\mathcal{N}(G)$-module, then $\mathbf{T}_{\mathcal{U}}M \subset \mathbf{T}_{\dim}M \subset \mathbf{T}M$;*

(2) If M is a finitely generated projective $\mathcal{N}(G)$-module, then $\mathbf{T}_{\mathcal{U}}M = \mathbf{T}_{\dim}M = \mathbf{T}M = 0$;

(3) If M is a finitely presented $\mathcal{N}(G)$-module, then $\mathbf{T}_{\mathcal{U}}M = \mathbf{T}_{\dim}M = \mathbf{T}M$;

(4) If M is a finitely generated $\mathcal{N}(G)$-module, then $\mathbf{T}_{\dim}M = \mathbf{T}M$.

Proof. (1) Since $\mathcal{U}(G)$ is flat over $\mathcal{N}(G)$, the $\mathcal{U}(G)$-map $\mathcal{U}(G) \otimes_{\mathcal{N}(G)} (\mathbf{T}_{\mathcal{U}}M) \rightarrow \mathcal{U}(G) \otimes_{\mathcal{N}(G)} M$ is injective. Since this map at the same time is the zero map, $\mathcal{U}(G) \otimes_{\mathcal{N}(G)} (\mathbf{T}_{\mathcal{U}}M)$ is trivial. We get $\dim_{\mathcal{N}(G)}(\mathbf{T}_{\mathcal{U}}M) = 0$ from Theorem 8.29. This implies $\mathbf{T}_{\mathcal{U}}M \subset \mathbf{T}_{\dim}M$. Let $f \colon M \rightarrow \mathcal{N}(G)$ be any $\mathcal{N}(G)$-map. Then $f(\mathbf{T}_{\dim}M)$ is an $\mathcal{N}(G)$-submodule of dimension zero in $\mathcal{N}(G)$ by Additivity (Theorem 6.7 (4b)). By Theorem 6.7 (1) and (4b) each finitely generated $\mathcal{N}(G)$-submodule of $f(\mathbf{T}_{\dim}M) \subset \mathcal{N}(G)$ is projective and has dimension zero. Any finitely generated projective $\mathcal{N}(G)$-module of dimension zero is trivial (Lemma 6.28 (3)). Hence $f(\mathbf{T}_{\dim}M)$ is trivial. This implies $\mathbf{T}_{\dim}M \subset \mathbf{T}M$.

(2) Obviously $\mathbf{T}P$ is trivial for a finitely generated projective $\mathcal{N}(G)$-module P. Now apply assertion (1).

(3) Let M be finitely presented. Then $M = \mathbf{P}M \oplus \mathbf{T}M$ with finitely generated projective $\mathbf{P}M$ by Theorem 6.7 (1). Because of assertions (1) and (2) it suffices to prove $\mathbf{T}_{\mathcal{U}}(\mathbf{T}M) = \mathbf{T}M$. This follows from Theorem 6.7 (4e) and Lemma 8.22 (4).

(4) This follows from Theorem 6.7 (4e). $\square$

Example 8.34. We want to construct a finitely generated non-trivial $\mathcal{N}(G)$-module M with the property that $\mathbf{T}_{\mathcal{U}}(M) = 0$ and $\mathbf{T}_{\dim}M = M$ and in particular $\mathbf{T}_{\mathcal{U}}(M) \neq \mathbf{T}_{\dim}M$. Let $I_1 \subset I_2 \subset \ldots \subset \mathcal{N}(G)$ be a nested sequence of ideals which are direct summands in $\mathcal{N}(G)$ such that $\dim_{\mathcal{N}(G)}(I_n) \neq 1$ and $\lim_{n \to \infty} \dim_{\mathcal{N}(G)}(I_n) = 1$. Let I be the ideal $\bigcup_{n=1}^{\infty} I_n$. Put $M = \mathcal{N}(G)/I$. Then $\dim_{\mathcal{N}(G)}(M) = 0$ by Theorem 6.7 (4b) and (4c). This implies $\mathbf{T}_{\dim}M = M$. By Lemma 8.33 (2) we get $\mathbf{T}_{\mathcal{U}}\mathcal{N}(G)/I_n = 0$ for $n \geq 1$. Lemma 8.36 (1) below implies that $\mathbf{T}_{\mathcal{U}}M = \operatorname{colim}_{n \geq 1} \mathbf{T}_{\mathcal{U}}\mathcal{N}(G)/I_n = 0$.

We get for instance for $G = \mathbb{Z}$ such a sequence of ideals $(I_n)_{n \geq 1}$ by the construction in Example 1.14.

Example 8.35. An example of an $\mathcal{N}(\mathbb{Z})$-module M with $M^* = 0$ and $\dim_{\mathcal{N}(G)}(M) = 1$ and hence with $\mathbf{T}_{\dim}M \neq \mathbf{T}M$ is given in the exercises of Chapter 6.

The notion $\mathbf{T}_{\mathcal{U}}$ has the best properties in comparison with $\mathbf{T}_{\dim}$ and $\mathbf{T}$.

Lemma 8.36. *(1) The functor $\mathbf{T}_{\mathcal{U}}$ is left exact and commutes with colimits over directed systems;*

(2) An $\mathcal{N}(G)$-module M is cofinal measurable in the sense of [343, Definition 2.1], i.e. all its finitely generated submodules are quotients of finitely presented $\mathcal{N}(G)$-modules of dimension zero, if and only if $\mathbf{T}_{\mathcal{U}}M = M$.

(1) Let $\{M_i \mid i \in I\}$ be a directed system of $\mathcal{N}(G)$-modules. Since the functor $\mathcal{U}(G) \otimes_{\mathcal{N}(G)} -$ is compatible with arbitrary colimits and the functor $\operatorname{colim}_{i \in I}$ is exact for directed systems, the canonical map $\operatorname{colim}_{i \in I} \mathbf{T}_{\mathcal{U}} M_i \to \mathbf{T}_{\mathcal{U}} (\operatorname{colim}_{i \in I} M_i)$ is an isomorphism. Since $\mathcal{U}(G)$ is flat over $\mathcal{N}(G)$ by Theorem 8.22 (2), $\mathbf{T}_{\mathcal{U}}$ is left exact.

(2) Obviously submodules of cofinal-measurable $\mathcal{N}(G)$-modules are cofinal-measurable again. Because of Lemma 8.33 (3) and assertion (1) it suffices to prove for a finitely generated $\mathcal{N}(G)$-module M that $\mathcal{U}(G) \otimes_{\mathcal{N}(G)} M = 0$ is true if and only if M is the quotient of a finitely presented $\mathcal{N}(G)$-module N with $\mathcal{U}(G) \otimes_{\mathcal{N}(G)} N = 0$.

The if-part is obvious, the only-if-part is proved as follows. Let M be a finitely generated $\mathcal{U}(G)$-module with $\mathcal{U}(G) \otimes_{\mathcal{N}(G)} M = 0$. Choose an epimorphism $f \colon \mathcal{N}(G)^n \to M$. Since $\mathcal{U}(G)$ is flat over $\mathcal{N}(G)$ by Theorem 8.22 (2), the inclusion $i \colon \ker(f) \to \mathcal{N}(G)^n$ induces an isomorphism $\mathcal{U}(G) \otimes_{\mathcal{N}(G)} \ker(f) \to \mathcal{U}(G) \otimes_{\mathcal{N}(G)} \mathcal{N}(G)^n$. Hence we can find an $\mathcal{A}$-map $g \colon \mathcal{N}(G)^n \to \ker(f)$ such that $\mathcal{U}(G) \otimes_{\mathcal{N}(G)} g$ is an isomorphism. Hence $\operatorname{coker}(i \circ g)$ is a finitely presented $\mathcal{N}(G)$-module with $\mathcal{U}(G) \otimes_{\mathcal{N}(G)} \operatorname{coker}(i \circ g) = 0$ which maps surjectively onto M. $\qquad\square$

8.5 Miscellaneous

Let R be a principal ideal domain with quotient field F. We have already mentioned in Example 6.12 that R together with the usual notion of the rank of a finitely generated free R-module satisfies Assumption 6.2 and that the extended dimension function of Definition 6.6 is given by $\dim_R(M) = \dim_F(F \otimes_R M)$. Notice that F is the Ore localization of R with respect to the set S of non-zero-divisors and that all the properties of $\mathcal{U}(G)$ as stated in Lemma 8.22 and Theorem 8.29 do also hold for F. Moreover, the condition $M = \mathbf{T}_{\mathcal{U}} M$ translates to the classical notion of a torsion module, namely that $F \otimes_R M = 0$, or, equivalently, that for each element $m \in M$ there is an element $r \in R$ with $r \neq 0$ and $rm = 0$.

More information about affiliated rings and their regularity properties can be found in [36], [37], [247]. A systematic study of $\mathcal{U}(G)$ and dimension functions is carried out in [435, chapters 2 and 3].

Exercises

8.1. A ring R with involution $*$ is called $*$-*regular* if it is von Neumann regular and for any $r \in R$ we have $r^* r = 0 \Leftrightarrow r = 0$. Show that $\mathcal{U}(G)$ is $*$-regular.

8.2. Construct a commutative diagram of $\mathbb{C}$-categories

$$\{\text{fin. gen. proj. } \mathcal{N}(G)\text{-mod.}\} \xrightarrow{\ \nu\ } \{\text{fin. gen. Hilb. } \mathcal{N}(G)\text{-mod.}\}$$

$$\downarrow \qquad\qquad\qquad\qquad\qquad\qquad \downarrow$$

$$\{\text{fin. gen. proj. } \mathcal{U}(G)\text{-mod.}\} \xrightarrow{\ \nu_{\mathcal{U}}\ } \{\text{fin. gen. Hilb. } \mathcal{N}(G)\text{-mod.}\}_{\mathcal{U}}$$

whose vertical arrows are the obvious inclusions and whose horizontal arrows are equivalences of $\mathbb{C}$-categories.

8.3. Let G be a group. Make sense out of the following chain of inclusions

$$G \subset \mathbb{Z}G \subset \mathbb{C}G \subset l^1(G) \subset C_r^*(G) \subset \mathcal{N}(G) \subset l^2(G) \subset \mathcal{U}(G),$$

where $l^1(G)$ is the Banach algebra of formal sums $\sum_{g \in G} \lambda_g \cdot g$ which satisfy $\sum_{g \in G} |\lambda_g| < \infty$.

8.4. Let $f \colon P \to Q$ be a homomorphism of finitely generated projective $\mathcal{U}(G)$-modules with $\dim_{\mathcal{U}(G)}(P) = \dim_{\mathcal{U}(G)}(Q)$. Show that the following assertions are equivalent: i.) f is injective, ii.) f is surjective, iii.) f is bijective.

8.5. Show that the functor $\mathbf{T}_{\dim}$ is left exact, but the functor $\mathbf{T}$ is not left exact.

8.6. Show that the functor $\mathbf{T}_{\dim}$ does not commute with colimits over directed systems, unless all structure maps are injective.

8.7. Give an example of a non-trivial $\mathcal{N}(\mathbb{Z})$-module M and a directed system $\{M_i \mid i \in I\}$ of submodules (directed by inclusion) with the properties that $M^* = 0$ and $M_i \cong \mathcal{N}(\mathbb{Z})$ for $i \in I$. Show that this implies $\mathbf{T}M \neq \bigcup_{i \in I} \mathbf{T}M_i$.

8.8. Given an $\mathcal{N}(G)$-module M, define $\mathbf{T}'_{\mathcal{U}}M$ by the cokernel of $i \colon M \to \mathcal{U}(G) \otimes_{\mathcal{N}(G)} M, \quad m \mapsto 1 \otimes m$. Construct for an exact sequence of $\mathcal{N}(G)$-modules $0 \to L \to M \to N \to 0$ a natural exact sequence $0 \to \mathbf{T}_{\mathcal{U}}L \to \mathbf{T}_{\mathcal{U}}M \to \mathbf{T}_{\mathcal{U}}N \to \mathbf{T}'_{\mathcal{U}}L \to \mathbf{T}'_{\mathcal{U}}M \to \mathbf{T}'_{\mathcal{U}}N \to 0$.

8.9. Show for an infinite locally finite group G that

$$b_p^{(2)}(G; \mathcal{N}(G)) = b_p^{(2)}(G; \mathcal{U}(G)) = 0$$

for all $p \geq 0$, but $H_0^G(EG; \mathcal{U}(G)) \neq 0$.

Show that we get an $\mathcal{N}(G)$-module M such that

$$\dim_{\mathcal{N}(G)}(M) = \dim_{\mathcal{U}(G)}(\mathcal{U}(G) \otimes_{\mathcal{U}(G)} M) = 0,$$

but $\mathcal{U}(G) \otimes_{\mathcal{U}(G)} M \neq \{0\}$.

9. Middle Algebraic K-Theory and L-Theory of von Neumann Algebras

Introduction

So far we have only dealt with the von Neumann algebra $\mathcal{N}(G)$ of a group G. We will introduce and study in Section 9.1 the general concept of a von Neumann algebra. We will explain the decomposition of a von Neumann algebra into different types. Any group von Neumann algebra is a finite von Neumann algebra. A lot of the material of the preceding chapters can be extended from group von Neumann algebras to finite von Neumann algebras as explained in Subsection 9.1.4. In Sections 9.2 and 9.3 we will compute $K_n(\mathcal{A})$ and $K_n(\mathcal{U})$ for $n = 0, 1$ in terms of the centers $\mathcal{Z}(\mathcal{A})$ and $\mathcal{Z}(\mathcal{U})$, where $\mathcal{U}$ is the algebra of operators which are affiliated to a finite von Neumann algebra $\mathcal{A}$. The quadratic L-groups $L_n^\epsilon(\mathcal{A})$ and $L_n^\epsilon(\mathcal{U})$ for $n \in \mathbb{Z}$ and the decorations $\epsilon = p, h, s$ are determined in Section 9.4. The symmetric L-groups $L_\epsilon^n(\mathcal{A})$ and $L_\epsilon^n(\mathcal{U})$ turn out to be isomorphic to their quadratic counterparts.

In Section 9.5 we will apply the results above to detect elements in the K- and G-theory of the group ring. We will show for a finite normal subgroup H of an arbitrary (discrete) group G that the map $\mathrm{Wh}(H)^G \to \mathrm{Wh}(G)$ induced by the inclusion of H into G has finite kernel, where the action of G on $\mathrm{Wh}(H)$ comes from the conjugation action of G on H (see Theorem 9.38). We will present some computations of the Grothendieck group $G_0(\mathbb{C}G)$ of finitely generated (not necessarily projective) $\mathbb{C}G$-modules. The main result, Theorem 9.65, says for an amenable group G that the rank of the abelian group $G_0(\mathbb{C}G)$ is greater or equal to the cardinality of the set $\mathrm{con}(G)_{f,cf}$ of conjugacy classes (g) of elements $g \in G$ for which $|g| < \infty$ and $|(g)| < \infty$ hold. The map detecting elements in $G_0(\mathbb{C}G)$ is based on the center valued dimension which is related to the Hattori-Stallings rank. We will review the Hattori-Stallings rank, the Isomorphism Conjecture for $K_0(\mathbb{C}G)$ and the Bass Conjecture in Subsection 9.5.2. This chapter needs only a small input from Chaper 8 and is independent of the other chapters.

9.1 Survey on von Neumann Algebras

So far we have only considered the von Neumann algebra $\mathcal{N}(G)$ of a group G (see Definition 1.1). In this section we introduce and study the notion of a von Neumann algebra in general.

9.1.1 Definition of a von Neumann Algebra

Let H be a Hilbert space and $\mathcal{B}(H)$ be the C^*-algebra of bounded (linear) operators from H to itself, where the norm is the operator norm. The *norm topology* is the topology on $\mathcal{B}(H)$ induced by the operator norm. The *strong topology, ultra-weak topology* or *weak topology* respectively is the unique topology such that a subset $A \subset \mathcal{B}(H)$ is closed if and only if for any net $(x_i)_{i \in I}$ of elements in A, which converges strongly, ultra-weakly or weakly respectively to an element $x \in \mathcal{B}(H)$, also x belongs to A (see Subsection 1.1.3 for the various notions of convergence). The norm topology contains both the strong topology and the ultra-weak topology, the strong topology contains the weak topology, the ultra-weak topology contains the weak topology. In general these inclusions are strict and there is no relation between the ultra-weak and strong topology [144, I.3.1 and I.3.2]. A map of topological spaces $h \colon X \to Y$ is continuous if and only if for any net $(x_i)_{i \in I}$ converging to x the net $(h(x_i))_{i \in I}$ converges to $h(x)$. Notice that this characterization of continuity is valid for all topological spaces. Only if X satisfies the first countability axiom (i.e. any point has a countable neighborhood basis) one can use sequences $(x_n)_{n \geq 0}$ instead of nets, but this axiom will not be satisfied for some of the topologies on $\mathcal{B}(H)$ introduced above.

Definition 9.1 (Von Neumann algebra). *A von Neumann algebra $\mathcal{A}$ is a sub-$*$-algebra of $\mathcal{B}(H)$ which is closed in the weak topology and contains* id$\colon H \to H$.

The condition weakly closed can be rephrased in a more algebraic fashion as follows. Given a subset $M \subset \mathcal{B}(H)$, its *commutant* M' is defined to be the subset $\{f \in \mathcal{B}(H) \mid fm = mf \text{ for all } m \in M\}$. If we apply this construction twice, we obtain the *double commutant* M''. The proof of the following so called Double Commutant Theorem can be found in [282, Theorem 5.3.1. on page 326].

Theorem 9.2 (Double Commutant Theorem). *Let $M \subset \mathcal{B}(H)$ be a sub-$*$-algebra, i.e. M is closed under addition, scalar multiplication, multiplication and under the involution $*$ and contains 0 and 1. Then the following assertions are equivalent.*

(1) M is closed in the weak topology;
(2) M is closed in the strong topology;
(3) $M = M''$.

In particular the closure of M in the weak topology as well as in the strong topology is M''.

9.1.2 Types and the Decomposition of von Neumann Algebras

Next we recall the various types of von Neumann algebras. A *projection* p in a von Neumann algebra $\mathcal{A}$ is an element satisfying $p^2 = p$ and $p^* = p$. It is called *abelian* if $p\mathcal{A}p$ is a commutative algebra. Two projections p and q are called *equivalent* $p \sim q$, if there is an element $u \in \mathcal{A}$ with $p = uu^*$ and $q = u^*u$. We write $p \leq q$ if $qp = p$. A projection p is *finite*, if $q \leq p$ and $q \sim p$ together imply $p = q$, and *infinite* otherwise. A projection p is *properly infinite* if p is infinite and cp is either zero or infinite for all central projections $c \in \mathcal{A}$. The *central carrier* c_p of a projection p is the smallest central projection $c_p \in \mathcal{A}$ satisfying $p \leq c_p$.

A von Neumann algebra $\mathcal{A}$ is *of type I* if it has an abelian projection whose central carrier is the identity 1. If $\mathcal{A}$ has no non-zero abelian projection but possesses a finite projection with central carrier 1, then $\mathcal{A}$ is *of type II*. If $\mathcal{A}$ has no non-zero finite projection, it is *of type III*. We call $\mathcal{A}$ *finite*, *infinite* or *properly infinite* respectively if 1 is a projection, which is finite, infinite or properly infinite respectively. A von Neumann algebra $\mathcal{A}$ is *of type I_f* or *of type II_1* respectively if $\mathcal{A}$ is finite and of type I or of type II respectively. It is *of type I_∞* or *of type II_∞* respectively if $\mathcal{A}$ is properly infinite and of type I or of type II respectively. All von Neumann algebras of type III are properly infinite. A von Neumann algebra can only be of at most one of the types I_f, I_∞, II_1, II_∞ and III. A von Neumann algebra $\mathcal{A}$ is called a *factor* if its center $\mathcal{Z}(\mathcal{A}) := \{a \in \mathcal{A} \mid ab = ba \text{ for all } b \in \mathcal{A}\}$ consists of $\{\lambda \cdot 1 \mid \lambda \in \mathbb{C}\}$. A factor is of precisely one of the types I_f, I_∞, II_1, II_∞ or III.

One has the following unique decomposition [283, Theorem 6.5.2 on page 422].

Theorem 9.3. *Given a von Neumann algebra $\mathcal{A}$, there is a natural unique decomposition*

$$\mathcal{A} = \mathcal{A}_{I_f} \times \mathcal{A}_{I_\infty} \times \mathcal{A}_{II_1} \times \mathcal{A}_{II_\infty} \times \mathcal{A}_{III}$$

into von Neumann algebras of type I_f, I_∞, II_1, II_∞ and III. In particular one obtains natural decompositions for the K-groups

$$K_n(\mathcal{A}) = K_n(\mathcal{A}_{I_f}) \times K_n(\mathcal{A}_{I_\infty}) \times K_n(\mathcal{A}_{II_1}) \times K_n(\mathcal{A}_{II_\infty}) \times K_n(\mathcal{A}_{III}).$$

Lemma 9.4. *Let G be a discrete group. Let G_f be the normal subgroup of G consisting of elements $g \in G$, whose centralizer has finite index (or, equivalently, whose conjugacy class (g) consists of finitely many elements). Then*

(1) The group von Neumann algebra $\mathcal{N}(G)$ is of type I if and only if G is virtually abelian;

(2) The group von Neumann algebra $\mathcal{N}(G)$ is of type II if and only if the index of G_f in G is infinite;

(3) Suppose that G is finitely generated. Then $\mathcal{N}(G)$ is of type I_f if G is virtually abelian, and of type II_1 if G is not virtually abelian;

(4) The group von Neumann algebra $\mathcal{N}(G)$ is a factor if and only if G_f is the trivial group.

Proof. (1) This is proved in [285], [490].

(2) This is proved in [285],[366].

(3) This follows from (1) and (2) since for finitely generated G the group G_f has finite index in G if and only if G is virtually abelian.

(4) This follows from [144, Proposition 5 in III.7.6 on page 319].   □

9.1.3 Finite von Neumann Algebras and Traces

One of the basic properties of finite von Neumann algebras is the existence of the center valued trace which turns out to be universal.

A *finite trace* $\mathrm{tr}\colon \mathcal{A} \to \mathbb{C}$ on a von Neumann algebra is a $\mathbb{C}$-linear mapping satisfying $\mathrm{tr}(ab) = \mathrm{tr}(ba)$ for $a, b \in \mathcal{A}$ and $\mathrm{tr}(a) \geq 0$ for $a \geq 0$ (i.e. $a = bb^*$ for some $b \in \mathcal{A}$). It is called *faithful*, if for $a \in \mathcal{A}$ with $a \geq 0$ we have $\mathrm{tr}(a) = 0 \Rightarrow a = 0$. It is called *normal* if for $f \in \mathcal{A}$, which is the supremum with respect to the usual ordering $\leq$ of positive elements (see 1.7) of some monotone increasing net $\{f_i \mid i \in I\}$ of positive elements in $\mathcal{A}$, we get $\mathrm{tr}(f) = \sup\{\mathrm{tr}(f_i) \mid i \in I\}$. The next result is taken from [283, Theorem 7.1.12 on page 462, Proposition 7.4.5 on page 483, Theorem 8.2.8 on page 517, Proposition 8.3.10 on page 525, Theorem 8.4.3 on page 532].

Theorem 9.5. *Let $\mathcal{A}$ be a finite von Neumann algebra on H. There is a map*

$$\mathrm{tr}^u = \mathrm{tr}^u_{\mathcal{A}}\colon \mathcal{A} \to \mathcal{Z}(\mathcal{A})$$

into the center $\mathcal{Z}(\mathcal{A})$ of $\mathcal{A}$ called the center valued trace or universal trace of $\mathcal{A}$, which is uniquely determined by the following properties:

(1) tr^u is a trace with values in the center, i.e. tr^u is $\mathbb{C}$-linear, for $a \in \mathcal{A}$ with $a \geq 0$ we have $\mathrm{tr}^u(a) \geq 0$ and $\mathrm{tr}^u(ab) = \mathrm{tr}^u(ba)$ for all $a, b \in \mathcal{A}$;
(2) $\mathrm{tr}^u(a) = a$ for all $a \in \mathcal{Z}(\mathcal{A})$.

The map tr^u has the following further properties:

(3) tr^u is faithful;
(4) tr^u is normal, or, equivalently, tr^u is continuous with respect to the ultra-weak topology on $\mathcal{A}$;
(5) $\| \mathrm{tr}^u(a) \| \leq \|a\|$ for $a \in \mathcal{A}$;
(6) $\mathrm{tr}^u(ab) = a\, \mathrm{tr}^u(b)$ for all $a \in \mathcal{Z}(\mathcal{A})$ and $b \in \mathcal{A}$;
(7) Let p and q be projections in $\mathcal{A}$. Then $p \sim q$, if and only if $\mathrm{tr}^u(p) = \mathrm{tr}^u(q)$;

(8) *Any linear functional* $f\colon \mathcal{A} \to \mathbb{C}$ *which is continuous with respect to the norm topology on* $\mathcal{A}$ *and which is central, i.e.* $f(ab) = f(ba)$ *for all* $a, b \in \mathcal{A}$, *factorizes as*

$$\mathcal{A} \xrightarrow{\mathrm{tr}^u} \mathcal{Z}(\mathcal{A}) \xrightarrow{f|_{\mathcal{Z}(\mathcal{A})}} \mathbb{C}.$$

Example 9.6. Let X be a compact space together with a finite measure ν on its Borel-σ-algebra. Let $L^\infty(X, \nu)$ be the Banach algebra of equivalence classes of essentially bounded measurable functions $X \to \mathbb{C}$, where two such functions are called equivalent if they only differ on a set of measure zero. It becomes a Banach algebra with the norm

$$||f||_\infty = \inf\{K > 0 \mid \nu(\{x \in X \mid |f(x)| \geq K\}) = 0\}$$

and the involution coming from complex conjugation. This turns out to be a commutative von Neumann algebra by the obvious embedding

$$L^\infty(X, \nu) \to \mathcal{B}(L^2(X, \nu))$$

coming from pointwise multiplication. Any commutative von Neumann algebra is isomorphic to $L^\infty(X, \nu)$ for appropriate X and ν [144, Theorem 1 and 2 in I.7.3 on page 132].

Example 9.7. Let G be a group. The *right regular representation* $\rho_r\colon \mathbb{C}G \to \mathcal{B}(l^2(G))$ sends $g \in G$ to the operator $r_{g^{-1}}\colon l^2(G) \to l^2(G)$, $u \mapsto ug$, whereas the *left regular representation* $\rho_l\colon \mathbb{C}G \to \mathcal{B}(l^2(G))$ sends $g \in G$ to the operator $l_g\colon l^2(G) \to l^2(G)$, $u \mapsto gu$. The left regular representation is a homomorphism of $\mathbb{C}$-algebras, whereas the right regular representation is an *anti-homomorphism of $\mathbb{C}$-algebras*, i.e. it respects the scalar multiplication and addition, but respects multiplication only up to changing the order. We get from [283, Theorem 6.7.2 on page 434]

$$\mathrm{im}(\rho_r)'' = \mathrm{im}(\rho_l)' = \mathcal{B}(l^2(G))^G.$$

Hence $\mathcal{N}(G)$ as introduced in Definition 1.1 is the closure of $\mathbb{C}G$, which we view as a $*$-subalgebra of $\mathcal{B}(l^2(G))$ by the right regular representation ρ_r, in $\mathcal{B}(l^2(G))$ with respect to the weak or strong topology. The closure of $\mathrm{im}(\rho_r)$ in $\mathcal{B}(l^2(G))$ with respect to the norm topology is called the *reduced C^*-algebra* $C_r^*(G)$ of G. In the special case $G = \mathbb{Z}^n$, we get $C_r^*(\mathbb{Z}^n) = C(T^n)$ (compare with Example 1.4), where $C(T^n)$ denotes the space of continuous functions from T^n to $\mathbb{C}$.

We will later need for the computation of $K_1(\mathcal{A})$ the following technical condition. It is always satisfied for a von Neumann algebra acting on a separable Hilbert space and in particular for the group von Neumann algebra $\mathcal{N}(G)$ of any countable group G.

Definition 9.8. *A von Neumann algebra is* countably composable *if every orthogonal family of non-zero projections is countable.*

9.1.4 Extending Results for Group von Neumann Algebras to Finite von Neumann Algebras

Let $\mathcal{A}$ be a finite von Neumann algebra with some faithful finite normal trace $\mathrm{tr}\colon \mathcal{A} \to \mathbb{C}$. Define a pre-Hilbert structure on $\mathcal{A}$ by $\langle a, b \rangle = tr(ab^*)$. Let $l^2(\mathcal{A})$ be the Hilbert completion of $\mathcal{A}$. Denote by $||a||$ the induced norm on $l^2(\mathcal{A})$. Given $a \in \mathcal{A}$, we obtain a linear operator $\mathcal{A} \to \mathcal{A}$ sending b to ab. This operator is bounded with operator norm $||a||$. Hence it extends uniquely to a bounded operator $\rho_l(a)\colon l^2(\mathcal{A}) \to l^2(\mathcal{A})$ satisfying $||\rho_l(a)||_\infty = ||a||$. This yields the *left regular representation*

$$\rho_l \colon \mathcal{A} \to \mathcal{B}(l^2(\mathcal{A})).$$

Thus we obtain a left $\mathcal{A}$-module structure on $l^2(\mathcal{A})$. Analogously we get $\rho_r(a)\colon l^2(\mathcal{A}) \to l^2(\mathcal{A})$ induced by $b \mapsto ba$. In particular we obtain the *right regular representation* $\rho_r \colon \mathcal{A} \to B_{\mathcal{A}}(l^2(\mathcal{A}))$ from $\mathcal{A}$ into the subalgebra $B_{\mathcal{A}}(l^2(\mathcal{A}))$ of linear bounded $\mathcal{A}$-operators of $B(l^2(\mathcal{A}))$. The following result is fundamental for the theory of Hilbert modules over a finite von Neumann algebra (see Dixmier [144, Theorem 1 in I.5.2 on page 80, Theorem 2 in I.6.2 on page 99]).

Theorem 9.9. *Let $\mathcal{A}$ be a finite von Neumann algebra. Then the right regular representation*

$$\rho_r \colon \mathcal{A} \to \mathcal{B}_{\mathcal{A}}(l^2(\mathcal{A}))$$

is an isometric anti-homomorphism of $\mathbb{C}$-algebras.

In the special case where $\mathcal{A}$ is the group von Neumann algebra $\mathcal{N}(G) := \mathcal{B}(l^2(G))^G$, the inclusion $\mathcal{N}(G) \to l^2(G)$ $f \to \overline{f^*(1)}$ induces an isometric isomorphism $l^2(\mathcal{N}(G)) \to l^2(G)$ and thus an identification of $\mathcal{B}_{\mathcal{N}(G)}(l^2(\mathcal{N}(G))$ with $\mathcal{B}(l^2(G))^G$. Under this identification the map ρ_r of Theorem 9.9 becomes the the anti-homomorphisms of $\mathbb{C}$-algebras $\mathcal{N}(G) \to \mathcal{N}(G)$ sending f to the operator $i \circ f^* \circ i$, where $i\colon l^2(G) \to l^2(G)$ sends $\sum_{g \in G} \lambda_g \cdot g$ to $\sum_{g \in G} \overline{\lambda_g} \cdot g$.

Remark 9.10. In view of Example 9.7 and Theorem 9.9 it is clear that a lot of the material of the preceding chapters extends from group von Neumann algebras (with the standard trace) to finite von Neumann algebras with a given finite faithful normal trace. For instance, there is an obvious notion of a finitely generated Hilbert $\mathcal{A}$-module and the corresponding category is equivalent to the category of finitely generated projective $\mathcal{A}$-modules as $\mathbb{C}$-category with involution (cf. Theorem 6.24). It is clear how to define the von Neumann dimension for a finitely generated $\mathcal{A}$-module and to extend for instance Theorem 6.5 and Theorem 6.7. Moreover, the definition of the L^2-Betti number $b_p^{(2)}(X; V)$ of a G-space X (see Definition 6.50) can be extended to the case where an $\mathcal{A}$-$\mathbb{C}G$-bimodule V, which is finitely generated projective over $\mathcal{A}$, is given. Then the definition of $b_p^{(2)}(X; \mathcal{N}(G))$ is the special case $V = \mathcal{N}(G)$, where the $\mathcal{N}(G)$-$\mathbb{C}G$-bimodule structure comes from the inclusion of rings $\mathbb{C}G \subset \mathcal{N}(G)$. Also the notion of the algebra $\mathcal{U}(G)$ of operators affiliated to $\mathcal{N}(G)$ extends to a finite von Neumann algebra $\mathcal{A}$. All the nice properties of $\mathcal{U}(G)$ carry over to $\mathcal{U} = \mathcal{U}(\mathcal{A})$, for instance $\mathcal{U}(\mathcal{A})$ is von Neumann regular and is the Ore localization of $\mathcal{A}$ with respect to the set of non-zero divisors.

For the rest of this chapter we will use some of the results, which we only have proved for group von Neumann algebras, also for finite von Neumann algebras.

9.2 Middle K-Theory of a Neumann Algebra

In this section we define and compute the K-groups $K_0(\mathcal{A})$, $K_1(\mathcal{A})$ and $K_1^{\mathrm{inj}}(\mathcal{A})$ for a von Neumann algebra $\mathcal{A}$.

9.2.1 K_0 of a von Neumann Algebra

Definition 9.11 (Projective class group $K_0(R)$). *Let R be an (associative) ring (with unit). Define its* projective class group $K_0(R)$ *to be the abelian group whose generators are isomorphism classes $[P]$ of finitely generated projective R-modules P and whose relations are $[P_0] + [P_2] = [P_1]$ for any exact sequence $0 \to P_0 \to P_1 \to P_2 \to 0$ of finitely generated projective R-modules. Define $G_0(R)$ analogously but replacing finitely generated projective by finitely generated.*

One should view $K_0(R)$ together with the assignment sending a finitely generated projective R-module P to its class $[P]$ in $K_0(R)$ as the universal dimension for finitely generated projective R-modules. Namely, suppose we are given an abelian group and an assignment d which associates to a finitely generated projective R-module an element $d(P) \in A$ such that $d(P_0) + d(P_2) = d(P_1)$ holds for any exact sequence $0 \to P_0 \to P_1 \to P_2 \to 0$ of finitely generated projective R-modules. Then there is precisely one homomorphism ϕ of abelian groups from $K_0(R)$ such that $\phi([P]) = d(P)$ holds for each finitely generated projective R-module. The analogous statement holds for $G_0(R)$ if we consider finitely generated R-modules instead of finitely generated projective R-modules.

Definition 9.12 (Center valued trace). *Let $\mathcal{A}$ be a finite von Neumann algebra and let tr^u be its center valued trace (see Theorem 9.5). For a finitely generated projective $\mathcal{A}$-module P define its center valued von Neumann dimension by*

$$\dim{}^u(P) := \mathrm{tr}^u(A) := \sum_{i=1}^{n} \mathrm{tr}^u(a_{i,i}) \quad \in \mathcal{Z}(\mathcal{A})^{\mathbb{Z}/2} = \{a \in \mathcal{Z}(\mathcal{A}) \mid a = a^*\}$$

for any matrix $A = (a_{i,j})_{i,j} \in M_n(\mathcal{A})$ with $A^2 = A$ such that $\mathrm{im}(R_A \colon \mathcal{A}^n \to \mathcal{A}^n)$ induced by right multiplication with A is $\mathcal{A}$-isomorphic to P.

The definition of $\dim^u(P)$ is independent of the choice of A (cf. (6.4)). The matrix A appearing in Definition 9.12 can be chosen to satisfy both $A^2 = A$ and $A^* = A$. This follows from Theorem 6.24. Therefore $\dim^u(P)$ is an element in $\mathcal{Z}(A)^{\mathbb{Z}/2}$ with respect to the $\mathbb{Z}/2$-action coming from taking the adjoint.

The next result follows from [283, Theorem 8.4.3 on page 532, Theorem 8.4.4 on page 533].

Theorem 9.13 (K_0 of finite von Neumann algebras). *Let A be a finite von Neumann algebra.*

(1) The following statements are equivalent for two finitely generated projective A-modules P and Q:

(a) P and Q are A-isomorphic;

(b) P and Q are stably A-isomorphic, i.e. $P \oplus V$ and $Q \oplus V$ are A-isomorphic for some finitely generated projective A-module V;

(c) $\dim^u(P) = \dim^u(Q)$;

(d) $[P] = [Q]$ in $K_0(A)$;

(2) The center valued dimension induces an injection

$$\dim^u \colon K_0(A) \to \mathcal{Z}(A)^{\mathbb{Z}/2} = \{a \in \mathcal{Z}(A) \mid a = a^*\},$$

where the group structure on $\mathcal{Z}(A)^{\mathbb{Z}/2}$ comes from the addition. If A is of type II_1, this map is an isomorphism.   □

Example 9.14. Let A be an abelian von Neumann algebra. Then it is of the shape $L^\infty(X, \nu)$ as explained in Example 9.6. Let $L^\infty(X, \nu, \mathbb{Z})$ be the abelian subgroup of $L^\infty(X, \nu)$ consisting of elements which can be represented by bounded measurable functions $f \colon X \to \mathbb{Z}$. We claim that the center-valued dimension induces an isomorphism

$$\dim^u \colon K_0(L^\infty(X, \nu)) \to L^\infty(X, \nu, \mathbb{Z}).$$

This follows from Theorem 9.13 (2) and the following result taken from [331, Lemma 4.1]. Namely, for an abelian von Neumann algebra A and an A-homomorphism $t \colon A^n \to A^n$, which is normal, i.e. t and t^* commute, there exists a unitary morphism $u \colon A^n \to A^n$ such that $u^* \circ t \circ u$ is diagonal.

Theorem 9.15. *Let A be a properly infinite von Neumann algebra. Then $K_0(A) = 0$.*

Proof. We firstly show that for a properly infinite projection in a von Neumann algebra A the class $[\text{im}(p)] \in K_0(A)$ of the finitely generated projective A-module $\text{im}(p)$ is zero. This follows essentially from [283, Lemma 6.3.3 on page 411] from which we get a projection q satisfying $q \leq p$ and $q \sim p - q \sim p$. Since equivalent projections have A-isomorphic images and $\text{im}(p)$ is A-isomorphic to $\text{im}(p - q) \oplus \text{im}(q)$, we get in $K_0(A)$

$$[\mathrm{im}(p)] = [\mathrm{im}(q)] + [\mathrm{im}(p-q)] = [\mathrm{im}(p)] + [\mathrm{im}(p)] \Rightarrow [\mathrm{im}(p)] = 0.$$

Given any projective $\mathcal{A}$-module P, choose a projection $p \in M_n(\mathcal{A})$ such that $\mathrm{im}(p\colon \mathcal{A}^n \to \mathcal{A}^n)$ is $\mathcal{A}$-isomorphic to P. Notice that $M_n(\mathcal{A})$ is again a von Neumann algebra.

Next we want to show that $p \oplus 1 \in M_{n+1}(\mathcal{A})$ is properly infinite. The center $\mathcal{Z}(M_{n+1}(\mathcal{A}))$ is $\{c \cdot I_{n+1} \mid c \in \mathcal{Z}(\mathcal{A})\}$, where I_{n+1} is the identity matrix in $M_{n+1}(\mathcal{A})$. This follows for instance from Theorem 9.2. Hence any central projection c in $M_{n+1}(\mathcal{A})$ is of the form $c = q \cdot I_{n+1}$ for some central projection $q \in \mathcal{A}$. If $c(p \oplus 1)$ is not zero, then $q \in \mathcal{A}$ is a non-zero central projection. Since $\mathcal{A}$ is by assumption properly infinite, q is infinite. But then also c is infinite by [283, Theorem 6.3.8 on page 414]. Hence $p \oplus 1$ is properly infinite.

The *Morita isomorphism* $\mu\colon K_0(M_{n+1}(\mathcal{A})) \to K_0(\mathcal{A})$ is defined by $\mu([P]) = [\mathcal{A}^{n+1} \otimes_{M_{n+1}(\mathcal{A})} P]$. In particular it maps the class of $\mathrm{im}(p \oplus 1)$ to the class of $\mathrm{im}(p) \oplus \mathrm{im}(1) = P \oplus \mathcal{A}$. Since we have already shown that properly infinite projections represent zero in K_0, we conclude $[\mathcal{A}] = 0 \in K_0(\mathcal{A})$ and $[\mathrm{im}(p \oplus 1)] = 0 \in K_0(M_{n+1}(\mathcal{A}))$. Hence we get in $K_0(\mathcal{A})$

$$[P] = [\mathrm{im}(p) \oplus \mathrm{im}(1)] - [\mathcal{A}] = 0 - 0 = 0. \square$$

In view of Theorem 9.3 and Theorem 9.15 we get for any von Neumann algebra $\mathcal{A}$ that $K_0(\mathcal{A}) = K_0(\mathcal{A}_{I_f}) \oplus K_0(\mathcal{A}_{II_1})$ and hence the computation of $K_0(\mathcal{A})$ follows from Theorem 9.13.

9.2.2 K_1 of a von Neumann Algebra

Definition 9.16 (K-group $K_1(R)$). *Let R be a ring. Define $K_1(R)$ to be the abelian group whose generators are conjugacy classes $[f]$ of automorphisms $f\colon P \to P$ of finitely generated projective R-modules with the following relations:*

i.) Given a commutative diagram of finitely generated projective R-modules

$$
\begin{array}{ccccccccc}
0 & \longrightarrow & P_1 & \xrightarrow{\ i\ } & P_2 & \xrightarrow{\ p\ } & P_3 & \longrightarrow & 0 \\
& & \downarrow{\scriptstyle f_1} & & \downarrow{\scriptstyle f_2} & & \downarrow{\scriptstyle f_3} & & \\
0 & \longrightarrow & P_1 & \xrightarrow{\ i\ } & P_2 & \xrightarrow{\ p\ } & P_3 & \longrightarrow & 0
\end{array}
$$

with exact rows and automorphisms as vertical arrows, we get $[f_1]+[f_3] = [f_2]$.

ii.) Given automorphisms $f, g\colon P \to P$ of a finitely generated projective R-module P, we get $[g \circ f] = [f] + [g]$.

Define $K_1^{\mathrm{inj}}(R)$ analogously by replacing automorphisms by injective endomorphisms everywhere.

We leave it to the reader to check that the definition of $K_1(R)$ by $K_1(R) := GL(R)/[GL(R), GL(R)]$ in Subsection 3.1.1 coincides with the one of Definition 9.16.

Let $\mathcal{Z}(\mathcal{A})^{\mathrm{inv}}$ be the abelian group of elements in the center of $\mathcal{A}$ which are invertible. The set of non-zero-divisors in $\mathcal{Z}(\mathcal{A})$ is an abelian monoid under multiplication and we denote by $\mathcal{Z}(\mathcal{A})^w$ the abelian group which is obtained from this abelian monoid by the Grothendieck construction. We can identify the abelian von Neumann algebra $\mathcal{Z}(\mathcal{A})$ with $L^\infty(X, \nu)$ (see Example 9.6). Then $\mathcal{Z}(\mathcal{A})^{\mathrm{inv}}$ becomes the space of equivalence classes of measurable functions $f\colon X \to \mathbb{C}$ which are essentially bounded from below and above, i.e. there are positive constants k and K such that $\{x \in X \mid |f(x)| \le k\}$ and $\{x \in X \mid |f(x)| \ge K\}$ have measure zero. The space $\mathcal{Z}(\mathcal{A})^w$ becomes the space of equivalence classes of measurable functions from $f\colon X \to \mathbb{C}$ for which $f^{-1}(0)$ has measure zero. Notice that $\mathcal{Z}(\mathcal{A})^w = \mathcal{U}(\mathcal{Z}(\mathcal{A}))^{\mathrm{inv}} = \mathcal{Z}(\mathcal{U}(\mathcal{A}))^{\mathrm{inv}}$. We get from [344, Theorem 2.1]

Theorem 9.17 (K_1 of von Neumann algebras of type I_f). *Let $\mathcal{A}$ be a von Neumann algebra of type I_f. Then there is a so called normalized determinant*

$$\mathrm{det}_{\mathrm{norm}}\colon M_k(\mathcal{A}) \to \mathcal{Z}(\mathcal{A})$$

with the following properties:

(1) If $A \in M_k(\mathcal{A})$ satisfies $\mathrm{det}_{\mathrm{norm}}(A) = 1$, then A is a product of two commutators in $GL_k(\mathcal{A})$;

(2) The normalized determinant induces isomorphisms

$$\mathrm{det}_{\mathrm{norm}}\colon K_1(\mathcal{A}) \xrightarrow{\cong} \mathcal{Z}(\mathcal{A})^{\mathrm{inv}};$$
$$\mathrm{det}_{\mathrm{norm}}\colon K_1^{\mathrm{inj}}(\mathcal{A}) \xrightarrow{\cong} \mathcal{Z}(\mathcal{A})^w,$$

which are compatible with the involutions.

We can use the center valued trace to define the center valued Fuglede-Kadison determinant

$$\mathrm{det}_{FK}\colon GL_k(\mathcal{A}) \to \mathcal{Z}(\mathcal{A})^{+,\mathrm{inv}}, \qquad A \mapsto \exp\left(\frac{1}{2} \cdot \mathrm{tr}^u(\ln(A^*A))\right),$$

where $\mathcal{Z}(\mathcal{A})^{+,\mathrm{inv}}$ is the (multiplicative) abelian group of elements in the center of $\mathcal{A}$ which are both positive and invertible. We get from [344, Theorem 3.3]

Theorem 9.18 (K_1 of von Neumann algebras of type II_1). *Let $\mathcal{A}$ be a countably composable von Neumann algebra of type II_1.*

(1) If $A \in GL_k(\mathcal{A})$ satisfies $\mathrm{det}_{FK}(A) = 1$, then A is a product of nine commutators in $GL_k(\mathcal{A})$;

(2) The center valued Fuglede-Kadison determinant induces an isomorphism

$$\mathrm{det}_{FK}\colon K_1(\mathcal{A}) \xrightarrow{\cong} \mathcal{Z}(\mathcal{A})^{+,\mathrm{inv}};$$

(3) $K_1^{\mathrm{inj}}(\mathcal{A}) = 0$.

Theorem 9.18 (2) has also been proved in [179], provided that $\mathcal{A}$ is a factor. We get from [344, Theorem 4.2]

Theorem 9.19. *Let $\mathcal{A}$ be a countably composable properly infinite von Neumann algebra. Then $K_1(\mathcal{A}) = K_1^{\mathrm{inj}}(\mathcal{A}) = 0$.*

In view of Theorem 9.3, Theorem 9.18 (3) and Theorem 9.19 we get for a countably composable von Neumann algebra $\mathcal{A}$ that $K_1(\mathcal{A}) = K_1(\mathcal{A}_{I_f}) \oplus K_1(\mathcal{A}_{II_1})$ and $K_1^{\mathrm{inj}}(\mathcal{A}) = K_1^{\mathrm{inj}}(\mathcal{A}_{I_f})$ and hence the computation of $K_1(\mathcal{A})$ and $K_1^{\mathrm{inj}}(\mathcal{A})$ follows from Theorem 9.17 (2) and Theorem 9.18 (2).

The condition countably composable appearing in Theorem 9.18 and Theorem 9.19 is purely technical, it may be possible that it can be dropped.

9.3 Middle K-Theory of the Algebra of Affiliated Operators

In this section we compute the K-groups $K_0(\mathcal{U})$ and $K_1(\mathcal{U})$ of the algebra $\mathcal{U}$ of operators affiliated to a finite von Neumann algebra $\mathcal{A}$ and deal with a part of the localization sequence associated to $\mathcal{A} \to \mathcal{U}$.

Theorem 9.20 (K-groups of $\mathcal{U}$). *Let $\mathcal{A}$ be a finite von Neumann algebra and $\mathcal{U}$ be the algebra of affiliated operators. Then*

(1) The map $i_*: K_0(\mathcal{A}) \to K_0(\mathcal{U})$ induced by the inclusion $i: \mathcal{A} \to \mathcal{U}$ is an isomorphism;

(2) There is a natural isomorphism

$$j: K_1^{\mathrm{inj}}(\mathcal{A}) \xrightarrow{\cong} K_1(\mathcal{U}).$$

Proof. (1) This follows from Theorem 8.22 (7) and (8).

(2) Since for an injective endomorphism $f: P \to P$ of a finitely generated projective $\mathcal{A}$-module P the induced map $\mathcal{U} \otimes_{\mathcal{A}} f: \mathcal{U} \otimes_{\mathcal{A}} P \to \mathcal{U} \otimes_{\mathcal{A}} P$ is a $\mathcal{U}$-isomorphism (see Lemma 6.28 (2) and Theorem 8.22 (5)), we obtain a natural map $j: K_1^{\mathrm{inj}}(\mathcal{A}) \to K_1(\mathcal{U})$. We define an inverse $k: K_1(\mathcal{U}) \to K_1^{\mathrm{inj}}(\mathcal{A})$ as follows. Let $\eta \in K_1^{\mathrm{inj}}(\mathcal{U})$ be an element for which there is a $\mathcal{U}$-automorphism $f: \mathcal{U}^n \to \mathcal{U}^n$ with $\eta = [f]$. By Lemma 8.8 we can choose injective endomorphisms $a, b: \mathcal{A}^n \to \mathcal{A}^n$ such that $\mathcal{U} \otimes_{\mathcal{A}} a = f \circ \mathcal{U} \otimes_{\mathcal{A}} b$ (identifying $\mathcal{U} \otimes_{\mathcal{A}} \mathcal{A}^n$ and $\mathcal{U}^n$). Define $k(\eta) = [a] - [b]$. We leave it to the reader to verify that k is well-defined and an inverse of j. $\square$

Notice that Theorem 9.20 together with the results of Subsections 9.2.1 and 9.2.2 give the complete computation of $K_n(\mathcal{U})$ for $n = 0, 1$, provided that $\mathcal{A}_{II}$ is countably composable.

Definition 9.21. *Let $K_0(\mathcal{A} \to \mathcal{U})$ be the abelian group whose generators are isomorphism classes of finitely presented $\mathcal{A}$-modules M with $\mathcal{U} \otimes_{\mathcal{A}} M = 0$ and whose relations are $[M_0] + [M_2] = [M_1]$ for any exact sequence $0 \to M_0 \to M_1 \to M_2 \to 0$ of such $\mathcal{A}$-modules.*

Recall that the following conditions for an $\mathcal{A}$-module M are equivalent by Lemma 6.28 (4) and Lemma 8.33 (3): i.) M is finitely presented with $\mathcal{U} \otimes_{\mathcal{A}} M = 0$, ii.) M is finitely presented with $\dim_{\mathcal{A}}(M) = 0$, iii.) M has a resolution $0 \to \mathcal{A}^n \to \mathcal{A}^n \to M \to 0$, iv.) M has a resolution $0 \to P_0 \to P_1 \to M \to 0$ for finitely generated projective $\mathcal{A}$-modules and $\mathcal{U} \otimes_{\mathcal{A}} M = 0$.

Let $S \subset R$ be a multiplicatively closed subset of the ring R satisfying the Ore condition (see Definition 8.14). Provided that S contains no zero-divisors, there is an exact localization sequence associated to an Ore localization $R \to RS^{-1}$ [39]

$$K_1(R) \xrightarrow{i_1} K_1(RS^{-1}) \xrightarrow{j} K_0(R \to RS^{-1}) \xrightarrow{k} K_0(R) \xrightarrow{i_0} K_0(RS^{-1}). \quad (9.22)$$

Here i_1 and i_0 are induced by the canonical map $i \colon R \to RS^{-1}$ and $K_0(R \to RS^{-1})$ is defined in terms of R-modules M which possess a resolution $0 \to P_1 \to P_0 \to M \to$ with finitely generated projective R-modules P_0 and P_1 and satisfy $RS^{-1} \otimes_R M = 0$. The map k sends the class of such an R-module M to $[P_0] - [P_1]$. The class of an automorphism $f \colon (RS^{-1})^n \to (RS^{-1})^n$ is sent by j to $[\mathrm{coker}(a)] - [\mathrm{coker}(b)]$ for any R-endomorphisms $a, b \colon R^n \to R^n$, for which $RS^{-1} \otimes_R a$ and $RS^{-1} \otimes_R b$ are bijective and satisfy $f = RS^{-1} \otimes_R a \circ (RS^{-1} \otimes_R b)^{-1}$.

Lemma 9.23. *Let $\mathcal{A}$ be a finite von Neumann algebra. Then*

(1) The localization sequence (9.22) yields for $\mathcal{A}$ and its Ore localization $\mathcal{U}$ (see Theorem 8.22 (1)) the exact sequence

$$K_1(\mathcal{A}) \xrightarrow{i_1} K_1(\mathcal{U}) \xrightarrow{j} K_0(\mathcal{A} \to \mathcal{U}) \xrightarrow{j} K_0(\mathcal{A}) \xrightarrow{i_1} K_0(\mathcal{U});$$

It splits into an exact sequence

$$K_1(\mathcal{A}) \xrightarrow{i_1} K_1(\mathcal{U}) \xrightarrow{j} K_0(\mathcal{A} \to \mathcal{U}) \to 0 \quad (9.24)$$

and an isomorphism

$$i_0 \colon K_0(\mathcal{A}) \xrightarrow{\cong} K_0(\mathcal{U});$$

(2) The localization sequence is the direct product of the localization sequences of the summands $\mathcal{A}_{I_f}$ and $\mathcal{A}_{II_1}$ appearing in the decomposition $\mathcal{A} = \mathcal{A}_{I_f} \times \mathcal{A}_{II_1}$ (see Theorem 9.3);

(3) If $\mathcal{A}$ is of type I_f, then the exact sequence (9.24) is isomorphic to the short exact sequence

$$0 \to \mathcal{Z}(\mathcal{A})^{\mathrm{inv}} \to \mathcal{Z}(\mathcal{A})^w \to \mathcal{Z}(\mathcal{A})^w / \mathcal{Z}(\mathcal{A})^{\mathrm{inv}} \to 0.$$

In particular $K_0(\mathcal{A} \to \mathcal{U}) = \mathcal{Z}(\mathcal{A})^w / \mathcal{Z}(\mathcal{A})^{\mathrm{inv}}$;

(4) If $\mathcal{A}$ is of type II_1 and countably composable, then the exact sequence (9.24) becomes

$$\mathcal{Z}(\mathcal{A})^{+,\mathrm{inv}} \to 0 \to 0 \to 0.$$

In particular $K_0(\mathcal{A} \to \mathcal{U}) = 0$.

Proof. (1) This follows from Theorem 9.20 (1).

(2) This is obvious.

(3) This follows from Theorem 9.17 and Theorem 9.20 (2).

(4) This follows from Theorem 9.18 and Theorem 9.20 (2). $\square$

9.4 L-Theory of a von Neumann Algebra and the Algebra of Affiliated Operators

In this section we give the computation of the L-groups of a von Neumann algebra $\mathcal{A}$ and of the algebra $\mathcal{U}$ of affiliated operators. The definitions of the various decorated quadratic and symmetric L-groups $L_n^\epsilon(R)$ and $L_\epsilon^n(R)$ for a ring R with involution $*: R \to R$ are given for instance in [429], [430]. The decoration $\epsilon = p$ or $\epsilon = h$ respectively means that the underlying modules are finitely generated projective or finitely generated free respectively. If we write $\epsilon = s$, we mean the L-groups with respect to the trivial subgroup in $\widetilde{K}_1$.

Before we state the result, we need some preparation. A non-singular symmetric form $a: P \to P^*$ on a finitely generated projective (left) R-module is an R-isomorphism $P \to P^*$ such that the composition $P \xrightarrow{i} (P^*)^* \xrightarrow{a^*} P^*$ is equal to a, where i is the canonical isomorphism and the involution on R is used to transform the canonical right module structure on P^* to a left module structure. Two symmetric non-singular forms $a: P \to P^*$ and $b: Q \to Q^*$ are isomorphic if there is an R-isomorphism $f: P \to Q$ satisfying $f^* \circ b \circ f = a$. Given a finitely generated projective R-module P, the associated hyperbolic non-singular symmetric form $h(P): P \otimes P^* \to (P \otimes P^*)^* = P^* \oplus P$ is given by the matrix $\begin{pmatrix} 0 & 1 \\ 1 & 0 \end{pmatrix}$. Two non-singular symmetric forms are called equivalent if they become isomorphic after adding hyperbolic forms. The *Witt group* of equivalence classes of non-singular symmetric forms with the addition coming from the orthogonal sum can be identified with $L^0(R)$ [429, Proposition 5.1 on page 160]. The analogous statement is true for $L_h^0(R)$ or $L_s^0(R)$ respectively if one considers only non-singular symmetric forms $a: F \to F^*$ for finitely generated free R-modules or non-singular symmetric forms $a: R^n \to (R^n)^*$ respectively such that the element in $\widetilde{K}_1(R)$ given by the composition of a with the standard isomorphism

$$i: (R^n)^* \to R^n, \qquad f \mapsto (f(e_1)^*, f(e_2)^*, \dots, f(e_n)^*)$$

vanishes. If $\mathcal{A}$ is a von Neumann algebra (actually it is enough to require $\mathcal{A}$ to be a C^*-algebra), there are maps

$$\text{sign}^{(2)} \colon L_p^0(\mathcal{A}) \to K_0(\mathcal{A}); \tag{9.25}$$

$$\iota \colon K_0(\mathcal{A}) \to L_p^0(\mathcal{A}), \tag{9.26}$$

which turn out to be inverse to one another [442, Theorem 1.6 on page 343]). The map ι of (9.26) above sends the class $[P] \in K_0(\mathcal{A})$ of a finitely generated projective $\mathcal{A}$-module P to the class of $\overline{\mu} \colon P \to P^*$ coming from some inner product μ on P (see Section 6.2). Such an inner product exists and the class of $\overline{\mu} \colon P \to P^*$ in $L_p^0(\mathcal{A})$ is independent of the choice of the inner product by Lemma 6.23.

Next we define $\text{sign}^{(2)}([a])$ for the class $[a] \in L^0(\mathcal{A})$ represented by a nonsingular symmetric form $a \colon P \to P^*$. Choose a finitely generated projective $\mathcal{A}$-module Q together with an isomorphism $u \colon \mathcal{A}^n \to P \oplus Q$. Let $i \colon (\mathcal{A}^n)^* \to \mathcal{A}^n$ be the standard isomorphism. Let $\overline{a} \colon \mathcal{A}^n \to \mathcal{A}^n$ be the endomorphism $i \circ u^* \circ (a + 0) \circ u$. We get by spectral theory projections $\chi_{(0,\infty)}(\overline{a}) \colon \mathcal{A}^n \to \mathcal{A}^n$ and $\chi_{(-\infty,0)}(\overline{a}) \colon \mathcal{A}^n \to \mathcal{A}^n$. Define P_+ and P_- to be image of $\chi_{(0,\infty)}(\overline{a})$ and $\chi_{(-\infty,0)}(\overline{a})$. Put $\text{sign}^{(2)}([a]) = [P_+] - [P_-]$. We leave it to the reader to check that this is well defined. The non-singular symmetric form $a \colon P \to P^*$ is isomorphic to the orthogonal sum of $a_+ \colon P_+ \to P_+^*$ and $a_- \colon P_- \to P_-^*$, where a_+ and $-a_-$ come from inner products. This implies that $\text{sign}^{(2)}$ and ι are inverse to one another. One can define analogously isomorphisms, inverse to one another,

$$\text{sign}^{(2)} \colon L_p^0(\mathcal{U}) \to K_0(\mathcal{U}); \tag{9.27}$$

$$\iota \colon K_0(\mathcal{U}) \to L_p^0(\mathcal{U}). \tag{9.28}$$

Let R be a ring with involution. The involution on R induces involutions on the reduced K-groups $\widetilde{K}_i(R)$ for $i = 0, 1$ which are defined as the cokernel of the homomorphisms $K_i(\mathbb{Z}) \to K_i(R)$ induced by the obvious ring homomorphism $\mathbb{Z} \to R$. Denote by $\widehat{H}^n(\mathbb{Z}/2; \widetilde{K}_i(R))$ the Tate cohomology of the group $\mathbb{Z}/2$ with coefficients in the $\mathbb{Z}[\mathbb{Z}/2]$-module $\widetilde{K}_i(R)$. For any $\mathbb{Z}[\mathbb{Z}/2]$-module M, $\widehat{H}^n(\mathbb{Z}/2; M)$ is 2-periodic with

$$\widehat{H}^0(\mathbb{Z}/2; M) = \ker(1 - t \colon M \to M) / \operatorname{im}(1 + t \colon M \to M);$$
$$\widehat{H}^1(\mathbb{Z}/2; M) = \ker(1 + t \colon M \to M) / \operatorname{im}(1 - t \colon M \to M),$$

where $t \in \mathbb{Z}/2$ is the generator. We get long exact *Rothenberg sequences* [429, Proposition 9.1 on page 181]

$$\ldots \to \widehat{H}^0(\mathbb{Z}/2; \widetilde{K}_0(R)) \to L_h^1(R) \to L_p^1(R) \to \widehat{H}^1(\mathbb{Z}/2; \widetilde{K}_0(R))$$

$$\to L_h^0(R) \to L_p^0(R) \xrightarrow{j} \widehat{H}^0(\mathbb{Z}/2; \widetilde{K}_0(R)) \to \ldots, \tag{9.29}$$

where j sends the class of a symmetric non-singular form $f \colon P \to P^*$ to the class of the element $[P] \in \widetilde{K}_0(R)$, and

$$\ldots \to \widehat{H}^0(\mathbb{Z}/2; \widetilde{K}_1(R)) \to L^1_s(R) \to L^1_h(R) \to \widehat{H}^1(\mathbb{Z}/2; \widetilde{K}_1(R))$$
$$\to L^0_s(R) \to L^0_h(R) \xrightarrow{k} \widehat{H}^0(\mathbb{Z}/2; \widetilde{K}_1(R)) \to \ldots, \tag{9.30}$$

where k sends the class of a symmetric non-singular form $a \colon F \to F^*$ to the class of the element $[i \circ h^* \circ a \circ h] \in \widetilde{K}_1(R)$ for any R-isomorphism $h \colon R^n \to F$ and $i \colon (R^n)^* \xrightarrow{\cong} R^n$ the standard isomorphism. There are also Rothenberg sequences for the quadratic L-groups, just replace the symmetric L-groups in (9.29) and (9.30) by the quadratic versions.

Theorem 9.31 (L-groups of von Neumann algebras). *Let $\mathcal{A}$ be a von Neumann algebra. If $\mathcal{A}$ is finite, let $\mathcal{U} = \mathcal{U}(\mathcal{A})$ be the algebra of affiliated operators. Then*

(1) The symmetrization maps

$$L^n_\epsilon(\mathcal{A}) \xrightarrow{\cong} L^\epsilon_n(\mathcal{A});$$
$$L^n_\epsilon(\mathcal{U}) \xrightarrow{\cong} L^\epsilon_n(\mathcal{U})$$

are isomorphisms for $n \in \mathbb{Z}$ and $\epsilon = p, h, s$;

(2) The quadratic L-groups are 2-periodic, i.e. there are natural isomorphisms

$$L^\epsilon_n(\mathcal{A}) \xrightarrow{\cong} L^\epsilon_{n+2}(\mathcal{A});$$
$$L^\epsilon_n(\mathcal{U}) \xrightarrow{\cong} L^\epsilon_{n+2}(\mathcal{U})$$

for $n \in \mathbb{Z}$ and $\epsilon = p, h, s$ and analogously for the symmetric L-groups;

(3) The L^2-signature maps $\mathrm{sign}^{(2)}$ defined in (9.25) and (9.27) and the maps ι defined in (9.26) and (9.28) are isomorphisms, inverse to one another, and yield a commutative square of isomorphisms

$$
\begin{array}{ccc}
L^0_p(\mathcal{A}) & \xrightarrow[\cong]{\mathrm{sign}^{(2)}} & K_0(\mathcal{A}) \\
\Big\downarrow{\cong} & & \Big\downarrow{\cong} \\
L^0_p(\mathcal{U}) & \xrightarrow[\cong]{\mathrm{sign}^{(2)}} & K_0(\mathcal{U})
\end{array}
$$

(4) We have $L^1_p(\mathcal{A}) = 0$ and $L^1_p(\mathcal{U}) = 0$;

(5) If $\mathcal{A}$ is of type II_1, then for $n \in \mathbb{Z}$ the diagram of natural maps

$$
\begin{array}{ccccccccc}
0 & \longrightarrow & \mathbb{Z}/2 & \longrightarrow & L^0_h(\mathcal{A}) & \longrightarrow & L^0_p(\mathcal{A}) & \longrightarrow & 0 \\
 & & \Big\downarrow{\mathrm{id}} & & \Big\downarrow{\cong} & & \Big\downarrow{\cong} & & \\
0 & \longrightarrow & \mathbb{Z}/2 & \longrightarrow & L^0_h(\mathcal{U}) & \longrightarrow & L^0_p(\mathcal{U}) & \longrightarrow & 0
\end{array}
$$

is commutative, the vertical maps are isomorphisms and the rows are exact. We have $L_h^1(\mathcal{A}) = L_h^1(\mathcal{U}) = 0$.

If $\mathcal{A}$ is countably composable and of type II_1, then the natural maps appearing in the commutative square

$$
\begin{array}{ccc}
L_s^n(\mathcal{A}) & \xrightarrow{\;\cong\;} & L_h^n(\mathcal{A}) \\[2pt]
\Big\downarrow{\scriptstyle\cong} & & \Big\downarrow{\scriptstyle\cong} \\[2pt]
L_s^n(\mathcal{U}) & \xrightarrow{\;\cong\;} & L_h^n(\mathcal{U})
\end{array}
$$

are isomorphisms for $n \in \mathbb{Z}$;

(6) *Suppose that $\mathcal{A}$ is of type I_f. Then the natural map $L_\epsilon^n(\mathcal{A}) \to L_\epsilon^n(\mathcal{U})$ is bijective for $n \in \mathbb{Z}$ and $\epsilon = p, h, s$. We have $L_h^1(\mathcal{A}) = 0$.*

Let $l \colon L_h^0(\mathcal{A}) \to K_0(\mathcal{A})$ be the composition of the natural map $i \colon L_h^0(\mathcal{A}) \to L_p^0(\mathcal{A})$ with the isomorphism $\mathrm{sign}^{(2)} \colon L_p^0(\mathcal{A}) \xrightarrow{\cong} K_0(\mathcal{A})$. If $[\mathcal{A}] \in 2 \cdot K_0(\mathcal{A})$, then we get an exact sequence

$$
0 \to \mathbb{Z}/2 \to L_h^0(\mathcal{A}) \xrightarrow{\;l_1\;} 2 \cdot K_0(\mathcal{A}) \to 0,
$$

where l_1 is induced by l. If $[\mathcal{A}] \notin 2 \cdot K_0(\mathcal{A})$, then l is injective and its image is generated by $2 \cdot K_0(\mathcal{A})$ and $[\mathcal{A}]$.

If $K_1(\mathbb{Z}) \to K_1(\mathcal{A})$ is trivial, then we get an exact sequence

$$
0 \to L_s^0(\mathcal{A}) \to L_h^0(\mathcal{A}) \to \{f \in \mathcal{Z}(\mathcal{A})^{\mathrm{inv}} \mid f^2 = 1\} \to 0
$$

and $L_1^s(\mathcal{A}) = 0$.

Suppose that $K_1(\mathbb{Z}) \to K_1(\mathcal{A})$ is not trivial. Let κ be the subgroup of $\{f \in \mathcal{Z}(\mathcal{A})^{\mathrm{inv}} \mid f^2 = 1\}$ generated by the image of $[-\mathrm{id} \colon \mathcal{A} \to \mathcal{A}]$ under the isomorphism $\det_{\mathrm{norm}} \colon K_1(\mathcal{A}) \xrightarrow{\cong} \mathcal{Z}(\mathcal{A})^{\mathrm{inv}}$ of Theorem 9.17. Then we obtain an exact sequence

$$
0 \to L_s^0(\mathcal{A}) \to L_h^0(\mathcal{A}) \to \{f \in \mathcal{Z}(\mathcal{A})^{\mathrm{inv}} \mid f^2 = 1\}/\kappa \to 0
$$

and $L_s^1(\mathcal{A}) \cong \mathbb{Z}/2$;

(7) *If $\mathcal{A}$ is properly infinite, then $L_p^n(\mathcal{A})$ and $L_h^n(\mathcal{A})$ vanish for $n \in \mathbb{Z}$. If $\mathcal{A}$ is countably composable and properly infinite, then $L_s^n(\mathcal{A})$ vanishes for $n \in \mathbb{Z}$.*

Proof. (1) This follows from the fact that 2 is invertible in $\mathcal{A}$ and $\mathcal{U}$ [429, Proposition 3.3 on page 139]. The proof is given there only for L^p but applies also to L^h and L^s.

(2) This follows from assertion (1) and from the fact that $i = \sqrt{-1}$ belongs to $\mathcal{Z}(\mathcal{A})$ and $\mathcal{Z}(\mathcal{U})$ and is sent under the involution to $-i$ (see [429, Proposition 4.3 on page 150]. The proof is given there only for L^p but applies also to L^h and L^s.

(3) This follows from the constructions and definitions and from Theorem 9.20 (1). The map $\mathrm{sign}^{(2)}\colon L_p^0(\mathcal{A}) \to K_0(\mathcal{A})$ is defined and bijective not only for a von Neumann algebra $\mathcal{A}$, but also for a C^*-algebra $\mathcal{A}$ (see [442, Theorem 1.6 on page 343]).

(4) The L-group $L_p^1(\mathcal{A})$ is isomorphic to the topological K-group $K_{\mathrm{top}}^1(\mathcal{A})$ (for any C^*-algebra $\mathcal{A}$) [442, Theorem 1.8 on page 347]. For a von Neumann algebra $K_1^{\mathrm{top}}(\mathcal{A})$ is trivial. [51, Example 8.1.2 on page 67], [514, Example 7.1.11 on page 134].

Since $\mathcal{U}(G)$ is von Neumann regular (see Theorem 8.22 (3)), any finitely generated submodule of a finitely generated projective module is a direct summand (see Lemma 8.18 (4)). Hence the argument that $L_1^p(R)$ vanishes for semisimple rings in [428] carries over to $\mathcal{U}$. One could also argue by doing surgery on the inclusion of $H_0(C_*) \to C_*$ in the sense of [429, Section 4].

(5) We conclude from Theorem 9.13 (2) and Theorem 9.18 (2) that the involution on $K_i(\mathcal{A})$ is trivial and $2 \cdot \mathrm{id}\colon K_i(\mathcal{A}) \xrightarrow{\cong} K_i(\mathcal{A})$ is bijective for $i = 0, 1$. Hence $\widehat{H}^n(\mathbb{Z}/2; K_i(\mathcal{A})) = 0$ for $i, n \in \{0, 1\}$. The natural map $K_1(\mathcal{A}) \to \widetilde{K}_1(\mathcal{A})$ is bijective by Theorem 9.18 (2). We get a short exact sequence $0 \to K_0(\mathbb{Z}) = \mathbb{Z} \to K_0(\mathcal{A}) \to \widetilde{K}_0(\mathcal{A}) \to 0$ from Theorem 9.13 (2). It induces a long exact sequence of Tate cohomology groups. This implies that $\widehat{H}^1(\mathbb{Z}/2, \widetilde{K}_0(\mathcal{A})) = \mathbb{Z}/2$, $\widehat{H}^0(\mathbb{Z}/2, \widetilde{K}_0(\mathcal{A})) = 0$ and $\widehat{H}^n(\mathbb{Z}/2, \widetilde{K}_1(\mathcal{A})) = 0$ for $n = 0, 1$. We conclude from Theorem 9.18 (3) and Theorem 9.20 that the natural map $\widehat{H}^n(\mathbb{Z}/2, \widetilde{K}_i(\mathcal{A})) \to \widehat{H}^n(\mathbb{Z}/2, \widetilde{K}_i(\mathcal{U}))$ is bijective for $n, i \in \{0, 1\}$. Now assertion (5) follows from the Rothenberg sequences (9.29) and (9.30).

(6) We have already proved in assertion (3) and (4) that $i\colon L_p^0(\mathcal{A}) \to L_p^0(\mathcal{U})$ is bijective for all n.

The involution on $K_0(\mathcal{A})$ is trivial and multiplication with 2 is injective by Theorem 9.13 (2). This implies

$$\widehat{H}^0(\mathbb{Z}/2; K_0(\mathcal{A})) = K_0(\mathcal{A})/2 \cdot K_0(\mathcal{A});$$
$$\widehat{H}^1(\mathbb{Z}/2; K_0(\mathcal{A})) = 0.$$

The long exact Tate cohomology sequence associated to $0 \to K_0(\mathbb{Z}) = \mathbb{Z} \to K_0(\mathcal{A}) \to \widetilde{K}_0(\mathcal{A}) \to 0$ yields the exact sequence

$$0 \to \widehat{H}^1(\mathbb{Z}/2; \widetilde{K}_0(\mathcal{A})) \to \mathbb{Z}/2 \xrightarrow{\eta} K_0(\mathcal{A})/2 \cdot K_0(\mathcal{A}) \to \widehat{H}^0(\mathbb{Z}/2; \widetilde{K}_0(\mathcal{A})) \to 0,$$

where η sends the generator of $\mathbb{Z}/2$ to the class of $[\mathcal{A}]$. The natural map $\widehat{H}^n(\mathbb{Z}/2; \widetilde{K}_0(\mathcal{A})) \to \widehat{H}^n(\mathbb{Z}/2; \widetilde{K}_0(\mathcal{U}))$ is bijective for $n = 0, 1$ by Theorem 9.20 (1). Hence the part of assertion (6) for L_h^n follows from assertions (3), (4) and from the Rothenberg sequence (9.29). It remains to treat L_s^n.

We conclude from Example 9.6 applied to $\mathcal{Z}(\mathcal{A})$, Theorem 9.17 (2) and Theorem 9.20 (2) that $\det_{\mathrm{norm}}$ induces isomorphisms

$$\widehat{H}^0(\mathbb{Z}/2; K_1(\mathcal{A})) = \{f \in \mathcal{Z}(\mathcal{A}) \mid f^2 = 1\};$$
$$\widehat{H}^1(\mathbb{Z}/2; K_1(\mathcal{A})) = 0,$$

and that the natural map $\widehat{H}^n(\mathbb{Z}/2; K_1(\mathcal{A})) \xrightarrow{\cong} \widehat{H}^n(\mathbb{Z}/2; K_1(\mathcal{U}))$ is bijective for $n = 0, 1$. We begin with the case where $K_1(\mathbb{Z}) \to K_1(\mathcal{A})$ is trivial. Then the natural maps $K_1(\mathcal{A}) \to \widetilde{K}_1(\mathcal{A})$ and $K_1(\mathcal{U}) \to \widetilde{K}_1(\mathcal{U})$ are bijective. The Rothenberg sequence (9.30) reduces to

$$0 \to L_s^0(\mathcal{A}) \to L_h^0(\mathcal{A}) \xrightarrow{k} \{f \in \mathcal{Z}(\mathcal{A}) \mid f^2 = 1\} \to L_s^1(\mathcal{A}) \to 0.$$

By inspecting the proof of Theorem 9.17 in [344, Theorem 2.1] one checks that any element in $x \in K_1(\mathcal{A})$ with $x^* = x$ can be represented by an element $a \in \mathcal{A}^{\mathrm{inv}}$ with $a^* = a$. (Notice that this is not completely obvious since the composition of the canonical map $\mathcal{Z}(\mathcal{A})^{\mathrm{inv}} \to K_1(\mathcal{A})$ with $\det_{\mathrm{norm}} \colon K_1(\mathcal{A}) \to \mathcal{Z}(\mathcal{A})^{\mathrm{inv}}$ is *not* the identity in general.) Hence the map k in the Rothenberg sequence above is surjective and assertion (6) follows.

It remains to treat the case where $K_1(\mathbb{Z}) \to K_1(\mathcal{A})$ is not trivial. The exact Tate cohomology sequence associated to the exact sequence $0 \to K_1(\mathbb{Z}) = \{\pm 1\} \to K_1(\mathcal{A}) \to \widetilde{K}_1(\mathcal{A}) \to 0$ and the computations above yield a short exact sequence

$$0 \to \{f \in \mathcal{Z}(\mathcal{A}) \mid f^2 = 1\}/\kappa \xrightarrow{i} \widehat{H}^0(\mathbb{Z}/2; \widetilde{K}_1(\mathcal{A})) \to \{\pm 1\} \to 0 \qquad (9.32)$$

and imply $\widehat{H}^1(\mathbb{Z}/2; \widetilde{K}_1(\mathcal{A})) = 0$ and that the natural map $\widehat{H}^n(\mathbb{Z}/2; \widetilde{K}_1(\mathcal{A})) \xrightarrow{\cong} \widehat{H}^n(\mathbb{Z}/2; \widetilde{K}_1(\mathcal{U}))$ is bijective for $n = 0, 1$. The Rothenberg sequence (9.30) reduces to

$$0 \to L_s^0(\mathcal{A}) \to L_h^0(\mathcal{A}) \xrightarrow{k} \widehat{H}^0(\mathbb{Z}/2; \widetilde{K}_1(\mathcal{A})) \to L_s^1(\mathcal{A}) \to 0.$$

Since any element in $x \in K_1(\mathcal{A})$ with $x^* = x$ can be represented by an element $a \in \mathcal{A}^{\mathrm{inv}}$ with $a^* = a$, the image of k contains the image of the map i appearing in the exact sequence (9.32). From the definition of k one concludes that $\mathrm{im}(k) \subset \mathrm{im}(i)$. Namely, for a symmetric non-degenerate symmetric form $f \colon P \to P^*$ or $a \colon F \to F^*$ respectively the element $[P] \in \widetilde{K}_0(R)$ or $[i \circ h^* \circ a \circ h] \in \widetilde{K}_1(R)$ respectively, which appears in the definition of the image under k of the element $[f] \in L_p^0(R)$ or $[a] \in L_h^0(R)$ respectively, lifts to an element in the unreduced K-group, which is fixed under the involution. Hence we get $\mathrm{im}(k) = \mathrm{im}(i)$. Now assertion (6) follows.

(7) We already know that $K_i(\mathcal{A}) = 0$ for $i = 0, 1$ (see Theorem 9.15 and Theorem 9.19). Now the claim follows for $\epsilon = p$ from assertions (2), (3) and (4). The claim for the other decorations $\epsilon = h, s$ is a direct consequence of the Rothenberg sequences (9.29) and (9.30). This finishes the proof of Theorem 9.31. $\square$

Remark 9.33. The canonical decomposition of $\mathcal{A}$ (see Theorem 9.3) induces an isomorphism

$$L_p^n(\mathcal{A}) \xrightarrow{\cong} L_p^n(\mathcal{A}_{I_f}) \times L_p^n(\mathcal{A}_{I_\infty}) \times L_p^n(\mathcal{A}_{II_1}) \times L_p^n(\mathcal{A}_{II_\infty}) \times L_p^n(\mathcal{A}_{III}).$$

If $\mathcal{A}$ is finite, the canonical decomposition of $\mathcal{A}$ (see Theorem 9.3) induces a splitting

$$L_p^n(\mathcal{U}(\mathcal{A})) \cong L_p^n(\mathcal{U}(\mathcal{A}_{I_f})) \times L_p^n(\mathcal{U}(\mathcal{A}_{II_1})).$$

This comes from a general splitting $L_p^n(R \times S) \xrightarrow{\cong} L_p^n(R) \times L_p^n(S)$ for rings with involution. This splitting is not available for the other decorations $\epsilon = h, s$ essentially because we only get a splitting $K_i(R \times S) \xrightarrow{\cong} K_i(R) \times K_i(S)$ for the unreduced K-groups, which does not carry over to the reduced K-groups.

Example 9.34. Let G be a finitely generated group which does not contain $\mathbb{Z}^n$ as subgroup of finite index. Then $\mathcal{N}(G)$ is of type II_1 and countably composable by Lemma 9.4. We conclude from Theorem 9.13, Theorem 9.18, Theorem 9.20 and Theorem 9.31 that

$$K_0(\mathcal{N}(G)) \cong \mathcal{Z}(\mathcal{N}(G))^{\mathbb{Z}/2};$$
$$K_1(\mathcal{N}(G)) \cong \mathcal{Z}(\mathcal{N}(G))^{+,\mathrm{inv}};$$
$$K_0(\mathcal{U}(G)) \cong \mathcal{Z}(\mathcal{N}(G))^{\mathbb{Z}/2};$$
$$K_1(\mathcal{U}(G)) = 0;$$
$$L_p^0(\mathcal{N}(G)) \cong \mathcal{Z}(\mathcal{N}(G))^{\mathbb{Z}/2};$$
$$L_h^0(\mathcal{N}(G)) \cong L_s^0(\mathcal{N}(G));$$

$$L_\epsilon^1(\mathcal{N}(G)) \cong 0 \qquad \text{for } \epsilon = p, h, s;$$
$$L_\epsilon^n(\mathcal{N}(G)) \cong L_\epsilon^n(\mathcal{U}(G)) \qquad \text{for } n \in \mathbb{Z}, \epsilon = p, h, s;$$
$$L_n^\epsilon(\mathcal{N}(G)) \cong L_\epsilon^n(\mathcal{N}(G)) \qquad \text{for } n \in \mathbb{Z}, \epsilon = p, h, s;$$
$$L_n^\epsilon(\mathcal{U}(G)) \cong L_\epsilon^n(\mathcal{U}(G)) \qquad \text{for } n \in \mathbb{Z}, \epsilon = p, h, s;$$

and that there is an exact sequence $0 \to \mathbb{Z}/2 \to L_h^0(\mathcal{N}(G)) \to \mathcal{Z}(\mathcal{N}(G))^{\mathbb{Z}/2} \to 0$. If G contains no element $g \in G$ with $g \neq 1$ and $|(g)| < \infty$, then we conclude from Lemma 9.4 (4)

$$\mathcal{Z}(\mathcal{N}(G))^{\mathbb{Z}/2} \cong \mathbb{R};$$
$$\mathcal{Z}(\mathcal{N}(G))^{+,\mathrm{inv}} \cong \mathbb{R}^{>0}.$$

Remark 9.35. Analogously to the localization sequence in K-theory (9.22) there is a long exact localization sequence in L-theory [430, Section 3.2]. Let $\mathcal{A}$ be a finite von Neumann algebra and $\mathcal{U}$ the algebra of affiliated operators such that $\mathcal{A}_{II_1}$ is countably composable. Then the maps $L_\epsilon^n(\mathcal{A}) \to L_\epsilon^n(\mathcal{U})$ and $L_n^\epsilon(\mathcal{A}) \to L_n^\epsilon(\mathcal{U})$ are bijective for all $n \in \mathbb{Z}$ and $\epsilon = p, h, s$ by Theorem 9.31. Hence the relative terms $L_\epsilon^n(\mathcal{A} \to \mathcal{U})$ and $L_n^\epsilon(\mathcal{A} \to \mathcal{U})$ must vanish for $n \in \mathbb{Z}$ and $\epsilon = p, h, s$. They are defined in terms of $\mathcal{U}$-acyclic Poincaré $\mathcal{A}$-chain complexes. Equivalently, they can be defined in terms of linking forms and formations on finitely presented $\mathcal{A}$-modules which are $\mathcal{U}$-torsion [430, Proposition 3.4.1 on page 228, Proposition 3.4.7 on page 274, Proposition 3.5.2 on page 292, Proposition 3.5.5 on page 361]. See also [184], [186].

9.5 Application to Middle K- and G-Theory of Group Rings

9.5.1 Detecting Elements in K_1 of a Complex Group Ring

We have introduced the Whitehead group $\mathrm{Wh}(G)$ of a group G in Subsection 3.1.1. Let $i\colon H \to G$ be the inclusion of a normal subgroup $H \subset G$. It induces a homomorphism $i_0\colon \mathrm{Wh}(H) \to \mathrm{Wh}(G)$. The conjugation action of G on H and on G induces a G-action on $\mathrm{Wh}(H)$ and on $\mathrm{Wh}(G)$ which turns out to be trivial on $\mathrm{Wh}(G)$. Hence i_0 induces homomorphisms

$$i_1\colon \mathbb{Z} \otimes_{\mathbb{Z}G} \mathrm{Wh}(H) \to \mathrm{Wh}(G); \tag{9.36}$$
$$i_2\colon \mathrm{Wh}(H)^G \to \mathrm{Wh}(G). \tag{9.37}$$

The main result of this subsection is the following theorem. We emphasize that it holds for all groups G.

Theorem 9.38 (Detecting elements in $\mathrm{Wh}(G)$). *Let $i\colon H \to G$ be the inclusion of a normal finite subgroup H into an arbitrary group G. Then the maps i_1 and i_2 defined in (9.36) and (9.37) have finite kernel.*

Proof. Since H is finite, $\mathrm{Wh}(H)$ is a finitely generated abelian group. Hence it suffices to show for $k = 1, 2$ that i_k is rationally injective, i.e. $\mathrm{id}_{\mathbb{Q}} \otimes_{\mathbb{Z}} i_k$ is injective. The G-action on H by conjugation $c\colon G \to \mathrm{aut}(H)$ factorizes as $G \xrightarrow{\mathrm{pr}} G/C_G H \xrightarrow{\overline{c}} \mathrm{aut}(H)$, where C_H is the centralizer of H, i.e. the kernel of c, and $\overline{c}$ is injective. Since H is finite, $\mathrm{aut}(H)$ and therefore $G/C_G H$ are finite. This implies that the natural map

$$b\colon \mathbb{Q} \otimes_{\mathbb{Z}} \mathrm{Wh}(H)^G = \mathbb{Q} \otimes_{\mathbb{Z}} \mathrm{Wh}(H)^{G/C_G H}$$
$$\to \mathbb{Q} \otimes_{\mathbb{Z}[G/C_G H]} \mathrm{Wh}(H) = \mathbb{Q} \otimes_{\mathbb{Z}G} \mathrm{Wh}(H)$$

sending $q \otimes_{\mathbb{Z}} x$ to $q \otimes_{\mathbb{Z}G} x$ is an isomorphism. Its composition with $\mathbb{Q} \otimes_{\mathbb{Z}G} i_1$ is $\mathbb{Q} \otimes_{\mathbb{Z}G} i_2$. Hence it suffices to show that i_2 is rationally injective.

Next we construct the following commutative diagram

$$\left(\mathcal{Z}(\mathbb{C}H)^{\mathrm{inv}}\right)^{G\times\mathbb{Z}/2} \xrightarrow{\ s\circ i_6\ } \mathcal{Z}(\mathcal{N}(G))^{+,\mathrm{inv}}$$

$$j_3\downarrow \qquad\qquad \uparrow \det{}_{FK}$$

$$K_1(\mathbb{C}H)^{G\times\mathbb{Z}/2} \xrightarrow{\ i_5\ } K_1(\mathcal{N}(G))$$

$$j_2\uparrow \qquad\qquad \uparrow k_2$$

$$K_1(\mathbb{Z}H)^{G\times\mathbb{Z}/2} \xrightarrow{\ i_4\ } K_1(\mathbb{Z}G)^{\mathbb{Z}/2}$$

$$p\downarrow \qquad\qquad \downarrow q$$

$$\mathrm{Wh}(H)^{G\times\mathbb{Z}/2} \xrightarrow{\ i_3\ } \mathrm{Wh}(G)^{\mathbb{Z}/2}$$

$$j_1\downarrow \qquad\qquad \downarrow k_1$$

$$\mathrm{Wh}(H)^G \xrightarrow{\ i_2\ } \mathrm{Wh}(G)$$

The G-actions on the various groups above come from the conjugation action of G on the normal subgroup H. The $\mathbb{Z}/2$-actions are given by the involutions coming from the involutions on the rings. Notice that these two actions commute. The maps j_1 and k_1 are the obvious inclusions. The homomorphisms p and q are induced by the canonical projections $K_1(\mathbb{Z}H) \to \mathrm{Wh}(H)$ and $K_1(\mathbb{Z}G) \to \mathrm{Wh}(G)$. The maps j_2 and k_2 come from the obvious ring homomorphisms. Let $j_3' \colon \mathcal{Z}(\mathbb{C}H)^{\mathrm{inv}} \to K_1(\mathbb{C}H)$ be the homomorphism sending $u \in \mathcal{Z}(\mathbb{C}H)^{\mathrm{inv}}$ to the class in $K_1(\mathbb{C}H)$ represented by the $\mathbb{C}H$-automorphism of $\mathbb{C}H$ given by multiplication with u. It induces the homomorphism j_3. We have indicated the definition of the center-valued Fuglede-Kadison determinant $K_1(\mathcal{N}(G)) \to \mathcal{Z}(\mathcal{N}(G))^{+,\mathrm{inv}}$ in Subsection 9.2.2. Notice that it is defined for any group G. The condition that $\mathcal{N}(G)$ is countably decomposable enters in the proof of Theorem 9.18, not in the construction of the Fuglede-Kadison determinant. The horizontal maps i_2, i_3 and i_4 are induced by the inclusion $i\colon H \to G$ in the obvious way. The map i_5 comes from the inclusion of $\mathbb{C}H$ into $\mathcal{N}(G)$. Let $i_6 \colon \left(\mathcal{Z}(\mathbb{C}H)^{\mathrm{inv}}\right)^{G\times\mathbb{Z}/2} \to \left(\mathcal{N}(G)^{\mathrm{inv}}\right)^{G\times\mathbb{Z}/2}$ be the injection induced by the inclusion of rings $\mathbb{C}H \to \mathcal{N}(G)$. We have $\mathcal{N}(G)^G = \mathcal{Z}(\mathcal{N}(G))$, since $\mathbb{C}G$ is dense in $\mathcal{N}(G)$ in the weak topology. We have $\mathcal{N}(G)^{\mathrm{inv}} \cap \mathcal{Z}(\mathcal{N}(G)) = \mathcal{Z}(\mathcal{N}(G))^{\mathrm{inv}}$. This implies $\left(\mathcal{N}(G)^{\mathrm{inv}}\right)^{G\times\mathbb{Z}/2} = \left(\mathcal{Z}(\mathcal{N}(G))^{\mathrm{inv}}\right)^{\mathbb{Z}/2}$. Hence we can define a map $s\colon \left(\mathcal{N}(G)^{\mathrm{inv}}\right)^{G\times\mathbb{Z}/2} \to \mathcal{Z}(\mathcal{N}(G))^{+,\mathrm{inv}}$ by sending a to $|a| = \sqrt{a^*a}$. For $u \in \mathcal{Z}(\mathcal{N}(G))^{\mathbb{Z}/2}$ the Fuglede-Kadison determinant of the $\mathcal{N}(G)$-automorphism of $\mathcal{N}(G)$ given by multiplication with u is $|u| = \sqrt{u^*u}$. Now one easily checks that the diagram above commutes. In order to show that i_2 is rationally injective, it suffices to prove the following assertions:

(1) s is rationally injective;
(2) i_6 is injective;
(3) j_3 is an isomorphism;
(4) j_2 is rationally injective;

(5) p is rationally an isomorphism;
(6) j_1 is rationally an isomorphism;
(7) k_1 is injective;
(8) $\det_{FK} \circ k_2(\ker(q)) = 1$.

(1) This follows from $s(a) = 1 \Leftrightarrow a^2 = 1$.

(2) This follows from the definition.

(3) Recall that j_3 is induced by a homomorphism $j_3' \colon \mathcal{Z}(\mathbb{C}H)^{\mathrm{inv}} \to K_1(\mathbb{C}H)$ by taking the $G \times \mathbb{Z}/2$- fixed point set. Since the $G \times \mathbb{Z}/2$-action factorizes through the finite group $G/C_G H \times \mathbb{Z}/2$, j_3 is a rational isomorphism if j_3' is a rational isomorphism. Since $\mathbb{C}H$ is semisimple, there is an isomorphism $\mathrm{pr} \colon \mathbb{C}H \to \prod_{a=1}^{b} M_{n_a}(\mathbb{C})$. Thus we obtain a commutative diagram

$$
\begin{array}{ccccc}
\mathcal{Z}(\mathbb{C}H)^{\mathrm{inv}} & \xrightarrow[\cong]{} & \prod_{a=1}^{b} \mathcal{Z}(M_{n_a}(\mathbb{C}))^{\mathrm{inv}} & \xleftarrow[\cong]{\prod_{a=1}^{b} d_a} & \prod_{a=1}^{b} \mathbb{C}^{\mathrm{inv}} \\
{\scriptstyle j_3'} \downarrow & & {\scriptstyle \prod_{a=1}^{b} k_a'} \downarrow & & {\scriptstyle \prod_{a=1}^{b} m_a} \downarrow \\
K_1(\mathbb{C}H) & \xrightarrow[\cong]{} & \prod_{a=1}^{b} K_1(M_{n_a}(\mathbb{C})) & \xrightarrow[\cong]{\prod_{a=1}^{b} \det_{\mathbb{C}} \circ \mu_a} & \prod_{a=1}^{b} \mathbb{C}^{\mathrm{inv}}
\end{array}
$$

Here the isomorphism $d_a \colon \mathbb{C}^{\mathrm{inv}} \to \mathcal{Z}(M_{n_a}(\mathbb{C}))^{\mathrm{inv}}$ sends λ to the diagonal (n_a, n_a)-matrix whose diagonal entries are all λ, the maps k_a' are defined analogously to j_3', the map $m_a \colon \mathbb{C}^{\mathrm{inv}} \to \mathbb{C}^{\mathrm{inv}}$ sends λ to λ^{n_a}, the Morita isomorphism $\mu_a \colon K_1(M_{n_a}(\mathbb{C})) \to K_1(\mathbb{C})$ is given by applying $\mathbb{C}^{n_a} \otimes_{M_{n_a}(\mathbb{C})} -$ and $\det_{\mathbb{C}} \colon K_1(\mathbb{C}) \to \mathbb{C}^{\mathrm{inv}}$ is the isomorphism induced by the determinant. Since all horizontal arrows and the maps m_a are rational isomorphisms, j_3' is rationally bijective.

(4) Wall [511] has shown for finite H that the kernel $SK_1(\mathbb{Z}H)$ of the change of rings map $K_1(\mathbb{Z}H) \to K_1(\mathbb{Q}H)$ is finite and maps under the canonical projection $K_1(\mathbb{Z}H) \to \mathrm{Wh}(H)$ bijectively onto the torsion subgroup $\mathrm{tors}(\mathrm{Wh}(H))$ of $\mathrm{Wh}(H)$ (see also [406, page 5 and page 180]). The change of rings map $K_1(\mathbb{Q}H) \to K_1(\mathbb{C}H)$ is injective [406, page 5 and page 43]. Hence the change of rings map $K_1(\mathbb{Z}H) \to K_1(\mathbb{C}H)$ is rationally injective. Since j_2 is obtained by taking the fixed point set under the action of the finite group $G/C_G H \times \mathbb{Z}/2$ from it, j_2 is rationally injective.

(5) The kernel of the projection $K_1(\mathbb{Z}H) \to \mathrm{Wh}(H)$ is finite as H is finite. Hence p is rationally bijective since it is obtained from a rational isomorphism by taking the fixed point set under the operation of the finite group $G/C_G H \times \mathbb{Z}/2$.

(6) The involution on $\mathrm{Wh}'(H) := \mathrm{Wh}(H)/\mathrm{tors}(\mathrm{Wh}(H)) = \mathrm{Wh}(H)/SK_1(\mathbb{Z}H)$ is trivial [511], [406, page 182]. This implies $\mathrm{Wh}'(H) = \mathrm{Wh}'(H)^{\mathbb{Z}/2}$. Since $SK_1(\mathbb{Z}H)$ is finite, we conclude that the inclusion $\mathrm{Wh}(H)^{\mathbb{Z}/2} \to \mathrm{Wh}(H)$ is a rational isomorphism. This implies that j_1 is rationally bijective.

(7) This is obvious.

(8) An element in the kernel of $K_1(\mathbb{Z}G) \to \mathrm{Wh}(G)$ is represented by a $\mathbb{Z}G$-automorphism $r_{\pm g} \colon \mathbb{Z}G \to \mathbb{Z}G$ which is given by right multiplication with an element $\pm g$ for $g \in G$. Since for the induced $\mathcal{N}(G)$-map $r_{\pm g} \colon \mathcal{N}(G) \to \mathcal{N}(G)$, we have $(r_{\pm g})^* \circ r_{\pm g} = 1$, we get $\det_{FK} \circ k_2([r_{\pm g}]) = 1$. This finishes the proof Theorem 9.38. $\square$

9.5.2 Survey on the Isomorphism Conjecture for K_0 of Complex Group Rings, the Bass Conjecture and the Hattori-Stallings Rank

In this section we explain the Isomorphism Conjecture for $K_0(\mathbb{C}G)$ and the Bass Conjecture and their relation. We introduce the Hattori-Stallings rank and study its relation to the center valued dimension of a group von Neumann algebra.

We begin with the Isomorphism Conjecture for $K_0(\mathbb{C}G)$. For more information about the Isomorphism Conjectures in algebraic K- and L-theory due to Farrell and Jones and the related *Baum-Connes Conjecture* for the topological K-theory of the reduced group C^*-algebra we refer for instance to [27, Conjecture 3.15 on page 254], [128, section 6], [194], [256] [257], [339], [340], and [501].

The *orbit category* $\mathrm{Or}(G)$ has as objects homogeneous spaces G/H and as morphisms G-maps. Let $\mathrm{Or}(G, \mathcal{FIN})$ be the full subcategory of $\mathrm{Or}(G)$ consisting of objects G/H with finite H. We define a covariant functor

$$\mathrm{Or}(G) \to \mathrm{ABEL}, \qquad G/H \mapsto K_0(\mathbb{C}H) \tag{9.39}$$

as follows. It sends G/H to the projective class group $K_0(\mathbb{C}H)$. Given a morphism $f \colon G/H \to G/K$ there is an element $g \in G$ with $g^{-1}Hg \subset K$ such that $f(g'H) = g'gK$. Define $f_* \colon K_0(\mathbb{C}H) \to K_0(\mathbb{C}K)$ as the map induced by induction with the group homomorphism $c_g \colon H \to K$, $h \mapsto g^{-1}hg$. This is independent of the choice of g since any inner automorphism of K induces the identity on $K_0(\mathbb{C}K)$.

Conjecture 9.40 (Isomorphism Conjecture). *The* Isomorphism Conjecture for $K_0(\mathbb{C}G)$ *says that the map*

$$a \colon \mathrm{colim}_{\mathrm{Or}(G,\mathcal{FIN})} K_0(\mathbb{C}H) \to K_0(\mathbb{C}G)$$

induced by the various inclusions $H \subset G$ is an isomorphism.

Notation 9.41. *Let $\mathrm{con}(G)$ be the set of conjugacy classes (g) of elements $g \in G$. Let $\mathrm{con}(G)_f$ be the subset of $\mathrm{con}(G)$ of conjugacy classes (g) for which each representative g has finite order. Let $\mathrm{con}(G)_{cf}$ be the subset of $\mathrm{con}(G)$ of conjugacy classes (g) which contain only finitely many elements. Put $\mathrm{con}(G)_{f,cf} = \mathrm{con}(G)_f \cap \mathrm{con}(G)_{cf}$. We denote by $\mathrm{class}_0(G)$, $\mathrm{class}_0(G)_f$, $\mathrm{class}_0(G)_{cf}$, and $\mathrm{class}_0(G)_{f,cf}$ respectively the complex vector space with the* set con(G), con(G)$_f$, con(G)$_{cf}$ and con(G)$_{f,cf}$ respectively as basis. We denote by class(G), class(G)$_f$, class(G)$_{cf}$, and class(G)$_{f,cf}$ respectively the complex vector space of functions from the set con(G), con(G)$_f$, con(G)$_{cf}$ and con(G)$_{f,cf}$ respectively to the complex numbers $\mathbb{C}$.

Notice that class$_0(G)$, class$_0(G)_f$, class$_0(G)_{cf}$ and class$_0(G)_{f,cf}$ can be identified with the complex vector space of functions with finite support from con(G), con(G)$_f$, con(G)$_{cf}$ and con(G)$_{f,cf}$ to $\mathbb{C}$. Recall that (g) is finite if and only if the centralizer C_g of g has finite index in G. We obtain an isomorphism of complex vector spaces

$$z\colon \text{class}_0(G)_{cf} \xrightarrow{\cong} \mathcal{Z}(\mathbb{C}G), \qquad (g) \mapsto \sum_{g' \in (g)} g'. \tag{9.42}$$

Define the *universal $\mathbb{C}G$-trace* of $\sum_{g \in G} \lambda_g g \in \mathbb{C}G$ by

$$\text{tr}^u_{\mathbb{C}G}\left(\sum_{g \in G} \lambda_g g\right) := \sum_{g \in G} \lambda_g \cdot (g). \qquad \in \text{class}_0(G). \tag{9.43}$$

Under the obvious identification of class$_0(G)$ with $\mathbb{C}G/[\mathbb{C}G, \mathbb{C}G]$, the trace $\text{tr}^u_{\mathbb{C}G}$ becomes the canonical projection. This extends to square matrices in the usual way

$$\text{tr}^u_{\mathbb{C}G}\colon M_n(\mathbb{C}G) \to \text{class}_0(G), \qquad A \mapsto \sum_{i=1}^n \text{tr}^u_{\mathbb{C}G}(a_{i,i}). \tag{9.44}$$

Let P be a finitely generated projective $\mathbb{C}G$-module. Define its *Hattori-Stallings rank* by

$$\text{HS}(P) := \text{tr}^u_{\mathbb{C}G}(A) \qquad \in \text{class}_0(G), \tag{9.45}$$

where A is any element in $M_n(\mathbb{C}G)$ with $A^2 = A$ such that the image of the map $\mathbb{C}G^n \to \mathbb{C}G^n$ given by right multiplication with A is $\mathbb{C}G$-isomorphic to P. This definition is independent of the choice of A (cf. (6.4)). The Hattori-Stallings rank defines homomorphisms

$$\text{HS}\colon K_0(\mathbb{C}G) \to \text{class}_0(G), \qquad [P] \mapsto \text{HS}(P); \tag{9.46}$$

$$\text{HS}_{\mathbb{C}}\colon K_0(\mathbb{C}G) \otimes_{\mathbb{Z}} \mathbb{C} \to \text{class}_0(G), \qquad [P] \otimes \lambda \mapsto \lambda \cdot \text{HS}(P). \tag{9.47}$$

Conjecture 9.48 (Bass Conjecture). *The* strong Bass Conjecture *for $\mathbb{C}G$ says that the image of the map* $\text{HS}_{\mathbb{C}}\colon K_0(\mathbb{C}G) \otimes_{\mathbb{Z}} \mathbb{C} \to \text{class}_0(G)$ *of (9.47) is* class$_0(G)_f$.

The strong Bass Conjecture *for $\mathbb{Z}G$ says that for a finitely generated projective $\mathbb{Z}G$-module P*

$$\text{HS}(\mathbb{C} \otimes_{\mathbb{Z}} P)(g) = \begin{cases} 0 & \text{if } g \neq 1; \\ \text{rk}_{\mathbb{Z}}(\mathbb{Z} \otimes_{\mathbb{Z}G} P) & \text{if } g = 1. \end{cases}$$

The weak Bass Conjecture *for $\mathbb{Z}G$ says that for a finitely generated projective $\mathbb{Z}G$-module P*

$$\mathrm{HS}(\mathbb{C} \otimes_{\mathbb{Z}} P)(1) = \sum_{(g) \in \mathrm{con}(G)} \mathrm{HS}(P)(g) = \mathrm{rk}_{\mathbb{Z}}(\mathbb{Z} \otimes_{\mathbb{Z}G} P).$$

Theorem 9.49. *For any group G there is a commutative diagram whose left vertical arrow is an isomorphism*

$$
\begin{array}{ccc}
\left(\mathrm{colim}_{\mathrm{Or}(G,\mathcal{FIN})} K_0(\mathbb{C}H)\right) \otimes_{\mathbb{Z}} \mathbb{C} & \xrightarrow{a \otimes_{\mathbb{Z}} \mathrm{id}_{\mathbb{C}}} & K_0(\mathbb{C}G) \otimes_{\mathbb{Z}} \mathbb{C} \\
h \downarrow \cong & & \downarrow \mathrm{HS}_{\mathbb{C}} \\
\mathrm{class}_0(G)_f & \xrightarrow{e} & \mathrm{class}_0(G)
\end{array}
$$

Proof. Firstly we explain the maps in the square. We have introduced the map a in Conjecture 9.40 and the map $\mathrm{HS}_{\mathbb{C}}$ in (9.47). The map e is given extending a function $\mathrm{con}(G)_f \to \mathbb{C}$ to a function $\mathrm{con}(G) \to \mathbb{C}$ by putting it to be zero on elements in $\mathrm{con}(G)$ which do not belong to $\mathrm{con}(G)_f$. Define for a group homomorphism $\psi \colon G \to G'$ a map $\psi_* \colon \mathrm{con}(G) \to \mathrm{con}(G')$ by sending (h) to $(\psi(h))$. It induces a homomorphism $\psi_* \colon \mathrm{class}_0(G) \to \mathrm{class}_0(G')$. One easily checks that the following diagram commutes

$$
\begin{array}{ccc}
K_0(\mathbb{C}G) & \xrightarrow{\psi_*} & K_0(\mathbb{C}G') \\
\mathrm{HS} \downarrow & & \downarrow \mathrm{HS} \\
\mathrm{class}_0(G) & \xrightarrow{\psi_*} & \mathrm{class}_0(G')
\end{array}
\tag{9.50}
$$

There is a canonical isomorphism

$$f_1 \colon \left(\mathrm{colim}_{\mathrm{Or}(G,\mathcal{FIN})} K_0(\mathbb{C}H)\right) \otimes_{\mathbb{Z}} \mathbb{C} \xrightarrow{\cong}$$
$$\mathrm{colim}_{\mathrm{Or}(G,\mathcal{FIN})} \left(K_0(\mathbb{C}H) \otimes_{\mathbb{Z}} \mathbb{C}\right). \tag{9.51}$$

The Hattori-Stallings rank induces for each finite subgroup H of G an isomorphism $K_0(\mathbb{C}H) \otimes_{\mathbb{Z}} \mathbb{C} \to \mathrm{class}(H)$ by elementary complex representation theory for finite groups [470, Theorem 6 in Chapter 2 on page 19]. Thus we get an isomorphism

$$f_2 \colon \mathrm{colim}_{\mathrm{Or}(G,\mathcal{FIN})} \left(K_0(\mathbb{C}H) \otimes_{\mathbb{Z}} \mathbb{C}\right) \xrightarrow{\cong} \mathrm{colim}_{\mathrm{Or}(G,\mathcal{FIN})} \mathrm{class}(H). \tag{9.52}$$

Let $f_3' \colon \mathrm{colim}_{\mathrm{Or}(G,\mathcal{FIN})} \mathrm{con}(H) \to \mathrm{con}(G)_f$ be the map induced by the inclusions of the finite subgroups H of G. Define a map $f_4' \colon \mathrm{con}(G)_f \to \mathrm{colim}_{\mathrm{Or}(G,\mathcal{FIN})} \mathrm{con}(H)$ by sending $(g) \in \mathrm{con}(G)_f$ to the image of $(g) \in \mathrm{con}(\langle g \rangle)$ under the structure map from $\mathrm{con}(\langle g \rangle)$ to $\mathrm{colim}_{\mathrm{Or}(G,\mathcal{FIN})} \mathrm{con}(H)$, where $\langle g \rangle$ is the cyclic subgroup generated by g. One easily checks that this is independent of the choice of the representative g of (g) and that f_3' and f_4' are inverse to one another. The bijection f_3' induces an isomorphism

$$f_3 \colon \operatorname{colim}_{\operatorname{Or}(G,\mathcal{FIN})} \operatorname{class}(H) \xrightarrow{\cong} \operatorname{class}_0(G)_f, \qquad (9.53)$$

because colimit and the functor sending a set to the complex vector space with this set as basis commute. Now the isomorphism h is defined as the composition of the isomorphisms f_1 of (9.51), f_2 of (9.52) and f_3 of (9.53). It remains to check that the square in Theorem 9.49 commutes. This follows from the commutativity of (9.50). This finishes the proof of Theorem 9.49.

$\square$

A group G is *poly-cyclic* if there is a finite sequence of subgroups $\{1\} = G_0 \subset G_1 \subset G_2 \subset \ldots G_r = G$ such that G_i is normal in G_{i+1} with cyclic quotient G_{i+1}/G_i for $i = 0, 1, 2, \ldots, r-1$. We call G *virtually poly-cyclic* if G contains a poly-cyclic subgroup of finite index.

Theorem 9.54. *(1) The map $a \otimes_{\mathbb{Z}} \operatorname{id}_{\mathbb{C}} \colon \left(\operatorname{colim}_{\operatorname{Or}(G,\mathcal{FIN})} K_0(\mathbb{C}H)\right) \otimes_{\mathbb{Z}} \mathbb{C} \to K_0(\mathbb{C}G) \otimes_{\mathbb{Z}} \mathbb{C}$ is injective;*

(2) The Isomorphism Conjecture for $K_0(\mathbb{C}G)$ 9.40 implies that the map $\operatorname{HS}_{\mathbb{C}} \colon K_0(\mathbb{C}G) \otimes_{\mathbb{Z}} \mathbb{C} \to \operatorname{class}_0(G)$ of (9.47) is injective with image $\operatorname{class}_0(G)_f$ and hence implies the strong Bass Conjecture for $K_0(\mathbb{C}G)$ 9.48;

(3) Let P be a finitely generated projective $\mathbb{C}G$-module and let $g \in G$ be an element of infinite order with finite (g). Then $\operatorname{HS}(P)(g) = 0$;

(4) Let P be a finitely generated projective $\mathbb{Z}G$-module. Then $\operatorname{HS}(\mathbb{C} \otimes_{\mathbb{Z}} P)(g) = 0$ for any element $g \in G$ with $g \neq 1$ for which $|g| < \infty$ or $|(g)| < \infty$;

(5) The strong Bass Conjecture for $\mathbb{Z}G$ is true for residually finite groups and linear groups;

(6) The strong Bass Conjecture for $\mathbb{C}G$ is true for amenable groups;

(7) The strong Bass Conjecture for $\mathbb{C}G$ is true for a countable group G, if the Bost Conjecture holds for G, which is the version of the Baum-Connes Conjecture with $C_r^*(G)$ replaced by $l^1(G)$. (A discussion of the Bost Conjecture and the class of groups for which it is known can be found in [38]. The class contains all countable groups which are a-T-menable);

(8) Suppose that G is virtually poly-cyclic. Then the map

$$a \colon \operatorname{colim}_{\operatorname{Or}(G,\mathcal{FIN})} K_0(\mathbb{C}H) \to K_0(\mathbb{C}G)$$

is surjective and induces an isomorphism

$$a \colon \left(\operatorname{colim}_{\operatorname{Or}(G,\mathcal{FIN})} K_0(\mathbb{C}H)\right) \otimes_{\mathbb{Z}} \mathbb{C} \xrightarrow{\cong} K_0(\mathbb{C}G) \otimes_{\mathbb{Z}} \mathbb{C},$$

and the forgetful map

$$f \colon K_0(\mathbb{C}G) \to G_0(\mathbb{C}G)$$

is an isomorphism;

(9) If P is a finitely generated projective $\mathbb{C}G$-module, then $\operatorname{HS}(P)(1) \in \mathbb{Q}$.

Proof. (1) and (2) These follow from Theorem 9.49.

(3) This follows from [23, Theorem 8.1 on page 180].

(4) and (5) These follow from [307, Theorem 4.1 on page 96].

(6) This is proved for elementary amenable groups in [197, Theorem 1.6]. The proof for amenable groups can be found in [38].

(7) This is proved in [38].

(8) Moody has shown [384] that the obvious map $\bigoplus_{H \in \mathcal{FIN}} G_0(\mathbb{C}H) \to G_0(\mathbb{C}G)$ given by induction is surjective. Since G is virtually poly-cyclic, the complex group ring $\mathbb{C}G$ is regular, i.e. Noetherian and any $\mathbb{C}G$-module has a finite dimensional projective resolution [447, Theorem 8.2.2 and Theorem 8.2.20]. This implies that $f \colon K_0(\mathbb{C}G) \to G_0(\mathbb{C}G)$ is bijective. The same is true for any finite subgroup $H \subset G$. Now the claim follows using assertion (1).

(9) is proved in [529]. See also [78]. $\square$

Next we investigate the relation between the center valued dimension $\dim^u_{\mathcal{N}(G)}$ (see Definition 9.12) and the Hattori-Stallings rank. Define a homomorphism

$$\phi \colon \mathcal{Z}(\mathcal{N}(G)) \to \text{class}(G)_{cf} \tag{9.55}$$

by assigning to $u \in \mathcal{Z}(\mathcal{N}(G))$

$$\phi(u) \colon \text{con}(G)_{cf} \to \mathbb{C}, \qquad (h) \mapsto \text{tr}_{\mathcal{N}(G)} \left(u \cdot \frac{1}{|(h)|} \cdot \sum_{\widetilde{h} \in (h)} \widetilde{h}^{-1} \right).$$

Equivalently, $\phi(u)$ can be described as follows. If we evaluate $u \in \mathcal{Z}(\mathcal{N}(G)) \subset \mathcal{N}(G) = \mathcal{B}(l^2(G))^G$ at the unit element $e \in G$, we obtain an element $u(e) = \sum_{g \in G} \lambda_g \cdot g \in l^2(G)$ with the property that λ_g depends only on the conjugacy class (g) of $g \in G$. This implies that $\lambda_g = 0$ if (g) is infinite and $\phi(u)(g) = \lambda_g$ for $g \in G$ with finite (g).

Lemma 9.56. *The map $\phi \colon \mathcal{Z}(\mathcal{N}(G)) \to \text{class}(G)_{cf}$ is injective. The composition of $k \colon \text{class}_0(G)_{cf} \to \mathcal{Z}(\mathcal{N}(G))$ with $\phi \colon \mathcal{Z}(\mathcal{N}(G)) \to \text{class}(G)_{cf}$ is the canonical inclusion $\text{class}_0(G)_{cf} \to \text{class}(G)_{cf}$, where k is given by*

$$k \left(\sum_{(g) \in \text{con}(G)} \lambda_{(g)} \cdot (g) \right) = \sum_{(g)} \lambda_{(g)} \cdot \left(\sum_{g' \in (g)} g' \right).$$

The following diagram commutes

$$
\begin{array}{ccccc}
K_0(\mathbb{C}G) & \xrightarrow{\text{HS}} & \text{class}_0(G) & \xrightarrow{r} & \text{class}_0(G)_{cf} \\
\downarrow{\scriptstyle i_*} & & & & \downarrow{\scriptstyle k} \\
K_0(\mathcal{N}(G)) & \xrightarrow{\dim^u_{\mathcal{N}(G)}} & \mathcal{Z}(\mathcal{N}(G))^{\mathbb{Z}/2} & \xrightarrow{j} & \mathcal{Z}(\mathcal{N}(G)),
\end{array}
$$

where r is given by restriction, i_* is induced by the inclusion $i \colon \mathbb{C}G \to \mathcal{N}(G)$, j is the inclusion and the other maps have been defined in Definition 9.12 and (9.46).

Proof. Obviously ϕ is injective and $\phi \circ k$ is the canonical inclusion. For the commutativity of the diagram above it is enough to show for an element $A \in M_n(\mathbb{C}G)$ and $h \in G$ with finite (h)

$$\mathrm{tr}^u_{\mathbb{C}G}(A)(h) = (\phi \circ \mathrm{tr}^u_{\mathcal{N}(G)}(A))(h), \tag{9.57}$$

since for $A \in M_n(\mathbb{C}G)$ with $A^2 = A$ we have

$$\phi \circ k \circ r \circ \mathrm{HS}(\mathrm{im}(A))(h) = \mathrm{tr}^u_{\mathbb{C}G}(A)(h);$$
$$\phi \circ j \circ \dim^u_{\mathcal{N}(G)} \circ i_*([\mathrm{im}(A)]) = (\phi \circ \mathrm{tr}^u_{\mathcal{N}(G)}(A))(h).$$

To prove (9.57) it suffices to show for $g \in G$ and $h \in G$ with finite (h)

$$\mathrm{tr}^u_{\mathbb{C}G}(g)(h) = (\phi \circ \mathrm{tr}^u_{\mathcal{N}(G)}(g))(h). \tag{9.58}$$

Obviously

$$\mathrm{tr}^u_{\mathbb{C}G}(g)(h) = \mathrm{tr}_{\mathcal{N}(G)}\left(g \cdot \sum_{\widetilde{h} \in (h)} \widetilde{h}^{-1} \right). \tag{9.59}$$

We conclude from the universal property of $\mathrm{tr}^u_{\mathcal{N}(G)}$ for all $x \in \mathcal{N}(G)$ (see Theorem 9.5)

$$\mathrm{tr}_{\mathcal{N}(G)}\left(x \cdot \sum_{h' \in (h)} (h')^{-1} \right) = (\phi \circ \mathrm{tr}^u_{\mathcal{N}(G)}(x))(h). \tag{9.60}$$

Now (9.58) and hence (9.57) follow from (9.59) and (9.60). This finishes the proof of Lemma 9.56.   □

We conclude from Theorem 9.54 (9) and Lemma 9.56

Corollary 9.61. *Let P be a finitely generated projective $\mathbb{C}G$-module. Then $\dim_{\mathcal{N}(G)}(\mathcal{N}(G) \otimes_{\mathbb{Z}G} P) = \mathrm{HS}(P)(1)$ and this is a non-negative rational number.*

The next result shows that one cannot detect elements in $\widetilde{K}_0(\mathbb{Z}G)$ by $\widetilde{K}_0(\mathcal{N}(G))$ in contrast to $\mathrm{Wh}(G)$ (cf. Theorem 9.38).

Theorem 9.62. *The change of rings map $\widetilde{K}_0(\mathbb{Z}G) \to \widetilde{K}_0(\mathcal{N}(G))$ is trivial.*

Proof. Because of Theorem 9.13 (2) and Lemma 9.56 it suffices to prove for a finitely generated projective $\mathbb{Z}G$-module P that $\mathrm{HS}(\mathbb{C} \otimes_{\mathbb{Z}} P)(g) = 0$ for $g \in G$ with $g \neq 1$ and $|(g)| < \infty$, and $\mathrm{HS}(\mathbb{C} \otimes_{\mathbb{Z}} P)(1) \in \mathbb{Z}$. This follows from Theorem 9.54 (4) and the fact that $P \cong_{\mathbb{Z}G} \mathrm{im}(A)$ for some $A \in M_n(\mathbb{Z}G)$ with $A^2 = A$.   □

Theorem 9.63. *The image of the composition*

$$K_0(\mathbb{C}G) \otimes_{\mathbb{Z}} \mathbb{C} \xrightarrow{\mathrm{HS}_{\mathbb{C}}} \mathrm{class}_0(G) \xrightarrow{r} \mathrm{class}_0(G)_{cf}$$

is $\mathrm{class}_0(G)_{f,cf}$.

Proof. Apply Theorem 9.49 and Theorem 9.54 (3). $\qquad\qquad\qquad\square$

9.5.3 G-Theory of Complex Group Rings

In this subsection we investigate $G_0(\mathbb{C}G)$ (see Definition 9.11). We have defined $\mathbf{T}\mathcal{N}(G) \otimes_{\mathbb{C}G} M$ and $\mathbf{P}\mathcal{N}(G) \otimes_{\mathbb{C}G} M$ in Definition 6.1 (see also Section 8.4). Recall from Theorem 6.5 and Theorem 6.7 (3) that $\mathbf{P}\mathcal{N}(G) \otimes_{\mathbb{C}G} M$ is a finitely generated projective $\mathcal{N}(G)$-module.

Theorem 9.64. *If G is amenable, the map*

$$l\colon G_0(\mathbb{C}G) \to K_0(\mathcal{N}(G)), \qquad [M] \mapsto [\mathbf{P}\mathcal{N}(G) \otimes_{\mathbb{C}G} M]$$

is a well-defined homomorphism. If $f\colon K_0(\mathbb{C}G) \to G_0(\mathbb{C}G)$ is the forgetful map sending $[P]$ to $[P]$ and $i_*\colon K_0(\mathbb{C}G) \to K_0(\mathcal{N}(G))$ is induced by the inclusion $i\colon \mathbb{C}G \to \mathcal{N}(G)$, then the composition $l \circ f$ agrees with i_*.

Proof. If $0 \to M_0 \xrightarrow{b} M_1 \xrightarrow{p} M_2 \to 0$ is an exact sequence of finitely generated $\mathbb{C}G$-modules we have to check in $K_0(\mathcal{N}(G))$

$$[\mathbf{P}\mathcal{N}(G) \otimes_{\mathbb{C}G} M_0] - [\mathbf{P}\mathcal{N}(G) \otimes_{\mathbb{C}G} M_1] + [\mathbf{P}\mathcal{N}(G) \otimes_{\mathbb{C}G} M_2] = 0.$$

Consider the induced sequence $\mathbf{P}\mathcal{N}(G) \otimes_{\mathbb{C}G} M_0 \xrightarrow{\bar{b}} \mathbf{P}\mathcal{N}(G) \otimes_{\mathbb{C}G} M_1 \xrightarrow{\bar{p}} \mathbf{P}\mathcal{N}(G) \otimes_{\mathbb{C}G} M_2$. Obviously $\bar{p}$ is surjective as p is surjective. We conclude from Theorem 6.7 (1) that $\ker(\bar{b})$ and $\ker(\bar{p})$ are finitely generated projective $\mathcal{N}(G)$-modules. Theorem 6.7 (4b) and (4e) and Theorem 6.37 imply

$$\dim_{\mathcal{N}(G)}\left(\ker(\bar{b})\right) = 0;$$
$$\dim_{\mathcal{N}(G)}\left(\mathbf{P}\mathcal{N}(G) \otimes_{\mathbb{C}G} M_0\right) = \dim_{\mathcal{N}(G)}\left(\ker(\bar{p})\right).$$

Lemma 6.28 shows that $\bar{b}$ is injective and $\mathbf{P}\mathcal{N}(G) \otimes_{\mathbb{C}G} M_0 \cong_{\mathcal{N}(G)} \ker(\bar{p})$. Obviously $\ker(\bar{p}) \oplus \mathbf{P}\mathcal{N}(G) \otimes_{\mathbb{C}G} M_2$ and $\mathbf{P}\mathcal{N}(G) \otimes_{\mathbb{C}G} M_1$ are $\mathcal{N}(G)$-isomorphic. $\qquad\square$

Now we can detect elements in $G_0(\mathbb{C}G)$ for amenable groups G.

Theorem 9.65. *Suppose that G is amenable. Then the image of the composition*

$$G_0(\mathbb{C}G) \otimes_{\mathbb{Z}} \mathbb{C} \xrightarrow{l \otimes_{\mathbb{Z}} \mathbb{C}} K_0(\mathcal{N}(G)) \otimes_{\mathbb{Z}} \mathbb{C} \xrightarrow{\dim_{\mathcal{N}(G)}^u} \mathcal{Z}(\mathcal{N}(G)) \xrightarrow{\phi} \mathrm{class}(G)_{cf}$$

contains the complex vector space $\mathrm{class}_0(G)_{f,cf}$. In particular

$$\mathrm{rk}_{\mathbb{Z}}(G_0(\mathbb{C}G)) \geq |\mathrm{con}(G)_{f,cf}|.$$

Proof. Apply Lemma 9.56, Theorem 9.63 and Theorem 9.64.   □

Theorem 9.66. *(1) If G is amenable, the class $[\mathbb{C}G]$ generates an infinite cyclic subgroup in $G_0(\mathbb{C}G)$ and is in particular not zero;*
(2) If G contains $\mathbb{Z} * \mathbb{Z}$ as subgroup, we get $[\mathbb{C}G] = 0$ in $G_0(\mathbb{C}G)$.

Proof. (1) The map $l\colon G_0(\mathbb{C}G) \to K_0(\mathcal{N}(G))$ of Theorem 9.64 sends $[\mathbb{C}G]$ to $[\mathcal{N}(G)]$ and $[\mathcal{N}(G)] \in K_0(\mathcal{N}(G))$ generates an infinite cyclic subgroup because of $\dim_{\mathcal{N}(G)}(\mathcal{N}(G)) = 1$.

(2) We abbreviate $F_2 = \mathbb{Z} * \mathbb{Z}$. Induction with the inclusion $F_2 \to G$ induces a homomorphism $G_0(\mathbb{C}F_2) \to G_0(\mathbb{C}G)$ which sends $[\mathbb{C}F_2]$ to $[\mathbb{C}G]$. Hence it suffices to show $[\mathbb{C}F_2] = 0$ in $G_0(\mathbb{C}F_2)$. The cellular chain complex of the universal covering of $S^1 \vee S^1$ yields an exact sequence of $\mathbb{C}F_2$-modules $0 \to (\mathbb{C}F_2)^2 \to \mathbb{C}F_2 \to \mathbb{C} \to 0$, where $\mathbb{C}$ is equipped with the trivial F_2-action. This implies $[\mathbb{C}F_2] = -[\mathbb{C}]$ in $G_0(\mathbb{C}F_2)$. Hence it suffices to show $[\mathbb{C}] = 0$ in $G_0(\mathbb{C}F_2)$. Choose an epimorphism $f\colon F_2 \to \mathbb{Z}$. Restriction with f defines a homomorphism $G_0(\mathbb{C}\mathbb{Z}) \to G_0(\mathbb{C}F_2)$. It sends $\mathbb{C}$ viewed as trivial $\mathbb{C}\mathbb{Z}$-module to $\mathbb{C}$ viewed as trivial $\mathbb{C}F_2$-module. Hence it remains to show $[\mathbb{C}] = 0$ in $G_0(\mathbb{C}\mathbb{Z})$. This follows from the exact sequence $0 \to \mathbb{C}\mathbb{Z} \xrightarrow{s-1} \mathbb{C}\mathbb{Z} \to \mathbb{C} \to 0$ for s a generator of $\mathbb{Z}$ which comes from the cellular $\mathbb{C}\mathbb{Z}$-chain complex of $\widetilde{S^1}$.   □

Theorem 9.66 gives evidence for

Conjecture 9.67. (Amenability and the regular representation in G-theory). *A group G is amenable if and only if $[\mathbb{C}G] \neq 0$ in $G_0(\mathbb{C}G)$.*

Example 9.68. Let D be the infinite dihedral group $D = \mathbb{Z}/2 * \mathbb{Z}/2$. Then we compute

$$\widetilde{K}_0(\mathbb{Z}D) = 0;$$
$$K_0(\mathbb{C}D) \cong \mathbb{Z}^3;$$
$$G_0(\mathbb{C}D) \cong K_0(\mathbb{C}D).$$

as follows. Since $\widetilde{K}_0(\mathbb{Z}[\mathbb{Z}/2]) = 0$ [125, Corollary 5.17] and the obvious map $\widetilde{K}_0(\mathbb{Z}[\mathbb{Z}/2]) \oplus \widetilde{K}_0(\mathbb{Z}[\mathbb{Z}/2]) \to \widetilde{K}_0(\mathbb{Z}D)$ is bijective [508, Corollary 11.5 and the following remark], we conclude $\widetilde{K}_0(\mathbb{Z}D) = 0$. If s_1 and s_2 are the generators of the subgroups $\mathbb{Z}/2 * 1$ and $1 * \mathbb{Z}/2$, then the set $\mathrm{con}_f(G)$ consists of the three elements (1), (s_1) and (s_2). Let $\mathbb{C}^-$ be the $\mathbb{C}[\mathbb{Z}/2]$-module with $\mathbb{C}$ as underlying complex vector space and the $\mathbb{Z}/2$-action given by $-\mathrm{id}$. Let P_1 and P_2 be the $\mathbb{C}D$-modules given by induction with $\mathbb{Z}/2 = \langle s_1 \rangle \subset \mathbb{Z}/2 * \mathbb{Z}/2$ and $\mathbb{Z}/2 = \langle s_2 \rangle \subset \mathbb{Z}/2 * \mathbb{Z}/2$ applied to $\mathbb{C}^-$. One easily checks that $\mathrm{HS}(P_i) = 1/2 \cdot ((e) - (s_i))$ for $i = 1, 2$ and $e \in D$ the unit element. Notice that D contains an infinite cyclic subgroup of index 2, namely $\langle s_1 s_2 \rangle$. Because of Theorem 9.54 (1) and (8) the forgetful map $f\colon K_0(\mathbb{C}G) \xrightarrow{\cong} G_0(\mathbb{C}G)$ is bijective and we obtain isomorphisms $u\colon \mathbb{Z}^3 \xrightarrow{\cong} K_0(\mathbb{C}D)$ and $v\colon K_0(\mathbb{C}D) \xrightarrow{\cong} \mathbb{Z}^3$, which are inverse to one another and defined by

$$u(n_0, n_1, n_2) = n_0 \cdot [\mathbb{Z}D] + n_1 \cdot [P_1] + n_2 \cdot [P_2];$$
$$v([P]) = (\mathrm{HS}(P)(1) + \mathrm{HS}(P)(s_1) + \mathrm{HS}(s_2), -2 \cdot \mathrm{HS}(P)(s_1), -2 \cdot \mathrm{HS}(s_2)).$$

The composition

$$\mathbb{Z}^3 \xrightarrow{u} K_0(\mathbb{C}D) \xrightarrow{i_*} K_0(\mathcal{N}(D)) \xrightarrow{\dim_{\mathcal{N}(D)}} \mathbb{R}$$

sends (n_0, n_1, n_2) to $n_0 + n_1/2 + n_2/2$. In particular the map $i_* \colon \widetilde{K}_0(\mathbb{C}D) \to \widetilde{K}_0(\mathcal{N}(D))$ is not trivial (compare with Theorem 9.62.)

Remark 9.69. The knowledge about $G_0(\mathbb{C}G)$ is very poor. At least there is Moody's result stated as Theorem 9.54 (8) and we will show in Example 10.13 that for the amenable locally finite group $A = \bigoplus_{n \in \mathbb{Z}} \mathbb{Z}/2$ the map $K_0(\mathbb{C}A) \to G_0(\mathbb{C}A)$ is not surjective. On the other hand we do not know a counterexample to the statement that for an amenable group with an upper bound on the orders of its finite subgroups both $a \colon \mathrm{colim}_{\mathrm{Or}(G, \mathcal{FIN})} K_0(\mathbb{C}H) \to K_0(\mathbb{C}G)$ and $K_0(\mathbb{C}G) \to G_0(\mathbb{C}G)$ are bijective. We also do not know a counterexample to the statement that for a group G, which is non amenable, $G_0(\mathbb{C}G) = \{0\}$. To our knowledge the latter equality is not even known for $G = \mathbb{Z} * \mathbb{Z}$.

9.6 Miscellaneous

We have emphasized the analogy between a finite von Neumann algebra $\mathcal{A}$ and a principal ideal domain R and their Ore localizations $\mathcal{U}$ and F with respect to the set of non-zero-divisors (see Example 6.12 and Section 8.5). Of course F is just the quotient field of R. Some of the results of this chapters are completely analogous for R and F. For instance $K_0(R) \to K_0(F)$ is bijective (compare Theorem 9.20 (1)). We have $K_1(F) = F^{\mathrm{inv}}$ and $K_0(R \to F) = F^{\mathrm{inv}}/R^{\mathrm{inv}}$. We also get $K_1(R) = R^{\mathrm{inv}}$, provided that R has a Euclidean algorithm, for instance if R is $\mathbb{Z}$ or $K[x]$ for a field K (compare Theorem 9.17, Theorem 9.20 (2) and Theorem 9.23 (3)).

The topological K-theory $K_p^{\mathrm{top}}(\mathcal{A})$ of a von Neumann algebra $\mathcal{A}$ equals by definition $K_0(\mathcal{A})$ for $p = 0$ and $\pi_0(GL(\mathcal{A}))$ for $p = 1$. We have computed $K_0^{\mathrm{top}}(\mathcal{A}) = K_0(\mathcal{A})$ in Subsection 9.2.1 and $K_1^{\mathrm{top}}(\mathcal{A}) = 0$ holds for any von Neumann algebra [51, Example 8.1.2 on page 67], [514, Example 7.1.11 on page 134].

More information about the Bass Conjecture can be found for instance in [23], [38], [159], [161], [165], [175], [176], [307], [458].

For an introduction to algebraic K-theory see for instance [378] and [441].

Computations of $G_0(AG)$ for Noetherian rings A and finite nilpotent groups G can be found in [246], [305] and [513]. In [246] a conjecture is stated which would give a computation for all finite groups G in terms of certain orders associated to the irreducible rational representations of G.

One can also consider instead of $G_0(R)$ the version $G_0^{FP_\infty}(R)$ of G-theory as suggested by Weibel. The group $G_0^{FP_\infty}(R)$ is defined in terms of R-modules which are of type FP_∞, i.e. possess a (not necessarily finite dimensional) finitely generated projective resolution (cf. Definition 9.11). This version is much closer to K-theory than G_0. For instance the forgetful map $f\colon K_0(AG) \to G_0^{FP_\infty}(AG)$ is bijective for a commutative ring A if the trivial AG-module A has a finite dimensional (not necessarily finitely generated) projective AG-resolution. The latter condition is always satisfied for a ring A containing $\mathbb{Q}$ as subring if there is a finite dimensional G-CW-model for the classifying space $E(G, \mathcal{FIN})$ for proper G-actions. This is the case for a word-hyperbolic group G and for a discrete subgroup G of a Lie group which has only finitely many path components (see [1, Corollary 4.14], [370]).

Exercises

9.1. Let $\mathcal{A}$ be a von Neumann algebra. Show that $\mathcal{A}$ is finite if and only if it possesses a finite normal faithful trace $\mathrm{tr}\colon \mathcal{A} \to \mathbb{C}$.

9.2. Let M be a closed hyperbolic manifold. Show that $\mathcal{N}(\pi_1(M))$ is of type II_1.

9.3. Let M be a compact connected orientable 3-manifold whose boundary does not contain S^2 as a path component. Show that $\mathcal{N}(\pi_1(M))$ is of type I_f if and only if M has a finite covering $\overline{M} \to M$ such that $\overline{M}$ is homotopy equivalent to S^3, $S^1 \times S^2$, $S^1 \times D^2$, $T^2 \times D^1$ or T^3, and that it is of type II_1 otherwise.

9.4. Construct for any ring R and natural number m natural isomorphisms $K_n(M_m(R)) \xrightarrow{\cong} K_n(R)$ for $n = 0, 1$.

9.5. Let $\mathcal{A}$ be an abelian von Neumann algebra. Show that $[\mathcal{A}] \in K_0(\mathcal{A})$ is not contained in $2 \cdot K_0(\mathcal{A})$ and the canonical projection $K_1(\mathcal{A}) \to \widetilde{K}_1(\mathcal{A})$ is not bijective but that $[M_2(\mathcal{A})] \in K_0(M_2(\mathcal{A}))$ is contained in $2 \cdot K_0(M_2(\mathcal{A}))$ and $K_1(M_2(\mathcal{A})) \to \widetilde{K}_1(M_2(\mathcal{A}))$ is bijective.

9.6. Show that the topological K-group $K_1^{\mathrm{top}}(\mathcal{A}) := \pi_0(GL(\mathcal{A}))$ is trivial for any von Neumann algebra $\mathcal{A}$.

9.7. Let G be a finite group. Let V and W be two finite dimensional unitary representations. Define the character of V by the function $\chi_V\colon G \to \mathbb{C}$ which sends $g \in G$ to the trace $\mathrm{tr}_\mathbb{C}(l_g\colon V \to V)$ of the endomorphism l_g of the finite dimensional $\mathbb{C}$-vector space V given by left multiplication with g. Show that V and W are unitarily $\mathbb{C}G$-isomorphic if and only if $\chi_V = \chi_W$.

9.8. Let R be a ring such that the set S of non-zero divisors satisfies the Ore condition. Let C_* be a finite free R-chain complex such that each homology group $H_n(C_*)$ has a 1-dimensional finitely generated free R-resolution and is S-torsion, i.e. $RS^{-1} \otimes_R H_n(C_*) = 0$. Define an element

$$\rho^h(C_*) := \sum_{n \in \mathbb{Z}} (-1)^n \cdot [H_n(C_*)] \quad \in K_0(R \to RS^{-1}).$$

Choose an R-basis for C_*. It induces an RS^{-1}-basis for $RS^{-1} \otimes_R C_*$. Show that $RS^{-1} \otimes_R C_*$ is an acyclic based free RS^{-1}-chain complex. Define its torsion

$$\rho(RS^{-1} \otimes_R C_*) \in K_1(RS^{-1})$$

as in (3.1). Show that the image of this element under the projection to the cokernel $\operatorname{coker}(i_1)$ of the map $i_1 \colon K_1(R) \to K_1(RS^{-1})$ appearing in the localization sequence (9.22) is independent of the choice of the R-basis and is an R-chain homotopy invariant of C_*. Show that the map $j \colon K_1(RS^{-1}) \to K_0(R \to RS^{-1})$ appearing in the localization sequence (9.22) maps $\rho(RS^{-1} \otimes_R C_*)$ to $\rho^h(C_*)$.

9.9. Compute explicitly for $R = \mathbb{Z}$ and $S \subset \mathbb{Z}$ the set of non-zero divisors that $RS^{-1} = \mathbb{Q}$ and that the localization sequence (9.22) becomes

$$\{\pm 1\} \to \mathbb{Q}^{\mathrm{inv}} \xrightarrow{p} \mathbb{Q}^{+,\mathrm{inv}} \xrightarrow{0} \mathbb{Z} \xrightarrow{\mathrm{id}} \mathbb{Z},$$

where p maps a rational number different from zero to its absolute value. Show for a finite free $\mathbb{Z}$-chain complex C_* that $H_n(C_*)$ has a 1-dimensional finitely generated free R-resolution and is S-torsion for all $n \in \mathbb{Z}$, if and only if $H_n(C_*)$ is a finite abelian group for all $n \in \mathbb{Z}$. Prove that in this case the invariant $\rho^h(C_*)$ introduced in the previous exercise becomes under these identifications

$$\rho^h(C_*) = \prod_{n \in \mathbb{Z}} |H_n(C_*)|^{(-1)^n} \quad \in \mathbb{Q}^{+,\mathrm{inv}}.$$

9.10. Let G be a finite group. Show that $\mathbb{C}G$ is semisimple, i.e. any submodule of a finitely generated projective module is a direct summand. Conclude that $\mathbb{C}G$ is a product of matrix algebras $M_n(\mathbb{C})$. Compute $K_n(\mathbb{C}G)$ for $n = 0, 1$ and $L_n^\epsilon(\mathbb{C}G)$ for $\epsilon = p, h, s$ and $n \in \mathbb{Z}$ as an abelian group.

9.11. Show that for a group G the following assertions are equivalent:

(1) $\mathcal{N}(G)$ is Noetherian;
(2) $\mathcal{U}(G)$ is Noetherian;
(3) $\mathcal{U}(G)$ is semisimple;
(4) $K_0(\mathcal{U}(G))$ is finitely generated;
(5) $K_0(\mathcal{N}(G))$ is finitely generated;
(6) G is finite.

9.12. For a ring R let $\mathrm{Nil}_0(R)$ be the abelian group whose generators $[P, f]$ are conjugacy classes of nilpotent endomorphisms $f\colon P \to P$ of finitely generated projective R-modules and whose relations are given by $[P_0, f_1] - [P_1, f_1] + [P_2, f_2] = 0$ for any exact sequence $0 \to (P_0, f_0) \to (P_1, f_1) \to (P_2, f_2) \to 0$ of such nilpotent endomorphisms. Nilpotent means that $f^n = 0$ for some natural number n. Show

(1) The natural map $i\colon K_0(R) \to \mathrm{Nil}_0(R)$ sending $[P]$ to $[P, 0]$ is split injective;

(2) Let $\widetilde{\mathrm{Nil}}_0(R)$ be the cokernel of the natural map $i\colon K_0(R) \to \mathrm{Nil}_0(R)$. If R is semihereditary, then $\widetilde{\mathrm{Nil}}_0(R) = 0$;

(3) Show for a group G that $\widetilde{\mathrm{Nil}}_0(\mathcal{N}(G)) = \widetilde{\mathrm{Nil}}_0(\mathcal{U}(G)) = 0$.

9.13. Let G be a group such that $\mathrm{Wh}(G)$ vanishes. Show that any finite subgroup of the center of G is isomorphic to a product of finitely many copies of $\mathbb{Z}/2$ and $\mathbb{Z}/3$ or to a product of finitely many copies of $\mathbb{Z}/2$ and $\mathbb{Z}/4$.

9.14. Suppose that the group G satisfies the weak Bass Conjecture for $\mathbb{Z}G$. Then the change of rings homomorphism $K_0(\mathbb{Z}G) \to K_0(\mathcal{N}(G))$ maps the class $[P]$ of a finitely generated projective $\mathbb{Z}G$-module P to $\mathrm{rk}_{\mathbb{Z}}(\mathbb{Z} \otimes_{\mathbb{Z}G} P) \cdot [\mathcal{N}(G)]$.

9.15. Show that G satisfies the strong Bass Conjecture for $\mathbb{C}G$ if and only if $\mathrm{HS}(P)(g) = 0$ holds for any finitely generated projective $\mathbb{C}G$-module P and any element $g \in G$ with $|g| = |(g)| = \infty$.

9.16. Show that the group G satisfies $\mathrm{con}(G)_{f,cf} = \{(1)\}$ if and only if the image of the change of rings map $\widetilde{K}_0(\mathbb{C}G) \to \widetilde{K}_0(\mathcal{N}(G))$ consists of torsion elements.

9.17. Let $\mathcal{A}$ be a finite von Neumann algebra. Construct homomorphisms $d_{\mathcal{A}}\colon G_0(\mathcal{A}) \to K_0(\mathcal{A})$ and $d_{\mathcal{U}}\colon G_0(\mathcal{U}) \to K_0(\mathcal{U})$ such that $d_{\mathcal{A}} \circ f_{\mathcal{A}} = \mathrm{id}$ and $d_{\mathcal{U}} \circ f_{\mathcal{U}} = \mathrm{id}$ holds for the forgetful maps $f_{\mathcal{A}}\colon K_0(\mathcal{A}) \to G_0(\mathcal{A})$ and $f_{\mathcal{U}}\colon K_0(\mathcal{U}) \to G_0(\mathcal{U})$.

9.18. Let F be a free group. Let V be a $\mathbb{C}F$-module whose underlying complex vector space is finite dimensional and let $H \subset F$ be a subgroup. Show that $V \otimes_{\mathbb{C}} \mathbb{C}[F/H]$ with the diagonal F-action is a finitely generated $\mathbb{C}F$-module whose class in $G_0(\mathbb{C}F)$ is zero.

9.19. Let G be finite. Show that the change of coefficients map $j\colon G_0(\mathbb{Z}G) \to K_0(\mathbb{Q}G)$ is well-defined and surjective.

10. The Atiyah Conjecture

Introduction

Atiyah [9, page 72] asked the question, whether the analytic L^2-Betti numbers $b_p^{(2)}(M)$ of a cocompact free proper Riemannian G-manifold with G-invariant Riemannian metric and without boundary, which are defined in terms of the heat kernel (see (1.60)), are always rational numbers. This is implied by the following conjecture which we call the (strong) Atiyah Conjecture in view of Atiyah's question above.

Given a group G, let $\mathcal{FIN}(G)$ be the set of finite subgroups of G. Denote by

$$\frac{1}{|\mathcal{FIN}(G)|} \mathbb{Z} \subset \mathbb{Q} \tag{10.1}$$

the additive subgroup of $\mathbb{R}$ generated by the set of rational numbers $\{\frac{1}{|H|} \mid H \in \mathcal{FIN}(G)\}$.

Conjecture 10.2 (Strong Atiyah Conjecture). *A group G satisfies the strong Atiyah Conjecture if for any matrix $A \in M(m, n, \mathbb{C}G)$ the von Neumann dimension of the kernel of the induced bounded G-operator*

$$r_A^{(2)} : l^2(G)^m \to l^2(G)^n, \quad x \mapsto xA$$

satisfies

$$\dim_{\mathcal{N}(G)} \left(\ker \left(r_A^{(2)} : l^2(G)^m \to l^2(G)^n \right) \right) \in \frac{1}{|\mathcal{FIN}(G)|} \mathbb{Z}.$$

We will explain in Subsection 10.1.4 that there are counterexamples to the strong Atiyah Conjecture 10.2, but no counterexample is known at the time of writing if one replaces $\frac{1}{|\mathcal{FIN}(G)|}\mathbb{Z}$ by $\mathbb{Q}$ or if one assumes that there is an upper bound for the orders of finite subgroups of G.

In Subsection 10.1.1 we will give various reformulations of the Atiyah Conjecture. In particular we will emphasize the module-theoretic point of view which seems to be the best one for proving the conjecture for certain classes of groups (see Lemma 10.7). This is rather surprising if one thinks about the original formulation of Atiyah's question in terms of heat kernels. On the other hand it gives some explanation for the strong Atiyah Conjecture 10.2. Roughly speaking, the strong Atiyah Conjecture 10.2 predicts that all L^2-Betti numbers for G are induced by L^2-Betti numbers of finite subgroups $H \subset G$. This is the same philosophy as in the Isomorphism Conjecture 9.40 for $K_0(\mathbb{C}G)$ or in the Baum-Connes Conjecture for $K^{top}_*(C^*_r(G))$. We discuss the relation of the Atiyah Conjecture to other conjectures like the Kaplansky Conjecture in Subsection 10.1.2. We will give a survey on positive results about the Atiyah Conjecture in Subsection 10.1.3.

In Section 10.2 we will discuss first a general strategy how to prove the strong Atiyah Conjecture 10.2. Then we will explain the concrete form in which the strategy will appear to prove the strong Atiyah Conjecture 10.2 for Linnell's class of groups $\mathcal{C}$.

In Section 10.3 we give the details of the proof of Linnell's Theorem 10.19. The proof is complicated, but along the way one can learn a lot of notions and techniques like transfinite induction, universal localization of rings, division closure and rational closure, crossed products and G- and K-theory. Knowledge of these concepts is not required for any other parts of the book. To get a first impression of the proof, one may only consider the proof in the case of a free group presented in Subsection 10.3.1 which can be read without knowing any other material from this Chapter 10.

To understand this chapter, the reader is only required to be familiar with Sections 1.1, 1.2 and 6.1.

10.1 Survey on the Atiyah Conjecture

In this section we formulate various versions of the Atiyah Conjecture, discuss its consequences and a strategy for the proof.

10.1.1 Various Formulations of the Atiyah Conjecture

Conjecture 10.3 (Atiyah Conjecture). *Let G be a group. Let F be a field $\mathbb{Q} \subset F \subset \mathbb{C}$ and let $\mathbb{Z} \subset \Lambda \subset \mathbb{R}$ be an additive subgroup of $\mathbb{R}$. We say that G satisfies the* Atiyah Conjecture of order Λ with coefficients in F *if for any matrix $A \in M(m, n, FG)$ the von Neumann dimension (see Definition 1.10) of the kernel of the induced bounded G-operator $r^{(2)}_A : l^2(G)^m \to l^2(G)^n$, $x \mapsto xA$ satisfies*

$$\dim_{\mathcal{N}(G)} \left(\ker \left(r^{(2)}_A : l^2(G)^m \to l^2(G)^n \right) \right) \in \Lambda.$$

Notice that the strong Atiyah Conjecture 10.2 is the special case $F = \mathbb{C}$ and $\Lambda = \frac{1}{|\mathcal{FIN}(G)|}\mathbb{Z}$, where the choices of F and Λ are the best possible ones. Sometimes we can only prove this weaker version for special choices of F and Λ but not the strong Atiyah Conjecture 10.2 itself.

Lemma 10.4. *Let G be the directed union of the directed system $\{G_i \mid i \in I\}$ of subgroups. Then G satisfies the Atiyah Conjecture 10.3 of order Λ with coefficients in F if and only if each G_i does. In particular a group G satisfies the Atiyah Conjecture 10.3 of order Λ with coefficients in F if and only if each finitely generated subgroup does.*

Proof. Let $A \in M(m, n, FG_i)$. Then we conclude from Lemma 1.24

$$\dim_{\mathcal{N}(G_i)} \left(\ker \left(r_A^{(2)} : l^2(G_i)^m \to l^2(G_i)^n \right) \right)$$
$$= \dim_{\mathcal{N}(G)} \left(\ker \left(r_A^{(2)} : l^2(G)^m \to l^2(G)^n \right) \right).$$

For any $A \in M(m, n, FG)$ there is G_i such that A already belongs to $M(m, n, FG_i)$ because there are only finitely many elements in G which appear with non-trivial coefficient in one of the entries of A. A group G is the directed union of its finitely generated subgroups.   □

We can reformulate the Atiyah Conjecture 10.3 for $F = \mathbb{Q}$ as follows.

Lemma 10.5. (Reformulation of the Atiyah Conjecture in terms of L^2-Betti numbers). *Let G be a group. Let $\mathbb{Z} \subset \Lambda \subset \mathbb{R}$ be an additive subgroup of $\mathbb{R}$. Then the following statements are equivalent:*

(1) For any cocompact free proper G-manifold M without boundary we have

$$b_p^{(2)}(M; \mathcal{N}(G)) \in \Lambda;$$

(2) For any cocompact free proper G-CW-complex X we have

$$b_p^{(2)}(X; \mathcal{N}(G)) \in \Lambda;$$

(3) The Atiyah Conjecture 10.3 of order Λ with coefficients in $\mathbb{Q}$ is true for G.

Proof. (3) $\Rightarrow$ (2) This follows from Additivity (see Theorem 1.12 (2)).

(2) $\Rightarrow$ (3) Because of Lemma 10.4 and the equality $b_p^{(2)}(G \times_H Z; \mathcal{N}(G)) = b_p^{(2)}(Z; \mathcal{N}(H))$ for $p \geq 0$, $H \subset G$ and Z an H-CW-complex (see Theorem 6.54 (7)), we can assume without loss of generality that G is finitely generated. Put $Y = \bigvee_{i=1}^g S^1$. Choose an epimorphism $f : \pi_1(Y) = *_{i=1}^g \mathbb{Z} \to G$ from the free group on g generators to G. Let $p : \overline{Y} \to Y$ be the G-covering associated to f. Let A be an (m,n)-matrix over $\mathbb{Q}G$ and let $d \geq 3$ be an integer. Choose $k \in \mathbb{Z}$ with $k \neq 0$ such that $k \cdot A$ has entries in $\mathbb{Z}G$. By attaching n copies of $G \times D^{d-1}$ to $\overline{Y}$ with attaching maps of the shape $G \times S^{d-2} \to G \to \overline{Y}$ and m d-cells $G \times D^d$ one can construct a pair of finite G-CW-complexes $(X, \overline{Y})$ such that the d-th differential of the cellular $\mathbb{Z}G$-chain complex $C_*(X)$ is given by $k \cdot A$ and the $(d+1)$-th differential is trivial. This follows from the observation that the composition of the Hurewicz map $\pi_{d-1}(X_{d-1}) \to H_{d-1}(X_{d-1})$ with the boundary map $H_{d-1}(X_{d-1}) \to H_{d-1}(X_{d-1}, X_{d-2})$ is surjective since each

element of the cellular $\mathbb{Z}G$-basis for $C_{d-1}(X) = H_{d-1}(X_{d-1}, X_{d-2})$ obviously lies in the image of the composition above. Hence $C_*^{(2)}(X)$ is a finite Hilbert $\mathcal{N}(G)$-chain complex such that $H_d^{(2)}(C_*^{(2)}(X)) = H_d^{(2)}(X; \mathcal{N}(G))$ is just the kernel of $r_{k \cdot A}^{(2)} : l^2(G)^m \to l^2(G)^n$. Obviously $r_{k \cdot A}^{(2)}$ and $r_A^{(2)}$ have the same kernel. Hence $\dim_{\mathcal{N}(G)}\left(\ker(r_A^{(2)})\right) = b_d^{(2)}(X; \mathcal{N}(G))$.

(1) $\Rightarrow$ (2) Given a cocompact free proper G-CW-complex X of dimension n, we can find an embedding of $G \backslash X$ in $\mathbb{R}^{2n+3}$ with regular neighborhood N with boundary ∂N [445, chapter 3]. Then there is an $(n+1)$-connected map $f : \partial N \longrightarrow G \backslash X$. Let $\overline{f} : \overline{\partial N} \to X$ be obtained by the pullback of $X \to G \backslash X$ with f. Then $\overline{f}$ is $(n+1)$-connected and $\overline{\partial N}$ is a cocompact free proper G-manifold without boundary. We get $b_p^{(2)}(X; \mathcal{N}(G)) = b_p^{(2)}(\overline{\partial N}; \mathcal{N}(G))$ for $p \le n$ from Theorem 1.35 (1).

(2) $\Rightarrow$ (1) This is obvious. This finishes the proof of Lemma 10.5. $\square$

Atiyah's question [9, page 72], whether the L^2-Betti numbers $b_p^{(2)}(M; \mathcal{N}(G))$ of a cocompact free proper Riemannian G-manifold M with G-invariant Riemannian metric and without boundary are always rational numbers, has a positive answer if and only if the Atiyah Conjecture 10.3 of order $\Lambda = \mathbb{Q}$ with coefficients in $F = \mathbb{Q}$ is true for G. This follows from the L^2-Hodge-de Rham Theorem 1.59, the expression of the L^2-Betti numbers in terms of the heat kernel (see (1.60) and Theorem 3.136 (1)) and Lemma 10.5 above.

All examples of L^2-Betti numbers of G-coverings of closed manifolds or of finite CW-complexes, which have explicitly been computed in this book, are consistent with the strong Atiyah Conjecture 10.2, provided that there is an upper bound on the orders of finite subgroups of G. Without the latter condition counterexamples do exist as we will explain in Remark 10.24.

One can rephrase the Atiyah Conjecture also in a more module-theoretic fashion, where one emphasizes the possible values $\dim_{\mathcal{N}(G)}(\mathcal{N}(G) \otimes_{FG} M)$ for FG-modules M. This point of view turns out to be the best one for proofs of the Atiyah Conjecture. Recall that we have defined the von Neumann dimension of any $\mathcal{N}(G)$-module in Definition 6.6 (using Theorem 6.5).

Notation 10.6. *Let G be a group and let F be a field with $\mathbb{Q} \subset F \subset \mathbb{C}$. Define $\Lambda(G, F)_{\text{fgp}}, \Lambda(G, F)_{\text{fp}}, \Lambda(G, F)_{\text{fg}}$ or $\Lambda(G, F)_{\text{all}}$ respectively to be the additive subgroup of $\mathbb{R}$ given by differences*

$$\dim_{\mathcal{N}(G)}(\mathcal{N}(G) \otimes_{FG} M_1) - \dim_{\mathcal{N}(G)}(\mathcal{N}(G) \otimes_{FG} M_0),$$

where M_0 and M_1 run through all finitely generated projective FG-modules, finitely presented FG-modules, finitely generated FG-modules or FG-modules with $\dim_{\mathcal{N}(G)}(\mathcal{N}(G) \otimes_{FG} M_i) < \infty$ for $i = 0, 1$ respectively.

Lemma 10.7. (Reformulation of the Atiyah Conjecture in terms of modules). *A group G satisfies the Atiyah Conjecture of order Λ with coefficients in F if and only if $\Lambda(G, F)_{\text{fp}} \subset \Lambda$, or equivalently, if and only if for any finitely presented FG-module M*

$$\dim_{\mathcal{N}(G)}(\mathcal{N}(G) \otimes_{FG} M) \in \Lambda.$$

Proof. Given a matrix $A \in M(m, n, FG)$, let $r_A \colon FG^m \to FG^n$ and $r_A^{\mathcal{N}(G)} \colon \mathcal{N}(G)^m \to \mathcal{N}(G)^n$ be the associated FG-homomorphism and $\mathcal{N}(G)$-homomorphism given by right multiplication with A. Since the tensor product $\mathcal{N}(G) \otimes_{FG} -$ is right exact, coker $\left(r_A^{\mathcal{N}(G)}\right)$ is $\mathcal{N}(G)$-isomorphic to $\mathcal{N}(G) \otimes_{FG}$ coker(r_A). We conclude from Additivity (see Theorem 6.7 (4b))

$$\dim_{\mathcal{N}(G)} \left(\ker \left(r_A^{\mathcal{N}(G)} \right) \right) = m - n + \dim_{\mathcal{N}(G)} \left(\mathcal{N}(G) \otimes_{FG} \operatorname{coker}(r_A) \right). \quad (10.8)$$

Since $\mathcal{N}(G)$ is semihereditary (see Theorem 6.5 and Theorem 6.7 (1)), $\ker \left(r_A^{\mathcal{N}(G)} \right)$ is finitely generated projective. We get from Theorem 6.24

$$\dim_{\mathcal{N}(G)} \left(\ker \left(r_A^{\mathcal{N}(G)} \right) \right) = \dim_{\mathcal{N}(G)} \left(\ker \left(r_A^{(2)} \right) \right). \quad (10.9)$$

We conclude from (10.8) and (10.9) that $\dim_{\mathcal{N}(G)} \left(\ker \left(r_A^{(2)} \right) \right)$ belongs to Λ if and only if $\dim_{\mathcal{N}(G)} \left(\mathcal{N}(G) \otimes_{FG} \operatorname{coker}(r_A) \right)$ belongs to Λ since $m - n \in \mathbb{Z} \subset \Lambda$. $\square$

Lemma 10.10. *Let G be a group and let F be a field with $\mathbb{Q} \subset F \subset \mathbb{C}$.*

(1) We have $\frac{1}{|\mathcal{FIN}(G)|}\mathbb{Z} \subset \Lambda(G, F)_{\mathrm{fgp}}$. If the Isomorphism Conjecture 9.40 for $K_0(\mathbb{C}G)$ is true, we have $\frac{1}{|\mathcal{FIN}(G)|}\mathbb{Z} = \Lambda(G, F)_{\mathrm{fgp}}$;
(2) $\Lambda(G, F)_{\mathrm{fgp}} \subset \Lambda(G, F)_{\mathrm{fp}} \subset \Lambda(G, F)_{\mathrm{fg}} \subset \Lambda(G, F)_{\mathrm{all}}$;
(3) $\Lambda(G, F)_{\mathrm{fg}} \subset \operatorname{clos}(\Lambda(G, F)_{\mathrm{fp}})$, where the closure is taken in $\mathbb{R}$;
(4) $\operatorname{clos}(\Lambda(G, F)_{\mathrm{fp}}) = \operatorname{clos}(\Lambda(G, F)_{\mathrm{fg}}) = \Lambda(G, F)_{\mathrm{all}}$.

Proof. (1) The image of the following composition

$$\operatorname{colim}_{\mathrm{Or}(G, \mathcal{FIN})} K_0(FH) \xrightarrow{a} K_0(FG) \xrightarrow{i} K_0(\mathcal{N}(G)) \xrightarrow{\dim_{\mathcal{N}(G)}} \mathbb{R}$$

is $\frac{1}{|\mathcal{FIN}(G)|}\mathbb{Z}$, where a is essentially given by the various inclusions of the finite subgroups H of G and is in the case $F = \mathbb{C}$ the assembly map appearing in the Isomorphism Conjecture 9.40 for $K_0(\mathbb{C}G)$, and i is a change of rings homomorphism. This follows from the compatibility of the dimension with induction (see Theorem 6.29 (2)) and the fact that $\dim_{\mathcal{N}(H)}(V) = \frac{1}{|H|} \cdot \dim_{\mathbb{C}}(V)$ holds for a finitely generated $\mathcal{N}(H)$-module for finite $H \subset G$. The image of the composition $K_0(FG) \xrightarrow{i} K_0(\mathcal{N}(G)) \xrightarrow{\dim_{\mathcal{N}(G)}} \mathbb{R}$ is by definition $\Lambda(G, F)_{\mathrm{fgp}}$.

(2) This is obvious.

(3) Let M be a finitely generated FG-module. Choose an exact sequence $\bigoplus_{i \in I} FG \xrightarrow{p} FG^n \to M \to 0$ for an appropriate $n \in \mathbb{N}$ and index set I.

For a finite subset $J \subset I$ let M_J be the cokernel of the restriction of p to $\bigoplus_{i \in J} FG$. We obtain a directed system of finitely presented FG-modules $\{M_J \mid J \subset I, |J| < \infty\}$ such that $\operatorname{colim}_{J \subset I, |J| < \infty} M_J = M$. We conclude $\operatorname{colim}_{J \subset I, |J| < \infty} \mathcal{N}(G) \otimes_{FG} M_J = \mathcal{N}(G) \otimes_{FG} M$ since $\mathcal{N}(G) \otimes_{FG} -$ has a right adjoint and hence respects colimits. Each of the structure maps $\mathcal{N}(G) \otimes_{FG} M_{J_1} \to \mathcal{N}(G) \otimes_{FG} M_{J_2}$ for $J_1 \subset J_2 \subset I$ with $|J_1|, |J_2| < \infty$ is surjective and each FG-module M_J is finitely presented. Hence the claim follows since the dimension is compatible with directed colimits (see Theorem 6.13 (2)).

(4) Let M be an FG-module. Let $\{M_i \mid i \in I\}$ be the directed system of its finitely generated submodules. Then M is $\operatorname{colim}_{i \in I} M_i$. Hence $\mathcal{N}(G) \otimes_{FG} M$ is $\operatorname{colim}_{i \in I} \mathcal{N}(G) \otimes_{FG} M_i$. Because of Theorem 6.13 (2) and assertion (3) it suffices to show that for any map $f \colon M_0 \to M_1$ of finitely generated FG-modules we have

$$\dim_{\mathcal{N}(G)} \left(\operatorname{im} (\mathrm{id} \otimes_{FG} f \colon \mathcal{N}(G) \otimes_{FG} M_0 \to \mathcal{N}(G) \otimes_{FG} M_1) \right) \in \Lambda(G, F)_{\mathrm{fg}}.$$

This follows from Additivity (see Theorem 6.7 (4b)) and the exact sequence

$$\mathcal{N}(G) \otimes_{FG} M_0 \xrightarrow{\mathrm{id} \otimes_{FG} f} \mathcal{N}(G) \otimes_{FG} M_1 \xrightarrow{\mathrm{id} \otimes_{FG} \mathrm{pr}} \mathcal{N}(G) \otimes_{FG} \operatorname{coker}(f) \to 0. \qquad \square$$

Remark 10.11. Let G be a group. Then $\frac{1}{|\mathcal{FIN}(G)|}\mathbb{Z}$ is closed in $\mathbb{R}$ if and only if there is an upper bound on the orders of finite subgroups. Suppose that there is such an upper bound. Then the least common multiple d of the orders of finite subgroups is defined and

$$\frac{1}{|\mathcal{FIN}(G)|}\mathbb{Z} = \{r \in \mathbb{R} \mid d \cdot r \in \mathbb{Z}\}$$

and by Lemma 10.7 and Lemma 10.10 the strong Atiyah Conjecture 10.2 is equivalent to the equality

$$\{r \in \mathbb{R} \mid d \cdot r \in \mathbb{Z}\} = \Lambda(G, \mathbb{C})_{\mathrm{fgp}} = \Lambda(G, \mathbb{C})_{\mathrm{fp}} = \Lambda(G, \mathbb{C})_{\mathrm{fg}} = \Lambda(G, \mathbb{C})_{\mathrm{all}}.$$

In particular the strong Atiyah Conjecture 10.2 for a torsionfree group G is equivalent to the equality

$$\mathbb{Z} = \Lambda(G, \mathbb{C})_{\mathrm{fgp}} = \Lambda(G, \mathbb{C})_{\mathrm{fp}} = \Lambda(G, \mathbb{C})_{\mathrm{fg}} = \Lambda(G, \mathbb{C})_{\mathrm{all}}.$$

Remark 10.12. Assume that G is amenable and that there is an upper bound on the orders of finite subgroups of G. Then G satisfies the strong Atiyah Conjecture 10.2 if and only if the image of the map

$$\dim_{\mathcal{N}(G)} \colon G_0(\mathbb{C}G) \to \mathbb{R}, \qquad [M] \mapsto \dim_{\mathcal{N}(G)}(\mathcal{N}(G) \otimes_{\mathbb{C}G} M)$$

(see Theorem 6.7 (4e) and Theorem 9.64) is contained in (and hence equal to) $\frac{1}{|\mathcal{FIN}(G)|}\mathbb{Z}$. Moreover, we conclude from Theorem 9.54 (8) and Lemma 10.10 (1) that the strong Atiyah Conjecture 10.2 holds for virtually poly-cyclic groups G.

Example 10.13. Let $A = \bigoplus_{n \in \mathbb{Z}} \mathbb{Z}/2$. This abelian group is locally finite and hence satisfies the Isomorphism Conjecture 9.40 for $K_0(\mathbb{C}G)$ and by Lemma 10.4 the strong Atiyah Conjecture 10.2 . This implies

$$\mathbb{Z}[1/2] = \frac{1}{|\mathcal{FIN}(A)|}\mathbb{Z} = \Lambda(A, \mathbb{C})_{\text{fgp}} = \Lambda(A, \mathbb{C})_{\text{fp}}.$$

We want to show that each real number $r \geq 0$ can be realized by $r = \dim_{\mathcal{N}(A)}(\mathcal{N}(A) \otimes_{\mathbb{Q}A} M)$ for a finitely generated $\mathbb{Q}G$-module M. This implies

$$\Lambda(A, \mathbb{C})_{\text{fg}} = \mathbb{R}$$

or, equivalently, that the map

$$\dim_{\mathcal{N}(A)} \colon G_0(\mathbb{C}A) \to \mathbb{R}, \quad [M] \mapsto \dim_{\mathcal{N}(A)}(\mathcal{N}(A) \otimes_{\mathbb{C}A} M)$$

(see Theorem 6.7 (4e) and Theorem 9.64) is surjective. The image of its composition with the forgetful map $f \colon K_0(\mathbb{C}A) \to G_0(\mathbb{C}A)$ is $\mathbb{Z}[1/2]$. In particular we conclude that f is not surjective and $G_0(\mathbb{C}A)$ is not countable.

Let $A_n \subset A$ be the finite subgroup $\bigoplus_{k=0}^{n-1} \mathbb{Z}/2$ of order 2^n for $n \geq 1$. Denote by $N_n \in \mathbb{Q}A$ the element $2^{-n} \cdot \sum_{a \in A_n} a$. Let I_n and (N_n) be the ideals of $\mathbb{Q}A$ generated by the elements $N_n - N_{n-1}$ and N_n. Since $N_m N_n = N_{\max\{m,n\}}$, we get a direct sum decomposition $(N_n) \oplus I_n = (N_{n-1})$ Since (N_n) is $\mathbb{Q}G$-isomorphic to $\mathbb{Q}G \otimes_{\mathbb{Q}A_n} \mathbb{Q}$ for $\mathbb{Q}$ with the trivial A_n-action and the dimension is compatible with induction (see Theorem 6.29 (2)), we get

$$\dim_{\mathcal{N}(A)}(\mathcal{N}(A) \otimes_{\mathbb{Q}A} I_n)$$
$$= \dim_{\mathcal{N}(A)}(\mathcal{N}(A) \otimes_{\mathbb{Q}A} (N_{n-1})) - \dim_{\mathcal{N}(A)}(\mathcal{N}(A) \otimes_{\mathbb{Q}A} (N_n))$$
$$= \frac{1}{|A_{n-1}|} - \frac{1}{|A_n|}$$
$$= 2^{-n}.$$

Fix any number $r \in [0, 1)$. Because it has a 2-adic expansion, we can find a (finite or infinite) sequence of positive integers $n_1 < n_2 < n_3 < \ldots$ such that

$$r = \sum_{i \geq 1} 2^{-n_i} = \sum_{i \geq 1} \dim_{\mathcal{N}(A)}(\mathcal{N}(A) \otimes_{\mathbb{Q}A} I_{n_i}).$$

Let $P_k \subset \mathbb{Q}A$ be the $\mathbb{Q}A$-submodule $\sum_{i=1}^{k} I_{n_i}$. By construction $P_k = \bigoplus_{i=1}^{k} I_{n_i}$ and P_k is a direct summand in $\mathbb{Q}A$. This implies

$$\dim_{\mathcal{N}(A)}(\mathcal{N}(A) \otimes_{\mathbb{Q}A} \mathbb{Q}A/P_k) = 1 - \sum_{i=1}^{k} 2^{-n_i}.$$

Let $P \subset \mathbb{Q}A$ be the union of the submodules P_k. Then $\mathcal{N}(A) \otimes_{\mathbb{Q}A} \mathbb{Q}A/P = \text{colim}_{k \to \infty} \mathcal{N}(A) \otimes_{\mathbb{Q}A} \mathbb{Q}A/P_k$. We conclude from Theorem 6.13 (2)

$$\dim_{\mathcal{N}(A)}(\mathcal{N}(A) \otimes_{\mathbb{Q}A} \mathbb{Q}A/P) = 1 - \sum_{i \geq 1} 2^{-n_i} = 1 - r.$$

Since $M = \mathbb{Q}A^n \oplus \mathbb{Q}A/P$ is a finitely generated $\mathbb{Q}A$-module and satisfies $\dim_{\mathcal{N}(A)}(\mathcal{N}(A) \otimes_{\mathbb{Q}A} M) = n + 1 - r$, the claim follows.

10.1.2 Relations of the Atiyah Conjecture to Other Conjectures

In this subsection we relate the Atiyah Conjecture to other conjectures.

Conjecture 10.14 (Kaplansky Conjecture). *The Kaplansky Conjecture for a torsionfree group G and a field F says that the group ring FG has no non-trivial zero-divisors.*

If G contains an element g of finite order $n > 1$, then $x = \sum_{i=1}^{n} g^i$ is a non-trivial zero-divisor, namely $x(n - x) = 0$. Hence the Kaplansky Conjecture is equivalent to the statement that, given a field F, a group G is torsionfree if and only if FG has no non-trivial zero-divisors.

If G is a right-ordered torsionfree group and F is a field, then the Kaplansky Conjecture is known to be true [302, Theorem 6.29 on page 101]), [310, Theorem 4.1]. Delzant [140] deals with group rings of word-hyperbolic groups and proves the Kaplansky Conjecture for certain word-hyperbolic groups. Given a torsionfree group G, a weaker version of the Kaplansky Conjecture predicts that FG has no non-trivial idempotents and a stronger version predicts that all units in FG are trivial, i.e. of the form $\lambda \cdot g$ for $\lambda \in F, \lambda \neq 0$ and $g \in G$ [302, (6.20) on page 95]). For more information about the Kaplansky Conjecture we refer for instance to [310, Section 4]. A proof of the next result can also be found in [172].

Lemma 10.15. *Let F be a field with $\mathbb{Z} \subset F \subset \mathbb{C}$ and let G be a torsionfree group. Then the Kaplansky Conjecture holds for G and the field F if the Atiyah Conjecture 10.3 of order $\Lambda = \mathbb{Z}$ with coefficients in F is true for G.*

Proof. Let $x \in FG$ be a zero-divisor. Let $r_x^{(2)} : l^2(G) \to l^2(G)$ be given by right multiplication with x. We get $0 < \dim_{\mathcal{N}(G)}\left(\ker(r_x^{(2)})\right) \leq 1$ from Theorem 1.12 (1) and (2). Since by assumption $\dim_{\mathcal{N}(G)}\left(\ker(r_x^{(2)})\right) \in \mathbb{Z}$, we conclude $\dim_{\mathcal{N}(G)}\left(\ker(r_x^{(2)})\right) = 1$. Since $\ker(r_x^{(2)})$ is closed in $l^2(G)$, Theorem 1.12 (1) and (2) imply $\ker(r_x^{(2)}) = l^2(G)$ and hence $x = 0$. $\square$

Lemma 10.16. *Let G be a torsionfree amenable group. Then the strong Atiyah Conjecture 10.2 for G is equivalent to the Kaplansky Conjecture 10.14 for G and the field $\mathbb{C}$.*

Proof. Because of Lemma 10.15 it suffices to prove the strong Atiyah Conjecture 10.2, provided that the Kaplansky Conjecture 10.14 holds for G

and $\mathbb{C}$. We first show for a $\mathbb{C}G$-module M which admits an exact sequence $\mathbb{C}G^n \xrightarrow{c} \mathbb{C}G \to M \to 0$ that $\dim_{\mathcal{N}(G)}(\mathcal{N}(G) \otimes_{\mathbb{C}G} M) \in \{0,1\}$. Let C_* be the chain complex concentrated in dimension 1 and 0 with c as first differential. We conclude from Lemma 1.18, Theorem 6.24 and the dimension-flatness of $\mathcal{N}(G)$ over $\mathbb{C}G$ for amenable G (see Theorem 6.37)

$$\dim_{\mathcal{N}(G)}(\mathcal{N}(G) \otimes_{\mathbb{C}G} M) = \dim_{\mathcal{N}(G)}(\operatorname{coker}(\mathcal{N}(G) \otimes_{\mathbb{C}G} c))$$
$$= \dim_{\mathcal{N}(G)}(H_0^{(2)}(l^2(G) \otimes_{\mathbb{C}G} C_*))$$
$$= \dim_{\mathcal{N}(G)}(\ker(\Delta_0 \colon l^2(G) \to l^2(G)))$$
$$= \dim_{\mathcal{N}(G)}(\ker(\mathcal{N}(G) \otimes_{\mathbb{C}G} c^*c))$$
$$= \dim_{\mathcal{N}(G)}(\mathcal{N}(G) \otimes_{\mathbb{C}G} \ker(c^*c)).$$

Since $c^*c \colon \mathbb{C}G \to \mathbb{C}G$ is $\mathbb{C}G$-linear and hence given by right multiplication with an element in $\mathbb{C}G$, its kernel is either trivial or $\mathbb{C}G$. This implies $\dim_{\mathcal{N}(G)}(\mathcal{N}(G) \otimes_{\mathbb{C}G} \ker(c^*c)) \in \{0,1\}$ and thus $\dim_{\mathcal{N}(G)}(\mathcal{N}(G) \otimes_{\mathbb{C}G} M) \in \{0,1\}$.

In order to prove the strong Atiyah Conjecture 10.2 it suffices to show for any finitely generated $\mathbb{C}G$-module M that $\dim_{\mathcal{N}(G)}(\mathcal{N}(G) \otimes_{\mathbb{C}G} M)$ is an integer. We proceed by induction over the number n of generators of M. In the induction beginning $n = 1$ we can assume the existence of an exact sequence $\bigoplus_{i \in I} \mathbb{C}G \xrightarrow{p} \mathbb{C}G \to M$. Given a finite subset $J \in I$, let M_J be the cokernel of the restriction of p to $\bigoplus_{i \in J} \mathbb{C}G$. Then $\dim_{\mathcal{N}(G)}(\mathcal{N}(G) \otimes_{\mathbb{C}G} M_J) \in \{0,1\}$ by the argument above. We conclude as in the proof of Lemma 10.10 (3) that $\dim_{\mathcal{N}(G)}(\mathcal{N}(G) \otimes_{\mathbb{C}G} M) \in \{0,1\}$. It remains to prove the induction step from n to $n+1$. Obviously we can find a short exact sequence of $\mathbb{C}G$-modules $0 \to L \to M \to N \to 0$ such that L is generated by one element and N by n elements. By induction hypothesis $\dim_{\mathcal{N}(G)}(\mathcal{N}(G) \otimes_{\mathbb{C}G} L)$ and $\dim_{\mathcal{N}(G)}(\mathcal{N}(G) \otimes_{\mathbb{C}G} N)$ are integers. We conclude from Additivity (see Theorem 6.7 (4b)) and dimension-flatness of $\mathcal{N}(G)$ over $\mathbb{C}G$ for amenable G (see Theorem 6.37)

$$\dim_{\mathcal{N}(G)}(\mathcal{N}(G) \otimes_{\mathbb{C}G} M) = \dim_{\mathcal{N}(G)}(\mathcal{N}(G) \otimes_{\mathbb{C}G} L) + \dim_{\mathcal{N}(G)}(\mathcal{N}(G) \otimes_{\mathbb{C}G} N).$$

Hence $\dim_{\mathcal{N}(G)}(\mathcal{N}(G) \otimes_{\mathbb{C}G} M)$ is an integer. $\square$

The result above follows also from [416, Theorem 2.2].

Suppose that there is an integer d such that the order of any finite subgroup of G divides d and that the strong Atiyah Conjecture 10.2 holds for G. Suppose furthermore that $h^{(2)}(G) < \infty$ and hence $\chi^{(2)}(G)$ is defined (see Definition 6.79). Theorem 6.99 (2) says that $d \cdot b_p^{(2)}(G) \in \mathbb{Z}$ holds for all p. This implies

$$d \cdot \chi^{(2)}(G) \in \mathbb{Z}.$$

This is consistent with the result of Brown [69, Theorem IX.9.3 on page 257] that $d \cdot \chi'(G) \in \mathbb{Z}$ holds for a group G of finite homological type, i.e.

a group G of finite virtual cohomological dimension such that for any $\mathbb{Z}G$-module M, which is finitely generated as abelian group, $H_i(G; M)$ is finitely generated for all $i \geq 0$. Here $\chi'(G)$ is defined by

$$\chi'(G) = \frac{1}{[G : G_0]} \cdot \sum_{p \geq 0} (-1)^p \operatorname{rk}_{\mathbb{Z}}(H_p(G_0; \mathbb{Z})) \tag{10.17}$$

for any torsionfree subgroup $G_0 \subset G$ of finite index $[G : G_0]$. Namely, if G contains a torsionfree subgroup G_0 of finite index, which has a finite model for BG, then both $\chi^{(2)}(G)$ and $\chi'(G)$ are defined and agree. This follows from Remark 6.81, where $\chi^{(2)}(G) = \chi_{\mathrm{virt}}(BG)$ is shown, and the fact that $\chi_{\mathrm{virt}}(BG) = \chi'(G)$ [69, page 247].

We will state, discuss and give some evidence for the Singer Conjecture in Chapter 11. Because of the Euler-Poincaré formula (see Theorem 1.35 (2)) the Singer Conjecture implies for a closed aspherical manifold M that all the L^2-Betti numbers $b_p^{(2)}(\widetilde{M})$ of its universal covering are integers, as predicted by the strong Atiyah Conjecture 10.2 in combination with Lemma 10.5.

A link between the Atiyah Conjecture and the Baum-Connes Conjecture will be discussed in Section 10.4. A connection between the Atiyah Conjecture and the Isomorphism Conjecture 9.40 for $K_0(\mathbb{C}G)$ has already been explained in Lemma 10.10 (1). Notice that the Atiyah Conjecture is harder than the Baum-Connes Conjecture and the Isomorphism Conjecture for $K_0(\mathbb{C}G)$ in the sense that it deals with finitely presented modules (see Lemma 10.7) and not only with finitely generated projective modules.

10.1.3 Survey on Positive Results about the Atiyah Conjecture

In this subsection we state some cases, where the Atiyah Conjecture is known to be true.

Definition 10.18 (Linnell's class of groups $\mathcal{C}$). *Let $\mathcal{C}$ be the smallest class of groups which contains all free groups and is closed under directed unions and extensions with elementary amenable quotients.*

We will extensively discuss the proof of the following theorem due to Linnell [309, Theorem 1.5] later in this chapter.

Theorem 10.19 (Linnell's Theorem). *Let G be a group such that there is an upper bound on the orders of finite subgroups and G belongs to $\mathcal{C}$. Then the strong Atiyah Conjecture 10.2 holds for G.*

The next result is a direct consequence of Theorem 13.3 (2), Theorem 13.31 (2) and Proposition 13.35 (1) which deal with the Approximation Conjecture 13.1.

Theorem 10.20. *Let $\mathbb{Z} \subset \Lambda \subset \mathbb{R}$ be an additive subgroup of $\mathbb{R}$ which is closed in $\mathbb{R}$. Let $\{G_i \mid i \in I\}$ be a directed system of groups such that each*

G_i belongs to the class $\mathcal{G}$ (see Definition 13.9) and satisfies the Atiyah Conjecture 10.3 of order Λ with coefficients in $\mathbb{Q}$. Then both its colimit (= direct limit) and its inverse limit satisfy the Atiyah Conjecture 10.3 of order Λ with coefficients in $\mathbb{Q}$.

The next definition and the next theorem, which is essentially a consequence of the two Theorems 10.19 and 10.20 above by transfinite induction, are taken from [463].

Definition 10.21. *Let $\mathcal{D}$ be the smallest non-empty class of groups such that*

(1) If $p\colon G \to A$ is an epimorphism of a torsionfree group G onto an elementary amenable group A and if $p^{-1}(B) \in \mathcal{D}$ for every finite group $B \subset A$, then $G \in \mathcal{D}$;
(2) $\mathcal{D}$ is closed under taking subgroups;
(3) $\mathcal{D}$ is closed under colimits and inverse limits over directed systems.

Theorem 10.22. *(1) If the group G belongs to $\mathcal{D}$, then G is torsionfree and the Atiyah Conjecture 10.3 of order $\Lambda = \mathbb{Z}$ with coefficients in $F = \mathbb{Q}$ is true for G;*
(2) The class $\mathcal{D}$ is closed under direct sums, direct products and free products. Every residually torsionfree elementary amenable group belongs to $\mathcal{D}$.

The fundamental group of compact 2-dimensional manifold M belongs to $\mathcal{C}$ since it maps onto the abelian group $H_1(M)$ with a free group as kernel. If $\pi_1(M)$ is torsionfree, it belongs also to $\mathcal{D}$. This follows from the fact that a finitely generated free group is residually torsionfree nilpotent [354, §2]. The pure braid groups belong to $\mathcal{D}$ since they are residually torsionfree nilpotent [180, Theorem 2.6]. The Atiyah Conjecture 10.3 of order $\Lambda = \mathbb{Z}$ with coefficients in $F = \mathbb{Q}$ is also true for the braid group (see [313]). Positive one-relator groups, i.e. one relator groups whose relation can be written as a word in positive multiples of the generators, belong to $\mathcal{D}$ because they are residually torsionfree solvable.

10.1.4 A Counterexample to the Strong Atiyah Conjecture

The *lamplighter group* L is defined by the semidirect product

$$L := \bigoplus_{n \in \mathbb{Z}} \mathbb{Z}/2 \rtimes \mathbb{Z}$$

with respect to the shift automorphism of $\bigoplus_{n \in \mathbb{Z}} \mathbb{Z}/2$, which sends $(x_n)_{n \in \mathbb{Z}}$ to $(x_{n-1})_{n \in \mathbb{Z}}$. Let $e_0 \in \bigoplus_{n \in \mathbb{Z}} \mathbb{Z}/2$ be the element whose entries are all zero except the entry at 0. Denote by $t \in \mathbb{Z}$ the standard generator of $\mathbb{Z}$ which we will also view as element of L. Then $\{e_0 t, t\}$ is a set of generators for L. The associated *Markov operator* $M\colon l^2(G) \to l^2(G)$ is given by right multiplication with $\frac{1}{4} \cdot (e_0 t + t + (e_0 t)^{-1} + t^{-1})$. It is related to the Laplace operator $\Delta_0\colon l^2(G) \to l^2(G)$ of the Cayley graph of G by $\Delta_0 = 4 \cdot \mathrm{id} - 4 \cdot M$. The following result is a special case of the main result in the paper of Grigorchuk and Žuk [230, Theorem 1 and Corollary 3] (see also [229]). An elementary proof can be found in [143].

Theorem 10.23 (Counterexample to the strong Atiyah Conjecture).
The von Neumann dimension of the kernel of the Markov operator M of the lamplighter group L associated to the set of generators $\{e_0 t, t\}$ is $1/3$. In particular L does not satisfy the strong Atiyah Conjecture 10.2.

Notice that each finite subgroup of the lamplighter group L is a 2-group and for any power 2^n of 2 a subgroup of order 2^n exists. Hence $\frac{1}{|\mathcal{FIN}(L)|} \cdot \mathbb{Z} = \mathbb{Z}[1/2]$ and we obtain a counterexample to the strong Atiyah Conjecture 10.2. At the time of writing the author does not know of a counterexample to the strong Atiyah Conjecture in the case, where one replaces $\frac{1}{|\mathcal{FIN}(L)|}\mathbb{Z}$ by $\mathbb{Q}$ or where one assumes the existence of an upper bound on the orders of finite subgroups.

Remark 10.24. The lamplighter group L is a subgroup of a finitely presented group G such that any finite subgroup of G is a 2-group. By Theorem 10.23 above and a slight variation of the proof of Lemma 10.5 we conclude the existence of a closed Riemannian manifold M with $G = \pi_1(M)$ such that not all the L^2-Betti numbers of its universal covering belong to $\frac{1}{|\mathcal{FIN}(\pi_1(M))|}\mathbb{Z} = \mathbb{Z}[1/2]$.

The group G is constructed as follows (see [229]). Let $\phi\colon L \to L$ be the injective endomorphism sending t to t and e_0 to $e_0 t^{-1} e_0 t$. Let G be the HHN-extension associated to the subgroups L and $\mathrm{im}(\phi)$ of L and the isomorphism $\phi\colon L \to \mathrm{im}(\phi)$. This group G has the finite presentation

$$G = \langle e_0, t, s \mid e_0^2 = 1, [t^{-1} e_0 t, e_0] = 1, [s, t] = 1, s^{-1} e_0 s = e_0 t^{-1} e_0 t \rangle.$$

The general theory of HHN-extensions implies that G contains L as a subgroup, namely, as the subgroup generated by e_0 and t and each finite subgroup of G is isomorphic to a subgroup of L [350, Theorem 2.1 on page 182, Theorem 2.4 on page 185].

Remark 10.25. For the lamplighter group L the Isomorphism Conjecture 9.40 is true, i.e. the assembly map

$$a\colon \mathrm{colim}_{L/H \subset \mathrm{Or}(L, \mathcal{FIN})} K_0(\mathbb{C}H) \to K_0(\mathbb{C}L)$$

is bijective.

We first prove surjectivity. A ring is called *regular coherent*, if any finitely presented module has a finite projective resolution. For a finite group H the rings $\mathbb{C}H$ and $\mathbb{C}[H \times \mathbb{Z}]$ are regular coherent. Given a commutative ring R, any finitely presented $R[\bigoplus_{n \in \mathbb{Z}} \mathbb{Z}/2]$-module or $R[(\bigoplus_{n \in \mathbb{Z}} \mathbb{Z}/2) \times \mathbb{Z}]$-module respectively is the induction of a finitely presented RH-module or $R[H \times \mathbb{Z}]$-module respectively for some finite subgroup $H \subset \bigoplus_{n \in \mathbb{Z}} \mathbb{Z}/2$. Hence the rings $\mathbb{C}[\bigoplus_{n \in \mathbb{Z}} \mathbb{Z}/2]$ and $\mathbb{C}[(\bigoplus_{n \in \mathbb{Z}} \mathbb{Z}/2) \times \mathbb{Z}]$ are regular coherent. For a commutative ring R and automorphism $\alpha \colon R \to R$ the change of rings homomorphism $K_0(R) \to K_0(R_\alpha[t, t^{-1}])$ is surjective, where $R_\alpha[t, t^{-1}]$ is the ring of α-twisted Laurent series over R, provided that R and $R[\mathbb{Z}]$ are regular coherent [508, Corollary 13.4 and the following sentence on page 221]. This implies that $K_0(\mathbb{C}[\bigoplus_{n \in \mathbb{Z}} \mathbb{Z}/2]) \to K_0(\mathbb{C}L)$ is surjective. Since $\bigoplus_{n \in \mathbb{Z}} \mathbb{Z}/2$ is locally finite, $\bigoplus_{n \in \mathbb{Z}} \mathbb{Z}/2$ satisfies the Isomorphism Conjecture 9.40. This proves the surjectivity of $a \colon \operatorname{colim}_{H \subset \mathcal{FIN}(L)} K_0(\mathbb{C}H) \to K_0(\mathbb{C}L)$.

Next we prove injectivity. For finite subgroups $H \subset K$ of L the inclusion $H \to K$ is split injective, because both H and K are $\mathbb{Z}/2$-vector spaces. Therefore $K_0(\mathbb{C}H) \to K_0(\mathbb{C}K)$ is a split injection of finitely generated free abelian groups. Hence $\operatorname{colim}_{H \subset \mathcal{FIN}(L)} K_0(\mathbb{C}H)$ is torsionfree. Since $a \otimes_{\mathbb{Z}} \mathbb{C}$ is injective by Theorem 9.54 (1), the assembly map a is injective.

We conclude for the lamplighter group L that $\Lambda(G, \mathbb{C})_{\mathrm{fgp}} \neq \Lambda(G, \mathbb{C})_{\mathrm{fp}}$.

10.2 A Strategy for the Proof of the Atiyah Conjecture

In this section we discuss a strategy for the proof of the strong Atiyah Conjecture 10.2. We begin with a general strategy. Although it cannot work in full generality because of the counterexample presented in Subsection 10.1.4, it is nevertheless very illuminating and is the basic guideline for proofs in special cases. We will give some basic facts about localization in order to formulate the result which we will prove and which implies Linnell's Theorem 10.19. After presenting some induction principles we will explain the concrete form in which the strategy will appear to prove the strong Atiyah Conjecture 10.2 for Linnell's class of groups $\mathcal{C}$. To get a first impression of the proof, one may only consider the proof in the case of a free group presented in Subsection 10.3.1 which can be read without knowing any other material from this Chapter 10.

Linnell proves his Theorem 10.19 in [309, Theorem 1.5]. Our presentation is based on the Ph. D. thesis of Reich [435].

10.2.1 The General Case

In this Subsection we discuss a general strategy for the proof of the strong Atiyah Conjecture 10.2. We have introduced and studied the notion of a von Neumann regular ring in Subsection 8.2.2 and of the algebra $\mathcal{U}(G)$ in Definition 8.9 and Subsection 8.2.3. It suffices to recall for the sequel the following facts. A ring is von Neumann regular if and only if any finitely presented R-module is finitely generated projective (see Lemma 8.18). The algebra $\mathcal{U}(G)$ is the Ore localization of $\mathcal{N}(G)$ with respect to the multiplicatively closed subset of non-zero-divisors in $\mathcal{N}(G)$ (see Theorem 8.22 (1)). The algebra $\mathcal{U}(G)$ is von Neumann regular (see Theorem 8.22 (3)). The explicit operator theoretic definition $\mathcal{U}(G)$ will not be needed. The dimension function $\dim_{\mathcal{N}(G)}$ for arbitrary $\mathcal{N}(G)$-modules extends to a dimension function $\dim_{\mathcal{U}(G)}$ for arbitrary $\mathcal{U}(G)$-modules (see Theorem 8.29).

Lemma 10.26 (General Strategy). *Let G be a group. Suppose that there is a ring $\mathcal{S}(G)$ with $\mathbb{C}G \subset \mathcal{S}(G) \subset \mathcal{U}(G)$ which has the following properties:*

(R') *The ring $\mathcal{S}(G)$ is von Neumann regular;*
(K') *The image of the composition*

$$K_0(\mathcal{S}(G)) \xrightarrow{i} K_0(\mathcal{U}(G)) \xrightarrow{\dim_{\mathcal{U}(G)}} \mathbb{R}$$

is contained in $\frac{1}{|\mathcal{FIN}(G)|}\mathbb{Z}$, where i denotes the change of rings homomorphism.

Then G satisfies the strong Atiyah Conjecture 10.2.

Proof. Because of Lemma 10.7 it suffices to show for a finitely presented $\mathbb{C}G$-module M that $\dim_{\mathcal{N}(G)}(\mathcal{N}(G) \otimes_{\mathbb{C}G} M)$ belongs to $\frac{1}{|\mathcal{FIN}(G)|}\mathbb{Z}$. Since $\mathcal{S}(G)$ is von Neumann regular, the finitely presented $\mathcal{S}(G)$-module $\mathcal{S}(G) \otimes_{\mathbb{C}G} M$ is finitely generated projective (see Lemma 8.18) and hence defines a class in $K_0(\mathcal{S}(G))$. The composition $\dim_{\mathcal{U}(G)} \circ i \colon K_0(\mathcal{S}(G)) \to \mathbb{R}$ has image $\frac{1}{|\mathcal{FIN}(G)|}\mathbb{Z}$ by assumption and sends the class of $\mathcal{S}(G) \otimes_{\mathbb{C}G} M$ to $\dim_{\mathcal{N}(G)}(\mathcal{N}(G) \otimes_{\mathbb{C}G} M)$ by Theorem 8.29. $\square$

Remark 10.27. The role of the ring-theoretic condition **(R')** is to reduce the problem from finitely presented modules to finitely generated projective modules. Thus one can use K-theory of finitely generated projective modules. The price to pay is that one has to enlarge $\mathbb{C}G$ to an appropriate ring $\mathcal{S}(G)$ which is nicer than $\mathbb{C}G$, namely, which is von Neumann regular. This will be done by a localization process, namely, we will choose $\mathcal{S}(G)$ to be the division closure of $\mathbb{C}G$ in $\mathcal{U}(G)$. This will be a minimal choice for $\mathcal{S}(G)$. Notice that localization usually improves the properties of a ring and that in general $\mathbb{C}G$ does not have nice ring theoretic properties. For instance $\mathbb{C}G$ is von Neumann regular if and only if G is locally finite (see Example 8.21). No counterexample is known to the conjecture that $\mathbb{C}G$ is Noetherian if and only if G is virtually poly-cyclic. The implications G virtually poly-cyclic $\Rightarrow$ $\mathbb{C}G$ regular $\Rightarrow$ $\mathbb{C}G$ Noetherian are proved in [447, Theorem 8.2.2 and Theorem 8.2.20]. In order to be able to treat dimensions, the algebra $\mathcal{U}(G)$ comes in, which in contrast to $\mathcal{N}(G)$ does contain $\mathcal{S}(G)$. The passage from $\mathcal{N}(G)$ to $\mathcal{U}(G)$ does not cause any problems since the dimension theory of $\mathcal{N}(G)$ extends to a dimension theory for $\mathcal{U}(G)$ and the change of rings map $K_0(\mathcal{N}(G)) \to K_0(\mathcal{U}(G))$ is bijective (see Theorem 8.29 and Theorem 9.20 (1)).

Lemma 10.28. *Let $\mathcal{S}(G)$ be a ring with $\mathbb{C}G \subset \mathcal{S}(G) \subset \mathcal{U}(G)$.*

(1) Suppose that $\mathcal{S}(G)$ is semisimple. Then there is an upper bound on the orders of finite subgroups of G;

(2) Suppose that $\mathcal{S}(G)$ satisfies the conditions (**R'**) and (**K'**) of Lemma 10.26 and that there is an upper bound on the orders of finite subgroups of G. Then $\mathcal{S}(G)$ is semisimple;

(3) Suppose that G is torsionfree. Then $\mathcal{S}(G)$ is a skew field if and only if $\mathcal{S}(G)$ satisfies the conditions (**R'**) and (**K'**) of Lemma 10.26. In this case we get for any $\mathbb{C}G$-module M

$$\dim_{\mathcal{N}(G)}(\mathcal{N}(G) \otimes_{\mathbb{C}G} M) = \dim_{\mathcal{U}(G)}(\mathcal{U}(G) \otimes_{\mathbb{C}G} M)$$
$$= \dim_{\mathcal{S}(G)}(\mathcal{S}(G) \otimes_{\mathbb{C}G} M),$$

where $\dim_{\mathcal{S}(G)}(V)$ for an $\mathcal{S}(G)$-module V is defined to be n if $V \cong_{\mathcal{S}(G)} \mathcal{S}(G)^n$ for a non-negative integer n and by ∞ otherwise.

Proof. (1) By Wedderburn's Theorem the semisimple ring $\mathcal{S}(G)$ is a finite product of matrix rings over skewfields. Since K_0 is compatible with products and $K_0(D) \cong \mathbb{Z}$ for a skewfield D, the Morita isomorphism implies that $K_0(\mathcal{S}(G))$ is a finitely generated free abelian group. Hence $\frac{1}{|\mathcal{FIN}(G)|}\mathbb{Z}$ is finitely generated since it is contained in the image of the composition $\dim_{\mathcal{U}(G)} \circ i\colon K_0(\mathcal{S}(G)) \to \mathbb{R}$. If $a_1, a_2, \ldots, a_r$ are generators for $\frac{1}{|\mathcal{FIN}(G)|}\mathbb{Z}$, we can find $l \in \mathbb{Z}$ such that $l \cdot a_i \in \mathbb{Z}$ for $i = 1, 2 \ldots, r$. Hence $l \cdot \frac{1}{|\mathcal{FIN}(G)|}\mathbb{Z} \subset \mathbb{Z}$. Therefore l is an upper bound on the orders of finite subgroups of G.

(2) Let l be the least common multiple of the orders of finite subgroups of G. By condition (**K'**) the image of the composition

$$K_0(\mathcal{S}(G)) \to K_0(\mathcal{U}(G)) \xrightarrow{\dim_{\mathcal{U}(G)}} \mathbb{R}$$

lies in $\frac{1}{l} \cdot \mathbb{Z} = \{r \in \mathbb{R} \mid l \cdot r \in \mathbb{Z}\}$. We want to show that for any chain of ideals $\{0\} = I_0 \subset I_1 \subset I_2 \subset \ldots \subset I_r = \mathcal{S}(G)$ of $\mathcal{S}(G)$ with $I_i \neq I_{i+1}$ we have $r \leq l$. Then $\mathcal{S}(G)$ is Noetherian and hence is semisimple by Lemma 8.20 (2).

Choose $x_i \in I_i$ with $x_i \notin I_{i-1}$ for $1 \leq i \leq r-1$. Let J_i be the ideal generated by $x_1, x_2, \ldots, x_i$ for $1 \leq i \leq r-1$. Then we obtain a sequence of finitely generated ideals of the same length $\{0\} = J_0 \subset J_1 \subset J_2 \subset \ldots \subset J_r = \mathcal{S}(G)$ of $\mathcal{S}(G)$ with $J_i \neq J_{i+1}$. Since $\mathcal{U}(G)$ is von Neumann regular, J_{i-1} is a direct summand in J_i (see Lemma 8.18 (4)). Hence we get direct sum decompositions $J_i = J_{i-1} \oplus K_i$ for $i = 1, 2, \ldots, r$ for finitely generated projective non-trivial $\mathcal{S}(G)$-modules $K_1, K_2, \ldots, K_r$. Choose an idempotent $p_i \in M_{n_i}(\mathcal{S}(G))$ representing K_i. Then p_i considered as an element $M_{n_i}(\mathcal{U}(G))$ represents $\mathcal{U}(G) \otimes_{\mathcal{S}(G)} K_i$ and is non-trivial. Hence $\mathcal{U}(G) \otimes_{\mathcal{S}(G)} K_i$ is a non-trivial finitely generated projective $\mathcal{U}(G)$-module. We conclude from Theorem 8.29 $\dim_{\mathcal{U}(G)}(\mathcal{U}(G) \otimes_{\mathcal{U}(S)} K_i) > 0$ for $i = 1, 2, \ldots, r$ and hence

$$0 < \dim_{\mathcal{U}(G)}(\mathcal{U}(G) \otimes_{\mathcal{S}(G)} J_1) < \dim_{\mathcal{U}(G)}(\mathcal{U}(G) \otimes_{\mathcal{S}(G)} J_2)$$
$$< \ldots < \dim_{\mathcal{U}(G)}(\mathcal{U}(G) \otimes_{\mathcal{S}(G)} J_{r-1}) < 1.$$

Since $l \cdot \dim_{\mathcal{U}(G)}(\mathcal{U}(G) \otimes_{\mathcal{S}(G)} J_i)$ is an integer for $i = 1, 2, \ldots, r - 1$, we get $r \leq l$.

(3) Suppose that $\mathcal{S}(G)$ satisfies conditions (**R'**) and (**K'**). By the argument above $\mathcal{S}(G)$ is semisimple and contains no non-trivial ideal. Hence it is a skew field. Recall that over a skew field $\mathcal{S}(G)$ any $\mathcal{S}(G)$-module V is isomorphic to $\bigoplus_{i \in I} \mathcal{S}(G)$ for some index set I. Now apply Theorem 8.29. Any skewfield satisfies the conditions (**R'**) and (**K'**). This finishes the proof of Lemma 10.28.

$\square$

Example 10.29. Consider the free abelian group $\mathbb{Z}^n$ of rank n. Let $\mathbb{C}[\mathbb{Z}^n]_{(0)}$ be the quotient field of the commutative integral domain $\mathbb{C}[\mathbb{Z}^n]$. By inspecting Example 8.11 we see that $\mathbb{C}[\mathbb{Z}^n] \subset \mathbb{C}[\mathbb{Z}^n]_{(0)} \subset \mathcal{U}(\mathbb{Z}^n)$. By Lemma 10.28 (3) the quotient field $\mathbb{C}[\mathbb{Z}^n]_{(0)}$ satisfies the conditions of Lemma 10.26. This proves the strong Atiyah Conjecture 10.2 for $\mathbb{Z}^n$ and that for any $\mathbb{C}[\mathbb{Z}^n]$-module M

$$\dim_{\mathcal{N}(\mathbb{Z}^n)}(\mathcal{N}(\mathbb{Z}^n) \otimes_{\mathbb{C}[\mathbb{Z}^n]} M) = \dim_{\mathbb{C}[\mathbb{Z}^n]_{(0)}}(\mathbb{C}[\mathbb{Z}^n]_{(0)} \otimes_{\mathbb{C}[\mathbb{Z}^n]} M)$$

holds. The reader should compare this with the explicit proof of Lemma 1.34, which corresponds to the case of a finitely presented $\mathbb{C}[\mathbb{Z}^n]$-module M.

Remark 10.30. Suppose that G is a torsionfree amenable group which satisfies the Kaplansky Conjecture 10.14. We have already shown in Example 8.16 that set S of non-zero-divisors of $\mathbb{C}G$ satisfies the Ore condition (see Definition 8.14) and that the Ore localization $S^{-1}\mathbb{C}G$ is a skew field. Recall from Theorem 8.22 (1) that $\mathcal{U}(G)$ is the Ore localization of the von Neumann algebra $\mathcal{N}(G)$ with respect to the set of all non-zero-divisors. A non-zero-divisor x of $\mathbb{C}G$ is still a non-zero-divisor when we regard it as an element in $\mathcal{N}(G)$ by the following argument. Denote by $r_x^{\mathcal{N}(G)}: \mathcal{N}(G) \to \mathcal{N}(G)$ and $r_x^{\mathbb{C}G}: \mathbb{C}G \to \mathbb{C}G$ the maps given by right multiplication with x. The kernel of $r_x^{\mathcal{N}(G)}$ is a finitely generated projective $\mathcal{N}(G)$-module since $\mathcal{N}(G)$ is semihereditary (see Theorem 6.5 and Theorem 6.7 (1)). We conclude $\dim_{\mathcal{N}(G)}(\ker(r_x^{\mathcal{N}(G)})) = \dim_{\mathcal{N}(G)}(\mathcal{N}(G) \otimes_{\mathbb{C}G} \ker(r_x^{\mathbb{C}G}))$ from the dimension-flatness of $\mathcal{N}(G)$ over $\mathbb{C}G$ for amenable G (see Theorem 6.37). Since $\ker(r_x) = \{0\}$, Lemma 6.28 (3) implies that $r_x^{\mathcal{N}(G)}$ has trivial kernel and hence x is a non-zero-divisor in $\mathcal{N}(G)$. We conclude $\mathbb{C}G \subset S^{-1}\mathbb{C}[G] \subset \mathcal{U}(G)$. Hence the strong Atiyah Conjecture 10.2 holds for G by Lemma 10.28 (3). Notice that this gives a different proof of Lemma 10.16.

Remark 10.31. We know already from Subsection 10.1.4 that the strong Atiyah Conjecture 10.2 does not hold for arbitrary groups G. We think that the extra condition that there is an upper bound on the order of finite subgroups of G is very important. Under this condition we see from Lemma 10.28 (2) that we can replace in Lemma 10.26 the condition (**R'**) that $S(G)$ is von Neumann regular by the stronger condition (**R**) that $S(G)$ is semisimple. Actually, the proof of Linnell's Theorem does only work if we use (**R**) instead of (**R'**).

$\square$

10.2.2 Survey on Universal Localizations and Division and Rational Closure

We give some basic facts about universal localization and division and rational closure.

Let R be a ring and Σ be a set of homomorphisms between (left) R-modules. A ring homomorphism $f \colon R \to S$ is called Σ-*inverting* if for every map $\alpha \colon M \to N \in \Sigma$ the induced map $S \otimes_R \alpha \colon S \otimes_R M \to S \otimes_R N$ is an isomorphism. A Σ-inverting ring homomorphism $i \colon R \to R_\Sigma$ is called *universal* Σ-*inverting* if for any Σ-inverting ring homomorphism $f \colon R \to S$ there is precisely one ring homomorphism $f_\Sigma \colon R_\Sigma \to S$ satisfying $f_\Sigma \circ i = f$. This generalizes Definition 8.13. If $f \colon R \to R_\Sigma$ and $f' \colon R \to R'_\Sigma$ are two universal Σ-inverting homomorphisms, then by the universal property there is precisely one isomorphism $g \colon R_\Sigma \to R'_\Sigma$ with $g \circ f = f'$. This shows the uniqueness of the universal Σ-inverting homomorphism. The universal Σ-inverting ring homomorphism exists if Σ is a set of homomorphisms of finitely generated projective modules [465, Section 4]. If Σ is a set of matrices, a model for R_Σ is given by considering the free R-ring generated by the set of symbols $\{\overline{a_{i,j}} \mid A = (a_{i,j}) \in \Sigma\}$ and dividing out the relations given in matrix form by $\overline{A}A = A\overline{A} = 1$, where $\overline{A}$ stands for $(\overline{a_{i,j}})$ for $A = (a_{i,j})$. The map $i \colon R \to R_\Sigma$ does not need to be injective and the functor $R_\Sigma \otimes_R -$ does not need to be exact in general.

Notation 10.32. *Let S be a ring and $R \subset S$ be a subring. Denote by $T(R \subset S)$ the set of all elements in R which become invertible in S. Denote by $\Sigma(R \subset S)$ the set of all square matrices A over R which become invertible in S.*

A subring $R \subset S$ is called *division closed* if $T(R \subset S) = R^{\mathrm{inv}}$, i.e. any element in R which is invertible in S is already invertible in R. It is called *rationally closed* if $\Sigma(R \subset S) = GL(R)$, i.e. any square matrix over R which is invertible over S is already invertible over R. Notice that the intersection of division closed subrings of S is again division closed, and analogously for rationally closed subrings. Hence the following definition makes sense.

Definition 10.33 (Division and rational closure). *Let S be a ring with subring $R \subset S$. The division closure $\mathcal{D}(R \subset S)$ or rational closure $\mathcal{R}(R \subset S)$ respectively is the smallest subring of S which contains R and is division closed or rationally closed respectively.*

Obviously $R \subset \mathcal{D}(R \subset S) \subset \mathcal{R}(R \subset S) \subset S$. One easily checks $\mathcal{D}(\mathcal{D}(R \subset S) \subset S) = \mathcal{D}(R \subset S)$ and $\mathcal{R}(\mathcal{R}(R \subset S) \subset S) = \mathcal{R}(R \subset S)$. The rational closure $\mathcal{R}(R \subset S)$ of R in S is the set of elements $s \in S$ for which there is a square matrix A over S and a matrix B over R such that $AB = BA = 1$ holds over S and s is an entry in A [118, section 7.1]. The easy proof of the next two results is left to the reader (or see [435, Proposition 13.17]).

Lemma 10.34. *Let S be a ring with subring $R \subset S$.*

(1) If R is von Neumann regular, then R is division closed and rationally closed in S;

(2) If $\mathcal{D}(R \subset S)$ is von Neumann regular, then $\mathcal{D}(R \subset S) = \mathcal{R}(R \subset S)$.

Lemma 10.35. *Let S be a ring with subring $R \subset S$.*

(1) The map $\phi\colon R_{T(R\subset S)} \to S$ given by the universal property factorizes as
$$\phi\colon R_{T(R\subset S)} \xrightarrow{\Phi} \mathcal{D}(R \subset S) \subset S;$$

(2) Suppose that the pair $(R, T(R \subset S))$ satisfies the (right) Ore condition (see Definition 8.14). Then
$$R_{T(R\subset S)} = RT(R \subset S)^{-1} = \mathcal{D}(R \subset S),$$

i.e. the map given by the universal property $R_{T(R\subset S)} \xrightarrow{j} RT(R \subset S)^{-1}$ is an isomorphism, and $\Phi\colon R_{T(R\subset S)} \to \mathcal{D}(R \subset S)$ is an isomorphism;

(3) The map $\psi\colon R_{\Sigma(R\subset S)} \to S$ given by the universal property factorizes as $\psi\colon R_{\Sigma(R\subset S)} \xrightarrow{\Psi} \mathcal{R}(R \subset S) \subset S$. The map $\Psi\colon R_{\Sigma(R\subset S)} \to \mathcal{R}(R \subset S)$ is always surjective.

The map $\Psi\colon R_{\Sigma(R\subset S)} \to \mathcal{R}(R \subset S)$ is not always injective. (An example will be given in the exercises.)

Lemma 10.36. *Let $R_i \subset S_i \subset S$ be subrings for $i \in I$ and $R = \bigcup_{i\in I} R_i$ and $S = \bigcup_{i\in I} S_i$ directed unions of rings. Suppose that all rings S_i are von Neumann regular. Then S is von Neumann regular and we obtain directed unions*

$$T(R \subset S) = \bigcup_{i\in I} T(R_i \subset S_i);$$

$$\Sigma(R \subset S) = \bigcup_{i\in I} \Sigma(R_i \subset S_i);$$

$$\mathcal{D}(R \subset S) = \bigcup_{i\in I} \mathcal{D}(R_i \subset S_i);$$

$$\mathcal{R}(R \subset S) = \bigcup_{i\in I} \mathcal{R}(R_i \subset S_i).$$

Proof. One easily checks that S satisfies Definition 8.17 of von Neumann regular.

Suppose that $x \in R$ becomes invertible in S. Choose $i \in I$ with $x \in R_i$. Obviously x is not a zero-divisor in S_i. Since S_i is von Neumann regular, x_i is invertible in S_i by Lemma 8.20 (1). This shows $T(R \subset S) = \bigcup_{i\in I} T(R_i \subset S_i)$. The proof of $\Sigma(R \subset S) = \bigcup_{i\in I} \Sigma(R_i \subset S_i)$ is similar using the fact that a matrix ring over a von Neumann ring is again von Neumann regular by Lemma 8.19 (1).

Since S_i is von Neumann regular, Lemma 10.34 (1) implies $\mathcal{D}(R_i \subset S_i) = \mathcal{D}(R_i \subset S) \subset \mathcal{D}(R \subset S)$ and hence $\bigcup_{i \in I} \mathcal{D}(R_i \subset S_i) \subset \mathcal{D}(R \subset S)$. Since $\bigcup_{i \in I} \mathcal{D}(R_i \subset S_i) = \bigcup_{i \in I} \mathcal{D}(R_i \subset S)$ is division closed in S, we get $\bigcup_{i \in I} \mathcal{D}(R_i \subset S_i) = \mathcal{D}(R \subset S)$. The proof for the rational closure is analogous. $\qquad\square$

10.2.3 The Strategy for the Proof of Linnell's Theorem

We will outline the strategy of proof for Linnell's Theorem 10.19.

Notation 10.37. *Let G be a group. Define $\mathcal{D}(G)$ to be the division closure $\mathcal{D}(\mathbb{C}G \subset \mathcal{U}(G))$ and $\mathcal{R}(G)$ to be the rational closure $\mathcal{R}(\mathbb{C}G \subset \mathcal{U}(G))$ of $\mathbb{C}G$ in $\mathcal{U}(G)$ (see Definition 10.33). We abbreviate $T(G) = T(\mathbb{C}G \subset \mathcal{U}(G))$ and $\Sigma(G) = \Sigma(\mathbb{C}G \subset \mathcal{U}(G))$ (see Notation 10.32).*

The ring $\mathcal{D}(G)$ will be our candidate for $\mathcal{S}(G)$ in Lemma 10.26. We conclude from Lemma 10.34 that any von Neumann regular ring $\mathcal{S}(G)$ with $\mathbb{C}G \subset \mathcal{S}(G) \subset \mathcal{U}(G)$ satisfies $\mathcal{D}(G) \subset \mathcal{R}(G) \subset \mathcal{S}(G)$ and that $\mathcal{D}(G) = \mathcal{R}(G)$ holds if $\mathcal{D}(G)$ is von Neumann regular. Hence $\mathcal{D}(G)$ is a minimal choice for $\mathcal{S}(G)$ in Lemma 10.26 and we should expect $\mathcal{D}(G) = \mathcal{R}(G)$. In view of Remark 10.31 we should also expect that $\mathcal{D}(G)$ is semisimple. The upshot of this discussion is that, in order to prove Linnell's Theorem 10.19, we will prove the following

Theorem 10.38 (Strong version of Linnell's Theorem). *Let G be a group in the class $\mathcal{C}$ (see Definition 10.18) such that there exists a bound on the orders of finite subgroups. Then*

(R) *The ring $\mathcal{D}(G)$ is semisimple;*
(K) *The composition*

$$\mathrm{colim}_{G/H \in \mathrm{Or}(G, \mathcal{FIN})} K_0(\mathbb{C}H) \xrightarrow{a} K_0(\mathbb{C}G) \xrightarrow{i} K_0(\mathcal{D}(G))$$

is surjective, where a is the assembly map appearing in the Isomorphism Conjecture 9.40 for $K_0(\mathbb{C}G)$ and essentially given by the various inclusions of the finite subgroups of G, and i is a change of rings homomorphism.

Notice that we have replaced the condition **(K')** appearing in Lemma 10.26 by the stronger condition **(K)** above. The stronger condition gives better insight in what is happening and also yields a connection to the Isomorphism Conjecture 9.40 for $K_0(\mathbb{C}G)$. Namely, the surjectivity of the change of rings map $K_0(\mathbb{C}G) \to K_0(\mathcal{D}(G))$ is equivalent to the condition **(K)** provided that the Isomorphism Conjecture 9.40 for $K_0(\mathbb{C}G)$ holds.

We will prove in the strong version of Linnell's Theorem 10.38 a statement which implies the strong Atiyah Conjecture 10.2 by Lemma 10.26 but which in general contains more information than the strong Atiyah Conjecture 10.2.

At least for torsionfree groups G the statement turns out to be equivalent to the strong Atiyah Conjecture 10.2 by the next lemma.

Lemma 10.39. *Let G be a torsionfree group. Then the strong Atiyah Conjecture 10.2 is true if and only if $\mathcal{D}(G)$ is a skewfield. In this case $\mathcal{D}(G) = \mathcal{R}(G)$ and the composition*

$$\mathrm{colim}_{\mathrm{Or}(G,\mathcal{FIN})}\, K_0(\mathbb{C}H) \xrightarrow{a} K_0(\mathbb{C}G) \xrightarrow{i} K_0(\mathcal{D}(G))$$

is surjective.

Proof. Suppose that $\mathcal{D}(G)$ is a skewfield. Then $\mathcal{D}(G) = \mathcal{R}(G)$ follows from Lemma 10.34 (2). Since any finitely generated module over a skewfield is finitely generated free, the composition $\mathrm{colim}_{\mathrm{Or}(G,\mathcal{FIN})}\, K_0(\mathbb{C}H) \xrightarrow{a} K_0(\mathbb{C}G) \xrightarrow{i} K_0(\mathcal{D}(G))$ is surjective. The strong Atiyah Conjecture 10.2 follows from Lemma 10.26.

Now suppose that the strong Atiyah Conjecture 10.2 holds. We want to show that $\mathcal{R}(G)$ is a skewfield. It suffices to show for any element $x \in \mathcal{R}(G)$ with $x \neq 0$ that x is invertible in $\mathcal{U}(G)$, because then x is already invertible in $\mathcal{R}(G)$ since $\mathcal{R}(G)$ is rationally closed in $\mathcal{U}(G)$. By Lemma 10.35 (3) there is $y \in \mathbb{C}G_{\Sigma(G)}$ which is mapped to x under the canonical ring homomorphism $\Psi \colon \mathbb{C}G_{\Sigma(G)} \to \mathcal{R}(G)$. Two square matrices A and B over a ring R are called *associated* if there are invertible square matrices U and V and non-negative integers m and n satisfying

$$U \begin{pmatrix} A & 0 \\ 0 & I_m \end{pmatrix} V = \begin{pmatrix} B & 0 \\ 0 & I_n \end{pmatrix}.$$

For a set of matrices Σ over R, the universal Σ-inverting homomorphism $f \colon R \to R_\Sigma$ and $s \in R_\Sigma$ we can find a square matrix A over R such that $f(A)$ and s are associated in R_Σ. This follows from Cramer's Rule (see [465, Theorem 4.3 on page 53]). If we apply this to $y \in \mathbb{C}G_{\Sigma(G)}$ regarded as $(1,1)$-matrix and push forward the result to $\mathcal{R}(G)$ by Ψ, we obtain invertible matrices U, V over $\mathcal{R}(G)$ and a matrix A over $\mathbb{C}G$ satisfying

$$U \begin{pmatrix} x & 0 \\ 0 & I_{m-1} \end{pmatrix} V = A.$$

If we consider this over $\mathcal{U}(G)$, we conclude for the maps r_x and r_A induced by right multiplication with x and A

$$\dim_{\mathcal{U}}(\mathrm{im}(r_x \colon \mathcal{U}(G) \to \mathcal{U}(G)) + m - 1 = \dim_{\mathcal{U}(G)}(\mathrm{im}(r_A \colon \mathcal{U}(G)^m \to \mathcal{U}(G)^m)).$$

Since by assumption the strong Atiyah Conjecture 10.2 holds, Theorem 8.29 and Lemma 10.7 imply that $\dim_{\mathcal{U}(G)}(\mathrm{im}\,(r_A \colon \mathcal{U}(G)^m \to \mathcal{U}(G)^m))$ must be an integer $\leq m$ and hence $\dim_{\mathcal{U}}(\ker(r_x \colon \mathcal{U}(G) \to \mathcal{U}(G))) \in \{0, 1\}$. Since x is by assumption non-trivial and $\mathcal{U}(G)$ is von Neumann regular by Theorem 8.22

(3), the kernel of r_x is trivial by Lemma 8.18 and Theorem 8.29. Hence x must be a unit by Lemma 8.20 (1). Hence $\mathcal{R}(G)$ is a skewfield. Since $\mathcal{D}(G)$ is division closed and contained in the skewfield $\mathcal{R}(G)$, $\mathcal{D}(G)$ is a skewfield. We conclude $\mathcal{D}(G) = \mathcal{R}(G)$ from Lemma 10.34 (2). This finishes the proof of Lemma 10.39.   □

The rest of this chapter is devoted to the proof of the strong version of Linnell's Theorem 10.38. First of all we explain the underlying induction technique.

Lemma 10.40 (Induction principle). *Suppose that (**P**) is a property for groups such that the following is true:*

*(1) (**P**) holds for any free group;*
*(2) Let $1 \to H \to G \to Q \to 1$ be an extension of groups such that (**P**) holds for H and Q is virtually finitely generated abelian, then (**P**) holds for G.*
*(3) If (**P**) holds for any finitely generated subgroup of G, then (**P**) holds for G;*

*Then (**P**) holds for any group G in $\mathcal{C}$.*

*If we replace condition (1) by the condition that (**P**) holds for the trivial group, then (**P**) holds for any elementary amenable group G.*

Proof. This follows from the following description of the class $\mathcal{C}$ and the class $\mathcal{EAM}$ of elementary amenable groups respectively by transfinite induction in [300, Lemma 3.1] and [309, Lemma 4.9]. Define for any ordinal α the class of groups $\mathcal{D}_\alpha$ as follows. Put $\mathcal{D}_0$ to be the class consisting of virtually free groups or of the trivial group respectively. If α is a successor ordinal, define $\mathcal{D}_\alpha$ to be the class of groups G which fit into an exact sequence $1 \to H \to G \to Q \to 1$ such that any finitely generated subgroup of H belongs to $\mathcal{D}_{\alpha-1}$ and Q is either finite or infinite cyclic. If α is a limit ordinal, then $\mathcal{D}_\alpha$ is the union of the $\mathcal{D}_\beta$ over all β with $\beta < \alpha$. Then $\mathcal{C}$ or $\mathcal{EAM}$ respectively is $\bigcup_{\alpha>0} \mathcal{D}_\alpha$.   □

Lemma 10.40 is the reason why it sometimes is easier to prove a property (**P**) for all elementary amenable groups than for all amenable groups. We conclude from Lemma 10.40.

Lemma 10.41 (Plan of proof). *In order to prove the strong version of Linnell's Theorem 10.38 and hence Linnell's Theorem 10.19 it suffices to show that the following statements are true:*

*(1) Any free group G has the property (**R**), i.e. $\mathcal{D}(G)$ is semisimple;*
*(2) If $1 \to H \to G \to Q \to 1$ is an extension of groups such that H has property (**R**) and Q is finite, then G has property (**R**);*
*(3) If $1 \to H \to G \to Q \to 1$ is an extension of groups such that H has property (**R**) and Q is infinite cyclic, then G has property (**R**);*
*(4) Any virtually free group G has property (**K**), i.e. the composition*

$$\mathrm{colim}_{G/H\in\mathrm{Or}(G,\mathcal{FIN})} K_0(\mathbb{C}H) \xrightarrow{a} K_0(\mathbb{C}G) \xrightarrow{i} K_0(\mathcal{D}(G))$$

is surjective;

(5) Let $1 \to H \to G \to Q \to 1$ be an extension of groups. Suppose that for any group H', which contains H as subgroup of finite index, properties (**K**) and (**R**) are true. Suppose that Q is virtually finitely generated abelian. Then (**K**) holds for G;

(6) Suppose that G is the directed union of subgroups $\{G_i \mid i \in I\}$ such that each G_i satisfies (**K**). Then G has property (**K**);

(7) Let G be a group such that there is an upper bound on the orders of finite subgroups. Suppose that G is the directed union of subgroups $\{G_i \mid i \in I\}$ such that each G_i satisfies both (**K**) and (**R**). Then G satisfies (**R**).

Proof. We call two groups G and G' *commensurable* if there exists subgroups $G_0 \subset G$ and $G'_0 \subset G'$ such that $[G : G_0] < \infty$ and $[G' : G'_0] < \infty$ holds and G_0 and G'_0 are isomorphic. In order to prove Lemma 10.41, it suffices to show for any $G \in \mathcal{C}$ the following property:

(**P**)   If there is an upper bound on the orders of finite subgroups of G and G_0 is commensurable to G, then G_0 satisfies (**R**) and (**K**).

This is done by proving the conditions in Lemma 10.40.

Since any group G which is commensurable with a free group is virtually free, condition (1) of Lemma 10.40 follows from assumptions (1), (2) and (4) in Lemma 10.41.

Let $1 \to H \to G \to Q \to 1$ be an extension such that Q is virtually finitely generated abelian, H has property (**P**) and there is an upper bound on the orders of finite subgroups of G. Let G_0 be any group which is commensurable to G. Then one easily constructs an extension $1 \to H_0 \to G_0 \to Q_0 \to 1$ such that H_0 is commensurable to H and Q_0 is commensurable to Q and hence virtually finitely generated abelian. Then G_0 satisfies (**K**) by assumption (5) of Lemma 10.41. Moreover one can find a filtration of G_0 by a sequence of normal subgroups $H_0 = G_n \subset G_{n-1} \subset \ldots \subset G_1 \subset G_0$ such that $G_i/G_{i-1} \cong \mathbb{Z}$ for $1 \leq i \leq n-1$ and G_0/G_1 is finite. The group G_0 has property (**R**) by assumptions (2) and (3) of Lemma 10.41. Hence G has property (**P**). This proves condition (2) of Lemma 10.40.

Finally we prove condition (3) of Lemma 10.40. Suppose that there is an upper bound on the orders of finite subgroups of G. Any finitely generated subgroup of G_0 is commensurable to a finitely generated subgroup of G. Hence G_0 has properties (**K**) and (**R**) by assumptions (6) and (7) of Lemma 10.41. Now Lemma 10.41 follows from Lemma 10.40.   □

Remark 10.42. Notice that in the formulation of condition (**P**) in the proof of Lemma 10.41 we have to build in the commensurable group G_0 since we cannot prove in general that a group G has property (**K**) if a subgroup $G_0 \subset G$ of finite index has property (**K**).

In the statements (5) and (7) of Lemma 10.41 we need both (**K**) and (**R**) to get a conclusion for (**K**) or (**R**) alone. Therefore we cannot separate the proof of (**K**) from the one of (**R**) and vice versa.

If we would replace $(\mathbf{R})$ by $(\mathbf{R}')$ in Lemma 10.41, statement (7) remains true if we omit the condition that there is an upper bound on the orders of finite subgroups and that each G_i satisfies $(\mathbf{K})$. We only need that each G_i satisfies $(\mathbf{R}')$ to prove that G satisfies $(\mathbf{R}')$. This looks promising. The problem is that we cannot prove statements (3) and (5) for $(\mathbf{R}')$ instead of $(\mathbf{R})$. Namely, the proofs rely heavily on the assumption that $\mathcal{D}(H)$ is Noetherian (see Goldie's Theorem 10.61 and Moody's Induction Theorem 10.67). Notice that a von Neumann regular ring is semisimple if and only if it is Noetherian (see Lemma 8.20 (2)). The counterexample to the strong Atiyah Conjecture given by the lamplighter group L (see Subsection 10.1.4) actually shows that at least one of the statements (3) and (5) applied to $1 \to \bigoplus_{n \in \mathbb{Z}} \mathbb{Z}/2 \to L \to \mathbb{Z} \to 0$ is wrong, if we replace $(\mathbf{R})$ by $(\mathbf{R}')$.

We can replace condition $(\mathbf{K})$ by $(\mathbf{K}')$ in Lemma 10.41 getting a weaker conclusion in the strong version of Linnell's Theorem 10.38. This still gives the strong Atiyah Conjecture and the proof of statement (4) simplifies considerably. But we prefer the stronger condition $(\mathbf{K})$ because we get a stronger result in strong version of Linnell's Theorem 10.38.

10.3 The Proof of Linnell's Theorem

This section is devoted to the proof that the conditions appearing in Lemma 10.41 are satisfied. Recall that then the strong version of Linnell's Theorem 10.38 and hence Linnell's Theorem 10.19 follow.

10.3.1 The Proof of Atiyah's Conjecture for Free Groups

In this subsection we prove

Lemma 10.43. *The strong Atiyah Conjecture 10.2 holds for any free group G, i.e. for any matrix $A \in M(m, n, \mathbb{C}G)$ the von Neumann dimension of the kernel of the induced bounded G-operator*

$$r_A^{(2)} : l^2(G)^m \to l^2(G)^n, \quad x \mapsto xA$$

satisfies

$$\dim_{\mathcal{N}(G)} \left(\ker \left(r_A^{(2)} : l^2(G)^m \to l^2(G)^n \right) \right) \in \mathbb{Z}.$$

The proof uses the notion of a Fredholm module and is designed along the new conceptual proof of the *Kadison Conjecture* for the free group F_2 on two generators due to Connes [120, IV.5] (see also [166]). The Kadison Conjecture says that there are no non-trivial idempotents in $C_r^*(G)$ for a torsionfree group G. We begin with introducing Fredholm modules.

Let H be a Hilbert space. Given a Hilbert basis $\{b_i \mid i \in I\}$ and $f \in \mathcal{B}(H)$, define $\text{tr}(f) \in \mathbb{C}$ to be $\sum_{i \in I} \langle f(b_i), b_i \rangle$ if this sum converges. We say that f is of *trace class* if $\operatorname{tr}(|f|) < \infty$ holds for one (and then automatically for all) Hilbert basis, where $|f|$ is the positive part in the polar decomposition of f. (Notice that this agrees with Definition 1.8 applied to $|f|$ in the case $G = \{1\}$.) If f is of trace class, then $\sum_{i \in I} \langle f(b_i), b_i \rangle$ converges and defines a number $\operatorname{tr}(f) \in \mathbb{C}$ which is independent of the choice of Hilbert basis. We call $f \in \mathcal{B}(H)$ *compact* if the closure of the image of the unit disk $\{x \in H \mid |x| \le 1\}$ is compact. We call $f \in \mathcal{B}(H)$ an *operator of finite rank* if the image of f is a finite dimensional vector space. We denote by $\mathcal{L}^0(H)$, $\mathcal{L}^1(H)$, and $\mathcal{K}(H)$ the (two-sided) ideal in $\mathcal{B}(H)$ of operators of finite rank, of operators of trace class and of compact operators. Denote by $\mathcal{L}^p(H) = \{f \in \mathcal{B}(H) \mid |f|^p \in \mathcal{L}^1(H)\}$ the *Schatten ideal* in $\mathcal{B}(H)$ for $p \in [1, \infty)$. Notice that $\mathcal{L}^0(H) \subset \mathcal{L}^1(H) \subset \mathcal{L}^q(H) \subset \mathcal{L}^{q'}(H) \subset \mathcal{K}(H)$ holds for $1 \le q \le q'$. A *$*$-algebra* or *algebra with involution* $\mathcal{B}$ is an algebra over $\mathbb{C}$ with an involution of rings $* \colon \mathcal{B} \to \mathcal{B}$, which is compatible with the $\mathbb{C}$-multiplication in the sense $*(\lambda \cdot b) = \overline{\lambda} \cdot *(b)$ for $\lambda \in \mathbb{C}$ and $b \in \mathcal{B}$. We call a $*$-homomorphism $\rho \colon \mathcal{B} \to \mathcal{B}(H)$ a *$\mathcal{B}$-representation*. Here we do *not* require that ρ sends $1 \in \mathcal{B}$ to $\operatorname{id}_H \in \mathcal{B}(H)$.

Definition 10.44. *Let $\mathcal{B} \subset \mathcal{A}$ be an arbitrary $*$-closed subalgebra of the C^*-algebra $\mathcal{A}$. For $p \in \{0\} \cup [1, \infty)$ a p-summable $(\mathcal{A}, \mathcal{B})$-Fredholm module consists of two $\mathcal{A}$-representations $\rho_\pm \colon \mathcal{A} \to \mathcal{B}(H)$ such that $\rho_+(a) - \rho_-(a) \in \mathcal{L}^p(H)$ holds for each $a \in \mathcal{B}$.*

We will later construct a 0-summable Fredholm $(\mathcal{N}(G), \mathbb{C}G)$-module. The next lemma follows using the equation

$$\rho_+(ab) - \rho_-(ab) = \rho_+(a)(\rho_+(b) - \rho_-(b)) + (\rho_+(a) - \rho_-(a))\rho_-(b).$$

Lemma 10.45. *Let $\rho_\pm \colon \mathcal{A} \to \mathcal{B}(H)$ be a p-summable $(\mathcal{A}, \mathcal{B})$-Fredholm module. Then*

(1) For $q \in \{0\} \cup [1, \infty)$ the set $\mathcal{A}^q = \{a \in \mathcal{A} \mid \rho_+(a) - \rho_-(a) \in \mathcal{L}^q(H)\}$ is a $*$-subalgebra of $\mathcal{A}$. For $q \le q'$ we have the inclusions $\mathcal{A}^0 \subset \mathcal{A}^1 \subset \mathcal{A}^q \subset \mathcal{A}^{q'} \subset \mathcal{A}$;

(2) The map

$$\tau \colon \mathcal{A}^1 \to \mathbb{C}, \quad a \mapsto \operatorname{tr}(\rho_+(a)) - \operatorname{tr}(\rho_-(a))$$

is linear and has the trace property $\tau(ab) = \tau(ba)$;

(3) Tensoring $\rho_\pm$ with the standard representation ρ_{st} of $M_n(\mathbb{C})$ on $\mathbb{C}^n$ yields a p-summable $(M_n(\mathcal{A}), M_n(\mathcal{B}))$-Fredholm module. We have $M_n(\mathcal{A}^p) \subset M_n(\mathcal{A})^p$.

To show that certain numbers are integers, the key ingredient will be the following lemma.

Lemma 10.46. *Let $p, q \in \mathcal{B}(H)$ be two projections with $p - q \in \mathcal{L}^1(H)$. Then $\operatorname{tr}(p - q)$ is an integer.*

Proof. The operator $(p - q)^2$ is a selfadjoint compact operator. We get a decomposition of H into the finite dimensional eigenspaces of $(p - q)^2$ [434, Theorem IV.16 on page 203]

$$H = \ker((p - q)^2) \oplus \bigoplus_{\lambda \neq 0} E_\lambda.$$

This decomposition is respected by both p and q since both p and q commute with $(p-q)^2$. Notice that $\ker(p-q) = \ker((p-q)^2)$. Next we compute the trace of $(p - q)$ with respect to a Hilbert basis which respects this decomposition

$$\text{tr}(p - q) \;=\; \sum_{\lambda \neq 0} \text{tr}((p - q)|_{E_\lambda}) \;=\; \sum_{\lambda \neq 0} \text{tr}(p|_{E_\lambda}) - \text{tr}(q|_{E_\lambda}).$$

Each difference $\text{tr}(p|_{E_\lambda}) - \text{tr}(q|_{E_\lambda})$ is an integer, namely, the difference of the dimensions of finite dimensional vector spaces $\dim_{\mathbb{C}}(p(E_\lambda)) - \dim_{\mathbb{C}}(q(E_\lambda))$. Since the sum of these integers converges to the real number $\text{tr}(p - q)$, the claim follows. □

Lemma 10.47. *Let $a, b \in \mathcal{B}(H)$ be elements with $a - b \in \mathcal{L}^0(H)$. Denote by $p_{\ker(a)}$ and $p_{\ker(b)}$ the orthogonal projections onto their kernels. Then $p_{\ker(a)} - p_{\ker(b)} \in \mathcal{L}^0(H)$.*

Proof. One easily checks that $p_{\ker(a)}$ and $p_{\ker(b)}$ agree on $\ker(a - b)$. Since the orthogonal complement of $\ker(a - b)$ is finite dimensional because of $a - b \in \mathcal{L}^0(H)$, we conclude $p_{\ker(a)} - p_{\ker(b)} \in \mathcal{L}^0(H)$. □

In order to construct the relevant Fredholm module, we need some basic properties of the free group F_2 in two generators s_1 and s_2. The associated Cayley graph is a tree with obvious F_2-action. The action is free and transitive on the set V of vertices. It is free on the set E of edges such that $F_2 \backslash E$ maps bijectively to the set of generators $\{s_1, s_2\}$. Hence we can choose bijections of F_2-sets $V \cong F_2$ and $E \cong F_2 \coprod F_2$. We equip the Cayley graph with the word-length metric. Then any to distinct points x and y on the Cayley graph can be joined by a unique geodesic which will be denoted by $x \to y$. On $x \to y$ there is a unique initial edge $\text{init}(x \to y)$ on $x \to y$ starting at x. Fix a base point $x_0 \in V$. Define a map

$$f \colon V \to E \coprod \{*\} \tag{10.48}$$

by sending x to $\text{init}(x \to x_0)$ if $x \neq x_0$ and to $*$ otherwise. Notice for $x \neq x_0$ that $g\,\text{init}(x \to x_0) = \text{init}(gx \to gx_0)$ is different from $\text{init}(gx \to x_0)$ if and only if gx lies on $gx_0 \to x_0$. This implies

Lemma 10.49. *The map f is bijective and almost equivariant in the following sense: For a fixed $g \in F_2$ there is only a finite number of vertices $x \neq x_0$ with $gf(x) \neq f(gx)$. The number of exceptions is equal to the distance from x_0 to gx_0.*

It is shown in [141] that free groups are the only groups which admit maps with properties similar to the one of f.

Let $l^2(V)$ and $l^2(E)$ be the Hilbert spaces with the sets V and E as Hilbert basis. We obtain isomorphisms of $\mathbb{C}[F_2]$-representations $l^2(V) \cong l^2(F_2)$ and $l^2(V) \cong l^2(F_2) \oplus l^2(F_2)$ from the bijections $V \cong F_2$ and $E \cong F_2 \coprod F_2$ above. Thus we can view $l^2(V)$ and $l^2(E)$ as $\mathcal{N}(F_2)$-representations. Denote by $\mathbb{C}$ the trivial $\mathcal{N}(F_2)$-representation for which $ax = 0$ for all $a \in \mathcal{N}(F_2)$ and $x \in \mathbb{C}$. Denote by $\rho_+ \colon \mathcal{N}(F_2) \to \mathcal{B}(l^2(V))$ and by $\rho \colon \mathcal{N}(F_2) \to \mathcal{B}(l^2(E) \oplus \mathbb{C})$ the corresponding representations. The map f from (10.48) induces an isometric isomorphism of Hilbert spaces $F \colon l^2(V) \xrightarrow{\cong} l^2(E) \oplus \mathbb{C}$. Let $\rho_- \colon \mathcal{N}(G) \to \mathcal{B}(l^2(V))$ be defined by $\rho_-(a) = F^{-1} \circ \rho(a) \circ F$.

Lemma 10.50. *(1) We have for $a \in \mathcal{N}(F_2)$*

$$\mathrm{tr}_{\mathcal{N}(F_2)}(a) = \sum_{x \in V} \langle (\rho_+(a) - \rho_-(a))(x), x \rangle;$$

(2) The representations ρ_+ and ρ_- define a 0-summable $(\mathcal{N}(F_2), \mathbb{C}[F_2])$-Fredholm module;

(3) We get $\mathrm{tr}_{\mathcal{N}(F_2)}(A) = \tau(A)$ for $A \in M_n(\mathcal{N}(F_2))^1$, if $\tau \colon M_n(\mathcal{N}(F_2))^1 \to \mathbb{R}$ is the map defined in Lemma 10.45 for the Fredholm module $\rho_\pm \otimes \rho_{\mathrm{st}}$.

Proof. (1) We get for $g \in F_2$ and $x \in V$ with $x \neq x_0$

$$\langle (\rho_+(g) - \rho_-(g))(x_0), x_0 \rangle = \mathrm{tr}_{\mathcal{N}(F_2)}(g);$$
$$\langle (\rho_+(g) - \rho_-(g))(x), x \rangle = 0.$$

By linearity we get the equations above for any $a \in \mathbb{C}[F_2]$. Recall that $\mathcal{N}(F_2)$ is the weak closure of $\mathbb{C}[F_2]$ in $\mathcal{B}(l^2(F_2))$. One easily checks that $\mathrm{tr}_{\mathcal{N}(F_2)}(a)$ and $\langle (\rho_+(a) - \rho_-(a))(x), x \rangle$ are weakly continuous functionals on $a \in \mathcal{N}(G)$ for each $x \in E$. This implies for $a \in \mathcal{N}(F_2)$ and $x \in V$ with $x \neq x_0$

$$\langle (\rho_+(a) - \rho_-(a))(x_0), x_0 \rangle = \mathrm{tr}_{\mathcal{N}(F_2)}(a);$$
$$\langle (\rho_+(a) - \rho_-(a))(x), x \rangle = 0.$$

Now assertion (1) follows by summing over $x \in E$.

(2) By Lemma 10.49 we have

$$\rho_+(g)(x) = gx = f^{-1}(gf(x)) = \rho_-(g)(x)$$

for all $x \in V$ with a finite number of exceptions. Hence $\rho_+(g) - \rho_-(g)$ has finite rank for $g \in F_2$. By linearity $\rho_+(a) - \rho_-(a)$ has finite rank for $a \in \mathbb{C}[F_2]$.

(3) This follows from (1) and Lemma 10.45 (3).   □

Now we are ready to prove Lemma 10.43.

Proof. Since any finitely generated free group occurs as a subgroup in F_2 and any finitely generated subgroup of a free group is a finitely generated free group, it suffices to prove the strong Atiyah Conjecture only for F_2 by Lemma 10.4. Since for any matrix $B \in M(m, n, \mathbb{C}[F_2])$ the matrix BB^* is a square matrix over $\mathbb{C}[F_2]$ and $\ker\left(r_{BB^*}^{(2)}\right) = \ker\left(r_B^{(2)}\right)$ holds, it suffices to prove for each square matrix $A \in M_n(\mathbb{C}[F_2])$

$$\dim_{\mathcal{N}(F_2)}\left(\ker\left(r_A^{(2)} : l^2(F_2)^n \to l^2(F_2)^n\right)\right) \in \mathbb{Z}.$$

Consider the 0-summable $(\mathcal{N}(F_2), \mathbb{C}F_2)$-Fredholm module of Lemma 10.50 (2). By 0-summability we know $\mathbb{C}[F_2] \subset \mathcal{N}(F_2)^0$. From Lemma 10.45 (3) we conclude $M_n(\mathbb{C}[F_2]) \subset M_n(\mathcal{N}(F_2)^0) \subset M_n(\mathcal{N}(F_2))^0$. Hence $\rho_+ \otimes \rho_{\mathrm{st}}(A) - \rho_- \otimes \rho_{\mathrm{st}}(A)$ is a finite rank operator. We conclude from Lemma 10.47 that $p_{\ker(\rho_+ \otimes \rho_{\mathrm{st}}(A))} - p_{\ker(\rho_- \otimes \rho_{\mathrm{st}}(A))}$ is of finite rank. Since up to the summand $\mathbb{C}$ we are dealing with sums of regular representations, we get $\rho_+ \otimes \rho_{\mathrm{st}}\left(p_{\ker(r_A^{(2)})}\right) = p_{\ker(\rho_+ \otimes \rho_{\mathrm{st}}(A))}$ and $\rho_- \otimes \rho_{\mathrm{st}}\left(p_{\ker(r_A^{(2)})}\right) \oplus p_{\mathbb{C}^n} = p_{\ker(\rho_- \otimes \rho_{\mathrm{st}}(A))}$, where $p_{\mathbb{C}^n}$ is the projection onto the summand $\mathbb{C}^n$. Hence $p_{\ker(r_A^{(2)})} \in M_n(\mathcal{N}(F_2))^0 \subset M_n(\mathcal{N}(F_2))^1$. Lemma 10.50 (3) implies

$$\dim_{\mathcal{N}(F_2)}\left(\ker\left(r_A^{(2)}\right)\right) = \mathrm{tr}_{\mathcal{N}(F_2)}\left(p_{\ker(r_A^{(2)})}\right) = \tau\left(p_{\ker(r_A^{(2)})}\right).$$

We conclude from Lemma 10.46 that $\tau\left(p_{\ker(r_A^{(2)})}\right)$ is an integer. This finishes the proof of Lemma 10.43. $\square$

Lemma 10.39 and Lemma 10.43 imply

Lemma 10.51. *Let G be a free group. Then $D(G)$ is a skewfield and in particular statement (1) in Lemma 10.41 holds.*

We will give a different proof for the Atiyah Conjecture 10.3 of order $\mathbb{Z}$ with coefficients in $\mathbb{Q}$ for any free group F in Example 13.5, which is based on approximation techniques.

Remark 10.52. The strong Atiyah Conjecture 10.2 for a virtually free group G follows if one knows it for free groups by the following argument. It suffices to treat the case where G is finitely generated by Lemma 10.4. If d is the least common multiple of the orders of finite subgroups of G, then one can find a finitely generated free subgroup $F \subset G$ of index d. For a proof of this well-known fact we refer for instance to [24, Theorem 8.3 and Theorem 8.4], [461, Theorem 5 in Section 7 on page 747]. For any matrix $A \in M(m, n, \mathbb{C}G)$ we get from Theorem 1.12 (6)

$$\dim_{\mathcal{N}(G)}(\ker(r_A^{(2)})) = \frac{1}{d} \cdot \dim_{\mathcal{N}(F)}(\ker(r_A^{(2)})).$$

10.3.2 Survey on Crossed Products

In this subsection we will introduce the concept of crossed product which we will need later.

Let R be a ring and let G be a group. Suppose that we are given maps of sets $c\colon G \to \mathrm{aut}(R)$, $g \mapsto c_g$ and $\tau\colon G \times G \to R^{\mathrm{inv}}$ satisfying

$$c_{\tau(g,g')} \circ c_{gg'} = c_g \circ c_{g'};$$
$$\tau(g,g') \cdot \tau(gg',g'') = c_g(\tau(g',g'')) \cdot \tau(g,g'g'')$$

for $g,g',g'' \in G$, where $c_{\tau(g,g')}\colon R \to R$ is conjugation with $\tau(g,g')$, i.e. it sends r to $\tau(g,g')r\tau(g,g')^{-1}$. Let $R*G = R*_{c,\tau} G$ be the free R-module with the set G as basis. It becomes a ring with the following multiplication

$$\left(\sum_{g \in G} \lambda_g g\right) \cdot \left(\sum_{h \in G} \mu_h h\right) = \sum_{g \in G}\left(\sum_{\substack{g',g'' \in G, \\ g'g''=g}} \lambda_{g'} c_{g'}(\mu_{g''})\tau(g',g'')\right) g.$$

This multiplication is uniquely determined by the properties $g \cdot r = c_g(r) \cdot g$ and $g \cdot g' = \tau(g,g') \cdot (gg')$. The two conditions above relating c and τ are equivalent to the condition that this multiplication is associative. We call $R*G = R*_{c,\tau} G$ the *crossed product* of R and G with respect to c and τ.

Example 10.53. Let $1 \to H \overset{i}{\to} G \overset{p}{\to} Q \to 1$ be an extension of groups. Let $s\colon Q \to G$ be a map satisfying $p \circ s = \mathrm{id}$. We do not require s to be a group homomorphism. Let R be a commutative ring. Define $c\colon Q \to \mathrm{aut}(RH)$ by $c_q(\sum_{h \in H} \lambda_h h) = \sum_{h \in H} \lambda_h s(q)hs(q)^{-1}$. Define $\tau\colon Q \times Q \to (RH)^{\mathrm{inv}}$ by $\tau(q,q') = s(q)s(q')s(qq')^{-1}$. Then we obtain a ring isomorphism $RH*Q \to RG$ by sending $\sum_{q \in Q} \lambda_q q$ to $\sum_{q \in Q} i(\lambda_q)s(q)$, where $i\colon RH \to RG$ is the ring homomorphism induced by $i\colon H \to G$. Notice that s is a group homomorphism if and only if τ is constant with value $1 \in R$.

Example 10.54. Let $\phi\colon R \to R$ be a ring automorphism. Fix a generator $t \in \mathbb{Z}$. Define $\bar{c}\colon \mathbb{Z} \to \mathrm{aut}(R)$ by $\bar{c}(t^n) = \phi^n$ and let $\bar{\tau}\colon \mathbb{Z} \times \mathbb{Z} \to R^{\mathrm{inv}}$ be the constant map with value 1. Then $R*_{\bar{c},\bar{\tau}} \mathbb{Z}$ is the *ϕ-twisted Laurent ring* and sometimes denoted by $R[t,t^{-1}]_\phi$. Given any other data $c\colon \mathbb{Z} \to \mathrm{aut}(R)$ and $\tau\colon \mathbb{Z} \times \mathbb{Z} \to R^{\mathrm{inv}}$ for a crossed product structure, there is an isomorphism of rings

$$R[t,t^{-1}]_\phi \overset{\cong}{\to} R*_{c,\tau} \mathbb{Z}$$

which sends $\sum_{n \in \mathbb{Z}} \lambda_n t^n$ to $\sum_{n \in \mathbb{Z}} (\lambda_n r_n) t^n$, where

$$
\begin{aligned}
r_n &= \tau(t,t)\tau(t^2,t)\ldots\tau(t^{n-1},t) && \text{for } n \geq 2; \\
r_n &= 1 && \text{for } n = 0,1; \\
r_n &= \tau(t^{-1},t)c_{t^{-1}}(\tau(t^{-1},t))\tau(t^{-1},t^{-1})\ldots \\
&\quad c_{t^{n+1}}(\tau(t^{-1},t))\tau(t^{n+1},t^{-1}) && \text{for } n \leq -2.
\end{aligned}
$$

The crossed product $R *_{c,\tau} G$ has the following universal property. Let S be a ring together with a map $\nu: G \to S^{\mathrm{inv}}$. Denote by $\tau_\nu: G \times G \to S^{\mathrm{inv}}$ the map which sends (g, g') to $\nu(g)\nu(g')\nu(gg')^{-1}$. Suppose that we are given a ring homomorphism $f: R \to S$ satisfying $c_{\nu(g)} \circ f = f \circ c_g$ and $\tau_\nu = f \circ \tau$, where $c_{\nu(g)}: S \to S$ sends s to $\nu(g)s\nu(g)^{-1}$. Then there is exactly one ring homomorphism called *crossed product homomorphism* $F = f *_\nu \mathrm{id}: R*_{c,\tau} G \to S$ with the properties that $F(r) = f(r)$ for $r \in R$ and $F(g) = \nu(g)$. It sends $\sum_{g \in G} \lambda_g g$ to $\sum_{g \in G} f(\lambda_g)\nu(g)$.

If $\overline{R}*_{\overline{c},\overline{\tau}}G$ is another crossed product and $f: R \to \overline{R}$ a ring homomorphism satisfying $\overline{c}_g \circ f = f \circ c_g$ and $\overline{\tau} = f \circ \tau$, then by the universal property above there is exactly one ring homomorphism, also called crossed product homomorphism, $F = f * \mathrm{id}: R*_{c,\tau} G \to \overline{R}*_{\overline{c},\overline{\tau}}G$ which in uniquely determined by the properties that $F(r) = f(r)$ for $r \in R$ and $F(g) = g$ for $g \in G$.

Next we collect the basic properties of crossed products which we will need later. Recall that a ring R is *(left) Artinian* if any descending sequence of finitely generated (left) R-modules $M_1 \supset M_2 \supset M_2 \supset \dots$ becomes stationary, i.e. there is an integer n with $M_m = M_{m+1}$ for $m \geq n$. A ring R is *(left) Noetherian* if any ascending sequence of finitely generated (left) R-modules becomes stationary. It suffices to check these condition for ideals. A ring R is (left) Noetherian if and only if any submodule of a finitely generated (left) R-module is again finitely generated. A ring R is *semiprime* if for any two-sided ideal $I \subset R$ the implication $I^2 = \{0\} \Rightarrow I = \{0\}$ holds. For more information about these notions we refer for instance to [448].

Lemma 10.55. *Let $R * G$ be a crossed product. Then*

(1) If R is Artinian and G is finite, then $R * G$ is Artinian;
(2) If R is Noetherian and G is virtually poly-cyclic, then $R*G$ is Noetherian;
(3) If R is semisimple of characteristic 0, then $R * G$ is semiprime;
(4) If R is semisimple of characteristic 0 and G is finite, then $R * G$ is semisimple.

Proof. (1) An ascending chain of finitely generated $R * G$-modules can be viewed as an ascending chain of finitely generated R-modules since G is by assumption finite. Hence it is stationary since R is Artinian by assumption.

(2) The same argument as in the proof of assertion (1) shows that $R * G$ is Noetherian if R is Noetherian and G is finite. A virtually poly-cyclic group can be obtained by iterated extensions with infinite cyclic groups or finite groups as quotient. If G is an extension $1 \to H \to G \to Q \to 1$, one gets $R * G \cong (R * H) * Q$. Hence it remains to treat the case where $Q = \mathbb{Z}$. By Example 10.54 it suffices to show that the twisted Laurent ring $R[t, t^{-1}]_\phi$ is Noetherian for an automorphism $\phi: R \to R$ of a Noetherian ring R. This follows from [446, Proposition 3.1.13 on page 354, Proposition 3.5.2 on page 395].

(3) This follows from [417, Theorem I].

(4) A ring is semiprime and Artinian if and only if it is semisimple [448, Theorem 2.3.10 on page 139]. Now apply assertions (2) and (3). $\qquad\square$

Lemma 10.56. *Let $R \subset S$ be an inclusion of rings and let $R * G \to S * G$ be an inclusion of crossed products.*

(1) *The subring of $S * G$ generated by $\mathcal{D}(R \subset S)$ and G in $S * G$ is a crossed product $\mathcal{D}(R \subset S) * G$ such that $R * G \subset \mathcal{D}(R \subset S) * G$ is an inclusion of crossed products. The analogous statement holds for $\mathcal{R}(R \subset S)$;*

(2) *Let Σ be a set of matrices of R which is invariant under the automorphisms $c_g \colon R \to R$ for all $g \in G$. Let $i \colon R \to R_\Sigma$ be universal Σ-inverting. Then $c \colon G \times G \to \operatorname{aut}(R)$ induces $c^\Sigma \colon G \to \operatorname{aut}(R_\Sigma)$ such that $i \circ c_g = c_g^\Sigma \circ i$ holds for all $g \in G$ and the map of crossed products $i * G \colon R * G \to R_\Sigma * G$ is universal Σ-inverting.*

Proof. (1) Since $R * G \subset S * G$ is by assumption an inclusion of crossed products, we get for the crossed product structure maps $c^R \colon G \to \operatorname{aut}(R)$, $\tau^R \colon G \times G \to R^{\operatorname{inv}}$, $c^S \colon G \to \operatorname{aut}(S)$ and $\tau^S \colon G \times G \to S^{\operatorname{inv}}$ that $c_g^S \colon S \to S$ induces $c_g^R \colon R \to R$ and $\tau^R = \tau^S$. By the universal property of the division closure c_g^S induces also an automorphism $c_g^{\mathcal{D}(R \subset S)} \colon \mathcal{D}(R \subset S) \to \mathcal{D}(R \subset S)$. Thus we can form the crossed product $\mathcal{D}(R \subset S) * G$ with respect to $c^{\mathcal{D}(R \subset S)} \colon G \times G \to \operatorname{aut}(\mathcal{D}(R \subset S))$ and $\tau^{\mathcal{D}(R \subset S)} \colon G \times G \to \mathcal{D}(R \subset S)^{\operatorname{inv}}$, where $\tau^{\mathcal{D}(R \subset S)}$ is given by τ^R. There is an obvious map from $\mathcal{D}(R \subset S) * G$ to $S * G$ whose image is the subring generated by $\mathcal{D}(R \subset S)$ and G. One easily checks that it is injective since $S * G$ is the free S-module with G as basis.

(2) The extension c^Σ exists by the universal property of $i \colon R \to R_\Sigma$ since $c_g(\Sigma) = \Sigma$ holds for $g \in G$ by assumption. From the universal property of the crossed product we deduce that $i * G$ is universal Σ-inverting, since i is universal Σ-inverting by assumption. $\qquad\square$

We conclude from Lemma 10.56 (using Notation 10.37)

Lemma 10.57. *Let $1 \to H \to G \to Q \to 1$ be an extension of groups. Let $s \colon Q \to G$ be a set-theoretic section.*

(1) *The subring of $\mathcal{U}(G)$ generated by $\mathcal{D}(H)$ and $s(Q)$ is a crossed product $\mathcal{D}(H) * Q$ such that*

$$\mathbb{C}G = \mathbb{C}H * Q \subset \mathcal{D}(H) * Q \subset \mathcal{U}(H) * Q \subset \mathcal{U}(G)$$

is given by inclusions of rings or crossed products. The analogous statement holds for $\mathcal{R}(H)$;

(2) *If $\mathbb{C}H \to \mathcal{R}(H)$ is universal $\Sigma(H)$-inverting, then $\mathbb{C}G = \mathbb{C}H * Q \to \mathcal{R}(H) * Q$ is universal $\Sigma(H)$-inverting;*

Lemma 10.58. *Let $1 \to H \to G \to Q \to 1$ be an extension of groups. Then*

$$\mathcal{D}(\mathcal{D}(H) * Q \subset \mathcal{U}(G)) = \mathcal{D}(G);$$
$$\mathcal{R}(\mathcal{R}(H) * Q \subset \mathcal{U}(G)) = \mathcal{R}(G);$$

Proof. Since $\mathcal{U}(H)$ is von Neumann regular (see Theorem 8.22 (3)) and hence division closed and rationally closed by Lemma 10.34 (1), we get an inclusion

$$\mathcal{D}(H) = \mathcal{D}(\mathbb{C}H \subset \mathcal{U}(H)) = \mathcal{D}(\mathbb{C}H \subset \mathcal{U}(G)) \subset \mathcal{D}(\mathbb{C}G \subset \mathcal{U}(G)) = \mathcal{D}(G).$$

Fix a set-theoretic section $s \colon Q \to G$. Since $\mathcal{D}(H) * Q$ can be viewed as the subring of $\mathcal{U}(G)$ generated by $\mathcal{D}(H)$ and $s(Q)$ by Lemma 10.57 (1) and $s(Q)$ lies in $\mathcal{D}(G)$, we get $\mathcal{D}(H) * Q \subset \mathcal{D}(G)$. This implies $\mathcal{D}(\mathcal{D}(H) * Q \subset \mathcal{U}(G)) \subset \mathcal{D}(\mathcal{D}(G) \subset \mathcal{U}(G)) = \mathcal{D}(G)$. Since $\mathcal{D}(G) = \mathcal{D}(\mathbb{C}G \subset \mathcal{U}(G)) = \mathcal{D}(\mathbb{C}H * Q \subset \mathcal{U}(G)) \subset \mathcal{D}(\mathcal{D}(H) * Q \subset \mathcal{U}(G))$ obviously holds, Lemma 10.58 follows for $\mathcal{D}(G)$. The proof for $\mathcal{R}(G)$ is analogous. $\square$

10.3.3 Property (R) Ascends to Finite Extensions

Lemma 10.59. *Statement (2) in Lemma 10.41 is true, i.e. if $1 \to H \to G \to Q \to 1$ is an extension of groups such that $\mathcal{D}(H)$ is semisimple and Q is finite, then $\mathcal{D}(G)$ is semisimple. Moreover, $\mathcal{D}(G)$ agrees with $\mathcal{D}(H) * Q$.*

Proof. We conclude from Lemma 10.55 (4) that $\mathcal{D}(H) * Q$ is semisimple. In particular $\mathcal{D}(H) * Q$ is von Neumann regular. From Lemma 10.34 (1) and Lemma 10.58 we conclude $\mathcal{D}(G) = \mathcal{D}(H) * Q$. $\square$

10.3.4 Property (R) Ascends to Extensions by Infinite Cyclic Groups

This subsection is devoted to the proof of

Lemma 10.60. *Statement (3) in Lemma 10.41 is true, i.e. if $1 \to H \to G \to \mathbb{Z} \to 1$ is an extension of groups such that $\mathcal{D}(H)$ is semisimple, then $\mathcal{D}(G)$ is semisimple. Moreover, $\mathrm{NZD}(\mathcal{D}(H) * \mathbb{Z}) = T(\mathcal{D}(H) * \mathbb{Z} \subset \mathcal{U}(G))$, the Ore localization $(\mathcal{D}(H) * \mathbb{Z}) \mathrm{NZD}(\mathcal{D}(H) * \mathbb{Z})^{-1}$ exists and is isomorphic to $\mathcal{D}(G)$.*

Here and elsewhere $\mathrm{NZD}(R)$ denotes the set of non-zero-divisors of a ring R. Its proof needs the following ingredients. We will fix throughout this subsection an extension $1 \to H \to G \to \mathbb{Z} \to 1$ together with a section $s \colon \mathbb{Z} \to G$ which is a group homomorphism.

Theorem 10.61 (Goldie's Theorem). *Let R be a Noetherian semiprime ring. Then $(R, \mathrm{NZD}(R))$ satisfies the Ore condition and the Ore localization $R \, \mathrm{NZD}(R)^{-1}$ is semisimple.*

Proof. [119, Section 9.4]. $\square$

We will need the following consequence

Lemma 10.62. *Let R be a semisimple ring. Let $R * \mathbb{Z}$ be a crossed product which is contained in the ring S. Suppose that $\mathrm{NZD}(R * \mathbb{Z}) = T(R * \mathbb{Z} \subset S)$. Then the Ore localization $(R * \mathbb{Z}) \, \mathrm{NZD}(R * \mathbb{Z})^{-1}$ exists, is a semisimple ring and embeds into S as the division closure $\mathcal{D}(R * \mathbb{Z} \subset S)$.*

Proof. Since $R * \mathbb{Z}$ is Noetherian and semiprime by Lemma 10.55, we conclude from Goldie's Theorem 10.61 that the Ore localization $(R * \mathbb{Z}) \, \mathrm{NZD}(R * \mathbb{Z})^{-1}$ exists and is semisimple. By Lemma 10.35 (2) $(R * \mathbb{Z}) \, \mathrm{NZD}(R * \mathbb{Z})^{-1}$ is isomorphic to $\mathcal{D}(R * \mathbb{Z} \subset S)$.   □

Theorem 10.63. *Suppose that the first non-vanishing coefficient of the Laurent polynomial $f(z) \in \mathcal{N}(H) * \mathbb{Z}$ is a non-zero-divisor in $\mathcal{N}(H)$. Then f is a non-zero-divisor in $\mathcal{N}(H) * \mathbb{Z}$.*

Proof. This is a special case of [308, Theorem 4].   □

We will need the following conclusion of Theorem 10.63 above.

Lemma 10.64. *Suppose that $f(z) \in \mathcal{U}(H) * \mathbb{Z}$ is of the form $1 + a_1 z + \ldots + a_n z^n$. Then f is invertible in $\mathcal{U}(G)$.*

Proof. Recall from Theorem 8.22 (1) that $\mathcal{U}(H)$ is the Ore localization of $\mathcal{N}(H)$ with respect to the set of non-zero-divisors in $\mathcal{N}(H)$. Hence we can find a non-zero-divisor $s \in \mathcal{N}(H)$ and $g(z) \in \mathcal{N}(H) * \mathbb{Z}$ such that $f(z) = s^{-1} g(z)$. The first non-vanishing coefficient of $g(z)$ is s. By Theorem 10.63 above $g(z)$ is a non-zero-divisor in $\mathcal{N}(G)$ and hence a unit in $\mathcal{U}(G)$. Therefore $f(z) = s^{-1} g(z)$ is a unit in $\mathcal{U}(G)$.   □

Finally we will need

Lemma 10.65. *Let R be a semisimple ring. Let $R * \mathbb{Z}$ be a crossed product which is contained in the ring S. Suppose that polynomials of the form $f(z) = 1 + a_1 z + \ldots + a_n z^n$ in $R * \mathbb{Z}$ become invertible in S. Then $\mathrm{NZD}(R * \mathbb{Z}) = T(R * \mathbb{Z} \subset S)$.*

Before we give the proof of Lemma 10.65, we explain how the main result of this subsection Lemma 10.60 follows from the results above. By assumption $\mathcal{D}(H)$ is semisimple. Lemma 10.64 ensures that the assumptions of Lemma 10.65 are satisfied for $R = \mathcal{D}(H)$ and $S = \mathcal{U}(G)$. Hence $\mathrm{NZD}(\mathcal{D}(H) * \mathbb{Z}) = T(\mathcal{D}(H) * \mathbb{Z} \subset \mathcal{U}(G))$. Lemma 10.62 implies that the Ore localization $(\mathcal{D}(H) * \mathbb{Z}) \, \mathrm{NZD}(\mathcal{D}(H) * \mathbb{Z})^{-1}$ exists, is semisimple and agrees with $\mathcal{D}(\mathcal{D}(H) * \mathbb{Z} \subset \mathcal{U}(G))$. Since $\mathcal{D}(\mathcal{D}(H) * \mathbb{Z} \subset \mathcal{U}(G))$ is $\mathcal{D}(G)$ by Lemma 10.58, Lemma 10.60 follows.

It remains to prove Lemma 10.65. Recall that by *Wedderburn's Theorem* [302, Theorem 3.5 in Chapter I on page 35] any semisimple ring R can be written as a finite product $\prod_{i=1}^{r} M_{n_i}(D_i)$ of matrix rings over skewfields D_i.

We begin with the case where R itself is a skewfield. Then any non-zero element in $R * \mathbb{Z}$ can be written up to multiplication with a unit in R and an element of the form z^n in the form $f(z) = 1 + a_1 z + \ldots + a_n z^n$. Hence it is invertible by assumption in S. This proves $\text{NZD}(R * \mathbb{Z}) = T(R * \mathbb{Z} \subset S)$ provided that R is a skewfield.

Next we treat the case where R looks like $M_n(D)$ for a skewfield D. We first change the given crossed product structure (c, τ) to a better one by a construction similar to the one in Example 10.54. We can write $c_t \colon M_n(D) \to M_n(D)$ as a composition $c_u \circ M_n(\theta)$ for some $u \in M_n(D)$ and automorphism $\theta \colon D \to D$, where c_u is conjugation with u and $M_n(\theta)$ is given by applying θ to each entry of a matrix [277, Theorem 8 on page 237]. Define a new crossed product structure $\bar{c} \colon \mathbb{Z} \to \text{aut}(M_n(D))$ and $\bar{\tau} \colon \mathbb{Z} \times \mathbb{Z} \to M_n(D)$ by $\bar{c}_{t^n} = (u^{-1} \circ c_t)^n$ and $\bar{\tau}(t^a, t^b) = 1$. Notice that $M_n(D) *_{c,\tau} \mathbb{Z}$ and $M_n(D) *_{\bar{c},\bar{\tau}} \mathbb{Z}$ are isomorphic. In the sequel we will use the new crossed product structure given by $\bar{c}$ and $\bar{\tau}$. Notice that the assumption about polynomials is invariant under this change of crossed product structure since $1 + a_1 t + \ldots + a_n t^n$ becomes with respect to the new crossed product structure $r_0 + a_1 r_1 t + \ldots + a_n r_n t^n$ for appropriate elements $r_0, r_1, \ldots, r_n$ in $M_n(D)^{\text{inv}}$.

In the sequel we consider $D \subset M_n(D)$ by the diagonal embedding which sends $d \in D$ to the diagonal matrix whose diagonal entries are all d. The crossed product structure on $M_n(D)$ induces one on D and we obtain an inclusion of crossed products $D * \mathbb{Z} \subset M_n(D) * \mathbb{Z}$. In particular $D * \mathbb{Z}$ becomes a subring of S because by assumption $M_n(D) * \mathbb{Z} \subset S$. Since D is a skewfield, we know already $\text{NZD}(D * \mathbb{Z}) = T(D * \mathbb{Z} \subset S)$. By Lemma 10.62 the Ore localization $(D * \mathbb{Z}) \text{NZD}(D * \mathbb{Z})^{-1}$ exists, is semisimple and embeds into S. In the sequel we identify $D * \mathbb{Z}$ as a subring of $M_n(D * \mathbb{Z})$ by the diagonal embedding. Since $\text{NZD}(D * \mathbb{Z}) \subset D * \mathbb{Z}$ satisfies the Ore condition, one easily checks that $\text{NZD}(D * \mathbb{Z}) \subset M_n(D * \mathbb{Z})$ satisfies the Ore condition and the inclusion $M_n(D * \mathbb{Z}) \subset S$ is $\text{NZD}(D * \mathbb{Z})$-inverting and hence induces a homomorphism $M_n(D * \mathbb{Z}) \text{NZD}(D * \mathbb{Z})^{-1} \to S$. It is injective since $M_n(D * \mathbb{Z}) \subset S$ and $(D * \mathbb{Z}) \text{NZD}(D * \mathbb{Z})^{-1} \subset S$ are injective. Hence we get an embedding $M_n(D * \mathbb{Z}) \text{NZD}(D * \mathbb{Z})^{-1} \subset S$. Consider $x \in \text{NZD}(M_n(D) * \mathbb{Z})$. It is a non-zero-divisor when considered in the Ore localization $M_n(D * \mathbb{Z}) \text{NZD}(D * \mathbb{Z})^{-1}$. Since $M_n(D * \mathbb{Z}) \text{NZD}(D * \mathbb{Z})^{-1} = M_n\left((D * \mathbb{Z}) \text{NZD}(D * \mathbb{Z})^{-1}\right)$ is von Neumann regular by Lemma 8.19 (1), any non-zero-divisor in $M_n(D * \mathbb{Z}) \text{NZD}(D * \mathbb{Z})^{-1}$ is invertible by Lemma 8.20 (1). Hence x becomes invertible in $M_n(D * \mathbb{Z}) \text{NZD}(D * \mathbb{Z})^{-1}$ and in particular in S. This shows $\text{NZD}(M_n(D) * \mathbb{Z}) = T(M_n(D) * \mathbb{Z} \subset S)$.

Finally it remains to treat the general case $R = \prod_{i=1}^r M_{n_i}(D_i)$. We will equip $R * \mathbb{Z}$ with the crossed product structure constructed in Example 10.54 for an appropriate automorphism $\phi \colon R \to R$. An automorphism of R permutes the factors $M_{n_i}(R)$. Hence we can find a positive integer m such that ϕ^m leaves the various factors invariant and can be written as $\prod_{i=1}^r \phi_i$ for appropriate automorphisms $\phi_i \colon M_{n_i}(D_i) \to M_{n_i}(D_i)$. This induces decompositions

$$R * m\mathbb{Z} = \prod_{i=1}^{r} \left(M_{n_i}(D_i) * \mathbb{Z} \right);$$

$$\mathrm{NZD}(R * m\mathbb{Z}) = \prod_{i=1}^{r} \mathrm{NZD}\left(M_{n_i}(D_i) * \mathbb{Z} \right);$$

$$T(R * m\mathbb{Z} \subset S) = \prod_{i=1}^{r} T\left(M_{n_i}(D_i) * \mathbb{Z} \subset S \right);$$

We have already proved $\mathrm{NZD}\left(M_{n_i}(D_i) * \mathbb{Z} \right) = T\left(M_{n_i}(D_i) * \mathbb{Z} \subset S \right)$ for $i = 1, 2 \ldots r$. This implies $\mathrm{NZD}(R * m\mathbb{Z}) = T(R * m\mathbb{Z} \subset S)$. We conclude from Lemma 10.62 that the Ore localization $(R * m\mathbb{Z}) \mathrm{NZD}(R * m\mathbb{Z})^{-1}$ exists, is semisimple and embeds as the division closure $\mathcal{D}(R * m\mathbb{Z} \subset S)$ into S. One easily checks that $\left((R * m\mathbb{Z}) \mathrm{NZD}(R * m\mathbb{Z})^{-1} \right) * \mathbb{Z}/m$ is isomorphic to $\left((R * m\mathbb{Z}) * \mathbb{Z}/m \right) \mathrm{NZD}(R * m\mathbb{Z})^{-1}$. We conclude from Lemma 10.55 (4) that $\left((R * m\mathbb{Z}) * \mathbb{Z}/m \right) \mathrm{NZD}(R * m\mathbb{Z})^{-1}$ is semisimple. Since $R * \mathbb{Z}$ agrees with $(R * m\mathbb{Z}) * \mathbb{Z}/m$ and $R * \mathbb{Z} \subset S$ by assumption, we get inclusions $R * \mathbb{Z} \subset (R * \mathbb{Z}) \mathrm{NZD}(R * m\mathbb{Z})^{-1} \subset S$ and $(R * \mathbb{Z}) \mathrm{NZD}(R * m\mathbb{Z})^{-1}$ is semisimple. Since any non-zero-divisor in $R * \mathbb{Z}$ is a non-zero-divisor in the Ore localization $(R * \mathbb{Z}) \mathrm{NZD}(R * m\mathbb{Z})^{-1}$ and any non-zero-divisor in a semisimple ring is a unit, we get $\mathrm{NZD}(R * Z) = T(R * \mathbb{Z} \subset S)$. This finishes the proof of Lemma 10.65 and hence of Lemma 10.60. $\qquad \square$

10.3.5 Property (K) and Extensions by Virtually Finitely Generated Abelian Groups

The main result of this section will be

Lemma 10.66. *The statement* (5) *in Lemma 10.41 is true.*

Again the proof will need some preparation. The main ingredient will be the next result. For its proof we refer to [383], [384], or to [112], [188] and [418, Chapter 8]. We have introduced the Grothendieck group of finitely generated R-modules $G_0(R)$ in Definition 9.11.

Theorem 10.67 (Moody's Induction Theorem). *Let R be a Noetherian ring and Q be a virtually finitely generated abelian group. Then the induction map*

$$\mathrm{colim}_{Q/K \in \mathrm{Or}(Q, \mathcal{FIN})} G_0(R * K) \rightarrow G_0(R * Q)$$

is surjective.

However, more input is needed. We begin with the following general result about iterated localization.

Lemma 10.68. *Let $R \subset S$ be a ring extension and $T \subset T(R \subset S)$ be a multiplicatively closed subset. Assume that the pairs (R, T) and $(RT^{-1}, T(RT^{-1} \subset S))$ satisfy the Ore condition. Then also $(R, T(R \subset S))$ satisfies the Ore condition and $RT(R \subset S)^{-1} = (RT^{-1})T(RT^{-1} \subset S)^{-1}$. We get inclusions*

$$R \subset (RT^{-1})T(RT^{-1} \subset S)^{-1} = RT(R \subset S)^{-1} \subset S.$$

If we assume in addition that $T(RT^{-1} \subset S) = \mathrm{NZD}(RT^{-1})$, then $T(R \subset S) = \mathrm{NZD}(R)$.

Proof. We begin with verifying the Ore condition for $(R, T(R \subset S))$. Given $r_1 \in R$ and $t_1 \in T$, we must find $r_2 \in R$ and $t_2 \in T$ with $t_1 r_2 = r_1 t_2$. Since $(RT^{-1}, T(RT^{-1} \subset S))$ satisfies the Ore condition, we can find $r_3, r_4 \in R$ and $t_3, t_4 \in T$ satisfying $t_1 r_3 t_3^{-1} = r_1 r_4 t_4^{-1}$ in RT^{-1} and $r_4 t_4^{-1} \in T(RT^{-1} \subset S)$. Since (R, T) satisfies the Ore condition, we can find $r_5 \in R$ and $t_5 \in T$ satisfying $t_3 t_5 = t_4 r_5$. This implies $t_1 r_3 t_5 = r_1 r_4 r_5$ and $r_5 \in T(R \subset S)$. Put $r_2 = r_3 t_5$ and $t_2 = r_4 r_5$. Since $r_4 t_4^{-1} \in T(RT^{-1} \subset S)$ and $r_5 \in T(R \subset S)$ holds, we get $t_2 \in T(R \subset S)$.

One easily checks that the composition $R \to RT^{-1} \to (RT^{-1})T(RT^{-1} \subset S)^{-1}$ is universally $T(R \subset S)$-inverting. Since the same holds for $R \to RT(R \subset S)^{-1}$, we conclude $RT(R \subset S)^{-1} = (RT^{-1})T(RT^{-1} \subset S)^{-1}$. Since a non-zero-divisor in R is a non-zero-divisor in RT^{-1}, we conclude from the assumption $T(RT^{-1} \subset S) = \mathrm{NZD}(RT^{-1})$ that $T(R \subset S) = \mathrm{NZD}(R)$ holds. $\square$

Lemma 10.69. *Let $1 \to H \to G \to Q \to 1$ be an extension of groups such that Q is virtually finitely generated abelian. Suppose that $\mathcal{D}(H)$ is semisimple. Then $\mathrm{NZD}(\mathcal{D}(H) * Q) = T(\mathcal{D}(H) * Q \subset \mathcal{U}(G))$, and the Ore localization $(\mathcal{D}(H) * Q) \mathrm{NZD}(\mathcal{D}(H) * Q)^{-1}$ exists, is semisimple and agrees with $\mathcal{D}(G)$.*

Proof. We begin with the special case where Q is finite. Then $\mathcal{D}(G) = \mathcal{D}(H) * Q$ and is semisimple by Lemma 10.59. Since in a semisimple ring any non-zero-divisor is a unit, we conclude

$$\mathrm{NZD}(\mathcal{D}(H) * Q) = (\mathcal{D}(H) * Q)^{\mathrm{inv}} = T(\mathcal{D}(H) * Q \subset \mathcal{U}(G)).$$

In particular $\mathcal{D}(H) * Q = (\mathcal{D}(H) * Q) \mathrm{NZD}(\mathcal{D}(H) * Q)^{-1}$. So the claim holds for finite Q.

If Q is infinite cyclic, the claim follows from Lemma 10.60.

Any extension by a virtually finitely generated abelian group can be replaced by an iterated extension by infinite cyclic groups and by finite groups. Hence we must ensure that we can iterate. But this follows from Lemma 10.68. $\square$

Lemma 10.70. *(1) Let $T \subset R$ be a set of non-zero-divisors. Suppose that (R, T) satisfies the Ore condition. Then the inclusion $R \to RT^{-1}$ induces an epimorphism $G_0(R) \to G_0(RT^{-1})$;*

(2) If R is semisimple, then the forgetful map $K_0(R) \to G_0(R)$ is bijective.

Proof. (1) Let M be a finitely generated RT^{-1}-module. Choose an epimorphism $f: (RT^{-1})^n \to M$. Let $i: R^n \to (RT^{-1})^n$ be the obvious injection. Now $N = f \circ i(R^n)$ is a finitely generated R-submodule of M. The inclusion $j: N \to M$ induces an injection $\mathrm{id} \otimes_R j: RT^{-1} \otimes_R N \to RT^{-1} \otimes_R M$ since $RT^{-1} \otimes_R -$ is exact by Lemma 8.15 (3). Since f is surjective, $\mathrm{id} \otimes_R j$ is bijective. Since for any RT^{-1}-module L we have $RT^{-1} \otimes_R L \cong_{RT^{-1}} L$, the RT^{-1}-modules M and $RT^{-1} \otimes_R N$ are isomorphic.

(2) Any finitely generated module over a semisimple ring is finitely generated projective. $\qquad \square$

Lemma 10.71. *Let $1 \to H \to G \xrightarrow{p} Q \to 1$ be an extension of groups such that Q is virtually finitely generated abelian and $\mathcal{D}(H)$ is semisimple. Then the map*

$$G_0(\mathcal{D}(H) * Q) \to G_0(\mathcal{D}(G))$$

is surjective.

Proof. This follows from Lemma 10.69 and Lemma 10.70 (1). $\qquad \square$

Now we are ready to prove the main result of this subsection Lemma 10.66. We have to show for an extension of groups $1 \to H \to G \to Q \to 1$ with virtually finitely generated abelian Q that the canonical map

$$\mathrm{colim}_{G/G_0 \in \mathrm{Or}(G, \mathcal{FIN})} K_0(\mathbb{C}G_0) \to K_0(\mathcal{D}(G))$$

is surjective, provided that for any group K with $H \subset K \subset G$ and $[K : H] < \infty$ the canonical map

$$\mathrm{colim}_{K/K_0 \in \mathrm{Or}(K, \mathcal{FIN})} K_0(\mathbb{C}K_0) \to K_0(\mathcal{D}(K))$$

is surjective and $\mathcal{D}(K)$ is semisimple. We will show that the map of interest can be written as the following composition of maps which are surjective or even bijective:

$$\mathrm{colim}_{G/G_0 \in \mathrm{Or}(G, \mathcal{FIN})} K_0(\mathbb{C}G_0)$$

$$\xrightarrow{f_1} \mathrm{colim}_{Q/Q_0 \in \mathrm{Or}(Q, \mathcal{FIN})} \mathrm{colim}_{p^{-1}(Q_0)/L \in \mathrm{Or}(p^{-1}(Q_0), \mathcal{FIN})} K_0(\mathbb{C}L)$$

$$\xrightarrow{f_2} \mathrm{colim}_{Q/Q_0 \in \mathrm{Or}(Q, \mathcal{FIN})} K_0(\mathcal{D}(p^{-1}(Q_0)))$$

$$\xrightarrow{f_3} \mathrm{colim}_{Q/Q_0 \in \mathrm{Or}(Q, \mathcal{FIN})} G_0(\mathcal{D}(p^{-1}(Q_0)))$$

$$\xrightarrow{f_4} \mathrm{colim}_{Q/Q_0 \in \mathrm{Or}(Q, \mathcal{FIN})} G_0(\mathcal{D}(H) * Q_0)$$

$$\xrightarrow{f_5} G_0(\mathcal{D}(H) * Q)$$

$$\xrightarrow{f_6} G_0(\mathcal{D}(G))$$

$$\xrightarrow{f_7} K_0(\mathcal{D}(G))$$

The isomorphism f_1 is due to the fact that any finite subgroup $G_0 \subset G$ occurs as finite subgroup in $p^{-1}(p(G_0))$ and $p(G_0)$ is finite. The map f_2 is bijective by assumption. The map f_3 is bijective by Lemma 10.70 (2) since $\mathcal{D}(p^{-1}(Q_0))$ is semisimple by assumption. The map f_4 is an isomorphism since $\mathcal{D}(H) * Q_0 = \mathcal{D}(p^{-1}(Q_0))$ holds by Lemma 10.59. The map f_5 is surjective by Moody's Induction Theorem 10.67. The map f_6 is surjective by Lemma 10.71. The map f_7 is an isomorphism by Lemma 10.70 (2) since $\mathcal{D}(G)$ is semisimple by Lemma 10.59 and Lemma 10.60. This finishes the proof of Lemma 10.66. $\qquad\square$

10.3.6 Property (K) holds for Virtually Free Groups

The main goal of this subsection is to prove

Lemma 10.72. *Statement (4) in Lemma 10.41 is true, i.e. for any virtually free group G the composition*

$$\mathrm{colim}_{\mathrm{Or}(G,\mathcal{FIN})} K_0(\mathbb{C}H) \xrightarrow{a} K_0(\mathbb{C}G) \xrightarrow{i} K_0(\mathcal{D}(G))$$

is surjective.

The main ingredients in the proof of Lemma 10.72 will be the following two results.

Theorem 10.73. *For a virtually free group the Isomorphism Conjecture 9.40 for $K_0(\mathbb{C}G)$ is true, i.e. the assembly map*

$$a\colon \mathrm{colim}_{G/H \in \mathrm{Or}(G,\mathcal{FIN})} K_0(\mathbb{C}H) \to K_0(\mathbb{C}G)$$

is surjective.

Proof. [309, Lemma 4.8]. $\qquad\square$

Theorem 10.74. *Let R be a hereditary ring with faithful projective rank function $\rho\colon K_0(R) \to \mathbb{R}$ and let Σ be a set of full matrices with respect to ρ. Then the universal Σ-inverting homomorphism $R \to R_\Sigma$ induces an epimorphism $K_0(R) \to K_0(R_\Sigma)$.*

Some explanations about Theorem 10.74 are in order. Recall that a ring is called *hereditary*, if any submodule of a projective module is again projective. A *projective rank function* for R is a function $\rho\colon K_0(R) \to \mathbb{R}$ such that for any finitely generated projective R-module P we have $\rho([P]) \geq 0$ and $\rho([R]) = 1$. It is called *faithful* if $\rho([P]) = 0$ implies $P = 0$. Our main example of a faithful projective rank function will be (see Theorem 8.29)

$$K_0(\mathbb{C}G) \to \mathbb{R}, \qquad [P] \mapsto \dim_{\mathcal{U}(G)}(\mathcal{U}(G) \otimes_{\mathbb{C}G} P).$$

Given an R-map $\alpha\colon P \to Q$ of finitely generated projective R-modules, define $\rho(\alpha)$ to be the infimum over all numbers $\rho(P')$, where P' runs through all finitely generated projective R-modules P' for which there exists a factorization $\alpha: P \to P' \to Q$. We call α *left full* or *right full* respectively if $\rho(\alpha) = \rho(P)$ or $\rho(\alpha) = \rho(Q)$ respectively holds. It is called *full* if it is both left full and right full. Theorem 10.74 is now a special case of [465, Theorem 5.2], since by [465, Lemma 1.1 and Theorem 1.11] a projective rank function is in particular a Sylvester projective function and by [465, Theorem 1.16] there are enough left and right full maps.

We want to apply Theorem 10.74 to the case $R = \mathbb{C}H$ for a virtually finitely generated free group H, the rank function ρ given by $\dim_{\mathcal{U}(H)}$ and $\Sigma = \Sigma(H) := \Sigma(\mathbb{C}H \subset \mathcal{U}(H))$. The trivial $\mathbb{C}H$-module $\mathbb{C}$ has a 1-dimensional projective resolution P_* since this is obviously true for a finitely generated free subgroup $F \subset H$ of finite index $[H : F]$ and $[H : F]$ is invertible in $\mathbb{C}$. Hence any $\mathbb{C}H$-module M has a 1-dimensional projective resolution, namely $P_* \otimes_{\mathbb{C}} M$ with the diagonal H-action. This implies that $\mathbb{C}H$ is hereditary.

Next we check that $\Sigma(H)$ consists of full maps. Suppose that $\alpha: P \to Q$ is a homomorphism of $\mathbb{C}H$-modules such that $\mathrm{id} \otimes_{\mathbb{C}H} \alpha: \mathcal{U}(H) \otimes_{\mathbb{C}H} P \to \mathcal{U}(H) \otimes_{\mathbb{C}H} Q$ is an isomorphism. If α factorizes over the finitely generated projective $\mathbb{C}H$-module P', then the isomorphism $\mathrm{id} \otimes_{\mathbb{C}H} \alpha$ factorizes over $\mathcal{U}(H) \otimes_{\mathbb{C}H} P'$. Theorem 8.29 implies

$$\rho(\alpha) \leq \rho(P) = \dim_{\mathcal{U}(H)}(\mathcal{U}(H) \otimes_{\mathbb{C}H} P) \leq \dim_{\mathcal{U}(H)}(\mathcal{U}(H) \otimes_{\mathbb{C}H} P').$$

Hence $\rho(\alpha) = \rho(P) = \rho(Q)$, i.e. α is full.

Now Theorem 10.74 implies that the map $K_0(\mathbb{C}H) \to K_0(\mathbb{C}H_{\Sigma(H)})$ is surjective for virtually finitely generated free group H. The third ingredient will be the next lemma

Lemma 10.75. *Let G be a virtually free group. Then the inclusion $\mathbb{C}G \to \mathcal{D}(G)$ is universally $\Sigma(G)$-inverting.*

Before we give its proof, we explain how we can derive Lemma 10.72 from it. Namely, we conclude that for any virtually finitely generated free group H the map $K_0(\mathbb{C}H) \to K_0(\mathcal{D}(H))$ is surjective. Let G be a virtually free group G. It is the directed union $G = \{G_i \mid i \in I\}$ of its finitely generated subgroups G_i. Each G_i is virtually finitely generated free. The ring $\mathbb{C}G$ is the directed union of the subrings $\mathbb{C}G_i$ and the ring $\mathcal{D}(G)$ is the directed union of the subrings $\mathcal{D}(G_i)$ (see Lemma 10.83). Since K_0 commutes with colimits, we get

$$K_0(\mathbb{C}G) = \mathrm{colim}_{i \in I} K_0(\mathbb{C}G_i);$$
$$K_0(\mathcal{D}(G)) = \mathrm{colim}_{i \in I} K_0(\mathcal{D}(G_i)).$$

Since a directed colimit of epimorphisms is again an epimorphism, we get

Lemma 10.76. *If G is a virtually free group, then $K_0(\mathbb{C}G) \to K_0(\mathcal{D}(G))$ is surjective.*

Theorem 10.73 and Lemma 10.76 together imply Lemma 10.72. It remains to give the proof of Lemma 10.75. This needs some preparation.

We will need the following notions. A ring homomorphism $f \colon R \to K$ from a ring into a skewfield is called an *R-field*. It is called *epic R-field* if K is generated as a skewfield by the image of f, or, equivalently, for any skewfield K' with $\mathrm{im}(f) \subset K' \subset K$ we have $K = K'$. An epic R-field is called *field of fractions for R* if f is injective. Unfortunately, this notion is not unique, a ring can have several non-isomorphic field of fractions. We have to develop this notion further in order to achieve uniqueness.

A *local homomorphism* from the R-field $f \colon R \to K$ to the R-field $g \colon R \to L$ is an R-algebra homomorphism $u \colon K_0 \to L$, where $K_0 \subset K$ is an R-subalgebra of K and $K_0 - \ker(u) \subset K_0^{\mathrm{inv}}$. The ring K_0 is a local ring with maximal ideal $\ker(u)$. Two local homomorphisms are equivalent if they restrict to a common homomorphism, which is again local. An equivalence class is called *specialization*. An initial object $f \colon R \to K$ in the category of epic R-fields is called a *universal R-field*. If f is a universal R-field with injective f, then it is called *universal R-field of fractions*. The universal R-field is unique up to isomorphism, but even if there exists a field of fractions for R, there need not exist a universal R-field. Moreover, if a universal R-field $f \colon R \to K$ exists, the map f is not necessarily injective, i.e. it is not necessarily a field of fractions for R.

The ring $R = \mathbb{Z}$ has the following epic $\mathbb{Z}$-fields $f_0 \colon \mathbb{Z} \to \mathbb{Q}$ and $f_p \colon \mathbb{Z} \to \mathbb{F}_p$, where $\mathbb{F}_p$ is the finite field of prime order p. Notice that for any R-field $f \colon R \to K$ there is an obvious factorization $f \colon R \to R_{\Sigma(f \colon R \to K)} \xrightarrow{\Psi} K$. In the case f_p it is given by $\mathbb{Z} \to \mathbb{Z}_{(p)} \to \mathbb{F}_p$. The next lemma will give a criterion when Ψ is bijective. A ring R is called *semifir* if every finitely generated submodule of a free module is free and for a free module any two basis have the same cardinality.

Lemma 10.77. *Let R be a semifir. Then there exists a universal field of fractions $f \colon R \to K$ and the canonical map $\Psi \colon R_{\Sigma(f \colon R \to K)} \to K$ is an isomorphism.*

Proof. By [118, Corollary 5.11 in Chapter 7 on page 417] there is a universal field of fractions $f \colon R \to K$ such that every full homomorphism $P \to Q$ of finitely generated projective R-modules becomes an isomorphism over K. Full is meant with respect to the faithful projective rank function given by the isomorphism $K_0(R) \xrightarrow{\cong} \mathbb{Z} \subset \mathbb{R}$. Hence the set of full matrices over R agrees with $\Sigma(R \subset K)$. Hence Ψ is bijective by [118, Proposition 5.7 in Chapter 7 on page 415]. $\square$

Given a free group F, there is a universal field of fractions $f \colon \mathbb{C}F \to K$ by Lemma 10.77 since $\mathbb{C}F$ is a semifir [117, Corollary on page 68], [142].

The next result is shown in [306, Proposition 6]. The notion of Hughes-free is explained below in the proof of Lemma 10.81.

Lemma 10.78. *The universal field of fractions of $\mathbb{C}F$ is Hughes-free.*

A special case of the main theorem in [267] says

Lemma 10.79. *Let F be a free group. Any two $\mathbb{C}F$-fields of fractions which are Hughes-free are isomorphic as $\mathbb{C}F$-fields.*

We conclude from Lemma 10.77, Lemma 10.78 and Lemma 10.79.

Lemma 10.80. *Let F be a free group. Let $\mathbb{C}F \to K$ be a Hughes-free $\mathbb{C}F$-field of fractions. Then it is a universal $\mathbb{C}F$-field of fractions and it is universal $\Sigma(\mathbb{C}F \to K)$-inverting.*

Now we are ready to show

Lemma 10.81. *Let F be a free group. Then $\mathcal{D}(F)$ is a skewfield and the inclusion $\mathbb{C}F \to \mathcal{D}(F)$ is universally $\Sigma(F)$-inverting.*

Proof. We know already that $\mathcal{D}(F)$ is a skewfield by Lemma 10.51. We have $\Sigma(F) = \Sigma(\mathbb{C}F \to \mathcal{D}(F))$. Because of Lemma 10.80 it suffices to show that $\mathbb{C}F \to \mathcal{D}(F)$ is *Hughes-free*. This means the following. Given any finitely generated subgroup $H \subset F$ and an element $t \in H$ together with a homomorphism $p_t \colon H \to \mathbb{Z}$ which maps t to a generator, the set $\{t^i \mid i \in \mathbb{Z}\}$ is $\mathcal{D}(\mathbb{C}[\ker(p_t)] \subset \mathcal{D}(F))$-linear independent in $\mathcal{D}(F)$. Since $\mathcal{U}(\ker(p_t))$ is von Neumann regular by Theorem 8.22 (3), Lemma 10.34 (1) implies

$$\mathcal{D}(\mathbb{C}[\ker(p_t)] \subset \mathcal{D}(F)) = \mathcal{D}(\mathbb{C}[\ker(p_t)] \subset \mathcal{U}(F))$$
$$= \mathcal{D}(\mathbb{C}[\ker(p_t)] \subset \mathcal{U}(\ker(p_t))) = \mathcal{D}(\ker(p_t)).$$

From the split exact exact sequence $1 \to \ker(p_t) \to H \xrightarrow{p_t} \mathbb{Z} \to 1$ and Lemma 10.57 (1) we conclude that the subring of $\mathcal{U}(\ker(p_t)) * \mathbb{Z}$ generated by $\mathcal{D}(\ker(p_t))$ and $\mathbb{Z}$ is itself a crossed product $\mathcal{D}(\ker(p_t)) * \mathbb{Z}$. In particular it is a free $\mathcal{D}(\ker(p_t))$-module with basis $\{t^i \mid i \in \mathbb{Z}\}$.   □

Now we are ready to prove Lemma 10.75. Choose an extension $1 \to F \to G \to Q \to 1$ for a free group F and a finite group Q. We conclude from Lemma 10.51 and Lemma 10.59 that $\mathcal{D}(G)$ is semisimple and $\mathcal{D}(G) = \mathcal{D}(F) * Q$. Since $\mathbb{C}F \to \mathcal{D}(F)$ is universally $\Sigma(F)$-inverting by Lemma 10.81, the inclusion $\mathbb{C}G = \mathbb{C}F * Q \subset \mathcal{D}(F) * Q = \mathcal{D}(G)$ is universally $\Sigma(F)$-inverting by Lemma 10.57 (2). Notice that $\mathcal{D}(G) = \mathcal{R}(G)$ by Lemma 10.34 (2). Hence $\mathbb{C}G \to \mathcal{D}(G)$ is both $\Sigma(G)$-inverting and universally $\Sigma(F)$-inverting. One easily checks that then it is automatically universally $\Sigma(G)$-inverting. This finishes the proof of Lemma 10.75 and hence of the main result of this subsection, Lemma 10.72.

10.3.7 The Induction Step for Directed Unions

In this subsection we want to show

Lemma 10.82. *The statements and (6) and (7) in Lemma 10.41 are true.*

We begin with the following result

Lemma 10.83. *Let the group G be the directed union of subgroups $G = \bigcup_{i \in I} G_i$. Then*

$$\mathcal{D}(G) = \bigcup_{i \in I} \mathcal{D}(G_i);$$

$$\mathcal{R}(G) = \bigcup_{i \in I} \mathcal{R}(G_i).$$

If each $\mathcal{D}(G_i)$ or $\mathcal{R}(G_i)$ respectively is von Neumann regular, then $\mathcal{D}(G)$ or $\mathcal{R}(G)$ respectively is von Neumann regular.

Proof. We have $\mathbb{C}G = \bigcup_{i \in I} \mathbb{C}G_i$. We conclude from Lemma 10.36 that

$$\mathcal{D}\left(\mathbb{C}G \subset \bigcup_{i \in I} \mathcal{U}(G_i) \right) = \bigcup_{i \in I} \mathcal{D}(G_i).$$

Since each $\mathcal{U}(G_i)$ is von Neumann regular (see Theorem 8.22 (3)), $\bigcup_{i \in I} \mathcal{U}(G_i)$ is von Neumann regular. Lemma 10.34 (1) implies

$$\mathcal{D}(G) = \mathcal{D}\left(\mathbb{C}G \subset \bigcup_{i \in I} \mathcal{U}(G_i) \right).$$

The proof for the rational closure is similar. $\qquad\square$

Now we can give the proof of Lemma 10.82. We begin with statement (6) We have to show the surjectivity of the map

$$\operatorname{colim}_{G/H \in \operatorname{Or}(G, \mathcal{FIN})} K_0(\mathbb{C}H) \to K_0(\mathcal{D}(G)),$$

provided that G is a directed union $G = \bigcup_{i \in I} G_i$ and each of the maps $\operatorname{colim}_{G_i/K \in \operatorname{Or}(G, \mathcal{FIN})} K_0(\mathbb{C}K) \to K_0(\mathcal{D}(G_i))$ is surjective. This follows from $\mathbb{C}G = \bigcup_{i \in I} \mathbb{C}G_i$ and $\mathcal{D}(G) = \bigcup_{i \in I} \mathcal{D}(G_i)$ (see Lemma 10.83) and the facts that K_0 commutes with colimits over directed systems and the colimit over a directed system of epimorphisms is again an epimorphism.

Next we prove statement (7). We have to show for G the directed union of subgroups $G = \bigcup_{i \in I} G_i$ that $\mathcal{D}(G)$ is semisimple provided that there is an upper bound on the orders of finite subgroups of G, each $\mathcal{D}(G_i)$ is semisimple and each map $\operatorname{colim}_{G_i/K \in \operatorname{Or}(G_i, \mathcal{FIN})} K_0(\mathbb{C}K) \to K_0(\mathcal{D}(G_i))$ is surjective. We have already shown that the latter condition implies that $\operatorname{colim}_{G/H \in \operatorname{Or}(G, \mathcal{FIN})} K_0(\mathbb{C}H) \to K_0(\mathcal{D}(G))$ is surjective. Because of Lemma 10.28 (2) it suffices to prove that $\mathcal{D}(G)$ is von Neumann regular. This follows from Lemma 10.83, since each $\mathcal{D}(G_i)$ is by assumption semisimple and hence von Neumann regular. This finishes the proof of Lemma 10.82. $\qquad\square$

Notice that we have proved the conditions appearing in Lemma 10.41, namely, statement (1) in Lemma 10.51, statement (2) in Lemma 10.59, statement (3) in Lemma 10.60, statement (5) in Lemma 10.66, statement (4) in Lemma 10.72 and statements (6) and (7) in Lemma 10.82. This proves the strong version of Linnell's Theorem 10.38 and hence Linnell's Theorem 10.19. $\qquad\square$

10.4 Miscellaneous

No counterexamples to the Atiyah Conjecture 10.3 of order $\Lambda = \mathbb{Q}$ with coefficients in $F = \mathbb{C}$ are known to the author at the time of writing. Recall that the Atiyah Conjecture 10.3 of order $\Lambda = \mathbb{Q}$ with coefficients in $F = \mathbb{Q}$ is equivalent to a positive answer to Atiyah's original question whether the L^2-Betti numbers of a G-covering of a closed Riemannian manifold are rational numbers. In [143, Example 5.4] the real number

$$\kappa(1/2, 1/2) = \sum_{k \geq 2} \frac{\phi(k)}{(2^k - 1)^2} = 0, 1659457149\ldots$$

is studied, where $\phi(k)$ denotes the number of primitive k-th roots of unity. It is shown that this number occurs as an L^2-Betti number of the universal covering of a closed manifold. Some evidence is given that this number may not be rational but no proof of the irrationality of this number is known to the author at the time of writing. So there are some doubts about this version of the Atiyah Conjecture. There is some hope that the strong Atiyah Conjecture 10.2 is true under the additional assumption that there is an upper bound on the orders of finite subgroups of G.

At the time of writing there is no proof known to the author that the strong Atiyah Conjecture 10.2 holds for a group G if it is true for a subgroup $H \subset G$ of finite index. Some special results in this direction can be found in [313].

Some of the results of this Chapter 10 have been extended from $\mathbb{Q}$ as coefficients to the field of algebraic numbers $\overline{\mathbb{Q}} \subset \mathbb{C}$. Namely, Theorem 10.20 and Theorem 10.22 have also been proved in the case where one replaces $\mathbb{Q}$ by $\overline{\mathbb{Q}}$ in [148, Theorem 1.6 and Proposition 1.9]. (There only torsionfree groups and $\Lambda = \mathbb{Z}$ is considered, but the same proof applies to the case where Λ is assumed to be closed in $\mathbb{R}$.) This is interesting because of Lemma 10.15 and the fact that a torsionfree group G satisfies the Kaplansky Conjecture for the field $\overline{\mathbb{Q}}$ if and only if is satisfies the Kaplansky Conjecture for the complex numbers $\mathbb{C}$. The well-known proof of this fact can be found for instance in [148, Proposition 4.1].

A group G has the *algebraic eigenvalue property* if for every matrix $A \in M_d(\overline{\mathbb{Q}}G)$ the eigenvalues of the operator $r_A : l^2(G)^d \to l^2(G)^d$ are algebraic. There is the conjecture that every group has this property [148, Conjecture 6.8]. It is based on the following results. If G is a free group, then the series

$$f(z) = \sum_{n=0}^{\infty} \mathrm{tr}_{\overline{\mathbb{Q}}G}(A^n) \cdot z^{-n-1}$$

is an algebraic function over $\overline{\mathbb{Q}}(z)$ and hence each eigenvalue of r_A is an algebraic number since it is a pole of $f(z)$ [456]. Hence a free group has the algebraic eigenvalue property. More generally any group in Linnell's class $\mathcal{C}$

(see Definition 10.18) has the algebraic eigenvalue property [148, Corollary 6.7].

So far we have studied the passage from modules over $\mathbb{C}G$ to $\mathcal{N}(G)$. Let $C_r^*(G)$ be the reduced C^*-algebra of G. We have $\mathbb{C}G \subset C_r^*(G) \subset \mathcal{N}(G)$. We have mentioned in Lemma 10.10 that $\frac{1}{|\mathcal{FIN}(G)|}\mathbb{Z} \subset \Lambda(G, \mathbb{C})_{\text{fgp}}$ and that the Isomorphism Conjecture 9.40 for $K_0(\mathbb{C}G)$ implies $\frac{1}{|\mathcal{FIN}(G)|}\mathbb{Z} = \Lambda(G, \mathbb{C})_{\text{fgp}}$, where $\Lambda(G, \mathbb{C})_{\text{fgp}}$ is the image of the composition

$$K_0(\mathbb{C}G) \xrightarrow{i} K_0(\mathcal{N}(G)) \xrightarrow{\dim_{\mathcal{N}(G)}} \mathbb{R}.$$

Let $\Lambda(G, C_r^*)_{\text{fgp}}$ be the image of the composition

$$K_0(C_r^*(G)) \xrightarrow{i} K_0(\mathcal{N}(G)) \xrightarrow{\dim_{\mathcal{N}(G)}} \mathbb{R}.$$

Obviously

$$\Lambda(G, \mathbb{C})_{\text{fgp}} \subset \Lambda(G, C_r^*)_{\text{fgp}}.$$

If equality would hold, this would suggest to study modules over $C_r^*(G)$ in order to attack the strong Atiyah Conjecture 10.2. This equality was conjectured in [26, page 21]. The question whether this equality holds is related to the Baum-Connes Conjecture (see [26], [27] [257]), which is the analog of the Isomorphism Conjecture 9.40 for $K_0(\mathbb{C}G)$ and known for a much larger class of groups. However, a counterexample to this equality has meanwhile been given by Roy [449]. There a group G is constructed which fits into an exact sequence $1 \to \pi_1(M) \to G \to \mathbb{Z}/3 \times \mathbb{Z}/3 \to 1$ for a closed aspherical 4-dimensional manifold M and whose non-trivial finite subgroups are isomorphic to $\mathbb{Z}/3$, but for which $-\frac{1105}{9}$ belongs to $\Lambda(G, C_r^*)_{\text{fgp}}$. It is not known to the author whether the Isomorphism Conjecture 9.40 for $K_0(\mathbb{C}G)$, the strong Atiyah Conjecture 10.2 or the Baum-Connes Conjecture is true or false for the group G above. The counterexample of Roy does not imply that one of these conjectures is false for this group G.

Let $\mathbb{Z} \subset \mathbb{Z}\left[\frac{1}{|\mathcal{FIN}(G)|}\right] \subset \mathbb{Q}$ be the subring of $\mathbb{Q}$ obtained from $\mathbb{Z}$ by inverting all the orders of finite subgroups of G. One should not confuse the abelian group $\frac{1}{|\mathcal{FIN}(G)|}\mathbb{Z}$ defined in (10.1) with the larger ring $\mathbb{Z}\left[\frac{1}{|\mathcal{FIN}(G)|}\right]$. Provided that G satisfies the Baum-Connes-Conjecture,

$$\Lambda(G, C_r^*)_{\text{fgp}} \subset \mathbb{Z}\left[\frac{1}{|\mathcal{FIN}(G)|}\right]$$

is shown in [339]. Notice that the lamplighter group L satisfies the Baum-Connes Conjecture by [257]) since it it amenable. Hence $\Lambda(L, C_r^*)_{\text{fgp}} \subset \mathbb{Z}\left[\frac{1}{2}\right]$ but $\Lambda(L, \mathbb{C})_{\text{fp}}$ contains $1/3$ (see Theorem 10.23) and hence is not contained in $\mathbb{Z}\left[\frac{1}{2}\right]$ or $\Lambda(L, C_r^*)_{\text{fgp}}$.

The next result is taken from [435, Theorem 9.1].

Theorem 10.84. *Let G be a group such that there is an upper bound on the orders of finite subgroups. Then*

(1) If G belongs to $\mathcal{C}$, then for any $\mathbb{C}G$-module M

$$\mathrm{Tor}_p^{\mathbb{C}G}(M; \mathcal{D}(G)) = 0 \qquad \text{for all } p \geq 2;$$

(2) If G is elementary amenable, then for any $\mathbb{C}G$-module M

$$\mathrm{Tor}_p^{\mathbb{C}G}(M; \mathcal{D}(G)) = 0 \qquad \text{for all } p \geq 1.$$

In other words, the functor $- \otimes_{\mathbb{C}G} \mathcal{D}(G)$ is exact.

Notice that $- \otimes_{\mathbb{Z}G} \mathbb{C}G$ is always exact and $- \otimes_{\mathcal{D}(G)} \mathcal{U}(G)$ is exact provided that $\mathcal{D}(G)$ is semisimple. Consider a group G, which belongs to $\mathcal{C}$ and has an upper bound on the orders of its finite subgroups, and a G-space X. We conclude from the strong version of Linnell's Theorem 10.38, Theorem 10.84 and the universal coefficients spectral sequence that there is an exact sequence

$$0 \to H_n(X; \mathbb{Z}) \otimes_{\mathbb{Z}G} \mathcal{U}(G) \to H_n^G(X; \mathcal{U}(G)) \to \mathrm{Tor}_1^{\mathbb{Z}G}(H_{n-1}(X; \mathbb{Z}), \mathcal{U}(G)) \to 0.$$

In particular we get in the special case $X = EG$

$$H_1^G(EG; \mathcal{U}(G)) = \mathrm{Tor}_1^{\mathbb{Z}G}(\mathbb{Z}, \mathcal{U}(G)); \tag{10.85}$$

$$H_n^G(EG; \mathcal{U}(G)) = 0 \qquad \text{for } n \geq 2. \tag{10.86}$$

Notice that for an infinite group G we have $b_0^{(2)}(G; \mathcal{N}(G)) = 0$ (see Theorem 6.54) (8b)) and hence $\dim_{\mathcal{U}(G)}(H_0^G(EG; \mathcal{U}(G))) = 0$ by Theorem 8.31. We conclude for an infinite group G, which belongs to $\mathcal{C}$ and has an upper bound on the orders of its finite subgroups, that $\chi^{(2)}(G)$ is defined if and only if $\dim_{\mathcal{U}(G)}(\mathrm{Tor}_1^{\mathbb{Z}G}(\mathbb{Z}, \mathcal{U}(G))) < \infty$, and in this case

$$\chi^{(2)}(G) = -\dim_{\mathcal{U}(G)}(\mathrm{Tor}_1^{\mathbb{Z}G}(\mathbb{Z}, \mathcal{U}(G))) \leq 0.$$

Let G be a group in the class $\mathcal{C}$ such that there exists a bound on the orders of finite subgroups. Then $\mathbb{C}G \to \mathcal{D}(G)$ is universally $\Sigma(G)$-inverting (see [435, Theorem 8.3]). The proof of this statement is along the lines of the one of the strong version of Linnell's Theorem 10.38. This is also true for the following statement (see [435, Theorem 8.4]).

Let G be an elementary amenable group such that there exists a bound on the orders of finite subgroups. Then the set of non-zero-divisors of $\mathbb{C}G$ equals $T(G)$, i.e. any non-zero-divisor in $\mathbb{C}G$ becomes invertible in $\mathcal{U}(G)$. The pair $(\mathbb{C}G, T(G))$ satisfies the Ore condition and the Ore localization is $\mathcal{D}(G)$. In particular $\mathbb{C}G \to \mathcal{D}(G)$ is both universally $\Sigma(G)$-inverting and universally $T(G)$-inverting.

Notice, however, that the lamplighter group L is amenable and neither $\mathcal{D}(L)$ nor $\mathcal{U}(L)$ is flat over $\mathbb{C}L$, and that $\mathbb{C}L$ does not have a classical ring of quotients, i.e. the set S of non-zero-divisors does not satisfies the left Ore condition [311].

Lemma 10.87. *Suppose for the group G that $\mathcal{D}(G)$ is semisimple. Then the change of rings map*

$$K_0(\mathcal{D}(G)) \to K_0(\mathcal{U}(G))$$

is injective.

Proof. We have a decomposition $\mathcal{D}(G) = \prod_{i=1}^{r} M_{n_i}(D_i)$ for skewfields D_1, D_2, ... , D_r by Wedderburn's Theorem. Let e_i be the projection onto $M_{n_i}(D_i)$. Then each e_i is central in $\mathcal{D}(G)$ and $e_i e_j = 0$ for $i \neq j$ and $K_0(\mathcal{D}(G))$ is the free abelian group with $\{\frac{1}{n_i} \cdot [\operatorname{im}(e_i)] \mid i = 1, 2 \ldots, r\}$ as basis. By Theorem 8.22 (6) we can choose a projection p_i in $\mathcal{N}(G)$ such that $p_i e_i = e_i$ and $e_i p_i = p_i$ holds in $\mathcal{U}(G)$.

The element $e_i \in \mathcal{U}(G)$ commutes with any element $f \in \mathcal{N}(G)$ by the following argument. Because of the Double Commutant Theorem 9.2 we can find a net $\{f_j \mid j \in J\}$ of elements in $\mathbb{C}G \subset \mathcal{N}(G) = \mathcal{B}(l^2(G))^G$ which converges to f in the strong topology. Since e_i commutes with any element in $\mathcal{D}(G)$, it commutes with any element f_j, i .e. we have the equality $e_i \circ f_j = f_j \circ e_i$ of affiliated unbounded operators from $l^2(G)$ to $l^2(G)$. Obviously $\operatorname{dom}(e_i) \subset \operatorname{dom}(f_j \circ e_i)$ and $\operatorname{dom}(e_i) \subset \operatorname{dom}(f \circ e_i)$ since each f_j and f are bounded. By definition $\operatorname{dom}(f_j \circ e_i) = \operatorname{dom}(e_i \circ f_j)$ and $\operatorname{dom}(f \circ e_i) = \operatorname{dom}(e_i \circ f)$. For $x \in \operatorname{dom}(e_i)$ we have

$$\lim_{j \in J} f_j(x) = f(x);$$

$$\lim_{j \in J} e_i \circ f_j(x) = \lim_{j \in J} f_j \circ e_i(x) = f \circ e_i(x).$$

Hence $f(x) \in \operatorname{dom}(e_i)$ and $e_i \circ f(x) = f \circ e_i(x)$. Lemma 8.3 (3) implies that $f \circ e_i = e_i \circ f$ holds in $\mathcal{U}(G)$ for each $f \in \mathcal{N}(G)$.

Since $\mathcal{U}(G)$ is the Ore localization of $\mathcal{N}(G)$ with respect to all non-zero-divisors (see Theorem 8.22 (1)), $e_i \in \mathcal{U}(G)$ is central in $\mathcal{U}(G)$. This implies $e_i = p_i$ in $\mathcal{U}(G)$ and $p_i \in \mathcal{Z}(\mathcal{N}(G))^{\mathbb{Z}/2}$. The composition

$$K_0(\mathcal{D}(G)) \to K_0(\mathcal{U}(G)) \overset{\cong}{\leftarrow} K_0(\mathcal{N}(G)) \xrightarrow{\dim_{\mathcal{N}(G)}} \mathcal{Z}(\mathcal{A})^{\mathbb{Z}/2}$$

(see Theorem 9.13 (2) and Theorem 9.20 (1)) sends $[\operatorname{im}(e_i)]$ to p_i. Since $p_i p_j = 0$ for $i \neq j$ and $p_i^2 = p_i$, the elements $p_1, p_2, \ldots, p_r$ are linearly independent over $\mathbb{R}$. Hence the composition above is injective. This finishes the proof of Lemma 10.87. $\qquad\square$

Exercises

10.1. Let G be a finitely presented group. Let $\mathbb{Z} \subset \Lambda \subset \mathbb{R}$ be an additive subgroup of $\mathbb{R}$. Show that the following statements are equivalent:

(1) For any closed manifold M with $G = \pi_1(M)$ we have $b_p^{(2)}(\widetilde{M}) \in \Lambda$;

(2) For any finite CW-complex X with $G = \pi_1(X)$ we have $b_p^{(2)}(\widetilde{X}) \in \Lambda$;

(3) The Atiyah Conjecture 10.3 of order Λ with coefficients in $\mathbb{Q}$ is true for G.

10.2. Suppose that there exists a cocompact free proper G-CW-complex X which has finitely many path components. Show that then G is finitely generated. On the other hand show for a finitely generated group G that there is a cocompact connected free proper G-manifold M without boundary.

10.3. Let $\Lambda \subset \mathbb{R}$ be an additive subgroup different from $\{0\}$. Let $d(\Lambda)$ be the infimum of the subset $\{\lambda \in \Lambda \mid \lambda > 0\}$ of $\mathbb{R}$. Show that $\Lambda \subset \mathbb{R}$ is closed if and only if

$$\Lambda = \{r \in \mathbb{R} \mid d(\Lambda) \cdot r \in \mathbb{Z}\}.$$

10.4. Let G be a discrete group, let F be a field with $\mathbb{Q} \subset F \subset \mathbb{C}$ and let $\mathbb{Z} \subset \Lambda \subset \mathbb{R}$ be an additive subgroup of $\mathbb{R}$. Suppose that $\Lambda \subset \mathbb{R}$ is closed and that G satisfies the Atiyah Conjecture 10.3 of order Λ with coefficients in F. Prove that then for any FG-chain complex C_*

$$\dim_{\mathcal{N}(G)}(H_p(\mathcal{N}(G) \otimes_{FG} C_*)) \in \Lambda \coprod \{\infty\}.$$

10.5. Let X be a G-CW-complex such that $h^{(2)}(X; \mathcal{N}(G)) < \infty$ and hence $\chi^{(2)}(X; \mathcal{N}(G))$ is defined (see Definition 6.79). Suppose that there is an integer d such that the order of any finite subgroup of G divides d. Assume that the strong Atiyah Conjecture 10.2 holds for G. Then show

$$d \cdot \chi^{(2)}(X; \mathcal{N}(G)) \in \mathbb{Z}.$$

10.6. Let G be a group. Suppose that there is a chain of finite subgroups $1 \subset H_1 \subset H_2 \subset \ldots$ and a prime number p such that the orders of the p-Sylow subgroups of H_n become arbitrary large. Show that there exists a chain of subgroups $1 \subset K_1 \subset K_2 \subset \ldots$ with $|K_n| = p^n$ and $\Lambda(A, \mathbb{C})_{\mathrm{fg}} = \mathbb{R}$ holds.

10.7. Prove for the class $\mathcal{C}$:

(1) $\mathcal{C}$ is closed under taking subgroups;
(2) If $\{G_i \mid i \in I\}$ is a collection of groups in $\mathcal{C}$, then $*_{i \in I} G_i$ belongs to $\mathcal{C}$;
(3) Let $1 \to H \to G \to K \to 1$ be an extension of groups such that H is finite or infinite cyclic and K belongs to $\mathcal{C}$, then G belongs to $\mathcal{C}$;
(4) If G belongs to $\mathcal{C}$ and $H \subset G$ is an elementary amenable normal subgroup of G, then G/H belongs to $\mathcal{C}$.

10.8. Let $(G_n)_{n\geq 1}$ be a sequence of groups and $(d_n)_{n\geq 1}$ be a sequence of positive integers such that G_n belongs $\mathcal{C}$ and d_n is an upper bound on the orders of finite subgroups of G_n. Show that $G = *_{n\geq 1}G_n$ satisfies the strong Atiyah Conjecture 10.2. Give an example of a group G such that G satisfies the strong Atiyah Conjecture 10.2, G contains only 2-groups as subgroups and has no upper bound on the orders of finite subgroups.

10.9. Show for any 2-dimensional manifold that $\pi_1(M)$ belongs to $\mathcal{C}$ and satisfies the strong Atiyah Conjecture 10.2.

10.10. Let M be a closed 3-manifold such that there is a finite covering $\overline{M} \to M$ and a fibration $F \to \overline{M} \to S^1$, where F is a closed connected orientable manifold. Show that $\pi_1(M)$ belongs to $\mathcal{C}$ and satisfies the strong Atiyah Conjecture 10.2.

10.11. Let G be a torsionfree group which satisfies the Atiyah Conjecture 10.3 of order $\Lambda = \mathbb{Z}$ with coefficients in $F = \mathbb{Q}$. Suppose that there is a model for BG of finite type. Show for any $p \geq 0$ that the Hilbert $\mathcal{N}(G)$-module $H_p^{(2)}(EG)$ is isomorphic to $l^2(G)^n$ for an appropriate integer $n \geq 0$.

10.12. Let M be a closed hyperbolic manifold of dimension $2n$. Show that $\pi_1(M)$ does belong to $\mathcal{C}$ if and only if $n = 1$.

10.13. Find an extension of groups $1 \to H \to G \to K \to 1$ such that H and K satisfy the strong Atiyah Conjecture 10.2 but G does not.

10.14. Let G be the directed union of the directed system of subgroups $\{G_i \mid i \in I\}$. Suppose that for any $i \in I$ there is a ring $\mathcal{S}(G_i)$ with $\mathbb{C}G_i \subset \mathcal{S}(G_i) \subset \mathcal{U}(G_i)$ which satisfies condition $(\mathbf{R'})$ or $(\mathbf{K'})$ respectively of Lemma 10.26. Suppose that $\mathcal{S}(G_i) \subset \mathcal{S}(G_j)$ holds for $i \leq j$. Put $\mathcal{S}(G) = \bigcup_{i\in I} \mathcal{S}(G_i)$. Show that $\mathbb{C}G \subset \mathcal{S}(G) \subset \mathcal{U}(G)$ and that $\mathcal{S}(G)$ satisfies conditions $(\mathbf{R'})$ or $(\mathbf{K'})$ respectively of Lemma 10.26. Give an example to show that the assertion above is not true in general for $(\mathbf{R})$.

10.15. Show that the map τ defined in Lemma 10.45 (2) and (3) induces a map
$$\tau \colon K_0(\mathcal{A}^1) \to \mathbb{R}$$
by sending an element in $K_0(\mathcal{A}^1)$ represented by an idempotent p to $\tau(p)$. Show that the image of this map is contained in $\mathbb{Z}$.

10.16.   Let $H \subset G$ be a subgroup of finite index d. Suppose that $\mathcal{D}(H)$ is semisimple. Show that the following diagram is well-defined and commutes

$$
\begin{array}{ccccc}
K_0(\mathcal{D}(G)) & \longrightarrow & K_0(\mathcal{U}(G)) & \xrightarrow{\dim_{\mathcal{U}(G)}} & \mathbb{R} \\
\text{res}\downarrow & & \text{res}\downarrow & & d\cdot\text{id}\downarrow \\
K_0(\mathcal{D}(H)) & \longrightarrow & K_0(\mathcal{U}(H)) & \xrightarrow{\dim_{\mathcal{U}(H)}} & \mathbb{R}
\end{array}
$$

Show that for all virtually free groups G the image of the composition

$$K_0(\mathcal{D}(G)) \xrightarrow{i} K_0(\mathcal{U}(G)) \xrightarrow{\dim_{\mathcal{U}(G)}} \mathbb{R}$$

is contained in $\frac{1}{|\mathcal{FIN}(G)|}\mathbb{Z}$.

10.17.   Let $\mathbb{C}\langle x, y, z\rangle$ and $\mathbb{C}\langle x, u\rangle$ be non-commutative polynomial rings. It is known that non-commutative polynomial rings are semifirs [118, Section 10.9]. Hence there are universal fields of fractions $\mathbb{C}\langle x, y, z\rangle \to K$ and $\mathbb{C}\langle x, u\rangle \to L$ by Lemma 10.77. Define a map $f \colon \mathbb{C}\langle x, y, z\rangle \to \mathbb{C}\langle x, u\rangle$ by $f(x) = x$, $f(y) = xu$ and $f(z) = xu^2$. Show that this map is injective. Let D be the division closure of the image of f in L. Show that f yields a field of fractions $\mathbb{C}\langle x, y, z\rangle \to D$. It factorizes by the universal property as the composition $\mathbb{C}\langle x, y, z\rangle \to \mathbb{C}\langle x, y, z\rangle_{\Sigma(\mathbb{C}\langle x,y,z\rangle \to D)} \xrightarrow{\Psi} D$. Show that Ψ maps $y^{-1}z - x^{-1}y$ to zero and deduce that Ψ in not injective. Prove that $\mathbb{C}\langle x, y, z\rangle \to D$ is not the universal field of fractions of $\mathbb{C}\langle x, y, z\rangle$ and that there are two non-isomorphic fields of fractions for $\mathbb{C}\langle x, y, z\rangle$.

10.18.   Let G be the directed union of subgroups $\bigcup_{i \in I} G_i$. Give an example to show that in general neither $\mathcal{N}(G) = \bigcup_{i \in I} \mathcal{N}(G_i)$ nor $\mathcal{U}(G) = \bigcup_{i \in I} \mathcal{U}(G_i)$ is true.

10.19.   Let G be a group in the class $\mathcal{C}$, which has an upper bound on the orders of finite subgroups. Show that the kernel of $K_0(\mathbb{C}G) \to K_0(\mathcal{D}(G))$ and of $K_0(\mathbb{C}G) \to K_0(\mathcal{N}(G))$ agree.

11. The Singer Conjecture

Introduction

This chapter is devoted to the following conjecture (see [477] and also [146, Conjecture 2]).

Conjecture 11.1 (Singer Conjecture). *If M is an aspherical closed manifold, then*

$$b_p^{(2)}(\widetilde{M}) \;=\; 0 \qquad if\; 2p \neq \dim(M).$$

If M is a closed connected Riemannian manifold with negative sectional curvature, then

$$b_p^{(2)}(\widetilde{M}) \begin{cases} = 0 & if\; 2p \neq \dim(M); \\ > 0 & if\; 2p = \dim(M). \end{cases}$$

Because of the Euler-Poincaré formula $\chi(M) = \sum_{p \geq 0}(-1)^p \cdot b_p^{(2)}(\widetilde{M})$ (see Theorem 1.35 (2)) the Singer Conjecture 11.1 implies the following conjecture in the cases where M is aspherical or has negative sectional curvature.

Conjecture 11.2 (Hopf Conjecture). *If M is an aspherical closed manifold of even dimension, then*

$$(-1)^{\dim(M)/2} \cdot \chi(M) \geq 0.$$

If M is a closed Riemannian manifold of even dimension with sectional curvature $\sec(M)$, then

$$\begin{aligned} (-1)^{\dim(M)/2} \cdot \chi(M) &> 0 & if\; \sec(M) < 0; \\ (-1)^{\dim(M)/2} \cdot \chi(M) &\geq 0 & if\; \sec(M) \leq 0; \\ \chi(M) &= 0 & if\; \sec(M) = 0; \\ \chi(M) &\geq 0 & if\; \sec(M) \geq 0; \\ \chi(M) &> 0 & if\; \sec(M) > 0. \end{aligned}$$

In original versions of the Singer Conjecture 11.1 and the Hopf Conjecture 11.2 the statements for aspherical manifolds did not appear. Notice that any Riemannian manifold with non-positive sectional curvature is aspherical by Hadamard's Theorem. We will also discuss

Conjecture 11.3 (L^2-torsion for aspherical manifolds). *If M is an aspherical closed manifold of odd dimension, then $\widetilde{M}$ is det-L^2-acyclic and*

$$(-1)^{\frac{\dim(M)-1}{2}} \cdot \rho^{(2)}(\widetilde{M}) \geq 0.$$

If M is a closed connected Riemannian manifold of odd dimension with negative sectional curvature, then $\widetilde{M}$ is det-L^2-acyclic and

$$(-1)^{\frac{\dim(M)-1}{2}} \cdot \rho^{(2)}(\widetilde{M}) > 0.$$

If M is an aspherical closed manifold whose fundamental group contains an amenable infinite normal subgroup, then $\widetilde{M}$ is det-L^2-acyclic and

$$\rho^{(2)}(\widetilde{M}) = 0.$$

Notice that for aspherical closed manifolds of odd dimension Conjecture 11.3 is stronger than the Singer Conjecture 11.1. Recall that the Euler characteristic of a closed manifold of odd dimension vanishes and that for a closed Riemannian manifold M of even dimension $\rho^{(2)}(\widetilde{M}) = 0$ holds. Therefore the Hopf Conjecture or Conjecture 11.3 respectively are not interesting for manifolds whose dimension is odd or even respectively.

We will discuss special cases for which these conjectures are known in Section 11.1. We give the proof of Gromov's result on Kähler manifolds in Section 11.2.

To understand this chapter, it is only required to be familiar with Sections 1.1 and 1.2 and to have some basic knowledge of Section 3.4. For Subsections 11.2.3 and 11.2.4 small input from Chapter 2 is needed.

11.1 Survey on Positive Results about the Singer Conjecture

In this section we discuss results which give some evidence for the Singer Conjecture 11.1, the Hopf Conjecture 11.2 and Conjecture 11.3. Notice that it suffices to consider closed connected manifolds which are orientable. If M is not orientable, one can pass to the orientation covering $\overline{M} \to M$ and use the fact that $b_p^{(2)}(\widetilde{M}; \mathcal{N}(\pi_1(M))) = 2 \cdot b_p^{(2)}(\widetilde{\overline{M}}; \mathcal{N}(\pi_1(\overline{M})))$ and $\rho^{(2)}(\widetilde{M}; \mathcal{N}(\pi_1(M))) = 2 \cdot \rho^{(2)}(\widetilde{\overline{M}}; \mathcal{N}(\pi_1(\overline{M})))$ holds (see Theorem 1.35 (9) and Theorem 3.96 (5)).

11.1.1 Low-Dimensional Manifolds

We begin with low-dimensional manifolds. Let F_g be the closed orientable surface of genus g. It is aspherical if and only if $g \geq 1$. It carries a hyperbolic structure if and only if $g \geq 2$, a flat Riemannian metric if and only if $g = 1$ and a Riemannian metric with positive constant sectional curvature if and only if $g = 0$. If F_g carries a Riemannian metric with sectional curvature $\sec(M) < 0$, $\sec(M) \leq 0$, $\sec(M) = 0$, $\sec(M) \geq 0$ or $\sec(M) > 0$ respectively, the Gauss-Bonnet Theorem [218, 3.20 on page 107 and 3.111 on page 147] implies $\chi(M) < 0$, $\chi(M) \leq 0$, $\chi(M) = 0$. $\chi(M) \geq 0$ or $\chi(M) > 0$ respectively. If $g \geq 1$, then $b_p^{(2)}(\widetilde{F_g}) = 0$ for $p \neq 1$ and $b_1^{(2)}(\widetilde{F_g}) = 2g - 2 = -\chi(M)$ holds by Example 1.36. Hence the Singer Conjecture 11.1 and the Hopf Conjecture 11.2 are true in dimension 2.

Next we consider dimension 3 and Conjecture 11.3. We have already explained in Section 4.1 that the Sphere Theorem [252, Theorem 4.3] implies that a compact connected orientable irreducible 3-manifold is aspherical if and only if it is a 3-disk or has infinite fundamental group. This implies that a closed connected orientable 3-manifold is aspherical if and only if its prime decomposition contains at most one factor which is not a homotopy sphere and this factor must be irreducible with infinite fundamental group. Hence a closed orientable 3-manifold M is aspherical if and only if it is diffeomorphic to $N \# \Sigma$ for a closed connected orientable irreducible 3-manifold N with infinite fundamental group and a homotopy sphere Σ. Choose a homotopy equivalence $g \colon \Sigma \to S^3$. We have briefly introduced the notion of Whitehead torsion and simple homotopy equivalence in Section 3.1.2. Since the Whitehead group of the trivial group is trivial and S^3 has trivial fundamental group, the Whitehad torsion of g is trivial. By homotopy invariance and the sum formula for Whitehead torsion we conclude that $\mathrm{id} \# g \colon N \# \Sigma \to N \# S^3$ has trivial Whitehead torsion. This shows that there is a simple homotopy equivalence $f \colon M \to N$. Hence a closed orientable 3-manifold M is aspherical if and only if it is simply homotopy equivalent to a closed connected orientable irreducible 3-manifold with infinite fundamental group. We conclude from Theorem 3.96 (1) that M is det-L^2-acyclic if and only if N is det-L^2-acyclic, and in this case $\rho^{(2)}(\widetilde{M}) = \rho^{(2)}(\widetilde{N})$.

Suppose from now on that N satisfies Thurston's Geometrization Conjecture. Then we conclude from Theorem 4.3 that $\widetilde{N}$ is det-L^2-acyclic and $\rho^{(2)}(\widetilde{N}) \leq 0$. Suppose that N carries a Riemannian metric of negative sectional curvature. Then its fundamental group is word-hyperbolic and hence cannot contain $\mathbb{Z} \oplus \mathbb{Z}$ as subgroup [65, Corollary 3.10 on page 462], Then N is already hyperbolic and $\rho^{(2)}(\widetilde{N}) < 0$. This is consistent with Conjecture 11.3. In other words, Conjecture 11.3 is true for 3-dimensional manifolds if Thurston's Geometrization Conjecture holds. Notice that there exist closed graph manifolds, i.e. closed irreducible 3-manifolds M whose decomposition along a family of disjoint, pairwise-nonisotopic incompressible tori in M contains only Seifert pieces, which do not admit a Riemannian metric of non-positive sectional curvature but are aspherical. [304, Example 4.2]. So it makes a difference whether one makes the assumption aspherical instead of the stronger condition that there is a Riemannian metric with non-positive sectional curvature. Also, weakening the assumption in the conjectures, gives them a different flavour because aspherical is a purely homotopy-theoretic condition.

The Hopf Conjecture 11.2 is known for a closed 4-dimensional Riemannian manifold with non-positive or non-negative curvature [59], [108], where the result is attributed to Milnor. Actually it is shown that the integrand appearing in the Gauss-Bonnet formula is non-negative. It follows from the example of Geroch [219] that one cannot deduce in higher even dimension the positivity of the Gauss-Bonnet integrand at a point x purely algebraically from the fact that the sectional curvature is positive. The Singer Conjecture 11.1 for a closed orientable 4-manifold M is equivalent to the statement that $b_1^{(2)}(\widetilde{M}) = 0$ because of Poincaré duality and the fact that infinite $\pi_1(M)$ implies $b_0^{(2)}(\widetilde{M}) = 0$ (see Theorem 1.35 (3) and (8)). We have given a lot of sufficient conditions on $\pi_1(M)$ to ensure $b_1^{(2)}(\widetilde{M}) = 0$ in Subsection 7.1.1. To the authors knowledge at the time of writing the Singer Conjecture 11.1 is not known for all closed 4-dimensional manifolds.

11.1.2 Pinched Curvature

The following two results are taken from the paper by Jost and Xin [280, Theorem 2.1 and Theorem 2.3].

Theorem 11.4. *Let M be a closed connected Riemannian manifold of dimension $\dim(M) \geq 3$. Suppose that there are real numbers $a > 0$ and $b > 0$ such that the sectional curvature satisfies $-a^2 \leq \sec(M) \leq 0$ and the Ricci curvature is bounded from above by $-b^2$. If the non-negative integer p satisfies $2p \neq \dim(M)$ and $2pa \leq b$, then*

$$b_p^{(2)}(\widetilde{M}) = 0.$$

Theorem 11.5. *Let M be a closed connected Riemannian manifold of dimension $\dim(M) \geq 4$. Suppose that there are real numbers $a > 0$ and $b > 0$ such that the sectional curvature satisfies $-a^2 \leq \sec(M) \leq -b^2$. If the non-negative integer p satisfies $2p \neq \dim(M)$ and $(2p - 1) \cdot a \leq (\dim(M) - 2) \cdot b$, then*

$$b_p^{(2)}(\widetilde{M}) = 0.$$

The next result is a consequence of a result of Ballmann and Brüning [18, Theorem B on page 594].

Theorem 11.6. *Let M be a closed connected Riemannian manifold. Suppose that there are real numbers $a > 0$ and $b > 0$ such that the sectional curvature satisfies $-a^2 \leq \sec(M) \leq -b^2$. If the non-negative integer p satisfies $2p < \dim(M) - 1$ and $p \cdot a < (\dim(M) - 1 - p) \cdot b$, then*

$$b_p^{(2)}(\widetilde{M}) = 0;$$
$$\alpha_p^{\Delta}(M) = \infty^+.$$

Notice that Theorem 11.5 and Theorem 11.6 are improvements of the older results by Donnelly and Xavier [155], where $b_p(\widetilde{M}) = 0$ for $p \notin \{\frac{\dim(M)-1}{2}, \dim(M), \frac{\dim(M)+1}{2}\}$ is shown provided that there is a real number ϵ satisfying $-1 \le \sec(M) \le -1 + \epsilon$ and $0 \le \epsilon < 1 - \frac{(\dim(M)-2)^2}{(\dim(M)-1)^2}$. The proof of Theorem 11.6 follows along the lines of [155], the decisive improvement is [18, Theorem 5.3 on page 619]. The Hopf Conjecture for pinched sectional curvature is also treated in [58].

11.1.3 Aspherical Manifolds and Locally Symmetric Spaces

Suppose that M is an aspherical closed Riemannian manifold. Then Corollary 5.16 shows that M satisfies the Singer Conjecture 11.1, the Hopf Conjecture 11.2 and Conjecture 11.3 provided that M carries the structure of a locally symmetric space. Suppose that M is a closed locally symmetric Riemannian manifold whose sectional curvature is negative. Then M satisfies the Singer Conjecture 11.1, the Hopf Conjecture 11.2 and Conjecture 11.3 by Corollary 5.16.

If the aspherical closed manifold M carries a non-trivial S^1-action, then it satisfies the Singer Conjecture 11.1, the Hopf Conjecture 11.2 and Conjecture 11.3 by Theorem 3.111.

If the aspherical closed manifold M appears in a fibration $F \to M \to S^1$ such that F is a connected CW-complex of finite type, then the Singer Conjecture 11.1 and the Hopf Conjecture 11.2 hold for M by Theorem 1.39. If additionally $\widetilde{F}$ is det-L^2-acyclic, M satisfies also Conjecture 11.3 by Theorem 3.106.

We have proved in Theorem 7.2 (1) and (2) that the Singer Conjecture 11.1 and the Hopf Conjecture 11.2 hold for an aspherical closed manifold M whose fundamental group contains an amenable infinite normal subgroup. We have proved in Theorem 3.113 that Conjecture 11.3 is true in the special case, where M is a closed aspherical manifold, whose fundamental group is of det $\ge$ 1-class and contains an elementary amenable infinite normal subgroup.

We will deal with the Singer Conjecture 11.1 for Kähler manifolds in Corollary 11.17.

11.2 The Singer Conjecture for Kähler Manifolds

Let M be a complex manifold (without boundary). Let h denote a Hermitian metric. Thus we have for any $x \in M$ a Hilbert space structure

$$h_x : T_x M \times T_x M \to \mathbb{C} \tag{11.7}$$

on the complex vector space $T_x M$. It induces a Riemannian metric g by

$$g_x : T_x M \times T_x M \to \mathbb{R}, \qquad (v, w) \mapsto \Re(h_x(v, w)). \tag{11.8}$$

The associated *fundamental form* is the 2-form ω defined by

$$\omega_x \colon T_x M \times T_x M \;\to\; \mathbb{R}, \qquad (v, w) \;\mapsto\; -\tfrac{1}{2} \cdot \Im(h_x(v, w)). \quad (11.9)$$

Obviously h contains the same information as g and ω together.

Definition 11.10 (Kähler manifold). *Let M be a complex manifold without boundary. A Hermitian metric h is called Kähler metric if (M, g) is a complete Riemannian manifold and the fundamental form $\omega \in \Omega^2(M)$ is closed, i.e. $d\omega = 0$. A Kähler manifold $M = (M, h)$ is a connected complex manifold M without boundary together with a Kähler metric h.*

Definition 11.11. *Let (M, g) be a connected Riemannian manifold. A $(p-1)$-form $\eta \in \Omega^{p-1}(M)$ is bounded if $||\eta||_\infty := \sup\{||\eta||_x \mid x \in M\} < \infty$ holds, where $||\eta||_x$ is the norm on $\mathrm{Alt}^{p-1}(T_x M)$ induced by g_x. A p-form $\omega \in \Omega^p(M)$ is called d(bounded) if $\omega = d(\eta)$ holds for some bounded $(p-1)$-form $\eta \in \Omega^{p-1}(M)$. A p-form $\omega \in \Omega^p(M)$ is called $\widetilde{d}$(bounded) if its lift $\widetilde{\omega} \in \Omega^p(\widetilde{M})$ to the universal covering $\widetilde{M}$ is d(bounded).*

The next definition is taken from [236, 0.3 on page 265].

Definition 11.12 (Kähler hyperbolic manifold). *A Kähler hyperbolic manifold is a closed connected Kähler manifold (M, h) whose fundamental form ω is $\widetilde{d}$(bounded).*

Example 11.13. The following list of examples of Kähler hyperbolic manifolds is taken from [236, Example 0.3]:

(1) M is a closed Kähler manifold which is homotopy equivalent to a Riemannian manifold with negative sectional curvature;
(2) M is a closed Kähler manifold such that $\pi_1(M)$ is word-hyperbolic in the sense of [234] and $\pi_2(M) = 0$;
(3) $\widetilde{M}$ is a symmetric Hermitian space of non-compact type;
(4) M is a complex submanifold of a Kähler hyperbolic manifold;
(5) M is a product of two Kähler hyperbolic manifolds.

The complex projective space $\mathbb{CP}^n$ equipped with the Fubini-Study metric, which is up to scaling with a constant the only $U(n+1)$-invariant Riemannian metric, is a closed Kähler manifold but is not Kähler hyperbolic. Namely, it is simply connected and its fundamental form is not of the form $d(\eta)$ since in $H^2(\mathbb{CP}^n; \mathbb{Z})$ it represents the non-trivial class $c_1(\gamma_1|_{\mathbb{CP}^n})$ given by the first Chern class of the restriction of the universal complex line bundle γ_1, which lives over $\mathbb{CP}^\infty$, to $\mathbb{CP}^n$ (see [520, Example 2.3 in VI.2 on page 224]).

The following result is due to Gromov [234, Theorem 1.2.B and Theorem 1.4.A on page 274].

Theorem 11.14. (L^2-Betti numbers and Novikov-Shubin invariants of Kähler hyperbolic manifolds). *Let M be a closed Kähler hyperbolic manifold of complex dimension m and real dimension $n = 2m$. Then*

$$b_p^{(2)}(\widetilde{M}) = 0 \qquad \text{if } p \neq m;$$
$$b_m^{(2)}(\widetilde{M}) > 0;$$
$$(-1)^m \cdot \chi(M) > 0;$$
$$\alpha_p(\widetilde{M}) = \infty^+ \qquad \text{for } p \geq 1.$$

This has interesting consequences in algebraic geometry.

Theorem 11.15. *Let M be a closed Kähler hyperbolic manifold of complex dimension m and real dimension $n = 2m$.*

(1) The canonical line bundle $L = \Lambda^m T^* M$ is quasi-ample, i.e. its Kodaira dimension is m;

(2) M satisfies all of the following four assertions (which are equivalent for closed Kähler manifolds):

(a) M is Moishezon, i.e. the transcendental degree of the field $\mathcal{M}(M)$ of meromorphic functions is equal to m;

(b) M is Hodge, i.e. the Kähler form represents a class in $H^2(M; \mathbb{C})$ which lies in the image of $H^2(M; \mathbb{Z}) \to H^2(M; \mathbb{C})$;

(c) M can be holomorphically embedded into $\mathbb{CP}^N$ for some N;

(d) M is a projective algebraic variety;

(3) The fundamental group is an infinite non-amenable group of deficiency ≤ 1. It cannot be a non-trivial free product.

Proof. (1) and (2a) We will see in Corollary 11.36 that there is a holomorphic L^2-integrable m-form on the universal covering $\widetilde{M}$ of any closed Kähler hyperbolic manifold. This is used by Gromov [236, Section 3] to construct meromorphic functions on M and leads to the result that the canonical line bundle L is quasi-ample and M is Moishezon. The equivalence of the assertions (2a), (2b), (2c) and (2d) for closed Kähler manifolds is due to results of Chow, Kodaira and Moishezon (see [249, Appendix B4 on page 445], [294], [382], [520, Remark on page 11, Theorem 4.1 in VI.4 on page 234]).

We conclude $\alpha_1(\widetilde{M}) = \infty^+$ and $b_0^{(2)}(\widetilde{M}) = 0$ from Theorem 11.14. Theorem 1.35 (8) and Theorem 2.55 (5b) imply that $\pi_1(M)$ is an infinite non-amenable group. If $m = 1$, then $\pi_1(M)$ is the fundamental group of a closed orientable surface with infinite fundamental group and hence has deficiency ≤ 1 and cannot be a non-trivial free product. Suppose $m \geq 2$. Then $b_1^{(2)}(\widetilde{M}) = 0$ by Theorem 11.14. This shows $b_1^{(2)}(\pi_1(M)) = 0$. Lemma 7.22 implies that $\pi_1(M)$ has deficiency ≤ 1. We conclude from Theorem 1.35 (5) that the only finitely presented group G, which satisfies $b_1^{(2)}(G) = 0$ and is a non-trivial free product, is $G = \mathbb{Z}/2 * \mathbb{Z}/2$. Since this group is amenable, $\pi_1(M)$ cannot be a non-trivial free product.   □

Gromov's notion of Kähler hyperbolic reflects the negative curvature case. The following notion due to Cao and Xavier [86] and Jost and Zuo [281] is weaker and modelled upon non-positive sectional curvature. A Kähler manifold M is called *Kähler non-elliptic manifold* if the lift $\widetilde{\omega}$ of the fundamental form ω to the universal covering $\widetilde{M}$ can be written as $d(\eta)$ for some 1-form η which satisfies for some point $x_0 \in \widetilde{M}$ and constants $C_1, C_2 > 0$, which depend on η but not on $x \in \widetilde{M}$,

$$||\eta||_x \leq C_1 \cdot d(x, x_0) + C_2.$$

Here $d(x, x_0)$ is the distance of x and x_0 with respect to the metric on $\widetilde{M}$ associated to its Riemannian metric. (Actually Jost and Zuo consider more generally arbitrary coverings and local systems.) Of course a Kähler hyperbolic manifold is in particular Kähler non-elliptic. If a closed Kähler manifold carries a Riemannian metric with non-positive sectional curvature, then it is Kähler non-elliptic. The following result is due to Cao and Xavier [86, Theorem 2 on page 485] and Jost and Xin [281, Theorem 3], where also applications to algebraic geometry are given.

Theorem 11.16. (L^2-Betti numbers of Kähler non-elliptic manifolds). *Let M be a closed Kähler non-elliptic manifold of complex dimension m and real dimension $n = 2m$. Then*

$$b_p^{(2)}(\widetilde{M}) = 0 \qquad \text{if } p \neq m;$$
$$(-1)^m \cdot \chi(M) \geq 0.$$

Theorem 11.14 and Theorem 11.16 imply

Corollary 11.17. *Let M be a closed Riemannian manifold whose sectional curvature is negative or non-positive respectively. Then M satisfies the Singer Conjecture 11.1 if M carries some Kähler structure. An aspherical closed Kähler manifold satisfies the Singer Conjecture 11.1 if its fundamental group is word-hyperbolic in the sense of [234].*

The remainder of this section is devoted to the proof of Theorem 11.14 and Theorem 11.16.

11.2.1 Hodge Theory on Kähler manifolds

We give a brief introduction to Hodge theory on a complex manifold M (without boundary). Let m be its complex and $n = 2m$ be its real dimension. Recall that each tangent space $T_x M$ is a complex vector space. We introduce the following notation for a complex vector space V

$$\mathbb{C} \otimes_{\mathbb{R}} V := \mathbb{C} \otimes_{\mathbb{R}} \mathrm{res}_{\mathbb{C}}^{\mathbb{R}} V;$$
$$\mathrm{Alt}^p(V \otimes_{\mathbb{R}} \mathbb{C}) := \mathrm{Alt}_{\mathbb{R}}^p\left(\mathrm{res}_{\mathbb{C}}^{\mathbb{R}} V, \mathrm{res}_{\mathbb{C}}^{\mathbb{R}} \mathbb{C}\right) = \mathrm{Alt}_{\mathbb{C}}^p\left(\mathbb{C} \otimes_{\mathbb{R}} V, \mathbb{C}\right).$$

Denote by $J : V \to V$ multiplication with i. It induces an involution of complex vector spaces $\mathbb{C} \otimes_{\mathbb{R}} J : \mathbb{C} \otimes_{\mathbb{R}} V \to \mathbb{C} \otimes_{\mathbb{R}} V$. Denote by $(\mathbb{C} \otimes_{\mathbb{R}} V)^+$ and $(\mathbb{C} \otimes_{\mathbb{R}} V)^-$ the eigenspace for $+i$ and $-i$. We get canonical identifications of complex vector spaces

$$\mathbb{C} \otimes_{\mathbb{R}} V = (\mathbb{C} \otimes_{\mathbb{R}} V)^+ \oplus (\mathbb{C} \otimes_{\mathbb{R}} V)^-;$$
$$V = (\mathbb{C} \otimes_{\mathbb{R}} V)^+;$$
$$\mathrm{Alt}^r(\mathbb{C} \otimes_{\mathbb{R}} V) = \bigoplus_{p+q=r} \mathrm{Alt}^p((\mathbb{C} \otimes_{\mathbb{R}} V)^+) \otimes_{\mathbb{C}} \mathrm{Alt}^q((\mathbb{C} \otimes_{\mathbb{R}} V)^-).$$

We have already defined the complex vector space $\Omega^p(M)$ of smooth p-forms on M in Section 1.3. We define the space of (p,q)-forms $\Omega^{p,q}(M)$ by

$$\mathrm{Alt}^{p,q}(\mathbb{C} \otimes_{\mathbb{R}} T_x M) := \mathrm{Alt}^p((\mathbb{C} \otimes_{\mathbb{R}} T_x M)^+) \otimes_{\mathbb{C}} \mathrm{Alt}^q((\mathbb{C} \otimes_{\mathbb{R}} T_x M)^-),$$
$$\Omega^{p,q}(M) := C^\infty(\mathrm{Alt}^{p,q}(\mathbb{C} \otimes_{\mathbb{R}} TM)). \tag{11.18}$$

We obtain a canonical decomposition

$$\Omega^r(M) = \bigoplus_{p+q=r} \Omega^{p,q}(M). \tag{11.19}$$

We have introduced the exterior differential d in (1.46), the Hodge-star operator $*$ in (1.48) and adjoint of the exterior differential δ in (1.50). The Hodge-star operator $*^r : \Omega^r(M) \to \Omega^{n-r}(M)$ decomposes as

$$*^{p,q} : \Omega^{p,q}(M) \to \Omega^{m-p,m-q}(M).$$

Define

$$\partial_{p,q} : \Omega^{p,q}(M) \to \Omega^{p+1,q}(M); \tag{11.20}$$
$$\overline{\partial}_{p,q} : \Omega^{p,q}(M) \to \Omega^{p,q+1}(M) \tag{11.21}$$

by the composition

$$\Omega^{p,q}(M) \xrightarrow{i} \Omega^{p+q}(M) \xrightarrow{d^{p+q}} \Omega^{p+q+1}(M) \xrightarrow{\mathrm{pr}} \begin{cases} \Omega^{p+1,q}(M) \\ \Omega^{p,q+1}(M) \end{cases}$$

where i is the inclusion and pr the projection. Define

$$\Delta_r = d^{r-1}\delta^r + \delta^{r+1}d^r : \Omega^r(M) \to \Omega^r(M); \tag{11.22}$$
$$\Box_{p,q} := \partial^{p-1,q}(\partial^{p-1,q})^* + (\partial^{p,q})^*\partial^{p,q} : \Omega^{p,q}(M) \to \Omega^{p,q}(M); \tag{11.23}$$
$$\Box_{p,q} := \overline{\partial}^{p,q-1}(\overline{\partial}^{p,q-1})^* + (\overline{\partial}^{p,q})^*\overline{\partial}^{p,q} : \Omega^{p,q}(M) \to \Omega^{p,q}(M). \tag{11.24}$$

The proof of the following lemma can be found in [520, Theorem 3.7 in Chapter I on page 34 and Theorem 4.7 in Chapter V on page 191].

Lemma 11.25. *If M is a Kähler manifold, then*

$$\partial \circ \partial = 0;$$
$$\overline{\partial} \circ \overline{\partial} = 0;$$
$$d = \partial + \overline{\partial};$$
$$\Delta = 2 \cdot \Box = 2 \cdot \overline{\Box}.$$

Definition 11.26. *Let M be a Kähler manifold. Define the space of harmonic L^2-integrable (p,q)-forms and harmonic r-forms*

$$\mathcal{H}^{p,q}_{(2)}(M) := \{\omega \in \Omega^{p,q}(M) \mid \Box(\omega) = 0, \int_M \omega \wedge *\omega < \infty\};$$

$$\mathcal{H}^r_{(2)}(M) := \{\omega \in \Omega^r(M) \mid \Delta(\omega) = 0, \int_M \omega \wedge *\omega < \infty\}.$$

We get the following extension of Theorem 1.57.

Theorem 11.27 (L^2-Hodge decomposition for Kähler manifolds). *If M is a Kähler manifold, then we get orthogonal decompositions*

$$L^2\Omega^r(M) = \mathcal{H}^r(M) \oplus \text{clos}(d^{r-1}(\Omega^{r-1}_c(M))) \oplus \text{clos}(\delta^{r+1}(\Omega^{r+1}_c(M)));$$
$$\mathcal{H}^r_{(2)}(M) = \bigoplus_{p+q=r} \mathcal{H}^{p,q}_{(2)}(M),$$

where $\Omega^p_c(M) \subset \Omega^p(M)$ is the subspace of p-forms with compact support.

11.2.2 The L^2-Lefschetz Theorem

Theorem 11.28 (L^2-Lefschetz Theorem). *Let M be a (not necessarily compact) Kähler manifold of complex dimension m and real dimension $n = 2m$. Let ω be its fundamental form. Then the linear map*

$$L^k \colon \Omega^r(M) \to \Omega^{r+2k}(M), \qquad \phi \mapsto \phi \wedge \omega^k$$

satisfies

(1) L commutes with Δ and d;
(2) L^k induces bounded operators denoted in the same way

$$L^k \colon L^2\Omega^r(M) \to L^2\Omega^{r+2k}(M);$$
$$L^k \colon \mathcal{H}^r_{(2)}(M) \to \mathcal{H}^{r+2k}_{(2)}(M);$$

(3) The operators L^k of assertion (2) are quasi-isometries, i.e. there is a positive constant C such that $C^{-1} \cdot ||\phi|| \leq ||L^k(\phi)|| \leq C \cdot ||\phi||$ holds, and in particular are injective, provided that k satisfies $2r + 2k \leq n$. These operators are surjective, if k satisfies $2r + 2k \geq n$.

Proof. (1) [520, Theorem 4.7 in Chapter V on page 191 and Theorem 4.8 in Chaper V on page 192].

(2) Notice that L^k is defined fiberwise, for each x we have a linear map $L_x^k\colon \mathrm{Alt}^r(T_xM \otimes_{\mathbb{R}} \mathbb{C}) \to \mathrm{Alt}^{r+2k}(T_xM \otimes_{\mathbb{R}} \mathbb{C})$. The Kähler condition implies that $\omega_x^m \neq 0$ for all $x \in M$ (see [520, (1.12) on page 158]). The fundamental form ω is parallel with respect to the Levi-Civita connection since the fiber transport $t_w\colon T_xM \to T_yM$ for any path w in M from x to y with respect to the Levi-Civita connection is an isometric $\mathbb{C}$-linear isomorphism. The following diagram commutes

$$
\begin{array}{ccc}
\mathrm{Alt}^r(T_xM \otimes_{\mathbb{R}} \mathbb{C}) & \xrightarrow{\;L_x^k\;} & \mathrm{Alt}^{r+2k}(T_xM \otimes_{\mathbb{R}} \mathbb{C}) \\
{\scriptstyle t_w}\downarrow & & {\scriptstyle t_w}\downarrow \\
\mathrm{Alt}^r(T_yM \otimes_{\mathbb{R}} \mathbb{C}) & \xrightarrow{\;L_y^k\;} & \mathrm{Alt}^{r+2k}(T_yM \otimes_{\mathbb{R}} \mathbb{C})
\end{array}
$$

This implies that L^k is bounded.

(3) Elementary linear algebra shows $L_x^k\colon \mathrm{Alt}^r(\mathbb{C}\otimes_{\mathbb{R}} T_xM) \to \mathrm{Alt}^{r+2k}(\mathbb{C}\otimes_{\mathbb{R}} T_xM)$ is injective for $2r+2k = n$ [520, Theorem 3.12 (c) in V.3 on page 182]. Since the source and target are complex vector spaces of the same finite dimension, it is bijective for $2r+2k = n$. Because of the identity $L_x^k \circ L_x^l = L_x^{k+l}$ we conclude that L_x^k is injective, if k satisfies $2r + 2k \leq n$, and surjective, if k satisfies $2r + 2k \geq n$. This implies that L^k is a quasi-isometry and injective if k satisfies $2r + 2k \leq n$. It remains to prove the claim about surjectivity. Because of the factorization

$$
L^{l+k}\colon \Omega^{r-2l}(M) \xrightarrow{\;L^l\;} \Omega^r(M) \xrightarrow{\;L^k\;} \Omega^{r+2k}(M)
$$

it suffices to prove surjectivity for L^k in the case $2r + 2k = n$.

Consider the adjoint $K_x\colon \mathrm{Alt}^{r+2k}(\mathbb{C} \otimes_{\mathbb{R}} T_xM) \to \mathrm{Alt}^r(\mathbb{C} \otimes_{\mathbb{R}} T_xM)$ of $L_x^k\colon \mathrm{Alt}^r(\mathbb{C} \otimes_{\mathbb{R}} T_xM) \to \mathrm{Alt}^{r+2k}(\mathbb{C} \otimes_{\mathbb{R}} T_xM)$. Since L_x^k is surjective, K_x^k is injective. We get a quasi-isometry $K\colon L^2\Omega^{r+2k}(M) \to L^2\Omega^r(M)$ which is the adjoint of L^k. Since K is injective, L^k has dense image. Since L^k is a quasi-isometry, L^k has closed image and hence is surjective. The proof for $\mathcal{H}_{(2)}^r(M)$ is analogous. This finishes the proof of Theorem 11.28.   □

Corollary 11.29. *Let M be a closed Kähler manifold of complex dimension m and real dimension $n = 2m$. Let $b_r(M) = \dim_{\mathbb{C}}\left(\mathcal{H}_{(2)}^r(M)\right) = \dim_{\mathbb{C}}(H_r(M;\mathbb{C}))$ be the ordinary r-th Betti number and define $h_{p,q}(M) := \dim_{\mathbb{C}}\left(\mathcal{H}_{(2)}^{p,q}(M)\right)$. Then*

(1) $b_r(M) = \sum_{p+q=r} h_{p,q}(M)$ for $r \geq 0$;
(2) $b_r(M) = b_{n-r}(M)$ and $h_{p,q}(M) = h_{m-p,m-q}(M)$ for $r, p, q \geq 0$;
(3) $h_{p,q}(M) = h_{q,p}(M)$ for $p, q \geq 0$;
(4) $b_r(M)$ is even for r odd;

(5) $h_{1,0}(M) = \frac{b_1(M)}{2}$ and depends only on $\pi_1(M)$;

(6) $b_r(M) \leq b_{r+2}(M)$ for $r < m$.

The corresponding statements except for (4) hold for the L^2-versions $b_r^{(2)}(\widetilde{M})$ and $h^{p,q}(\widetilde{M})$.

Now we can give the proof of Theorem 11.16.

Proof. Because of Poincaré duality (see Theorem 1.35 (3)) it suffices to show $\mathcal{H}^r_{(2)}(\widetilde{M}) = 0$ for $r < m$. Let $\widetilde{\omega}$ be the lift of the fundamental form on M to the universal covering $\widetilde{M}$. Then

$$L^1 \colon \mathcal{H}^r_{(2)}(\widetilde{M}) \to \mathcal{H}^{r+2}_{(2)}(\widetilde{M}), \qquad \phi \mapsto \phi \wedge \widetilde{\omega}$$

is injective by Theorem 11.28 (3). Consider $\phi \in \mathcal{H}^r_{(2)}(\widetilde{M})$. By assumption there is $\eta \in \Omega^1(\widetilde{M})$ which satisfies

$$\widetilde{\omega} = d(\eta);$$
$$||\eta||_x \leq C_1 \cdot d(x, x_0) + C_2$$

for positive constants $C_1, C_2 > 0$ and some point $x_0 \in \widetilde{M}$. By the Hodge Decomposition Theorem 11.27 it suffices to show $\phi \wedge \widetilde{\omega} \in \mathrm{clos}\left(d^{r+1}(\Omega_c^{r+1}(\widetilde{M}))\right)$ because then $L^1(\phi) = \phi \wedge \widetilde{\omega} \in \mathcal{H}^{r+2}(\widetilde{M})$ implies $L^1(\phi) = 0$ and thus $\phi = 0$.

Let B_r be the closed ball of radius r around x_0. Choose a smooth function $a_r \colon \widetilde{M} \to [0,1]$ together with a constant $C_3 > 0$ such that $a_r(x) = 1$ for $x \in B_r$, $a_r(x) = 0$ for $x \notin B_{2r}$ and $||da_r||_x \leq \frac{C_3}{d(x,x_0)}$ for $x \in B_{2r} - B_r$. Since $\phi \wedge a_r \eta$ lies in $\Omega_c^{r+1}(\widetilde{M})$, it suffices to prove that $(-1)^r \cdot d(\phi \wedge a_r \eta)$ converges to $\phi \wedge \widetilde{\omega}$ in the L^2-norm.

We have

$$||\phi \wedge \widetilde{\omega} - \phi \wedge a_r \widetilde{\omega}||_{L^2}^2 = \int_{\widetilde{M}} ||\phi \wedge \widetilde{\omega} - \phi \wedge a_r \widetilde{\omega}||_x^2 \, d\mathrm{vol}$$

$$= \int_{\widetilde{M}} (1 - a_r(x))^2 \cdot ||\phi \wedge \widetilde{\omega}||_x^2 \, d\mathrm{vol}$$

$$\leq \int_{\widetilde{M} - B_r} ||\phi \wedge \widetilde{\omega}||_x^2 \, d\mathrm{vol}.$$

Since $\phi \wedge \widetilde{\omega}$ is L^2-integrable, we get $\lim_{r \to \infty} \int_{\widetilde{M} - B_r} ||\phi \wedge \widetilde{\omega}||_x^2 \, d\mathrm{vol} = 0$ from Lebesgue's Theorem of majorized convergence. This shows that $\phi \wedge a_r \widetilde{\omega}$ converges to $\phi \wedge \widetilde{\omega}$ in the L^2-norm. Since

$$(-1)^r \cdot d(\phi \wedge a_r \eta) = \phi \wedge da_r \wedge \eta + \phi \wedge a_r \widetilde{\omega},$$

it remains to show that $\phi \wedge da_r \wedge \eta$ converges to zero in the L^2-norm. We get for appropriate constants $C_4, C_5 > 0$ and $r \geq 1$

$$\int_{\widetilde{M}} ||\phi \wedge da_r \wedge \eta||_x^2 \, d\mathrm{vol}$$

$$\leq \int_{B_{2r}-B_r} ||\phi \wedge da_r \wedge \eta||_x^2 \, d\mathrm{vol}$$

$$\leq \int_{B_{2r}-B_r} C_4 \cdot ||\phi||_x^2 \cdot ||da_r||_x^2 \cdot ||\eta||_x^2 \, d\mathrm{vol}$$

$$\leq \int_{B_{2r}-B_r} C_4 \cdot ||\phi||_x^2 \cdot \frac{C_3^2}{d(x,x_0)^2} \cdot (C_1 \cdot d(x,x_0) + C_2)^2 \, d\mathrm{vol}$$

$$\leq C_5 \cdot \int_{B_{2r}-B_r} ||\phi||_x^2 \, d\mathrm{vol}.$$

Since ϕ is L^2-integrable, $\lim_{r\to\infty} \int_{B_{2r}-B_r} ||\phi||_x^2 \, d\mathrm{vol} = 0$. Hence $\phi \wedge da_r \wedge \eta$ converges to zero in the L^2-norm. This finishes the proof of Theorem 11.16.
□

11.2.3 Novikov-Shubin Invariants for Kähler Hyperbolic Manifolds

In this subsection we will prove

Theorem 11.30. *Let M be a closed Kähler hyperbolic manifold. Then*

$$\alpha_p(\widetilde{M}) = \infty^+ \qquad \text{for } p \geq 1.$$

Proof. Because of Poincaré duality (see Theorem 2.55 (2)) it suffices to treat the case $p > m$, if the real dimension of M is $2m$.

We will use the analytic definition of the Novikov-Shubin invariants in terms of the analytic spectral density function $F_p(\widetilde{M})$ (see Definition 2.64). Recall that the analytic and cellular versions agree by Theorem 2.68. We have to find $\epsilon > 0$ such that $F_p(\widetilde{M})(\lambda) = F_p(\widetilde{M})(0)$ holds for $0 \leq \lambda \leq \epsilon$. We conclude from the Hodge Decomposition Theorem 1.57 that the orthogonal complement of the kernel of $d_{\min}^p \colon \mathrm{dom}(d_{\min}^p) \subset L^2\Omega^p(\widetilde{M}) \to L^2\Omega^{p+1}(\widetilde{M})$ is $\mathrm{clos}\left(\delta^{p+1}(\Omega_c^{p+1}(\widetilde{M}))\right)$. Because of Definition 2.1 and Lemma 2.2 (1) is suffices to find $\epsilon > 0$ such that for all elements ψ in $\mathrm{dom}(d_{\min}^p) \cap \mathrm{clos}\left(\delta^{p+1}(\Omega_c^{p+1}(\widetilde{M}))\right)$ with $d(\psi) := d_{\min}^p(\psi) \in \mathrm{dom}(\delta^{p+1})$ we have

$$||\psi||_{L^2} \leq \epsilon \cdot ||d(\psi)||_{L^2}. \tag{11.31}$$

Put $k = p - m$. In the sequel $C_1, C_2, \ldots$ are positive constants which depend only on M (but not on ψ). Let $\widetilde{\omega} \in \Omega^2(\widetilde{M})$ be the lift of the fundamental form ω of M to the universal covering $\widetilde{M}$. The map $L^k \colon L^2\Omega^{p-2k}(\widetilde{M}) \to L^2\Omega^p(\widetilde{M})$ sending ϕ to $\phi \wedge \widetilde{\omega}^k$ is bijective, commutes with Δ and satisfies

$$C_1 \cdot ||\phi|| \leq ||L^k(\phi)||_{L^2} \leq \frac{1}{C_1} \cdot ||\phi||$$

by Theorem 11.28. Consider ψ in $\mathrm{dom}(d^p_{\min}) \cap \mathrm{clos}\left(\delta^{p+1}(\Omega^{p+1}_c(\widetilde{M}))\right)$ with $d(\psi) \in \mathrm{dom}(\delta^{p+1})$. We can choose ϕ with

$$\phi \in \mathrm{dom}((\Delta_{p-2k})_{\min});$$
$$\psi = L^k(\phi).$$

Notice that some complication in the proof comes from the fact that $d(\phi)$ (in contrast to $\delta \circ d(\phi) = \Delta(\phi)$) does not lie in the range, where we know that L^k is a quasi-isometry. Since L^k commutes with Δ and hence with $\Delta^{1/2}$ we conclude using partial integration (see Lemma 1.56)

$$||d(\psi)||^2_{L^2} = \langle d(\psi), d(\psi)\rangle = \langle \delta \circ d(\psi), \psi\rangle = \langle \Delta(\psi), \psi\rangle = \langle \Delta^{1/2}(\psi), \Delta^{1/2}(\psi)\rangle$$

$$= \langle \Delta^{1/2} \circ L^k(\phi), \Delta^{1/2} \circ L^k(\phi)\rangle = \langle L^k \circ \Delta^{1/2}(\phi), L^k \circ \Delta^{1/2}(\phi)\rangle =$$

$$||L^k \circ \Delta^{1/2}(\phi)||^2_{L^2} \geq C^2_1 \cdot ||\Delta^{1/2}(\phi)||^2_{L^2} = C^2_1 \cdot \langle \Delta^{1/2}(\phi), \Delta^{1/2}(\phi)\rangle =$$

$$C^2_1 \cdot \langle \Delta(\phi), \phi\rangle = C^2_1 \cdot (\langle \delta \circ d(\phi), \phi\rangle + \langle d \circ \delta(\phi), \phi\rangle)$$

$$= C^2_1 \cdot (\langle d(\phi), d(\phi)\rangle + \langle \delta(\phi), \delta(\phi)\rangle) = C^2_1 \cdot (||d(\phi)||^2_{L^2} + ||\delta(\phi)||^2_{L^2})$$

$$\geq C^2_1 \cdot ||d(\phi)||^2_{L^2}.$$

This shows

$$||d(\phi)||_{L^2} \leq \frac{1}{C_1} \cdot ||d(\psi)||_{L^2}. \tag{11.32}$$

Since M is Kähler hyperbolic by assumption, we can find $\eta \in \Omega^1(\widetilde{M})$ with $||\eta||_\infty < \infty$ and $\widetilde{\omega} = d(\eta)$. We have $||\widetilde{\omega}||_{L^2} < \infty$. Put

$$\theta := (-1)^p \cdot \phi \wedge \widetilde{\omega}^{k-1} \wedge \eta;$$
$$\mu := (-1)^p \cdot d\phi \wedge \widetilde{\omega}^{k-1} \wedge \eta.$$

Both forms θ and μ are L^2-integrable and

$$||\mu||_{L^2} \leq C_2 \cdot ||d\phi||_{L^2} \cdot ||\widetilde{\omega}^{k-1}||_\infty \cdot ||\eta||_\infty; \tag{11.33}$$
$$\psi = d\theta - \mu.$$

We conclude using partial integration (see Lemma 1.56)

$$||\psi||^2_{L^2} = \langle\psi, \psi\rangle_{L^2} = \langle\psi, d\theta - \mu\rangle_{L^2} = \langle\psi, d\theta\rangle_{L^2} - \langle\psi, \mu\rangle_{L^2}$$

$$= \langle\delta\psi, \theta\rangle_{L^2} - \langle\psi, \mu\rangle_{L^2} = -\langle\psi, \mu\rangle_{L^2} \leq ||\psi||_{L^2} \cdot ||\mu||_{L^2}.$$

This implies

$$||\psi||_{L^2} \leq ||\mu||_{L^2}. \tag{11.34}$$

If we put $\epsilon = \frac{C_2}{C_1} \cdot ||\widetilde{\omega}^{k-1}||_\infty \cdot ||\eta||_\infty$, we get the desired inequality (11.31) from the inequalities (11.32), (11.33) and (11.34). This finishes the proof of Theorem 11.30. $\qquad\square$

11.2.4 Non-Vanishing of the Middle L^2-Betti Number for Kähler Hyperbolic Manifolds

In this subsection we sketch the proof of the following result. The proof is due to Gromov [236, Section 2] and inspired by [499]. Notice that Theorem 11.14 then follows from Theorem 11.16, Theorem 11.30 and Theorem 11.35 below.

Theorem 11.35. *Let M be a closed Kähler hyperbolic manifold of complex dimension m and real dimension $n = 2m$. Then*

$$\mathcal{H}^{p,q}_{(2)}(\widetilde{M}) \neq 0 \qquad \text{for } p + q = m.$$

Proof. Fix p with $0 \leq p \leq m$. Consider the following elliptic cochain complex of differential operators of order 1

$$\cdots \xrightarrow{\overline{\partial}_{p,q-2}} \Omega^{p,q-1}(\widetilde{M}) \xrightarrow{\overline{\partial}_{p,q-1}} \Omega^{p,q}(\widetilde{M}) \xrightarrow{\overline{\partial}_{p,q}} \cdots$$

We define its (reduced) L^2-cohomology as in Subsection 1.4.2 by

$$Z^{p,q}(\widetilde{M}) := \ker\left(\left(\overline{\partial}^{p,q} : \Omega^{p,q}_c(\widetilde{M}) \to L^2\Omega^{p,q+1}(\widetilde{M})\right)_{\min}\right);$$

$$B^{p,q}(\widetilde{M}) := \operatorname{im}\left(\left(\overline{\partial}^{p,q-1} : \Omega^{p,q-1}_c(\widetilde{M}) \to L^2\Omega^{p,q}(\widetilde{M})\right)_{\min}\right);$$

$$H^{p,q}_{(2)}(\widetilde{M}) := Z^{p,q}(\widetilde{M})/\operatorname{clos}(B^{p,q}(\widetilde{M})).$$

The Hilbert spaces $H^{p,q}_{(2)}(\widetilde{M})$ are finitely generated Hilbert $\mathcal{N}(\pi_1(M))$-modules and isomorphic to $\mathcal{H}^{p,q}_{(2)}(\widetilde{M})$ since the Laplacian associated to the elliptic complex $(\Omega^{p,*}(\widetilde{M}), \overline{\partial}_{p,*})$ is by definition $\square_{p,q}$ introduced in (11.24). The L^2-index of the elliptic complex $(\Omega^{p,*}(\widetilde{M}), \overline{\partial}_{p,*})$ is by definition

$$\operatorname{index}^{(2)}\left(\Omega^{p,*}(\widetilde{M}), \overline{\partial}_{p,*}\right) = \sum_{q \geq 0}(-1)^q \cdot \dim_{\mathcal{N}(\pi_1(M))}\left(H^{p,q}_{(2)}(\widetilde{M})\right).$$

Using the lift $\widetilde{\omega}$ of the fundamental form ω of M to $\widetilde{M}$, Gromov [236, Section 2] constructs a family of elliptic complexes $(\Omega^{p,*}(\widetilde{M}), \overline{\partial}_{p,*}[t])$ with $t \in \mathbb{R}$, which depends continuously on t and agrees with $(\Omega^{p,*}(\widetilde{M}), \overline{\partial}_{p,*})$ for $t = 0$. Moreover, he shows using a twisted L^2-index theorem (see also [361, Theorem 3.6 on page 223]) that for some non-constant polynomial $p(t)$

$$\operatorname{index}^{(2)}\left(\Omega^{p,*}(\widetilde{M}), \overline{\partial}_{p,*}[t]\right) = p(t).$$

This implies that there exist arbitrary small $t > 0$ with the property that at least one of the Hilbert modules $H^{p,q}_{(2)}\left(\Omega^{p,*}(\widetilde{M}), \overline{\partial}_{p,*}[t]\right) = \ker\left((\square_{p,q}[t])_{\min}\right)$ is different from zero.

We want to use this fact to show that the assumption $\mathcal{H}^{p,m-p}_{(2)}(\widetilde{M}) = 0$ leads to a contradiction. We already know that $\mathcal{H}^{p,q}_{(2)}(\widetilde{M}) = 0$ for $q \neq m - p$ from Theorem 11.16 and Theorem 11.27. In Theorem 11.30 we have already proved that $\alpha_r(\widetilde{M}) = \infty^+$ for all $r \geq 1$. Lemma 2.66 (2) implies that $\alpha((\Delta_r)_{\min} \colon \mathrm{dom}((\Delta_r)_{\min}) \subset L^2\Omega^r(\widetilde{M}) \to L^2\Omega^r(\widetilde{M})) = \infty^+$ for all $r \geq 0$. We conclude from Lemma 11.25 that

$$\alpha\left((\Box_{p,q})_{\min} \colon \mathrm{dom}((\Box_{p,q})_{\min}) \subset L^2\Omega^{p,q}(\widetilde{M}) \to L^2\Omega^{p,q}(\widetilde{M})\right) = \infty^+$$

for $q \geq 0$. Hence there is $\epsilon > 0$ such that for the spectral density function $F\left((\Box_{p,q})_{\min}\right)(\lambda) = 0$ holds for all $q \geq 0$ and $\lambda \leq \epsilon$. Since by construction $\Box_{p,q}[t] - \Box_{p,q}$ is a bounded operator with operator norm $\leq C \cdot t$ for some constant C, this contradicts the existence of arbitrary small $t > 0$ such that $\ker\left((\Box_{p,q}[t])_{\min}\right) \neq 0$ for some q. This finishes the proof of Theorem 11.35.
$\quad\square$

For applications in algebraic geometry the following consequence is crucial.

Corollary 11.36. *Let M be a closed Kähler hyperbolic manifold of complex dimension m and real dimension $n = 2m$. Then there exists a non-trivial holomorphic L^2-integrable m-form on $\widetilde{M}$.*

Proof. Choose a non-trivial element ϕ in $\mathcal{H}^{0,m}(\widetilde{M})$ whose existence follows from Theorem 11.35. Obviously $\Box_{0,m}(\phi) = 0$ implies $\overline{\partial}_{0,m}(\phi)$, i.e. ϕ is holomorphic. $\quad\square$

11.3 Miscellaneous

One may consider the following more general version of the Singer Conjecture.

Conjecture 11.37. (Singer Conjecture for contractible proper cocompact Poincaré G-CW-complexes).
Let X be a finite proper G-CW-complex. Suppose that X is contractible (after forgetting the group action). Suppose that there is an element $[X] \in H^G_{\dim(X)}(X;\mathbb{Q})$ for $\mathbb{Q}$ with the trivial action such that the $\mathbb{Q}G$-chain map $\cap[X]\colon C^{\dim(X)-*}(X) \to C_*(X)$, which is uniquely defined up to $\mathbb{Q}G$-chain homotopy, is a $\mathbb{Q}G$-chain homotopy equivalence. Then

$$b^{(2)}_p(X;\mathcal{N}(G)) = b^{(2)}_p(G) = 0 \qquad \textit{if } 2p \neq \dim(X).$$

This version seems to cut down the assumptions to the absolutely necessary and decisive properties, namely contractibility, Poincaré duality, properness and cocompactness. If M is an aspherical closed orientable manifold, then $\widetilde{M}$ with the canonical $\pi_1(M)$-action satisfies the assumptions of Conjecture 11.37. Hence Conjecture 11.37 implies the Singer Conjecture 11.1 for aspherical closed orientable manifolds. Notice that in the situation of Conjecture 11.37 the equality

$$b_p^{(2)}(X; \mathcal{N}(G)) = b_p^{(2)}(G)$$

follows from Theorem 6.54 (3). If a group G acts properly on a smooth contractible manifold M by orientation preserving diffeomorphisms, the assumption of Conjecture 11.37 are satisfied because there exist smooth equivariant triangulations [273]. Poincaré duality is explained for instance in [345, page 245].

Anderson [5], [6] has constructed simply connected complete Riemannian manifolds with negative sectional curvature such that the space of harmonic forms $\mathcal{H}_{(2)}^p(M)$ is non-trivial for some p with $2p \neq \dim(M)$. This manifold M does not admit a cocompact free proper action of a group G by isometries. Hence it yields a counterexample neither to the Singer Conjecture 11.1 nor to Conjecture 11.37. But this example shows the necessity of the condition of cocompactness for the Singer Conjecture 11.1 to be true.

The author does not know a counterexample to the conjecture that for a closed even-dimensional Riemannian manifold M with negative sectional curvature $\alpha_p(\widetilde{M}) = \infty^+$ holds for $p \geq 1$. This is true for locally symmetric spaces (see Corollary 5.16 (3)), Kähler manifolds (see Theorem 11.14) and for pinched curvature $-1 \leq \sec(M) < -(1 - \frac{2}{\dim(M)})^2$ (see Theorem 11.6).

Next we mention the work of Davis and Okun [130]. A simplicial complex L is called a *flag complex* if each finite non-empty set of vertices, which pairwise are connected by edges, spans a simplex of L. To such a flag complex they associate a right-angled Coxeter group W_L defined by the following presentation [130, Definition 5.1]. Generators are the vertices v of L. Each generator v satisfies $v^2 = 1$. If two vertices v and w span an edge, there is the relation $(vw)^2 = 1$. Given a finite flag complex L, Davis and Okun associate to it a finite proper W_L-CW-complex Σ_L, which turns out to be a model for the classifying space of the family of finite subgroups $E(G, \mathcal{FIN})$ [130, 6.1, 6.1.1 and 6.1.2]. Equipped with a specific metric, Σ_L turns out to be non-positive curved in a combinatorial sense, namely, it is a CAT(0)-space [130, 6.5.3]. If L is a generalized rational homology $(n-1)$-sphere, i.e. a homology $(n-1)$-manifold with the same rational homology as S^{n-1}, then Σ_L is a polyhedral homology n-manifold with rational coefficients [130, 7.4]. So Σ_L is a reminiscence of the universal covering of a closed n-dimensional manifold with non-positive sectional curvature and fundamental group G. In view of the Singer Conjecture 11.1 the conjecture makes sense that $b_p^{(2)}(\Sigma_L; \mathcal{N}(W_L)) = 0$ for $2p \neq n$ provided that the underlying topological space of L is S^{n-1} (or, more generally, that it is a homology $(n-1)$-sphere) [130, Conjecture 0.4 and 8.1]. Davis and Okun show that the conjecture is true in dimension $n \leq 4$ and that it is true in dimension $(n+1)$ if it holds in dimension n and n is odd [130, Theorem 9.3.1 and Theorem 10.4.1].

A group G is called *Kähler group* if it is the fundamental group of a closed Kähler manifold. A lot of information about Kähler groups can be found in [3]. For instance if G is a Kähler group with $b_1^{(2)}(G) \neq 0$, then it contains a subgroup G_0 of finite index for which there is an epimorphism $G_0 \to \pi_1(F_g)$ with finite kernel for some closed orientable surface F_g of genus $g \geq 2$ [235, 3.1 on page 69]. Thus a Kähler group has $b_1^{(2)}(G) = 0$ or looks like a surface group of genus ≥ 2.

Exercises

11.1. Construct a closed connected Riemannian manifold M of even dimension, whose sectional curvature is non-positive but not identically zero, such that $b_p^{(2)}(\widetilde{M}) = 0$ holds for all $p \geq 0$.

11.2. Show for a flat closed connected Riemannian manifold M that it is det-L^2-acyclic and $\rho^{(2)}(\widetilde{M}) = 0$.

11.3. Let M be a closed connected 4-manifold. Suppose that there is a fibration $F \to M \to B$ for closed connected manifolds F and B of dimension 1, 2 or 3. Prove:

(1) $b_0^{(2)}(\widetilde{M}) \neq 0$ if and only if F and B have finite fundamental groups. In this case

$$b_p^{(2)}(\widetilde{M}) = \frac{1}{|\pi_1(M)|} \qquad \text{for } p = 0, 4;$$
$$b_p^{(2)}(\widetilde{M}) = 0 \qquad \text{for } p \neq 0, 2, 4;$$
$$b_2^{(2)}(\widetilde{M}) = \chi(F) \cdot \chi(B) - \frac{2}{|\pi_1(M)|};$$
$$\chi(M) \quad > 0;$$

(2) Suppose that $\pi_1(F)$ is finite and $\pi_1(B)$ is infinite or that $\pi_1(F)$ is infinite and $\pi_1(B)$ is finite. Then

$$b_p^{(2)}(\widetilde{M}) = \frac{\chi(F) \cdot \chi(B)}{2} \qquad \text{for } p = 1, 3;$$
$$b_p^{(2)}(\widetilde{M}) = 0 \qquad \text{for } p \neq 1, 3;$$
$$\chi(M) \quad \leq 0;$$

(3) Suppose that F and B have infinite fundamental groups. Then

$$b_p^{(2)}(\widetilde{M}) = 0 \qquad \text{for } p \neq 2;$$
$$b_p^{(2)}(\widetilde{M}) = \chi(F) \cdot \chi(B) \qquad \text{for } p = 2;$$
$$\chi(M) \quad \geq 0;$$

(4) Suppose that M is aspherical. Show that $\chi(M) = \chi(B) \cdot \chi(F) \geq 0$ and that we have $\chi(M) > 0$ if and only if both F and B carry a hyperbolic structure.

11.4. Let M be a closed connected Riemannian manifold of even dimension such that its sectional curvature satisfies $-1 \leq \sec(M) < -(1 - \frac{2}{\dim(M)})^2$. Show that $b_p^{(2)}(\widetilde{M}) = 0$ for $2p \neq \dim(M)$ and $\alpha_p(\widetilde{M}) = \infty^+$ for $p \geq 1$ holds.

11.5. Let M be a Kähler manifold and $N \subset M$ be a complex submanifold. Show that N inherits the structure of a Kähler manifold. Show the analogous statement for Kähler hyperbolic instead of Kähler.

11.6. Let M be a Kähler manifold of complex dimension m. Let ω be its fundamental form and $d\mathrm{vol}$ be its volume form. Show

$$m! \cdot d\mathrm{vol} = \omega^m,$$

where ω^m denotes the m-fold wedge product of ω with itself.

11.7. Let M be a Kähler hyperbolic manifold. Show that its universal covering does not contain a closed complex submanifold of positive dimension.

11.8. Show that the torus T^{2n} carries the structure of a Kähler manifold but not of a Kähler hyperbolic manifold.

11.9. Let M be a Kähler manifold. Show that there is an isometric $\mathbb{C}$-isomorphism from $\mathcal{H}^{p,q}(\widetilde{M})$ to the complex dual of $\mathcal{H}^{q,p}(\widetilde{M})$.

11.10. Suppose that the group G contains a subgroup G_0 of finite index for which there is an epimorphism $G_0 \to \pi_1(F_g)$ with finite kernel for some closed connected orientable surface F_g of genus $g \geq 2$. Show that $b_1^{(2)}(G) > 0$.

11.11. A symplectic form ω on a smooth manifold of even dimension $n = 2m$ is a 2-form $\omega \in \Omega^2(M)$ such that $\omega_x^m \neq 0$ for all $x \in M$ and $d(\omega) = 0$. Suppose that M comes with a complex structure. We call ω compatible with the complex structure if $\omega(Jv, Jw) = \omega(v, w)$ holds, where J is multiplication with i. Show that a Kähler structure on a complex manifold is the same as a symplectic structure which is compatible with the complex structure.

12. The Zero-in-the-Spectrum Conjecture

Introduction

In this section we deal with the zero-in-the-spectrum Conjecture which to the author's knowledge appears for the first time in Gromov's article [233, page 120].

Conjecture 12.1 (Zero-in-the-spectrum Conjecture). *Let $\widetilde{M}$ be a complete Riemannian manifold. Suppose that $\widetilde{M}$ is the universal covering of an aspherical closed Riemannian manifold M (with the Riemannian metric coming from M). Then for some $p \geq 0$ zero is in the spectrum of the minimal closure*

$$(\Delta_p)_{\min} \colon \operatorname{dom}((\Delta_p)_{\min}) \subset L^2\Omega^p(\widetilde{M}) \to L^2\Omega^p(\widetilde{M})$$

of the Laplacian acting on smooth p-forms on $\widetilde{M}$.

Remark 12.2. Lott [318, page 347] gives five versions of this conjecture, stated as a question, where the conditions on $\widetilde{M}$ are varied. Namely,

(1) $\widetilde{M}$ is a complete Riemannian manifold;
(2) $\widetilde{M}$ is a complete Riemannian manifold with bounded geometry, i.e. the injectivity radius is positive and the sectional curvature is pinched between -1 and 1;
(3) $\widetilde{M}$ is a complete Riemannian manifold, which is *uniformly contractible*, i.e. for all $r > 0$ there is an $R(r) > 0$ such that for all $m \in M$ the open ball $B_r(m)$ around m of radius r is contractible within $B_{R(r)}(m)$;
(4) $\widetilde{M}$ is the universal covering of a closed Riemannian manifold M (with the Riemannian metric coming from M);
(5) $\widetilde{M}$ is the universal covering of an aspherical closed Riemannian manifold M (with the Riemannian metric coming from M).

Notice that version (5) agrees with Conjecture 12.1. Meanwhile there are counterexamples to version (4) due to Farber and Weinberger [187] which also yield counterexamples to versions (1) and (2). To the authors knowledge there are no counterexamples known to versions (3) and (5) (at the time of writing).

We will give a purely algebraic reformulation of the zero-in-the-spectrum Conjecture 12.1 in Section 12.1. Some evidence for the zero-in-the-spectrum Conjecture 12.1 is presented in Section 12.2. Most of the positive results are based on direct computations or on appropriate index theorems applied to pertubations of invertible operators. We deal with the counterexamples of Farber and Weinberger in the non-aspherical case in Section 12.3.

To understand this chapter some knowledge about Chapters 1 and 2 is required.

12.1 An Algebraic Formulation of the Zero-in-the-Spectrum Conjecture

Next we reformulate the zero-in-the-spectrum Conjecture 12.1 into a purely algebraic statement, namely, that for an aspherical closed manifold M with fundamental group $\pi = \pi_1(M)$ at least one of the homology groups $H_p^\pi(\widetilde{M}; \mathcal{N}(\pi))$ is different from zero. Recall that for a G-CW-complex X the $\mathcal{N}(G)$-module $H_p^G(X; \mathcal{N}(G))$ is defined by $H_p(\mathcal{N}(G) \otimes_{\mathbb{Z}G} C_*(X))$, where $\mathcal{N}(G)$ is just viewed as a ring, no topology enters. Analogously define $H_p^G(X; C_r^*(G))$, where $C_r^*(G)$ is the reduced C^*-algebra of G.

Lemma 12.3. *Let X be a free G-CW-complex of finite type. Let n be a non-negative integer or ∞. Then the following assertions are equivalent, where in assertion (5) we assume that X is a cocompact free proper G-manifold with G-invariant Riemannian metric.*

(1) We have

$$b_p^{(2)}(X; \mathcal{N}(G)) = 0 \qquad \text{for } p \leq n;$$
$$\alpha_p(X; \mathcal{N}(G)) = \infty^+ \qquad \text{for } p \leq n+1;$$

(2) $H_p^G(X; \mathcal{N}(G))$ vanishes for $p \leq n$;
(3) $H_p^G(X; C_r^*(G))$ vanishes for $p \leq n$;
(4) For each $p \leq n$ zero is not in the spectrum of the combinatorial Laplacian $\Delta_p \colon C_p^{(2)}(X) \to C_p^{(2)}(X)$, or, equivalently, Δ_p is invertible;
(5) For each $p \leq n$ zero is not in the spectrum of the minimal closure

$$(\Delta_p)_{\min} \colon \operatorname{dom}((\Delta_p)_{\min}) \subset L^2\Omega^p(X) \to L^2\Omega^p(X)$$

of the Laplacian acting on smooth p-forms on X.

Proof. (1) $\Rightarrow$ (5) We conclude from Lemma 2.66 (2) and Theorem 2.68 that assertion (1) is equivalent to the condition that for each $p \leq n$ there exists $\epsilon_p > 0$ such that the spectral density function F_p^Δ of $(\Delta_p)_{\min} \colon \operatorname{dom}((\Delta_p)_{\min}) \subset L^2\Omega^p(X) \to L^2\Omega^p(X)$ satisfies $F_p^\Delta(\lambda) = 0$ for $\lambda \leq \epsilon_p$. This is equivalent to the existence of $\epsilon_p > 0$ such that

$$\epsilon_p \cdot ||\omega||_{L^2} \leq ||(\Delta_p)_{\min}(\omega)||_{L^2} \qquad \text{for } \omega \in \text{dom}\,((\Delta_p)_{\min})\,.$$

We conclude that $(\Delta_p)_{\min}$ is injective and has closed image. Since $(\Delta_p)_{\min}$ is selfadjoint (see Lemma 1.70 (1)), we have $\text{im}\,((\Delta_p)_{\min})^{\perp} = \ker\,((\Delta_p)_{\min}) = 0$. Hence $(\Delta_p)_{\min}\colon \text{dom}\,((\Delta_p)_{\min}) \to L^2\Omega^p(X)$ is bijective. One easily checks that the inverse is a bounded operator. Hence zero is not in the spectrum of $(\Delta_p)_{\min}$ for $p \leq n$.

(5) $\Rightarrow$ (1) By assumption we can find for each $p \leq n$ a bounded operator $S_p\colon L^2\Omega^p(X) \to L^2\Omega^p(X)$ with $\text{im}(S_p) = \text{dom}\,((\Delta_p)_{\min})$ such that $S_p \circ (\Delta_p)_{\min} = \text{id}_{\text{dom}((\Delta_p)_{\min})}$ and $(\Delta_p)_{\min} \circ S_p = \text{id}_{L^2\Omega^p(X)}$. Consider $\omega \in \text{dom}\,((\Delta_p)_{\min})$. Choose $\phi \in L^2\Omega^p(X)$ with $S_p(\phi) = \omega$. Denote by $||S_p||_\infty$ the operator norm of S_p. We get

$$||(\Delta_p)_{\min}(\omega)||_{L^2} = ||(\Delta_p)_{\min} \circ S_p(\phi)||_{L^2} = ||\phi||_{L^2}$$
$$\geq ||S_p||_\infty^{-1} \cdot ||S_p(\phi)||_{L^2} = ||S_p||_\infty^{-1} \cdot ||\omega||_{L^2}.$$

This implies $F_p^\Delta(\lambda) = 0$ for $\lambda < ||S_p||_{L^2}^{-1}$. Hence (1) is true.

(1) $\Leftrightarrow$ (4) Its proof is analogous but simpler because the combinatorial Laplacian is a bounded everywhere defined operator.

(1) $\Leftrightarrow$ (2) This follows from Lemma 6.98.

(4) $\Rightarrow$ (3) Let $f\colon \mathbb{C}G^m \to \mathbb{C}G^n$ be a $\mathbb{C}G$-linear map. Let A be the (m, n)-matrix describing f. Let A^* be obtained from A by transposing and applying to each entry the standard involution $\mathbb{C}G \to \mathbb{C}G$ which sends $\sum_{g \in G} \lambda_g \cdot g$ to $\sum_{g \in G} \overline{\lambda_g} \cdot g^{-1}$. Define the adjoint $f^*\colon \mathbb{C}G^n \to \mathbb{C}G^m$ of f to be the $\mathbb{C}G$-linear map given by A^*. Recall that $C_*(X)$ is the cellular $\mathbb{Z}G$-chain complex and the finitely generated Hilbert $\mathcal{N}(G)$-chain complex $C_*^{(2)}(X)$ is defined by $l^2(G) \otimes_{\mathbb{Z}G} C_*(X)$. Define $\Delta_p^c\colon C_p(X) \to C_p(X)$ to be the $\mathbb{C}G$-linear map $c_{p+1}c_{p+1}^* + c_p^*c_p$, where the adjoints of c_p and c_{p+1} are taken with respect to a cellular basis. (The cellular basis is not quite unique but the adjoints are). Then the combinatorial Laplacian $\Delta_p\colon C_p^{(2)}(X) \to C_p^{(2)}(X)$ is $\text{id}_{l^2(G)} \otimes_{\mathbb{Z}G} \Delta_p^c$. We conclude from (6.22) that Δ_p is invertible if and only if $\text{id}_{\mathcal{N}(G)} \otimes_{\mathbb{Z}G} \Delta_p^c\colon \mathcal{N}(G) \otimes_{\mathbb{Z}G} C_p(X) \to \mathcal{N}(G) \otimes_{\mathbb{Z}G} C_p(X)$ is bijective. We will need the following standard result about C^*-algebras. If $A \subset B$ is an inclusion of C^*-algebras and $a \in A$ is invertible in B, then a is already invertible in A [282, Proposition 4.1.5 on page 241]. If we apply this to $A = M_l(C_r^*(G))$ and $B = M_l(\mathcal{N}(G))$, we get that $\text{id}_{C_r^*(G)} \otimes_{\mathbb{Z}G} \Delta_p^c\colon C_r^*(G) \otimes_{\mathbb{Z}G} C_p(X) \to C_r^*(G) \otimes_{\mathbb{Z}} C_p(X)$ is invertible for $p \leq n$. One easily checks $c_p \circ \Delta_p^c = \Delta_{p-1}^c \circ c_p$ for all p. This implies that $\text{id}_{C_r^*(G)} \otimes_{\mathbb{Z}G} c_p \circ \left(\text{id}_{C_r^*(G)} \otimes_{\mathbb{Z}G} \Delta_p^c\right)^{-1} = \left(\text{id}_{C_r^*(G)} \otimes_{\mathbb{Z}G} \Delta_{p-1}^c\right)^{-1} \circ \text{id}_{C_r^*(G)} \otimes_{\mathbb{Z}G} c_p$ holds for $p \leq n$. For $p \leq n$ consider $x \in \ker\left(\text{id}_{C_r^*(G)} \otimes_{\mathbb{Z}G} c_p\right)$. Then x lies in the image of $\text{id}_{C_r^*(G)} \otimes_{\mathbb{Z}G} c_{p+1}$ because of the easily verified equation

$$\mathrm{id}_{C_r^*(G)} \otimes_{\mathbb{Z}G} c_{p+1} \circ \mathrm{id}_{C_r^*(G)} \otimes_{\mathbb{Z}G} c_{p+1}^* \circ \left(\mathrm{id}_{C_r^*(G)} \otimes_{\mathbb{Z}G} \Delta_p^c \right)^{-1} (x) = x.$$

This proves $H_p^G(X; C_r^*(G)) = 0$ for $p \leq n$.

$(3) \Rightarrow (2)$ Since $C_r^*(G) \otimes_{\mathbb{Z}G} C_*(X)$ is a projective $C_r^*(G)$-chain complex whose modules vanish in negative dimensions, (3) implies that it is $C_r^*(G)$-chain homotopy equivalent to a projective $C_r^*(G)$-chain complex P_* with $P_p = 0$ for $p \leq n$. Since $\mathcal{N}(G) \otimes_{\mathbb{Z}G} C_*(X)$ is $\mathcal{N}(G)$-chain homotopy equivalent to $\mathcal{N}(G) \otimes_{C_r^*(G)} P_*$, assertion (2) follows. This finishes the proof of Lemma 12.3. $\square$

Remark 12.4. Let M be an aspherical closed Riemannian manifold with fundamental group π. We conclude from Lemma 12.3 that $\widetilde{M}$ satisfies the zero-in-the-spectrum Conjecture 12.1 if and only if $H_p^\pi(E\pi; \mathcal{N}(\pi)) \neq 0$ for at least one $p \geq 0$. In particular the answer to the problem, whether the universal covering of a closed connected Riemannian manifold M satisfies the zero-in-the-spectrum Conjecture 12.1, depends only on the homotopy type of M and not on its Riemannian metric.

So one may ask the question for which groups G, whose classifying space BG possesses a finite CW-model, finite dimensional CW-model or CW-model of finite type, $H_p^\pi(E\pi; \mathcal{N}(\pi)) = 0$ holds for $p \geq 0$. In comparison with the zero-in-the-spectrum Conjecture 12.1, the condition Poincaré duality is dropped but a finiteness condition and the contractibility of $E\pi$ remain. Without the finiteness assumption such groups G exists. For example the group $G = \prod_{n=1}^{\infty} (\mathbb{Z} * \mathbb{Z})$ satisfies $H_p^G(EG; \mathcal{N}(G)) = 0$ for all $p \geq 0$ (see Lemma 12.11 (5))).

12.2 Survey on Positive Results about the Zero-in-the-Spectrum Conjecture

In this section we present some positive results on the zero-in-the-spectrum Conjecture 12.1. Notice that it suffices to consider closed connected manifolds which are orientable. If M is not orientable, one can pass to the orientation covering $\overline{M} \to M$ and use the fact that $b_p^{(2)}(\widetilde{M}; \mathcal{N}(\pi_1(M))) = 2 \cdot b_p^{(2)}(\widetilde{M}; \mathcal{N}(\pi_1(\overline{M})))$ and $\alpha(\widetilde{M}; \mathcal{N}(\pi_1(M))) = \alpha(\widetilde{M}; \mathcal{N}(\pi_1(\overline{M})))$ holds (see Theorem 1.35 (9) and Theorem 2.55 (6)).

12.2.1 The Zero-in-the-Spectrum Conjecture for Low-Dimensional Manifolds

The zero-in-the-spectrum Conjecture 12.1 is true for universal coverings of closed 2-dimensional manifolds by Example 1.36, actually zero is contained in the spectrum of the Laplacian in dimension 1.

We have already explained in Section 4.1 that the Sphere Theorem [252, Theorem 4.3] implies that a closed connected orientable 3-manifold is aspherical if and only if it is homotopy equivalent to a closed connected orientable irreducible 3-manifold M with infinite fundamental group. If M is non-exceptional, then $\widetilde{M}$ satisfies the zero-in-the-spectrum Conjecture 12.1 by Theorem 4.2 (2) and (4). More precisely, zero is contained in the spectrum of the Laplacian in dimension 1 and 2. In other words, if Thurston's Geometrization Conjecture is true, then the zero-in-the-spectrum Conjecture 12.1 holds for universal coverings of aspherical closed 3-manifolds.

Lemma 12.5. *Let M be a closed connected 4-dimensional manifold with fundamental group π. Then zero is not the spectrum of $(\Delta_p)_{\min}\colon \operatorname{dom}((\Delta_p)_{\min}) \subset L^2\Omega^p(\widetilde{M}) \to L^2\Omega^p(\widetilde{M})$ for all $p \geq 0$ if and only if $H_p^\pi(E\pi; \mathcal{N}(\pi)) = 0$ for $p = 0, 1$ and $\chi(M) = 0$.*

Proof. We conclude from Lemma 12.3 that zero is not the spectrum of $(\Delta_p)_{\min}\colon \operatorname{dom}((\Delta_p)_{\min}) \subset L^2\Omega^p(\widetilde{M}) \to L^2\Omega^p(\widetilde{M})$ for all $p \geq 0$ if and only if $b_p^{(2)}(\widetilde{M}) = 0$ and $\alpha_{p+1}(\widetilde{M}) = \infty^+$ for $p \geq 0$. Since the classifying map $f\colon M \to B\pi$ for $\pi = \pi_1(M)$ is 2-connected, we have $b_p^{(2)}(\widetilde{M}) = b_p^{(2)}(\pi)$ for $p = 0, 1$ and $\alpha_p(\widetilde{M}) = \alpha_p(\pi)$ for $p \leq 2$ (see Theorem 1.35 (1) and Theorem 2.55 (1)). Now the claim follows from Lemma 12.3, Poincaré duality (see Theorem 1.35 (3) and Theorem 2.55 (2)) and the Euler-Poincaré formula (see Theorem 1.35 (2)). $\square$

The possible geometries on 4-manifolds have been classified by Wall [512]. If a closed connected 4-manifold possesses a geometry, one can compute the L^2-Betti numbers and get enough information about their Novikov-Shubin invariants to conclude that M satisfies the zero-in-the-spectrum Conjecture 12.1 [329, Theorem 4.3] and [318, Proposition 18]. Let M be a 4-manifold with a complex structure. Then its universal covering satisfies the zero-in-the-spectrum Conjecture 12.1. Otherwise we would get a contradiction. Namely, a closed 4-manifold with complex structure, which satisfies $b_p^{(2)}(\widetilde{M}) = 0$ for $p \geq 0$ and hence $\chi(M) = \operatorname{sign}(M) = 0$ by Theorem 1.35 (2) and Lemma 7.22, possesses a geometry [512, page 148-149].

12.2.2 The Zero-in-the-Spectrum Conjecture for Locally Symmetric Spaces

The zero-in-the-spectrum Conjecture 12.1 holds for the universal covering $\widetilde{M}$ of aspherical closed locally symmetric spaces M by Corollary 5.16 (2). Actually zero is in the spectrum of the Laplacian in dimension p, if $\frac{\dim(M)-1}{2} \leq p \leq \frac{\dim(M)+1}{2}$.

12.2.3 The Zero-in-the-Spectrum Conjecture for Kähler Hyperbolic Manifolds

If M is a closed Kähler hyperbolic manifold (see Definition 11.12), then zero is in the spectrum of the Laplacian on the universal covering in dimension $\dim(M)/2$ and in particular M satisfies the zero-in-the-spectrum Conjecture 12.1 by Theorem 11.14.

12.2.4 The Zero-in-the-Spectrum Conjecture for HyperEuclidean Manifolds

A complete Riemannian manifold N is called *hyperEuclidean* if there is a proper distance non-increasing map $F \colon N \to \mathbb{R}^n$ of nonzero degree for $n = \dim(N)$. Proper means that the preimage of a compact set is again compact. The *degree* of a proper map $f \colon N \to P$ of oriented manifolds of the same dimension n is the integer $\deg(f)$ for which the map induced by f on the n-th homology with compact support sends the fundamental class $[N]$ to the fundamental class $[P]$ multiplied by $\deg(f)$. Suppose that $f \colon N \to P$ is smooth and proper and y is a regular value of f, i.e. $T_x f \colon T_x N \to T_y P$ is bijective for all $x \in f^{-1}(y)$. Then $f^{-1}(y)$ consists of finitely many points. Define $\epsilon(x)$ for $x \in f^{-1}(y)$ to be 1 if $T_x f$ is orientation preserving and to be -1 otherwise. Then $\deg(f) = \sum_{x \in f^{-1}(y)} \epsilon(x)$.

Any simply connected Riemannian manifold with non-positive sectional curvature is hyperEuclidean, namely, take f to be the inverse of the exponential map $\exp_y \colon T_y M \to M$ for some point $y \in N$.

Theorem 12.6. *(1) Let N be a complete Riemannian manifold which is hyperEuclidean. Then zero is in the spectrum of $(\Delta_p)_{\min}$ if $\frac{\dim(N)-1}{2} \le p \le \frac{\dim(N)+1}{2}$;*

(2) Let M be a closed Riemannian manifold with non-positive sectional curvature. Then the universal covering $\widetilde{M}$ satisfies the zero-in-the-spectrum Conjecture 12.1.

Proof. (1) We only sketch the proof, more details can be found in [233] and [318, Proposition 7]. Suppose that $n = \dim(N)$ is even and (1) is not true. This leads to a contradiction by the following argument. Take a non-trivial complex vector bundle over $\mathbb{R}^n$ with fixed trivialization at infinity and pull it back via the composition of f with $\epsilon \cdot \mathrm{id} \colon \mathbb{R}^n \to \mathbb{R}^n$. Then the pullback F_ϵ carries an ϵ-flat connection which is trivial at infinity. The signature operator is invertible, since we assume that zero is not in the spectrum of $(\Delta_p)_{\min}$ for all p. If we twist the signature operator with this ϵ-flat connection, then for small enough ϵ this pertubation is still invertible and hence its index is zero. On the other hand a relative index theorem [239, Proposition 4.13 on page 121] shows that the index of the pertubated operator is non-zero for small ϵ. Here one needs the fact that the vector bundle F is non-trivial and f has degree different from zero. The case, where n is odd, is reduced to the case, where n is even, by considering $N \times \mathbb{R}$.

(2) This follows from (1) since $\widetilde{M}$ is hyperEuclidean.   □

12.2.5 The Zero-in-the-Spectrum Conjecture and the Strong Novikov Conjecture

The *strong Novikov Conjecture* says for a group G that the so called assembly map $a \colon K_p(BG) \to K_p^{\text{top}}(C_r^*(G))$ from the topological K-homology of BG to the topological K-theory of the reduced group C^*-algebra $C_r^*(G)$ of G is rationally injective for all $p \in \mathbb{Z}$. For more information about this conjecture we refer for instance to [199]. Notice that the Baum-Connes Conjecture implies the strong Novikov Conjecture.

Theorem 12.7. *Suppose that M is an aspherical closed Riemannian manifold. If its fundamental group $\pi = \pi_1(M)$ satisfies the strong Novikov Conjecture, then $\widetilde{M}$ satisfies the zero-in-the-spectrum Conjectur 12.1. Actually zero is in the spectrum of the Laplacian on $\widetilde{M}$ in dimension p if $\frac{\dim(M)-1}{2} \le p \le \frac{\dim(M)+1}{2}$.*

Proof. We give a sketch of the proof. More details can be found in [318, Corollary 4]. We only explain that the assumption that in every dimension zero is not in the spectrum of the Laplacian on $\widetilde{M}$, yields a contradiction in the case that $n = \dim(M)$ is even. Namely, this assumption implies that the index of the signature operator twisted with the flat bundle $\widetilde{M} \times_\pi C_r^*(\pi) \to M$ in $K_0(C_r^*(\pi))$ is zero. This index is the image of the class $[D]$ defined by the signature operator in $K_0(B\pi)$ under the assembly $a \colon K_0(B\pi) \to K_0(C_r^*(\pi))$. Since by assumption the assembly map is injective, this implies $[D] = 0$ in $K_0(B\pi)$. Notice that M is aspherical by assumption and hence $M = B\pi$. The homological Chern character defines an isomorphism

$$\mathbb{Q} \otimes_{\mathbb{Z}\pi} K_0(B\pi) = \mathbb{Q} \otimes_{\mathbb{Z}\pi} K_0(M) \to H_{\text{ev}}(M; \mathbb{Q}) = \bigoplus_{p \ge 0} H_{2p}(M; \mathbb{Q}),$$

which sends D to the Poincaré dual $L(M) \cap [M]$ of the Hirzebruch L-class $L(M) \in H^{\text{ev}}(M; \mathbb{Q})$. This implies that $L(M) \cap [M] = 0$ and hence $L(M) = 0$. This contradicts the fact that the component of $L(M)$ in $H^0(M; \mathbb{Q})$ is 1.   □

12.2.6 The Zero-in-the-Spectrum Conjecture and Finite Asymptotic Dimension

A metric space X has *finite asymptotic dimension* if there exists an integer n such that for any $r > 0$ there is a covering $X = \bigcup_{i \in I} C_i$ together with a constant D such that the diameter of the subset $C_i \subset X$ is bounded by D for all $i \in I$ and a ball of radius r intersects at most $(n+1)$ of the sets C_i non-trivially. The smallest such n is called the *asymptotic dimension*. This is the coarse geometric analog of the notion of finite covering dimenson in topology. The following theorem is due to Yu [528, Theorem 1.1 and Corollary 7.4]. Its proof is based on the fact that the coarse Baum-Connes Conjecture holds for proper metric spaces with finite asymptotic dimension [528, Corollary 7.1] and coarse index theory.

Theorem 12.8. *(1) Let M be a uniformly contractible Riemannian mani-fold without boundary which has finite asymptotic dimension. Then zero is in the spectrum of $(\Delta_p)_{\min}$ for at least one $p \geq 0$;*

(2) Let G be a group such that there is a finite model for BG and G re-garded as a metric space with respect to the word-length metric has finite asymptotic dimension. Then G satisfies the strong Novikov Conjecture;

(3) Let M be an aspherical closed manifold. Suppose that $\pi_1(M)$ regarded as a metric space with respect to the word-length metric has finite asymp-totic dimension. Then the universal covering $\widetilde{M}$ satisfies the zero-in-the-spectrum Conjecture 12.1.

If G is a finitely generated group which is word-hyperbolic in the sense of [234], then G regarded as a metric space with respect to the word-length met-ric has finite asymptotic dimension [237, Remark on page 31]. A torsionfree word-hyperbolic group has a finite model for BG given by the Rips complex [65, Corollary 3.26 in III. Γ on page 470] and hence satisfies the assumptions of Theorem 12.8 (2).

If the finitely generated group G regarded as a metric space with respect to the word-length metric has finite asymptotic dimension, then any finitely generated subgroup $H \subset G$ has the same property [528, Proposition 6.2].

Meanwhile Gromov has announced that he can construct groups with finite models for BG such that G as a metric space with respect to the word-length metric does not have finite asymptotic dimension.

12.3 Counterexamples to the Zero-in-the-Spectrum Conjecture in the Non-Aspherical Case

For some time it was not known whether the zero-in-the-spectrum Conjecture 12.1 is true if one drops the condition aspherical. Farber and Weinberger [187] gave the first example of a closed Riemannian manifold for which zero is not in the spectrum of the minimal closure $(\Delta_p)_{\min} \colon \operatorname{dom}((\Delta_p)_{\min}) \subset L^2 \Omega^p(\widetilde{M}) \to L^2 \Omega^p(\widetilde{M})$ of the Laplacian acting on smooth p-forms on $\widetilde{M}$ for each $p \geq 0$. We will present a short proof of a more general result following ideas of Higson, Roe and Schick [258] which also were obtained independently by the author. We want to emphasize that the key idea is due to Farber and Weinberger [187]. Namely, they considered $H_p^G(\widetilde{X}; C_r^*(G; \mathbb{R}))$ instead of $H_p^G(\widetilde{X}; \mathcal{N}(G))$, where $C_r^*(G;\mathbb{R})$ is the reduced real group C^*-algebra. This is possible in the context of the zero-in-the-spectrum Conjecture 12.1 since $M_l(C_r^*(G;\mathbb{R}))$ is division closed in $M_l(\mathcal{N}(G;\mathbb{R}))$ (see [282, Proposition 4.1.5 on page 241]). The crucial observation is that for any invertible element in $M_l(C_r^*(G;\mathbb{R}))$ we can find an element $B \in M_l(\mathbb{Q}G)$, which is arbitrary close to A in the C^*-norm and hence invertible in $M_l(C_r^*(G;\mathbb{R}))$. (This is not true for $\mathcal{N}(G;\mathbb{R})$ with its C^*-norm because $C_r^*(G;\mathbb{R}) \neq \mathcal{N}(G;\mathbb{R})$ in general). We have used this fact at several places before when we applied index theorems to small pertubations of invertible operators of Hilbert modules over C^*-algebras.

Theorem 12.9. *Let G be a finitely presented group. Then the following assertions are equivalent:*

(1) We have $H_p^G(EG;\mathcal{N}(G)) = 0$ for $p = 0,1,2$;
(2) We have $H_p^G(EG;C_r^*(G)) = 0$ for $p = 0,1,2$;
(3) There is a finite 3-dimensional CW-complex X with fundamental group $\pi_1(X) = G$ such that $H_p^G(\widetilde{X};C_r^*(G)) = 0$ for $p \geq 0$;
(4) There is a finite 3-dimensional CW-complex X with fundamental group $\pi_1(X) = G$ such that $H_p^G(\widetilde{X};\mathcal{N}(G)) = 0$ for $p \geq 0$;
(5) For any integer $n \geq 6$ there is a connected closed orientable Riemannian manifold M with fundamental group $\pi_1(M) = G$ of dimension n such that $H_p^G(\widetilde{M};\mathcal{N}(G)) = 0$ for $p \geq 0$;
(6) For any integer $n \geq 6$ there is a connected closed orientable Riemannian manifold M of dimension n with fundamental group $\pi_1(M) = G$ such that for each $p \geq 0$ zero is not in the spectrum of $(\Delta_p)_{\min}\colon \mathrm{dom}\,((\Delta_p)_{\min}) \subset L^2\Omega^p(\widetilde{M}) \to L^2\Omega^p(\widetilde{M})$.

Proof. (1) $\Rightarrow$ (2) This follows from Lemma 12.3.

(2) $\Rightarrow$ (3) Since G is finitely presented, we can find a CW-model for BG whose 2-skeleton is finite. By assumption $H_p^G(EG;C_r^*(G)) = 0$ for $p = 0,1,2$. Let $C_r^*(G;\mathbb{R})$ be the reduced real group C^*-algebra of G. Since $C_r^*(G) = \mathbb{C} \otimes_\mathbb{R} C_r^*(G;\mathbb{R})$, we conclude $H_p^G(EG;C_r^*(G;\mathbb{R})) = 0$ for $p = 0,1,2$. Let Z be the 2-skeleton of BG. Then we obtain an exact sequence of $C_r^*(G;\mathbb{R})$-modules $0 \to H_2^G(\widetilde{Z};C_r^*(G;\mathbb{R})) \to C_r^*(G;\mathbb{R}) \otimes_{\mathbb{Z}G} C_2(\widetilde{Z}) \to C_r^*(G;\mathbb{R}) \otimes_{\mathbb{Z}G} C_1(\widetilde{Z}) \to C_r^*(G;\mathbb{R}) \otimes_{\mathbb{Z}G} C_0(\widetilde{Z}) \to 0$. Since the $C_r^*(G;\mathbb{R})$-module $C_r^*(G;\mathbb{R}) \otimes_{\mathbb{Z}G} C_p(\widetilde{Z})$ is finitely generated free for $p = 0,1,2$, the $C_r^*(G;\mathbb{R})$-module $H_2^G(\widetilde{Z};C_r^*(G;\mathbb{R}))$ becomes finitely generated free after taking the direct sum with $C_r^*(G;\mathbb{R})^l$ for an appropriate integer $l \geq 0$. Put $Y = Z \vee \bigvee_{i=1}^l S^2$. Then Y is a finite 2-dimensional CW-complex with $\pi_1(Y) = G$ such that $H_p^G(\widetilde{Y};C_r^*(G))$ vanishes for $p = 0,1$ and is a finitely generated free $C_r^*(G)$-module for $p = 2$.

The universal coefficient spectral sequence converges to $H_{p+q}^G(\widetilde{Y};C_r^*(G;\mathbb{R}))$ and its E^2-term is $E_{p,q}^2 = \mathrm{Tor}_p^{\mathbb{Z}G}(H_q(\widetilde{Y});C_r^*(G;\mathbb{R}))$. Since $H_1(\widetilde{Y}) = 0$, it gives an exact sequence of $C_r^*(G;\mathbb{R})$-modules

$$C_r^*(G;\mathbb{R}) \otimes_{\mathbb{Z}G} H_2(\widetilde{Y}) \xrightarrow{\psi} H_2^G(\widetilde{Y}; C_r^*(G;\mathbb{R})) \to H_2^G(EG; C_r^*(G;\mathbb{R})) \to 0,$$

where ψ is the obvious map. Since $H_2^G(EG; C_r^*(G;\mathbb{R})) = 0$ by assumption, $\psi \colon C_r^*(G;\mathbb{R}) \otimes_{\mathbb{Z}G} H_2(\widetilde{Y}) \to H_2^G(\widetilde{Y}; C_r^*(G;\mathbb{R}))$ is surjective.

Let $\phi \colon H_2(\widetilde{Y}) \to C_r^*(G;\mathbb{R}) \otimes_{\mathbb{Z}G} H_2(\widetilde{Y})$ be the $\mathbb{Z}G$-map sending u to $1 \otimes_{\mathbb{Z}G} u$. Next we show that the image of $\psi \circ \phi \colon H_2(\widetilde{Y}) \to H_2^G(\widetilde{Y}; C_r^*(G;\mathbb{R}))$ contains a $C_r^*(G;\mathbb{R})$-basis. Fix a $C_r^*(G;\mathbb{R})$-basis $\{b_1, b_2, \ldots, b_l\}$ of $H_2^G(\widetilde{Y}; C_r^*(G;\mathbb{R}))$. Since ψ is surjective, we can find elements $c_{i,j} \in C_r^*(G;\mathbb{R})$ and $h_{i,j} \in H_2(\widetilde{Y})$ such that

$$b_i = \sum_{j_i=1}^{s_i} c_{i,j_i} \cdot \psi \circ \phi(h_{i,j_i})$$

holds for $i = 1, 2, \ldots, l$.

Let $\mu \colon \bigoplus_{i=1}^{l} C_r^*(G;\mathbb{R})^{s_i} \to M_l(C_r^*(G;\mathbb{R}))$ be the $C_r^*(G;\mathbb{R})$-linear map which is uniquely determined by the property

$$\sum_{k=1}^{l} (\mu(x_1, \ldots, x_l))_{i,k} \cdot b_k = \sum_{j_i=1}^{s_i} x_{i,j_i} \cdot \psi \circ \phi(h_{i,j_i}) \qquad \text{for } i = 1, 2, \ldots, l.$$

We equip $C_r^*(G;\mathbb{R})^m$ with the norm $||(x_1, \ldots, x_m)||_1 = \sum_{i=1}^{m} ||x_i||_{C_r^*(G;\mathbb{R})}$ and $M_l(C_r^*(G;\mathbb{R}))$ with the standard C^*-norm. Since μ is $C_r^*(G;\mathbb{R})$-linear, it is continuous with respect to these norms by the following calculation. Here e_{i,j_i} is the obvious element of the standard basis for the source of μ and $C := \max\{||\mu(e_{i,j_i})||_{M_l(C_r^*(G;\mathbb{R}))} \mid 1 \leq j_i \leq s_i, 1 \leq i \leq l\}$

$$||\mu(x)||_{M_l(C_r^*(G;\mathbb{R}))} \leq \left|\left|\sum_{i=1}^{l}\sum_{j_i=1}^{s_i} x_{i,j_i} \cdot \mu(e_{i,j_i})\right|\right|_{M_l(C_r^*(G;\mathbb{R}))}$$

$$\leq \sum_{i=1}^{l}\sum_{j_i=1}^{s_i} ||x_{i,j_i} \cdot \mu(e_{i,j_i})||_{M_l(C_r^*(G;\mathbb{R}))}$$

$$\leq \sum_{i=1}^{l}\sum_{j_i=1}^{s_i} ||x_{i,j_i}||_{C_r^*(G;\mathbb{R})} \cdot ||\mu(e_{i,j_i})||_{M_l(C_r^*(G;\mathbb{R}))}$$

$$\leq \sum_{i=1}^{l}\sum_{j_i=1}^{s_i} ||x_{i,j_i}||_{C_r^*(G;\mathbb{R})} \cdot C$$

$$= C \cdot ||x||_1.$$

The set of invertible matrices $Gl_l(C_r^*(G;\mathbb{R})) \subset M_l^*(C_r^*(G;\mathbb{R}))$ is open. This standard fact for C^*-algebras follows from the observation that for an element A in a C^*-algebra with norm $||A|| < 1$ the sequence $\sum_{n \geq 0} A^n$ converges to an inverse of $1 - A$. With respect to the C^*-norm $\mathbb{Q}G \subset C_r^*(G;\mathbb{R})$ is dense. Since μ sends $\left((c_{1,j_1})_{j_1=1}^{s_1}, \ldots, (c_{l,j_l})_{j_l=1}^{s_l}\right)$ to the identity matrix, we can find $q := \left((q_{1,j_1})_{j_1=1}^{s_1}, \ldots, (q_{l,j_l})_{j_l=1}^{s_l}\right)$ in $\bigoplus_{i=1}^{l} \mathbb{Q}G^{s_i}$ such that $\mu(q)$ lies in $GL_l(C_r^*(G;\mathbb{R}))$. Choose an integer N such that each number $n_{i,j_i} := N \cdot q_{i,j_i}$ is an integer. Obviously $\mu(N \cdot q) = N \cdot \mu(q)$ lies in $GL_l(C_r^*(G;\mathbb{R}))$. Define $u_i \in H_2(\widetilde{Y})$ for $i = 1, 2, \ldots, l$ by $u_i = \sum_{j_i=1}^{s_i} n_{i,j} \cdot h_{i,j}$. Then $\{\psi \circ \phi(u_i) \mid i = 1, 2, \ldots, l\}$ is a $C_r^*(G;\mathbb{R})$-basis for $H_2^G(\widetilde{Y}; C_r^*(G;\mathbb{R}))$.

Since $\widetilde{Y}$ is simply connected, the Hurewicz homomorphism $h \colon \pi_2(\widetilde{Y}) \to H_2(\widetilde{Y})$ is surjective [521, Corollary 7.8 in Chapter IV.7 on page 180]. Hence we can find maps $a_i \colon S^2 \to \widetilde{Y}$ for $i = 1, 2, \ldots, l$ such that the composition $\psi \circ \phi \circ h \colon \pi_2(\widetilde{Y}) \to H_2^G(\widetilde{Y}, C_r^*(G;\mathbb{R}))$ sends $\{[a_1], \ldots, [a_l]\}$ to a $C_r^*(G;\mathbb{R})$-basis. Now attach to Z a 3-cell with attaching map $p \circ a_i \colon S^2 \to Y$ for each $i = 1, 2, \ldots, l$, where $p \colon \widetilde{Y} \to Y$ is the projection. We obtain a CW-complex X such that the inclusion $Y \to X$ is 2-connected. In particular $\pi_1(X) = \pi_1(Y) = G$ and $H_p^G(\widetilde{X}; C_r^*(G;\mathbb{R})) = H_p^G(\widetilde{Y}; C_r^*(G;\mathbb{R}))$ for $p = 0, 1$. Moreover, there is an exact sequence of $C_r^*(G;\mathbb{R})$-modules

$$0 \to H_3^G(\widetilde{X}; C_r^*(G;\mathbb{R})) \to C_r^*(G;\mathbb{R}) \otimes_{\mathbb{Z}G} C_3(\widetilde{X}) \xrightarrow{\nu} H_2^G(\widetilde{Y}; C_r^*(G;\mathbb{R}))$$

$$\to H_2^G(\widetilde{X}; C_r^*(G;\mathbb{R})) \to 0,$$

where ν sends $1 \otimes v_i$ for v_i the i-th element of the cellular basis of $C_3(\widetilde{X}) = \mathbb{Z}G^l$ to $\psi \circ \phi \circ h([a_i])$. Hence ν is an isomorphism. This implies $H_p^G(\widetilde{X}, C_r^*(G;\mathbb{R})) = 0$ for $p = 2, 3$ and hence $H_p^G(\widetilde{X}, C_r^*(G)) = 0$ for $p \geq 0$.

(3) $\Leftrightarrow$ (4) This follows from Lemma 12.3.

(4) $\Rightarrow$ (5) We can find an embedding of X in $\mathbb{R}^{n+1}$ for $n \geq 6$ with regular neighborhood N with boundary ∂N [445, chapter 3]. Then X and N are homotopy equivalent and the inclusion $\partial N \to N$ is 2-connected. In particular we get identifications $G = \pi_1(X) = \pi_1(N) = \pi_1(\partial N)$ and $H_p^G(\widetilde{N}; \mathcal{N}(G)) = H_p^G(\widetilde{X}; \mathcal{N}(G)) = 0$ for $p \geq 0$. By Poincaré duality $H_p^G(\widetilde{N}, \widetilde{\partial N}; \mathcal{N}(G))$ is $\mathcal{N}(G)$-isomorphic to $H^{n+1-p}(\widetilde{N}; \mathcal{N}(G))$, where the latter is defined by the cohomology of $\hom_{\mathbb{Z}G}(C_*(\widetilde{N}), \mathcal{N}(G))$. There is a canonical $\mathcal{N}(G)$-isomorphism $\hom_{\mathbb{Z}G}(C_*(\widetilde{N}), \mathcal{N}(G)) \cong \hom_{\mathcal{N}(G)}(\mathcal{N}(G) \otimes_{\mathbb{Z}G} C_*(\widetilde{N}), \mathcal{N}(G))$. Since $\mathcal{N}(G) \otimes_{\mathbb{Z}G} C_*(\widetilde{N})$ is a free $\mathcal{N}(G)$-chain complex with trivial homology, it is contractible. This implies that $\hom_{\mathcal{N}(G)}(\mathcal{N}(G) \otimes_{\mathbb{Z}G} C_*(\widetilde{N}), \mathcal{N}(G))$ is contractible. Hence $H_p^G(\widetilde{N}, \widetilde{\partial N}; \mathcal{N}(G))$ vanishes for $p \geq 0$. The long exact homology sequence

$$\ldots \to H_{p+1}^G(\widetilde{N}, \widetilde{\partial N}; \mathcal{N}(G)) \to H_p^G(\widetilde{\partial N}; \mathcal{N}(G)) \to H_p^G(\widetilde{N}; \mathcal{N}(G))$$

$$\to H_p^G(\widetilde{N}, \widetilde{\partial N}; \mathcal{N}(G)) \to \ldots$$

implies that ∂N is a closed connected orientable Riemannian manifold with $G = \pi_1(\widetilde{\partial N})$ and $H_p^G(\widetilde{\partial N}; \mathcal{N}(G)) = 0$ for $p \geq 0$.

$(5) \Rightarrow (1)$ This follows from the fact that the classifying map $M \to BG$ is 2-connected and hence induces epimorphisms $H_p^G(\widetilde{M}; \mathcal{N}(G)) \to H_p^G(EG; \mathcal{N}(G))$ for $p = 0, 1, 2$.

$(5) \Leftrightarrow (6)$ This follows from Lemma 12.3. This finishes the proof of Theorem 12.9. □

It remains to investigate the class of groups for which Theorem 12.9 applies.

Definition 12.10. *Let d be a non-negative integer or $d = \infty$. Define $\mathcal{Z}_d$ to be the class of groups for which $H_p^G(EG; \mathcal{N}(G)) = 0$ holds for $p \leq d$.*

Lemma 12.11. *Let d be a non-negative integer or $d = \infty$. Then*

(1) Let G be the directed union $\bigcup_{i \in} G_i$ of subgroups $G_i \subset G$. Suppose that $G_i \in \mathcal{Z}_d$ for each $i \in I$. Then $G \in \mathcal{Z}_d$;
(2) If G contains a normal subgroup $H \subset G$ with $H \in \mathcal{Z}_d$, then $G \in \mathcal{Z}_d$;
(3) If $G \in \mathcal{Z}_d$ and $H \in \mathcal{Z}_e$, then $G \times H \in \mathcal{Z}_{d+e+1}$;
(4) $\mathcal{Z}_0$ is the class of non-amenable groups;
(5) Let G be a non-amenable group, for instance a non-abelian free group. Then $\prod_{i=0}^{d} G$ lies in $\mathcal{Z}_d$.

Proof. (1) Inspecting for instance the bar-resolution or the infinite join model for EG, one sees that EG is the colimit of a directed system of G-CW-subcomplexes of the form $G \times_{G_i} EG_i$ directed by I. Hence

$$H_p^G(EG; \mathcal{N}(G)) = \text{colim}_{i \in I} H_p^G(G \times_{G_i} EG_i; \mathcal{N}(G)).$$

We get $H_p^G(G \times_{G_i} EG_i; \mathcal{N}(G)) = \mathcal{N}(G) \otimes_{\mathcal{N}(G_i)} H_p^{G_i}(EG_i, \mathcal{N}(G_i)) = 0$ from Theorem 6.29 (1).

(2) The Serre spectral sequence applied to $BH \to BG \to B(G/H)$ with coefficients in the $\mathbb{Z}G$-module $\mathcal{N}(G)$ converges to $H_{p+q}^G(EG; \mathcal{N}(G))$ and has as E^2-term $E_{p,q}^2 = H_p^{G/H}(E(G/H); H_q^H(EH; \mathcal{N}(G)))$. Theorem 6.29 (1) implies $H_q^H(EH; \mathcal{N}(G)) = \mathcal{N}(G) \otimes_{\mathcal{N}(H)} H_q^H(EH; \mathcal{N}(H)) = 0$ for $q \leq d$. Hence $H_n^G(EG; \mathcal{N}(G)) = 0$ for $n \leq d$.

(3) The condition $G \in \mathcal{Z}_d$ implies that $\mathcal{N}(G) \otimes_{\mathbb{Z}G} C_*(EG)$ is $\mathcal{N}(G)$-chain homotopy equivalent to an $\mathcal{N}(G)$-chain complex P_* such that $P_p = 0$ holds for $p \leq d$, and analogously for the condition $H \in \mathcal{Z}_e$. Notice that $C_*(E(G \times H))$ is $\mathbb{Z}[G \times H]$-isomorphic to $C_*(EG) \otimes_{\mathbb{Z}} C_*(EH)$. Hence $\mathcal{N}(G \times H) \otimes_{\mathbb{Z}[G \times H]} C_*(E(G \times H))$ is $\mathcal{N}(G \times H)$-chain homotopy equivalent to the $\mathcal{N}(G \times H)$-chain complex $\mathcal{N}(G \times H) \otimes_{\mathcal{N}(G) \otimes_c \mathcal{N}(H)} (P_* \otimes_{\mathbb{C}} Q_*)$, whose chain modules in dimension $n \leq d + e + 1$ are zero. Hence $G \times H \in \mathcal{Z}_{d+e+1}$.

(4) This follows from Lemma 6.36.

(5) This follows from assertions (3) and (4) for $d < \infty$. The case $d = \infty$ follows from (2). This finishes the proof of Lemma 12.10. □

12.4 Miscellaneous

The Singer Conjecture 11.1 and the results presented in Section 12.2 suggest to consider the following stronger version of the zero-in-the-spectrum Conjecture 12.1. Let M be an aspherical closed manifold. Then zero is in the spectrum of $(\Delta_p)_{\min}$: dom$\,((\Delta_p)_{\min}) \subset L^2\Omega^p(\widetilde{M}) \to L^2\Omega^p(\widetilde{M})$ for p with $\frac{\dim(M)-1}{2} \leq p \leq \frac{\dim(M)+1}{2}$. Roughly speaking, zero is always in the spectrum of the Laplacian in the middle dimension.

If M is a 2-dimensional complete Riemannian manifold, then zero is in the spectrum of $(\Delta_p)_{\min}$ for some $p \geq 0$ by an argument due to Dodziuk and Lott [318, Proposition 10].

Let M be a complete Riemannian manifold. We have introduced its reduced L^2-cohomology $H^p_{(2)}(M)$ and unreduced L^2-cohomology $H^p_{(2),\mathrm{unr}}(M)$ in Definition 1.71. The unreduced cohomology $H^p_{(2),\mathrm{unr}}(M)$ vanishes if and only if zero is not in the spectrum of $(\Delta_p)_{\min}$: dom$\,((\Delta_p)_{\min}) \subset L^2\Omega^p(M) \to L^2\Omega^p(M)$. The question whether the unreduced and reduced L^2-cohomology are quasi-isometry invariants is investigated in [237, page 291] and [410].

We have seen for an integer n and a free G-CW-complex X of finite type in Lemma 12.3 that $H^G_p(X;\mathcal{N}(G)) = 0$ for $p \leq n$ is equivalent to $H^G_p(X; C^*_r(G))) = 0$ for $p \leq n$. The proof was essentially based on the fact that $C^*_r(G)$ is rationally closed in $\mathcal{N}(G)$, i.e. a matrix in $M_l(C^*_r(G))$ which is invertible in $M_l(\mathcal{N}(G))$ is already invertible in $M_l(C^*_r(G))$. This does not imply that $\mathcal{N}(G)$ is flat over $C^*_r(G)$ as the following elementary example illustrates.

Example 12.12. We want to show that $\mathcal{N}(\mathbb{Z})$ is not flat over $C^*_r(\mathbb{Z})$. Choose a numeration by the positive integers of $\mathbb{Q} \cap [0,1] = \{q_n \mid n = 1,2,\ldots\}$. Fix a sequence of positive real numbers $(\epsilon_n)_{n\geq 1}$ with $\sum_{n\geq 1} \epsilon_n < 1/2$. Let $A \subset S^1$ be the complement of the open subset $\bigcup_{n\geq 0}\{\exp(2\pi i x) \mid x \in (q_n - \epsilon_n, q_n + \epsilon_n)\}$ of S^1. Obviously A is closed and its volume is greater than zero. Choose a continuous function $f \colon S^1 \to \mathbb{C}$ such that $f^{-1}(0) = A$. (Actually one can choose f to be smooth [67, Satz 14.1 on page 153].) Then multiplication with $f \in C^*_r(\mathbb{Z}) = C(S^1)$ induces an injective $C^*_r(\mathbb{Z})$-homomorphism $r_f \colon C(S^1) \to C(S^1)$ since the complement of A contains the dense subset $\{\exp(2\pi i q) \mid q \in \mathbb{Q} \cap [0,1]\}$. Multiplication with $f \in \mathcal{N}(\mathbb{Z}) = L^\infty(S^1)$ (see Example 1.4) induces a $\mathcal{N}(\mathbb{Z})$-homomorphism $L^\infty(S^1) \to L^\infty(S^1)$ which is not injective since its kernel contains the non-zero element χ_A given by the characteristic function of A. We mention that both $C^*_r(\mathbb{Z})$ and $\mathcal{N}(\mathbb{Z})$ are flat over $\mathbb{C}\mathbb{Z}$ (see Conjecture 6.49).

A result similar to Theorem 12.9 has been proved by Kervaire [289].

Theorem 12.13. *Let G be a finitely presented group. Suppose $H_p(BG;\mathbb{Z}) = 0$ for $p = 1,2$. Then there is for each $n \geq 5$ a closed n-dimensional manifold M with $H_p(M;\mathbb{Z}) \cong H_p(S^n;\mathbb{Z})$ for $p \geq 0$ and $\pi_1(M) \cong G$.*

More information about the zero-in-the-spectrum Conjecture 12.1 can be found in [318] and in [237, section 8.A5].

Exercises

12.1. Let $T\colon \operatorname{dom}(T) \subset H \to H$ be a densely defined closed operator. Suppose that T is selfadjoint. Prove

$$\operatorname{resolv}(T) \;=\; \{\lambda \in \mathbb{C} \mid (T - \lambda\operatorname{id})(\operatorname{dom}(T)) = H\}.$$

12.2. Let M be a compact connected orientable 3-manifold such that each prime factor is non-exceptional. Show that then $H_p^\pi(E\pi; \mathcal{N}(\pi))$ is non-zero for $p = 0$ or $p = 1$.

12.3. Let M be an aspherical closed locally symmetric space with fundamental group π. Then $H_p^\pi(E\pi; \mathcal{N}(\pi)) = 0$ for $p < \frac{\dim(M_{\mathrm{ncp}}) - \mathrm{f\text{-}rk}(M_{\mathrm{ncp}})}{2}$ and $H_p^\pi(E\pi; \mathcal{N}(\pi)) \neq 0$ for $p = \frac{\dim(M_{\mathrm{ncp}}) - \mathrm{f\text{-}rk}(M_{\mathrm{ncp}})}{2}$.

12.4. Show that the universal covering of an aspherical closed Riemannian manifold is uniformly contractible.

12.5. Let $f\colon M \to N$ be a map of closed connected orientable manifolds of the same dimension. Suppose that f has non-zero degree and induces an isomorphism on the fundamental groups. Assume that zero is not in the spectrum of $(\Delta_p)_{\min}\colon \operatorname{dom}((\Delta_p)_{\min}) \subset L^2\Omega^p(\widetilde{M}) \to L^2\Omega^p(\widetilde{M})$ for $p \geq 0$. Show that then zero is not in the spectrum of $(\Delta_p)_{\min}\colon \operatorname{dom}((\Delta_p)_{\min}) \subset L^2\Omega^p(\widetilde{N}) \to L^2\Omega^p(\widetilde{N})$ for $p \geq 0$.

12.6. Let $F \to E \to B$ be a fibration of closed connected manifolds such that $\pi_1(F) \to \pi_1(E)$ is injective. Suppose that zero is not in the spectrum of $(\Delta_p)_{\min}\colon \operatorname{dom}((\Delta_p)_{\min}) \subset L^2\Omega^p(\widetilde{F}) \to L^2\Omega^p(\widetilde{F})$ for $p \geq 0$. Show that then zero is not in the spectrum of $(\Delta_p)_{\min}\colon \operatorname{dom}((\Delta_p)_{\min}) \subset L^2\Omega^p(\widetilde{E}) \to L^2\Omega^p(\widetilde{E})$ for $p \geq 0$.

12.7. Let X be a finite 2-dimensional CW-complex with fundamental group π. Suppose that π is not amenable, $H_1^\pi(E\pi; \mathcal{N}(\pi)) = 0$ and $\chi(X) = 0$. Show that $H_p^\pi(X; \mathcal{N}(\pi)) = 0$ for $p \geq 0$ and X is aspherical.

12.8. Let G be a finitely presented group G such that $H_p^G(EG; \mathcal{N}(G)) = 0$ for $p \leq 1$. Show

$$\operatorname{def}(G) \leq 1.$$

Prove that $\operatorname{def}(G) = 1$ implies the existence of an aspherical finite CW-complex X of dimension 2 such that $G = \pi_1(X)$ and $H_p^G(\widetilde{X}; \mathcal{N}(G)) = 0$ for $p \geq 0$.

12.9. Let G be a finitely presented group G such that $H_p^G(EG; \mathcal{N}(G)) = 0$ for $p \leq 1$. Define $q(G)$ to be the infimum over the Euler characteristics $\chi(M)$ of all closed connected orientable 4-manifolds with $\pi_1(M) \cong G$. Show

$$q(G) \geq 0.$$

Suppose $q(G) = 0$. Then there exists a closed connected orientable 4-manifold M with $G \cong \pi_1(M)$ such that zero is not in the spectrum of $(\Delta_p)_{\min} \colon \operatorname{dom}((\Delta_p)_{\min}) \subset L^2\Omega^p(\widetilde{M}) \to L^2\Omega^p(\widetilde{M})$ for $p \geq 0$.

12.10. Let G and H be groups with models for BG and BH of finite type. Let d and e be integers or be ∞. Suppose that

$$
\begin{aligned}
H_p^G(EG; \mathcal{N}(G)) &= 0 & p &\leq d; \\
H_p^G(EG; \mathcal{N}(G)) &\neq 0 & p &= d+1; \\
H_p^H(EH; \mathcal{N}(H)) &= 0 & q &\leq e; \\
H_p^H(EH; \mathcal{N}(H)) &\neq 0 & q &= e+1.
\end{aligned}
$$

Then $H_n^{G \times H}(E(G \times H); \mathcal{N}(G \times H))$ is zero for $n \leq d + e + 1$ and is not zero for $p = d + e + 2$.

12.11. Determine for the following groups G the supremum s over the set of integers n for which $H_p^G(EG; \mathcal{N}(G)) = 0$ for $p \leq n$ holds: $G = \mathbb{Z}/2$, $\mathbb{Z}^k$, $*_{i=1}^r \mathbb{Z}$, $\prod_{i=1}^r \mathbb{Z} * \mathbb{Z}$, $\mathbb{Z} \times (\mathbb{Z} * \mathbb{Z})$, $\prod_{i=1}^\infty (\mathbb{Z}/2 * \mathbb{Z}/2)$, $\prod_{i=1}^\infty (\mathbb{Z}/2 * \mathbb{Z}/3)$.

12.12. Let N be a compact connected oriented manifold of dimension $n = 4k$ for some integer k. Let M be a complete Riemannian manifold which is diffeomorphic to the interior $N - \partial N$. Show

(1) The image of $H^{2k}(N; \partial N; \mathbb{C}) \to H^{2k}(N; \mathbb{C})$ is isomorphic to the image of $H_c^{2k}(M; \mathbb{C}) \to H^{2k}(M; \mathbb{C})$, where $H_c^{2k}(M; \mathbb{C})$ denotes cohomology with compact support;

(2) Suppose that the signature of N is different from zero. Conclude that then the image of $H_c^{2k}(M; \mathbb{C}) \to H^{2k}(M; \mathbb{C})$ is non-trivial and zero is in the spectrum of $(\Delta_{2k})_{\min} \colon \operatorname{dom}((\Delta_{2k})_{\min}) \subset L^2\Omega^{2k}(M) \to L^2\Omega^{2k}(M)$.

13. The Approximation Conjecture and the Determinant Conjecture

Introduction

This chapter is devoted to the following two conjectures.

Conjecture 13.1 (Approximation Conjecture). *A group G satisfies the* Approximation Conjecture *if the following holds:*

Let $\{G_i \mid i \in I\}$ be an inverse system of normal subgroups of G directed by inclusion over the directed set I. Suppose that $\bigcap_{i \in I} G_i = \{1\}$. Let X be a G-CW-complex of finite type. Then $G_i \backslash X$ is a G/G_i-CW-complex of finite type and

$$b_p^{(2)}(X; \mathcal{N}(G)) = \lim_{i \in I} b_p^{(2)}(G_i \backslash X; \mathcal{N}(G/G_i)).$$

Let us consider the special case where the inverse system $\{G_i \mid i \in I\}$ is given by a nested sequence of normal subgroups of finite index

$$G = G_0 \supset G_1 \supset G_2 \supset G_3 \supset \dots.$$

Notice that then $b_p^{(2)}(G_i \backslash X; \mathcal{N}(G/G_i)) = \frac{b_p(G_i \backslash X)}{[G:G_i]}$, where $b_p(G_i \backslash X)$ is the ordinary p-th Betti number of the finite CW- complex $G_i \backslash X$ (see Example 1.32). The inequality

$$\limsup_{i \to \infty} \frac{b_p(G_i \backslash X)}{[G : G_i]} \leq b_p^{(2)}(X; \mathcal{N}(G))$$

is discussed by Gromov [237, pages 20, 231] and is essentially due to Kazhdan [284]. In this special case Conjecture 13.1 was formulated by Gromov [237, pages 20, 231] and proved in [328, Theorem 0.1]. Thus we get an asymptotic relation between the L^2-Betti numbers and Betti numbers, namely

$$\lim_{i \to \infty} \frac{b_p(G_i \backslash X)}{[G : G_i]} = b_p^{(2)}(X; \mathcal{N}(G)),$$

although the L^2-Betti numbers of the universal covering $\widetilde{Y}$ and the Betti numbers of a finite CW-complex Y have nothing in common except the fact that their alternating sum gives $\chi(Y)$ (see Example 1.38).

The Approximation Conjecture 13.1 in its present form for $I = \{0, 1, 2, \ldots\}$ appears in [462, Conjecture 1.10].

The second conjecture is

Conjecture 13.2 (Determinant Conjecture). *Every group G is of* det $\geq$ *1-class (see Definition 3.112), i.e. for any $A \in M(m, n, \mathbb{Z}G)$ the Fuglede-Kadison determinant (see Definition 3.11) of the morphism $r_A^{(2)} : l^2(G)^m \to l^2(G)^n$ given by right multiplication with A satisfies*

$$\det(r_A^{(2)}) \geq 1.$$

We will see that the Determinant Conjecture 13.2 implies both the Approximation Conjecture 13.1 and the Conjecture 3.94 about the homotopy invariance of L^2-torsion (see Lemma 13.6 and Theorem 13.3 (1)).

In Section 13.1 we will explain the relevance of the two conjectures above and formulate the main Theorem 13.3 of this chapter, which says that the two conjectures above are true for a certain class $\mathcal{G}$ of groups. This theorem is proved in Section 13.2. Variations of the approximation results are presented in Section 13.3.

To understand this chapter, it is only required to be familiar with Sections 1.1, 1.2 and to have some basic knowledge of Section 3.2.

13.1 Survey on the Approximation Conjecture and Determinant Conjecture

In this section we discuss the Approximation Conjecture 13.1 and the Determinant Conjecture 13.2 and introduce the class of groups for which they are known.

13.1.1 Survey on Positive Results about the Approximation Conjecture and the Determinant Conjecture

The following theorem is the main result of this chapter. It is proved in [462, Theorem 1.14, Theorem 1.19 and Theorem 1.21] for a class of groups which is slightly smaller than our class $\mathcal{G}$ which will be introduced in Definition 13.9 and studied in Subsection 13.1.3. The general strategy is based on [328], where the case that each quotient G/G_i is finite is carried out. See also [82, Theorem 5.1 on page 71] for a treatment of determinant class in this special case. The treatment of amenable exhaustions uses ideas of [149]. We will present the proof of Theorem 13.3 in Section 13.2.

Theorem 13.3. *(1) The Approximation Conjecture 13.1 for G and the inverse system $\{G_i \mid i \in I\}$ is true if each group G_i is of* det $\geq$ *1-class;*
(2) If the group G belongs $\mathcal{G}$, then G is of det $\geq$ *1-class and satisfies the Approximation Conjecture 13.1;*

(3) *The class of groups, which are of* det ≥ 1-*class, is closed under the operations* (1), (2), (3), (4) *and* (5) *appearing in Definition 13.9;*

(4) *The class of groups, for which assertion* (1) *of Conjecture 3.94 is true, is closed under the operations* (2), (3) *and* (4) *appearing in Definition 13.9.*

13.1.2 Relations to Other Conjectures

We leave the proof of the next lemma, which is similar to the one of Lemma 10.5, to the reader.

Lemma 13.4. *Let G be a group. Let $\{G_i \mid i \in I\}$ be an inverse system of normal subgroups directed by inclusion over the directed set I. Suppose that $\bigcap_{i \in I} G_i = \{1\}$. Then the following assertions are equivalent:*

(1) *The group G satisfies the Approximation Conjecture 13.1 for the inverse system $\{G_i \mid i \in I\}$;*

(2) *The group G satisfies the Approximation Conjecture 13.1 for the inverse system $\{G_i \mid i \in I\}$ for all cocompact free proper G-manifolds X without boundary;*

(3) *Let $A \in M(m, n, \mathbb{Q}G)$ be a matrix. Let $A_i \in M(m, n, \mathbb{Q}[G/G_i])$ be the matrix obtained from A by applying elementwise the ring homomorphism $\mathbb{Q}G \to \mathbb{Q}[G/G_i]$ induced by the projection $G \to G/G_i$. Then*

$$\dim_{\mathcal{N}(G)} \left(\ker \left(r_A^{(2)} : l^2(G)^m \to l^2(G)^n \right) \right)$$

$$= \lim_{i \to \infty} \dim_{\mathcal{N}(G/G_i)} \left(\ker \left(r_{A_i}^{(2)} : l^2(G/G_i)^m \to l^2(G/G_i)^n \right) \right),$$

where $r_A^{(2)}$ and $r_{A_i}^{(2)}$ are given by right multiplication.

The Approximation Conjecture is interesting in connection with the Atiyah Conjecture. The main application has already been stated in Theorem 10.20.

Example 13.5. We can give another proof that the Atiyah Conjecture 10.3 of order $\mathbb{Z}$ with coefficients in $\mathbb{Q}$ holds for any free group F. Namely, we can assume that F is finitely generated free by Lemma 10.4. The descending central series of a finitely generated free group F yields a sequence of in F normal subgroups $F = F_0 \supset F_1 \supset F_2 \supset \ldots$ such that $\bigcap_{i \geq 1} F_i = \{1\}$ and each quotient F/F_i is torsionfree and nilpotent [354, §2]. Since finitely generated nilpotent groups are poly-cyclic groups, each quotient F/F_i satisfies the Atiyah Conjecture of order $\mathbb{Z}$ with coefficients in $\mathbb{Q}$ by Moody's induction theorem (see Remark 10.12). Theorem 13.3 (2) and Theorem 10.20 imply that $\lim_{i \in I} F/F_i$ satisfies the Atiyah Conjecture of order $\mathbb{Z}$ with coefficients in $\mathbb{Q}$. Since F is a subgroup of $\lim_{i \in I} F/F_i$, Lemma 10.4 shows that F satisfies Atiyah Conjecture of order $\mathbb{Z}$ with coefficients in $\mathbb{Q}$. Recall that we have already proved the strong Atiyah Conjecture 10.2 for a free group F_2 of rank two and thus for all free groups using Fredholm modules in Subsection 10.3.1.

The Determinant Conjecture is important in connection with L^2-torsion because of the next result.

Lemma 13.6. *If the group G satisfies the Determinant Conjecture 13.2 (for instance if G belongs to the class $\mathcal{G}$ introduced in Subsection 13.1.3), then G satisfies Conjecture 3.94.*

Proof. Suppose that G satisfies the Determinant Conjecture 13.2. Then assertion (3) of Conjecture 3.94 follows directly from the definitions. Assertion (1) of Conjecture 3.94 follows from the following calculation based on Theorem 3.14 (1)

$$1 \leq \det(r_A^{(2)}) = \frac{1}{\mathrm{dot}(r_A^{(2)}{}_1)} \leq 1.$$

Theorem 3.93 (1) implies that assertion (1) of Conjecture 3.94 implies assertion (2) of Conjecture 3.94. □

We conclude from Theorem 2.68 and Theorem 3.28

Theorem 13.7. (Logarithmic estimate for spectral density functions). *If G satisfies Conjecture 3.94 (for instance if G belongs to the class $\mathcal{G}$ introduced in Subsection 13.1.3), then we get for any cocompact free proper G-manifold M with G-invariant Riemannian metric and $p \geq 1$ that its p-th analytic spectral density function F_p (see Definition 2.64) satisfies for appropriate $\epsilon > 0$ and $C > 0$*

$$F_p(\lambda) \leq \frac{C}{-\ln(\lambda)} \qquad \text{for } 0 < \lambda < \epsilon.$$

13.1.3 A Class of Groups

Now we define the class $\mathcal{G}$ of groups for which we can prove the Approximation Conjecture 13.1 and the Determinant Conjecture 13.2.

Definition 13.8. *Let $H \subset G$ be a subgroup. The discrete homogeneous space G/H is called amenable if we can find a G-invariant metric $d: G/H \times G/H \to \mathbb{R}$ such that any set with finite diameter consists of finitely many elements and for all $K > 0$ and $\epsilon > 0$ there is a finite non-empty subset $A \subset G/H$ with*

$$|B_K(A) \cap B_K(G/H - A)| \leq \epsilon \cdot |A|.$$

Here and elsewhere $B_K(A)$ denotes for a subset A of a metric space (X, d) the set $\{x \in X \mid d(x, A) < K\}$. If $H \subset G$ is a normal subgroup with amenable quotient, then G/H is an amenable discrete homogeneous space by the following argument. Consider $K > 0$ and $\epsilon > 0$. Equip G/H with the word length metric. Because of Lemma 6.35 we can choose a set $B \subset G/H$ such that for $S = B_{2K}(eH)$ we get $|\partial_S B| \leq \frac{\epsilon}{1+\epsilon} \cdot |B|$. Put $A = B - (B \cap B_K(G/H - B))$. Then $B_K(A) \cap B_K(G/H - A) \subset \partial_S B$. This implies

$$|B_K(A) \cap B_K(G/H - A)| \leq \frac{\epsilon}{1+\epsilon} \cdot |B|.$$

Since

$$(1 - \frac{\epsilon}{1+\epsilon}) \cdot |B| \leq |B| - |\partial_S B| = |B| - |B \cap B_{2K}(G/H - B)|$$
$$\leq |B| - |B \cap B_K(G/H - B)| \leq |A|,$$

we get

$$|B_K(A) \cap B_K(G/H - A)| \leq \epsilon \cdot |A|.$$

Definition 13.9. *Let $\mathcal{G}$ be the smallest class of groups which contains the trivial group and is closed under the following operations:*

(1) Amenable quotient
   Let $H \subset G$ be a (not necessarily normal) subgroup. Suppose that $H \in \mathcal{G}$ and the quotient G/H is an amenable discrete homogeneous space. Then $G \in \mathcal{G}$ (In particular $G \in \mathcal{G}$ if G contains a normal amenable subgroup $H \subset G$ with $G/H \in \mathcal{G}$) ;
(2) Colimits
   If $G = \mathrm{colim}_{i \in I} G_i$ is the colimit of the directed system $\{G_i \mid i \in I\}$ of groups indexed by the directed set I and each G_i belongs to $\mathcal{G}$, then G belongs to $\mathcal{G}$;
(3) Inverse limits
   If $G = \lim_{i \in I} G_i$ is the limit of the inverse system $\{G_i \mid i \in I\}$ of groups indexed by the directed set I and each G_i belongs to $\mathcal{G}$, then G belongs to $\mathcal{G}$;
(4) Subgroups
   If H is isomorphic to a subgroup of the group G with $G \in \mathcal{G}$, then $H \in \mathcal{G}$;
(5) Quotients with finite kernel
   Let $1 \to K \to G \to Q \to 1$ be an exact sequence of groups. If K is finite and G belongs to $\mathcal{G}$, then Q belongs to $\mathcal{G}$.

Next we provide some information about the class $\mathcal{G}$. Notice that Schick defines in [462, Definition 1.12] a smaller class, also denoted by $\mathcal{G}$, by requiring that his class contains the trivial subgroup and is closed under operations (1), (2), (3) and (4), but not necessarily under operation (5).

Let (P) be a property of groups, e.g. being finite or being amenable. Then a group is *residually (P)* if for any $g \in G$ with $g \neq 1$ there is an epimorphism $p \colon G \to G'$ to a group having property (P) such that $p(g) \neq 1$. If G is countable and (P) is closed under taking subgroups and extensions, this is equivalent to the condition that there is a nested sequence of subgroups

$G = G_0 \supset G_1 \supset G_2 \supset \ldots$ such that each G_i is normal in G and each quotient G/G_i has property (P). In particular we get the notions of a *residually finite* and *residually amenable* group. For the reader's convenience we record some basic properties of residually finite groups. For more information we refer to the survey article [355].

Theorem 13.10 (Survey on residually finite groups).

(1) *The free product of two residually finite groups is again residually finite [241], [114, page 27];*

(2) *A finitely generated residually finite group has a solvable word problem [389];*

(3) *The automorphism group of a finitely generated residually finite group is residually finite [29];*

(4) *A finitely generated residually finite group is Hopfian, i.e, any surjective endomorphism is an automorphism [356], [399, Corollary 41.44];*

(5) *Let G be a finitely generated group possessing a faithful representation into $GL(n, F)$ for F a field. Then G is residually finite [356], [517, Theorem 4.2];*

(6) *Let G be a finitely generated group. Let G^{rf} be the quotient of G by the normal subgroup which is the intersection of all normal subgroups of G of finite index. The group G^{rf} is residually finite and any finite dimensional representation of G over a field factorizes over the canonical projection $G \to G^{rf}$;*

(7) *The fundamental group of a compact 3-manifold whose prime decomposition consists of non-exceptional manifolds (i.e., which are finitely covered by a manifold which is homotopy equivalent to a Haken, Seifert or hyperbolic manifold) is residually finite (see [253, page 380]);*

(8) *There is an infinite group with four generators and four relations which has no finite quotient except the trivial one [255].*

Lemma 13.11. (1) *A group G belongs to $\mathcal{G}$ if and only if each finitely generated subgroup of G belongs to $\mathcal{G}$;*

(2) *The class $\mathcal{G}$ is residually closed, i.e. if there is a nested sequence of subgroups $G = G_0 \supset G_1 \supset G_2 \supset \ldots$ such that $\bigcap_{i \geq 0} G_i = \{1\}$ and each G_i belongs to $\mathcal{G}$, then G belongs to $\mathcal{G}$;*

(3) *Any residually amenable and in particular any residually finite group belongs to $\mathcal{G}$;*

(4) *Suppose that G belongs to $\mathcal{G}$ and $f: G \to G$ is an endomorphism. Define the "mapping torus group" G_f to be the quotient of $G * \mathbb{Z}$ obtained by introducing the relations $t^{-1}gt = f(g)$ for $g \in G$ and $t \in \mathbb{Z}$ a fixed generator. Then G_f belongs to $\mathcal{G}$;*

(5) *Let $\{G_j \mid j \in J\}$ be a set of groups with $G_j \in \mathcal{G}$. Then the direct sum $\bigoplus_{j \in J} G_j$ and the direct product $\prod_{j \in J} G_j$ belong to $\mathcal{G}$.*

Proof. (1) Any group G is the directed union of its finitely generated subgroups and a directed union is a special case of a colimit of a directed system.

(2) There is an obvious group homomorphism $G \to \lim_{i \in I} G/G_i$. It is injective because of $\bigcap_{i \geq 0} G_i = \{1\}$.

(3) This follows from (2).

(4) There is an exact sequence $1 \to K \to G_f \to \mathbb{Z} \to 1$, where K is the colimit of the directed system $\ldots \to G \xrightarrow{f} G \xrightarrow{f} G \to \ldots$ indexed by the integers.

(5) We first show that $G \times H$ belongs to $\mathcal{G}$, provided that G and H belong to $\mathcal{G}$. Let $\mathcal{G}_G^\times$ be the class of groups $\{H \mid G \times H \in \mathcal{G}\}$. It contains the trivial group because G belongs to $\mathcal{G}$ by assumption. It suffices to show that $\mathcal{G}_G^\times$ is closed under the operation appearing in Definition 13.9 because then $\mathcal{G} \subset \mathcal{G}_G^\times$ holds by definition of $\mathcal{G}$. This is obvious for operations (1), (4) and (5). For operations (2) and (3) one uses the fact that for a directed system or inverse system $\{G_i \mid i \in I\}$ over the directed set I the canonical maps induce isomorphisms

$$\operatorname{colim}_{i \in I}(G \times G_i) \cong G \times \operatorname{colim}_{i \in I} G_i;$$
$$G \times \lim_{i \in I} G_i \cong \lim_{i \in I}(G \times G_i).$$

Since $\prod_{j \in J} G_j$ is the limit of the inverse system $\{\prod_{j \in I} G_j \mid I \subset J, I \text{ finite}\}$ and $\bigoplus_{i \in I} G_i$ is by definition a subgroup of $\prod_{j \in J} G_i$, assertion (5) follows. This finishes the proof of Lemma 13.11. $\square$

13.2 The Proof of the Approximation Conjecture and the Determinant Conjecture in Special Cases

In this section we give the proof of Theorem 13.3. As a warm-up we prove

Lemma 13.12. *The trivial group is of* $\det \geq 1$-*class.*

Proof. Let $A \in M_n(\mathbb{Z})$ be a matrix. Let $\lambda_1, \lambda_2, \ldots, \lambda_r$ be the eigenvalues of AA^* (listed with multiplicity), which are different from zero. Then we get from Example 3.12

$$\det(r_A^{(2)}) = \sqrt{\prod_{i=1}^{r} \lambda_i}.$$

Let $p(t) = \det_{\mathbb{C}}(t - AA^*)$ be the characteristic polynomial of AA^*. It can be written as $p(t) = t^a \cdot q(t)$ for some polynomial $q(t)$ with integer coefficients and $q(0) \neq 0$. One easily checks

$$|q(0)| = \prod_{i=1}^{r} \lambda_i.$$

Since q has integer coefficients, we conclude $\det(r_A^{(2)}) \geq 1$. $\qquad\qquad\square$

13.2.1 The General Strategy

In this subsection we prove in Theorem 13.19 a general result about approximations. It will be applied to the various cases, we will have to deal with, when we will for instance show assertion (3) of Theorem 13.3 that the class of groups, for which the Determinant Conjecture 13.2 is true, is closed under the operations (1), (2), (3), (4) and (5) appearing in Definition 13.9.

Throughout this subsection we will consider the following data:

- R is a ring which satisfies $\mathbb{Z} \subset R \subset \mathbb{C}$ and is closed under complex conjugation;
- G is a group, A is a matrix in $M_d(RG)$ and $\operatorname{tr}_{\mathcal{N}(G)} \colon M_d(\mathcal{N}(G)) \to \mathbb{C}$ is the von Neumann trace (see Definition 1.2). Suppose that $r_A^{(2)} \colon l^2(G)^d \to l^2(G)^d$, which is given by right multiplication with A, is positive;
- I is a directed set. For each $i \in I$ we have a group G_i, $A_i \in M_d(RG_i)$ and a finite normal trace $\operatorname{tr}_i \colon M_d(\mathcal{N}(G_i)) \to \mathbb{C}$ (see Subsection 9.1.3) such that $\operatorname{tr}_i(I_d) = d$ holds for the unit matrix $I_d \in M_d(\mathcal{N}(G_i))$. Suppose that $r_{A_i}^{(2)} \colon l^2(G_i)^d \to l^2(G_i)^d$ is positive.

Let $F \colon [0, \infty) \to [0, \infty)$ be the spectral density function of the morphism $r_A^{(2)} \colon l^2(G)^d \to l^2(G)^d$ (see Definition 2.1). Let $F_i \colon [0, \infty) \to [0, \infty)$ be the spectral density function of the morphism $r_{A_i}^{(2)} \colon l^2(G_i)^d \to l^2(G_i)^d$ defined with respect to the trace tr_i (which may or may not be $\operatorname{tr}_{\mathcal{N}(G_i)}$). Since $r_A^{(2)}$ is positive, we get $F(\lambda) = \operatorname{tr}_{\mathcal{N}(G)}(E_\lambda)$ for $\{E_\lambda \mid \lambda \in \mathbb{R}\}$ the family of spectral projections of $r_A^{(2)}$ (see Lemma 2.3 and Lemma 2.11 (11)). The analogous statement holds for F_i. Recall that for a directed set I and a net $(x_i)_{i \in I}$ of real numbers one defines

$$\liminf_{i \in I} x_i := \sup\{\inf\{x_j \mid j \in I, j \geq i\} \mid i \in I\};$$
$$\limsup_{i \in I} x_i := \inf\{\sup\{x_j \mid j \in I, j \geq i\} \mid i \in I\}.$$

Put

$$\underline{F}(\lambda) := \liminf_{i \in I} F_i(\lambda); \qquad\qquad (13.13)$$
$$\overline{F}(\lambda) := \limsup_{i \in I} F_i(\lambda). \qquad\qquad (13.14)$$

Notice that F and F_i are density functions in the sense of Definition 2.7, i.e. they are monotone non-decreasing and right-continuous. The functions $\overline{F}$ and $\underline{F}$ are monotone non-decreasing, but it is not clear whether they are right-continuous. To any monotone non-decreasing function $f \colon [0, \infty) \to [0, \infty)$ we can assign a density function

$$f^+ : [0, \infty) \to [0, \infty), \qquad \lambda \mapsto \lim_{\epsilon \to 0+} f(\lambda + \epsilon). \qquad (13.15)$$

We always have

$$\underline{F}(\lambda) \leq \overline{F}(\lambda); \qquad (13.16)$$
$$\underline{F}(\lambda) \leq \underline{F}^+(\lambda); \qquad (13.17)$$
$$\overline{F}(\lambda) \leq \overline{F}^+(\lambda). \qquad (13.18)$$

Define $\det(r_A^{(2)})$ as in Definition 3.11 and analogously define $\det_i(r_{A_i}^{(2)})$ by $\exp\left(\int_{0+}^{\infty} \ln(\lambda)\, dF_i \right)$ if $\int_{0+}^{\infty} \ln(\lambda)\, dF_i > -\infty$ and by 0 if $\int_{0+}^{\infty} \ln(\lambda)\, dF_i = -\infty$. The main technical result of this subsection is

Theorem 13.19. *Suppose that there is a constant $K > 0$ such that for the operator norms $||r_A^{(2)}|| \leq K$ and $||r_{A_i}^{(2)}|| \leq K$ hold for $i \in I$. Suppose that for any polynomial p with real coefficients we have*

$$\mathrm{tr}_{\mathcal{N}(G)}(p(A)) = \lim_{i \in I} \mathrm{tr}_i(p(A_i)).$$

Then

(1) For every $\lambda \in \mathbb{R}$
$$F(\lambda) = \underline{F}^+(\lambda) = \overline{F}^+(\lambda);$$

(2) If $\det_i(r_{A_i}^{(2)}) \geq 1$ for each $i \in I$, then

$$F(0) = \lim_{i \in I} F_i(0);$$

$$\det(r_A^{(2)}) \geq 1.$$

Proof. (1) Fix $\lambda \geq 0$. Choose a sequence of polynomials $p_n(x)$ with real coefficients such that

$$\chi_{[0,\lambda]}(x) \leq p_n(x) \leq \chi_{[0,\lambda+1/n]}(x) + \tfrac{1}{n} \cdot \chi_{[0,K]}(x) \qquad (13.20)$$

holds for $0 \leq x \leq K$. Recall that χ_S is the characteristic function of a set S. Recall that the von Neumann trace is normal and satisfies $\mathrm{tr}_{\mathcal{N}(G)}(f) \leq \mathrm{tr}_{\mathcal{N}(G)}(g)$ for $0 \leq f \leq g$, and each tr_i has the same properties by assumption. We conclude from (13.20) and the assumptions $||r_{A_i}^{(2)}|| \leq K$ and $\mathrm{tr}_i(\mathrm{Id}) = d$

$$F_i(\lambda) \leq \mathrm{tr}_i(r_{p_n(A_i)}^{(2)}) \leq F_i(\lambda + \frac{1}{n}) + \frac{d}{n}.$$

Taking the limit inferior or the limit superior we conclude using the assumption $\mathrm{tr}_{\mathcal{N}(G)}(p_n(A)) = \lim_{i \in I} \mathrm{tr}_i(p(A_i))$

$$\overline{F}(\lambda) \leq \mathrm{tr}_{\mathcal{N}(G)}(r_{p_n(A)}^{(2)}) \leq \underline{F}(\lambda + \tfrac{1}{n}) + \tfrac{d}{n}. \qquad (13.21)$$

We conclude from (13.20) and the assumption $||r_A^{(2)}|| \leq K$

$$F(\lambda) \leq \operatorname{tr}_{\mathcal{N}(G)}(r_{p_n(A)}^{(2)}) \leq F(\lambda + \frac{1}{n}) + \frac{d}{n}.$$

This implies

$$\lim_{n \to \infty} \operatorname{tr}_{\mathcal{N}(G)}(r_{p_n(A)}^{(2)}) = F(\lambda).$$

We get by applying $\lim_{n \to \infty}$ to (13.21)

$$\overline{F}(\lambda) \leq F(\lambda) \leq \underline{F}^+(\lambda).$$

Since $\underline{F}$ is monotone non-decreasing, we get for $\epsilon > 0$ using (13.16)

$$F(\lambda) \leq \underline{F}(\lambda + \epsilon) \leq \overline{F}(\lambda + \epsilon) \leq F(\lambda + \epsilon).$$

Since F is right continuous, we get by taking $\lim_{\epsilon \to 0+}$

$$F(\lambda) - \underline{F}^+(\lambda) - \overline{F}^+(\lambda),$$

This finishes the proof of assertion (1).

(2) We conclude from Lemma 3.15 (1) (the proof for $\det_i$ is analogous)

$$\ln(\det(r_A^{(2)})) = \ln(K) \cdot (F(K) - F(0)) - \int_{0+}^K \frac{F(\lambda) - F(0)}{\lambda} \, d\lambda; \quad (13.22)$$

$$\ln(\det_i(r_{A_i}^{(2)})) = \ln(K) \cdot (F_i(K) - F_i(0)) - \int_{0+}^K \frac{F_i(\lambda) - F_i(0)}{\lambda} \, d\lambda. \quad (13.23)$$

We get from the assumptions $\ln(\det_i(r_{A_i}^{(2)})) \geq 0$ and $\operatorname{tr}_i(I_d) = d$

$$\int_{0+}^K \frac{F_i(\lambda) - F_i(0)}{\lambda} \, d\lambda \leq \ln(K) \cdot (F_i(K) - F_i(0)) \leq \ln(K) \cdot d. \quad (13.24)$$

We conclude

$$\int_{\epsilon}^K \frac{\underline{F}(\lambda) - F(0)}{\lambda} \, d\lambda = \int_{\epsilon}^K \frac{\underline{F}^+(\lambda) - F(0)}{\lambda} \, d\lambda. \quad (13.25)$$

from the following calculation based on Lebesgue's Theorem of majorized convergence and the fact that $F(\lambda) = F(K) = d$ for $\lambda \geq K$

$$\left| \int_{\epsilon}^K \frac{\underline{F}^+(\lambda) - F(0)}{\lambda} \, d\lambda - \int_{\epsilon}^K \frac{\underline{F}(\lambda) - F(0)}{\lambda} \, d\lambda \right|$$

$$= \left| \int_{\epsilon}^K \frac{\lim_{n \to \infty} \underline{F}(\lambda + 1/n) - F(0)}{\lambda} \, d\lambda - \int_{\epsilon}^K \frac{\underline{F}(\lambda) - F(0)}{\lambda} \, d\lambda \right|$$

$$= \left| \int_{\epsilon}^K \lim_{n \to \infty} \frac{\underline{F}(\lambda + 1/n) - F(0)}{\lambda + 1/n} \, d\lambda - \int_{\epsilon}^K \frac{\underline{F}(\lambda) - F(0)}{\lambda} \, d\lambda \right|$$

$$= \left| \lim_{n \to \infty} \int_{\epsilon}^K \frac{\underline{F}(\lambda + 1/n) - F(0)}{\lambda + 1/n} \, d\lambda - \int_{\epsilon}^K \frac{\underline{F}(\lambda) - F(0)}{\lambda} \, d\lambda \right|$$

$$= \lim_{n\to\infty} \left| \int_\epsilon^K \frac{F(\lambda + 1/n) - F(0)}{\lambda + 1/n} \, d\lambda - \int_\epsilon^K \frac{F(\lambda) - F(0)}{\lambda} \, d\lambda \right|$$

$$= \lim_{n\to\infty} \left| \int_{\epsilon+1/n}^{K+1/n} \frac{F(\lambda) - F(0)}{\lambda} \, d\lambda - \int_\epsilon^K \frac{F(\lambda) - F(0)}{\lambda} \, d\lambda \right|$$

$$\leq \lim_{n\to\infty} \left(\left| \int_\epsilon^{\epsilon+1/n} \frac{F(\lambda) - F(0)}{\lambda} \, d\lambda \right| + \left| \int_K^{K+1/n} \frac{F(\lambda) - F(0)}{\lambda} \, d\lambda \right| \right)$$

$$\leq \lim_{n\to\infty} \left(\frac{F(K) - F(0)}{\epsilon \cdot n} + \frac{F(K) - F(0)}{K \cdot n} \right)$$

$$= 0.$$

Since (13.25) holds for all $\epsilon > 0$, we conclude from Levi's Theorem of monotone convergence

$$\int_{0+}^K \frac{F(\lambda) - F(0)}{\lambda} \, d\lambda = \int_{0+}^K \frac{F^+(\lambda) - F(0)}{\lambda} \, d\lambda. \qquad (13.26)$$

Next we show

$$\int_\epsilon^K \frac{\liminf_{i\in I}(F_i(\lambda) - F_i(0))}{\lambda} \, d\lambda \leq \liminf_{i\in I} \int_\epsilon^K \frac{F_i(\lambda) - F_i(0)}{\lambda} \, d\lambda; \quad (13.27)$$

$$\int_{0+}^K \frac{\liminf_{i\in I}(F_i(\lambda) - F_i(0))}{\lambda} \, d\lambda \leq \liminf_{i\in I} \int_{0+}^K \frac{F_i(\lambda) - F_i(0)}{\lambda} \, d\lambda. \quad (13.28)$$

These follow from the following calculation for $S = [\epsilon, K]$ or $S = (0, K]$. Since $\sup \{\inf \{ F_j(\lambda) - F_j(0) \mid j \in I, i \leq j \} \mid i \in I\}$ is a monotone non-decreasing function in $\lambda \in S$, we can find a sequence of step function $t_1 \leq t_2 \leq t_3 \leq \ldots$ which converges pointwise on S to it. Hence we can find a sequence $i_1 \leq i_2 \leq \ldots$ of elements in I such that $t_n - \frac{1}{n} \leq \inf \{ F_j(\lambda) - F_j(0) \mid j \in I, i_n \leq j\}$ holds for $n \geq 0$ and hence

$$\sup \{\inf \{ F_j(\lambda) - F_j(0) \mid j \in I, i \leq j \} \mid i \in I\}$$
$$= \lim_{n\to\infty} (\inf \{ F_j(\lambda) - F_j(0) \mid j \in I, i_n \leq j\}).$$

We conclude from Levi's Theorem of monotone convergence applied to the sequence $\left(\inf \left\{ \frac{F_j(\lambda) - F_j(0)}{\lambda} \mid j \in I, i_n \leq j\right\}\right)_n$

$$\int_A \frac{\liminf_{i\in I}(F_i(\lambda) - F_i(0))}{\lambda} \, d\lambda$$

$$= \int_A \sup \left\{\inf \left\{ \frac{F_j(\lambda) - F_j(0)}{\lambda} \mid j \in I, i \leq j \right\} \mid i \in I\right\} \, d\lambda$$

$$= \int_A \lim_{n\to\infty} \left(\inf \left\{ \frac{F_j(\lambda) - F_j(0)}{\lambda} \mid j \in I, i_n \leq j \right\}\right) \, d\lambda$$

$$= \lim_{n\to\infty} \int_A \inf \left\{ \frac{F_j(\lambda) - F_j(0)}{\lambda} \mid j \in I, i_n \leq j\right\} \, d\lambda$$

$$\leq \sup\left\{ \int_A \inf\left\{ \left. \frac{F_j(\lambda) - F_j(0)}{\lambda} \,\right|\, j \in I, i \leq j \right\} d\lambda \,\right|\, i \in I \right\}$$

$$\leq \sup\left\{ \inf\left\{ \left. \int_A \frac{F_j(\lambda) - F_j(0)}{\lambda} \, d\lambda \,\right|\, j \in I, i \leq j \right\} \,\right|\, i \in I \right\}$$

$$= \liminf_{i \in I} \int_A \frac{F_j(\lambda) - F_j(0)}{\lambda} \, d\lambda.$$

We get for $\epsilon > 0$ from assertion (1) and equations (13.18), (13.24), (13.25) and (13.27)

$$\int_\epsilon^K \frac{F(\lambda) - F(0)}{\lambda} \, d\lambda = \int_\epsilon^K \frac{\underline{F}^+(\lambda) - F(0)}{\lambda} \, d\lambda$$

$$= \int_\epsilon^K \frac{\underline{F}(\lambda) - F(0)}{\lambda} \, d\lambda$$

$$\leq \int_\epsilon^K \frac{\underline{F}(\lambda) - \overline{F}(0)}{\lambda} \, d\lambda$$

$$= \int_\epsilon^K \frac{\liminf_{i \in I} F_i(\lambda) - \limsup_{i \in I} F_i(0)}{\lambda} \, d\lambda$$

$$\leq \int_\epsilon^K \frac{\liminf_{i \in I}(F_i(\lambda) - F_i(0))}{\lambda} \, d\lambda$$

$$\leq \liminf_{i \in I} \int_\epsilon^K \frac{F_i(\lambda) - F_i(0)}{\lambda} \, d\lambda$$

$$\leq \liminf_{i \in I} \int_{0+}^K \frac{F_i(\lambda) - F_i(0)}{\lambda} \, d\lambda$$

$$\leq \ln(K) \cdot d.$$

Since this holds for all $\epsilon > 0$, we conclude from Levi's Theorem of monotone convergence

$$\int_{0+}^K \frac{F(\lambda) - F(0)}{\lambda} \, d\lambda \leq \int_{0+}^K \frac{\underline{F}(\lambda) - \overline{F}(0)}{\lambda} \, d\lambda \leq \ln(K) \cdot d.$$

In particular we get $\int_{0+}^K \frac{\underline{F}(\lambda) - \overline{F}(0)}{\lambda} \, d\lambda < \infty$ which is only possible if

$$\underline{F}^+(0) - \overline{F}(0) = \lim_{\lambda \to 0+} \underline{F}(\lambda) - \overline{F}(0) = 0.$$

This and assertion (1) implies $\overline{F}(0) = F(0)$. The same argument applies to any directed subset $J \subset I$, so that we get for any directed subset $J \subset I$

$$\limsup_{i \in J} F_i(0) = F(0). \tag{13.29}$$

This implies

$$\lim_{i \in I} F_i(0) = F(0), \qquad (13.30)$$

because otherwise we could construct $\epsilon > 0$ and a sequence $i_1 \leq i_2 \leq i_3 \leq \ldots$ of elements in I with $F_{i_n}(0) \leq F(0) - \epsilon$ for $n \geq 1$, what would contradict (13.29).

We conclude using $F(K) = d = F_i(K)$ from assertion (1) and equations (13.22), (13.23), (13.26), (13.28) and (13.30)

$$\ln(\det(r_A^{(2)}))$$

$$= \ln(K) \cdot (F(K) - F(0)) - \int_{0+}^{K} \frac{F(\lambda) - F(0)}{\lambda} \, d\lambda$$

$$= \ln(K) \cdot (F(K) - F(0)) - \int_{0+}^{K} \frac{F^{+}(\lambda) - F(0)}{\lambda} \, d\lambda$$

$$= \ln(K) \cdot (F(K) - F(0)) - \int_{0+}^{K} \frac{\underline{F}(\lambda) - F(0)}{\lambda} \, d\lambda$$

$$= \ln(K) \cdot (F(K) - \lim_{i \in I} F_i(0)) - \int_{0+}^{K} \frac{\liminf_{i \in I} F_i(\lambda) - \lim_{i \in I} F_i(0)}{\lambda} \, d\lambda$$

$$= \ln(K) \cdot (F(K) - \lim_{i \in I} F_i(0)) - \int_{0+}^{K} \frac{\liminf_{i \in I}(F_i(\lambda) - F_i(0))}{\lambda} \, d\lambda$$

$$\geq \ln(K) \cdot (F(K) - \lim_{i \in I} F_i(0)) - \liminf_{i \in I} \int_{0+}^{K} \frac{F_i(\lambda) - F_i(0)}{\lambda} \, d\lambda$$

$$= \limsup_{i \in I} \left(\ln(K) \cdot (F(K) - F_i(0)) - \int_{0+}^{K} \frac{F_i(\lambda) - F_i(0)}{\lambda} \, d\lambda \right)$$

$$= \limsup_{i \in I} \left(\ln(K) \cdot (F_i(K) - F_i(0)) - \int_{0+}^{K} \frac{F_i(\lambda) - F_i(0)}{\lambda} \, d\lambda \right)$$

$$= \limsup_{i \in I} \ln(\det_i(r_{A_i}^{(2)}))$$

$$\geq 0.$$

This finishes the proof of Theorem 13.19. $\qquad\qquad\qquad\qquad\qquad\qquad$ $\square$

13.2.2 Limits of Inverse Systems

Throughout this subsection we consider an inverse system of groups $\{G_i \mid i \in I\}$ over the directed set I. Denote its limit by $G = \lim_{i \in I} G_i$. Let $\psi_i \colon G \to G_i$ for $i \in I$ and $\phi_{i,j} \colon G_j \to G_i$ for $i, j \in I, i \leq j$ be the structure maps. Let $\mathbb{Z} \subset R \subset \mathbb{C}$ be a ring closed under complex conjugation. Let $A \in M(m, n, RG)$ be a matrix. Denote by $A_i \in M(m, n, RG)$ the matrix obtained by applying elementwise the ring homomorphism $RG \to RG_i$ induced by ψ_i for $i \in I$. Let F be the spectral density function and $\det(r_A^{(2)})$ be the Fuglede-Kadison determinant of the morphism of finitely generated Hilbert $\mathcal{N}(G)$ modules $r_A^{(2)}: l^2(G)^m \to l^2(G)^n$. Let F_i be the spectral density function and $\det(r_{A_i}^{(2)})$ be the Fuglede-Kadison determinant of the morphism of finitely generated Hilbert $\mathcal{N}(G_i)$-modules $r_{A_i}^{(2)}: l^2(G_i)^m \to l^2(G_i)^n$. Define $\underline{F}$, $\overline{F}$, $\underline{F}^+$ and $\overline{F}^+$ as in (13.13), (13.14) and (13.15). We want to prove

Theorem 13.31. *We get in the notation above:*

(1) For every $\lambda \in \mathbb{R}$

$$F(\lambda) = \underline{F}^+(\lambda) = \overline{F}^+(\lambda);$$

(2) If $\det_i(r_{A_i}^{(2)}) \geq 1$ for each $i \in I$, then

$$\dim_{\mathcal{N}(G)}(\ker(r_A^{(2)})) = \lim_{i \in I} \dim_{\mathcal{N}(G_i)}(\ker(r_{A_i}^{(2)}));$$

$$\det(r_A^{(2)}) \geq 1.$$

In order to prove this we want to apply Theorem 13.19, but we first need the following preliminaries. Define for a matrix $B \in M(m, n, \mathbb{C}G)$ a (very secret Russian) real number

$$K^G(B) := mn \cdot \max\{||b_{i,j}||_1 \mid 1 \leq i \leq m, 1 \leq j \leq n\}, \qquad (13.32)$$

where for $a = \sum_{g \in G} \lambda_g g \in \mathbb{C}G$ its L^1-norm $||a||_1$ is defined by $\sum_{g \in G} |\lambda_g|$.

Lemma 13.33. *We get for any $B \in M(m, n, RG)$*

$$||r_B^{(2)} : l^2(G)^m \to l^2(G)^n|| \leq K^G(B).$$

Proof. We get for $u \in l^2(G)$ and $b = \sum_{g \in G} \lambda_g \cdot g \in \mathbb{C}G$

$$||ub||_{l^2} = \left|\left|\sum_{g \in G} \lambda_g \cdot ug\right|\right|_{l^2} \leq \sum_{g \in G} ||\lambda_g \cdot ug||_{l^2} \leq \sum_{g \in G} |\lambda_g| \cdot ||u||_{l^2} = ||b||_1 \cdot ||u||_{l^2}.$$

This implies for $x \in l^2(G)^m$

$$\left|\left|r_B^{(2)}(x)\right|\right|_{l^2}^2 = \sum_{j=1}^{n} \left|\left|\sum_{i=1}^{m} x_i B_{i,j}\right|\right|_{l^2}^2$$

$$\leq \sum_{j=1}^{n} \left(\sum_{i=1}^{m} ||x_i B_{i,j}||_{l^2}\right)^2$$

$$\leq \sum_{j=1}^{n} \left(\sum_{i=1}^{m} ||x_i||_{l^2} \cdot ||B_{i,j}||_1\right)^2$$

$$\leq \sum_{j=1}^{n} \left(\sum_{i=1}^{m} ||x_i||_{l^2} \cdot \max\{||B_{i,j}||_1 \mid 1 \leq i \leq m\}\right)^2$$

$$= \sum_{j=1}^{n} \max\{||B_{i,j}||_1 \mid 1 \le i \le m\}^2 \cdot \left(\sum_{i=1}^{m} ||x_i||_{l^2}\right)^2$$

$$\le \sum_{j=1}^{n} \max\{||B_{i,j}||_1 \mid 1 \le i \le m\}^2 \cdot m^2 \cdot \sum_{i=1}^{m} ||x_i||_{l^2}^2$$

$$\le n \cdot \max\{||B_{i,j}||_1 \mid 1 \le i \le m, 1 \le j \le n\}^2 \cdot m^2 \cdot ||x||_{l^2}^2$$

$$\le (nm \cdot \max\{||B_{i,j}||_1 \mid 1 \le i \le m, 1 \le j \le n\} \cdot ||x||_{l^2})^2$$

$$= \left(K^G(B) \cdot ||x||_{l^2}\right)^2.$$

This finishes the proof of Lemma 13.33.   □

Lemma 13.34. *Let $p(x)$ be a polynomial with real coefficients and $B \in M_m(\mathbb{C}G)$. Then there is $i_0 \in I$ with*

$$\mathrm{tr}_{\mathcal{N}(G)}(p(B)) = \mathrm{tr}_{\mathcal{N}(G_i)}(p(B_i)) \qquad \text{for } i \ge i_0.$$

Proof. Let $p(B)$ be the matrix $\left(\sum_{g \in G} \lambda_g(i,j) \cdot g\right)_{i,j}$. Then

$$\mathrm{tr}_{\mathcal{N}(G)}(p(B)) = \sum_{i=1}^{m} \lambda_1(i,i);$$

$$\mathrm{tr}_{\mathcal{N}(G_i)}(p(B_i)) = \sum_{i=1}^{m} \sum_{g \in G, \psi_i(g)=1} \lambda_g(i,i).$$

Only finitely many of the numbers $\lambda_g(i,j)$ are different from zero. Since G is $\lim_{i \in I} G_i$, there is an index i_0 such that $\psi_i(g) = 1$ together with $\lambda_g(i,i) \ne 0$ implies $g = 1$ for all $i \ge i_0$. Now Lemma 13.34 follows.   □

We want to apply Theorem 13.19 to AA^*, where A^* is obtained by transposing A and applying elementwise the involution of RG, which sends $\sum_{g \in G} \lambda_g \cdot g$ to $\sum_{g \in G} \overline{\lambda_g} \cdot g^{-1}$. The conditions appearing in Theorem 13.19 are satisfied, where we take the constant K appearing in Theorem 13.19 to be $K^G(A^*A)$ (see (13.32)). This follows from Lemma 13.33 and Lemma 13.34, since $K^{G_i}(\psi_i(B)) \le K^G(B)$ holds for any matrix $B \in M(m,n,\mathbb{C}G)$.

Recall that $r_{A^*}^{(2)} = \left(r_A^{(2)}\right)^*$. Hence $r_{AA^*}^{(2)}$ is positive, the spectral density functions of $r_{AA^*}^{(2)}$ and $r_A^{(2)}$ differ by a factor 2 (see Lemma 2.11 (11)) and $\det(r_{AA^*}^{(2)}) = \det(r_A^{(2)})^2$ (see Theorem 3.14 (1) and (3)). The analogous statement holds for A_i for $i \in I$. Hence Theorem 13.31 follows.   □

13.2.3 Colimits of Directed Systems

Throughout this subsection we consider a directed system of groups $\{G_i \mid i \in I\}$ over the directed set I. Denote its colimit by $G = \mathrm{colim}_{i \in I} G_i$. Let $\psi_i \colon G_i \to G$ for $i \in I$ and $\phi_{i,j} \colon G_i \to G_j$ for $i, j \in I, i \le j$ be the structure maps. Let $A \in M(m,n,\mathbb{Z}G)$ be a matrix. We want to show

Proposition 13.35. *We get in the notation above:*

(1) Suppose that each G_i is of det ≥ 1-class. Then the same is true for G and there are matrices $A_i \in M(m, n, \mathbb{Z}G_i)$ such that

$$\dim_{\mathcal{N}(G)} \left(\ker \left(r_A^{(2)} : l^2(G)^m \to l^2(G)^n \right) \right)$$
$$= \lim_{i \in I} \dim_{\mathcal{N}(G)} \left(\ker \left(r_{A_i} : l^2(G_i)^m \to l^2(G_i)^n \right) \right);$$

(2) If $\phi^{G_i} \colon \mathrm{Wh}(G_i) \to \mathbb{R}$, $[A] \mapsto \ln(\det(r_A^{(2)}))$ (see (3.92)) is trivial for each $i \in I$, then $\phi^G \colon \mathrm{Wh}(G) \to \mathbb{R}$ is trivial.

Proof. (1) Consider a matrix $A \in M(m, n, \mathbb{Z}G)$. Write $A_{i,j} = \sum_{g \in G} \lambda_g(i, j) \cdot g$. Let V be the set of elements $g \in G$ for which $\lambda_g(i, j) \neq 0$ holds for some (l, j). The set V is finite. Since G is $\mathrm{colim}_{i \in I} G_i$, we can find $i_0 \in I$ with $V \subset \mathrm{im}(\psi_{i_0})$. Choose for each $g \in V$ an element $\overline{g} \in G_{i_0}$ with $\psi_{i_0}(\overline{g})$. Define a matrix $A_{i_0} \in M_m(\mathbb{Z}G_{i_0})$ by requiring that its (i, j)-th entry is $\sum_{g \in V} \lambda_g(i, j) \cdot \overline{g}$. Then $A = \psi_{i_0}(A_{i_0})$, where we denote by the same letter ψ_{i_0} the ring homomorphism $M_m(\mathbb{Z}G_i) \to M_m(\mathbb{Z}G)$ induced by ψ_{i_0} Notice that this lift A_{i_0} is not unique. Let A_i for $i \geq i_0$ be the matrix in $M_m(\mathbb{Z}G_i)$ obtained from $\phi_{i_0,i}$ and A_{i_0}. Put $B = AA^*$, $B_{i_0} = A_{i_0}A_{i_0}^*$ and $B_i = A_iA_i^*$ for $i \geq i_0$.

By construction we get $K^{G_i}(B_i) \leq K^{G_{i_0}}(B_{i_0})$ for all $i \geq i_0$ and $K^G(B) \leq K^{G_{i_0}}(B_{i_0})$. If we put $K = K^{G_{i_0}}(B_{i_0})$, we conclude from Lemma 13.33

$$||r_B^{(2)}|| \leq K; \tag{13.36}$$
$$||r_{B_i}^{(2)}|| \leq K \qquad \text{for } i \geq i_0. \tag{13.37}$$

Let $p(x)$ be a polynomial with real coefficients. Then $\psi_{i_0}(p(B_{i_0})) = p(B)$. Choose $i_1 \in I$ with $i_0 \leq i_1$ such that for any (of the finitely many) $g \in G_{i_1}$, for which for some entry in $p(B_{i_1})$ the coefficient of g is non-trivial, the implication $\psi_{i_1}(g) = 1 \Rightarrow g = 1$ holds. Then we get

$$\mathrm{tr}_{\mathcal{N}(G_i)}(p(B_i)) = \mathrm{tr}_{\mathcal{N}(G)}(p(B)) \qquad \text{for } i \geq i_1. \tag{13.38}$$

Because of (13.36), (13.37) and (13.38) the conditions of Theorem 13.19 are satisfied. Since by assumption $\det(r_{B_i}^{(2)}) \geq 1$ for all $i \in I$ and r_B and r_{B_i} for $i \in I$ are positive, we conclude $\det(r_B^{(2)}) \geq 1$ and

$$\dim_{\mathcal{N}(G)} \left(\ker \left(r_B^{(2)} : l^2(G)^m \to l^2(G)^n \right) \right)$$
$$= \lim_{i \in I} \dim_{\mathcal{N}(G)}(r_{B_i}^{(2)} : l^2(G_i)^m \to l^2(G_i)^n).$$

Since $\det(r_A^{(2)}) = \det(r_B^{(2)})^2$ holds by Theorem 3.14 (1) and (3) and the kernels of $r_A^{(2)}$ and $r_B^{(2)}$ and the kernels of $r_{B_i}^{(2)}$ and $r_{A_i}^{(2)}$ for $i \in I$ agree, assertion (1) follows.

(2) Suppose that $A \in M_m(\mathbb{Z}G)$ is invertible. We use the same notation and construction as before and apply it to $B = AA^*$ and to $B^{-1} = (A^{-1})^* A^{-1}$. The lifts B_{i_0} and $B_{i_0}^{-1}$ are not inverse to one another. But we can find an index $i_1 \geq i_0$ such that B_i and B_i^{-1} are inverse to one another for all $i \geq i_1$. By assumption $\det(r_{B_i}^{(2)}) = \det(r_{B_i^{-1}}^{(2)}) = 1$. We conclude from Theorem 13.19 as explained above that $\det(r_B^{(2)}) \geq 1$ and $\det(r_{B^{-1}}^{(2)}) \geq 1$ hold. Theorem 3.14 (1) and (3) implies $\det(r_B^{(2)}) = 1$ and hence $\det(r_A^{(2)}) = 1$. This finishes the proof of Theorem 13.35. □

13.2.4 Amenable Extensions

This subsection is devoted to the proof of

Proposition 13.39. *Let $H \subset G$ be a subgroup such that H is of* det ≥ 1-*class and G/H is an amenable discrete homogeneous space. Then G is of* det ≥ 1-*class.*

Let (X, d) be a metric space such that any bounded set consists of finitely many elements. Define for $K \geq 0$ and $A \subset X$ the set $N_K(A) = B_K(A) \cap B_K(X - A)$. A nested sequence of finite subsets $E_1 \subset E_2 \subset E_2 \subset \ldots$ of X is called an *amenable exhaustion* if $X = \bigcup_{n \geq 1} E_n$ and for all $K > 0$ and $\epsilon > 0$ there is an integer $n(K, \epsilon)$ such that $|N_K(E_n)| \leq \epsilon \cdot |E_n|$ holds for $n \geq n(K, \epsilon)$.

Lemma 13.40. *Let $H \subset G$ be a subgroup such that G/H is an amenable discrete homogeneous space. Then it possesses an amenable exhaustion.*

Proof. By assumption we can find for positive integers n, K a subset $A_{n,K}$ with $|N_K(A_{n,K})| \leq \frac{1}{n} \cdot |A_{n,K}|$. Since G acts transitively on G/H and the metric is G-invariant, we may assume after translation that the element $1H \in G/H$ belongs to $A_{n,K}$. We construct the desired exhaustion $(E_n)_{n \geq 1}$ inductively over n. Put $E_1 = A_{1,1}$. Suppose that we have already constructed $E_1, E_2, \ldots, E_n$ with $|N_k(E_k)| \leq \frac{1}{k} \cdot |E_k|$ for $k = 1, 2, \ldots, n$. Choose a positive integer d such that $E_n \subset B_d(1H)$ and $d \geq n + 1$ holds. Define $E_{n+1} := E_n \cup B_n(1H) \cup A_{n+1,2d}$. We conclude $B_d(E_{n+1}) \subset B_{2d}(A_{n+1,2d})$ from the triangle inequality and $1H \in A_{n+1,2d}$. We have $B_d(G/H - E_{n+1}) \subset B_{2d}(G/H - A_{n+1,2d})$. This implies $N_{n+1}(E_{n+1}) \subset N_d(E_{n+1}) \subset N_{2d}(A_{n+1,2d})$. We conclude

$$|N_{n+1}(E_{n+1})| \leq |N_{2d}(A_{n+1,2d})| \leq \frac{1}{n+1} \cdot |A_{n+1,2d}| \leq \frac{1}{n+1} \cdot |E_{n+1}|.$$

Since $B_n(1G) \subset E_{n+1}$ by construction, we have $G/H = \bigcup_{n \geq 1} E_n$. □

Fix an amenable exhaustion $E_1 \subset E_2 \subset E_2 \subset \ldots$ of G/H. For $B \in M_d(\mathcal{N}(H))$ put

$$\operatorname{tr}_n(B) := \frac{\operatorname{tr}_{\mathcal{N}(H)}(B)}{|E_n|}.$$

Consider $A \in M_d(\mathcal{N}(G))$ such that the associated morphism $r_A \colon l^2(G)^d \to l^2(G)^d$ of Hilbert $\mathcal{N}(G)$-modules is positive. Let $q_n \colon l^2(G)^d \to l^2(G)^d$ be $\bigoplus_{i=1}^d q'_n$, where $q'_n \colon l^2(G) \to l^2(G)$ is the orthogonal projection onto the subspace of $l^2(G)$ generated by the preimage of E_n under the canonical projection $\mathrm{pr} \colon G \to G/H$. Notice that q_n is not G-equivariant in general, but is H-equivariant. The image of q_n is a finitely generated Hilbert $\mathcal{N}(H)$-module, namely, after a choice of a preimage $\bar{x}$ of each element x in E_n under the projection $\mathrm{pr} \colon G \to G/H$ we get an obvious isometric H-isomorphism $u_n \colon \mathrm{im}(q_n) \xrightarrow{\cong} l^2(H)^{d|E_n|}$. Let $A_n \in M_{d|E_n|}(\mathcal{N}(H))$ be the matrix uniquely determined by the property that $r_{A_n} \colon l^2(H)^{d|E_n|} \to l^2(H)^{d|E_n|}$ is the morphism $u \circ q_n \circ r_A \circ u^{-1}$. In the sequel we identify $\mathrm{im}(q_n)$ and $l^2(H)^{d|E_n|}$ by u. In particular $r_{A_n} = q_n r_A q_n$ and r_{A_n} is positive. Since $||q_n|| \leq 1$ holds for any projection, we conclude

Lemma 13.41. *We have $||r_{A_n}|| \leq ||r_A||$ for all $n \geq 1$.*

Lemma 13.42. *Consider a matrix $A \in M_d(\mathcal{N}(G))$. Let p be a polynomial with real coefficients. Then*

$$\mathrm{tr}_{\mathcal{N}(G)}(p(A)) = \lim_{n \to \infty} \mathrm{tr}_n(p(A_n)).$$

Proof. By linearity it suffices to treat the case $p(x) = x^s$ for some positive integer s. For $k = 1, 2, \ldots, d$ let $e_k \in l^2(G)^d$ be the element, whose components are zero except for the k-th entry, which is the unit element of G considered as element in $l^2(G)$. Recall that $\mathrm{pr} \colon G \to G/H$ denotes the canonical projection.

Fix $g \in G$ with $\mathrm{pr}(g) \in E_n$ and $k \in \{1, 2, \ldots, d\}$. Since $q_n(ge_k) = ge_k$ and q_n is selfadjoint, we get

$$\langle r_{A_n}^s(ge_k), ge_k \rangle = \langle (q_n r_A q_n)^s(ge_k), ge_k \rangle = \langle r_A q_n r_A q_n \cdots q_n r_A(ge_k), ge_k \rangle.$$

We have the telescope sum

$$r_A q_n r_A q_n \cdots q_n r_A = r_A^s - r_A(1 - q_n)r_A^{s-1} - r_A q_n r_A(1 - q_n)r_A^{s-2}$$
$$- \ldots - r_A q_n r_A q_n \cdots r_A(1 - q_n)r_A.$$

This implies

$$|\langle r_A^s(ge_k), ge_k \rangle - \langle r_{A_n}^s(ge_k), ge_k \rangle|$$
$$\leq \sum_{i=1}^{s-1} |\langle (1 - q_n)r_A^i(ge_k), (r_A^* q_n)^{s-i}(ge_k) \rangle|$$
$$\leq \sum_{i=1}^{s-1} ||(1 - q_n)r_A^i(ge_k)||_{l^2} \cdot ||r_A^*||^{s-i}. \tag{13.43}$$

Fix $\epsilon > 0$ and $i \in \{1, 2, \ldots, s\}$. For $k = 1, 2, \ldots, d$ we write

$$r_A^i(e_k) = \left(\sum_{g \in G} \lambda_g(k,l) \cdot g \right)^d_{l=1} \in l^2(G)^d.$$

Since this element belongs to $l^2(G)^d$, we can choose $R = R(\epsilon) \geq 0$ such that

$$\left\| r_A^i(e_k) - \left(\sum_{g \in G, \mathrm{pr}(g) \in B_R(1H)} \lambda_g(k,l) \cdot g \right)^d_{l=1} \right\|_{l^2} \leq \epsilon.$$

Suppose that $g_0 \in G$ satisfies $B_R(\mathrm{pr}(g_0)) \subset E_n$. Then

$$\{g \in G, \mathrm{pr}(g_0 g) \notin E_n\} \subset \{g \in G, \mathrm{pr}(g) \notin B_R(1H)\}.$$

We conclude

$$\|(1 - q_n) r_A^i(g_0 e_k)\|_{l^2}$$

$$= \left\| (1 - q_n) \left(\sum_{g \in G} \lambda_g(k,l) \cdot g_0 g \right)^d_{l=1} \right\|_{l^2}$$

$$= \left\| \left(\sum_{g \in G, \mathrm{pr}(g_0 g) \notin E_n} \lambda_g(k,l) \cdot g_0 g \right)^d_{l=1} \right\|_{l^2}$$

$$\leq \left\| \left(\sum_{g \in G, \mathrm{pr}(g) \notin B_R(1H)} \lambda_g(k,l) \cdot g_0 g \right)^d_{l=1} \right\|_{l^2}$$

$$= \left\| r_A^i(g_0 e_k) - \left(\sum_{g \in G, \mathrm{pr}(g) \in B_R(1H)} \lambda_g(k,l) \cdot g_0 g \right)^d_{l=1} \right\|_{l^2}$$

$$= \left\| r_A^i(e_k) - \left(\sum_{g \in G, \mathrm{pr}(g) \in B_R(1H)} \lambda_g(k,l) \cdot g \right)^d_{l=1} \right\|_{l^2}$$

$$\leq \epsilon. \tag{13.44}$$

Choose for any $x \in E_n$ an element $\overline{x} \in G$ with $\mathrm{pr}(\overline{x}) = x$. We conclude from (13.43) and (13.44)

$$\left| \mathrm{tr}_{\mathcal{N}(G)}(A^s) - \mathrm{tr}_n(A_n^s) \right|$$

$$= \left| \frac{1}{|E_n|} \sum_{k=1}^{d} \sum_{x \in E_n} \langle r_A^s(\overline{x}e_k), \overline{x}e_k \rangle - \langle r_{A_n}^s(\overline{x}e_k), \overline{x}e_k \rangle \right|$$

$$\leq \frac{1}{|E_n|} \sum_{k=1}^{d} \sum_{x \in E_n} \left| \langle r_A^s(\overline{x}e_k), \overline{x}e_k \rangle - \langle r_{A_n}^s(\overline{x}e_k), \overline{x}e_k \rangle \right|$$

$$\leq \frac{1}{|E_n|} \sum_{k=1}^{d} \sum_{x \in E_n} \sum_{j=1}^{s-1} \left\| (1 - q_n) r_A^j(\overline{x}e_k) \right\|_{l^2} \cdot \| r_A^* \|^{s-j}$$

$$\leq \frac{1}{|E_n|} \sum_{k=1}^{d} \sum_{x \in E_n, x \notin N_R(E_n)} \sum_{j=1}^{s-1} \left\| (1 - q_n) r_A^j(\overline{x}e_k) \right\|_{l^2} \cdot \| r_A^* \|^{s-j}$$

$$+ \frac{1}{|E_n|} \sum_{k=1}^{d} \sum_{x \in E_n, x \in N_R(E_n)} \sum_{j=1}^{s-1} \left\| (1 - q_n) r_A^j(\overline{x}e_k) \right\|_{l^2} \cdot \| r_A^* \|^{s-j}$$

$$\leq \frac{1}{|E_n|} \sum_{k=1}^{d} \sum_{x \in E_n, x \notin N_R(E_n)} \sum_{j=1}^{s-1} \epsilon \cdot \| r_A^* \|^{s-j}$$

$$+ \frac{|N_R(E_n)|}{|E_n|} \cdot ds \cdot \|1 - q_n\| \cdot \max\{ \|r_A\|^j \cdot \|r_A^*\|^{s-j} \mid j = 1, 2 \ldots, s-1 \}$$

$$\leq \epsilon \cdot ds \cdot \max\{ \|r_A^*\|^j \mid j = 1, 2, \ldots, s \}$$

$$+ \frac{|N_R(E_n)|}{|E_n|} \cdot ds \cdot 2 \cdot \max\{ \|r_A\|^j \cdot \|r_A^*\|^{s-j} \mid j = 1, 2 \ldots, s-1 \}.$$

Put

$$C_1 := ds \cdot \max\{ \|r_A^*\|^j \mid j = 1, 2, \ldots, s \};$$
$$C_2 := ds \cdot 2 \cdot \max\{ \|r_A\|^j \cdot \|r_A^*\|^{s-j} \mid j = 1, 2 \ldots, s-1 \}.$$

Then C_1 and C_2 are independent of ϵ and n and we get

$$\left| \mathrm{tr}_{\mathcal{N}(G)}(A^s) - \mathrm{tr}_n(A_n^s) \right| \leq C_1 \cdot \epsilon + C_2 \cdot \frac{|N_R(E_n)|}{|E_n|}.$$

Since $(E_n)_{n \geq 1}$ is an amenable exhaustion, we can find $n(\epsilon)$ such that $\frac{|N_R(E_n)|}{|E_n|} \leq \epsilon$ for $n \geq n(\epsilon)$ holds. Hence we get for $n \geq n(\epsilon)$

$$\left| \mathrm{tr}_{\mathcal{N}(G)}(A^s) - \mathrm{tr}_n(A_n^s) \right| \leq C_1 \cdot \epsilon + C_2 \cdot \epsilon.$$

Since $\epsilon > 0$ was arbitrary, Lemma 13.42 follows. $\qquad\qquad\square$

Now we can finish the proof of Proposition 13.39. Namely, because of Lemma 13.41 and Lemma 13.42 the assumptions of Theorem 13.19 are satisfied and Proposition 13.39 follows.

13.2.5 Quotients with Finite Kernels

This subsection is devoted to the proof of

Lemma 13.45. *Let* $1 \to K \to G \xrightarrow{p} Q \to 1$ *be a group extension such that K is finite. Let $f \colon M \to N$ be a morphism of finitely generated Hilbert $\mathcal{N}(Q)$-modules.*

(1) The Hilbert space $\mathrm{res}_p M$ together with the G-action obtained from the given Q-action and $p \colon G \to Q$ is a finitely generated Hilbert $\mathcal{N}(G)$-module. The morphism f yields a morphism of finitely generated Hilbert $\mathcal{N}(G)$-modules $\mathrm{res}_p f \colon \mathrm{res}_p M \to \mathrm{res}_p N$;

(2) If $M = N$, then the von Neumann traces of f and $\mathrm{res}_p f$ satisfy
$$\mathrm{tr}_{\mathcal{N}(G)}(\mathrm{res}_p f) = \tfrac{1}{|K|} \cdot \mathrm{tr}_{\mathcal{N}(Q)}(f);$$

(3) We get for the spectral density functions of f and $\mathrm{res}_p f$
$$F^G(\mathrm{res}_p f) = \frac{1}{|K|} \cdot F^Q(f);$$

(4) $\dim_{\mathcal{N}(G)}(\ker(\mathrm{res}_p f)) = \tfrac{1}{|K|} \cdot \dim_{\mathcal{N}(Q)}(\ker(f));$

(5) The Novikov-Shubin invariants of f and $\mathrm{res}_p f$ satisfy
$$\alpha(\mathrm{res}_p f) = \alpha(f);$$

(6) The Fuglede-Kadison determinants of f and res_f satisfy
$$\mathrm{det}_{\mathcal{N}(G)}(\mathrm{res}_p f) = \left(\mathrm{det}_{\mathcal{N}(Q)}(f)\right)^{\frac{1}{|K|}};$$

(7) The group Q is of $\det \geq 1$-class, if G is of $\det \geq 1$-class.

Proof. (1) Obviously it suffices to check this for $M = l^2(Q)$. Let $N_K \in \mathbb{C}G$ be the element $\sum_{k \in K} k$. Then right multiplication with $|K|^{-1} \cdot N_K$ yields an orthogonal projection $r_{|K|^{-1} \cdot N_K} \colon l^2(G) \to l^2(G)$. We get an isometric G-isomorphism $v \colon \mathrm{res}_p l^2(Q) \to \mathrm{im}(r_{|K|^{-1} \cdot N_K})$ by sending $\sum_{q \in Q} \lambda_q \cdot q$ to $\sum_{q \in Q} \lambda_q \cdot |K|^{-1/2} \cdot \sum_{g \in p^{-1}(q)} g$.

(2) Obviously it suffices to treat the case $M = N = l^2(Q)$. Denote by $e_Q \in Q$ and $e_G \in G$ the unit element. Let $f(e_Q) = \sum_{q \in Q} \lambda_q \cdot q$. If $i \colon \mathrm{im}(r_{|K|^{-1} \cdot N_K}) \to l^2(G)$ is the inclusion, we get
$$i \circ v \circ \mathrm{res}_p f \circ v^{-1} \circ r_{|K|^{-1} \cdot N_K}(e_G) = \sum_{q \in Q} \frac{\lambda_q}{|K|} \cdot \sum_{g \in p^{-1}(q)} g.$$

This implies
$$
\begin{aligned}
\mathrm{tr}_{\mathcal{N}(Q)}(f) &= \langle f(e_Q), e_Q \rangle_{l^2(Q)} \\
&= \lambda_{e_Q} \\
&= |K| \cdot \langle i \circ v \circ f \circ v^{-1} \circ r_{|K|^{-1} \cdot N_K}(e_G), e_G \rangle_{l^2(G)} \\
&= |K| \cdot \mathrm{tr}_{\mathcal{N}(G)}(\mathrm{res}_p f).
\end{aligned}
$$

(3) This follows from assertions (1) and (2).

(4), (5) and (6) follow from assertions (1) and (3).

(7) Let $A \in M_d(\mathbb{Z}Q)$ be a matrix. It induces a morphism $r_A^{(2)} \colon l^2(Q)^d \to l^2(Q)^d$ of finitely generated Hilbert $\mathcal{N}(Q)$-modules by right multiplication. We have to show $\det(r_A^{(2)}) \geq 1$ provided that G is of det $\geq$ 1-class.

Let n be any positive integer. We get a morphism $\mathrm{res}_p \, r_{A^n}^{(2)} \colon \mathrm{res}_p \, l^2(Q)^d \to \mathrm{res}_p \, l^2(Q)^d$ of finitely generated Hilbert $\mathcal{N}(G)$-modules. Notice that we have the orthogonal sum decomposition

$$l^2(G) = \mathrm{im}(r_{|K|^{-1} \cdot N_K}) \oplus \mathrm{im}(r_{1 - |K|^{-1} \cdot N_K}).$$

Consider the morphism

$$\left(v^d \circ r_{A^n}^{(2)} \circ (v^{-1})^d \right) \oplus \mathrm{id}_{\mathrm{im}(r_{1 - |K|^{-1} \cdot N_K})}^d \colon l^2(G)^d \to l^2(G)^d,$$

where the isomorphism v and the orthogonal projection $r_{|K|^{-1} \cdot N_K}$ are taken from the proof of assertion (1). We conclude from Assertion (6), Theorem 3.14 (1) and Lemma 3.15 (7)

$$\left(\det_{\mathcal{N}(Q)}(r_A^{(2)}) \right)^n$$

$$= \det_{\mathcal{N}(Q)}(r_{A^n}^{(2)})$$

$$= \frac{1}{|K|} \cdot \det_{\mathcal{N}(G)}(\mathrm{res}_p \, r_{A^n}^{(2)})$$

$$= \frac{1}{|K|} \cdot \det_{\mathcal{N}(G)} \left(v^d \circ r_{A^n}^{(2)} \circ (v^{-1})^d \right) \cdot \det_{\mathcal{N}(G)} \left((\mathrm{id}_{\mathrm{im}(r_{1 - |K|^{-1} \cdot N_K})})^d \right)$$

$$= \frac{1}{|K|} \cdot \det_{\mathcal{N}(G)} \left(v^d \circ r_{A^n}^{(2)} \circ (v^{-1})^d \oplus (\mathrm{id}_{\mathrm{im}(r_{1 - |K|^{-1} \cdot N_K})})^d \right). \quad (13.46)$$

For $u = \sum_{q \in Q} \lambda_q \cdot q$ in $\mathbb{Z}Q$ let $\bar{u} \in \mathbb{Z}G$ be the element $\sum_{q \in Q} \lambda_q \cdot \sum_{g \in p^{-1}(q)} g$. Define $B = (b_{i,j})_{i,j} \in M_d(\mathbb{Z}G)$ to be the matrix obtained from $A^n = (a_{i,j})_{i,j}$ by putting $b_{i,j} = \overline{a_{i,j}} - 1$. One easily checks

$$\frac{1}{|K|} \cdot r_{|K| \cdot I_d + B}^{(2)} = v^d \circ r_{A^n}^{(2)} \circ (v^{-1})^d \oplus (\mathrm{id}_{\mathrm{im}(r_{1 - |K|^{-1} \cdot N_K})})^d,$$

where I_d is the identity matrix in $M_d(\mathbb{Z}G)$. Notice that $|K| \cdot I_d + B$ lies in $M_d(\mathbb{Z}G)$ so that by assumption $\det_{\mathcal{N}(G)} \left(r_{|K| \cdot I_d + B}^{(2)} \right) \geq 1$ holds. Theorem 3.14 (1) implies

$$\det\nolimits_{\mathcal{N}(G)}\left(v^d \circ r^{(2)}_{A^n} \circ (v^{-1})^d \;\oplus\; (\mathrm{id}_{\mathrm{im}(r_{1-|K|^{-1}\cdot N_K})})^d\right)$$

$$= \det\nolimits_{\mathcal{N}(G)}\left(\frac{1}{|K|}\cdot r^{(2)}_{|K|\cdot I_d+B}\right)$$

$$= \frac{1}{|K|^d}\cdot \det\nolimits_{\mathcal{N}(G)}\left(r^{(2)}_{|K|\cdot I_d+B}\right)$$

$$\geq \frac{1}{|K|^d}. \tag{13.47}$$

We conclude from (13.46) and (13.47) that for any positive integer n

$$\det\nolimits_{\mathcal{N}(Q)}(r^{(2)}_A) \;\geq\; |K|^{-(d+1)/n}$$

holds. Hence $\det_{\mathcal{N}(Q)}(r^{(2)}_A) \geq 1$. This finishes the proof of Lemma 13.45. $\quad\square$
  Finally we can give the proof of Theorem 13.3.

Proof. (1) Because of Lemma 13.4 it suffices to prove assertion (3) of Lemma 13.4. The natural homomorphism $i\colon G \to \lim_{i\in I} G/G_i$ is injective because of $\bigcap_{i\geq 1} G_i = \{1\}$. We conclude

$$\dim\nolimits_{\mathcal{N}(G)}\left(\ker\left(r^{(2)}_A\right)\right) = \dim\nolimits_{\mathcal{N}(\lim_{i\in I} G/G_i)}\left(\ker\left(r^{(2)}_{i_*A}\right)\right)$$

from Lemma 1.24. Since the composition of $i\colon G \to \lim_{i\in I} G/G_i$ with the structure map $\psi_i\colon \lim_{i\in I} G/G_i \to G/G_i$ is the canonical projection $G \to G/G_i$, the claim follows from Theorem 13.31 (2).

(3) This follows from Theorem 3.14 (6), Theorem 13.31 (2), Proposition 13.35 Proposition 13.39 and Lemma 13.45 (7).

(2) follows from assertions (1) and (3) as soon as we have shown that the trivial group is of det ≥ 1-class. This has already been done in Lemma 13.12.

(4) This follows from Theorem 3.14 (6), Theorem 13.31 (2) and Proposition 13.35. This finishes the proof of the main result of this chapter, namely, Theorem 13.3. $\quad\square$

13.3 Variations of the Approximation Results

In this sections we discuss some variations of the approximation results above (Theorem 13.48, Theorem 13.49 and Theorem 13.50) whose proofs are essentially modifications of the proof of Theorem 13.3.
  We begin with a result of Dodziuk and Mathai [149, Theorem 0.1]. We will state it as presented by Eckmann [163], where a different proof is given.
  Let G be an amenable group and let X be a finite free simplicial G-complex. We get a closed fundamental domain $\mathcal{F}$ as follows. For each simplex e of $G\backslash X$ choose a simplex $\widehat{e}$ in X, which lies in the preimage of e under the

projection $p \colon X \to G \backslash X$ and is mapped bijectively onto e under p. Let $\mathcal{F}$ be the union $\bigcup_e \hat{e}$, where e runs through the simplices of $G \backslash X$. This is a sub-CW-complex of X. Let $E_m \subset X$ for $m = 1, 2, \ldots$ be a sub-CW-complex of X, which is the union of N_m translates (by elements of G) of $\mathcal{F}$. Let $\partial \mathcal{F}$ be the topological boundary of $\mathcal{F}$. Denote by ∂N_m the number of translates of $\mathcal{F}$ which meet ∂E_m. We call $(E_m)_{m \geq 1}$ a Følner exhaustion if $E_1 \subset E_2 \subset \ldots$, $X = \bigcup_{m \geq 1} E_m$ and $\lim_{m \to \infty} \frac{\partial N_m}{N_m} = 0$. Since G is assumed to be amenable, such a Følner exhaustion exists. Let $b_p(E_m)$ be the (ordinary) Betti number of the finite CW-complex E_m.

Theorem 13.48. *We get for the amenable group G and the finite free simplicial G-complex X with Følner exhaustion $(E_m)_{m \geq 1}$*

$$b_p^{(2)}(X; \mathcal{N}(G)) = \lim_{m \to \infty} \frac{b_p(E_m)}{N_m}.$$

The proof of Theorem 13.48 is analogous to the one of Proposition 13.39. The modified versions of Lemma 13.41 and Lemma 13.42 are proved in [149, Lemma 2.2 and Lemma 2.3].

Next we explain two results of Farber [183, Theorem 0.3 and Theorem 0.4].

Theorem 13.49. *Let G be a group with a sequence of (not necessarily normal) subgroups of finite index $[G : G_i]$*

$$G = G_0 \supset G_1 \supset G_2 \supset G_3 \supset \ldots$$

such that $\bigcap_{i \geq 0} G_i = \{1\}$. Let n_i be the number of subgroups in G which are conjugate to G_i. For a given $g \in G$ let $n_i(g)$ be the number of subgroups of G which are conjugate to G_i and contain g. Suppose that for any $g \in G$ with $g \neq 1$

$$\lim_{i \to \infty} \frac{n_i(g)}{n_i} = 0.$$

Let X be a free G-CW-complex of finite type. Then

$$b_p^{(2)}(X; \mathcal{N}(G)) = \lim_{i \to \infty} \frac{b_p(G_i \backslash X)}{[G : G_i]}.$$

If each G_i is normal in G, then the condition $\lim_{i \to \infty} \frac{n_i(g)}{n_i} = 0$ is obviously satisfied and Theorem 13.49 reduces to [328, Theorem 0.1] and can be viewed as a special case of Theorem 13.3.

Notice that one can construct out of the given nested sequence $G = G_0 \supset G_1 \supset G_2 \supset \ldots$ a new one $G = G_0' \supset G_1' \supset G_2' \supset \ldots$ such that each $G_i' \subset G$ is a normal subgroup of finite index and $\bigcap_{i \in I} G_i' = \{1\}$. Namely, take $G_i' = \bigcap_{g \in G} g^{-1} G_i g$. So the group G appearing in Theorem 13.49 is residually finite.

The proof of Theorem 13.49 is an application of Theorem 13.19. One has to prove the relevant versions of Lemma 13.33 and Lemma 13.34. Lemma 13.33 carries over directly. For Lemma 13.34 one has to use the following easily verified equation

$$\frac{1}{[G : G_i]} \cdot \text{tr}_{\mathbb{C}}(r_g) = \frac{n_i(g)}{n_i},$$

where $g \in G$ and $r_g \colon \mathbb{C}[G_i\backslash G] \to \mathbb{C}[G_i\backslash G]$ is given by right multiplication with g. Notice that $\det(r_{A_i}) \geq 1$ by the argument appearing in the proof of Theorem 13.3 (2).

The proof of the next result (and of some generalizations of it) can be found in [183, page 360].

Theorem 13.50. *Let G be a group with a sequence of normal subgroups of finite index $[G : G_i]$*

$$G = G_0 \supset G_1 \supset G_2 \supset G_3 \supset \dots$$

such that $\bigcap_{i \geq 0} G_i = \{1\}$. Let $K \subset \mathbb{C}$ be an algebraic number field which is closed under complex conjugation. Let $R \subset K$ be the ring of algebraic integers in K. Consider a unitary G-representation $\rho \colon G \to M_m(R)$. Let X be a free G-CW-complex of finite type. Denote by V^i the flat vector bundle over $G_i\backslash X$ given by the representation ρ restricted to G_i. Then

$$b_p^{(2)}(X; \mathcal{N}(G)) = \lim_{i \to \infty} \frac{\dim_{\mathbb{C}}(H_p(G_i\backslash X; V^i))}{\dim_{\mathbb{C}}(V^i) \cdot [G : G_i]}.$$

If one takes V to be $\mathbb{C}$ with the trivial G-action, Theorem 13.50 reduces to [328, Theorem 0.1].

One can also deal with the signature and the L^2-signature instead of the Betti numbers and the L^2-Betti numbers. The next result is taken from [347], where also the (rather obvious) notions of a finite Poincaré pair (X, Y), of the L^2-signature of a G-covering $(\overline{X}, \overline{Y})$ and of the signature of (X, Y) are explained.

Theorem 13.51. *Let G be a group with a sequence of normal subgroups of finite index $[G : G_i]$*

$$G = G_0 \supset G_1 \supset G_2 \supset G_3 \supset \dots$$

such that $\bigcap_{i \geq 0} G_i = \{1\}$. Let (X, Y) be a 4n-dimensional finite Poincaré pair. Let $p \colon \overline{X} \to X$ be a G-covering of X and $\overline{Y}$ be the preimage of Y under p. Denote by (X_i, Y_i) the 4n-dimensional finite Poincaré pair given by $(G_i\backslash\overline{X}, G_i\backslash\overline{Y})$. Then

$$\text{sign}^{(2)}(\overline{X}, \overline{Y}) = \lim_{i \to \infty} \frac{\text{sign}(X_i, Y_i)}{[G : G_i]}.$$

Some generalizations and variations of Theorem 13.51 and a version for signatures of Theorem 13.48 can be found in [347]. From the Atiyah-Patodi-Singer index theorem [11, Theorem 4.14] and its L^2-version due to Cheeger and Gromov [105, (0.9)] it is clear that Theorem 13.51 implies a similar result for eta-invariants and L^2-eta-invariants of the spaces $\overline{Y}$ and Y_k. Such a convergence result has been proved without the assumption that Y appears in a pair (X, Y) in [500, Theorem 3.12]. Namely, if M is a closed $(4n - 1)$-dimensional manifold and $p\colon \overline{M} \to M$ is a G-covering, then we get for $M_i = G_i \backslash M$

$$\eta^{(2)}(\overline{M}) \;=\; \lim_{i \to \infty} \frac{\eta(M_i)}{[G : G_i]}.$$

Question 13.52. (Approximating Fuglede-Kadison determinants by determinants).

Is there also an approximation results for the (generalized) Fuglede-Kadison determinant?

This seems to be a difficult question and to involve more input. At least we can show the following very special result.

Lemma 13.53. *Let $p \in \mathbb{Q}[\mathbb{Z}]$ be an element different from zero and let $f\colon \mathbb{Q}[\mathbb{Z}] \longrightarrow \mathbb{Q}[\mathbb{Z}]$ be the $\mathbb{Q}[\mathbb{Z}]$-map given by multiplication with p. Write p as a product*

$$p = c \cdot t^k \cdot \prod_{i=1}^{r}(t - a_i)$$

for non-zero complex numbers c, a_1, ..., a_r and an integer k. Then we get

$$\ln(\det(f^{(2)})) \;=\; \lim_{n \to \infty} \frac{\ln(\det(f[n]))}{n} \;=\; \ln(|c|) + \sum_{\substack{i=1,2,\ldots,r \\ |a_i|>1}} \ln(|a_i|),$$

where $\det(f^{(2)})$ is the Fuglede-Kadison-determinant of the morphism of Hilbert $\mathcal{N}(\mathbb{Z})$-modules $f^{(2)}\colon l^2(\mathbb{Z}) \to l^2(\mathbb{Z})$ obtained from f by tensoring with $l^2(\mathbb{Z})$ over $\mathbb{Q}[\mathbb{Z}]$, and $\det(f[n])$ is the Fuglede-Kadison determinant of the morphism of Hilbert $\mathcal{N}(1)$-modules $f[n]\colon \mathbb{C}[\mathbb{Z}/n] \to \mathbb{C}[\mathbb{Z}/n]$ obtained from f by taking the tensor product with $\mathbb{C}[\mathbb{Z}/n]$ over $\mathbb{Q}[\mathbb{Z}]$.

Proof. We obtain from (3.23)

$$\ln(\det(f^{(2)})) = \ln(|c|) + \sum_{\substack{i=1,2,\ldots,r \\ |a_i|>1}} \ln(|a_i|). \tag{13.54}$$

Next we compute $\det(f[n])$. The regular $\mathbb{Z}/n$-representation $\mathbb{C}[\mathbb{Z}/n]$ is the orthogonal direct sum $\bigoplus_{j=0}^{n-1} V_j$ of one-dimensional $\mathbb{Z}/n$-representations V_j, where the generator of $\mathbb{Z}/n$ acts on $V_j = \mathbb{C}$ by multiplication with ζ_n^j for $\zeta_n =$

$\exp(2\pi i/n)$. Then $f[n]$ is the direct sum $\bigoplus_{j=0}^{n-1} f[n]_j$, where $f[n]_j \colon V_j \longrightarrow V_j$ is multiplication with $p(\zeta_n^j)$. We get from Example 3.12 and Lemma 3.15 (7)

$$\det(f[n]) = \prod_{j=0}^{n-1} \det(f[n]_j);$$

$$\det(f[n]_j) = \begin{cases} |c| \cdot \prod_{i=1}^{r} |\zeta_n^j - a_i| & \text{if } \zeta_n^j \notin \{a_1, a_2, \ldots, a_r\} \\ 1 & \text{otherwise} \end{cases};$$

$$\det(f[n]) = \prod_{\substack{j=0,\ldots,(n-1) \\ \zeta_n^j \notin \{a_1,a_2,\ldots,a_r\}}} |c| \cdot \prod_{i=1}^{r} |\zeta_n^j - a_i|. \tag{13.55}$$

Since $n - r \le \#\{j \in \{0, 1, \ldots, (n-1)\} \mid \zeta_n^j \notin \{a_1, a_2, \ldots, a_r\}\} \le n$ and $\lim_{n \to \infty} \frac{n-r}{n} = \lim_{n \to \infty} \frac{n}{n} = 1$ holds, we get

$$\lim_{n \to \infty} \frac{1}{n} \cdot \ln \left(\prod_{\substack{j=0,\ldots,(n-1) \\ \zeta_n^j \notin \{a_1,a_2,\ldots,a_r\}}} |c| \right) = \ln(|c|). \tag{13.56}$$

Notice that c is the highest coefficient of p and hence a rational number. We conclude from (13.54), (13.55) and (13.56) that we can assume without loss of generality $c = 1$.

Equation (13.55) implies

$$\det(f[n]) = \left(\prod_{\substack{j=0,1,\ldots,(n-1) \\ \zeta_n^j \in \{a_1,a_2,\ldots a_r\}}} \prod_{\substack{i=1,\ldots,r \\ \zeta_n^j \ne a_i}} |\zeta_n^j - a_i| \right)^{-1} \cdot \prod_{j=0}^{n-1} \prod_{\substack{i=1,\ldots,r \\ a_i^n \ne 1}} |\zeta_n^j - a_i|$$

$$\cdot \prod_{j=0}^{n-1} \prod_{\substack{i=1,\ldots,r \\ a_i^n=1, a_i \ne \zeta_n^j}} |\zeta_n^j - a_i|.$$

$$= \left(\prod_{\substack{l=1,\ldots,r \\ a_l^n=1}} \prod_{\substack{i=1,\ldots,r \\ a_l \ne a_i}} |a_l - a_i| \right)^{-1} \cdot \prod_{\substack{i=1,\ldots,r \\ a_i^n \ne 1}} \prod_{j=0}^{n-1} |\zeta_n^j - a_i|$$

$$\cdot \prod_{\substack{i=1,\ldots,r \\ a_i^n=1}} \prod_{\substack{j=0,1,\ldots,(n-1) \\ a_i \ne \zeta_n^j}} |\zeta_n^j - a_i|. \tag{13.57}$$

From the identities of polynomials with complex coefficients

$$\prod_{j=0}^{n-1}(t-\zeta_n^j)=t^n-1;$$

$$\prod_{j=1}^{n-1}(t-\zeta_n^j)=\sum_{j=0}^{n-1}t^j,$$

we conclude

$$\prod_{\substack{i=1,\ldots,r\\a_i^n\neq1}}\prod_{j=0}^{n-1}\left|\zeta_n^j-a_i\right|=\prod_{\substack{i=1,\ldots,r\\a_i^n\neq1}}\left|a_i^n-1\right|;\tag{13.58}$$

$$\prod_{\substack{i=1,\ldots,r\\a_i^n-1}}\prod_{\substack{j=0,1,\ldots,(n-1)\\a_i\neq\zeta_n^j}}\left|\zeta_n^j-a_i\right|=\prod_{\substack{i=1,\ldots,r\\a_i^n=1}}n.\tag{13.59}$$

We conclude from (13.57), (13.58) and (13.59)

$$\frac{\ln(\det(f[n]))}{n}=\sum_{\substack{i=1,\ldots,r\\a_i^n\neq1}}\frac{\ln(|a_i^n-1|)}{n}+\sum_{\substack{i=1,\ldots,r\\a_i^n=1}}\frac{\ln(n)}{n}$$

$$-\sum_{\substack{l=1,\ldots,r\\a_l^n=1}}\sum_{\substack{i=1,\ldots,r\\a_l\neq a_i}}\frac{\ln(|a_l-a_i|)}{n}.\tag{13.60}$$

We have

$$\lim_{n\to\infty}\sum_{\substack{l=1,\ldots,r\\a_l^n=1}}\sum_{\substack{i=1,\ldots,r\\a_l\neq a_i}}\frac{\ln(|a_l-a_i|)}{n}=0;\tag{13.61}$$

$$\lim_{n\to\infty}\frac{\ln(|a_i^n-1|)}{n}=\ln(|a_i|)\quad\text{if }|a_i|>1;\tag{13.62}$$

$$\lim_{n\to\infty}\frac{\ln(|a_i^n-1|)}{n}=0\qquad\text{if }|a_i|<1;\tag{13.63}$$

$$\lim_{n\to\infty}\frac{\ln(n)}{n}=0.\tag{13.64}$$

Notice that p is by assumption a polynomial with rational coefficients and hence each root a_i is an algebraic integer. The main input, which is not needed when dealing with Betti numbers or signatures, is the following result taken from [472, Corollary B1 on page 30]. Namely, there is a constant $D>0$ (depending only on p but not on n) such that for $n\geq2$ and a_i with $a_i^n\neq1$ the equation

$$|a_i^n-1|\geq n^{-D}\tag{13.65}$$

holds. Hence we get for $n\geq2$ from (13.65) for a_i with $a_i^n\neq1$ and $|a_i|=1$

$$\frac{-D \cdot \ln(n)}{n} \leq \frac{\ln(|a_i^n - 1|)}{n} \leq \frac{\ln(2)}{n}. \tag{13.66}$$

Equations (13.64) and (13.66) imply

$$\lim_{n \to \infty} \sum_{\substack{i=1,\dots,r \\ a_i^n \neq 1, |a_i|=1}} \frac{\ln(|a_i^n - 1|)}{n} = 0. \tag{13.67}$$

We conclude from equations (13.60), (13.61), (13.62), (13.63), (13.64) and (13.67)

$$\lim_{n \to \infty} \frac{\det(f[n])}{n} = \sum_{i=1,2,\dots r, |a_i|>1} \ln(|a_i|). \tag{13.68}$$

Now Lemma 13.53 follows from (13.54) and (13.68).   □

Next we show that Lemma 13.53 is not true in general for polynomials $p \in \mathbb{C}[\mathbb{Z}]$ with complex coefficients. Hence the condition that the coefficients are rational (or at least algebraic) is essential.

Example 13.69. Fix a sequence of positive integers $2 = n_1 < n_2 < n_3 < \dots$ such that for each $k \geq 1$

$$\frac{1}{n_{k+1}} \leq \frac{1}{2^k n_k \exp(n_k)}. \tag{13.70}$$

Choose a sequence of positive integers $(m_k)_k$ with $m_1 = 1$ such that for all $k \geq 1$

$$0 < \frac{m_{k+1}}{n_{k+1}} - \frac{m_k}{n_k} \leq \frac{1}{n_{k+1}}. \tag{13.71}$$

Namely, define m_{k+1} inductively as the smallest integer for which $0 < \frac{m_{k+1}}{n_{k+1}} - \frac{m_k}{n_k}$ holds. We get from (13.70) and (13.71)

$$\sum_{l=k}^{\infty} \left(\frac{m_{l+1}}{n_{l+1}} - \frac{m_l}{n_l} \right) \leq \sum_{l=k+1}^{\infty} \frac{1}{2^l n_l \exp(n_l)}$$

$$\leq \frac{\exp(-n_k)}{n_k} \cdot \sum_{l=k+1}^{\infty} 2^{-l}$$

$$\leq \frac{\exp(-n_k)}{n_k}.$$

This shows that we can define a real number

$$s = \lim_{k \to \infty} \frac{m_k}{n_k}$$

and that the following holds

$$0 \le n_k s - m_k \le \exp(-n_k). \tag{13.72}$$

Put $a = \exp(2\pi i s)$. The number s is not rational (in fact it is not even algebraic) and hence $a^n \ne 1$ for all $n \ge 1$, since otherwise (13.72) would imply that $\frac{m_k}{n_k}$ becomes stationary for large k, a contradiction to (13.71). We conclude from the estimate $|\exp(iu) - 1| \le 2 \cdot \sqrt{|u|}$ and (13.60) and (13.72)

$$
\begin{aligned}
\frac{\ln(|\det((t - a)\colon \mathbb{C}[\mathbb{Z}/n_k] \to \mathbb{C}[\mathbb{Z}/n_k])|)}{n_k} &= \frac{\ln(|a^{n_k} - 1|)}{n_k} \\
&= \frac{\ln(|\exp(2\pi i n_k s) - 1|)}{n_k} \\
&= \frac{\ln(|\exp(2\pi i(n_k s - m_k)) - 1|)}{n_k} \\
&\le \frac{\ln(2 \cdot \sqrt{2\pi(n_k s - m_k)})}{n_k} \\
&\le \frac{\ln(2 \cdot \sqrt{2\pi \exp(-n_k)})}{n_k} \\
&= -\frac{1}{2} + \frac{\ln(2) + \ln(\sqrt{2\pi})}{n_k}.
\end{aligned}
$$

Hence the sequence $\frac{\ln(|\det((t-a)\colon \mathbb{C}[\mathbb{Z}/n_k] \to \mathbb{C}[\mathbb{Z}/n_k])|)}{n_k}$ cannot converge to zero. On the other hand we get $\det\big((t - a)\colon l^2(\mathbb{Z}) \longrightarrow l^2(\mathbb{Z})\big) = 0$ from (3.23).

We mention without proof that Lemma 13.53 can be extended to matrices $A \in M(m, n, \mathbb{Q}[\mathbb{Z}])$.

13.4 Miscellaneous

Some special cases of the results of this chapter have been proved in [110], [153] and [527]. Further variations of the approximation results can be found in Farber [183], where von Neumann categories and Dixmier traces are investigated. Generalizations of some of the results of this chapter from $\mathbb{Q}$ as coefficients to the field of algebraic numbers $\overline{\mathbb{Q}} \subset \mathbb{C}$ can be found in [148]. Let G be a group with a sequence of normal subgroups of finite index $[G : G_i]$

$$G = G_0 \supset G_1 \supset G_2 \supset G_3 \supset \dots$$

such that $\bigcap_{i \ge 0} G_i = \{1\}$. Estimates for the individual Betti numbers $b_p(G_i \backslash X)$ are given in [111] in the case that $b_p^{(2)}(X) = 0$ or that $b_p^{(2)}(X) = 0$ and $\alpha(X) = \infty^+$.

Lemma 13.53 motivates the following question, which is linked to Question 13.52.

Question 13.73 (Approximating L^2-torsion by torsion). *Let M be a closed Riemannian manifold and let*

$$\pi_1(M) = G_0 \supset G_1 \supset G_2 \supset \ldots$$

be a nested sequences of normal subgroups of finite index with $\bigcap_{i \geq 0} G_i = \{1\}$. Equip $\widetilde{M}$ and $G_i \backslash \widetilde{M}$ with the induced Riemannian metric. Let $\rho_{\mathrm{an}}^{(2)}(\widetilde{M})$ be the analytic L^2-torsion of $\widetilde{M}$ (see Definition 3.128) and let $\rho_{\mathrm{an}}(G_i \backslash \widetilde{M})$ be the analytic Ray-Singer torsion of $G_i \backslash \widetilde{M}$ with respect to the trivial representation $\mathbb{C}$ (see (3.9)).

Under which circumstances does

$$\rho_{\mathrm{an}}^{(2)}(\widetilde{M}) \;=\; \lim_{i \to \infty} \frac{\rho_{\mathrm{an}}(G_i \backslash \widetilde{M})}{[\pi_1(M) : G_i]}$$

hold?

Maybe there is a link between the Volume Conjecture 4.8 and Question 13.73 above.

Exercises

13.1. Show that the maps $\Phi^G \colon \mathrm{Wh}(G) \to \mathbb{R}$ and $\Phi^{G \times \mathbb{Z}} \colon \mathrm{Wh}(G \times \mathbb{Z}) \to \mathbb{R}$ have the same image.

13.2. Let $f \colon U \to U$ be a positive endomorphism of a finitely generated Hilbert $\mathcal{N}(G)$-module. Show for the spectral density function F of f

$$\dim_{\mathcal{N}(G)}(\{u \in U \mid f(u) = \lambda u\}) \;=\; F(\lambda) - \lim_{\epsilon \to 0+} F(\lambda - \epsilon).$$

13.3. Consider the situation described in the beginning of Subsection 13.2.1 for $R = \mathbb{C}$. Suppose that there is a constant $K > 0$ such that for the operator norms $||r_A^{(2)}||_\infty \leq K$ and $||r_{A_i}^{(2)}||_\infty \leq K$ hold for $i \in I$. Suppose that for any polynomial p with real coefficients we have

$$\mathrm{tr}_{\mathcal{N}(G)}(p(A)) \;=\; \lim_{i \in I} \mathrm{tr}_i(p(A_i)).$$

Let $\lambda \geq 0$ be a number such that there is no $x \neq 0$ with $r_A^{(2)}(x) = \lambda \cdot x$. Prove

$$F(\lambda) \;=\; \lim_{i \in I} F_i(\lambda).$$

13.4. Let $\{G_i \mid i \in I\}$ be an inverse system of normal subgroups of the group G directed by inclusion over the directed set I such that $\bigcap_{i \in I} G_i = \{1\}$.

Suppose that G is countable. Show that there is a sequence $i_1 \leq i_2 \leq i_3 \leq \ldots$ of elements in I such that $\bigcap_{n=1}^{\infty} G_{i_n} = \{1\}$.

13.5. Let G be a group which is of det ≥ 1-class. Show for any finite subgroup $H \subset G$ that its Weyl group $WH = NH/H$ is of det ≥ 1-class. Show that the homomorphism

$$K_1(WH) \to \mathbb{R},$$

which sends the class of a matrix $A \in GL_n(\mathbb{Z}[WH])$ to the Fuglede-Kadison determinant of the morphism of finitely generated Hilbert $\mathcal{N}(G)$-modules $r_A^{(2)} : l^2(G/H)^n \to l^2(G/H)^n$ given by right multiplication with A, is trivial.

13.6.   If G contains a residually finite or residually amenable subgroup of finite index, then G is residually finite or residually amenable.

13.7. Let F_g be a closed orientable surface of genus $g \geq 2$. Put $X = S^1 \times F_g$. Construct a sequence $\pi_1(X) = G_0 \supset G_1 \supset G_2 \supset \ldots$, of normal subgroups of finite index with $\bigcap_{i=0}^{\infty} G_i = \{1\}$ such that

$$\lim_{i \to \infty} \frac{b_1(G_i \backslash \widetilde{X})}{[\pi_1(X) : G_i]} = 0,$$

but for any $\epsilon > 0$

$$\lim_{i \to \infty} \frac{b_1(G_i \backslash \widetilde{X})}{[\pi_1(X) : G_i]^{1-\epsilon}} = \infty$$

holds.

13.8.   Let N be a closed Riemannian manifold. Consider nested sequences of normal subgroups of finite index $\pi_1(S^1) \supset G_0 \supset G_1 \supset G_2 \supset \ldots$ and $\pi_1(N) \supset H_0 \supset H_1 \supset H_2 \supset \ldots$ with $\bigcap_{i \geq 0} H_i = \bigcap_{i \geq 0} G_i = \{1\}$. Fix a Riemannian metric on S^1 and equip $S^1 \times N$ with the product Riemannian metric. Equip each covering of $S^1 \times N$ with the induced Riemannian metric. Show

$$\rho_{an}^{(2)}(\widetilde{S^1 \times N}) = \lim_{i \to \infty} \frac{\rho_{an}(\widetilde{(G_i \times H_i) \backslash (S^1 \times N)})}{[\pi_1(S^1 \times N) : (G_i \times H_i)]}.$$

14. L^2-Invariants and the Simplicial Volume

Introduction

In this chapter we give a survey on bounded cohomology and Gromov's notion of the simplicial volume $||M|| \in \mathbb{R}$ of a closed connected orientable manifold and discuss the following conjecture.

Conjecture 14.1 (Simplicial volume and L^2-invariants). *Let M be an aspherical closed orientable manifold of dimension ≥ 1. Suppose that its simplicial volume $||M||$ vanishes. Then $\widetilde{M}$ is of determinant class and*

$$b_p^{(2)}(\widetilde{M}) = 0 \quad \text{for } p \geq 0;$$
$$\rho^{(2)}(\widetilde{M}) = 0.$$

The part about the L^2-Betti numbers is taken from [237, section 8A on page 232], whereas the part about the L^2-torsion appears in [330, Conjecture 3.2].

In Section 14.1 we give a survey on the basic definitions and properties of bounded cohomology and simplicial volume. In Section 14.2 we discuss Conjecture 14.1 and give some evidence for it. Conjecture 14.1 suggests an interesting connection between two rather different notions. It is illuminating to draw conclusions from it, which partially have already been proved or whose direct proof should be easier than a possible proof of Conjecture 14.1.

To understand this chapter, it is only required to be familiar with Sections 1.2 and 3.4.

14.1 Survey on Simplicial Volume

In this section we define the notions of simplicial volume and bounded cohomology and give a survey about their basic properties.

14.1.1 Basic Definitions

Let X be a topological space and let $C_*^{\text{sing}}(X; \mathbb{R})$ be its singular chain complex with real coefficients. Recall that a singular p-simplex of X is a continuous map $\sigma\colon \Delta_p \to X$, where here Δ_p denotes the standard p-simplex (and not the Laplace operator). Let $S_p(X)$ be the set of all singular p-simplices. Then $C_p^{\mathrm{sing}}(X;\mathbb{R})$ is the real vector space with $S_p(X)$ as basis. The p-th differential ∂_p sends the element σ given by a p-simplex $\sigma\colon \Delta_p \to X$ to $\sum_{i=0}^{p}(-1)^i\cdot\sigma\circ s_i$, where $s_i\colon \Delta_{p-1} \to \Delta_p$ is the i-th face map. Define the L^1-*norm* of an element $x \in C_p^{\mathrm{sing}}(X;\mathbb{R})$, which is given by the (finite) sum $\sum_{\sigma\in S_p(X)}\lambda_\sigma\cdot\sigma$, by

$$||x||_1 := \sum_\sigma |\lambda_\sigma|. \tag{14.2}$$

Define the L^1-*seminorm* of an element y in the p-th singular homology $H_p^{\mathrm{sing}}(X;\mathbb{R}) := H_p(C_*^{\mathrm{sing}}(X;\mathbb{R}))$ by

$$||y||_1 := \inf\{||x||_1 \mid x \in C_p^{\mathrm{sing}}(X;\mathbb{R}), \partial_p(x) = 0, y = [x]\}. \tag{14.3}$$

Notice that $||y||_1$ defines only a seminorm on $H_p^{\mathrm{sing}}(X,\mathbb{R})$, it is possible that $||y||_1 = 0$ but $y \neq 0$. The next definition is taken from [232, page 8].

Definition 14.4 (Simplicial volume). *Let M be a closed connected orientable manifold of dimension n. Define its* simplicial volume *to be the nonnegative real number*

$$||M|| := ||j([M])||_1 \qquad \in [0, \infty)$$

for any choice of fundamental class $[M] \in H_n^{\mathrm{sing}}(M;\mathbb{Z})$ and $j\colon H_n^{\mathrm{sing}}(M;\mathbb{Z}) \to H_n^{\mathrm{sing}}(M;\mathbb{R})$ the change of coefficients map associated to the inclusion $\mathbb{Z} \to \mathbb{R}$.

Let $\phi \in C_{\mathrm{sing}}^p(X;\mathbb{R}) := \hom_{\mathbb{R}}(C_p^{\mathrm{sing}}(X;\mathbb{R}),\mathbb{R}) = \mathrm{map}(S_p(X),\mathbb{R})$ be a singular cochain. Define its L^∞-*norm* by

$$||\phi||_\infty := \sup\{|\phi(\sigma)| \mid \sigma \in S_p(X)\}. \tag{14.5}$$

Define the L^∞-*seminorm* of an element ψ in the p-th singular cohomology $H_{\mathrm{sing}}^p(X;\mathbb{R}) := H^p(C_{\mathrm{sing}}^*(X;\mathbb{R}))$ by

$$||\psi||_\infty := \inf\{||\phi||_\infty \mid \phi \in C_{\mathrm{sing}}^p(X;\mathbb{R}), \delta^p(\phi) = 0, \psi = [\phi]\}, \tag{14.6}$$

where $\delta^p\colon C_{\mathrm{sing}}^p(X;\mathbb{R}) \to C_{\mathrm{sing}}^{p+1}(X;\mathbb{R})$ is the p-th differential. Notice that $||\phi||_\infty$ and $||\psi||_\infty$ can take the value ∞.

Definition 14.7 (Bounded cohomology). *Let $\widehat{C}^*(X)$ be the subcochain complex of $C_{\mathrm{sing}}^*(X;\mathbb{R})$ consisting of singular cochains ϕ with $||\phi||_\infty < \infty$. The* bounded cohomology $\widehat{H}^p(X)$ *is defined to be the cohomology $H^p(\widehat{C}^*(X))$ of $\widehat{C}^*(X)$. For a group G we put $\widehat{H}^*(G) := \widehat{H}^*(BG)$.*

Notice that $\widehat{H}^*(BG)$ is independent of the choice of model of BG because of homotopy invariance of bounded cohomology. Sometimes the bounded cohomology of a group is defined in terms of the obvious bounded cochain complex of the real cochain complex given by the bar-resolution. These two notions agree (see [232, page 49], [275, (3.4) and Corollary 3.6.1]).

In some situations we want to extend the definition of simplicial volume to the case where M is not compact. The *locally finite homology* $H_p^{\mathrm{lf}}(X; \mathbb{Z})$ of a topological space X is the homology of the chain complex $C_*^{\mathrm{lf}}(X)$ of (formal possibly infinite) sums $\sum_\sigma \lambda_\sigma \cdot \sigma$ of singular simplices with coefficients $\lambda_\sigma \in \mathbb{Z}$ which are locally finite, i.e. for any compact subset $C \subset X$ the number of singular simplices σ, which meet C and satisfy $\lambda_\sigma \neq 0$, is finite. The notion of L^1-norm carries over to elements in the locally finite homology if one also allows the value ∞. Recall that for an orientable manifold M without boundary of dimension n the n-th locally finite homology $H_n^{\mathrm{lf}}(M; \mathbb{Z})$ is isomorphic to $\mathbb{Z}$ and that a choice of an orientation corresponds to a choice of a generator $[M]$ called fundamental class.

Definition 14.8 (Simplicial volume). *Let M be a connected (not necessarily compact) orientable manifold of dimension n without boundary. Define its simplicial volume by*

$$\|M\| := \|j([M])\|_1 \qquad \in [0, \infty],$$

where $j \colon H_n^{\mathrm{lf}}(M; \mathbb{Z}) \to H_n^{\mathrm{lf}}(M; \mathbb{R})$ is the change of coefficients map associated to the inclusion $\mathbb{Z} \to \mathbb{R}$ and $[M] \in H_n^{\mathrm{lf}}(M; \mathbb{Z})$ is any choice of fundamental class.

14.1.2 Elementary Properties

Bounded cohomology is not a cohomology theory. It is natural, i.e. a continuous map $f \colon X \to Y$ induces a homomorphism $\widehat{H}^p(f) \colon \widehat{H}^p(Y) \to \widehat{H}^p(X)$ such that $\widehat{H}^p(g \circ f) = \widehat{H}^p(f) \circ \widehat{H}^p(g)$ and $\widehat{H}^p(\mathrm{id}) = \mathrm{id}$ hold. It satisfies homotopy invariance, i.e. two homotopic maps induce the same homomorphism on bounded cohomology, and the dimension axiom, i.e. $\widehat{H}^p(\{*\})$ is zero for $p \geq 1$ and $\mathbb{R}$ for $p = 0$. But excision does not hold and there is no long exact sequence associated to a pair and no Mayer-Vietoris sequence. Therefore bounded cohomology is much harder to compute than ordinary cohomology. Moreover, we will see that the bounded cohomology $\widehat{H}^p(X)$ of a finite CW-complex X can be non-trivial for some $p > \dim(X)$.

Notice that the inclusion $i_* \colon \widehat{C}^*(X) \to C_{\mathrm{sing}}^*(C; \mathbb{R})$ induces homomorphisms

$$i^p \colon \widehat{H}^p(X) \to H_{\mathrm{sing}}^p(X; \mathbb{R}) . \tag{14.9}$$

The *Kronecker product* $\langle \psi, y \rangle$ of $\psi \in H_{\mathrm{sing}}^p(X; \mathbb{R})$ and $y \in H_p^{\mathrm{sing}}(X; \mathbb{R})$ is defined by $\phi(x)$ for any choice of elements $\phi \in \ker(\delta^p)$ and $x \in \ker(\partial_p)$ with $\psi = [\phi]$ and $y = [x]$. For $\psi \in \widehat{H}^p(X)$ and $x \in H_p^{\mathrm{sing}}(X; \mathbb{R})$ we define $\langle \psi, y \rangle$ by $\langle i^p(\psi), y \rangle$.

Lemma 14.10. *(1) Let X be a topological space. Consider $x \in H_p^{\mathrm{sing}}(X; \mathbb{R})$. We have $||x||_1 = 0$ if and only if for any $\psi \in \widehat{H}^p(X)$ the Kronecker product satisfies $\langle \psi, x \rangle = 0$. Otherwise $||x||_1 > 0$ and*

$$\frac{1}{||x||_1} = \inf\{||\psi||_\infty \mid \psi \in \widehat{H}^p(X), \langle \psi, x \rangle = 1\};$$

(2) Let M be a closed connected orientable manifold of dimension n. Then the simplicial volume $||M||$ vanishes if and only if $i^n : \widehat{H}^n(M) \to H_{\mathrm{sing}}^n(M; \mathbb{R})$ is trivial. Otherwise we get

$$\frac{1}{||M||} = ||\alpha||_\infty,$$

where $\alpha \subset H_{\mathrm{sing}}^n(M; \mathbb{R})$ is the image of the fundamental cohomology class of M under the change of rings homomorphism $H_{\mathrm{sing}}^n(M; \mathbb{Z}) \to H_{\mathrm{sing}}^n(M; \mathbb{R})$.

Proof. (1) This is proved for instance in [35, Proposition F.2.2 on page 278].
(2) This follows from (1). □

Lemma 14.11. *Let $f : M \to N$ be a map of closed connected oriented manifolds of the same dimension n. Let $\deg(f)$ be the degree of f. Then*

$$||M|| \geq |\deg(f)| \cdot ||N||.$$

In particular the simplicial volume is a homotopy invariant.

Proof. For any $x \in C_n^{\mathrm{sing}}(X; \mathbb{R})$ we get $||C_n^{\mathrm{sing}}(f)(x)||_1 = ||x||_1$. □

Corollary 14.12. *If $f : M \to M$ is a selfmap of a closed connected orientable manifold of degree different from -1, 0 and 1, then $||M|| = 0$.*

Lemma 14.13. *Let $p : M \to N$ be a d-sheeted covering of closed connected orientable manifolds. Then*

$$||M|| = d \cdot ||N||.$$

Proof. Since p is a map of degree d, it suffices to prove because of Lemma 14.11 that $||M|| \leq d \cdot ||N||$. For a singular simplex $\sigma : \Delta_n \to N$ let $L(\sigma)$ be the set of singular simplices $\widetilde{\sigma} : \Delta_n \to M$ with $p \circ \widetilde{\sigma} = \sigma$. It consists of precisely d elements. If the singular chain $u = \sum_\sigma \lambda_\sigma \cdot \sigma$ represents $[N]$, then $\widetilde{u} = \sum_\sigma \lambda_\sigma \cdot \sum_{\widetilde{\sigma} \in L(\sigma)} \widetilde{\sigma}$ represents $[M]$ and $||\widetilde{u}||_1 = d \cdot ||u||_1$. □

Lemma 14.14. *Let X be a topological space. We have $||y||_1 = 0$ for any $y \in H_1(X; \mathbb{R})$. Moreover, $\widehat{H}^1(X) = 0$.*

Proof. Any element $y \in H_1(X; \mathbb{R})$ can be represented by an $\mathbb{R}$-linear combination of elements of the shape $[\sigma \colon S^1 \to X]$ for continuous maps $\sigma \colon S^1 \to X$. If $\sigma_k \colon S^1 \to X$ is the composition of σ with the map $S^1 \to S^1$ sending z to z^k for some integer $k > 0$, then σ and $\frac{1}{k} \cdot \sigma_k$ represent the same element in $H_1(X; \mathbb{R})$ and $||\frac{1}{k} \cdot \sigma_k||_1 = \frac{1}{k} \cdot ||\sigma||_1$. This implies $||y||_1 = 0$.

Consider an element $\psi \in \widehat{H}^1(X)$. Fix a representative $\phi \in \ker(\widehat{\delta}^1)$. We conclude from Lemma 14.10 (1) that $i^1 \colon \widehat{H}^1(X) \to H^1_{\mathrm{sing}}(X; \mathbb{R})$ is trivial. Hence there is $u \in C^0_{\mathrm{sing}}(X)$ with $\delta^0(u) = \phi$. We have $|u(\sigma_0) - u(\sigma_1)| \leq ||\phi||_\infty$ for any two singular 0-simplices σ_0 and σ_1 whose images lie in the same path component of X. Since $\phi \in \widehat{C}^1(X)$, we can arrange $u \in \widehat{C}^0(X)$. Hence $\psi = 0$. $\qquad \square$

14.1.3 Bounded Cohomology and Amenable Groups

The next result is due to Gromov [232, page 40], [275, Theorem 4.3].

Theorem 14.15. (Mapping theorem for bounded cohomology).
Let $f \colon X \to Y$ be a map of path-connected topological spaces which induces an epimorphism $\pi_1(X) \to \pi_1(Y)$ with an amenable group as kernel. Then the induced map $\widehat{H}^p(f) \colon \widehat{H}^p(Y) \to \widehat{H}^p(X)$ is an isometric isomorphism.

Theorem 14.15 and Lemma 14.10 (1) imply

Corollary 14.16. *(1) Let X be a path-connected topological space with fundamental group π and classifying map $f \colon X \to B\pi$. Then f induces for $p \geq 0$ an isometric isomorphism $\widehat{H}^p(f) \colon \widehat{H}^p(B\pi) \xrightarrow{\cong} \widehat{H}^p(X)$;*
(2) Let X be a path-connected topological space with amenable fundamental group. Then $\widehat{H}^p(X)$ is zero for $p \geq 1$ and $\mathbb{R}$ for $p = 0$;
(3) Let M be a connected closed orientable manifold of dimension n with fundamental group π and classifying map $f \colon M \to B\pi$. Then

$$||M|| = ||H^{\mathrm{sing}}_n(f)([M])||_1;$$

(4) Let M be a closed connected orientable manifold of dimension ≥ 1 with amenable fundamental group. Then

$$||M|| = 0.$$

14.1.4 The Simplicial Volume of Hyperbolic and Low-Dimensional Manifolds

Define the positive real number v_n to be the supremum of the volumes of all n-dimensional geodesic simplices, i.e. the convex hull of $(n + 1)$ points in general position, in the n-dimensional hyperbolic space $\mathbb{H}^n$. This is the same as the maximum over the volumes of all simplices in $\overline{\mathbb{H}}^n$, which is obtained from $\mathbb{H}^n$ by adding its natural boundary. Any regular ideal simplex in $\overline{\mathbb{H}}^n$ has volume v_n and any simplex in $\overline{\mathbb{H}}^n$ of volume v_n is a regular ideal simplex [243]. We have (see [243])

$$v_2 = \pi;$$

$$v_3 = \frac{3}{2} \cdot \sum_{j \geq 1} \frac{\sin(2\pi j/3)}{j^2} \simeq 1.10149;$$

$$v_4 = \frac{10\pi}{3} \cdot \arcsin(1/3) - \frac{\pi^2}{3} \simeq 0.26889;$$

$$v_n \leq \frac{\pi}{(n-1)!};$$

$$\lim_{n \to \infty} \frac{n! \cdot v_n}{\sqrt{n} \cdot e} = 1.$$

Theorem 14.17 (Simplicial volume of hyperbolic manifolds). *Let M be a closed hyperbolic orientable manifold of dimension n. Then*

$$||M|| = \frac{\mathrm{vol}(M)}{v_n}.$$

If $n = 2r + 1$ is odd and $C_n > 0$ is the constant introduced in (3.151), then $\widetilde{M}$ is L^2-det-acyclic and

$$||M|| = \frac{(-1)^r}{C_n \cdot v_n} \cdot \rho^{(2)}(\widetilde{M}).$$

Proof. The equation $||M|| = \frac{\mathrm{vol}(M)}{v_n}$ follows from work of Gromov and Thurston (see [232, page 11]). The second equation follows from Theorem 3.152. ☐

Theorem 14.18 (Simplicial volume of low-dimensional manifolds).

(1) We have $||S^1|| = 0$;

(2) We have $||S^2|| = ||T^2|| = 0$. Let F_g be the closed connected orientable surface of genus $g \geq 1$. Then

$$||F_g|| = 2 \cdot |\chi(F_g)| = 4g - 4;$$

(3) Let M be a compact connected orientable irreducible 3-manifold with infinite fundamental group such that the boundary of M is empty or a disjoint union of incompressible tori. Suppose that M satisfies Thurston's Geometrization Conjecture, i.e. there is a decomposition along a minimal family of pairwise non-isotopic not boundary-parallel incompressible 2-sided tori in M whose pieces are Seifert manifolds or hyperbolic manifolds. Let $M_1, M_2, \ldots, M_r$ be the hyperbolic pieces. They all have finite volume [385, Theorem B on page 52]. Then M is det-L^2-acyclic and

$$||M|| = \frac{1}{v_3} \cdot \sum_{i=1}^{r} \mathrm{vol}(M_i) = \frac{-6\pi}{v_3} \cdot \rho^{(2)}(\widetilde{M}).$$

In particular, $\rho^{(2)}(\widetilde{M}) = 0$ if and and only if $||M|| = 0$.

Proof. (1) This follows from Corollary 14.12 or Lemma 14.14.

(2) This follows for S^2 and T^2 from Corollary 14.12. The other cases are proved in [232, page 9].

(3) The claim for $||M||$ is proved in [479], [491]. The claim for $\rho^{(2)}(\widetilde{M})$ has already been proved in Theorem 4.3. □

14.1.5 Volume and Simplicial Volume

In this subsection we discuss the relation of the simplicial volume and the volume of a closed Riemannian manifold.

The next result is due to Thurston. Its proof can be found for instance in [232, Section 1.2]. (See also [274], [491].)

Theorem 14.19 (Thurston's estimate on the simplicial volume). *Let M be a complete connected orientable Riemannian manifold of dimension n, which has finite volume and whose sectional curvature satisfies $-k \leq \sec(M) \leq -1$ for some real number $k \geq 1$. Then there is a constant $C_n > 0$, which depends only on $n = \dim(M)$ but not on M such that*

$$\mathrm{vol}(M) \leq C_n \cdot ||M||.$$

In particular $||M|| > 0$.

One knows that $C_n \leq \frac{\pi}{(n-1)!}$. Certainly $C_n \geq v_n$ by Theorem 14.17. It is unknown whether one can take $C_n = v_n$.

Next we want to get reverse estimates, i.e. we want to estimate the volume by the simplicial volume from below. This is motivated by the following result due to Cheeger [102].

Theorem 14.20 (Cheeger's finiteness theorem). *For any given numbers $D > 0$ and $v > 0$ there are only finitely many diffeomorphism classes of closed Riemannian manifolds M of fixed dimension n such that for the sectional curvature $|\sec(M)| \leq 1$, for the diameter $\mathrm{diam}(M) \leq D$ and for the volume $\mathrm{vol}(M) \geq v$ hold.*

Definition 14.21 (Minimal volume). *Let M be a smooth manifold. Define its* minimal volume $\mathrm{minvol}(M)$ *to be the infimum over all volumes $\mathrm{vol}(M, g)$, where g runs though all complete Riemannian metrics on M, for which the sectional curvature satisfies $|\sec(M, g)| \leq 1$.*

Let M be a closed connected orientable Riemannian manifold of dimension n. Then for any $\tilde{x} \in \widetilde{M}$ the following limit exists

$$h_{\mathrm{vol}}(M) := \lim_{R \to \infty} \frac{\ln(\mathrm{vol}(B_R(\tilde{x}))}{R}, \tag{14.22}$$

where $B_R(\tilde{x})$ is the geodesic ball in the universal covering $\widetilde{M}$ with center $\tilde{x}$ of radius R. This limit is independent of the choice of $\tilde{x}$ and called the *(volume) entropy* of M. One always has the inequality $h_{\mathrm{vol}(M)}(M) \le (\dim(M) - 1)$ by Bishop's inequality if $\sec(M) \ge -1$.

The entropy $h(G, S)$ of a finitely generated group G with respect to a finite set of generators S is defined by

$$h(G, S) := \liminf_{R \to \infty} \frac{\ln(N(R))}{R},$$

where $N(R)$ is the number of elements of G which can be written as a word of length $\le R$ in the generators of S or their inverses. The *entropy* $h(G)$ of a finitely generated group G is the infimum over all numbers $h(G, S)$, where S runs through all finite sets of generators. If M is a closed Riemannian manifold of diameter $\mathrm{diam}(M)$, then $h(\pi_1(M)) \le 2 \cdot \mathrm{diam}(M) \cdot h_{\mathrm{vol}}(M)$ (see [238, Theorem 5.16 on page 282] or [377] for a proof).

The next result is taken from [232, page 12 and page 37], [238, 5.39 on page 307]. For a real number k and a complete Riemannian manifold M of dimension n the bound on the sectional curvature $\sec(M) \ge -k^2$ implies the bound on its Ricci curvature

$$\mathrm{Ricci}_x(v, v) \ge -(n - 1) \cdot k^2 \cdot \langle v, v \rangle_x \tag{14.23}$$

for all $x \in M$ and $v \in T_x M$.

Theorem 14.24 (Gromov's estimate on the simplicial volume).

(1) Let M be a closed connected orientable Riemannian manifold of dimension n whose Ricci curvature satisfies for all $x \in M$ and $v \in T_x M$

$$\mathrm{Ricci}_x(v, v) \ge -\frac{1}{n - 1} \cdot \langle v, v \rangle_x.$$

Then there exists a constant C_n with $0 < C_n < n!$, which depends on n but not on M, such that

$$\|M\| \le C_n \cdot \mathrm{vol}(M);$$

(2) Let M be a closed connected orientable Riemannian manifold of dimension n. Then there exists a constant C_n with $0 < C_n < n!$, which depends on n but not on M, such that

$$\|M\| \le C_n \cdot h_{\mathrm{vol}}(M)^n \cdot \mathrm{vol}(M);$$

(3) Let M be a closed connected orientable Riemannian manifold of dimension n. Then

$$||M|| \leq (n-1)^n \cdot n! \cdot \text{minvol}(M).$$

Example 14.25. Obviously any closed flat Riemannian manifold has vanishing minimal volume. Hence we get

$$\text{minvol}(T^n) = ||T^n|| = 0.$$

Let F_g be the closed orientable surface of genus g, then

$$\text{minvol}(F_g) = 2\pi \cdot |\chi(F_g)| = 2\pi \cdot |2 - 2g| = \pi \cdot ||F_g||$$

by the following argument. The Gauss-Bonnet formula implies for any Riemannian metric on F_g whose sectional curvature satisfies $|\sec| \leq 1$

$$\text{vol}(F_g) \geq \int_{F_g} |\sec| \, d\text{vol} \geq \left| \int_{F_g} \sec \, d\text{vol} \right| = |2\pi \cdot \chi(F_g)|.$$

If $g \neq 1$ and we take the Riemannian metric whose sectional curvature is constant 1 or -1, then the Gauss-Bonnet Theorem shows

$$|2\pi \cdot \chi(F_g)| = \left| \int_{F_g} \sec \, d\text{vol} \right| = \text{vol}(F_g).$$

Now the claim follows.

Notice that $||S^2|| = 0$ and $\text{minvol}(S^2) \neq 0$. We have (see [32], [232, page 93])

$$\text{minvol}(\mathbb{R}^2) = 2\pi(1 + \sqrt{2});$$
$$\text{minvol}(\mathbb{R}^n) = 0 \qquad \text{for } n \geq 3.$$

The next result is taken from [232, page 6].

Theorem 14.26 (Minimal volume and characteristic classes). *Let M be a closed connected orientable manifold of dimension n. Denote by $p_I(M)$ the Pontryagin number of M for a fixed partition I of n (see [379, page 185]). Then there is a constant $C(I,n)$, which depends on I and n but not on M, such that*

$$|p_I(M)| \leq C(I,n) \cdot \text{minvol}(M).$$

There is also a dimension constant E_n satisfying

$$|\chi(M)| \leq E_n \cdot \text{minvol}(M).$$

Proof. The Chern-Weil-Theorem (see [379, Corollary 1 on page 308]) implies that for an appropriate polynomial $P_I(x)$

$$|p_I(M)| = \left| \int_M P_I(\Omega) \right| \leq \sup\{\|P_I(\Omega)\|_x \mid x \in M\} \cdot \mathrm{vol}(M),$$

where Ω is the curvature tensor. If the sectional curvature satisfies $|\sec(M)| \leq 1$, then one can find a constant $C(I, n)$, which depends only on I and n but not on M, such that $\sup\{\|P_I(\Omega)\|_x \mid x \in M\} \leq C(I, n)$. The proof for the Euler characteristic is similar and based on the Gauss-Bonnet Theorem. □

A discussion of examples of Riemannian manifolds with vanishing minimal volume is given in [232, Appendix 2].

14.1.6 Simplicial Volume and Betti Numbers

The next theorem is taken from [232, page 12].

Theorem 14.27 (Betti numbers and simplicial volume). *Let M be a complete connected orientable Riemannian manifold of dimension n with finite volume. Let $k_1 \geq k_2 > 0$ be positive constants such that the sectional curvature satisfies $-k_1 \leq \sec(M) \leq -k_2$. Then there is a constant $C(n, k_1/k_2)$, which depends only on n and the ratio k_1/k_2 but not on M, such that*

$$\sum_{i \geq 0} b_p(M) \leq C(n, k_1/k_2) \cdot \|M\|.$$

Notice that both Theorem 14.19 and Theorem 14.27 imply that $\|M\| > 0$ holds for a closed connected orientable Riemannian manifold with negative sectional curvature.

The next example is illuminating since it shows how information about the fundamental class gives bounds on the sum of the Betti numbers.

Example 14.28. Let X be a simplicial complex with N simplices. Then

$$\sum_{p \geq 0} b_p(X) \leq \sum_{p \geq 0} \dim_{\mathbb{Z}} C_p(X) \leq N.$$

Let M be a closed manifold of dimension n. Suppose that M admits a triangulation with t simplices of dimension n. Then the total number of simplices in this triangulation is bounded by 2^t since any p-simplex σ is the intersection of the n-simplices which contain σ. This implies

$$\sum_{p \geq 0} b_p(M) \leq 2^t.$$

One can also get an estimate in terms of singular simplices. Let M be a closed oriented manifold M of dimension n. Suppose that $[M] \in H_n(M; \mathbb{Z})$ can be represented by a $\mathbb{Z}$-linear combination $\sum_{i=1}^{t} \lambda_i \cdot \sigma_i$ of t singular simplices σ_i. Then we get

$$\sum_{p \geq 0} b_p(M) \leq t \cdot 2^n.$$

This follows from Poincaré duality by the following argument. Let $S_p(M)$ be the set of singular p-simplices. Let $S_p(M)'$ be the subset of $S_p(M)$ consisting of those singular p-simplices which are obtained from one of the singular n-simplices $\sigma_i \colon \Delta_n \to M$ above by composition with some face map $\Delta_p \to \Delta_n$. Let $\psi \in H^p(M; \mathbb{R})$ be any element which can be represented by a cocyle $\phi \in C^p_{\mathrm{sing}}(M; \mathbb{R}) = \mathrm{map}(S_p(M), \mathbb{R})$ which vanishes on any singular simplex in $S_p(M)'$. Then

$$\psi \cap [M] = \left[\phi \cap \left(\sum_{i=1}^{t} \lambda_i \cdot \sigma_i \right) \right] = \left[\sum_{i=1}^{t} \lambda_i \cdot \phi \cap \sigma_i \right] = \left[\sum_{i=1}^{t} \lambda_i \cdot 0 \right] = 0$$

follows from the definition of the cap-product [226, 24.19 on page 152]. Since $? \cap [M] \colon H^p(M; \mathbb{R}) \to H_{n-p}(M, \mathbb{R})$ is bijective by Poincaré duality, this implies $\psi = 0$. We conclude that the dimension of $H^p(M; \mathbb{R})$ is at most the cardinality of $S_p(M)'$. Now the claim follows.

Remark 14.29. Notice, however, that in general the minimum over all the numbers t for which $[M] \in H_n(M; \mathbb{Z})$ can be represented by a $\mathbb{Z}$-linear combination $\sum_{i=1}^{t} \lambda_i \cdot \sigma_i$ of t singular simplices (see Example 14.28 above) is not related to $||M||$. It is bounded from above by the infimum of the L^1-seminorms of all *integral* singular cycles representing $[M]$ but this infimum has not the nice properties as $||M||$ and is very hard to handle. A better version is defined by Gromov [238, page 306] using ideas of Thurston. Gromov introduces an extension of the simplicial volume to foliations with transverse measures and defines a new invariant $||[M]_{\mathbb{Z}}||_{\Delta}^{\mathcal{F}}$ for a closed orientable manifold M. It satisfies

$$||M|| \leq ||[M]_{\mathbb{Z}}||_{\Delta}^{\mathcal{F}}. \tag{14.30}$$

There seems to be no known counterexample to the equality $||M|| = ||[M]_{\mathbb{Z}}||_{\Delta}^{\mathcal{F}}$.

The next result appears as an exercise in [238, page 307].

Theorem 14.31. *Let M be an aspherical closed orientable manifold of dimension n. Then*

$$\sum_{p \geq 0} b_p^{(2)}(\widetilde{M}) \leq 2^n \cdot ||[M]_{\mathbb{Z}}||_{\Delta}^{\mathcal{F}}.$$

14.1.7 Simplicial Volume and S^1-Actions

Theorem 14.32. *Let M be a connected closed orientable manifold. If M carries a non-trivial S^1-action, then*

$$||M|| = 0.$$

If M carries an S^1-action with $M^{S^1} = \emptyset$, then

$$\mathrm{minvol}(M) = 0.$$

Proof. The claim for the simplicial volume is proved in [232, Section 3.1, page 41], [526], the one for the minimal volume is taken from [232, page 7]. $\square$

14.1.8 Some Information about the Second Bounded Cohomology of Groups

We mention the following two results due to Epstein-Fujiwara [178], [209] and Bestvina-Fujiwara [41].

Theorem 14.33. *The second bounded cohomology $\widehat{H}^2(BG)$ is a real vector space of infinite dimension if G is a non-elementary word-hyperbolic group or if G is a subgroup, which is not virtually abelian, of the mapping class group of a compact orientable surface of genus g with p punctures.*

This shows in particular that the second bounded cohomology of $B(\mathbb{Z} * \mathbb{Z}) = S^1 \vee S^1$ is non-trivial despite the fact that $S^1 \vee S^1$ is a 1-dimensional finite CW-complex. We see that there are no Mayer-Vietoris sequences for bounded cohomology in general. The behaviour of the second bounded cohomology of a group under free amalgamated products and HNN-extensions is analysed in [210]. Burger and Monod [76], [77, Theorem 20 and Theorem 21] show that the natural map from the second bounded cohomology group to the second cohomology group of G is injective if G is an irreducible lattice in a connected semisimple Lie group with finite center, no compact factors and of rank ≥ 2.

14.1.9 Further Properties of the Simplicial Volume

We begin by stating a proportionality principle for the simplicial volume [232, page 11] which is analogous to the one for L^2-invariants (see Theorem 3.183).

Theorem 14.34 (Proportionality principle for the simplicial volume). *Let M and N be closed connected orientable Riemannian manifolds, whose universal coverings are isometrically diffeomorphic. Then*

$$\frac{||M||}{\mathrm{vol}(M)} = \frac{||N||}{\mathrm{vol}(N)}.$$

The proof of the next result can be found for instance in [35, Theorem F.2.5. on page 279] and [232, page 10].

Lemma 14.35. *Let M and N be closed connected orientable manifolds of dimensions m and n. There exists for each integer k a constant $C(k)$ which depends on k but not on M and N, such that*

$$||M \times N|| \leq C(m+n) \cdot ||M|| \cdot ||N||;$$
$$||M|| \cdot ||N|| \leq ||M \times N||.$$

The next result is taken from [232, page 10].

Theorem 14.36. *Let M and N be closed connected orientable manifolds of the same dimension $n \geq 3$. Then we get for their connected sum $M\#N$*

$$||M\#N|| = ||M|| + ||N||.$$

14.2 Simplicial Volume and L^2-Invariants of Universal Coverings of Closed Manifolds

In this section we give some evidence for Conjecture 14.1 and discuss some consequences.

Remark 14.37. Conjecture 14.1 is definitely false if one drops the condition "aspherical" since the simplicial volume of a simply connected closed orientable manifold M is zero (see Corollary 14.16 (4)), but $b_0^{(2)}(\widetilde{M}) = b_0(M) = 1$. One may ask whether for a (not necessarily aspherical) det-L^2-acyclic closed orientable manifold M with vanishing simplicial volume the L^2-torsion $\rho^{(2)}(\widetilde{M})$ must be trivial. The answer is again negative. Namely, let M be a 3-dimensional hyperbolic manifold. Put $N = S^2 \times M$. Then $||N|| = 0$ by Corollary 14.12, the universal covering is det-L^2-acyclic and $\rho^{(2)}(\widetilde{N}) \neq 0$ by Theorem 3.96 (4) and Theorem 3.152.

14.2.1 Hyperbolic Manifolds and 3-Manifolds

Let M be a hyperbolic orientable manifold of dimension n or a closed connected orientable manifold of dimension 3 satisfying the assumptions of Theorem 14.18 (3). Then M is aspherical and of determinant class and we get for a dimension constant D_n which is different from zero for n odd and zero for n even (see Theorem 4.1, Theorem 3.96 (3), Theorem 14.17 and Theorem 14.18 (3))

$$b_p^{(2)}(\widetilde{M}) = 0 \qquad \text{for } p \geq 0 \text{ if } ||M|| = 0;$$
$$\rho^{(2)}(\widetilde{M}) = D_n \cdot ||M||.$$

Hence for such manifolds Conjecture 14.1 is true.

The considerations above lead to the question whether the relation $\rho^{(2)}(\widetilde{M}) = D_n \cdot ||M||$ may hold for all aspherical closed orientable manifolds of dimension n for a dimension constant D_n. This is not the case as the following example shows.

Example 14.38. Let M be a 3-dimensional orientable closed hyperbolic 3-manifold. Let N be $M \times M \times M$ and F be an orientable closed hyperbolic surface. Let F^d be the d-fold cartesian product of F with itself. We get from Theorem 3.96 (4), Theorem 14.17 and Theorem 14.35

$$\rho^{(2)}(\widetilde{N \times F^d}) = 0;$$
$$\rho^{(2)}(\widetilde{M \times F^{d+3}}) \neq 0;$$
$$||N \times F^d|| \neq 0.$$

All the manifolds appearing in the list above are orientable closed aspherical manifolds of dimension $9 + 2d$. If $\rho^{(2)}(\widetilde{M}) = D_{9+2d} \cdot ||M||$ holds for all of them, we get $D_{9+2d} = 0$ from the first and third equation and $D_{9+2d} \neq 0$ from the second, a contradiction.

14.2.2 S^1-Actions

Let M be an aspherical closed orientable manifold with non-trivial S^1-action. Then $||M|| = 0$ by Theorem 14.32. Hence Conjecture 14.1 predicts that $\widetilde{M}$ is det-L^2-acyclic and $\rho^{(2)}(\widetilde{M}) = 0$. This has already been proved in Theorem 3.111.

14.2.3 Amenable Fundamental Groups

Let M be an aspherical closed orientable manifold with amenable fundamental group. Then $||M|| = 0$ by Corollary 14.16 (4). Hence Conjecture 14.1 predicts that $\widetilde{M}$ is det-L^2-acyclic and $\rho^{(2)}(\widetilde{M}) = 0$. This has already been proved in Theorem 3.113 in the special case where the fundamental group is elementary amenable.

Actually, it suffices in Theorem 3.113 to require that $\pi_1(M)$ contains a normal infinite elementary amenable subgroup. This raises

Question 14.39. (Simplicial volume and normal infinite amenable subgroups).
For which closed connected orientable manifolds, whose fundamental group contains an amenable infinite normal subgroup, does the simplicial volume vanish?

14.2.4 Selfmaps of Degree Different from −1, 0 and 1

Let $f \colon M \to M$ be a selfmap of an aspherical closed orientable manifold M whose degree is different from −1, 0 and 1. Then $\|M\| = 0$ by Corollary 14.12. Hence Conjecture 14.1 predicts that $b_p^{(2)}(\widetilde{M}) = 0$ for all $p \geq 0$, $\widetilde{M}$ is of determinant class and $\rho^{(2)}(\widetilde{M}) = 0$. There is the conjecture that f is homotopic to a covering with $\deg(f)$ sheets and then this conclusion would follow from Theorem 1.35 (9) and Theorem 3.96 (5) provided that $\widetilde{M}$ is of determinant class. Without this conjecture we can at least prove the claim for the L^2-Betti numbers under certain assumptions about the fundamental group.

A group G is *Hopfian* if any surjective group homomorphism $f \colon G \to G$ is an isomorphism. Examples of Hopfian groups are residually finite groups [356], [399, Corollary 41.44]. It is not true that a subgroup of finite index of a Hopfian group is again Hopfian [31, Theorem 2]. At least any subgroup of finite index in a residually finite group is again residually finite and hence Hopfian.

Theorem 14.40. *Let M be an aspherical closed orientable manifold. Suppose that any normal subgroup of finite index of its fundamental group is Hopfian and there is a selfmap $f \colon M \to M$ of degree $\deg(f)$ different from −1, 0, and 1. Then*

$$b_p^{(2)}(\widetilde{M}) = 0 \qquad \text{for } p \geq 0.$$

Proof. Fix an integer $n \geq 1$. Let $p \colon \overline{M} \to M$ be the covering of M associated to the image of $\pi_1(f^n) \colon \pi_1(M) \to \pi_1(M)$. By elementary covering theory there is a map $\overline{f^n} \colon M \to \overline{M}$ satisfying $p \circ \overline{f^n} = f^n$. Since $\deg(f^n) = \deg(\overline{f^n}) \cdot \deg(p)$ and $\deg(p) = [\pi_1(M) \colon \operatorname{im}(\pi_1(f))]$, we conclude

$$[\pi_1(M) \colon \operatorname{im}(\pi_1(f^n))] \leq |\deg(f^n)| = |\deg(f)|^n. \qquad (14.41)$$

Consider a map $g \colon N_1 \to N_2$ of two closed connected oriented n-dimensional manifolds of degree $\deg(g) \neq 0$. Abbreviate $G_1 = \pi_1(N_1)$ and $G_2 = \pi_1(N_2)$. Then there is a diagram of $\mathbb{Z}G_2$-chain complexes which commutes up to $\mathbb{Z}G_2$-homotopy

$$
\begin{array}{ccc}
\hom_{\mathbb{Z}G_1}(C_{n-*}(\widetilde{N_1}), \mathbb{Z}G_1) \otimes_{\mathbb{Z}[\pi_1(g)]} \mathbb{Z}G_2 & \xleftarrow{\ g^*\ } & \hom_{\mathbb{Z}G_2}(C_{n-*}(\widetilde{N_2}), \mathbb{Z}G_2) \\[4pt]
{\scriptstyle (?\cap[N_1])\otimes_{\mathbb{Z}[\pi_1(g)]}\mathrm{id}}\Big\downarrow & & \Big\downarrow{\scriptstyle ?\cap g_*([N_1])} \\[4pt]
C_*(\widetilde{N_1}) \otimes_{\mathbb{Z}[\pi_1(g)]} \mathbb{Z}G_2 & \xrightarrow{\ g_*\ } & C_*(\widetilde{N_2})
\end{array}
$$

Tensoring with $l^2(G_2)$ and then applying (reduced) L^2-homology yields a commutative diagram of finitely generated Hilbert $\mathcal{N}(G_2)$-modules

$$H_{(2)}^{n-p}\left(\hom_{\mathbb{Z}G_1}(C_*(\widetilde{N_1}),\mathbb{Z}G_1)\otimes_{\mathbb{Z}[\pi_1(g)]}l^2(G_2)\right) \xleftarrow{\;g^*\;} H_{(2)}^{n-p}(\widetilde{N_2};\mathcal{N}(G_2))$$

$$\left.H_*^{(2)}((?\cap[N_1])\otimes_{\mathbb{Z}[\pi_1(g)]}\mathrm{id})\right\downarrow \qquad\qquad\qquad \left.H_*^{(2)}(?\cap g_*([N_1]))\right\downarrow$$

$$H_p^{(2)}\left(C_*(\widetilde{N_1})\otimes_{\mathbb{Z}[\pi_1(g)]}l^2(G_2)\right) \qquad \xrightarrow{\;g_*\;} H_p^{(2)}(\widetilde{N_2};\mathcal{N}(G_2))$$

The right vertical arrow is bijective since $\deg(g)$ is invertible in $\mathbb{R}$, $g_*([N_1]) = \deg(g)\cdot[N_2]$ and $?\cap[N_2]\colon \hom_{\mathbb{Z}G_2}(C_{n-*}(\widetilde{N_2}),\mathbb{Z}G_2)\to C_*(\widetilde{N_2})$ is a $\mathbb{Z}[G_2]$-chain homotopy equivalence by Poincaré duality. Hence the morphism of finitely generated Hilbert $\mathcal{N}(G_2)$-modules

$$g_*\colon H_p^{(2)}\left(C_*(\widetilde{N_1})\otimes_{\mathbb{Z}[\pi_1(g)]}l^2(G_2)\right) \to H_p^{(2)}(\widetilde{N_2};\mathcal{N}(G_2))$$

is surjective. Theorem 1.12 (2) implies

$$\dim_{\mathcal{N}(G_2)}\left(H_p^{(2)}\left(C_*(\widetilde{N_1})\otimes_{\mathbb{Z}[\pi_1(g)]}l^2(G_2)\right)\right) \;\geq\; b_p^{(2)}(\widetilde{N_2}).$$

Let $e_p(N_1)$ be the number of p-cells in a fixed triangulation of N_1. We conclude from Theorem 1.12 (2)

$$\dim_{\mathcal{N}(G_2)}\left(H_p^{(2)}\left(C_*(\widetilde{N_1})\otimes_{\mathbb{Z}[\pi_1(g)]}l^2(G_2)\right)\right)$$
$$\leq \dim_{\mathcal{N}(G_2)}\left(C_p(\widetilde{N_1})\otimes_{\mathbb{Z}[\pi_1(g)]}l^2(G_2)\right)$$
$$\leq e_p(N_1).$$

This shows

$$b_p^{(2)}(\widetilde{N_2}) \leq e_p(N_1) \qquad\qquad \text{for } p \geq 0. \tag{14.42}$$

If we apply (14.42) to $g = \overline{f^n}$, we get

$$b_p^{(2)}(\widetilde{\overline{M}}) \;\leq\; e_p(M).$$

Since $b_p^{(2)}(\widetilde{\overline{M}}) = [\pi_1(M) : \mathrm{im}(\pi_1(f^n))]\cdot b_p^{(2)}(\widetilde{M})$ holds by Theorem 1.35 (9), we get

$$b_p^{(2)}(\widetilde{M}) \leq \frac{e_p(M)}{[\pi_1(M) : \mathrm{im}(\pi_1(f^n))]}. \tag{14.43}$$

Hence it suffices to show that there is no integer n such that $\mathrm{im}(\pi_1(f^n)) = \mathrm{im}(\pi_1(f^k))$ for all $k \geq n$ since then the limit for $n \to \infty$ of the right-hand side of (14.43) is zero. Suppose that such n exists. Then the composition $\overline{f^n}\circ p\colon \overline{M}\to \overline{M}$ induces an epimorphism on $\pi_1(\overline{M})$. Since $\pi_1(\overline{M})$ is Hopfian by assumption, $\overline{f^n}\circ p\colon \overline{M}\to \overline{M}$ is an isomorphism. Since $\overline{M}$ is aspherical, $\overline{f^n}\circ p$ is a homotopy equivalence. This implies

$$1 = \deg(\overline{f^n} \circ p) = \deg(\overline{f^n}) \cdot \deg(p).$$

This shows that $|\deg(p)| = 1$ and hence $\pi_1(f^n)$ is surjective. Since $\pi_1(M)$ is Hopfian and M aspherical, we conclude $|\deg(f^n)| = |\deg(f)^n| = 1$. This contradicts the assumption $\deg(f) \notin \{-1, 0, 1\}$. This finishes the proof of Theorem 14.40. $\square$

14.2.5 Negative Sectional Curvature and Locally Symmetric Spaces

There is the following conjecture which is attributed to Gromov in [457, page 239].

Conjecture 14.44. (Gromov's Conjecture about positivity of the simplicial volume for non-negative sectional and negative Ricci curvature). *Let M be a closed connected orientable Riemannian manifold whose sectional curvature is non-negative and whose Ricci curvature is everywhere negative. Then*
$$||M|| > 0.$$

Suppose that M is a closed connected orientable manifold with negative sectional curvature. Then the Singer Conjecture 11.1 and Conjecture 11.3 predict that either one of the L^2-Betti numbers $b_p^{(2)}(\widetilde{M})$ or $\rho^{(2)}(\widetilde{M})$ is different from zero. Hence together with Conjecture 14.1 they predict $||M|| > 0$. Indeed $||M|| > 0$ follows from Theorem 14.19 or from Theorem 14.27 for such M.

Conjecture 14.44 implies the following

Conjecture 14.45. (Positivity of the simplicial volume of locally symmetric spaces of non-compact type). *Let M be a closed connected Riemannian manifold whose universal covering is a symmetric space of non-compact type. Then*
$$||M|| > 0.$$

Conjecture 14.45 has been proved by Savage [457, page 239] in the case where the universal covering $\widetilde{M}$ is $SL(n, \mathbb{R})/SO(n)$.

Thus Conjecture 14.1 is true for aspherical locally symmetric spaces if Conjecture 14.45 is true (see Lemma 5.10 and equations (5.13) and (5.15)).

14.2.6 Simplicial Volume and L^2-Invariants

The following result follows from Gromov [238, page 300].

Theorem 14.46 (L^2-Betti numbers and volume). *Let M be an aspherical closed Riemannian manifold of dimension n. Suppose that for some $\delta > 0$ the Ricci curvature satisfies for all $x \in M$ and $v \in T_x M$*

$$\mathrm{Ricci}_x(v, v) \geq -\delta \cdot \langle v, v \rangle_x.$$

Then there is a constant $C(n, \delta)$, which depends on n and on δ but not on M, such that

$$\sum_{p \geq 0} b_p^{(2)}(\widetilde{M}) \leq C(n, \delta) \cdot \mathrm{vol}(M).$$

Theorem 14.46 implies together with (14.23)

Theorem 14.47 (L^2-Betti numbers and minimal volume). *Let M be an aspherical closed Riemannian manifold with $\mathrm{minvol}(M) = 0$. Then for $p \geq 0$*

$$b_p^{(2)}(\widetilde{M}) = 0.$$

The essential property of Riemannian manifolds with $\mathrm{Ricci}_x(v, v) \geq -\delta \cdot \langle v, v \rangle_x$ comes from packing inequalities as explained in [238, 5.31 on page 294]. A metric space X has packing type P for a function $P(R, r)$ with values in the positive integers if every ball of radius R contains at most $P(R, r)$ pairwise disjoint balls of radius r. Bishop's inequality implies for a Riemannian manifold M of dimension n with $\mathrm{Ricci}_x(v, v) \geq -(n-1) \cdot \langle v, v \rangle_x$ that $P(R, r) \leq C \cdot (R/r)^n$ for $0 \leq r \leq R \leq 1$ and $P(R, r) \leq \exp(\alpha_n \cdot R)$ for $1/2 \leq r \leq R$ and $1 \leq R$, where C is a universal constant and α_n a dimension constant. These packing inequalities are used to describe the manifold M in question as a homotopy retract of a simplicial complex Q which comes from the nerve of a covering of M and whose number of simplices is controlled.

Some evidence for Conjecture 14.1 comes from Theorem 14.24 (3) and Theorem 14.47. It raises the following question

Question 14.48. (Minimal volume and L^2-torsion for aspherical manifolds).
Let M be an aspherical closed Riemannian manifold with $\mathrm{minvol}(M) = 0$. Is then $\widetilde{M}$ of determinant class and

$$\rho^{(2)}(\widetilde{M}) = 0 ?$$

Some further evidence for Conjecture 14.1 comes from Remark 14.29 and Theorem 14.31.

14.3 Miscellaneous

More information about the simplicial volume can be found for instance in [76], [77] [178], [227], [232], [238], [275], [412] and [480].

One may ask

Question 14.49 (Simplicial volume and S^1-foliations). *Does the simplicial volume $||M||$ of a connected closed orientable manifold M vanish, if M carries an S^1-foliation?*

This question has a positive answer if one considers special S^1-foliations of certain 3-manifolds [52]. In view of of Conjecture 14.1 this is related to Question 3.186.

We mention the following result due to Besson-Courtois-Gallot [40, page 733].

Theorem 14.50. *Let M be a closed connected manifold M of dimension $n \geq 3$ with Riemannian metric g, whose Ricci curvature satisfies $\mathrm{Ricci}_x(v,v) \geq -(n-1)\langle v,v \rangle_x$ for $x \in M$ and $v \in T_x M$ and whose volume $\mathrm{vol}(M,g)$ coincides with its minimal volume $\mathrm{minvol}(M)$. Then (M,g) is hyperbolic, i.e. $\sec(M,g) = -1$.*

We mention the following improvement by Smillie and Gromov (see [232, page 23], [35, Proposition F.4.12 on page 293], [478]) of a result of Milnor [372] and Sullivan [487].

Theorem 14.51 (Flat bundles and simplicial volume). *Let $E \to B$ be a flat orientable n-dimensional vector bundle over a topological space B. Then its Euler class $e(E) \in H^n(B;\mathbb{R})$ (see [379, page 98]) satisfies*

$$\|e(E)\|_\infty \leq 2^{-n}.$$

In particular one gets for a flat n-dimensional bundle $E \to M$ over a closed connected orientable manifold M of dimension n

$$\chi(E) \leq 2^{-n} \cdot \|M\|,$$

where $\chi(E) = \langle e(E), [M] \rangle$ is the Euler number of E.

We mention the following conjecture due to Sullivan [487, page 187]. A manifold M of dimension n has an *affine structure* if there is an atlas whose change of coordinates maps are restrictions of affine isomorphisms $\mathbb{R}^n \to \mathbb{R}^n$.

Conjecture 14.52. (Sullivan's Conjecture on affine manifolds and Euler characteristic). *Let M be an n-dimensional closed connected orientable manifold which admits an affine structure. Then $\chi(M) = 0$.*

Notice that the tangent bundle TM of a manifold M with affine structure is flat and that the Euler number $\chi(TM)$ of the tangent bundle of a closed manifold M is equal to its Euler characteristic $\chi(M)$ [379, Corollary 11.12 on page 130]. Hence for an n-dimensional closed connected orientable manifold M, which admits an affine structure, the vanishing of $\|M\|$ implies $\chi(M) = 0$ by Theorem 14.51. This raises

Question 14.53. (Vanishing of the simplicial volume for affine manifolds).
For which closed connected orientable manifolds M, which admit an affine structure, does $\|M\| = 0$ hold?

A proof of the following result using bounded cohomology is presented in [35, Theorem F.6.5 on page 307]. The original proof appeared in [261].

Theorem 14.54. *Let M be an n-dimensional closed connected orientable manifold which admits an affine structure. Suppose that the range of the holonomy $\pi_1(M) \to GL(n, \mathbb{R})^+$ of the flat tangent bundle TM is amenable. Then $\chi(M) = 0$.*

The key idea is to show that the projection $p\colon STM \to M$ induces an injection $\widehat{H}^n(M) \to \widehat{H}^n(STM)$ and then use the observation that the pullback p^*TM has a non-vanishing section and hence $p^*e(TM) = e(p^*(TM)) = 0$.

Because of the Euler-Poincaré formula (see Theorem 1.35 (2)) Sullivan's Conjecture 14.52 follows from the following conjecture which is posed as a question in [237, page 232].

Conjecture 14.55. (Gromov's Conjecture on affine structures and L^2-Betti numbers). *Let M be an n-dimensional closed connected orientable manifold which admits an affine structure. Then $b_p^{(2)}(\widetilde{M}) = 0$ for $p \geq 0$.*

We have discussed the Volume Conjecture 4.8 already in Section 4.3 which links the simplicial volume $||S^3 - K||$ (or, equivalently, $\rho^{(2)}(K)$ of Definition 4.5 by Theorem 14.18 (3)) for a knot $K \subset S^3$ to the colored Jones polynomial.

Exercises

14.1.   Let M be a closed connected orientable manifold of dimension n. Let m be the minimal number of n-dimensional simplices appearing in a smooth triangulation of M. Show $||M|| \leq m$.

14.2.   Prove $||\mathbb{R}|| = \infty$.

14.3.   Let M and N be oriented manifolds without boundary. Suppose that there is a proper map $f\colon M \to N$ of degree d. Show that then $||M|| \geq d \cdot ||N||$. Conclude $||\mathbb{R}^n|| = 0$ for $n \geq 2$.

14.4.   Show for a path-connected topological space X that $\widehat{H}^0(X) \cong \mathbb{R}$.

14.5.   Let X be a simplicial complex. Let $C^*(X; \mathbb{R})$ be the cellular cochain complex with coefficients in $\mathbb{R}$. Let I_p be a cellular basis for $C^p(X)$. Then $C^p(X) = \mathrm{map}(I_p, \mathbb{R})$. Define for $\phi \in C^p(X)$ its L^∞-norm by

$$||\phi||_\infty := \sup\{\phi(b) \mid b \in I_p\}.$$

Show that this is independent of the choice of cellular basis. Let $l^\infty C^p(X) \subset C^p(X)$ be the submodule given by elements ϕ with $||\phi||_\infty < \infty$. Show that this defines a subchain complex $l^\infty C^*(X)$ of $C^*(X; \mathbb{R})$ so that $H^*(l^\infty C^*(X))$ is defined. Show that in general $H^*(l^\infty C^*(X))$ is not isomorphic to the bounded cohomology $\widehat{H}^p(X)$.

14.6. Let M be a closed connected orientable Riemannian manifold of dimension n with residually finite fundamental group. Let $k_1 \geq k_2 > 0$ be positive constants such that the sectional curvature satisfies $-k_1 \leq \sec(M) \leq -k_2$. Then there is a constant $C(n, k_1/k_2)$, which depends only on n and the ratio k_1/k_2 but not on M, such that

$$\sum_{i \geq 0} b_p^{(2)}(\widetilde{M}) \ \leq \ C(n, k_1/k_2) \cdot ||M||.$$

14.7. Let M be a closed connected oriented Riemannian manifold of dimension n. Let $dvol_M$ be its volume form. Show that the cochain $\phi \in C^n_{\text{sing}, C^\infty}(M; \mathbb{R})$ sending a singular smooth simplex $\sigma: \Delta_n \to M$ to $\frac{1}{\text{vol}(M)} \cdot \int_M \sigma^* dvol_M$ is a cocycle and represents an element in $H^n_{\text{sing}}(M; \mathbb{R})$ which is the image of the cohomological fundamental class of M under the change of rings map $H^n(M; \mathbb{Z}) \to H^n(M; \mathbb{R})$. Prove that $||\phi||_\infty = \infty$.

14.8. Let $f: X \to Y$ be a map of topological spaces. Show for any $y \in H_p^{\text{sing}}(X; \mathbb{R})$ and $\psi \in H^p_{\text{sing}}(Y; \mathbb{R})$

$$||H_p^{\text{sing}}(f)(y)||_1 \leq ||y||_1;$$
$$||H^p_{\text{sing}}(f)(\psi)||_\infty \leq ||\psi||_\infty.$$

14.9. Let M be an aspherical closed orientable Riemannian manifold of dimension n. Suppose that for some $\delta > 0$ the Ricci curvature satisfies for all $x \in M$ and $v \in T_x M$

$$\text{Ricci}_x(v, v) \geq -\delta \cdot \langle v, v \rangle_x.$$

Show that there is a constant $C(n, \delta)$, which depends on n and on δ but not on M, such that

$$\chi(M) \leq C(n, \delta) \cdot \text{vol}(M);$$
$$\text{sign}(M) \leq C(n, \delta) \cdot \text{vol}(M).$$

14.10. Show for the free group F_g on g generators

$$h(F_g) \ = \ \ln(2g - 1).$$

14.11. Let M and N be closed connected oriented manifolds. Let $f: M \to N$ be a map of degree 1. Show that $\pi_1(f)$ is surjective. Prove that $||M|| = ||N||$ if the kernel of $\pi_1(f)$ is amenable.

14.12. Let M be an aspherical closed connected orientable manifold which carries a non-trivial S^1-action. Show $\text{minvol}(M) = 0$.

14.13. Let M be a closed connected orientable manifold. Suppose that it carries an S^1-action such that M^{S^1} is empty. Show that then all Pontryagin numbers vanish.

14.14. Give an example of a closed connected orientable manifold M with infinite fundamental group whose minimal volume minvol(M) is zero and at least one L^2-Betti number of its universal covering is positive.

14.15. Let $F \to E \overset{p}{\to} B$ be a fibration of closed connected orientable manifolds with $\dim(F) \geq 1$. Suppose that $\pi_1(F)$ is amenable. Show $||E|| = 0$.

14.16. Show that the torus T^2 is the only closed connected orientable 2-dimensional manifold which admits an affine structure.

14.17. Let M be a closed connected manifold of dimension n. Suppose that for any $c > 0$ there are positive integers d_ϵ and t_ϵ satisfying $\frac{t_\epsilon}{d_\epsilon} \leq \epsilon$ together with a d_ϵ-sheeted covering $p_\epsilon \colon M_\epsilon \to M$ such that M_ϵ has a smooth triangulation with t_ϵ simplices of dimension n. Prove

$$||M|| = 0;$$
$$b_p^{(2)}(\widetilde{M}) = 0.$$

Assume furthermore that the fundamental group $\pi_1(M)$ is of det ≥ 1-class. Then show that $\widetilde{M}$ is det-L^2-acyclic and

$$\rho^{(2)}(\widetilde{M}) \ = \ 0.$$

15. Survey on Other Topics Related to L^2-Invariants

Introduction

In this chapter we discuss very briefly some topics which are related to L^2-invariants but cannot be covered in this books either because the author does not feel competent enough to write more about them or either because it would require to write another book to treat them in detail. The interested reader should look at the references cited below. Neither the list of references nor the list of topics is complete.

15.1 L^2-Index Theorems

The starting point for investigating L^2-invariants was the L^2-version of the index theorem by Atiyah [9]. Let P be an elliptic differential operator on a closed Riemannian manifold M and let $\overline{M} \to M$ be a regular covering of M with G as group of deck transformations. Then we can lift the Riemannian metric and the operator P to $\overline{M}$. The operator P is Fredholm and its *index* is defined by

$$\mathrm{ind}(P) = \dim_{\mathbb{C}}(\ker(P)) - \dim_{\mathbb{C}}(\ker(P^*)). \tag{15.1}$$

Using the von Neumann trace one can define the L^2-*index* of the lifted operator $\overline{P}$ analogously

$$\mathrm{ind}_{\mathcal{N}(G)}(\overline{P}) = \dim_{\mathcal{N}(G)}(\ker(\overline{P})) - \dim_{\mathcal{N}(G)}(\ker(\overline{P}^*)). \tag{15.2}$$

Then the L^2-index theorem says

$$\mathrm{ind}_{\mathcal{N}(G)}(\overline{P}) = \mathrm{ind}(P). \tag{15.3}$$

If one puts elliptic boundary conditions on the operator, this result was generalized to the case where M is compact and has a boundary by Schick [459]. This generalization is the L^2-version of the index theorem in [10].

These boundary conditions are local. Atiyah, Patodi and Singer [11], [12], [13] prove versions of the index theorem for manifolds with boundary using global boundary conditions, which, in contrast to the local conditions, apply to important geometrically defined operators such as the signature operator. This index theorem involves as a correction term the eta-invariant. The L^2-version of the eta-invariant has been defined and studied by Cheeger and Gromov [105], [106]. The L^2-version of this index theorem for manifolds with boundary using global boundary conditions has been proved by Ramachandran [427] for Dirac type operators.

The difference $\rho^{(2)}(M) := \eta^{(2)}(\widetilde{M}) - \eta(M)$ is the *relative von Neumann eta-invariant*. This invariant depends only on the smooth structure of M, not on the Riemannian metric [105]. If $\pi_1(M)$ is torsionfree and the Baum-Connes Conjecture holds for the *maximal* group C^*-algebra of $\pi_1(M)$, then Keswani [291] shows that $\rho^{(2)}(M)$ is a homotopy invariant. See also [359]. Chang and Weinberger [98] use $\rho^{(2)}(M)$ to prove their Theorem 0.10.

The L^2-index Theorem of Atiyah is a consequence of the index theorem of Mishchenko-Fomenko [380], where an index is defined which takes values in $K_0(C_r^*(G))$. If one applies the composition of the change of rings homomorphism $K_0(C_r^*(G)) \to K_0(\mathcal{N}(G))$ and the dimension homomorphism $\dim_{\mathcal{N}(G)} \colon K_0(\mathcal{N}(G)) \to \mathbb{R}$ to the index of Mishchenko and Fomenko, one obtains the L^2-index of Atiyah.

Some further references on L^2-index theory and related topics are [7], [21], [71], [72], [120], [121], [122], [154], [339], [386], [391], [392], [394], [437], [438], [439], [440], [460], [474], [485], [486].

Connections between L^2-cohomology and discrete series of representations of Lie groups are investigated, for instance, in [14], [123], [131], [135], [464].

15.2 L^p-Cohomology

One can also define and investigate L^p-cohomology for $p \in [1, \infty]$, as done for instance by Gromov [237, section 8] and Pansu [409], [410], [413]. One systematically replaces the L^2-condition by a L^p-condition and $l^2(G)$ by $l^p(G)$. Thus one gets L^p-cohomology for each $p \in [1, \infty]$. However, only for $p = 2$ the reduced L^p-cohomology is a Hilbert space and one can associate to the L^p-cohomology a real number, its dimension. There exists a L^p-version of the de Rham Theorem. The question whether for an element $\alpha \in \mathbb{C}G$ with $\alpha \neq 0$ there is $\beta \in l^p(G)$ with $\beta \neq 0$ and $\alpha * \beta = 0$ is studied in [312], [422]. Notice that the Atiyah-Conjecture predicts that for $p = 2$ and G torsionfree such an element β does not exists, but for $p > 2$ such elements are constructed in the case of finitely generated free and finitely generated free abelian groups in [312], [422].

Further references on L^p-cohomology are [2], [101], [170], [222], [223], [535].

15.3 Intersection Cohomology

When we have investigated L^2-invariants of manifolds M, we mostly have considered the case where there is a proper and cocompact action of a group G on M. Of course L^2-cohomology can also be interesting and has been investigated for complete non-necessarily compact Riemannian manifolds without such an action. For instance algebraic and arithmetic varieties have been studied. In particular, the Cheeger-Goresky-MacPherson Conjecture [104] and the Zucker Conjecture [531] have created a lot of activity. They link the L^2-cohomology of the regular part with the intersection homology of an algebraic variety. Some references concerning these topics are [53], [55], [95], [174], [266], [315], [319], [367], [397], [398], [402], [403], [414], [415], [451], [452], [453], [454], [455], [530], [531], [532], [533], [534].

15.4 Knot Concordance and L^2-signature

A knot in the 3-sphere is *slice* if there exists a locally flat topological embedding of the 2-disk into D^4 whose restriction to the boundary is the given knot. For a long time Casson-Gordon invariants [97] have been the only known obstructions for a knot to be slice. Cochran, Orr and Teichner give in [113] new obstructions for a knot to be slice using L^2-signatures. They construct a hierarchy of obstructions depending on one another and on certain choices. The obstructions take values in relative L-groups associated to Ore localizations of certain group rings. These rings map to the von Neumann algebra $\mathcal{N}(G)$ of certain groups G and the Ore localizations map to the algebra $\mathcal{U}(G)$ of affiliated operators. Recall that $\mathcal{U}(G)$ is the Ore localization of $\mathcal{N}(G)$ with respect to the multiplicative set of non-zero-divisors in $\mathcal{N}(G)$. Therefore they can use L^2-signatures to detect these obstructions in certain cases. Thus they can construct an explicit knot, which is not slice, but whose Casson-Gordon invariants are all trivial.

16. Solutions of the Exercises

Chapter 1

1.1. If H is finite, then use the projection $l^2(G) \to l^2(G)$ given by right multiplication with $\frac{1}{|H|} \cdot \sum_{h \in H} h$. Suppose that $l^2(G/H)$ is a Hilbert $\mathcal{N}(G)$-module. Since $l^2(G/H)^H$ is non-trivial, there must be a non-trivial Hilbert space V such that $(V \otimes l^2(G))^H \neq 0$. Hence $l^2(G)^H \neq 0$. This implies $|H| < \infty$.

1.2. Because of Theorem 1.9 (5) it suffices to prove for a weak isomorphism of finite dimensional Hilbert $\mathcal{N}(G)$-modules $u: V \to W$ and positive endomorphisms $f: V \to V$ and $g: W \to W$ with $u \circ f = g \circ u$ that $\mathrm{tr}_{\mathcal{N}(G)}(f) = \mathrm{tr}_{\mathcal{N}(G)}(g)$. By polar decomposition of u one can achieve $U = V$ and that u is positive. Let $\{E_\lambda \mid \lambda \in [0, \infty)\}$ be the spectral family of u. Consider the decomposition $V = \mathrm{im}(E_\lambda) \oplus \mathrm{im}(1 - E_\lambda)$. One writes f as a matrix with respect to this decomposition $\begin{pmatrix} f_{1,1} & f_{1,2} \\ f_{2,1} & f_{2,2} \end{pmatrix}$ and similar for g. Since u respects this decomposition, the morphisms $f_{2,2}$ and $g_{2,2}$ are conjugated by the automorphism of $\mathrm{im}(1 - E_\lambda)$ induced by u. Hence $\mathrm{tr}_{\mathcal{N}(G)}(f_{2,2}) = \mathrm{tr}_{\mathcal{N}(G)}(g_{2,2})$. We have $\mathrm{tr}_{\mathcal{N}(G)}(f) - \mathrm{tr}_{\mathcal{N}(G)}(f_{2,2}) = \mathrm{tr}_{\mathcal{N}(G)}(f_{1,1})$ and $\mathrm{tr}_{\mathcal{N}(G)}(g) - \mathrm{tr}_{\mathcal{N}(G)}(g_{2,2}) = \mathrm{tr}_{\mathcal{N}(G)}(g_{1,1})$. Since $\mathrm{tr}_{\mathcal{N}(G)}(f_{1,1})$ and $\mathrm{tr}_{\mathcal{N}(G)}(g_{1,1})$ are bounded by $\max\{||f||, ||g||\} \cdot \dim_{\mathcal{N}(G)}(\mathrm{im}(E_\lambda))$ and Theorem 1.12 (4) implies

$$\lim_{\lambda \to 0} \dim_{\mathcal{N}(G)}(\mathrm{im}(E_\lambda)) = \dim_{\mathcal{N}(G)}(\ker(u)) = 0,$$

the claim follows.

1.3. Put $I = \{1, 2, \ldots\}$. Let V_n be the Hilbert $\mathcal{N}(G)$-module given by the Hilbert sum $\bigoplus_{m=n}^{\infty} l^2(G)$. Then $\dim_{\mathcal{N}(G)}(V_n) = \infty$ for all n but $\bigcap_{n \geq 0} V_n = \{0\}$.

1.4. Replace U_1 and V_1 by $U_1/\ker(u_1)$ and $V_1/\ker(v_1)$ and replace U_5 and V_5 by $\mathrm{clos}(\mathrm{im}(u_4))$ and $\mathrm{clos}(\mathrm{im}(v_4))$. Then split the problem into three diagrams whose rows are short weakly exact sequences. It suffices to prove the claim for these diagrams. One reduces the claim to the case, where the rows are short exact sequences, by replacing a weakly exact sequence $0 \to V_0 \to V_1 \xrightarrow{q} V_2 \to 0$ by the exact sequence $0 \to \ker(q) \to V_1 \xrightarrow{\mathrm{pr}} \ker(q)^\perp \to 0$. Then the claim follows from the long weakly exact homology sequence (see Theorem 1.21).

1.5. Use $(f \circ g)^* = g^* \circ f^*$ for composable bounded operators of Hilbert spaces.

1.6.   Existence of finite self-covering (see Example 1.37), computation in terms of the quotient field of the group ring (see Lemma 1.34), S^1-action (see Theorem 1.40), mapping torus (see Theorem 1.39).

1.7.   One can construct a fibration $S^1 \to X \to B$. The main difficulty is to show that B is up to homotopy of finite type (see [331, Lemma 7.2]). Then one can apply Theorem 1.41. A more general statement will be proved in Theorem 7.2 (1) and (2).

1.8.   See [434, Theorem VIII.3 on page 256].

1.9.   One easily checks that this is a spectral family and then use the characterization of $\int_{-\infty}^{\infty} g(\lambda)dE_\lambda$ given in (1.65).

1.10.   The graph norm is the L^2-norm on the kernel of an operator

1.11.   Take $M = \mathbb{R} - \{0\}$ with the Riemannian metric coming from $\mathbb{R}$. Take $\omega = \chi_{(0,\infty)}$ and $\eta = \exp(-x^2) \cdot dx$.

1.12.   Write $\mathbb{C} \otimes_{\mathbb{Z}} C_*(X)$ as a direct sum of suspensions of 1-dimensional chain complexes of the shape $\mathbb{C}[\mathbb{Z}]^n \to \mathbb{C}[\mathbb{Z}]^n$ with non-trivial differentials and of zero-dimensional chain complexes with $\mathbb{C}[\mathbb{Z}]^n$ in dimension zero (see the proof of Lemma 2.58). The 1-dimensional chain complexes do not contribute to the L^2-Betti number (see Lemma 1.34 (1)). The contribution of the zero-dimensional chain complexes to the L^2-Betti numbers and to the Betti numbers are the same.

1.13.   The zero-th and first L^2-Betti numbers of $E(\mathbb{Z}/2 * \mathbb{Z}/2)$ vanish since $\mathbb{Z}/2 * \mathbb{Z}/2$ contains $\mathbb{Z}$ as subgroup of finite index (see Lemma 1.34 (2) and Theorem 1.35 (9)). Hence the zero-th and first L^2-Betti numbers of $\widetilde{M}$ vanish by Theorem 1.35 (1). From the Euler-Poincaré formula and Poincaré duality (see Theorem 1.35 (2) and (3)) we get $\chi(M) = b_2^{(2)}(\widetilde{M})$. Since $b_1(B(\mathbb{Z}/2 * \mathbb{Z}/2)) = 0$, one shows analogously $\chi(M) = 2 + b_2(M)$.

1.14.   Use Example 1.38 and truncate the relevant CW-complexes of finite type BG and X occuring there above dimension $n + 1$. Notice that $b_p^{(2)}(\widetilde{X \vee Y}) = b_p^{(2)}(\widetilde{X}) + b_p^{(2)}(\widetilde{Y})$ holds for $p \geq 1$ if $\pi_1(X)$ and $\pi_1(Y)$ are infinite (see Theorem 1.35 (5) and (8)) and that $b_p(X \vee Y) = b_p(X) + b_p(Y)$ holds for $p \geq 1$.

Chapter 2

2.1.   Assertion i.) is obvious. Assertion ii.) is proved as follows. If f and g are Fredholm and weak isomorphisms, $g \circ f$ is a weak isomorphism and Fredholm by Lemma 2.13 (3). Suppose that $g \circ f$ and f are Fredholm and weak isomorphisms. Injectivity and Fredholmness of g follow from Lemma 2.13 (2), density of the image of g is obvious. Suppose that g and $g \circ f$ are Fredholm and weak isomorphisms. Then f is a weak isomorphism and Fredholm since f^* is a weak isomorphism and Fredholm by the argument above applied to $f^* \circ g^*$ and g^*. Notice that f is Fredholm and a weak isomorphism if and only if f^* is Fredholm and a weak isomorphism (see Lemma 2.4). The desired counterexample is constructed in Example 1.19

2.2.   For $G = \mathbb{Z}$ we find u by Lemma 2.58. By induction (see (2.57)) we get the desired u for each group G with $\mathbb{Z} \subset G$. Now construct X as follows. Start with the finite 2-skeleton of EG. Then attach for $p = 3, 4, \ldots$ trivially a p-cell and then a $(p+1)$-cell, where the attaching map is given by an element u_p corresponding to α_p and involves only the trivially attached p-cell.

2.3.   The matrix A lives already over a finitely generated subgroup H of G. Now use induction (see (2.57)) and the fact that over a finite group all Novikov-Shubin invariants are ∞^+ (see Example 2.5).

2.4.   The condition $\alpha_1(\widetilde{X}) = 1$ implies that π is virtually nilpotent with growth rate precisely 1 (see Theorem 2.55 (5)). Hence $\pi_1(X)$ contains $\mathbb{Z}$ as subgroup of finite index by the formula for the growth rate appearing in Subsection 2.1.4. We conclude from Theorem 2.55 (1) and (6) that $\alpha_2(\widetilde{X}) = \alpha_2(E\pi) = \alpha_2(E\mathbb{Z}) = \infty^+$.

2.5.   The fundamental group π of M is finitely presented. Hence there is a model for $B\pi$ with finite 2-skeleton. Let $f \colon M \to B\pi$ be the classifying map. If we apply Theorem 2.55 (1) to its restriction to the 2-skeletons $M_2 \to B\pi_2$ we conclude that π determines $\alpha_p(\widetilde{M}) = \alpha$ for $p = 1, 2$. Now apply Poincaré duality Theorem 2.55 (2) and the fact that a compact 3-manifold M with non-empty boundary is homotopy equivalent to a finite 2-dimensional CW-complex.

2.6.   If $\pi_2(B) \to \pi_1(S^1)$ is not trivial, then $\pi_1(E)$ is finite and all Novikov-Shubin invariants must be ∞^+ (see Example 2.5). Suppose that $\pi_2(B) \to \pi_1(S^1)$ is trivial. Then $\pi_1(S^1) \to \pi_1(E)$ is an isomorphism. Now apply Lemma 2.58 and Theorem 2.61.

2.7.   Suppose that the $\mathbb{C}[\mathbb{Z}^n]$-module $l^2(\mathbb{Z}^n)$ is flat. Then $\operatorname{im}(c_2^{(2)}) = \ker(c_1^{(2)})$ holds for the differentials of $C_*^{(2)}(\widetilde{T^n})$ and $\operatorname{im}(c_2^{(2)})$ must be closed. We conclude from Lemma 2.11 (8) and (9) that $\alpha_2(\widetilde{T^n}) = \infty^+$. Example 2.59 implies $n = 1$.

It remains to show for each $\mathbb{C}[\mathbb{Z}]$-module M that $\text{Tor}_p^{\mathbb{C}[\mathbb{Z}]}(M, l^2(\mathbb{Z})) = 0$ for $p \geq 1$. The functor Tor commutes in both variables with colimits over directed systems [94, Proposition VI.1.3. on page 107]. Hence we can assume without loss of generality that M is a finitely generated $\mathbb{C}[\mathbb{Z}]$-module since each module is the directed union of its finitely generated submodules. Recall that $\mathbb{C}[\mathbb{Z}]$ is a principal ideal domain (see [15, Proposition V.5.8 on page 151 and Corollary V.8.7 on page 162]). Hence M is a finite sum of principal ideals. Therefore it suffices to show that for any non-trivial $p \in \mathbb{C}[\mathbb{Z}]$ multiplication with p induces an injection $l^2(\mathbb{Z}) \to l^2(\mathbb{Z})$. Since the von Neumann dimension of this kernel is zero by Lemma 2.58), the kernel is trivial by Theorem 1.12 (1).

2.8.   Use the product formula (see Theorem 1.35 (4) and Theorem 2.55 (3)) and Examples 2.59 and 2.69.

2.9.   Use the product formula (see Theorem 1.35 (4) and Theorem 2.55 (3)) and Examples 2.59 and 2.70.

2.10.   Use the product formula (see Theorem 2.55 (3)).

2.11.   See Theorem 3.100 and Remark 3.184.

2.12.   There exist up to isomorphism only countably many finitely presented groups. Moreover, for a given finitely presented group G and $m, n \geq 0$ there are only countably many elements in $M(m, n, \mathbb{Z}G)$. Hence A is countable. We get $\{r \mid r \in \mathbb{Q}, r \geq 0\} \subset A$ from the product formula (see Theorem 2.55 (3)), Lemma 2.58 and Example 2.59.

Chapter 3

3.1.   First show for a chain homotopy equivalence $f : C_* \to D_*$ of finite based free $\mathbb{Z}$-chain complexes that

$$\rho^{\mathbb{Z}}(D_*) - \rho^{\mathbb{Z}}(C_*) = \tau(f) - \sum_{p \geq 0}(-1)^p \cdot \ln\left(\det_{\mathbb{Z}}(H_p(f))\right) = 0.$$

The proof is an easy version of the one of Theorem 3.35 (5). Then prove that both sides of the equation are homotopy invariants and additive under direct sums of $\mathbb{Z}$-chain complexes. Construct a $\mathbb{Z}$-chain complex D_* which is $\mathbb{Z}$-homotopy equivalent to C_* and which is a direct sum of iterated suspensions of 0-dimensional $\mathbb{Z}$-chain complexes with $\mathbb{Z}$ as zero-th chain module and of 1-dimensional $\mathbb{Z}$-chain complexes of the shape $\mathbb{Z} \xrightarrow{n} \mathbb{Z}$ for some $n \in \mathbb{Z}, n \neq 0$.

This reduces the claim to these special chain complexes, where the proof is trivial.

3.2. Follow the proof of Theorem 3.35 (2) or see [116, (23.1)].

3.3. Follow the proof of Theorem 3.35 (4).

3.4. Follow the proof of Theorem 3.35 (6c) or see [116, (23.2)].

3.5. See [327, Corollary 5.5].

3.6. See [327, Example 1.15].

3.7. Use Lemma 3.15 (1) and $\int_\epsilon^a \frac{1}{\lambda \cdot (-\ln(-\ln(\lambda)) \cdot \ln(\lambda))} \, d\lambda = -\ln(\ln(-\ln(a))) + \ln(\ln(-\ln(\epsilon)))$ for $0 < \epsilon < a < e^{-e}$.

3.8. Since $\mathbb{C}[\mathbb{Z}^n]_{(0)}$ is a field, we can transform A over $\mathbb{C}[\mathbb{Z}^n]_{(0)}$ by elementary row and column operations into a diagonal matrix. Hence we can find over $\mathbb{C}[\mathbb{Z}^n]$ matrices $E_1, E_2, \ldots, E_{m+n}$ and a diagonal matrix D such that each E_i is a triangular matrix with non-zero diagonal entries or a permutation matrix and

$$E_1 \cdot E_2 \cdot \ldots \cdot E_m \cdot A \cdot E_{m+1} \cdot E_2 \cdot \ldots \cdot E_{m+n} = D.$$

The claim is easily verified for each matrix E_i and the matrix D and then proved for A using Lemma 1.34 (1), Theorem 3.14 (1) and (2) and Lemma 3.37 (1) and (2).

3.9. All claims except for $\rho^{(2)}(X)$ have already been proved in Lemma 2.58.

Notice that the determinant of $R_n \colon l^2(\mathbb{Z}) \to l^2(\mathbb{Z})$ for $n \in \mathbb{Z}, n > 0$ is n and the homology of the 1-dimensional $\mathbb{C}[\mathbb{Z}]$-chain complex $\mathbb{C}[\mathbb{Z}] \xrightarrow{R_n} \mathbb{C}[\mathbb{Z}]$ is zero for $n \in \mathbb{Z}, n > 0$. For each $n \in \mathbb{Z}, n > 0$ one constructs a free $\mathbb{Z}$-CW-complex which is obtained from $\widetilde{S^1}$ by attaching trivially an $\mathbb{Z}$-equivariant 3-cell and then a $\mathbb{Z}$-equivariant 4-cell such that the 4-th differential of $C_*(X_n)$ looks like R_n. Then X_m and X_n for $m \neq n$ leads to the example of two finite $\mathbb{Z}$-CW-complexes whose homology with $\mathbb{C}$-coefficients vanishes in all dimensions exept in dimension zero and hence is isomorphic in all dimensions but whose L^2-torsion are different.

Let C_* be a finite free $\mathbb{Z}[\mathbb{Z}]$-chain complex such that $\mathbb{C} \otimes_{\mathbb{Z}} H_p(C_*)$ is a torsion module over $\mathbb{C}[\mathbb{Z}]$ for all $p \in \mathbb{Z}$. Then $l^2(\mathbb{Z}) \otimes_{\mathbb{Z}[\mathbb{Z}]} C_*$ is det-L^2-acyclic. After a choice of a $\mathbb{Z}[\mathbb{Z}]$-basis for C_* we can define $\rho^{(2)}(l^2(\mathbb{Z}) \otimes_{\mathbb{Z}[\mathbb{Z}]} C_*) \in \mathbb{R}$.

Let $f_* \colon C_* \to D_*$ be a $\mathbb{Z}[\mathbb{Z}]$-chain homotopy equivalence of finite based free $\mathbb{Z}[\mathbb{Z}]$-chain complexes for which $\mathbb{C} \otimes_{\mathbb{Z}} H_p(C_*)$ and $\mathbb{C} \otimes_{\mathbb{Z}} H_p(D_*)$ are torsion modules over $\mathbb{C}[\mathbb{Z}]$ for $p \in \mathbb{Z}$. We get

$$\rho^{(2)}(l^2(\mathbb{Z}) \otimes_{\mathbb{Z}[\mathbb{Z}]} C_*) - \rho^{(2)}(l^2(\mathbb{Z}) \otimes_{\mathbb{Z}[\mathbb{Z}]} D_*) = \Phi^{\mathbb{Z}}(\tau(f_*))$$

for the homomorphism $\Phi^{\mathbb{Z}}\colon \mathrm{Wh}(\mathbb{Z}) \to \mathbb{R}$ defined in (3.92). This follows from the chain complex version of Theorem 3.93 (1). Since $\mathrm{Wh}(\mathbb{Z}) = \{1\}$, we conclude

$$\rho^{(2)}(l^2(\mathbb{Z}) \otimes_{\mathbb{Z}[\mathbb{Z}]} C_*) = \rho^{(2)}(l^2(\mathbb{Z}) \otimes_{\mathbb{Z}[\mathbb{Z}]} D_*).$$

Hence $\rho^{(2)}(l^2(\mathbb{Z}) \otimes_{\mathbb{Z}[\mathbb{Z}]} C_*)$ depends only on the $\mathbb{Z}[\mathbb{Z}]$-chain homotopy type of C_* and is in particular independent of the choice of the $\mathbb{Z}[\mathbb{Z}]$-basis. Let $0 \to C_* \to D_* \to E_* \to 0$ be an exact sequence of finite free $\mathbb{Z}[\mathbb{Z}]$-chain complexes for which $\mathbb{C} \otimes_{\mathbb{Z}} H_p(C_*)$, $\mathbb{C} \otimes_{\mathbb{Z}} H_p(D_*)$ and $\mathbb{C} \otimes_{\mathbb{Z}} H_p(E_*)$ are torsion modules over $\mathbb{C}[\mathbb{Z}]$ for $p \geq 0$. From Additivity of L^2-torsion (see Theorem 3.35 (1)) we conclude

$$\rho^{(2)}(l^2(\mathbb{Z}) \otimes_{\mathbb{Z}[\mathbb{Z}]} C_*) - \rho^{(2)}(l^2(\mathbb{Z}) \otimes_{\mathbb{Z}[\mathbb{Z}]} D_*) + \rho^{(2)}(l^2(\mathbb{Z}) \otimes_{\mathbb{Z}[\mathbb{Z}]} E_*) = 0.$$

Let M be a finitely generated $\mathbb{Z}[\mathbb{Z}]$-module such that $\mathbb{C} \otimes_{\mathbb{Z}} M$ is a torsion $\mathbb{C}[\mathbb{Z}]$-module. The ring $\mathbb{Z}[\mathbb{Z}]$ is Noetherian. Any finitely generated $\mathbb{Z}[\mathbb{Z}]$-module M has a finite free $\mathbb{Z}[\mathbb{Z}]$-resolution F_*. Define

$$\rho^{(2)}(M) = \rho^{(2)}(l^2(\mathbb{Z}) \otimes_{\mathbb{Z}[\mathbb{Z}]} F_*)$$

for any finite free $\mathbb{Z}[\mathbb{Z}]$-resolution F_* of M. The choice of F_* does not matter since two such resolutions are $\mathbb{Z}[\mathbb{Z}]$-chain homotopy equivalent.

Let C_* be any finite based free $\mathbb{Z}[\mathbb{Z}]$-chain complex such that $\mathbb{C} \otimes_{\mathbb{Z}} H_p(C_*)$ is a torsion $\mathbb{C}[\mathbb{Z}]$-module for $p \in \mathbb{Z}$. Without loss of generality we can assume $C_p = 0$ for $p < 0$. Then one can show by induction over the number n for which $H_p(C_*) = 0$ for $p \geq n$

$$\rho^{(2)}(l^2(\mathbb{Z}) \otimes_{\mathbb{Z}[\mathbb{Z}]} C_*) = \sum_{p \geq 0} (-1)^p \cdot \rho^{(2)}(H_p(C_*)).$$

The induction beginning $n = 0$ follows from the fact that the vanishing of $H_p(C_*)$ for all $p \geq 0$ implies that C_* is $\mathbb{Z}[\mathbb{Z}]$-chain homotopy equivalent to the trivial $\mathbb{Z}[\mathbb{Z}]$-chain complex. In the induction step from $(n-1)$ to n choose a finite free resolution F_* of $H_n(C_*)$ and construct a $\mathbb{Z}[\mathbb{Z}]$-chain map $\Sigma^n F_* \to C_*$ which induces an isomorphism on the n-th homology. We get an exact sequence of finite free $\mathbb{Z}[\mathbb{Z}]$-chain complexes $0 \to C_* \to \mathrm{cone}(f_*) \to \Sigma^{n+1} F_* \to 0$. We conclude that $H_p(\mathrm{cone}(f_*)) = H_p(C_*)$ for $p \leq n-1$ and $H_p(\mathrm{cone}(f_*)) = 0$ for $p \geq n-1$. Moreover, we get

$$\rho^{(2)}(l^2(\mathbb{Z}) \otimes_{\mathbb{Z}[\mathbb{Z}]} C_*) = \rho^{(2)}(l^2(\mathbb{Z}) \otimes_{\mathbb{Z}[\mathbb{Z}]} \mathrm{cone}(f_*)) - \rho^{(2)}(l^2(\mathbb{Z}) \otimes_{\mathbb{Z}[\mathbb{Z}]} \Sigma^{n+1} P_*).$$

From the definitions we get

$$-\rho^{(2)}(l^2(\mathbb{Z}) \otimes_{\mathbb{Z}[\mathbb{Z}]} \Sigma^{n+1} P_*) = (-1)^n \cdot \rho^{(2)}(H_n(C_*)).$$

This finishes the induction step.

We can write

$$\rho^{(2)}(l^2(\mathbb{Z}) \otimes_{\mathbb{Z}[\mathbb{Z}]} C_*) = -\frac{1}{2} \cdot \sum_{p \geq 0} (-1)^p \cdot p \cdot$$

$$\ln\left(\det\left(\operatorname{id}_{l^2(\mathbb{Z})} \otimes r_{A_p} : l^2(\mathbb{Z}) \otimes_{\mathbb{Z}[\mathbb{Z}]} \mathbb{Z}[\mathbb{Z}]^{s_p} \to l^2(\mathbb{Z}) \otimes_{\mathbb{Z}[\mathbb{Z}]} \mathbb{Z}[\mathbb{Z}]^{s_p}\right)\right)$$

for appropriate matrices $A_p \in M_{s_p}(\mathbb{Z}[\mathbb{Z}])$. If $d_p = \det_{[\mathbb{Z}[\mathbb{Z}]]}(A_p)$, we get from the previous Exercise 3.8

$$\det\left(\operatorname{id}_{l^2(\mathbb{Z})} \otimes r_{A_p} : l^2(\mathbb{Z}) \otimes_{\mathbb{Z}[\mathbb{Z}]} \mathbb{Z}[\mathbb{Z}]^{s_p} \to l^2(\mathbb{Z}) \otimes_{\mathbb{Z}[\mathbb{Z}]} \mathbb{Z}[\mathbb{Z}]^{s_p}\right)$$
$$= \det\left(\operatorname{id}_{l^2(\mathbb{Z})} \otimes r_{d_p} : l^2(\mathbb{Z}) \otimes_{\mathbb{Z}[\mathbb{Z}]} \mathbb{Z}[\mathbb{Z}] \to l^2(\mathbb{Z}) \otimes_{\mathbb{Z}[\mathbb{Z}]} \mathbb{Z}[\mathbb{Z}]\right).$$

We can write $\ln\left(\det\left(\operatorname{id}_{l^2(\mathbb{Z})} \otimes r_{d_p} : l^2(\mathbb{Z}) \otimes_{\mathbb{Z}[\mathbb{Z}]} \mathbb{Z}[\mathbb{Z}] \to l^2(\mathbb{Z}) \otimes_{\mathbb{Z}[\mathbb{Z}]} \mathbb{Z}[\mathbb{Z}]\right)\right)$ as $\ln(|a|) - \ln(|b|)$ for algebraic integers a, b with $|a|, |b| \geq 1$ because of Example 3.22. Hence the same is true for $\rho^{(2)}(l^2(\mathbb{Z}) \otimes_{\mathbb{Z}[\mathbb{Z}]} C_*)$.

3.10.   Consider an endomorphism $a: V \to V$ of a finitely generated Hilbert $\mathcal{N}(G)$-module which is a weak isomorphism and not of determinant class. Let C_* be the 1-dimensional $\mathcal{N}(G)$-chain complex $V \oplus V \xrightarrow{\operatorname{pr}_1} V$ and D_* be the $\mathcal{N}(G)$-chain complex with $D_1 = V$ and $D_p = 0$ for $p \neq 1$. Let $f_*: C_* \to D_*$ be the chain map given by $f_1 = \operatorname{id} \oplus a$ and $f_0 = 0$. Then $H_1^{(2)}(f_*)$ can be identified with a and f_p is of determinant class for all $p \in \mathbb{Z}$. Hence $H_1^{(2)}(f_*)$ is not of determinant class and f_* cannot be of determinant class by Theorem 3.35 (5).

3.11.   Without loss of generality the double chain complex is in the first quadrant. It suffices to prove the claim for the alternating sum of the rows, otherwise skip the double complex. Use induction over the width n, i.e. the smallest number $n \geq -1$ such that $C_{p,q} = 0$ for $p > n$. The induction beginning $n \leq 1$ follows from Theorem 3.35 (1) applied to the obvious exact sequence $0 \to C_{0,*} \to T_* \to \Sigma C_{1,*} \to 0$. In the induction step from $n - 1 \geq 1$ to n we can consider the double subcomplex $C'_{*,*} \subset C_{*,*}$ with $C'_{n,*} = C_{n,*}$, $C'_{n-1,*} = \operatorname{clos}(\operatorname{im}(C_{n,*} \to C_{n-1,*}))$ and $C'_{p,*} = 0$ for $p \neq n - 1, n$ and define $C''_{*,*}$ to be $C_{*,*}/C'_{*,*}$. We have an exact sequence of double complexes $0 \to C'_{*,*} \to C_{*,*} \to C''_{*,*} \to 0$ which induces also an exact sequence of the associated total complexes. Then for $p \in \mathbb{Z}$ the columns and rows of $C'_{*,*}$ and $C''_{*,*}$ are det-L^2-acyclic (see Lemma 3.14 (2), Lemma 3.15 (3) and Theorem 3.35 (1)). Now apply Theorem 3.35 (1) and the induction hypothesis to $C'_{*,*}$ and $C''_{*,*}$.

3.12.   Statement (1) is equivalent to the corresponding statement about finitely generated groups since any matrix over $\mathbb{Z}G$ comes by induction of a matrix over a finitely generated subgroup and Theorem 3.14 (6) holds. Because of Theorem 3.93 (1) it suffices to prove for a finitely generated group G that for any element in $x \in \operatorname{Wh}(G)$ we can find a G-homotopy equivalence $f: M \to N$ of cocompact free proper G-manifolds without boundary such that $\tau^G(f) = x + *(x)$ since $\phi^G(x + *(x)) = 2 \cdot \phi^G(x)$. One can find a cocompact free proper G-manifold W of even dimension such that $\partial W = \partial_0 W \coprod \partial_1 W$, the inclusion $i_k \colon \partial_k W \to W$ is a G-homotopy equivalence for $k = 0, 1$ and $\tau^G(i_0) = x$. Poincaré duality implies $\tau^G(i_1^{-1} \circ i_0) = x + *(x)$.

3.13.   This follows from Example 3.12 and Theorem 3.14 (6) and the obvious fact that the (classical) determinant of an invertible matrix over $\mathbb{Z}$ takes values in $\{\pm 1\}$.

3.14.   The set of matrices over $\mathbb{Z}G$ of finite size is a countable set. Hence the three sets are countable.

The first set is closed under addition because of $b_p^{(2)}(X \cup_G Y; \mathcal{N}(G)) = b_p^{(2)}(X; \mathcal{N}(G)) + b_p^{(2)}(Y; \mathcal{N}(G))$ for $p \geq 1$ and $|G| = \infty$ (see Theorem 1.35 (5) and (8)) and the first set is $\{n \cdot |G|^{-1} \mid n \in \mathbb{Z}, n \geq 0\}$ for $|G| < \infty$ by Example 1.32.

The third set can be identified with the subgroup of $\mathbb{R}$

$$\{\ln(\det(r_A)) - \ln(\det(r_B))\}$$

where A and B run through all matrices $A \in M_k(\mathbb{Z}G)$ and $B \in M_l(\mathbb{Z}G)$ for $k, l \geq 1$ such that $r_A \colon l^2(G)^k \to l^2(G)^k$ and $r_B \colon l^2(G)^l \to l^2(G)^l$ are weak isomorphisms of determinant class. Namely, fix a det-L^2-acyclic finite free proper G-CW-complex X. Then $X \times S^3$ is a finite, free, proper and det-L^2-acyclic with $\rho^{(2)}(X \times S^3) = 0$ by Theorem 3.93 (4). Given such a matrix A one can attach to $X \times S^3$ free G-cells in two consecutive dimensions such that the resulting G-CW-complex Z_A or Z_B respectively is finite, free, proper and det-L^2-acyclic and $\rho^{(2)}(Z_A; \mathcal{N}(G)) = \ln(\det(r_A))$ or $\rho^{(2)}(Z_B; \mathcal{N}(G)) = -\ln(\det(r_B))$ respectively (use Theorem 3.35 (1)). Now consider $Z = Z_A \cup_{X \times S^3} Z_B$. It is a det-$L^2$-acyclic finite free proper G-CW-complex with $\rho^{(2)}(Z; \mathcal{N}(G)) = \ln(\det(r_A)) - \ln(\det(r_B))$ by Theorem 3.93 (2).

3.15.   Use the exercise before and the facts that there are only countably many finitely presented groups and the fundamental group of a connected finite CW-complex is finitely presented.

3.16.   Consider a connected finite CW-complex X with fundamental group π and $A \in M_k(\mathbb{Z}\pi)$ such that $[A] \in \text{Wh}(\pi)$ is different from zero. Attach trivially k 4-cells to X and call the result Y. Construct an extension of id$\colon X \to X$ to a self map $f \colon Y \to Y$ such that $C_*(\widetilde{f}, \widetilde{\text{id}})$ is given by A. Then $\tau(f) = [A]$ holds in $\text{Wh}(G)$ and f induces the identity on π. We obtain a fibration $Y \to T_f \to S^1$ such that $\theta(p) \in H^1(S^1; \text{Wh}(\pi_1(T_f))) = \text{Wh}(\pi_1(T_f))$ defined in (3.98) is the image of $\tau(f)$ under the homomorphism $i_* \colon \text{Wh}(\pi) \to \text{Wh}(\pi_1(T_f))$ induced by the inclusion $i \colon Y \to T_f$. Since i induces the obvious split injection on the fundamental groups $\pi \to \pi \times \mathbb{Z} = \pi_1(T_f)$, $\theta(p)$ is non-trivial.

3.17.   The fundamental group $\pi_1(T_f)$ has a presentation as an HNN-extension

$$\langle \pi, s \mid sws^{-1} = f_*(w) \text{ for } w \in \pi \rangle.$$

This follows from the Theorem of Seifert and van Kampen applied to an appropriate decomposition of the mapping torus into two connected pieces with connected intersection.

3.18.   This follows from Example 3.107.

3.19.   Use the exercise before.

3.20.   (1) We get from Lemma 3.139 (2)

$$\theta_F(t) = t \cdot \int_0^\infty e^{-t\lambda} \cdot F(\lambda) \, d\lambda$$

$$= t \cdot \int_0^K e^{-t\lambda} \cdot F(\lambda) \, d\lambda + F(K) \cdot t \cdot \int_K^\infty e^{-t\lambda} \, d\lambda$$

$$= t \cdot \int_0^K e^{-t\lambda} \cdot F(\lambda) \, d\lambda + F(K) \cdot e^{-tK}$$

$$= t \cdot \int_0^K \sum_{n \geq 0} \frac{(-\lambda)^n}{n!} \cdot t^n \cdot F(\lambda) \, d\lambda + F(K) \cdot \sum_{n \geq 0} \frac{(-K)^n}{n!} \cdot t^n$$

$$= \sum_{n \geq 0} \frac{(-1)^n}{n!} \cdot \left(- \int_0^K n \cdot \lambda^{n-1} \cdot F(\lambda) \, d\lambda + F(K) \cdot K^n \right) \cdot t^n.$$

We get from the definitions

$$a_0 = \theta_F(0) = \int_0^\infty e^{0 \cdot \lambda} \, dF(\lambda) = F(K) - F(0) = F(K).$$

(2) We can write $\theta_F(t) = \sum_{n=0}^N a_n \cdot t^n + R(t)$ for a function $R(t)$ satisfying $\lim_{t \to 0} \frac{R(t)}{t^N} = 0$. We get

$$\frac{1}{\Gamma(s)} \cdot \int_0^\epsilon t^{s-1} \cdot \theta_F(t) \, dt = \sum_{n=0}^N a_n \cdot \frac{1}{\Gamma(s)} \cdot \int_0^\epsilon t^{s+n-1} \, dt$$

$$+ \frac{1}{\Gamma(s)} \cdot \int_0^\epsilon t^{s-1} \cdot R(t) \, dt$$

$$= \sum_{n=0}^N a_n \cdot \frac{1}{\Gamma(s+1)} \cdot \frac{s}{s+n} \cdot \epsilon^{s+n}$$

$$+ \frac{1}{\Gamma(s)} \cdot \int_0^\epsilon t^{s-1} \cdot R(t) \, dt.$$

The function $\frac{1}{\Gamma(s)} \cdot \int_0^\epsilon t^{s-1} \cdot R(t)\, dt$ is holomorphic for $\Re(s) > -N$.

(3) For any α with $0 < \alpha < \alpha(F)$ we can find t_0 such that $\theta_F(t) \le t^{-\alpha}$ holds for $t \ge t_0$. Now go through the calculation (3.131) in the case $\delta = \infty$.

(4) We get from partial integration (3.16)

$$\int_{\epsilon+}^K \lambda^{-s}\, dF(\lambda) = -s \cdot \int_\epsilon^K \lambda^{-s-1} \cdot F(\lambda)\, d\lambda + K^{-s} \cdot F(K) - \epsilon^{-s-1} \cdot F(\epsilon).$$

For any α with $0 < \alpha < \alpha(F)$ there is $\lambda(\alpha)$ such that $F(\lambda) \le \lambda^\alpha$ holds for $0 \le \lambda \le \lambda(\alpha)$. We conclude from Lebesgue's Theorem of majorized convergence

$$\int_0^\infty \lambda^{-s}\, dF(\lambda) = -s \cdot \int_0^K \lambda^{-s-1} \cdot F(\lambda)\, d\lambda + K^{-s} \cdot F(K).$$

This implies using Lemma 3.15 (1)

$$
\begin{aligned}
\frac{d}{ds} \int_0^\infty \lambda^{-s}\, dF(\lambda)\Big|_0 &= \Big(-s \cdot \frac{d}{ds} \int_0^K \lambda^{-s-1} \cdot F(\lambda)\, d\lambda \\
&\quad - \int_0^K \lambda^{-s-1} \cdot F(\lambda)\, d\lambda + \ln(K) \cdot K^{-s} \cdot F(K) \Big)\Big|_{s=0} \\
&= -\int_0^K \frac{F(\lambda)}{\lambda}\, d\lambda + \ln(K) \cdot F(K) \\
&= \int_0^\infty \ln(\lambda)\, dF(\lambda).
\end{aligned}
$$

(5) This is analogous to (3.133) using (3.132).

(6) This follows from the other assertions.

3.21. The heat kernel of Δ_0 on $\mathbb{R}$ is given by

$$e^{-t\Delta_0}(x,y) = \frac{1}{4\pi t} \cdot e^{-(x-y)^2/4t}.$$

Since $[0,1]$ is a fundamental domain for the $\mathbb{Z}$-action on $\mathbb{R} = \widetilde{S^1}$, we get for $t > 0$

$$\theta_0(\widetilde{S^1})(t) = \int_0^1 \frac{1}{4\pi t} \cdot e^{-(x-x)^2/4t}\, dx = \frac{1}{4\pi t}.$$

From $* \circ \Delta_1 = \Delta_0 \circ *$ we get $\theta_1(\widetilde{S^1}) = \theta_0(\widetilde{S^1})$. We have for $\Re(s) > 1$

$$\frac{1}{\Gamma(s)} \cdot \int_0^1 t^{s-1} \cdot \theta_1(\widetilde{S^1})\, dt = \frac{1}{4\pi} \cdot s \cdot \frac{1}{(s-1) \cdot \Gamma(s+1)}.$$

This implies

$$\frac{d}{ds}\frac{1}{\Gamma(s)}\cdot\int_0^1 t^{s-1}\cdot\theta_1(\widetilde{S^1})\,dt\,\bigg|_{s=0} = -\frac{1}{4\pi}.$$

A direct computation shows

$$\int_1^\infty t^{-1}\cdot\theta_1(\widetilde{S^1})\,dt = \frac{1}{4\pi}.$$

3.22. See [327, Example 1.18].

3.23. From Poincaré duality (see Theorem 1.35 (3)) and from the long exact weak homology sequence (see Theorem 1.21) we conclude

$$b_p^{(2)}(M) = b_p^{(2)}(M,\partial M) = b_p^{(2)}(\partial M).$$

From the version of Poincaré duality for manifolds with boundary (cf. Theorem 3.93 (3) and its proof) we conclude that $(M,\partial M)$ is det-L^2-acyclic and satisfies

$$\rho^{(2)}(M) = -\rho^{(2)}(M,\partial M).$$

We conclude from Theorem 3.35 (1) that ∂M is det-L^2-acyclic and we have

$$\rho^{(2)}(M) = \rho^{(2)}(M,\partial M) + \rho^{(2)}(\partial M).$$

3.24. Let A be the $(1,1)$-matrix over $\mathbb{C}[\mathbb{Z}]$ with entry $(z-\lambda)$ for $\lambda\in\mathbb{R}^{\geq 0}$. The operator norm of R_A is $1+\lambda$. We compute:

$$\operatorname{tr}\left(\left(1-\frac{AA^*}{(1+\lambda)^2}\right)^p\right)$$

$$= \operatorname{tr}\left(\left(1-\frac{(z-\lambda)(z^{-1}-\lambda)}{(1+\lambda)^2}\right)^p\right)$$

$$= \operatorname{tr}\left(\left(1-\frac{1+\lambda^2-\lambda\cdot(z+z^{-1})}{(1+\lambda)^2}\right)^p\right)$$

$$= \operatorname{tr}\left(\left(\frac{2\cdot\lambda}{(1+\lambda)^2}+\frac{\lambda}{(1+\lambda)^2}\cdot(z+z^{-1})\right)^p\right)$$

$$= \frac{\lambda^p}{(1+\lambda)^{2p}}\cdot\operatorname{tr}\left((2+(z+z^{-1}))^p\right)$$

$$= \frac{\lambda^p}{(1+\lambda)^{2p}}\cdot\operatorname{tr}\left(\sum_{k=0}^p\binom{p}{k}\cdot 2^{p-k}\cdot(z+z^{-1})^k\right)$$

$$= \frac{(2\cdot\lambda)^p}{(1+\lambda)^{2p}}\cdot\sum_{k=0}^p\binom{p}{k}\cdot 2^{-k}\cdot\operatorname{tr}\left((z+z^{-1})^k\right)$$

$$= \frac{(2 \cdot \lambda)^p}{(1+\lambda)^{2p}} \cdot \sum_{k=0}^{p} \binom{p}{k} \cdot 2^{-k} \cdot \mathrm{tr}\left(\sum_{j=0}^{k} \binom{k}{j} \cdot z^j \cdot z^{-(k-j)}\right)$$

$$= \frac{(2 \cdot \lambda)^p}{(1+\lambda)^{2p}} \cdot \sum_{k=0}^{p} \binom{p}{k} \cdot 2^{-k} \cdot \sum_{j=0}^{k} \binom{k}{j} \cdot \mathrm{tr}\left(z^{2j-k}\right)$$

$$= \frac{(2 \cdot \lambda)^p}{(1+\lambda)^{2p}} \cdot \sum_{k=0,k \text{ even}}^{p} \binom{p}{k} \cdot 2^{-k} \cdot \binom{k}{k/2}$$

$$= \frac{(2 \cdot \lambda)^p}{(1+\lambda)^{2p}} \cdot \sum_{k=0}^{[p/2]} \binom{p}{2k} \cdot 4^{-k} \cdot \binom{2k}{k}$$

$$- \frac{(2 \cdot \lambda)^p}{(1+\lambda)^{2p}} \cdot \sum_{k=0}^{[p/2]} 4^{-k} \cdot \frac{p!}{(p \quad 2k)! \cdot k! \cdot k!}$$

Now apply (3.24) and Theorem 3.172.

Chapter 4

4.1. Suppose that M has a prime decomposition whose factors are homotopy spheres with precisely one exception which is D^3 or an irreducible manifold with infinite fundamental group. By the Sphere Theorem [252, Theorem 4.3] an irreducible 3-manifold is aspherical if and only if it is a 3-disk or has infinite fundamental group. Hence M is aspherical.

Suppose that M is aspherical. Then its fundamental group is torsionfree. Hence each summand in its prime decomposition $M = M_1 \# M_2 \# \ldots \# M_r$ has trivial or torsionfree infinite fundamental group. We can assume without loss of generality that no M_i is homotopy equivalent to S^3 since the connected sum with such manifolds does not affect the homotopy type. Since S^3 is not aspherical, we have $r \geq 1$. There is an obvious 2-connected map $\#_{i=1}^r M_i \to \bigvee_{i=1}^r M_i$. Hence $\pi_2\left(\bigvee_{i=1}^r M_i\right) \cong \pi_2\left(\widetilde{\bigvee_{i=1}^r M_i}\right) \cong H_2\left(\widetilde{\bigvee_{i=1}^r M_i}\right)$ is trivial. A Mayer Vietoris argument implies $H_2\left(\widetilde{\bigvee_{i=1}^r M_i}|_{M_i}\right) = 0$ for $i = 1, 2, \ldots, r$. Since the inclusion of M_i into $\bigvee_{i=1}^r M_i$ is injective, we conclude $H_2(\widetilde{M_i}) = 0$ for $i = 1, 2, \ldots, r$. This shows that each M_i is different from $S^1 \times S^2$. Hence each M_i is either an irreducible manifold with infinite fundamental group or D^3. In particular each M_i is aspherical. The obvious map $\#_{i=1}^r M_i \to \bigvee_{i=1}^r M_i$ is a map of aspherical spaces inducing an isomorphism on the fundamental groups. Hence it is a homotopy equivalence. It induces a homeomorphism $\partial M \to \coprod_{i=1}^r \partial M_i$. Hence it induces an isomorphism

$$H_3(M, \partial M)) \cong \mathbb{Z} \to H_3\left(\bigvee_{i=1}^r M_i, \coprod_{i=1}^r \partial M_i\right) \cong \bigoplus_{i=1}^r H_3(M_i, \partial M_i) \cong \mathbb{Z}^r.$$

This implies $r = 1$.

4.2. Theorem 3.183 implies for two closed connected 3-manifolds M and N with the same geometry that $\alpha_p(\widetilde{M}) = \alpha_p(\widetilde{N})$ for $p \geq 1$ and that $b_p^{(2)}(\widetilde{M}) = 0 \Leftrightarrow b_p^{(2)}(\widetilde{N}) = 0$ for $p \geq 0$. If M has a geometry, it is either aspherical or the geometry is S^3 or $S^2 \times \mathbb{R}$. Since we assume $\pi_1(M)$ to be infinite, the geometry S^3 cannot occur. If M is aspherical, it is an irreducible manifold with infinite fundamental group by Exercise 4.1 and hence $b_p^{(2)}(\widetilde{M}) = 0$ for all $p \geq 0$ by Theorem 4.1. If M has the geometry $S^2 \times \mathbb{R}$, we conclude $b_p^{(2)}(\widetilde{M}) = 0$ for all $p \geq 0$, since $S^1 \times S^2$ has geometry $S^2 \times \mathbb{R}$ and $b_p^{(2)}(\widetilde{S^1 \times S^2}) = 0$ for all $p \geq 0$. We conclude from Theorem 4.2 that one can distinguish the geometries from one another by the Novikov-Shubin invariants of the universal coverings of closed 3-manifolds except $\mathbb{H}^2 \times \mathbb{R}$ and $\widetilde{SL_2(\mathbb{R})}$.

4.3.

(1) $\mathbb{H}^3$: The fundamental group of a closed hyperbolic manifold does not contain $\mathbb{Z} \oplus \mathbb{Z}$ as subgroup [466, Corollary 4.6 on page 449]. It cannot be virtually cyclic because this would imply that a finite covering of M is homotopy equivalent to S^1 or because this would imply $\alpha_1(\widetilde{M}) = 1$ by Theorem 2.55 (5a) contradicting Theorem 4.2 (2) which says $\alpha_1(\widetilde{M}) = \infty^+$;

(2) S^3: π is finite if and only if $\widetilde{M}$ is compact;

(3) $S^2 \times \mathbb{R}$: M is finitely covered by $S^1 \times S^2$ and hence has infinite virtually cyclic fundamental group. Since $\alpha_1(M) = 1$ is only true for the geometry $S^2 \times \mathbb{R}$ by Theorem 4.2 (2), the fundamental groups for all the other geometries cannot be infinite virtually cyclic by Theorem 2.55 (5a);

(4) $\mathbb{R}^3$: M is finitely covered by T^3 and hence contains $\mathbb{Z}^3$ as subgroup of finite index. Since $\alpha_1(M) = 3$ is only true for the geometry $\mathbb{R}^3$ by Theorem 4.2 (2), the fundamental groups of all the other geometries cannot contain $\mathbb{Z}^3$ as subgroup of finite index by Theorem 2.55 (5a);

(5) Nil: M is finitely covered by a closed manifold $\overline{M}$ which is a S^1-fibration $S^1 \to \overline{M} \to T^2$. Hence π contains a subgroup of finite index G which can be written as an extension $1 \to \mathbb{Z} \to G \to \mathbb{Z}^2 \to 1$;

(6) $\mathbb{H}^2 \times \mathbb{R}$: M is finitely covered by $S^1 \times F_g$ for a closed surface of genus $g \geq 2$. Hence π contains a subgroup of finite index which is isomorphic to $\mathbb{Z} \times G$ for some group G which is not solvable. Hence π is not solvable. The fundamental groups of all geometries except $\mathbb{H}^3$, $\mathbb{H}^2 \times \mathbb{R}$ and $\widetilde{Sl_2(\mathbb{R})}$ are solvable;

(7) $\widetilde{Sl_2(\mathbb{R})}$: M is finitely covered by a closed manifold $\overline{M}$ which is a S^1-prinicipal bundle $S^1 \to \overline{M} \to F_g$ for a closed surface of genus $g \geq 2$. This shows that π is not solvable and contains $\mathbb{Z} \oplus \mathbb{Z}$ as subgroup. Next we show that π does not contain a subgroup of finite index of the shape $\mathbb{Z} \times G$ for a group G. Suppose the contrary. Then we can arrange by possibly passing

to a finite covering over $\overline{M}$ that $\pi_1(\overline{M}) = \mathbb{Z} \times G$. Since $S^1 \to \overline{M} \to F_g$ is a S^1-principal bundle, the center of $\pi_1(\overline{M})$ contains the image of $\pi_1(S^1) \to \pi_1(\overline{M})$. Since $\mathcal{Z}(\pi_1(F_g))$ does not contain $\mathbb{Z} \oplus \mathbb{Z}$ and hence has trivial center, the center of $\pi_1(\overline{M})$ is the image of $\pi_1(S^1) \to \pi_1(\overline{M})$. The center of $\mathbb{Z} \times G$ is $\mathbb{Z} \times \mathcal{Z}(G)$. This implies that the map $\pi_1(S^1) \to \pi_1(\overline{M})$ has a retraction which comes from the projection $\mathbb{Z} \times G \to \mathbb{Z}$. Hence the S^1-bundle $S^1 \to \overline{M} \to F_g$ is trivial. This implies that $\overline{M}$ is diffeomorphic to $S^1 \times F_g$ and hence carries the geometry $\mathbb{H}^2 \times \mathbb{R}$, a contradiction;

(8) *Sol*: Any *Sol*-manifold M has a finite covering $\overline{M}$ which can be written as a bundle $T^2 \to \overline{M} \to S^1$ [466, Theorem 5.3 on page 447]. Hence we get a subgroup $G \subset \pi$ of finite index which is an extension $0 \to \mathbb{Z}^2 \to G \to \mathbb{Z} \to 0$. The group π cannot be virtually abelian, since otherwise π contains $\mathbb{Z}^3$ as subgroup of finite index and carries the geometry $\mathbb{R}^3$.

4.4.   All Novikov-Shubin invariants are ∞^+ (see Example 2.5) and we have $b_p^{(2)}(\widetilde{M}) = |\pi|^{-1} \cdot b_p(\widetilde{M})$ (see Theorem 1.35 (9)). If M is closed, $\widetilde{M}$ is homotopy equivalent to S^3. If M has boundary, then $H_p(\widetilde{M}) = 0$ for $p = 1, 3$ and we get $\dim_{\mathbb{C}}(H_2(\widetilde{M};\mathbb{C})) = \chi(\widetilde{M}) - 1$. Hence

$$
\begin{aligned}
b_0^{(2)}(\widetilde{M}) &= |\pi|^{-1}; \\
b_1^{(2)}(\widetilde{M}) &= 0; \\
b_2^{(2)}(\widetilde{M}) &= \chi(M) - |\pi|^{-1} \quad \text{if } \partial M \neq \emptyset; \\
b_2^{(2)}(\widetilde{M}) &= 0 \quad\qquad\qquad \text{if } \partial M = \emptyset; \\
b_3^{(2)}(\widetilde{M}) &= 0 \quad\qquad\qquad \text{if } \partial M \neq \emptyset; \\
b_3^{(2)}(\widetilde{M}) &= |\pi|^{-1} \quad\qquad \text{if } \partial M = \emptyset.
\end{aligned}
$$

4.5.   We have to check that it is not true that $b_p^{(2)}(\widetilde{M}) = 0$ for all $p \geq 0$ and $\alpha(\widetilde{M}) = \infty^+$ for all $p \geq 1$ holds. Suppose that it is true. Then we conclude from Theorem 4.1 and Theorem 4.2 (2) and (4) that M is an irreducible 3-manifold with non-empty compressible torus boundary. Now use the construction in the proof of [322, Lemma 6.4 on page 52] based on the Loop Theorem [252, Theorem 4.2 on page 39] and the formula for the second Novikov-Shubin invariants under wedges (cf. Theorem 2.55 (4)) to get a contradiction also in this remaining case.

4.6.   See [322, Proposition 6.5 on page 54].

4.7.   This follows from Theorem 3.106. Notice that $\pi_1(T_f)$ belongs to the class $\mathcal{G}$ (cf. the paragraph after the proof of Theorem 3.106).

Chapter 5

5.1.   See [251, Lemma 3.1 in IV.3 on page 205].

5.2.   See see [293, page 4 and 5].

5.3.   See [293, page 7].

5.4.   See [293, page 8].

5.5.   Since $GL(n, \mathbb{C}) \times GL(m, \mathbb{C})$ is a subgroup of $GL(m+n, \mathbb{C})$ closed under taking conjugate transpose and similar for $\mathbb{R}$, we get that the product $G \times H$ of two connected linear reductive Lie groups G and H is again a connected linear reductive Lie group. Since we get for the center $\mathcal{Z}(G \times H) = \mathcal{Z}(G) \times \mathcal{Z}(H)$, the claim follows by the previous Exercise 5.4.

5.6.   Since the spaces in question are homogeneous, it suffices to construct the global symmetry t_x in some model at one point x.

5.7.   One can identify $G_p(\mathbb{R}^{p+q})$ with $SO(p+q)/SO(p) \times SO(q)$. Now apply Example 5.11.

5.8.   By Theorem 5.12 (2) one can read off $\dim(M) - \text{f-rk}(\widetilde{M})$ from $\alpha_p(\widetilde{M})$ for $p \geq 1$ which depends only on the homotopy type (see Theorem 2.55 (1)). Notice that M and N are aspherical by Lemma 5.10 and hence homotopic if they have isomorphic fundamental groups.

5.9.   Amenability of $\pi_1(M)$ is equivalent to $\alpha_1(\widetilde{M}) \neq \infty^+$ by Theorem 2.55 (5b) provided that $\pi_1(M)$ is infinite. We conclude from (5.14) and the fact that $\frac{\dim(\widetilde{M_{\text{ncp}}}) - \text{f-rk}(\widetilde{M_{\text{ncp}}})}{2} \geq 1$ if $\widetilde{M_{\text{ncp}}} \neq \{*\}$ (see Theorem 5.12 (2)) that $\widetilde{M_{\text{ncp}}} = \{*\}$ if $\pi_1(M)$ is amenable.

   Suppose that $\widetilde{M_{\text{ncp}}} = \{*\}$. Then M carries a Riemannian metric with non-negative sectional curvature (see Theorem 5.9) and hence with non-negative Ricci curvature. Then $\pi_1(M)$ has polynomial growth (see [218, Theorem 3.106 on page 144]). Hence $\pi_1(M)$ is virtually nilpotent and in particular amenable.

5.10.   Let X^n for $n = 2, 3, 5$ be a closed hyperbolic manifold of dimension n. Put $M = X^5 \times X^5$ and $N = X^3 \times X^3 \times X^2 \times X^2$. We have $\mathbb{Z}^4 \subset \pi_1(N)$, but $\mathbb{Z}^4$ cannot be a subgroup of $\pi_1(M)$ since any abelian subgroup of $\pi_1(X^5)$ is trivial or $\mathbb{Z}$. Hence M and N have non-isomorphic fundamental groups. All L^2-Betti numbers, all Novikov-Shubin invariants and the L^2-torsion of $\widetilde{M}$ and $\widetilde{N}$ with respect to the $\pi_1(M)$- and $\pi_1(N)$-action agree. This follows from

the various product formulas and the known values for hyperbolic manifolds
or directly from Theorem 5.12.

5.11. Since M is aspherical, $\widetilde{M} = \widetilde{M}_{\mathrm{Eucl}} \times \widetilde{M}_{\mathrm{ncp}}$ by Lemma 5.10. Now apply
Theorem 1.35 (2) and (5.15).

5.12. Suppose that $b_p^{(2)}(\widetilde{M}) = 0$ for all $p \geq 0$. Since M carries a Riemannian
metric with non-positive sectional curvature (see Theorem 5.9 and Lemma
5.10), the Whitehead group of its fundamental group vanishes [192, page 61].
Hence also $\widetilde{N}$ is det-L^2-acyclic and $\rho^{(2)}(\widetilde{M}) = \rho^{(2)}(\widetilde{N})$ by Theorem 3.96 (1).
    Suppose that $b_p^{(2)}(\widetilde{M}) \neq 0$ for some $p \geq 0$. Then M has even dimen-
sion by Theorem 5.12 (1). But $\rho^{(2)}(\widetilde{N}) = 0$ for any closed even-dimensional
Riemannian manifold (see Subsection 3.4.3).

Chapter 6

6.1. Since M is the directed union of its finitely generated submodules, we
can assume without loss of generality because of Theorem 6.7 (1) and (4c)
that M is finitely generated projective. Now apply Lemma 6.28 (3).

6.2.   Choose an exact sequence $0 \to K \to \mathcal{N}(G)^n \to M \to 0$. We conclude
from Theorem 6.7 (4b) that $\dim_{\mathcal{N}(G)}(K) + \dim_{\mathcal{N}(G)}(M) = n$. Since K is
the directed union of its finitely generated submodules, we get from Theo-
rem 6.7 (1) and (4c) the existence of a finitely generated projective $\mathcal{N}(G)$-
module $P \subset K$ with $\dim_{\mathcal{N}(G)}(P) \geq \dim_{\mathcal{N}(G)}(K) - \epsilon/2$. Hence $\mathcal{N}(G)^n/P$
is a finitely presented $\mathcal{N}(G)$-module which satisfies $\dim_{\mathcal{N}(G)}(\mathcal{N}(G)^n/P) \leq$
$\dim_{\mathcal{N}(G)}(M) + \epsilon/2$ because of Theorem 6.7 (4b) and maps surjectively onto
M. Therefore we can assume without loss of generality that M is finitely
presented.
    Because of Theorem 6.7 (3) we may assume $M = \mathbf{T}M$.
    From Lemma 6.28 (4), we get an exact sequence $0 \to \mathcal{N}(G)^n \xrightarrow{f} \mathcal{N}(G)^n \to$
$M \to 0$ for some positive $\mathcal{N}(G)$-map f. Let $\{E_\lambda \mid \lambda \geq 0\}$ be the spectral
family of $\nu(f)$ (see (6.22)). Then $\nu(f)$ splits as the orthogonal sum of a weak
isomorphism $g_\lambda \colon \mathrm{im}(E_\lambda) \to \mathrm{im}(E_\lambda)$ and an automorphism $g_\lambda^\perp \colon \mathrm{im}(E_\lambda)^\perp \to$
$\mathrm{im}(E_\lambda)^\perp$. Thus we obtain an exact sequence $0 \to \nu^{-1}(\mathrm{im}(E_\lambda)) \xrightarrow{\nu(g_\lambda)}$
$\nu^{-1}(\mathrm{im}(E_\lambda)) \to M \to 0$ and $\dim_{\mathcal{N}(G)}(\nu^{-1}(\mathrm{im}(E_\lambda))) = \dim_{\mathcal{N}(G)}(\mathrm{im}(E_\lambda))$
from Theorem 6.24. Since $\ker(\nu(f)) = 0$ by Theorem 6.24 (3) we conclude
from Theorem 1.12 (4) that $\lim_{\lambda \to 0} \dim_{\mathcal{N}(G)}(\mathrm{im}(E_\lambda)) = 0$. Hence we can find
$\lambda > 0$ with $\dim_{\mathcal{N}(G)}(\nu^{-1}(\mathrm{im}(E_\lambda))) \leq \epsilon$. Since M is a quotient of $\nu^{-1}(\mathrm{im}(E_\lambda))$,
the claim follows.

6.3.   From the previous Exercise 6.2 we get a finitely generated projective
$\mathcal{N}(G)$-module Q with an epimorphism $f \colon Q \to P/M$ such that $\dim_{\mathcal{N}(G)}(Q) \leq$

$\dim_{\mathcal{N}(G)}(P/M) + \epsilon$ holds. Choose an $\mathcal{N}(G)$-map $\overline{p}\colon P \to Q$ such that $f \circ \overline{p} = p$ holds for the canonical projection $p\colon P \to P/M$. Because of Theorem 6.7 (1) $P/\ker(\overline{p})$ is a finitely generated projective $\mathcal{N}(G)$-module and hence $\ker(\overline{p})$ is a direct summand in P. Theorem 6.7 (4b) implies $\dim_{\mathcal{N}(G)}(M) \le \dim_{\mathcal{N}(G)}(\ker(\overline{p})) + \epsilon$. Take $P' = \ker(\overline{p})$.

6.4.   If M is finitely generated and projective, M and M^* are isomorphic, an isomorphism is given by any inner product (see Lemma 6.23). If M is finitely generated, the claim follows from Theorem 6.7 (3) and (4e). The countably generated case is reduced to the finitely generated case as follows.

Since M is countably generated, we can find a family $\{M_n \mid n = 0, 1, 2, \ldots\}$ of finitely generated submodules of M such that $M_m \subset M_n$ holds for $m \le n$ and M is the union $\bigcup_{n \ge 0} M_n$. Then

$$M = \mathrm{colim}_{n \to \infty} M_n;$$
$$M^* = \lim_{n \to \infty} M_n^*.$$

Theorem 6.13 (1) and Theorem 6.18 imply

$$\dim(M) = \sup\{\dim(M_n) \mid n \ge 0\};$$
$$\dim(M^*) = \sup\{\inf\{\dim(\mathrm{im}(M_n^* \to M_m^*)) \mid n = m, m+1, \ldots\} \mid m \ge 0\}.$$

Hence it suffices to show that $\dim(\mathrm{im}(M_n^* \to M_m^*)) = \dim(M_m^*)$ for $m \le n$. As the canonical projection induces an isomorphism $(\mathbf{P}M_n)^* \to M_n^*$ it suffices to show for an inclusion $P \subset Q$ of finitely generated projective $\mathcal{N}(G)$-modules that $\dim(\mathrm{im}(Q^* \to P^*)) = \dim(P^*)$.

Since $\overline{P}$ is a direct summand in Q, it suffices to treat the case $Q = \overline{P}$. Theorem 6.7 (4d) implies $\dim(P) = \dim(\overline{P})$ and hence $\dim(P^*) = \dim(\overline{P}^*)$. We get an exact sequence $0 \to (\overline{P}/P)^* \to \overline{P}^* \xrightarrow{i^*} P^* \to \mathrm{coker}(i^*) \to 0$. Since $(\overline{P}/P)^*$ is zero, we conclude $\dim(\mathrm{coker}(i^*)) = 0$ and hence the claim from Theorem 6.7 (4b).

6.5.   Each map $\phi_{J,K}$ is injective and hence its image has dimension 1 by Theorem 6.7 (4b). Theorem 6.13 (2) implies $\dim_{\mathcal{N}(\mathbb{Z})}(M) = 1$. Notice that $\phi_{J,K}^*$ is given by multiplication with $\prod_{u \in K-J}(z - u)$. Since M^* is the limit over the inverse system given by the maps $\phi_{J,K}^*$, we get an isomorphism

$$M^* = \bigcap_{u \in S^1}(z - u),$$

where $(z - u)$ is the ideal generated by $(z - u)$. We have already shown in Example 6.19 that $\bigcap_{u \in S^1}(z - u) = \{0\}$.

6.6.   We first show that $\mathcal{N}(\mathbb{Z})$ is not Noetherian. Identify $\mathcal{N}(\mathbb{Z}) = L^\infty(S^1)$ (see Example 1.4). Consider the sequence of $L^\infty(S^1)$-submodules $P_1 \subset P_2 \subset$

$P_3 \subset \ldots \subset L^\infty(S^1)$ such that P_n is the ideal generated by the characteristic function of $\{\exp(2\pi it) \mid t \in [1/n, 1]\}$. One easily checks that $P = \bigcup_{n \geq 1} P_n$ is a submodule of $L^\infty(S^1)$ which can be written as the infinite sum of non-trivial $L^\infty(S^1)$-modules Q_n for $n \geq 1$, where Q_n is the ideal generated by the characteristic function of $\{\exp(2\pi it) \mid t \in [1/(n+1), 1/n]\}$. Hence P cannot be finitely generated.

If $i \colon \mathbb{Z} \to G$ is an inclusion, i_*P is a submodule of $i_*\mathcal{N}(\mathbb{Z}) = \mathcal{N}(G)$ and is the infinite sum of non-trivial $\mathcal{N}(G)$-modules i_*Q_n by Theorem 6.29 (1).

We will later show in an exercise of Chapter 9 that $\mathcal{N}(G)$ is Noetherian if and only if G is finite.

6.7.   Choose a free AG-module F and AG-maps $i \colon P \to F$ and $r \colon F \to P$ with $r \circ i = \mathrm{id}$. Since $M \subset P$ is finitely generated, there is a finitely generated free direct summand $F_0 \subset F$ with $i(M) \subset F_0$ and $F_1 = F/F_0$ free. Hence i induces a map $f \colon P/M \to F_1$. It suffices to show that f is trivial because then $i(P) \subset F_0$ and the restriction of r to F_0 yields an epimorphism $F_0 \twoheadrightarrow P$.

Let $g \colon AG \to P/M$ be any AG-map. Since the von Neumann algebra $\mathcal{N}(G)$ is semihereditary (see Theorem 6.7 (1)), the image of $\mathcal{N}(G) \otimes_{AG} (f \circ g)$ is a finitely generated projective $\mathcal{N}(G)$-module. Its von Neumann dimension is zero by Theorem 6.7 (4b) since $\mathcal{N}(G) \otimes_{AG} P/M$ has trivial dimension by assumption. We conclude from Lemma 6.28 (3) that the image of $\mathcal{N}(G) \otimes_{AG} (f \circ g)$ is trivial. Hence $\mathcal{N}(G) \otimes_{AG} (f \circ g)$ is the zero map. Since $AG \to \mathcal{N}(G)$ is injective, $f \circ g$ is trivial. This implies that f is trivial since g was arbitrary.

6.8.   We conclude from Theorem 6.7 (4b) that $\mathbf{T}_{\dim} M$ is the largest $\mathcal{N}(G)$-submodule of M with vanishing von Neumann dimension.

This definition of $\mathbf{T}_{\dim} M$ coincides with the one of Definition 6.1 of $\mathbf{T}M$ by Theorem 6.7 (3), (4b) and (4e) and Lemma 6.28 (3), provided that M is finitely generated.

A counterexample for not finitely generated M comes from Exercise 6.5, where a module M was constructed with $\dim(M) = 1$ and $M^* = 0$.

6.9.   Let M be a $\mathbb{C}H$-module. If P_* is a projective $\mathbb{C}H$-resolution for M, then $\mathbb{C}G \otimes_{\mathbb{C}H} P_*$ is a projective $\mathbb{C}G$-resolution for $\mathbb{C}G \otimes_{\mathbb{C}H} M$. We have $\mathcal{N}(G) \otimes_{\mathbb{C}G} \mathbb{C}G \otimes_{\mathbb{C}H} P_* \cong_{\mathcal{N}(G)} \mathcal{N}(G) \otimes_{\mathcal{N}(H)} \mathcal{N}(H) \otimes_{\mathbb{C}H} P_*$. Theorem 6.29 implies

$$\dim_{\mathcal{N}(H)}\left(\mathrm{Tor}_p^{\mathbb{C}H}(\mathcal{N}(H), M)\right) = \dim_{\mathcal{N}(G)}\left(\mathcal{N}(G) \otimes_{\mathcal{N}(H)} \mathrm{Tor}_p^{\mathbb{C}H}(\mathcal{N}(H), M)\right)$$
$$= \dim_{\mathcal{N}(G)}\left(\mathrm{Tor}_p^{\mathbb{C}G}(\mathcal{N}(G), \mathbb{C}G \otimes_{\mathbb{C}H} M)\right).$$

6.10.   Let $\mathbb{Z} \subset G$ be an infinite cyclic subgroup of finite index. Let M be a $\mathbb{C}G$-module. Consider the canonical $\mathbb{C}G$-map $p \colon \mathbb{C}G \otimes_{\mathbb{C}[\mathbb{Z}]} \mathrm{res}_G^{\mathbb{Z}} M \to M$

sending $g \otimes m$ to gm, where $\operatorname{res}_G^{\mathbb{Z}} M$ is the restriction of the $\mathbb{C}G$-module M to a $\mathbb{C}[\mathbb{Z}]$-module. Define a $\mathbb{C}G$-map $i\colon M \to \mathbb{C}G \otimes_{\mathbb{C}[\mathbb{Z}]} \operatorname{res}_G^{\mathbb{Z}} M$ by sending m to $\frac{1}{[G:\mathbb{Z}]} \cdot \sum_{g\mathbb{Z} \in G/\mathbb{Z}} g \otimes g^{-1}m$. One easily checks that it is independent of the choice of representative g for $g\mathbb{Z}$ and satisfies $p \circ i = \mathrm{id}$. Since $[G : \mathbb{Z}]$ is finite, $\mathcal{N}(G)$ is isomorphic as $\mathcal{N}(\mathbb{Z})$-module to $\bigoplus_{[G:\mathbb{Z}]} \mathcal{N}(\mathbb{Z})$. This implies that $\operatorname{Tor}_p^{\mathbb{C}G}(\mathcal{N}(G); M)$ is a direct summand in

$$\operatorname{Tor}_p^{\mathbb{C}G}(\mathcal{N}(G), \mathbb{C}G \otimes_{\mathbb{C}[\mathbb{Z}]} \operatorname{res}_G^{\mathbb{Z}} M) \cong \bigoplus_{[G:\mathbb{Z}]} \operatorname{Tor}_p^{\mathbb{C}[\mathbb{Z}]}(\mathcal{N}(\mathbb{Z}), \operatorname{res}_G^{\mathbb{Z}} M).$$

Hence it suffices to prove the claim for $G = \mathbb{Z}$.

The functor Tor commutes in both variables with colimits over directed systems [94, Proposition VI.1.3. on page 107]. Hence it suffices to prove for a finitely generated $\mathbb{C}\mathbb{Z}$-module M that $\operatorname{Tor}_p^{\mathbb{C}[\mathbb{Z}]}(\mathcal{N}(\mathbb{Z}), M) = 0$ for $p \geq 1$. Since $\mathbb{C}[\mathbb{Z}]$ is a principal ideal domain [15, Proposition V.5.8 on page 151 and Corollary V.8.7 on page 162], we can write

$$M = \mathbb{C}[\mathbb{Z}]^{n_p} \oplus \left(\bigoplus_{i_p=1}^{s_p} \mathbb{C}[\mathbb{Z}]/((z - a_{p,i_p})^{r_{p,i_p}}) \right)$$

for $a_{p,i_p} \in \mathbb{C}$ and $n_p, s_p, r_{p,i_p} \in \mathbb{Z}$ with $n_p, s_p \geq 0$ and $r_{p,i_p} \geq 1$, where z is a fixed generator of $\mathbb{Z}$. Hence we can assume without loss of generality that $M = \mathbb{C}[\mathbb{Z}]/(z-a)^n$ for appropriate $a \in \mathbb{C}$ and $n \geq 1$. Using the obvious $\mathbb{C}[\mathbb{Z}]$-resolution one reduces the claim to the assertion that $\mathcal{N}(\mathbb{Z}) \xrightarrow{r_{(z-a)^n}} \mathcal{N}(\mathbb{Z})$ is injective which is obvious in view of the identification $\mathcal{N}(\mathbb{Z}) = L^\infty(S^1)$ (see Example 1.4).

6.11. (1) This is a modification of the corresponding argument for dimension-flatness appearing in the solution to Exercise 6.9.

(2) Show that $\operatorname{Tor}_p^{\mathbb{C}K}(\mathcal{N}(K), M)$ is a direct summand in

$$\operatorname{Tor}_p^{\mathbb{C}K}(\mathcal{N}(K), \mathbb{C}K \otimes_{\mathbb{C}G} \operatorname{res}_K^G M) \cong \bigoplus_{[K:G]} \operatorname{Tor}_p^{\mathbb{C}G}(\mathcal{N}(G), \operatorname{res}_K^G M)$$

for any $\mathbb{C}K$-module M. The argument is analogous to the one appearing in the solution of the previous Exercise 6.10.

(3) Since $\mathcal{N}(G)$ is flat over $\mathbb{C}G$, we get for $p \geq 1$

$$H_p^G(EG; \mathcal{N}(G)) = \mathcal{N}(G) \otimes_{\mathbb{C}G} H_p(EG) = 0.$$

(4) This follows from (3).

(5) This follows from (3) and Lemma 6.98.

(6), (7) (8) These follow from (1), (4) and (5) and the following statements. We get from Theorem 1.35 (2) and (8) and Example 2.59

$$\alpha_1(E\mathbb{Z}^n; \mathcal{N}(\mathbb{Z}^n)) = n;$$
$$b_1^{(2)}(E(\mathbb{Z} * \mathbb{Z}); \mathcal{N}(\mathbb{Z} * \mathbb{Z})) = \chi(S^1 \vee S^1) - b_0^{(2)}(E(\mathbb{Z} * \mathbb{Z}); \mathcal{N}(\mathbb{Z} * \mathbb{Z})) = 1.$$

If M is an aspherical manifold whose universal covering is a symmetric space, then we get for $G = \pi_1(M)$ from Corollary 5.16 (2) that one of the L^2-Betti numbers $b_p^{(2)}(G)$ is different from zero or one of the Novikov Shubin invariants $\alpha_p(EG; \mathcal{N}(G))$ is different from ∞^+.

(9) Because of (1) and (3) it suffices to prove for the fundamental group π of a connected sum $M = M_1 \# \ldots \# M_r$ of (compact connected orientable) non-exceptional prime 3-manifolds M_j that $H_p^G(E\pi; \mathcal{N}(\pi))$ is different from zero for at least one $p \in \{0, 1, 2\}$. We can assume without loss of generality that ∂M contains no S^2, otherwise glue D^3-s to M. It suffices to treat the case $|\pi| = \infty$ because $H_0(EG; \mathcal{N}(G))$ is non-zero for finite G (see Theorem 6.54 (8b)). We conclude from Theorem 4.1 that either $b_1^{(2)}(\widetilde{M}) \neq 0$ or M is $\mathbb{RP}^3 \# \mathbb{RP}^3$ or M is a prime manifold with infinite fundamental group whose boundary is empty or a disjoint union of tori. Since the classifying map $M \to B\pi$ is 2-connected we conclude from Lemma 6.98 that $H_1(E\pi; \mathcal{N}(\pi)) \neq 0$ if $b_1^{(2)}(\widetilde{M}) \neq 0$ or $\alpha_1(\widetilde{M}) \neq 0$. Since $\alpha_1(\widetilde{\mathbb{RP}^3 * \mathbb{RP}^3}) = \alpha_1(\widetilde{S^1 \times S^2}) = 1$ by Theorem 4.2 (2), only the case of an non-exceptional irreducible 3-manifold with infinite fundamental group whose boundary is empty or a disjoint union of tori is left. Notice that M is aspherical by Exercise 4.1. Hence it suffices to show that $\alpha_2(\widetilde{M}) \neq \infty^+$ because of Lemma 6.98. This follows in the case, where the boundary is empty or a union of incompressible tori from Theorem 4.2. Now reduce the case, where the boundary contains a compressible torus, to the case, where the boundary is a union of incompressible tori, using the construction in the proof of [322, Lemma 6.4 on page 52] based on the Loop Theorem [252, Theorem 4.2 on page 39] and the formula for the second Novikov-Shubin invariants under wedges (cf. Theorem 2.55 (4)).

6.12.   See [334, Theorem 5.3 on page 233].

6.13.   See [107, Section 4].

6.14.   Fix a natural number n. Then the number of cells with isotropy group whose order is less or equal to n must be finite since otherwise

$$m(X; \mathcal{N}(G)) \geq \sum_{c \in I(X), |G_c| \leq n} |G_c|^{-1} \geq \sum_{c \in I(X), |G_c| \leq n} n^{-1} \geq \infty.$$

6.15.   Use induction over the cells of B. In the induction step apply the sum formula (see Theorem 6.80 (2)) to the pushout on the level of total spaces which is obtained from a pushout on the base space level, which is given by attaching a cell, by the pullback construction.

6.16.   Let K be the kernel of the projection $p \colon \mathbb{Z} * \mathbb{Z} \to \mathbb{Z}$ which sends the two generators of $\mathbb{Z} * \mathbb{Z}$ to the generator of $\mathbb{Z}$. We obtain a fibration $BK \to B(\mathbb{Z} * \mathbb{Z}) \to B\mathbb{Z}$. Hence $B(\mathbb{Z} * \mathbb{Z})$ is homotopy equivalent to the mapping torus of an appropriate selfhomotopy equivalence $BK \to BK$. But $b_1^{(2)}(\mathbb{Z} * \mathbb{Z}) = 1$.

6.17.   We can assume without loss of generality that the boundary of M contains no S^2-s, otherwise attach D^3-s to M.

   If M is a prime manifold with infinite fundamental group, then either M is irreducible and hence aspherical by Exercise 4.1 and we get $\chi^{(2)}(\pi_1(M)) = \chi(M)$ from Theorem 1.35 (2) or $M = S^1 \times S^2$ and hence $\chi^{(2)}(\pi_1(S^1 \times S^2)) = \chi^{(2)}(\widetilde{S^1}) = \chi(S^1) = \chi(S^1 \times S^2) = 0$ by Theorem 1.35 (2). Notice that $\pi_1(M) = *_{i=1}^r \pi_1(M_i)$ and $\chi(M) = \sum_{i=1}^r \chi(M_i)$. We conclude $\chi^{(2)}(\widetilde{M}) = 1 - r + \sum_{i=1}^r \chi^{(2)}(\widetilde{M_i})$ from Theorem 6.80 (2) and (8). We have $\chi^{(2)}(EG; \mathcal{N}(G)) = |G|^{-1}$ for a finite group G by Theorem 6.80 (7). Now the claim follows from Theorem 4.1.

6.18. $b_p^{(2)}(G) = b_p^{(2)}(E(G; \mathcal{FIN}); \mathcal{N}(G))$ follows from Theorem 6.54 (2) since $EG \times E(G; \mathcal{FIN})$ is a model for EG.

6.19. $\mathbb{C} \otimes_{\mathbb{Z}} C_*(E(G; \mathcal{FIN}))$ is a projective $\mathbb{C}G$-resolution of $\mathbb{C}$ equipped with the trivial G-action.

6.20.   This follows from Theorem 6.54 (2) since $EG \times X$ is a model for EG.

6.21.   There is an obvious exact sequence of $\mathbb{Z}[\mathbb{Z}/5]$ modules $0 \to \mathbb{Z} \to \mathbb{Z}[\mathbb{Z}/5] \to \mathbb{Z}[\mathbb{Z}/5]/(N) \to 0$, where $\mathbb{Z}$ carries the trivial $\mathbb{Z}/5$-action. The associated long cohomology sequence implies

$$H^1(\mathbb{Z}/5; \mathbb{Z}[\mathbb{Z}/5]/(N)) \cong H^2(\mathbb{Z}/5; \mathbb{Z}) \cong \mathbb{Z}/5.$$

Notice that the obvious $\mathbb{Z}/5$-action on $\mathbb{Z}[\mathbb{Z}/5]/(N)$ has no non-trivial fixed point. Now apply Example 6.94.

6.22.   Since $\mathrm{im}(c_p)$ and $\ker(c_p)$ are submodules of free $\mathbb{C}[F_g]$-modules, they are free. Since $0 \to \ker(c_p) \to C_p \to \mathrm{im}(c_p) \to 0$ is exact, we conclude that $\mathrm{im}(c_p)$ and $\ker(c_p)$ are finitely generated free $\mathbb{C}[F_g]$-modules. In particular we see that $H_p(C_*(X))$ has a finite free 1-dimensional resolution, for instance $0 \to \mathrm{im}(c_{p+1}) \to \ker(c_p) \to H_p(C_*(X)) \to 0$. Now fix any finite based free $\mathbb{C}[F_g]$-chain complex $F[p]_*$ which is concentrated in dimension p and $p+1$ and satisfies $H_n(F[p]_*) = 0$ for $n \neq p$ and $H_p(F[p]_*) \cong_{\mathbb{C}[F_g]} H_p(C_*(X))$. Notice that the $\mathbb{C}[F_g]$-chain homotopy type of $F[p]_*$ depends only on the $\mathbb{C}[F_g]$-isomorphism class of $H_p(C_*(X))$. One easily constructs a $\mathbb{C}[F_g]$-chain map $f_* \colon \bigoplus_{p \geq 0} F[p]_* \to C_*(X)$ which induces an isomorphism on homology and hence is a $\mathbb{C}[G]$-chain equivalence. We conclude from the homotopy invariance of the Novikov-Shubin invariants and the L^2-Betti numbers (see Theorem 2.19)

$$b_p^{(2)}(X; \mathcal{N}(G)) = b_p(l^2(G) \otimes_{\mathbb{C}[F_g]} F[p]_*) + b_p(l^2(G) \otimes_{\mathbb{C}[F_g]} F[p-1]_*);$$
$$\alpha_p(X; \mathcal{N}(G)) = \alpha_p(l^2(G) \otimes_{\mathbb{C}[F_g]} F[p-1]_*)$$

and that $b_p(l^2(G) \otimes_{\mathbb{C}F_g} F[p]_*)$ and $\alpha_{p+1}(l^2(G) \otimes_{\mathbb{C}F_g} F[p]_*)$ depend only on the $\mathbb{C}[F_g]$-isomorphism class of $H_p(C_*(X))$.

6.23. We know $b_0^{(2)}(\mathbb{Z} * \mathbb{Z}) = 0$ and $\alpha_1(\mathbb{Z} * \mathbb{Z}) = \infty^+$ from Theorem 1.35 (8) and Theorem 2.55 (5). Let $G_r \subset G$ be the subgroup $\prod_{i=1}^r \mathbb{Z} * \mathbb{Z}$. We conclude from the product formulas Theorem 1.35 (4) and Theorem 2.55 (3) that $b_p^{(2)}(G_r) = 0$ for $p \leq r-1$ and $\alpha_p(G_r) = \infty^+$ for $p \leq r$ holds. Lemma 6.98 implies $H_p^{G_r}(EG_r; \mathcal{N}(G_r)) = 0$ for $p \leq r-1$.

Let G_r' be the subgroup $\prod_{i=r+1}^\infty \mathbb{Z} * \mathbb{Z}$ of G. Obviously $G = G_r \times G_r'$. Hence $EG_r \times EG_r'$ with the obvious $G = G_r \times G_r'$-operation is a model for EG. Hence we get an $\mathcal{N}(G)$-chain isomorphism

$$\mathcal{N}(G) \otimes_{\mathcal{N}(G_r) \otimes \mathcal{N}(G_r')} \left((\mathcal{N}(G_r) \otimes_{\mathbb{C}G_r} C_*(EG_r)) \otimes_\mathbb{C} (\mathcal{N}(G_r') \otimes_{\mathbb{C}G_r'} C_*(EG_r')) \right)$$
$$\cong \mathcal{N}(G) \otimes_{\mathbb{C}G} C_*(EG).$$

Since $H_p(\mathcal{N}(G_r) \otimes_{\mathbb{C}G_r} C_*(EG_r)) = 0$ for $p \leq r-1$, and $\mathcal{N}(G_r) \otimes_{\mathbb{C}G_r} C_*(EG_r)$ is a free $\mathcal{N}(G_r)$-chain complex, $\mathcal{N}(G_r) \otimes_{\mathbb{C}G_r} C_*(EG_r)$ is $\mathcal{N}(G_r)$-chain homotopy equivalent to a projective $\mathcal{N}(G_r)$-chain complex P_* with $P_p = 0$ for $p \leq r-1$. Hence $\mathcal{N}(G) \otimes_{\mathbb{C}G} C_*(EG)$ is $\mathcal{N}(G)$-chain homotopy equivalent to an $\mathcal{N}(G)$-chain complex Q_* with $Q_p = 0$ for $p \leq r-1$. This implies $H_n(\mathcal{N}(G) \otimes_{\mathbb{C}G} C_*(EG)) = 0$ for $p \leq r-1$. Since r can be chosen arbitrarily large, the assertion follows.

A more general statement will be proved in Lemma 12.11 (5).

6.24. We get $H_p^\mathbb{Z}(\widetilde{S^1}; \mathcal{N}(\mathbb{Z})) = \{0\}$ for $p \geq 1$ and $H_1^{\mathbb{Z}^2}(\widetilde{T^2}; \mathcal{N}(\mathbb{Z}^2)) \neq \{0\}$ from Lemma 6.98 since $b_p^{(2)}(\widetilde{S^1}) = 0$ for $p \geq 1$, $\alpha_p(\widetilde{S^1}) = \infty^+$ for $p \geq 2$ and $\alpha_2(\widetilde{T^2}) = 2$ holds by Example 2.59.

6.25. If $N = \mathcal{N}(H)$, Theorem 6.104 follows from Theorem 6.29 applied to the inclusion $G \to G \times H$. Now Theorem 6.104 follows for any projective $\mathcal{N}(H)$-module N by Theorem 6.7 (4b) and (4c). Since the statement in Theorem 6.104 (2) is symmetric in M and N, it is now proved in the case, where M or N is projective.

If N is a finitely generated $\mathcal{N}(H)$-module, it splits as $\mathbf{P}N \oplus \mathbf{T}N$, where $\mathbf{P}N$ is finitely generated projective and $\dim_{\mathcal{N}(H)}(\mathbf{T}N) = 0$ (see Theorem 6.7 (3) and (4e)). Since $\dim_{\mathcal{N}(H)}(\mathbf{T}N) = 0$ and each $\mathcal{N}(G)$-module M is a quotient of a free $\mathcal{N}(G)$-module, we conclude $\dim_{\mathcal{N}(G\times H)}(\mathcal{N}(G \times H) \otimes_{\mathcal{N}(G)\otimes_{\mathbb{C}}\mathcal{N}(H)} M \otimes_{\mathbb{C}} \mathbf{T}N) = 0$ for any $\mathcal{N}(G)$-module M from Theorem 6.7 (4b). Hence Theorem 6.104 follows for finitely generated N.

Since N is the directed colimit of its finitely generated submodules and $\mathcal{N}(G\times H)\otimes_{\mathcal{N}(G)\otimes_{\mathbb{C}}\mathcal{N}(H)} M \otimes_{\mathbb{C}} -$ commutes with directed colimits and directed colimits are exact, Theorem 6.104 follows from Additivity (see Theorem 6.7 (4b)) and Theorem 6.13.

6.26.   Let C_* and D_* be positive $\mathcal{N}(G)$ and $\mathcal{N}(H)$-chain complexes. There are canonical isomorphisms of $\mathcal{N}(G) \otimes_{\mathbb{C}} \mathcal{N}(H)$-modules

$$\alpha: \bigoplus_{p+q=n} H_p(C_*) \otimes_{\mathbb{C}} H_q(D_*) \xrightarrow{\cong} H_n(C_* \otimes D_*).$$

Next we want to show that the composition of the canonical map

$$\mathcal{N}(G\times H)\otimes_{\mathcal{N}(G)\otimes\mathcal{N}(H)}H_p(C_*\otimes_{\mathbb{C}}D_*) \to H_p(\mathcal{N}(G\times H)\otimes_{\mathcal{N}(G)\otimes\mathcal{N}(H)}(C_*\otimes_{\mathbb{C}}D_*))$$

with $\mathcal{N}(G \times H) \otimes_{\mathcal{N}(G)\otimes\mathcal{N}(H)} \alpha$ is a dim-isomorphism, i.e. its kernel and cokernel have zero dimension. Obviously we may assume that D_* is finite dimensional. We use induction over the dimension d of D_* If D_* is zero-dimensional, this follows from Theorem 6.104 (1). In the induction step use the long dim-exact sequences associated to $0 \to D_*|_{d-1} \to D_* \to d[D_d] \to 0$ and a Five-Lemma for dim-isomorphisms for dim-exact sequences whose proof is an easy consequence of Theorem 6.7 (4b). Here dim-exact has the obvious meaning and follows from Theorem 6.104 (1).

As explained in the proof of 6.54 (5), we can assume that X is a G-CW-complex and Y is an H-CW-complex. Now apply the results above to $C_* = \mathcal{N}(G) \otimes_{\mathbb{Z}G} C_*(X)$ and $D_* = \mathcal{N}(H) \otimes_{\mathbb{Z}H} C_*(Y)$. The claim follows from Theorem 6.104 (2) since there is a $\mathbb{Z}[G \times H]$-chain isomorphism $C_*(X) \otimes_{\mathbb{C}} C_*(Y) \to C_*(X \times Y)$.

Chapter 7

7.1.   This follows from an iterated application of Theorem 7.2 (6).

7.2.   This follows from the previous Exercise 7.1.

7.3.   We can write BG as the colimit of the directed system

$$BG_1 \subset BG_1 \vee BG_2 \subset BG_1 \vee BG_2 \vee BG_3 \subset \dots.$$

Notice that each of the structure maps has a retraction. If $p: EG \to BG$ is the universal covering, we can write EG as the colimit of the directed system of free G-CW-complexes

$$EG|_{BG_1} \subset EG|_{BG_1 \vee BG_2} \subset EG|_{BG_1 \vee BG_2 \vee BG_3} \subset \cdots.$$

We conclude from Theorem 6.7 (4c) and Theorem 6.29

$$b_p^{(2)}(G) = \lim_{r \to \infty} b_p^{(2)} \left(\bigvee_{n=1}^{r} G_i \right).$$

We have for $p \geq 2$

$$b_p^{(2)} \left(\bigvee_{n=1}^{r} G_i \right) = \sum_{n=1}^{r} b_p^{(2)}(G_i)$$

and

$$b_1^{(2)} \left(\bigvee_{n=1}^{r} C_i \right) - b_0^{(2)} \left(\bigvee_{n=1}^{r} G_i \right) = r - 1 + \sum_{n=1}^{r} (b_1^{(2)}(G_i) - b_0^{(2)}(G_i)).$$

It suffices to prove this in the case $r = 2$, where it follows from the long exact homology sequence $\cdots \to H_p^G(G; \mathcal{N}(G)) \to H_p^G(G \times_{G_1} EG_1; \mathcal{N}(G)) \oplus H_p^G(G \times_{G_2} EG_2; \mathcal{N}(G)) \to H_p^G(EG; \mathcal{N}(G)) \to H_{p-1}^G(G; \mathcal{N}(G)) \to \cdots$, Theorem 6.7 (4b) and Theorem 6.29. Notice that $b_0^{(2)}(G_i; \mathcal{N}(G_i)) \leq 1/2$ holds for all $i \in I$ by Theorem 6.54 (8b).

7.4.   This follows from an iterated application of Theorem 6.54 (5).

7.5.   It suffices to show that $G \in \mathcal{B}_d$ for each $d \geq 0$. Since $\mathcal{B}_0$ is the class of infinite groups by Theorem 6.54 (8b), this is clear for $d = 0$. The induction step from d to $d+1$ is done as follows. We can write $G = G_1 \times G_2$, where G_1 and G_2 are of the shape $\prod_{j \in J} H_j$ for an infinite index set J and non-trivial groups H_j. Hence $G_1 \in \mathcal{B}_d$ and $G_2 \in \mathcal{B}_0$ by induction hypothesis. Then $G = G_1 \times G_2 \in \mathcal{B}_{d+1}$ by Theorem 6.54 (5).

As $G = \prod_{i \in I} \mathbb{Z}$ is abelian and hence amenable, we get $H_0^G(EG; \mathcal{N}(G)) \neq 0$ from Theorem 6.54 (8c).

7.6.   An infinite locally finite group is an infinite amenable group. Hence $H_0^G(EG; \mathcal{N}(G)) \neq 0$ by Theorem 6.54 (8c) and $b_p^{(2)}(G) = 0$ for all $p \geq 0$ by Theorem 7.2 (1). We have $H_p^H(EH; \mathcal{N}(H)) = H_p^G(G \times_H EH; \mathcal{N}(G)) = 0$ for $p \geq 1$ for any finite subgroup $H \subset G$ by Theorem 6.29 (1). Since G is the colimit of the system of its finite subgroups, $H_p^G(EG; \mathcal{N}(G)) = 0$ for $p \geq 1$ by a colimit argument.

7.7.   Suppose that K is finite. Then the claim follows from Theorem 6.54 (6b).

Suppose that H is finite. It suffices to show for any fibration $BH \to E_0 \xrightarrow{f} E_1$ of connected CW-complexes that

$$b_p^{(2)}(\widetilde{E_1}) = |H| \cdot b_p^{(2)}(\widetilde{E_0}).$$

The map h induces a $\mathbb{C}[\pi_1(E_0)]$-chain homotopy equivalence $f_* \colon C_*(\widetilde{E_0}) \to$ res $C_*(\widetilde{E_1})$, where res denotes the restriction coming from the epimorphism $\pi_1(f)$. It yields an $\mathcal{N}(\pi_1(E_0))$-chain equivalence

$$\mathcal{N}(\pi_1(E_0)) \otimes_{\mathbb{C}[\pi_1(E_0)]} C_*(\widetilde{E_0}) \xrightarrow{\simeq} \text{res}\left(\mathcal{N}(\pi_1(E_1)) \otimes_{\mathbb{C}[\pi_1(E_1)]} C_*(\widetilde{E_1})\right).$$

Since for any $\mathcal{N}(\pi_1(E_1))$-module M we have (see Lemma 13.45)

$$|H| \cdot \dim_{\mathcal{N}(\pi_1(E_0))}(\text{res } M) = \dim_{\mathcal{N}(\pi_1(E_1))}(M),$$

the claim follows.

7.8.   Let $f \colon B \to B(\pi_1(B))$ be the classifying map of B. We can find a fibration $F_0 \xrightarrow{j} E_0 \xrightarrow{q} B(\pi_1(B))$ together with a homotopy equivalence $u \colon E_0 \to E$ such that $f \circ p \circ u = q$ (see [521, Theoren 7.30 in I.7 on page 42]). Since the L^2-Betti numbers are homotopy invariants (see Theorem 6.54 (1)) and $\pi_1(B) \in \mathcal{B}_d$ by assumption, it suffices to prove that $b_p^{(2)}(\widetilde{F_0}) < \infty$ for $p \le d$. There is also a fibration $F \xrightarrow{k} F_1 \xrightarrow{r} \widetilde{B}$ and a homotopy equivalence $v \colon F_1 \to F_0$ such that $u \circ j \circ v \circ k \simeq i$. It remains to prove that $b_p^{(2)}(\widetilde{F_1}) < \infty$ for $p \le d$.

The Serre spectral sequence converges to $H_{p+q}^{\pi_1(F_1)}(\widetilde{F_1}; \mathcal{N}(\pi_1(F_1)))$ and has as E^2-term

$$E_{p,q}^2 = H_p(\widetilde{B}) \otimes_{\mathbb{C}} H_q\left(\mathcal{N}(\pi_1(F_1)) \otimes_{\mathbb{C}[\pi_1(F_1)]} \mathbb{C}[\pi_1(F_1)] \otimes_{\mathbb{C}[\pi_1(k)]} C_*(\widetilde{F})\right)$$

since $\widetilde{B}$ is simply connected. Because of Additivity (see Theorem 6.7 (4b)) and the assumption that $H_p(\widetilde{B}; \mathbb{C})$ is finite dimensional for $p \le d$, it suffices to show that for $q \le d$

$$\dim_{\mathcal{N}(\pi_1(F_1))}\left(H_q\left(\mathcal{N}(\pi_1(F_1)) \otimes_{\mathbb{C}[\pi_1(F_1)]} \mathbb{C}[\pi_1(F_1)] \otimes_{\mathbb{C}[\pi_1(k)]} C_*(\widetilde{F})\right)\right) < \infty.$$

But this is $b_q^{(2)}(\pi_1(E) \times_{\pi_1(i)} \widetilde{F}; \mathcal{N}(\pi_1(E)))$ by Theorem 6.29.

7.9.   We have already explained in the solution of Exercise 7.3 that

$$b_1^{(2)}(G * H) - b_0^{(2)}(G * H) = 1 + b_1^{(2)}(G) - b_0^{(2)}(G) + b_1^{(2)}(H) - b_0^{(2)}(H).$$

Since $b_0^{(2)}(G) = |G|^{-1}$ by Theorem 6.54 (8), the claim follows.

7.10.   Theorem 6.80 (4) and Lemma 6.87 show

$$\text{ch}_{\{1\}}^g(\chi^G(E(G; \mathcal{FIN}))) = \chi^{(2)}(E(G; \mathcal{FIN}); \mathcal{N}(G)) = \chi^{(2)}(G).$$

By definition $\mathrm{ch}^g_{\{1\}} \subset \{r \in \mathbb{R} \mid d \cdot r \in \mathbb{Z}\}$.

7.11. $SL(2,\mathbb{Z})$ is isomorphic to $\mathbb{Z}/6 *_{\mathbb{Z}/2} \mathbb{Z}/4$. We have $\chi^{(2)}(G) = |G|^{-1}$ for finite G. Now apply (7.13).

7.12.   The proof is analogous to the one of Lemma 7.16.

7.13.   (1) The Seifert-van Kampen theorem applied to an appropriate decomposition of T_{Bf} into two connected pieces with connected intersection implies that $\pi_1(T_{Bf})$ is isomorphic to the quotient of the group $G * \mathbb{Z}$, which is obtained by introducing for a fixed generator $t \in \mathbb{Z}$ and each $g \in G$ the relation $tgt^{-1} = f(g)$. This group is isomorphic to $K \rtimes \mathbb{Z}$. Analogous to the proof in Theorem 7.10 one shows that the universal covering of T_{Bf} has trivial homology and hence T_{Bf} is a model for $B(K \rtimes \mathbb{Z})$.

(2) We conclude from Theorem 1.39 that $b_p(B(\widetilde{K \rtimes \mathbb{Z}})) = b_p(\widetilde{T_{Bf}}) = 0$ for $p \geq 0$. Since G is of det $\geq$ 1-class by assumption, $K \rtimes \mathbb{Z}$ is of det $\geq$ 1-class by Theorem 13.3 (3). The invariant is well-defined because of Lemma 13.6.

(3) obvious.

(4) The proofs of assertion (1), (2), (3) and (4) of Theorem 7.27 carry directly over to the case of an endomorphism. The proofs of assertions (5) of Theorem 7.27 carries over since there is a canonical bijection $K/K_0 \xrightarrow{\cong} G/G_0$ if we assume that f induces a bijection $G/G_0 \xrightarrow{\cong} G/G_0$. The proofs of assertions (6) and (7) of Theorem 7.27 carry directly over to the case of an endomorphism if one reformulates them as stated.

Chapter 8

8.1.   We have to show for a Hilbert space H and a densely defined closed operator $f\colon \mathrm{dom}(f) \subset H \to H$ that $f^*f = 0$ implies $f = 0$. This follows from Lemma 2.2 (1) and the equation $\langle f^*f(x), x\rangle = \langle f(x), f(x)\rangle$ for $x \in \mathrm{dom}(f^*f)$.

8.2.   We get a commutative diagram of $\mathbb{C}$-categories

$$
\begin{array}{ccc}
\{\mathcal{N}(G)^n\} & \xrightarrow{\ \nu\ } & \{l^2(G)^n\} \\
\downarrow & & \downarrow \\
\{\mathcal{U}(G)^n\} & \xrightarrow{\ \nu_\mathcal{U}\ } & \{l^2(G)^n\}_\mathcal{U}
\end{array}
$$

whose vertical arrows are the obvious inclusions and whose horizontal arrows are isomorphisms of $\mathbb{C}$-categories. The map ν has already been introduced in

(6.22). The categories $\{\mathcal{U}(G)^n\}$ and $\{l^2(G)^n\}_{\mathcal{U}}$ and the functor $\nu_{\mathcal{U}}$ are defined analogously. Now one can apply the functor idempotent completion Idem to it. This yields a commutative diagram with isomorphisms of $\mathbb{C}$-categories as horizontal maps. Now identify this diagram with the desired diagram by equivalences of $\mathbb{C}$-categories. This has been done for the upper horizontal arrow in the proof of Theorem 6.24 and is done for the lower horizontal arrow analogously using Lemma 8.3 (4).

8.3. The inclusions $G \subset \mathbb{Z}G \subset \mathbb{C}G \subset l^1(G)$ are obvious. Given $u = \sum_{g \in G} \lambda_g \cdot g \in l^1(G)$, we obtain a bounded G-operator $r_u : l^2(G) \to l^2(G)$ which is uniquely determined by $g_0 \mapsto \sum_{g \in G} \lambda_g \cdot g_0 g$. The operator norm of r_u satisfies $||r_u||_\infty \le ||u||_1$. Since $C_r^*(G)$ is the closure in the norm topology of the image of the map $\mathbb{C}G \to \mathcal{B}(l^2(G))$, $u \mapsto r_u$, we get an inclusion $l^1(G) \subset C_r^*(G)$. Since $\mathcal{N}(G)$ is defined as $\mathcal{B}(l^2(G))^G$ and $\mathcal{B}(l^2(G))^G$ is closed in the norm topology in $\mathcal{B}(l^2(G))$, we get an inclusion $C_r^*(G) \subset \mathcal{N}(G)$. We get an inclusion $\mathcal{N}(G) \subset l^2(G)$ by $f \mapsto f(e)$ for $e \in G$ the unit element. For $v \in l^2(G)$ we get a densely defined operator $r_v : \mathrm{dom}(r_v) = \mathbb{C}G \subset l^2(G) \to l^2(G)$ which maps $u = \sum_{i=1}^n \lambda_i \cdot g_i$ to $\sum_{i=1}^n \lambda_i \cdot g_i \cdot v$. Since its adjoint is densely defined, this operator is closable. Its minimal closure is an element in $\mathcal{U}(G)$. This gives the inclusion $l^2(G) \subset \mathcal{U}(G)$.

8.4. Since $\mathcal{U}(G)$ is regular (see Theorem 8.22 (3)), $\mathrm{im}(f) \subset Q$ and $\ker(f) \subset P$ are direct summands by Lemma 8.18. We get from Theorem 8.29

$$\ker(f) = 0 \Leftrightarrow \dim_{\mathcal{U}(G)}(\ker(f)) = 0$$
$$\Leftrightarrow \dim_{\mathcal{U}(G)}(\mathrm{im}(f)) = \dim_{\mathcal{U}(G)}(Q) \Leftrightarrow \mathrm{im}(f) = Q.$$

8.5.   Let $0 \to M_0 \xrightarrow{i} M_1 \xrightarrow{p} M_2 \to 0$ be an exact sequence of $\mathcal{N}(G)$-modules. Then $0 \to \mathbf{T}_{\dim} M_0 \xrightarrow{\mathbf{T}_{\dim} i} \mathbf{T}_{\dim} M_1 \xrightarrow{\mathbf{T}_{\dim} p} \mathbf{T}_{\dim} M_2$ is exact since $\mathrm{im}(i) \cap \mathbf{T}_{\dim} M_1 = i(\mathbf{T}_{\dim} M_0)$ follows from Additivity (see Theorem 6.7 (4b)). Hence $\mathbf{T}_{\dim}$ is left exact.

In Exercise 6.5 we have constructed an $\mathcal{N}(\mathbb{Z})$-module $M = \mathrm{colim}_{j \in I} N_j$ satisfying

$$\dim_{\mathcal{N}(\mathbb{Z})}(M) = 1;$$
$$M^* = 0.$$

The structure map of N_j for $j = \emptyset \in I$ induces an injection $i : \mathcal{N}(\mathbb{Z}) \to M$. Thus we obtain an exact sequence $0 \to \mathcal{N}(\mathbb{Z}) \xrightarrow{i} M \xrightarrow{p} \mathrm{coker}(i) \to 0$. Obviously $\mathbf{T}\mathcal{N}(\mathbb{Z}) = 0$. We conclude $\mathbf{T}M = M$ from $M^* = 0$ and hence $\dim(\mathbf{T}M) = 1$. We conclude from Additivity (see Theorem 6.7 (4b)) that $\dim_{\mathcal{N}(G)}(\mathrm{coker}(i)) = 0$ and hence $\dim_{\mathcal{N}(G)}(\mathbf{T}\,\mathrm{coker}(i)) = 0$ and that the sequence $\mathbf{T}\mathcal{N}(\mathbb{Z}) \xrightarrow{\mathbf{T}i} \mathbf{T}M \xrightarrow{\mathbf{T}p} \mathbf{T}\operatorname{coker}(i)$ cannot be exact at $\mathbf{T}M$. Hence $\mathbf{T}$ is not left exact.

8.6.   The following is taken from Example 8.34. Let $I_1 \subset I_2 \subset \ldots \subset \mathcal{N}(G)$ be a nested sequence of ideals which are direct summands in $\mathcal{N}(\mathbb{Z})$ such that $\dim_{\mathcal{N}(\mathbb{Z})}(I_n) \neq 1$ and $\lim_{n \to \infty} \dim_{\mathcal{N}(\mathbb{Z})}(I_n) = 1$. Let I be the ideal $\bigcup_{n=1}^{\infty} I_n$. Put $M = \mathcal{N}(\mathbb{Z})/I = \operatorname{colim}_{n \to \infty} \mathcal{N}(G)/I_n$. Then Additivity (see Theorem 6.7 (4b)) and Theorem 6.13 (1) imply $\dim(M) = 0$ and hence $M = \mathbf{T}_{\dim}M$. Since $\mathcal{N}(\mathbb{Z})/I_n$ is projective and $\mathcal{N}(\mathbb{Z})$ is semihereditary (see Theorem 6.5 and Theorem 6.7 (1)), we conclude from Theorem 6.13 (1) and Lemma 6.28 (3) that $\mathbf{T}_{\dim}(\mathcal{N}(G)/I_n) = 0$. Hence $\mathbf{T}_{\dim}(\operatorname{colim}_{n \to \infty} \mathcal{N}(G)/I_n) = M \neq \{0\}$ is different from $\operatorname{colim}_{n \to \infty} \mathbf{T}_{\dim}(\mathcal{N}(G)/I_n) = \{0\}$. This shows that $\mathbf{T}_{\dim}$ does not commute with colimits over directed systems in general.

However, it does commute with colimits over directed systems with injective maps as structure maps since for any submodule $N \subset M$ we get $\mathbf{T}_{\dim}N = \mathbf{T}_{\dim}M \cap N$ from Additivity (see Theorem 6.7 (4b)).

8.7.   In Exercise 6.5 we have constructed an $\mathcal{N}(\mathbb{Z})$-module $M = \operatorname{colim}_{i \in I} N_i$ satisfying

$$\dim_{\mathcal{N}(\mathbb{Z})}(M) = 1;$$
$$M^* = 0;$$

Take M_i to be the image of the injective structure map $\psi_i \colon N_i \to M$. Then $M = \mathbf{T}M$ follows from $M^* = \{0\}$ and $\mathbf{T}M_i = 0$ from $M_i \cong \mathcal{N}(\mathbb{Z})$ since $\mathcal{N}(\mathbb{Z})$ is semihereditary (see Theorem 6.5 and Theorem 6.7 (1)) and Theorem 6.13 (1) and Lemma 6.28 (3) hold.

8.8.   Consider an exact sequence $0 \to M_0 \to M_1 \to M_2 \to 0$ of $\mathcal{N}(G)$-modules. It stays exact after applying $\mathcal{U}(G) \otimes_{\mathcal{N}(G)} -$ by Theorem 8.22 (2) and we obtain by the various maps j_{M_i} a map from the first to the second exact sequence. Now apply the snake lemma to get the desired six-term sequence.

8.9.   Let G be a locally finite group G which is infinite. Then G is an infinite amenable group and we get $b_p^{(2)}(G; \mathcal{N}(G)) = b_p^{(2)}(G; \mathcal{U}(G)) = 0$ for all $p \geq 0$ from Theorem 7.2 (1) and Theorem 8.31.

Let I be the directed set of finite subgroups of G. Consider the directed system over I which assigns to $H, K \in I$ with $H \subset K$ the canonical projection $\mathbb{C}[G/H] \to \mathbb{C}[G/K]$. We obtain an obvious epimorphism of $\mathbb{C}[G]$-modules $\alpha \colon \operatorname{colim}_{H \in I} \mathbb{C}[G/H] \to \mathbb{C}$. Since G is locally finite, I is the same as the directed set of finitely generated subgroups and hence α is bijective. As $\mathcal{U}(G) \otimes_{\mathbb{C}[G]} -$ commutes with colimits, we get an isomorphism

$$\beta \colon \operatorname{colim}_{H \in I} \mathcal{U}(G) \otimes_{\mathbb{C}[G]} \mathbb{C}[G/H] \xrightarrow{\cong} \mathcal{U}(G) \otimes_{\mathbb{C}G} \mathbb{C}.$$

Since $\dim_{\mathcal{U}(G)}(\mathcal{U}(G) \otimes_{\mathbb{C}[G]} \mathbb{C}[G/K]) = |K|^{-1}$ is different from zero for finite $K \subset G$, the element $1_{\mathcal{U}(G)} \otimes_{\mathbb{C}G} 1K$ in $\mathcal{U}(G) \otimes_{\mathbb{C}[G]} \mathbb{C}[G/K]$ is different from zero for each finite subgroup $K \subset G$. Hence the source of β is non-trivial and therefore $\mathcal{U}(G) \otimes_{\mathbb{C}G} \mathbb{C} \neq \{0\}$. This shows

$$H_0(EG;\mathcal{U}(G)) = \mathcal{U}(G) \otimes_{\mathbb{C}G} \mathbb{C} \neq \{0\}.$$

Chapter 9

9.1. Suppose that tr is a faithful finite normal trace on the von Neumann algebra $\mathcal{A}$. We want to show that $\mathcal{A}$ is finite, i.e. for any projection $p \in \mathcal{A}$ with $p \sim 1$ we have $p = 1$. Choose $u \in \mathcal{A}$ with $p = uu^*$ and $1 = u^*u$. Then

$$\operatorname{tr}(1 - p) = \operatorname{tr}(1) - \operatorname{tr}(p) = \operatorname{tr}(u^*u) - \operatorname{tr}(uu^*) = 0.$$

Since tr is faithful, we conclude $p = 1$.

Suppose that $\mathcal{A}$ is finite. Then there is the universal trace $\operatorname{tr}^u \colon \mathcal{A} \to \mathcal{Z}(\mathcal{A})$ (see Theorem 9.5). Since $\mathcal{Z}(\mathcal{A})$ is an abelian von Neumann algebra, we get from Example 9.6 an identification $\mathcal{Z}(\mathcal{A}) = L^\infty(X, \nu)$. We obtain a faithful finite normal trace

$$\operatorname{tr} \colon \mathcal{Z}(\mathcal{A}) = L^\infty(X, \nu) \to \mathbb{C}, \quad f \mapsto \int_X f d\nu.$$

Then $\operatorname{tr} \circ \operatorname{tr}^u$ is a faithful finite normal trace on $\mathcal{A}$.

9.2. Since M is hyperbolic, $\mathbb{Z} \oplus \mathbb{Z}$ is not a subgroup of $\pi_1(M)$ (see [65, Corollary 3.10 on page 462]). Since M is aspherical of dimension ≥ 2, its fundamental group cannot be virtually cyclic. Hence $\pi_1(M)$ is not virtually abelian. Since M is closed, $\pi_1(M)$ is finitely generated. Now apply Lemma 9.4 (3).

9.3. Since M is closed, $\pi_1(M)$ is finitely generated. We conclude from Lemma 9.4 (3) that either $\mathcal{N}(\pi_1(M))$ is of type II_1 or $\pi_1(M)$ is virtually abelian and $\mathcal{N}(\pi_1(M))$ is of type I_f. Suppose that $\pi_1(M)$ is virtually abelian. Then there is a finite covering $\overline{M} \to M$ such that $\pi_1(\overline{M}) \cong \mathbb{Z}^d$ for some integer $d \geq 0$. Up to homotopy we can assume that the prime decomposition of $\overline{M}$ contains no D^3 and no homotopy sphere. Hence any factor in the prime decomposition of $\overline{M}$ has non-trivial fundamental group. Since $\pi_1(\overline{M})$ is abelian, this implies that $\overline{M}$ is prime. Hence either $\overline{M}$ is $S^1 \times S^2$ or an irreducible manifold with fundamental group $\mathbb{Z}^d$ for some integer $d \geq 1$. This implies that $\overline{M}$ is aspherical. Hence it is homotopy equivalent to T^d. Since $H_n(M) = 0$ for $n \geq 4$, we must have $d = 1, 2, 3$.

9.4.   The two isomorphisms, inverse to one another, are given by tensoring with the R-$M_n(R)$-bimodule $_R R^n{}_{M_n(R)}$ and the $M_n(R)$-R-bimodule $_{M_n(R)} R^n{}_R$. Notice that there is an isomorphism of $M_n(R)$-$M_n(R)$-bimodules $(_{M_n(R)} R^n{}_R) \otimes_R (_R R^n{}_{M_n(R)}) \cong M_n(R)$ and an isomorphism of R-R-bimodules $(_R R^n{}_{M_n(R)}) \otimes_{M_n(R)} (_{M_n(R)} R^n{}_R) \cong R$.

9.5.   From Example 9.14 we get an isomorphism

$$\dim^u \colon K_0(L^\infty(X,\nu)) \to L^\infty(X,\nu,\mathbb{Z}).$$

Obviously $\dim^u(\mathcal{A})$ is the constant function with value 1 which cannot be written as 2-times an element in $L^\infty(X,\nu,\mathbb{Z})$. If we compose the canonical map $K_1(\mathbb{Z}) \to K_1(\mathcal{A})$ with the map given by the ordinary determinant of commutative rings $\det \colon K_1(\mathcal{A}) \to \mathcal{A}^{\mathrm{inv}}$, we see that the element in $K_1(\mathbb{Z})$ given by (-1) is mapped to the non-trivial element -1 in $\mathcal{A}$. Hence $K_1(\mathcal{A}) \to \widetilde{K}_1(\mathcal{A})$ is not injective.

   Consider the matrices

$$P_1 = \begin{pmatrix} 1 & 0 \\ 0 & 0 \end{pmatrix}, \qquad P_2 = \begin{pmatrix} 0 & 0 \\ 0 & 1 \end{pmatrix}, \qquad U = \begin{pmatrix} 0 & 1 \\ 1 & 0 \end{pmatrix}.$$

Then $P_i^2 = P_i$ for $i = 1,2$, $P_1 + P_2 = I$ and $UP_1U^{-1} = P_2$. Hence we get for the $M_2(\mathcal{A})$-module $\mathcal{A}^2$

$$\mathcal{A}^2 \oplus \mathcal{A}^2 \cong (P_1) \oplus (P_1) \cong (P_1) \oplus (P_2) \cong (I_2) = M_2(\mathcal{A}).$$

Hence $\mathcal{A}^2$ is finitely generated projective and we get $[M_2(\mathcal{A})] = 2 \cdot [\mathcal{A}^2]$ in $K_0(\mathcal{A})$. The generator of $K_1(\mathbb{Z})$ is sent under the canonical map $K_1(\mathbb{Z}) \to K_1(M_2(\mathcal{A}))$ to zero because of

$$\begin{pmatrix} -1 & 0 \\ 0 & -1 \end{pmatrix} = \begin{pmatrix} -1 & 0 \\ 0 & 1 \end{pmatrix} \cdot U \cdot \begin{pmatrix} -1 & 0 \\ 0 & 1 \end{pmatrix}^{-1} \cdot U^{-1}.$$

9.6.   See [51, Example 8.1.2 on page 67], [514, Example 7.1.11 on page 134].

9.7.   This follows from Theorem 9.13 since $\mathcal{N}(G) = \mathbb{C}G$ for a finite group G. See also [470, Corollary 2 in Chapter 2 on page 16].

9.8.   Let $f_* \colon C_* \to D_*$ be an R-chain homotopy equivalence of finite based free R-chain complexes such that $H_n(C_*)$ and $H_n(D_*)$ are S-torsion for each n. Since RS^{-1} is flat as R-module, the RS^{-1}-chain complexes $RS^{-1} \otimes_R C_*$ and $RS^{-1} \otimes_R D_*$ are finite based free acyclic RS^{-1}-chain complexes. Hence $\rho(RS^{-1} \otimes_R C_*)$ and $\rho(RS^{-1} \otimes_R D_*)$ are defined and we get in $K_1(RS^{-1})$

$$\rho(RS^{-1} \otimes_R D_*) - \rho(RS^{-1} \otimes_R C_*) = \tau(RS^{-1} \otimes_R f_*).$$

Since $K_1(R) \to K_1(RS^{-1})$ sends $\tau(f_*)$ to $\tau(RS^{-1} \otimes_R f_*)$, we see that the class of $\rho(RS^{-1} \otimes_R C_*)$ in $\mathrm{coker}(i_1)$ depends only on the R-chain homotopy type of C_* and is in particular independent of the choice of R-basis. Notice that there is a canonical isomorphism from $\mathrm{coker}(i_1)$ to the image of $K_1(RS^{-1}) \to K_0(R \to S)$.

Fix based free 1-dimensional chain complexes $F_*[n]$ such that $H_i(F_*[n]) = 0$ for $i \neq 0$ and $H_0(F_*[n])$ is R-isomorphic to $H_n(C_*)$. Let d be the dimension of C_*. One easily constructs an R-chain homotopy equivalence from $\bigoplus_{n=0}^{d} \Sigma^n F_*[n]$ to C_*. This implies that $\rho(F_*[n])$ and $\sum_{n=0}^{d}(-1)^n \cdot \rho(F_*[n])$ have the same image under the map $K_1(RS^{-1}) \to K_0(R \to S)$. One easily checks that $\rho(F_*[n])$ is mapped to $[H_n(C_*)]$ under the map $K_1(RS^{-1}) \to K_0(R \to S)$.

9.9.   Obviously $RS^{-1} = \mathbb{Q}$. The determinant of commutative rings induces isomorphisms $K_1(\mathbb{Z}) \xrightarrow{\cong} \{\pm 1\}$ and $K_1(\mathbb{Q}) \xrightarrow{\cong} \mathbb{Q}^{\mathrm{inv}}$ and the rank of an abelian group and the dimension of a rational vector space induce isomorphisms $K_0(\mathbb{Z}) \xrightarrow{\cong} \mathbb{Z}$ and $K_0(\mathbb{Q}) \xrightarrow{\cong} \mathbb{Z}$. This follows from the fact that $\mathbb{Z}$ has a Euclidean algorithm and that $\mathbb{Q}$ is a field. A $\mathbb{Z}$-module M is $\mathbb{Q}$-torsion and has a 1-dimensional finite free resolution if and only if it is a finite abelian group. This follows from the fact that a finitely generated abelian group is a direct sum of finitely many copies of cyclic groups. The isomorphism $K_0(\mathbb{Z} \to \mathbb{Q}) \xrightarrow{\cong} \mathbb{Q}^{+,\mathrm{inv}}$ sends a finite abelian group M to its order.

9.10.   Since $\mathbb{C}G = l^2(G)$ any $\mathbb{C}G$-submodule of $\mathbb{C}G$ is a direct summand. Hence $\mathbb{C}G$ is semisimple. By Wedderburn's Theorem we have $\mathbb{C}G = \prod_{i=1}^{r} M_{n_i}(\mathbb{C})$ since any finite-dimensional skew field over $\mathbb{C}$ is $\mathbb{C}$ itself. We have

$$\mathcal{Z}(\mathbb{C}G) \cong \prod_{i=1}^{r} \mathcal{Z}(M_{n_i}(\mathbb{C})) \cong \prod_{i=1}^{r} \mathbb{C}.$$

This shows that r is the number of conjugacy classes of elements in G. We have

$$K_0(\mathbb{C}G) \cong K_0\left(\prod_{i=1}^{r} M_{n_i}(\mathbb{C}) \right) \cong \prod_{i=1}^{r} K_0(\mathbb{C}) \cong \mathbb{Z}^r.$$

Analogously we get

$$K_1(\mathbb{C}G) \cong K_1\left(\prod_{i=1}^{r} M_{n_i}(\mathbb{C}) \right) \cong \prod_{i=1}^{r} K_1(\mathbb{C}) \cong \prod_{i=1}^{r} \mathbb{C}^{\mathrm{inv}}.$$

We conclude from Theorem 9.31 that

$$L_0^p(\mathbb{C}G) = K_0(\mathbb{C}G) = \mathbb{Z}^r;$$
$$L_1^p(\mathbb{C}G) = 0;$$
$$L_1^h(\mathbb{C}G) = 0;$$
$$L_1^s(\mathbb{C}G) = \mathbb{Z}/2;$$

and that $L_0^s(\mathbb{C}G)$ is a subgroup of finite index in $L_0^h(\mathbb{C}G)$, and $L_0^h(\mathbb{C}G)$ is a subgroup of finite index in $K_0(\mathbb{C}G)$. This implies

$$L_0^h(\mathbb{C}G) = \mathbb{Z}^r;$$
$$L_0^s(\mathbb{C}G) = \mathbb{Z}^r.$$

9.11.   An Ore localization of a Noetherian ring is again Noetherian [448, Poposition 3.1.13 on page 278]. Hence $\mathcal{U}(G)$ is Noetherian if $\mathcal{N}(G)$ is (see Theorem 8.22 (1)). If $\mathcal{U}(G)$ is Noetherian, it must be semisimple since it is von Neumann regular (see Lemma 8.20 (2) and Theorem 8.22 (3)). For any semisimple ring its K_0-group is finitely generated. We conclude $K_0(\mathcal{N}(G)) = K_0(\mathcal{U}(G))$ from Theorem 9.20 (1).

Suppose that $K_0(\mathcal{N}(G))$ is finitely generated. We can split $\mathcal{N}(G) = \mathcal{N}(G)_I \times \mathcal{N}(G)_{II}$. Since $K_0(\mathcal{N}(G)_{II}) \to \mathcal{Z}(\mathcal{N}(G)_{II})^{\mathbb{Z}/2}$ is surjective by Theorem 9.13 (2), $\mathcal{N}(G)_{II}$ must be trivial. Hence $\mathcal{N}(G)$ is of type I. Lemma 9.4 (1) implies that G is virtually abelian. Since the image of $\dim_{\mathcal{N}(G)} \colon K_0(\mathcal{N}(G)) \to \mathbb{R}$ is finitely generated and contains $|H|^{-1}$ for any finite subgroup $H \subset G$, there must be a bound on the orders of finite subgroups of G. Suppose that $\mathbb{Z} \subset G$. Since $\dim_{\mathcal{N}(\mathbb{Z})} \colon K_0(\mathcal{N}(\mathbb{Z})) \to \mathbb{R}$ is surjective and dim is compatible with induction, $\dim_{\mathcal{N}(G)} \colon K_0(\mathcal{N}(G)) \to \mathbb{R}$ is surjective and therefore $K_0(\mathcal{N}(G))$ is not finitely generated. Hence G is a virtually abelian group which does not contain $\mathbb{Z}$ as subgroup and has a bound on the orders of its finite subgroups. This implies that G is finite.

9.12.   Suppose that R is semihereditary. Let $f \colon P \to P$ be a nilpotent R-endomorphism of a finitely generated projective R-module. Then $\mathrm{im}(f) \subset P$ is a finitely generated submodule of a finitely generated projective one and hence projective. We have exact sequences $0 \to (\mathrm{im}(f), f) \to (P, f) \xrightarrow{\mathrm{pr}} (P/\mathrm{im}(f), 0) \to 0$ and $0 \to (\mathrm{im}(f), 0) \to (P, 0) \xrightarrow{\mathrm{pr}} (P/\mathrm{im}(f), 0) \to 0$. Let (M, g) be the pullback of $(P, f) \xrightarrow{\mathrm{pr}} (P/\mathrm{im}(f), 0)) \xleftarrow{\mathrm{pr}} (P, 0)$. We obtain exact sequences $0 \to (\mathrm{im}(f), f) \to (M, g) \to (P, 0) \to 0$ and $0 \to (\mathrm{im}(f), 0) \to (M, g) \to (P, f) \to 0$. Notice that M is finitely generated projective (what is not necessarily true for $P/\mathrm{im}(f)$) and g is nilpotent as f is nilpotent. We get in $\mathrm{Nil}(R)$

$$[\mathrm{im}(f), f] + [P, 0] = [M, g] = [\mathrm{im}(f), 0] + [P, f].$$

This implies $[P, f] = [\mathrm{im}(f), f] \in \widetilde{\mathrm{Nil}}(R)$. We can iterate this argument and get $[P, \mathrm{im}(f)] = [\mathrm{im}(f^n), f] \in \widetilde{\mathrm{Nil}}(R)$ for all natural numbers n. Choose n large enough that $f^n = 0$ and we conclude $[P, f] = 0 \in \widetilde{\mathrm{Nil}}(R)$.

Both $\mathcal{N}(G)$ and $\mathcal{U}(G)$ are semihereditary (see Theorem 6.5, Theorem 6.7 (1), Lemma 8.18 and Theorem 8.22 (3).

9.13.   Suppose $\mathrm{Wh}(G)$ is trivial. Consider a finite subgroup H of the center of G. We derive from Theorem 9.38 that $\mathbb{Q} \otimes_{\mathbb{Z}} \mathrm{Wh}(H)$ is trivial. The rank of the finitely generated abelian group $\text{Wh}(H)$ is the difference $r - q$, where r is the number of $\mathbb{R}$-conjugacy classes and q the number of $\mathbb{Q}$-conjugacy classes in G (see Subsection 3.1.1). We conclude $r = q$. A finite abelian group H satisfies $r = q$ if and only if any cyclic subgroup has at most two generators. A non-trivial cyclic group $\mathbb{Z}/n$ has at most two generators if and only if $n = 2, 3, 4, 6$. Hence H is a finite product of copies of $\mathbb{Z}/2$, $\mathbb{Z}/3$ or of $\mathbb{Z}/2$ and $\mathbb{Z}/4$.

9.14.   We conclude from Theorem 9.62 that the image of $K_0(\mathcal{N}(\mathbb{Z}G)) \to K_0(\mathcal{N}(G))$ is generated by $[\mathcal{N}(G)]$. We conclude from Lemma 9.56 for any finitely generated projective $\mathbb{Z}G$-module P that the image of $[P]$ under $K_0(\mathbb{Z}G) \to K_0(\mathcal{N}(G))$ is $\text{HS}(\mathbb{C} \otimes_{\mathbb{Z}} P)(1) \cdot [\mathcal{N}(G)]$.

9.15.   This follows from Theorem 9.54 (3).

9.16.   This follows from Theorem 9.13 (2), Lemma 9.56 and Theorem 9.63.

9.17.   The map $d_{\mathcal{A}}$ sends $[M]$ to $[\mathbf{P}M]$. The proof that it is well-defined is similar to the one of Theorem 9.64. The argument for $\mathcal{U}$ is analogous using Lemma 8.27.

9.18.   The map $G_0(\mathbb{C}F) \to G_0(\mathbb{C}F)$ given by $V \otimes_{\mathbb{C}} -$ and the diagonal F-action is well-defined and sends the class of $\mathbb{C}[F/H]$ to the class of $V \otimes_{\mathbb{C}} \mathbb{C}[F/H]$. Hence it suffices to show that $[\mathbb{C}[F/H]] = 0$ in $G_0(\mathbb{C}F)$. If H is trivial, this has already been proved in Theorem 9.66 (2). Since any subgroup of a free group is free, it remains to treat the case, where H is a non-trivial free group.

The map given by induction $G_0(\mathbb{C}H) \to G_0(\mathbb{C}F)$ sends $[\mathbb{C}]$ to $[\mathbb{C}[G/H]]$, where $\mathbb{C}$ is viewed as $\mathbb{C}H$-module by the trivial action. Let $f \colon H \to \mathbb{Z}$ be an epimorphism. Restriction with f induces a homomorphism $G_0(\mathbb{C}\mathbb{Z}) \to G_0(\mathbb{C}H)$ which sends $[\mathbb{C}]$ to $[\mathbb{C}]$. We have already shown in the proof of Theorem 9.66 (2) that $[\mathbb{C}] = 0$ in $G_0(\mathbb{C}\mathbb{Z})$.

9.19.   Since G is finite, $\mathbb{Q}G$ is semisimple and hence any finitely generated $\mathbb{Q}G$-module is finitely generated projective. The functor sending a $\mathbb{Z}G$-module M to the $\mathbb{Q}G$-module $\mathbb{Q}G \otimes_{\mathbb{Z}G} M \cong_{\mathbb{Q}G} \mathbb{Q} \otimes_{\mathbb{Z}} M$ is exact. Hence j is well-defined.

Consider a finitely generated projective $\mathbb{Q}G$-module P. Choose an idempotent matrix $A \in M_n(\mathbb{Q}G)$ such that the image of the map $r_A \colon \mathbb{Q}G^n \to \mathbb{Q}G^n$ given by right multiplication with A is $\mathbb{Q}G$-isomorphic to P. Choose an integer $l \geq 1$ such that $B := l \cdot A$ is a matrix over $\mathbb{Z}G$. Let M be the cokernel of the map $r_B \colon \mathbb{Z}G^n \to \mathbb{Z}G^n$. Then $G_0(\mathbb{Z}G) \to K_0(\mathbb{Q}G)$ maps $[M]$ to $[P]$. Hence $G_0(\mathbb{Z}G) \to K_0(\mathbb{Q}G)$ is surjective.

A more general result is proved in Lemma 10.70 (1).

Chapter 10

10.1. Analogous to the proof of Lemma 10.5, where here Y should be a finite 2-dimensional CW-complex with $\pi_1(Y) \cong G$ and $p \colon \overline{Y} \to Y$ be its universal covering.

10.2. We get a G-covering $X \to X/G$ with finite $G\backslash X$. We get an exact sequence $\pi_1(G\backslash X) \to G \to \pi_0(G\backslash X)$. Since $\pi_1(G\backslash X)$ is finitely generated and $\pi_0(X)$ is a finite set, G contains a finitely generated subgroup of finite index. Hence G is finitely generated.

Suppose that there is an epimorphism $f' \colon F_g \to G$ from the free group of rank g to G. Choose a map $f \colon \#_{i=1}^g S^1 \times S^2 \to BG$ which induces f' on the fundamental groups. Take M to be the pullback of $EG \to BG$ with f. Then M is a cocompact free proper G-manifold. The long exact homotopy sequence implies that M is connected.

10.3. Suppose that $d(\Lambda) = 0$. Consider $r \in \mathbb{R}$ and $\epsilon > 0$. We can find $g \in \Lambda$ with $0 < g < \epsilon$. Choose $n \in \mathbb{Z}$ with $n \cdot g \le r \le (n+1) \cdot g$. Hence $0 \le r - n \cdot g \le \epsilon$. Hence Λ is dense in $\mathbb{R}$. This shows that $\Lambda = \mathbb{R}$ if and only if Λ is closed and satisfies $d(\Lambda) = 0$.

Suppose that $d(\Lambda) > 0$ and Λ is closed. Then $d(\Lambda) \in \Lambda$. Consider $g \in \Lambda$. Choose $n \in \mathbb{Z}$ with $n \cdot d(\Lambda) \le g < (n+1) \cdot d(\Lambda)$. Since $0 \le g - n \cdot d(\Lambda) < d(\Lambda)$ and $g - n \cdot d(\Lambda) \in \Lambda$, we conclude $g = n \cdot d(\Lambda)$.

10.4. Without loss of generality we can assume that C_* is finite dimensional. Since any module is the colimit of its finitely generated submodules, we can find a directed system of finite FG-subchain complexes $C_*[i] \subset C_*$ such that $C_* = \operatorname{colim}_{i \in I} C_*[i]$. Since the functor colimit over directed systems is exact and $\mathcal{N}(G) \otimes_{FG} -$ commutes with colimits, we conclude

$$H_p(\mathcal{N}(G) \otimes_{FG} H_p(C_*)) = \operatorname{colim}_{i \in I} H_p(\mathcal{N}(G) \otimes C_*[i]).$$

Because of Theorem 6.13 (2) it suffices to show for any FG-chain map $f_* \colon D_* \to E_*$ of finite FG-chain complexes

$$\dim_{\mathcal{N}(G)} \left(\operatorname{im} \left(H_p \left(\operatorname{id} \otimes_{FG} f_* \colon \mathcal{N}(G) \otimes_{FG} D_* \to \mathcal{N}(G) \otimes_{FG} E_* \right) \right) \right) \in \Lambda.$$

Since we have the long exact homology sequence associated to the exact sequence $0 \to D_* \to \operatorname{cone}(f_*) \to \Sigma C_* \to 0$, we conclude from Additivity (see Theorem 6.7 (4b)) that it suffices to show for any finite FG-chain complex F_* and $p \in \mathbb{Z}$ that

$$\dim_{\mathcal{N}(G)} \left(H_p \left(\mathcal{N}(G) \otimes_{FG} F_* \right) \right) \in \Lambda.$$

Again from Additivity (see Theorem 6.7 (4b)) we conclude that it suffices to show for any FG-map $g \colon M_0 \to M_1$ of finitely generated FG-modules that

$$\dim_{\mathcal{N}(G)} \left(\mathrm{im} \left(\mathrm{id} \otimes_{FG} g \colon \mathcal{N}(G) \otimes_{FG} M_0 \to \mathcal{N}(G) \otimes_{FG} M_1 \right) \right) \ \in \ \Lambda.$$

Since $\mathcal{N}(G) \otimes_{FG} M_0 \xrightarrow{\mathrm{id} \otimes_{FG} g} \mathcal{N}(G) \otimes_{FG} M_1 \xrightarrow{\mathrm{pr}} \mathcal{N}(G) \otimes_{FG} \mathrm{coker}(g) \to 0$ is exact, Additivity (see Theorem 6.7 (4b)) implies

$$\begin{aligned}
&\dim_{\mathcal{N}(G)} \left(\mathrm{im} \left(\mathrm{id} \otimes_{FG} g \colon \mathcal{N}(G) \otimes_{FG} M_0 \to \mathcal{N}(G) \otimes_{FG} M_1 \right) \right) \\
&\quad = \ \dim_{\mathcal{N}(G)} \left(\mathcal{N}(G) \otimes_{FG} M_1 \right) - \dim_{\mathcal{N}(G)} \left(\mathcal{N}(G) \otimes_{FG} \mathrm{coker}(g) \right).
\end{aligned}$$

But $\dim_{\mathcal{N}(G)} \left(\mathcal{N}(G) \otimes_{FG} M_1 \right)$ and $\dim_{\mathcal{N}(G)} \left(\mathcal{N}(G) \otimes_{FG} \mathrm{coker}(g) \right)$ belong to Λ because of Lemma 10.7 and Lemma 10.10 (3) since G satisfies by assumption the Atiyah Conjecture 10.3 of order Λ with coefficients in F.

10.5.   This follows from the previous Exercise 10.4 applied to $\mathcal{N}(G) \otimes_{\mathbb{Z}G} C_*(EG)$ or from Theorem 6.99.

10.6.   Pass to the sequence of p-Sylow subgroups. Now use the fact that a maximal proper subgroup of a p-group has index p and any proper subgroup is contained in a maximal proper subgroup to construct the desired sequence $K_1 \subset K_2 \subset K_3 \subset \dots$. Now the argument is a variation of the argument appearing in Example 10.13. Notice that $|K_n|^{-1} - |K_{n-1}|^{-1}|$ is $(p-1)/p$. Use the p-adic expansion of $r/(p-1)$ and replace $\mathbb{Q}G$ by $(\mathbb{Q}G)^{p-1}$. If the n-th coefficient of the p-adic expansion is k_n, use $(I_n)^{k_n} \subset (\mathbb{Q}G)^{k_n} \subset (\mathbb{Q}G)^n$.

10.7.   This is done by transfinite induction (see Lemma 10.40). Details can be found in [435, Theorem 7.7].

10.8.   By the previous Exercise 10.7 $\mathcal{C}$ if closed under free products. For any positive integer k the group $*_{n=1}^k G_n$ belongs to Linnell's class $\mathcal{C}$. Since any finite subgroup of $*_{n=1}^k G_n$ is conjugate to a subgroup of one of the summands G_i (see [471, Theorem 8 in Section I.4 on page 36]), there is an upper bound on the orders of finite subgroups of $*_{n=1}^k G_n$. Hence $*_{n=1}^k G_n$ satisfies the strong Atiyah Conjecture 10.2 by Theorem 10.19. Since G is the directed union of the subgroups $*_{n=1}^k G_n$, we conclude from Lemma 10.4 that G satisfies the strong Atiyah Conjecture 10.2. Now take $G = *_{n=1}^\infty \mathbb{Z}/2^n$. We could also take $G = \bigoplus_{n \in \mathbb{Z}} \mathbb{Z}/2$ which satisfies the strong Atiyah Conjecture 10.2 by Lemma 10.4 since it is locally finite.

10.9.   Any manifold M which has a boundary or which is not compact is homotopy equivalent to a CW-complex of dimension $\dim(M) - 1$. Any 1-dimensional CW-complex has a free group as fundamental group. Hence $\pi_1(M)$ is free and hence belongs to $\mathcal{C}$ if M is a 2-dimensional manifold which has boundary or which is not compact.

Let M be a closed 2-dimensional manifold. If $H_1(M)/\mathrm{tors}(M)$ is trivial, then $\pi_1(M)$ is finite and hence belongs to $\mathcal{C}$. Suppose $H_1(M)/\mathrm{tors}(M)$ is not trivial. Then $H_1(M)/\text{tors}(M)$ is $\mathbb{Z}^n$ for some integer $n \geq 1$. Let $\overline{M} \to M$ be the covering associated to the epimorphism $\pi_1(M) \to H_1(M)/\text{tors}(H_1(M))$. Then $\overline{M}$ is a 2-dimensional non-compact manifold and hence has a free fundamental group. Because there is an exact sequence $1 \to \pi_1(\overline{M}) \to \pi_1(M) \to \mathbb{Z}^n \to 1$, the fundamental group $\pi_1(M)$ belongs to $\mathcal{C}$. Now apply Theorem 10.19.

10.10.   The fundamental group of F belongs to $\mathcal{C}$ by the previous Exercise 10.9. Since $\mathcal{C}$ is closed under extensions with amenable quotient and $\mathbb{Z}$ and each finite group are amenable, $\pi_1(M)$ belongs to $\mathcal{C}$. Now apply Theorem 10.19.

10.11.   Suppose that G contains a non-trivial element g such that the centralizer $C_G(g)$ has finite index in G. Then $b_p^{(2)}(G; \mathcal{N}(G)) = 0$ for all $p \geq 0$ by Theorem 6.54 (6b) and Theorem 7.2 (1) and (2). We conclude $H_p^{(2)}(EG; \mathcal{N}(G)) = 0$ from Theorem 6.7 (4e), Lemma 6.28 (3) and Lemma 6.52.

   Suppose that G contains no non-trivial element g such that the centralizer $C_G(g)$ has finite index in G. Then $\mathcal{N}(G)$ is a factor by Lemma 9.4 (4) and $\dim_{\mathcal{N}(G)}$ and the universal center valued dimension $\dim_{\mathcal{N}(G)}^u$ agree. By assumption $b_p^{(2)}(EG; \mathcal{N}(G)) = \dim_{\mathcal{N}(G)}(H_p^G(EG; \mathcal{N}(G))$ is a non-negative integer n. We conclude from Theorem 6.7 (4e) and Theorem 9.13 (1) that $\mathbf{P}H_p^G(EG; \mathcal{N}(G))$ is $\mathcal{N}(G)$-isomorphic to $\mathcal{N}(G)^n$. Lemma 6.52 implies that the Hilbert $\mathcal{N}(G)$-modules $H_p^{(2)}(EG)$ and $l^2(G)^n$ are isomorphic.

10.12.   Suppose that $\pi_1(M)$ belongs to $\mathcal{C}$. We conclude $b_p^{(2)}(\widetilde{M}) = 0$ for $p \geq 2$ from Theorem 8.31 and (10.86). Since $b_n^{(2)}(\widetilde{M}) \neq 0$ by Theorem 1.62, we conclude $n = 1$. If $n = 1$, we get $\pi_1(M) \in \mathcal{C}$ from Exercise 10.9.

10.13.   Let $0 \to \bigoplus_{n \in \mathbb{Z}} \mathbb{Z}/2 \to L \to \mathbb{Z} \to 0$ be the canonical exact sequence for the lamplighter group L. The lamplighter group does not satisfy the strong Atiyah Conjecture 10.2 by Theorem 10.23. Since $\bigoplus_{\mathbb{Z}} \mathbb{Z}/2$ is locally finite, it satisfies the strong Atiyah Conjecture 10.2 because of Lemma 10.4. The group $\mathbb{Z}$ satisfies the strong Atiyah Conjecture 10.2 by Lemma 1.34 (1).

10.14.   Since each $\mathcal{S}(G_i)$ is von Neumann regular by assumption, $\mathcal{S}(G)$ is von Neumann regular. Since K_0 commutes with directed unions of rings and each $\mathcal{S}(G_i)$ satisfies $\mathbf{K}'$, the union $\mathcal{S}(G)$ satisfies $\mathbf{K}'$ by Theorem 3.14 (6). Let G be the locally finite group $\bigoplus_{n \in \mathbb{Z}} \mathbb{Z}/2$ and $\{G_i \mid i \in I\}$ be the directed system of its finitely generated subgroups. Each G_i is finite and hence $\mathcal{D}(G_i) = \mathbb{C}G_i$ is semisimple for each $i \in I$. We get $\mathcal{D}(G) = \bigcup_{i \in I} \mathcal{D}(G_i)$ from Lemma 10.83. But $\mathcal{D}(G)$ cannot be semisimple by Lemma 10.28 (1).

10.15.   The proof that the map is well-defined is analogous to (6.4). The integrality of its image follows from Lemma 10.46.

10.16.   Fix a set-theoretic section $s\colon G/H \to G$ of the projection $G \to G/H$. Then a $\mathbb{C}H$-basis for $\mathbb{C}G$ considered as $\mathbb{C}H$-module is given by $\{s(gH) \mid gH \in G/H\}$. The image of this basis under the obvious inclusion $\mathbb{C}G \subset \mathcal{D}(G)$ is a $\mathcal{D}(H)$-basis for $\mathcal{D}(G)$ considered as $\mathcal{D}(H)$-module (see Lemma 10.59). The image of this $\mathcal{D}(H)$-basis under the inclusion $\mathcal{D}(G) \to \mathcal{U}(G)$ is a $\mathcal{U}(H)$-basis for $\mathcal{U}(G)$ considered as $\mathcal{U}(H)$-module. This shows that the left and middle vertical arrow are well-defined and that the diagram commutes.

Let G be a virtually finitely generated free group G. If d is the least common multiple of the orders of finite subgroups of G, then one can find a finitely generated free subgroup $F \subset G$ of index d (see for instance [24, Theorem 8.3 and Theorem 8.4], [461, Theorem 5 in Section 7 on page 747]). Then $\mathcal{D}(H)$ is semisimple by Lemma 10.39 and Theorem 10.43. Hence the image of the composition

$$K_0(\mathcal{D}(G)) \xrightarrow{i} K_0(\mathcal{U}(G)) \xrightarrow{\dim_{\mathcal{U}(G)}} \mathbb{R}$$

is contained in $\frac{1}{|\mathcal{FIN}(G)|}\mathbb{Z}$, since this is true for F by Lemma 10.51 and the diagram appearing in the exercise commutes. Since any virtually free group G is the directed union of its finitely generated subgroups $\{G_i \mid i \in I\}$, each G_i is again virtually finitely generated free, $\mathcal{D}(G) = \bigcup_{i \in I} \mathcal{D}(G_i)$ by Lemma 10.83 and K_0 commutes with directed unions, the claim follows.

10.17.   One easily checks that f is injective. We have

$$\psi(y^{-1}z - x^{-1}y) = (xu)^{-1}xu^2 - x^{-1}xu = u - u = 0.$$

Since L is a skewfield, the division closure D of $\operatorname{im}(f)$ is itself a skewfield. Since the skewfield generated by $\operatorname{im}(f)$ is division closed, it agrees with D, i.e. f yields a field of fractions $f\colon \mathbb{C}\langle x, y, z\rangle \to D$. It cannot be a universal field of fractions by Lemma 10.77. Moreover, $f\colon \mathbb{C}\langle x, y, z\rangle \to D$ and $\mathbb{C}\langle x, y, z\rangle \to K$ cannot be isomorphic fields of fractions.

10.18.   Let G be a countable group which is not finitely generated, for instance the free group on a countable infinite set of generators. Let $\{G_i \mid i \in I\}$ be the directed system of its finitely generated subgroups. Then $G = \bigcup_{i \in I} G_i$ but $G \neq G_i$ for all $i \in I$. Choose a numeration $G = \{g_1, g_2, g_3, \ldots\}$. We get an element $u = \sum_{n=1}^{\infty} \frac{1}{n^2} \cdot g_n$ in $l^1(G)$. We have already seen in Exercise 8.3 that $l^1(G) \subset \mathcal{N}(G)$, namely, right multiplication with u defines a bounded G-equivariant operator $r_u\colon l^2(G) \to l^2(G)$. If this element lies in $\mathcal{N}(G_i)$ for a finitely generated subgroup G_i, then $r_u(1) \in l^2(G_i)$. This yields a contradiction because $G_i \neq G$. This shows $\mathcal{N}(G) \neq \bigcup_{i \in I} \mathcal{N}(G_i)$. This implies also $\mathcal{U}(G) \neq \bigcup_{i \in I} \mathcal{U}(G_i)$ since $\mathcal{N}(G) \cap \mathcal{U}(G_i) = \mathcal{N}(G_i)$ by Lemma 8.3 (4).

10.19.   This follows from Theorem 9.20 (1), Theorem 10.38 and Theorem 10.87.

Chapter 11

11.1.   Take $M = T^2 \times F_g$ for a closed connnected orientable surface F_g of genus $g \geq 2$ and equip M with the product of the flat Riemannian metric on T^2 and the hyperbolic Riemannian metric on F_g. Theorem 3.105 implies $b_p^{(2)}(\widetilde{M}) = 0$ for $p \geq 0$.

11.2.   The universal covering of a flat closed manifold M is isometrically diffeomorphic to $\mathbb{R}^n$. Recall that $\widetilde{T^n}$ is det-L^2-acyclic and $\rho^{(2)}(\widetilde{T^n}) = 0$ (see Theorem 3.105). Now apply Theorem 3.183.

11.3.   Suppose that $\pi_1(F)$ and $\pi_1(B)$ are finite. Then $\pi_1(E)$ is finite. The values of the zero-th and first L^2-Betti number of $\widetilde{E}$ can be derived from Theorem 1.35 (1) and (8). Now apply the Euler-Poincaré formula and Poincaré duality (see Theorem 1.35 (2) and (3)).

Suppose that exactly one of the groups $\pi_1(F)$ and $\pi_1(B)$ is finite. Then $\pi_1(E)$ is infinite and hence $b_0^{(2)}(\widetilde{E}) = 0$ by Theorem 1.35 (8). A spectral sequence argument shows $b_2^{(2)}(\widetilde{M}) = 0$, the corresponding entries on the E_2-term have vanishing von Neumann dimension. We get $b_1^{(2)}(\widetilde{M}) = b_3^{(2)}(\widetilde{M})$ and $b_0^{(2)}(\widetilde{M}) = b_4^{(2)}(\widetilde{M}) = 0$ from Poincaré duality (see Theorem 1.35 (3)). Now apply the Euler-Poincaré formula (see Theorem 1.35 (2)) and the product formula for the Euler characteristic $\chi(M) = \chi(F) \cdot \chi(B)$.

Suppose that both $\pi_1(B)$ and $\pi_1(F)$ are infinite. We conclude $b_p^{(2)}(\widetilde{M}) = 0$ for $p = 0, 1$ from Theorem 6.67. Now apply Euler-Poincaré formula and Poincaré duality (see Theorem 1.35 (2) and (3)).

If M is aspherical, either F and B are closed surfaces of genus $g \geq 1$ or one of them is S^1.

11.4.   Suppose that $\dim(M) = 2$. Then M is a closed surface of genus ≥ 2 by the Gauss-Bonnet formula and the claim follows from Example 1.36 and Example 2.70.

Suppose $\dim(M) \geq 4$. We conclude from Theorem 11.6 that $b_p^{(2)}(\widetilde{M}) = 0$ and $\alpha_p^{\triangle}(\widetilde{M}) = \infty^+$ holds for $p < \dim(M)/2$ since for some b with $1 - \frac{\dim(M)}{2} < b < 1$ and $a = 1$ we have $-a^2 \leq \sec(M) \leq -b^2$. Now apply Poincaré duality (see Theorem 1.35 (3) and Theorem 2.55 (2)) and Theorem 2.66 (2).

11.5.   See [520, Proposition 4.6 on page 190] for M a Kähler manifold. Kähler hyperbolicity is proved analogously.

11.6.   See [520, (1.12) on page 158].

11.7.   Suppose such $N \subset \widetilde{M}$ exists. Then N is closed Kähler with Kähler form $\omega|_N$ if ω is the Kähler form of M. Then $(\omega|_N)^{\dim(N)/2}$ is a multiple of the volume form of N [520, (1.12) on page 158] and cannot be exact although ω is exact.

11.8.   The Kähler structure is given in [520, Example 4.4 on page 189]. Since $b_n^{(2)}(\widetilde{T^{2n}}) = 0$, Theorem 11.14 implies that there is no Kähler hyperbolic structure on T^{2n}.

11.9.   The desired isomorphism is induced by complex conjugation (see [520, Theorem 5.1 in V.5 on page 197]).

11.10.   Because of Theorem 6.54 (6b) it suffices to treat the case $G = G_0$. Obviously G acts properly on the contractible space $E\pi$ for $\pi = \pi_1(F)$. We conclude from Example 1.36, Theorem 6.54 (2) and Theorem 13.45 (4)

$$b_1^{(2)}(G) = b_1^{(2)}(E\pi; \mathcal{N}(G)) = k \cdot b_1^{(2)}(E\pi; \mathcal{N}(\pi)) = k \cdot b_1^{(2)}(\pi) \neq 0$$

if k is the order of the kernel of $G_0 \to \pi$.

11.11.   See [3, page 3]).

Chapter 12

12.1.   See [496, Satz 18.2 in §18 on page 223].

12.2.   Because of Lemma 12.3 and homotopy invariance (see Theorem 1.35 (1) and Theorem 2.55 (1)) it suffices to show that it is not possible that both $b_p^{(2)}(\widetilde{M}) = b_p^{(2)}(\pi)$ vanishes for $p = 0, 1$ and $\alpha_p(\widetilde{M}) = \alpha_p(\pi)$ is ∞^+ for $p = 1, 2$. Suppose the contrary.

We can assume without loss of generality that ∂M has no boundary component which is homotopy equivalent to S^2. Otherwise we can attach finitely many copies of D^3 to M, what does not affect the fundamental group of π of M. Furthermore we can assume that the prime decomposition contains no homotopy sphere because removing such factor from the prime decomposition does not alter the homotopy type.

Since $\chi(\partial M) = 2 \cdot \chi(M)$ implies $\chi(M) \leq 0$, we conclude from Theorem 4.1 that $b_p^{(2)}(\widetilde{M}) = 0$ for $p = 0, 1, 2, 3$ and hence M is homotopy equivalent to $RP^3 \# RP^3$ or a prime 3-manifold with infinite fundamental group whose boundary is empty or a union of tori. Theorem 4.2 (2) implies that M is not homotopy equivalent to $\mathbb{RP}^3 \# \mathbb{RP}^3$ or $S^1 \times S^2$. Hence we are left with the case that M is irreducible and non-exceptional and has infinite fundamental group. The case that M is closed is excluded by Theorem 4.2 (2) and (4). The case that ∂M contains an incompressible torus is ruled out by Theorem 4.2 (4). The case, where ∂M contains an compressible torus is reduced to the case, where ∂M contains no compressible torus by the construction appearing in the proof of [322, Lemma 6.4 on page 52] based on the Loop Theorem [252, Theorem 4.2 on page 39] and the formula for first L^2-Betti number under wedges (see Theorem 1.35 (5)).

12.3.   By Lemma 5.10 we have $\widetilde{M} = \widetilde{M}_{\mathrm{Eucl}} \times \widetilde{M}_{\mathrm{ncp}}$. Now the claim follows from (5.14) and (5.15) and Lemma 12.3.

12.4.   Consider $r > 0$. Choose $r_0 > 0$ and $\widetilde{x}_0 \in \widetilde{M}$ such that $\widetilde{M} = \bigcup_{w \in \pi_1(M)} B_{r_0}(w \cdot \widetilde{x}_0)$ and $r_0 \geq r$. This is possible since M is compact. Since $\widetilde{M}$ is contractible, we can find $R > 2r_0$ such that $B_{2r_0}(\widetilde{x}_0)$ is contractible in $B_R(\widetilde{x}_0)$. Then $B_{2r_0}(w \cdot \widetilde{x}_0)$ is contractible in $B_R(w \cdot \widetilde{x}_0)$. For each $\widetilde{x} \in \widetilde{M}$ we can find $w \in \pi_1(M)$ such that $\widetilde{x} \in B_{r_0}(w \cdot \widetilde{x}_0)$. Hence we get for appropriate $w \in \pi_1(M)$ that $B_r(\widetilde{x}) \subset B_{2r_0}(w \cdot \widetilde{x}_0)$. This shows that $B_r(\widetilde{x})$ is contractible in $B_R(w \cdot \widetilde{x}_0)$. Since $B_R(w \cdot \widetilde{x}_0) \subset B_{2R}(\widetilde{x})$, we conclude that $B_r(\widetilde{x})$ is contractible in $B_{2R}(\widetilde{x})$ for a number $2R$ which does not depend on $\widetilde{x}$.

12.5.   See [341, Lemma 11.6 (5) on page 806].

12.6.   This follows from a spectral sequence argument applied to homology with coefficients in $\mathcal{N}(\pi_1(E))$ and Lemma 12.3.

12.7.   Since π is not amenable, $H_0^\pi(E\pi; \mathcal{N}(\pi)) = 0$ by Lemma 12.11 (4). The condition $\chi(X) = 0$ and $H_p^\pi(\widetilde{X}; \mathcal{N}(\pi)) = H_p^\pi(E\pi; \mathcal{N}(\pi)) = 0$ for $p = 0, 1$ imply $b_2^{(2)}(\widetilde{X}) = 0$ by the Euler-Poincaré formula (see Theorem 1.35 (2)). Since $\mathcal{N}(\pi)$ is semihereditary (see Theorem 6.5 and Theorem 6.7 (1)) and $H_2(\widetilde{X}; \mathcal{N}(\pi))$ is the kernel of the second differential $c_2^{\mathcal{N}(\pi)}$ of $\mathcal{N}(\pi) \otimes_{\mathbb{Z}[\pi]} C_*(\widetilde{X})$, $H_2(\widetilde{X}; \mathcal{N}(\pi))$ is direct mmand in $\mathcal{N}(\pi) \otimes_{\mathbb{Z}[\pi]} C_2(\widetilde{X}) \cong \mathcal{N}(\pi)^s$. Since $H_2(\widetilde{X}; \mathcal{N}(\pi))$ has trivial von Neumann dimension, it must be zero by Lemma 6.28 (3). Since $c_2^{\mathcal{N}(\pi)}$ is injective, c_2 is injective. Hence $H_p(\widetilde{X}) = 0$ for $p \geq 2$ and $\widetilde{X}$ must be contractible.

12.8.   We conclude $\mathrm{def}(G) \leq 1$ from Lemma 7.22.
    Suppose that $\mathrm{def}(G) = 1$. Then we can find a connected finite 2-dimensional CW-complex X with $G \cong \pi_1(X)$ and $\chi(X) = 0$. Lemma 12.11 (4) implies that G is non-amenable. Now apply the previous Exercise 12.7 to X.

12.9. We get $q(G) \geq 0$ from the Euler-Poincaré formula (see Theorem 1.35 (2)) and Poincaré duality (Theorem 1.35 (3)). The claim in the case $q(G) = 0$ follows from Lemma 12.5.

12.10. We get $H_n^{G \times H}(E(G \times H); \mathcal{N}(G \times H)) = 0$ for $n \leq d + e + 1$ from Lemma 12.11 (3). We conclude $H_{l+e+2}^{G \times H}(E(G \times H); \mathcal{N}(G \times H)) \neq 0$ from Lemma 12.3 and the product formulas Lemma 2.35 (1) and Theorem 6.54 (5) using Lemma 2.17 (1). (Notice that in the case of the Novikov-Shubin invariants we do not need the assumption that the spectral density functions have the limit property. This enters only when we want to get numerical values but not when we want to decide whether the Novikov-Shubin invariant is ∞^+.)

12.11. $G = \mathbb{Z}/2, \mathbb{Z}^n, \prod_{i=1}^{\infty}(\mathbb{Z}/2 * \mathbb{Z}/2)$: Since G is in all cases an extension of two abelian groups and hence amenable, we conclude $s = -1$ from Lemma 12.11 (4).

$G = *_{i=1}^r \mathbb{Z}$ for $r \geq 2$: This group is non-amenable and $b_1^{(2)}(G) = r - 1 \neq 0$. We conclude $s = 0$ from Lemma 12.3 and Lemma 12.11 (4).

$G = \prod_{i=1}^r \mathbb{Z} * \mathbb{Z}$: Since $\mathbb{Z} * \mathbb{Z}$ is non-amenable and $b_r^{(2)}(G) \neq 0$ by the product formula (see Theorem 6.54 (5)), we conclude $s = r - 1$ from Lemma 12.3 and Lemma 12.11 (3).

$G = \mathbb{Z} \times (\mathbb{Z} * \mathbb{Z})$: We have already computed s for $\mathbb{Z}$ and $\mathbb{Z} * \mathbb{Z}$. Now $s = 0$ follows from the previous Exercise 12.10.

$\prod_{i=1}^{\infty}(\mathbb{Z}/2 * \mathbb{Z}/3)$: The group $\mathbb{Z}/2 * \mathbb{Z}/3$ is not amenable since it contains a non-abelian free group of finite index. Hence $s = \infty$ by Lemma 12.11 (5).

12.12. (1) is clear when one works with differential forms and uses a collar of ∂N in N.

(2) Since the signature of N is non-zero, the image of $H^{2k}(N; \partial N; \mathbb{C}) \to H^{2k}(N; \mathbb{C})$ and hence the image of $H_c^{2k}(M; \mathbb{C}) \to H^{2k}(M; \mathbb{C})$ are non-trivial. Lemma 1.92 implies that $H_{(2)}^p(M)$ is different from zero. From Theorem 1.72 we conclude that the kernel of $(\Delta_{2k})_{\min} \colon \text{dom}((\Delta_{2k})_{\min}) \subset L^2\Omega^{2k}(M) \to L^2\Omega^{2k}(M)$ is non-trivial. Hence this operator contains zero in its spectrum.

Chapter 13

13.1. The Bass-Heller Swan decomposition for $(\mathbb{Z}[G])[\mathbb{Z}] = \mathbb{Z}[G \times \mathbb{Z}]$ yields an isomorphisms

$$K_0(\mathbb{Z}G) \oplus K_1(\mathbb{Z}G) \oplus \text{Nil}(\mathbb{Z}G) \oplus \text{Nil}(\mathbb{Z}G) \xrightarrow{\cong} K_1(\mathbb{Z}[G \times \mathbb{Z}]).$$

Its reduced version is the isomorphism

$$\tilde{K}_0(\mathbb{Z}G) \oplus \mathrm{Wh}(G) \oplus \widetilde{\mathrm{Nil}}(\mathbb{Z}G) \oplus \widetilde{\mathrm{Nil}}(\mathbb{Z}G) \xrightarrow{\cong} \mathrm{Wh}(G \times \mathbb{Z}).$$

We have analogous Bass-Heller-Swan isomorphisms for $\mathcal{N}(G)[\mathbb{Z}]$. These are natural with respect to the change of rings map $\mathbb{Z}G \to \mathcal{N}(G)$.

The map $\tilde{K}_0(\mathbb{Z}G) \to \tilde{K}_0(\mathcal{N}(G))$ is trivial (see Theorem 9.62). We get $\widetilde{\mathrm{Nil}}(\mathcal{N}(G)) = 0$ from Exercise 9.12. The map $K_1(\mathbb{Z}[G \times \mathbb{Z}]) \to K_1(\mathcal{N}(G \times \mathbb{Z}))$ factorizes through $K_1(\mathcal{N}(G)[\mathbb{Z}])$. The composition $\mathrm{Wh}(G) \to \mathrm{Wh}(G \times \mathbb{Z}) \xrightarrow{\Phi^{G \times \mathbb{Z}}} \mathbb{R}$ agrees with ϕ^G by Theorem 3.14 (6).

13.2.   One checks for $x \in \mathrm{im}(E_\lambda) \cap \bigcap_{\mu < \lambda} \mathrm{im}(E_\mu)^\perp$ using (1.65) and Lemma 2.2 (2)

$$
\begin{aligned}
\langle f(x) - \lambda x, f(x) - \lambda x \rangle &= ||f(x)||^2 - 2\lambda \cdot \langle f(x), x \rangle + \lambda^2 \cdot ||x||^2 \\
&= \lambda^2 ||x||^2 - 2\lambda^2 \cdot ||x||^2 + \lambda^2 \cdot ||x||^2 = 0.
\end{aligned}
$$

This implies

$$\mathrm{im}(E_\lambda) \cap \bigcap_{\mu < \lambda} \mathrm{im}(E_\mu)^\perp = \{u \in U \mid f(u) = \lambda\}.$$

Now apply Additivity and Continuity of $\dim_{\mathcal{N}(G)}$ (see Theorem 1.12 (2) and (4)).

13.3.   We conclude from the previous Exercise 13.2 that F is continuous at λ. Theorem 13.19 (1) implies

$$F(\lambda) = \underline{F}^+(\lambda) = \overline{F}^+(\lambda).$$

If $\underline{F}(\lambda) = F(\lambda) - \epsilon$, we get $F(\mu) = \underline{F}^+(\mu) \leq \underline{F}(\lambda) \leq F(\lambda) - \epsilon$ for $\mu < \lambda$. Continuity of F at λ implies $\epsilon = 0$. Hence $\underline{F}(\lambda) = \overline{F}(\lambda) = F(\lambda)$.

13.4.   Choose a numeration $\{g_n \mid n = 0, 1, 2, \ldots\}$ of G with $g_0 = 1$. We construct inductively a sequence $i_0 \leq i_1 \leq i_2 \leq \ldots$ such that $\{g_0, g_2, \ldots, g_n\} \cap \bigcap_{k=1}^n G_{i_k} = \{1\}$. The induction beginning $n = 0$ is trivial. In the induction step we can find an index $i'_n \in I$ such that $g_n \notin G_{i'_n}$. Choose $i_n \in I$ such that $i'_n \leq i_n$ and $i_{n-1} \leq i_n$.

13.5.   Since G is of det $\geq$ 1-class, NH and WH are of det $\geq$ 1-class by Theorem 13.3 (3). Hence $\Phi^{WH} : \mathrm{Wh}(WH) \to \mathbb{R}$ is trivial by Lemma 13.6. The following diagram commutes by Theorem 3.14 (6) and Theorem 13.45 (6).

$$
\begin{array}{ccccc}
K_1(\mathbb{Z}WH) & \xrightarrow{\mathrm{res}} & K_1(\mathbb{Q}NH) & \xrightarrow{\mathrm{ind}} & K_1(\mathbb{Q}G) \\
\downarrow & & \downarrow & & \downarrow \\
K_1(\mathcal{N}(WH)) & \xrightarrow{\mathrm{res}} & K_1(\mathcal{N}(NH)) & \xrightarrow{\mathrm{ind}} & K_1(\mathcal{N}(G)) \\
\downarrow & & \downarrow & & \downarrow \\
\mathbb{R} & \xrightarrow{\cdot\frac{1}{|H|}} & \mathbb{R} & \xrightarrow{\mathrm{id}} & \mathbb{R}
\end{array}
$$

The map in question is obtained by starting at the left upper corner, going to the right upper corner and then going down to the right lower corner. Hence it factorizes through Φ^{WH} which agrees with the composition of the map, which is obtained by going from the upper left corner down to the left lower corner, and the projection $K_1(\mathbb{Z}WH) \to \mathrm{Wh}(WH)$.

13.6.   Let $H \subset G$ be a subgroup of finite index which is residually finite or residually amenable respectively. Then $H' = \bigcap_{g \in G} gHg^{-1}$ is a normal subgroup of G of finite index in both H and G. Consider $g \in G$. If g does not lie in H', then the projection $G \to G/H'$ maps g to an element different from 1 and G/H' is finite and amenable. Suppose $g \in H'$. Then we can find an epimorphism $f\colon H \to L$ for some group L, which is finite or amenable respectively, such that g is not mapped to $1 \in L$. Let K be the kernel of f which is a normal subgroup in H. Then $K' \cap \bigcap_{g \in G} gKg^{-1} \subset G$ has finite index and G/K' is finite or amenable respectively and the projection $G \to G/K'$ sends g to an element different from 1 in G/K'.

13.7.   Choose a sequence of normal subgroups of finite index

$$
\pi_1(F_g) = H_0 \supset H_1 \supset H_2 \supset \ldots
$$

with $\bigcap_{i=0}^{\infty} H_i = \{1\}$ and a sequence of integers $n_1 \leq n_2 \leq \ldots$ such that $[\pi_1(F_g) : H_i] \geq \exp(i \cdot n_i)$ holds for $i \geq 1$. Put $G_i = 2^{n_i} \cdot \mathbb{Z} \times H_i$. Then

$$
\begin{aligned}
\frac{b_1(G_i \backslash \widetilde{X})}{[\pi_1(X) : G_i]^{1-\epsilon}} &= \frac{(2g-2) \cdot [\pi_1(F_g) : H_i] + 3}{[\pi_1(F_g) : H_i]^{1-\epsilon} \cdot 2^{n_i \cdot (1-\epsilon)}} \\
&= \frac{(2g-2) \cdot [\pi_1(F_g) : H_i]^{\epsilon}}{2^{n_i \cdot (1-\epsilon)}} + \frac{3}{[\pi_1(F_g) : H_i]^{1-\epsilon} \cdot 2^{n_i \cdot (1-\epsilon)}}.
\end{aligned}
$$

If $\epsilon = 0$, this converges for $i \to \infty$ to zero. Suppose $\epsilon > 0$, we get

$$
\frac{b_1(G_i \backslash \widetilde{X})}{[\pi_1(X) : G_i]^{1-\epsilon}} \geq \frac{(2g-2) \cdot \exp(i \cdot n_i \cdot \epsilon)}{2^{n_i \cdot (1-\epsilon)}}
$$

and hence $\frac{b_1(G_i \backslash \widetilde{X})}{[\pi_1(X):G_i]^{1-\epsilon}}$ converges to ∞ for $i \to \infty$.

13.8.   We get from Exercise 3.5 and the product formula for analytic torsion (which can be viewed as a special case of (3.126) in combination with

Theorem 3.149) and Theorem 3.105

$$\rho_{\mathrm{an}}(G_i\backslash\widetilde{S^1}) = \ln(\mathrm{vol}(G_i\backslash S^1));$$
$$\rho_{\mathrm{an}}((G_i\times H_i)\backslash(\widetilde{S^1\times N})) = \rho_{\mathrm{an}}(G_i\backslash\widetilde{S^1})\cdot\chi(H_i\backslash\tilde{N});$$
$$\rho_{\mathrm{an}}^{(2)}(\widetilde{S^1\times N}) = 0.$$

Since $\lim_{i\to\infty}\frac{\ln(\mathrm{vol}(G_i\backslash S^1))}{[\pi_1(S^1):G_i]} = 0$ and $\frac{\chi(H_i\backslash N)}{[\pi_1(N):H_i]} = \chi(N)$ hold the claim follows.

Chapter 14

14.1. Let σ_1, σ_2, ..., σ_m be the n-simplices in an oriented triangulation. Denote by σ'_1, σ'_2, ..., σ'_m the associated singular n-simplices. Then $||M||$ is represented by $\sum_{i=1}^m \sigma'_i$ and $||M|| \le ||\sum_{i=1}^m \sigma'_i||_1 = m$.

14.2. Let $u = \sum_{\sigma\in I}\lambda_\sigma\cdot\sigma$ for some subset $I\subset S_1(\mathbb{R})$ be a locally finite singular 1-chain which is in the kernel of the first differential and represents $[\mathbb{R}]\in H_1^{\mathrm{lf}}(\mathbb{R};\mathbb{Z})$. We can arrange without changing the properties above and its L^1-norm that each singular simplex $\sigma\colon[0,1]\to\mathbb{R}$ appearing in I is an affine orientation preserving embedding and $\lambda_\sigma\ne 0$ holds for $\sigma\in I$. Notice that I must be countable. Since u represents $[\mathbb{R}]$, we get for any $x\in\mathbb{R}$ such that x does not meet $\sigma(\{0,1\})$ for $\sigma\in I$

$$\sum_{\substack{\sigma,\\x\in\sigma(0,1)}}\lambda_\sigma = 1.$$

Since u is locally finite, we can choose a sequence $(x_n)_{n\ge 0}$ of real numbers such that for any $\sigma\in I$ and $n\ge 0$ we have $x_n\notin\sigma(\{0,1\})$ and for any $\sigma\in I$ the implication $x_m, x_n\in\mathrm{im}(\sigma)\Rightarrow n = m$ holds . Then we get for an integer $k > 0$

$$\left\|\sum_{\sigma\in I}\lambda_\sigma\cdot\sigma\right\|_1 = \sum_{\sigma\in I}|\lambda_\sigma| \ge \sum_{i=1}^k\sum_{\substack{\sigma\in I,\\x_i\in\sigma(0,1)}}|\lambda_\sigma| \ge \sum_{i=1}^k 1 = k.$$

14.3. Since f is proper it induces a chain map $C_*^{\mathrm{lf}}(f)\colon C_*^{\mathrm{lf}}(M;\mathbb{R})\to C_*^{\mathrm{lf}}(N;\mathbb{R})$. This chain map preserves the L^1-norm. Since f has degree d, we get $H_n^{\mathrm{lf}}(f)([M]) = d\cdot[N]$. This implies $||M||\ge d\cdot||N||$.

There exists a proper selfmap $\mathbb{R}^n\to\mathbb{R}^n$ of degree $d\ge 2$ provided that $n\ge 2$.

14.4.   Consider $\phi \in \widehat{C^0}(X)$ with $\delta^0(\phi) = 0$. Then $\phi: S_0(X) \to \mathbb{R}$ is the constant function since X is path-connected by assumption.

14.5.   The condition that X is a simplicial complex (and not only a CW-complex) ensures that $l^\infty C^*(X)$ is indeed a subcomplex.

If X is finite, $l^\infty C^*(X) = C^*(X)$, but in general $H^*(X; \mathbb{R})$ and $\widehat{H}^*(X)$ are different for an finite simplicial complex X.

14.6.   This follows from 13.3 (1) and (2), Lemma 14.13 and Theorem 14.27.

14.7.   The cochain ϕ is the image under the de Rham cochain map of the element which is given by the volume form in the de Rham complex multiplied with $\mathrm{vol}(M)^{-1}$. Hence it is a cocycle which represents the image of the cohomological fundamental class of M under the change of rings map $H^n(M; \mathbb{Z}) \to H^n(M; \mathbb{R})$ (see Theorem 1.47).

Choose an orientation preserving embedding $i: S^1 \times [0,3]^{n-1} \to M$. Let $p_n : S^1 \times [0,3]^{n-1} \to S^1 \times [0,3]^{n-1}$ be the map which sends $(z, t_1, \ldots, t_{n-1})$ to $(z^n, t_1, \ldots, t_{n-1})$. Let $\sigma: \Delta_n \to S^1 \times [0,3]^{n-1}$ be an orientation preserving embedding whose image contains $S^1 \times [1,2]^{n-1}$. Then

$$|\phi(i \circ p_n \circ \sigma)| = \int_{\Delta_p} (i \circ p_n \circ \sigma)^* \, d\mathrm{vol}_M$$

$$\geq \int_{S^1 \times [1,2]^{n-1}} (i \circ p_n)^* \, d\mathrm{vol}_M$$

$$= n \cdot \int_{S^1 \times [1,2]^{n-1}} i^* \, d\mathrm{vol}_M .$$

Since $\int_{S^1 \times [1,2]^{n-1}} i^* \, d\mathrm{vol}_M > 0$ and n can be chosen arbitrary large, we conclude $||\phi||_\infty = \infty$.

14.8.   The chain map $C_p^{\mathrm{sing}}(f): C_p^{\mathrm{sing}}(X; \mathbb{R}) \to C_p^{\mathrm{sing}}(Y; \mathbb{R})$ is isometric with respect to the L^1-norm. The cochain map $C_{\mathrm{sing}}^p(f): C_{\mathrm{sing}}^p(Y; \mathbb{R}) \to C_{\mathrm{sing}}^p(X; \mathbb{R})$ is norm decreasing with respect to the L^∞-norm.

14.9.   This follows from Theorem 14.46 using the Euler Poincaré formula (see Theorem 1.35 (2)), the obvious inequality

$$\mathrm{sign}^{(2)}(\widetilde{M}) \leq b_{2m}^{(2)}(\widetilde{M})$$

for $\dim(M) = 4m$ and the equality

$$\mathrm{sign}^{(2)}(\widetilde{M}) = \mathrm{sign}(M)$$

which is a conclusion of Atiyah's L^2-index theorem (see [9]) applied to the signature operator.

14.10.   See [238, Example 5.13 on page 281].

14.11.   Denote by $p\colon \overline{N} \to N$ the covering associated to the subgroup $\mathrm{im}(\pi_1(f))$ of $\pi_1(N)$. Then we can find a map $\overline{f}\colon M \to \overline{N}$ satisfying $f = p \circ \overline{f}$. Since f has degree one, $H_n(p)\colon H_n(\overline{N}) \to H_n(N)$ must be surjective for $n = \dim(M)$. Since $H_n(\overline{N}) = 0$ for non-compact $\overline{N}$, the space $\overline{N}$ must be compact and hence the index of $\mathrm{im}(\pi_1(f))$ in $\pi_1(N)$ is finite. Since this index is the degree of p, it must be 1 and hence $\pi_1(f)$ must be surjective. Now apply Lemma 14.10 (1) and Theorem 14.15.

14.12.   This follows from Corollary 1.43 and Theorem 14.32.

14.13.   This follows from Theorem 14.26 and Theorem 14.32.

14.14.   Let F_g be a closed orientable surface with $g \geq 2$. Then $b_p^{(2)}(\widetilde{S^3 \times F_g}) \neq 0$ for $p = 1, 4$ by Example 1.32, Theorem 1.35 (4) and Example 1.36. Since S^3 and hence $S^3 \times F_g$ carries a free S^1-action, the minimal volume of $S^3 \times F_g$ vanishes by Theorem 14.32.

14.15.   Since the quotient of an amenable group is amenable again, the map $\pi_1(E) \to \pi_1(B)$ is surjective with an amenable group as kernel. Hence p induces an isometric isomorphism $\widehat{H}^{\dim(E)}(B) \xrightarrow{\cong} \widehat{H}^{\dim(E)}(E)$ by Theorem 14.15.

Recall that for any closed oriented manifold M of dimension n the n-th homology $H_n(M; \mathbb{Z})$ is $\mathbb{Z}$. Suppose that the $\pi_1(B)$-action on $H_{\dim(F)}(F; \mathbb{Z})$ given by the fiber transport is trivial. A spectral sequence argument shows that $H_n(E; \mathbb{Z}) = 0$ for $n > \dim(F) + \dim(B)$ and $H_n(E; \mathbb{Z}) = \mathbb{Z}$ for $n = \dim(F) + \dim(B)$. This implies $\dim(E) = \dim(F) + \dim(B)$. If the $\pi_1(B)$-action is non-trivial, pass to the 2-sheeted covering $\overline{B} \to B$ given by the subgroup of elements in $\pi_1(B)$ which act trivially on $H_{\dim(F)}(F; \mathbb{Z})$ and apply the same argument to the pullback fibration $F \to \overline{E} \to \overline{B}$. Hence we get in all cases $\dim(B) < \dim(E)$ since we assume $\dim(F) \geq 1$. We conclude that the map $H_{\dim(E)}(p) : H_{\dim(E)}(E; \mathbb{Z}) \to H_{\dim(E)}(B; \mathbb{Z})$ is trivial and maps in particular $[E]$ to 0.

Now $\|E\| = 0$ follows from Lemma 14.10 (1).

14.16.   Suppose that F_g carries an affine structure. Then its tangent bundle is flat. We conclude from Theorem 14.51 that $4 \cdot |\chi(F_g)| \leq \|F_g\|$. We get $\|F_g\| = 2 \cdot |\chi(F_g)|$ for $g \geq 1$ and $\|S^2\| = 0$ from Theorem 14.18 (2). Hence $\chi(F_g) = 2 - 2g$ must be zero. This shows $F_g = T^2$. Obviously T^2 carries an affine structure.

14.17.   Fix $\epsilon > 0$. Let $p_\epsilon\colon M_\epsilon \to M$ be a d_ϵ-sheeted covering and let K_ϵ be a triangulation for M_ϵ with t_ϵ simplices of dimension n such that $t_\epsilon \leq d_\epsilon \cdot \epsilon$

holds. The number of p-simplices in K_ϵ is bounded by $\binom{n+1}{p} \cdot t_\epsilon$. Since each simplex has at most $(n+1)$ faces, the norm of the p-th differential $c_p^{(2)}[\epsilon]$ in $C_*^{(2)}(\widetilde{K_\epsilon})$ is bounded by $(n+1)$ for all $p \geq 0$. Let $\Delta_p[\epsilon] \colon C_p^{(2)}(\widetilde{K_\epsilon}) \to C_p^{(2)}(\widetilde{K_\epsilon})$ be the Laplacian. We conclude for all $p \geq 0$

$$\|\Delta_p[\epsilon]\| \leq 2 \cdot (n+1)^2;$$
$$\dim_{\mathcal{N}(\pi_1(K_\epsilon))}(C_p^{(2)}(\widetilde{K_\epsilon})) \leq \binom{n+1}{p} \cdot t_\epsilon.$$

This implies for all $p \geq 0$ (see Lemma 3.15 (3) and (6))

$$\dim_{\mathcal{N}(\pi_1(K_\epsilon))}(\ker(\Delta_p[\epsilon])) \leq \binom{n+1}{p} \cdot t_\epsilon;$$
$$\det{}_{\mathcal{N}(\pi_1(K_\epsilon))}(\Delta_p[\epsilon]) \leq \binom{n+1}{p} \cdot t_\epsilon \cdot 2 \cdot (n+1)^2.$$

Hence we get for all $p \geq 0$

$$b_p^{(2)}(\widetilde{K_\epsilon}) \leq \binom{n+1}{p} \cdot t_\epsilon;$$
$$\rho^{(2)}(\widetilde{K_\epsilon}) \leq \left(\sum_{p \geq 1} p \right) \cdot \binom{n+1}{p} \cdot t_\epsilon \cdot 2 \cdot (n+1)^2.$$

Recall from Exercise 14.1 that $\|M_\epsilon\| \leq t_\epsilon$. Since $b_p^{(2)}(\widetilde{M_\epsilon}) = d_\epsilon \cdot b_p^{(2)}(\widetilde{M})$ holds by Theorem 1.35 (9) and $\|M_\epsilon\| = d_\epsilon \cdot \|M\|$ holds by Lemma 14.13, we conclude

$$b_p^{(2)}(\widetilde{M}) \leq \binom{n+1}{p} \cdot \epsilon;$$
$$\|M\| \leq \epsilon.$$

Since $\epsilon > 0$ was arbitrary, we conclude

$$b_p^{(2)}(\widetilde{M}) = 0;$$
$$\|M\| = 0.$$

If we assume that $\pi_1(M)$ is of det $\geq$ 1-class, then $\widetilde{M}$ is det-L^2-acyclic and $\rho^{(2)}(\widetilde{M})$ is given by $\rho^{(2)}(\widetilde{K})$ for any choice of triangulation (see Lemma 13.6). Because of Theorem 3.96 (5) the same is true for $\widetilde{M_\epsilon}$ and we get

$$\rho^{(2)}(\widetilde{M}) = \frac{\rho^{(2)}(\widetilde{K_\epsilon})}{d_\epsilon} \leq \left(\sum_{p \geq 1} p \right) \cdot \binom{n+1}{p} \cdot \epsilon \cdot 2 \cdot (n+1)^2.$$

Since this holds for all $\epsilon > 0$ we get $\rho^{(2)}(\widetilde{M}) = 0$.

References

1. H. Abels. A universal proper G-space. *Math. Z.*, 159(2):143–158, 1978.
2. D. Alexandru-Rugină. Harmonic forms and L_p-cohomology. *Tensor (N.S.)*, 57(2):176–191, 1996.
3. J. Amorós, M. Burger, K. Corlette, D. Kotschick, and D. Toledo. *Fundamental groups of compact Kähler manifolds*. American Mathematical Society, Providence, RI, 1996.
4. D. R. Anderson and H. J. Munkholm. *Boundedly controlled topology*. Springer-Verlag, Berlin, 1988. Foundations of algebraic topology and simple homotopy theory.
5. M. T. Anderson. L^2 harmonic forms and a conjecture of Dodziuk-Singer. *Bull. Amer. Math. Soc. (N.S.)*, 13(2):163–165, 1985.
6. M. T. Anderson. L^2 harmonic forms on complete Riemannian manifolds. In *Geometry and analysis on manifolds (Katata/Kyoto, 1987)*, pages 1–19. Springer-Verlag, Berlin, 1988.
7. N. Anghel. L^2-index formulae for perturbed Dirac operators. *Comm. Math. Phys.*, 128(1):77–97, 1990.
8. P. Ara and D. Goldstein. A solution of the matrix problem for Rickart C^*-algebras. *Math. Nachr.*, 164:259–270, 1993.
9. M. F. Atiyah. Elliptic operators, discrete groups and von Neumann algebras. *Astérisque*, 32-33:43–72, 1976.
10. M. F. Atiyah and R. Bott. The index problem for manifolds with boundary. In *Differential Analysis, Bombay Colloq., 1964*, pages 175–186. Oxford Univ. Press, London, 1964.
11. M. F. Atiyah, V. K. Patodi, and I. M. Singer. Spectral asymmetry and Riemannian geometry. I. *Math. Proc. Cambridge Philos. Soc.*, 77:43–69, 1975.
12. M. F. Atiyah, V. K. Patodi, and I. M. Singer. Spectral asymmetry and Riemannian geometry. II. *Math. Proc. Cambridge Philos. Soc.*, 78(3):405–432, 1975.
13. M. F. Atiyah, V. K. Patodi, and I. M. Singer. Spectral asymmetry and Riemannian geometry. III. *Math. Proc. Cambridge Philos. Soc.*, 79(1):71–99, 1976.
14. M. F. Atiyah and W. Schmid. A geometric construction of the discrete series for semisimple Lie groups. *Invent. Math.*, 42:1–62, 1977.
15. M. Auslander and D. A. Buchsbaum. *Groups, rings, modules*. Harper & Row Publishers, New York, 1974. Harper's Series in Modern Mathematics.
16. A. Avez. Variétés riemanniennes sans points focaux. *C. R. Acad. Sci. Paris Sér. A-B*, 270:A188–A191, 1970.
17. W. Ballmann. *Lectures on spaces of nonpositive curvature*. Birkhäuser Verlag, Basel, 1995. With an appendix by M. Brin.
18. W. Ballmann and J. Brüning. On the spectral theory of manifolds with cusps. *J. Math. Pures Appl. (9)*, 80(6):593–625, 2001.

19. W. Ballmann and P. B. Eberlein. Fundamental groups of manifolds of non-positive curvature. *J. Differential Geom.*, 25(1):1–22, 1987.

20. W. Ballmann, M. Gromov, and V. Schroeder. *Manifolds of nonpositive curvature*. Birkhäuser Boston Inc., Boston, MA, 1985.

21. D. Barbasch and H. Moscovici. L^2-index and the Selberg trace formula. *J. Funct. Anal.*, 53(2):151–201, 1983.

22. H. Bass. The degree of polynomial growth of finitely generated nilpotent groups. *Proc. London Math. Soc. (3)*, 25:603–614, 1972.

23. H. Bass. Euler characteristics and characters of discrete groups. *Invent. Math.*, 35:155–196, 1976.

24. H. Bass. Covering theory for graphs of groups. *J. Pure Appl. Algebra*, 89(1-2):3–47, 1993.

25. H. Bass, A. Heller, and R. G. Swan. The Whitehead group of a polynomial extension. *Inst. Hautes Études Sci. Publ. Math.*, (22):61–79, 1964.

26. P. Baum and A. Connes. Geometric K-theory for Lie groups and foliations. *Enseign. Math. (2)*, 46(1-2):3–42, 2000.

27. P. Baum, A. Connes, and N. Higson. Classifying space for proper actions and K-theory of group C^*-algebras. In *C^*-algebras: 1943–1993 (San Antonio, TX, 1993)*, pages 240–291. Amer. Math. Soc., Providence, RI, 1994.

28. B. Baumslag and S. J. Pride. Groups with one more generator than relators. *Math. Z.*, 167(3):279–281, 1979.

29. G. Baumslag. Automorphism groups of residually finite groups. *J. London Math. Soc.*, 38:117–118, 1963.

30. G. Baumslag and P. B. Shalen. Amalgamated products and finitely presented groups. *Comment. Math. Helv.*, 65(2):243–254, 1990.

31. G. Baumslag and D. Solitar. Some two-generator one-relator non-Hopfian groups. *Bull. Amer. Math. Soc.*, 68:199–201, 1962.

32. C. Bavard and P. Pansu. Sur le volume minimal de $\mathbb{R}^2$. *Ann. Sci. École Norm. Sup. (4)*, 19(4):479–490, 1986.

33. M. E. B. Bekka, P.-A. Cherix, and A. Valette. Proper affine isometric actions of amenable groups. In *Novikov conjectures, index theorems and rigidity, Vol. 2 (Oberwolfach, 1993)*, pages 1–4. Cambridge Univ. Press, Cambridge, 1995.

34. M. E. B. Bekka and A. Valette. Group cohomology, harmonic functions and the first L^2-Betti number. *Potential Anal.*, 6(4):313–326, 1997.

35. R. Benedetti and C. Petronio. *Lectures on hyperbolic geometry*. Springer-Verlag, Berlin, 1992.

36. S. K. Berberian. *Baer *-rings*. Springer-Verlag, New York, 1972. Die Grundlehren der mathematischen Wissenschaften, Band 195.

37. S. K. Berberian. The maximal ring of quotients of a finite von Neumann algebra. *Rocky Mountain J. Math.*, 12(1):149–164, 1982.

38. A. J. Berrick, I. Chatterji, and G. Mislin. From acyclic groups to the Bass conjecture for amenable groups. Preprint, 2001.

39. A. J. Berrick and M. E. Keating. The localization sequence in K-theory. *K-Theory*, 9(6):577–589, 1995.

40. G. Besson, G. Courtois, and S. Gallot. Entropies et rigidités des espaces localement symétriques de courbure strictement négative. *Geom. Funct. Anal.*, 5(5):731–799, 1995.

41. M. Bestvina and K. Fujiwara. Bounded cohomology of subgroups of mapping class groups. Preprint, 2000.

42. J.-M. Bismut. From Quillen metrics to Reidemeister metrics: some aspects of the Ray-Singer analytic torsion. In *Topological methods in modern mathematics (Stony Brook, NY, 1991)*, pages 273–324. Publish or Perish, Houston, TX, 1993.

43. J.-M. Bismut and D. S. Freed. The analysis of elliptic families. I. Metrics and connections on determinant bundles. *Comm. Math. Phys.*, 106(1):159–176, 1986.

44. J.-M. Bismut and D. S. Freed. The analysis of elliptic families. II. Dirac operators, eta invariants, and the holonomy theorem. *Comm. Math. Phys.*, 107(1):103–163, 1986.

45. J.-M. Bismut, H. Gillet, and C. Soulé. Analytic torsion and holomorphic determinant bundles. I. Bott-Chern forms and analytic torsion. *Comm. Math. Phys.*, 115(1):49–78, 1988.

46. J.-M. Bismut, H. Gillet, and C. Soulé. Analytic torsion and holomorphic determinant bundles. II. Direct images and Bott-Chern forms. *Comm. Math. Phys.*, 115(1):79–126, 1988.

47. J.-M. Bismut, H. Gillet, and C. Soulé. Analytic torsion and holomorphic determinant bundles. III. Quillen metrics on holomorphic determinants. *Comm. Math. Phys.*, 115(2):301–351, 1988.

48. J.-M. Bismut and J. Lott. Flat vector bundles, direct images and higher real analytic torsion. *J. Amer. Math. Soc.*, 8(2):291–363, 1995.

49. J.-M. Bismut and W. Zhang. An extension of a theorem by Cheeger and Müller. *Astérisque*, (205):235, 1992. With an appendix by F. Laudenbach.

50. J.-M. Bismut and W. Zhang. Milnor and Ray-Singer metrics on the equivariant determinant of a flat vector bundle. *Geom. Funct. Anal.*, 4(2):136–212, 1994.

51. B. Blackadar. *K-theory for operator algebras.* Springer-Verlag, New York, 1986.

52. M. Boileau, S. Druck, and E. Vogt. A vanishing theorem for the Gromov volume of 3-manifolds with an application to circle foliations. *Math. Ann.*, 322(3):493–524, 2002.

53. A. Borel. L^2-cohomology and intersection cohomology of certain arithmetic varieties. In *Emmy Noether in Bryn Mawr (Bryn Mawr, Pa., 1982)*, pages 119–131. Springer-Verlag, New York, 1983.

54. A. Borel. The L^2-cohomology of negatively curved Riemannian symmetric spaces. *Ann. Acad. Sci. Fenn. Ser. A I Math.*, 10:95–105, 1985.

55. A. Borel and W. Casselman. Cohomologie d'intersection et L^2-cohomologie de variétés arithmétiques de rang rationnel 2. *C. R. Acad. Sci. Paris Sér. I Math.*, 301(7):369–373, 1985.

56. A. Borel and N. R. Wallach. *Continuous cohomology, discrete subgroups, and representations of reductive groups.* Princeton University Press, Princeton, N.J., 1980.

57. R. Bott and L. W. Tu. *Differential forms in algebraic topology.* Springer-Verlag, New York, 1982.

58. J.-P. Bourguignon and H. Karcher. Curvature operators: pinching estimates and geometric examples. *Ann. Sci. École Norm. Sup. (4)*, 11(1):71–92, 1978.

59. J.-P. Bourguignon and A. Polombo. Intégrands des nombres caractéristiques et courbure: rien ne va plus dès la dimension 6. *J. Differential Geom.*, 16(4):537–550 (1982), 1981.

60. T. P. Branson and P. B. Gilkey. The asymptotics of the Laplacian on a manifold with boundary. *Comm. Partial Differential Equations*, 15(2):245–272, 1990.

61. C. Bratzler. L^2-*Betti Zahlen und Faserungen.* Diplomarbeit, Johannes Gutenberg-Universität Mainz, 1997. http://www.math.uni-muenster.de/u/lueck/publ/diplome/bratzler.dvi.

62. M. Braverman. Witten deformation of analytic torsion and the spectral sequence of a filtration. *Geom. Funct. Anal.*, 6(1):28–50, 1996.

63. G. Bredon. *Topology and geometry*, volume 139 of *Graduate texts in Mathematics*. Springer, 1993.

64. G. E. Bredon. *Introduction to compact transformation groups*. Academic Press, New York, 1972. Pure and Applied Mathematics, Vol. 46.

65. M. R. Bridson and A. Haefliger. *Metric spaces of non-positive curvature*. Springer-Verlag, Berlin, 1999. Die Grundlehren der mathematischen Wissenschaften, Band 319.

66. M. G. Brin and C. C. Squier. Groups of piecewise linear homeomorphisms of the real line. *Invent. Math.*, 79(3):485–498, 1985.

67. T. Bröcker and K. Jänich. *Einführung in die Differentialtopologie*. Springer-Verlag, Berlin, 1973. Heidelberger Taschenbücher, Band 143.

68. R. Brooks. The fundamental group and the spectrum of the Laplacian. *Comment. Math. Helv.*, 56(4):581–598, 1981.

69. K. Brown. *Cohomology of groups*, volume 87 of *Graduate texts in Mathematics*. Springer, 1982.

70. K. S. Brown and R. Geoghegan. An infinite-dimensional torsion-free fp_∞ group. *Invent. Math.*, 77(2):367–381, 1984.

71. J. Brüning. L^2-index theorems on certain complete manifolds. *J. Differential Geom.*, 32(2):491–532, 1990.

72. J. Brüning. On L^2-index theorems for complete manifolds of rank-one-type. *Duke Math. J.*, 66(2):257–309, 1992.

73. J. Brüning and M. Lesch. Hilbert complexes. *J. Funct. Anal.*, 108(1):88–132, 1992.

74. U. Bunke. Equivariant torsion and G-CW-complexes. *Geom. Funct. Anal.*, 9(1):67–89, 1999.

75. G. Burde and H. Zieschang. *Knots*. Walter de Gruyter & Co., Berlin, 1985.

76. M. Burger and N. Monod. Bounded cohomology of lattices in higher rank Lie groups. *J. Eur. Math. Soc. (JEMS)*, 1(2):199–235, 1999.

77. M. Burger and N. Monod. Continuous bounded cohomology and applications to rigidity theory. preprint, to appear in GAFA, 2001.

78. M. Burger and A. Valette. Idempotents in complex group rings: theorems of Zalesskii and Bass revisited. *J. Lie Theory*, 8(2):219–228, 1998.

79. D. Burghelea, L. Friedlander, and T. Kappeler. Meyer-Vietoris type formula for determinants of elliptic differential operators. *J. Funct. Anal.*, 107(1):34–65, 1992.

80. D. Burghelea, L. Friedlander, and T. Kappeler. Asymptotic expansion of the Witten deformation of the analytic torsion. *J. Funct. Anal.*, 137(2):320–363, 1996.

81. D. Burghelea, L. Friedlander, and T. Kappeler. Witten deformation of the analytic torsion and the Reidemeister torsion. In *Voronezh Winter Mathematical Schools*, pages 23–39. Amer. Math. Soc., Providence, RI, 1998. Trans. Ser. 2 184 (1998).

82. D. Burghelea, L. Friedlander, and T. Kappeler. Torsions for manifolds with boundary and glueing formulas. *Math. Nachr.*, 208:31–91, 1999.

83. D. Burghelea, L. Friedlander, and T. Kappeler. Relative torsion. *Commun. Contemp. Math.*, 3(1):15–85, 2001.

84. D. Burghelea, L. Friedlander, T. Kappeler, and P. McDonald. Analytic and Reidemeister torsion for representations in finite type Hilbert modules. *Geom. Funct. Anal.*, 6(5):751–859, 1996.

85. J. W. Cannon, W. J. Floyd, and W. R. Parry. Introductory notes on Richard Thompson's groups. *Enseign. Math. (2)*, 42(3-4):215–256, 1996.

86. J. Cao and F. Xavier. Kähler parabolicity and the Euler number of compact manifolds of non-positive sectional curvature. *Math. Ann.*, 319(3):483–491, 2001.

87. S. E. Cappell. On connected sums of manifolds. *Topology*, 13:395–400, 1974.

88. S. E. Cappell. A splitting theorem for manifolds. *Invent. Math.*, 33(2):69–170, 1976.

89. S. E. Cappell and J. L. Shaneson. Nonlinear similarity. *Ann. of Math. (2)*, 113(2):315–355, 1981.

90. S. E. Cappell and J. L. Shaneson. On 4-dimensional s-cobordisms. *J. Differential Geom.*, 22(1):97–115, 1985.

91. S. E. Cappell, J. L. Shaneson, M. Steinberger, S. Weinberger, and J. E. West. The classification of nonlinear similarities over Z_{2^r}. *Bull. Amer. Math. Soc. (N.S.)*, 22(1):51–57, 1990.

92. A. L. Carey, M. Farber, and V. Mathai. Determinant lines, von Neumann algebras and L^2 torsion. *J. Reine Angew. Math.*, 484:153–181, 1997.

93. A. L. Carey and V. Mathai. L^2-torsion invariants. *J. Funct. Anal.*, 110(2):377–409, 1992.

94. H. Cartan and S. Eilenberg. *Homological algebra*. Princeton University Press, Princeton, N. J., 1956.

95. W. Casselman. Introduction to the L^2-cohomology of arithmetic quotients of bounded symmetric domains. In *Complex analytic singularities*, pages 69–93. North-Holland, Amsterdam, 1987.

96. A. J. Casson and S. A. Bleiler. *Automorphisms of surfaces after Nielsen and Thurston*. Cambridge University Press, Cambridge, 1988.

97. A. J. Casson and C. M. Gordon. On slice knots in dimension three. In *Algebraic and geometric topology (Proc. Sympos. Pure Math., Stanford Univ., Stanford, Calif., 1976), Part 2*, pages 39–53. Amer. Math. Soc., Providence, R.I., 1978.

98. S. Chang and S. Weinberger. On invariants of Hirzebruch and Cheeger-Gromov. preprint, 2001.

99. T. A. Chapman. Compact Hilbert cube manifolds and the invariance of White-head torsion. *Bull. Amer. Math. Soc.*, 79:52–56, 1973.

100. T. A. Chapman. Topological invariance of Whitehead torsion. *Amer. J. Math.*, 96:488–497, 1974.

101. M. Chayet and N. Lohoue. Sur la cohomologie L^p des variétés. *C. R. Acad. Sci. Paris Sér. I Math.*, 324(2):211–213, 1997.

102. J. Cheeger. Finiteness theorems for Riemannian manifolds. *Amer. J. Math.*, 92:61–74, 1970.

103. J. Cheeger. Analytic torsion and the heat equation. *Ann. of Math. (2)*, 109(2):259–322, 1979.

104. J. Cheeger, M. Goresky, and R. MacPherson. L^2-cohomology and intersection homology of singular algebraic varieties. In *Seminar on Differential Geometry*, pages 303–340. Princeton Univ. Press, Princeton, N.J., 1982.

105. J. Cheeger and M. Gromov. Bounds on the von Neumann dimension of L^2-cohomology and the Gauss-Bonnet theorem for open manifolds. *J. Differential Geom.*, 21(1):1–34, 1985.

106. J. Cheeger and M. Gromov. On the characteristic numbers of complete manifolds of bounded curvature and finite volume. In *Differential geometry and complex analysis*, pages 115–154. Springer-Verlag, Berlin, 1985.

107. J. Cheeger and M. Gromov. L_2-cohomology and group cohomology. *Topology*, 25(2):189–215, 1986.

108. S.-s. Chern. On curvature and characteristic classes of a Riemann manifold. *Abh. Math. Sem. Univ. Hamburg*, 20:117–126, 1955.

109. P. R. Chernoff. Essential self-adjointness of powers of generators of hyperbolic equations. *J. Functional Analysis*, 12:401–414, 1973.
110. B. Clair. Residual amenability and the approximation of L^2-invariants. *Michigan Math. J.*, 46(2):331–346, 1999.
111. B. Clair and K. Whyte. Growth of Betti numbers. Preprint, 2001.
112. G. Cliff and A. Weiss. Moody's induction theorem. *Illinois J. Math.*, 32(3):489–500, 1988.
113. T. Cochran, K. Orr, and P. Teichner. Knot concordance, Whitney towers and L^2-signatures. Preprint, to appear in *Ann. of Math.*, 1999.
114. D. E. Cohen. *Combinatorial group theory: a topological approach*. Cambridge University Press, Cambridge, 1989.
115. J. M. Cohen. von Neumann dimension and the homology of covering spaces. *Quart. J. Math. Oxford Ser. (2)*, 30(118):133–142, 1979.
116. M. M. Cohen. *A course in simple-homotopy theory*. Springer-Verlag, New York, 1973. Graduate Texts in Mathematics, Vol. 10.
117. P. M. Cohn. Free ideal rings. *J. Algebra*, 1:47–69, 1964.
118. P. M. Cohn. *Free rings and their relations*. Academic Press Inc. [Harcourt Brace Jovanovich Publishers], London, second edition, 1985.
119. P. M. Cohn. *Algebra. Vol. 3*. John Wiley & Sons Ltd., Chichester, second edition, 1991.
120. A. Connes. *Noncommutative geometry*. Academic Press Inc., San Diego, CA, 1994.
121. A. Connes and H. Moscovici. The L^2-index theorem for homogeneous spaces. *Bull. Amer. Math. Soc. (N.S.)*, 1(4):688–690, 1979.
122. A. Connes and H. Moscovici. The L^2-index theorem for homogeneous spaces of Lie groups. *Ann. of Math. (2)*, 115(2):291–330, 1982.
123. A. Connes and H. Moscovici. L^2-index theory on homogeneous spaces and discrete series representations. In *Operator algebras and applications, Part I (Kingston, Ont., 1980)*, pages 419–433. Amer. Math. Soc., Providence, R.I., 1982.
124. P. Connor and F. Raymond. Actions of compact Lie groups on aspherical manifolds. In *Topology of manifolds*, pages 227–264. Markham, Chicago, 1970. Cantrell and Edwards.
125. C. W. Curtis and I. Reiner. *Methods of representation theory. Vol. II*. John Wiley & Sons Inc., New York, 1987. With applications to finite groups and orders, A Wiley-Interscience Publication.
126. X. Dai and H. Fang. Analytic torsion and R-torsion for manifolds with boundary. *Asian J. Math.*, 4(3):695–714, 2000.
127. X. Dai and R. Melrose. Adiabatic limit of the analytic torsion. Preprint.
128. J. F. Davis and W. Lück. Spaces over a category and assembly maps in isomorphism conjectures in K- and L-theory. *K-Theory*, 15(3):201–252, 1998.
129. M. Davis and I. Leary. The L^2-cohomology of Artin groups. Preprint, 2001.
130. M. W. Davis and B. Okun. Vanishing theorems and conjectures for the ℓ^2-homology of right-angled Coxeter groups. *Geom. Topol.*, 5:7–74 (electronic), 2001.
131. D. L. de George and N. R. Wallach. Limit formulas for multiplicities in $L^2(\gamma \backslash G)$. *Ann. Math. (2)*, 107(1):133–150, 1978.
132. P. de la Harpe. *Topics in geometric group theory*. University of Chicago Press, Chicago, IL, 2000.
133. G. de Rham. Reidemeister's torsion invariant and rotations of S^n. In *Differential Analysis, Bombay Colloq.*, pages 27–36. Oxford Univ. Press, London, 1964.

134. G. de Rham. *Differentiable manifolds*. Springer-Verlag, Berlin, 1984. Forms, currents, harmonic forms, Translated from the French by F. R. Smith, With an introduction by S. S. Chern.

135. D. L. DeGeorge and N. R. Wallach. Limit formulas for multiplicities in $L^2(\gamma\backslash G)$. II. The tempered spectrum. *Ann. of Math. (2)*, 109(3):477–495, 1979.

136. A. Deitmar. Higher torsion zeta functions. *Adv. Math.*, 110(1):109–128, 1995.

137. A. Deitmar. A determinant formula for the generalized Selberg zeta function. *Quart. J. Math. Oxford Ser. (2)*, 47(188):435–453, 1996.

138. A. Deitmar. Equivariant torsion of locally symmetric spaces. *Pacific J. Math.*, 182(2):205–227, 1998.

139. A. Deitmar. Regularized and L^2-determinants. *Proc. London Math. Soc. (3)*, 76(1):150–174, 1998.

140. T. Delzant. Sur l'anneau d'un groupe hyperbolique. *C. R. Acad. Sci. Paris Sér. I Math.*, 324(4):381–384, 1997.

141. W. Dicks and P. H. Kropholler. Free groups and almost equivariant maps. *Bull. London Math. Soc.*, 27(4):319–326, 1995.

142. W. Dicks and P. Menal. The group rings that are semifirs. *J. London Math. Soc. (2)*, 19(2):288–290, 1979.

143. W. Dicks and T. Schick. The spectral measure of certain elements of the complex group ring of a wreath product. Preprintreihe SFB 478 — Geometrische Strukturen in der Mathematik, Heft 159, Münster, to appear in Geometriae Dedicata, 2001.

144. J. Dixmier. *von Neumann algebras*. North-Holland Publishing Co., Amsterdam, 1981. With a preface by E. C. Lance, Translated from the second French edition by F. Jellett.

145. J. Dodziuk. de Rham-Hodge theory for L^2-cohomology of infinite coverings. *Topology*, 16(2):157–165, 1977.

146. J. Dodziuk. L^2 harmonic forms on rotationally symmetric Riemannian manifolds. *Proc. Amer. Math. Soc.*, 77(3):395–400, 1979.

147. J. Dodziuk. Sobolev spaces of differential forms and de Rham-Hodge isomorphism. *J. Differential Geom.*, 16(1):63–73, 1981.

148. J. Dodziuk, P. A. Linnell, V. Mathai, T. Schick, and S. Yates. Approximating L^2-invariants, and the Atiyah conjecture. Preprintreihe SFB 478 — Geometrische Strukturen in der Mathematik, Heft 170, Münster, 2001.

149. J. Dodziuk and V. Mathai. Approximating L^2 invariants of amenable covering spaces: a combinatorial approach. *J. Funct. Anal.*, 154(2):359–378, 1998.

150. J. Dodziuk and V. K. Patodi. Riemannian structures and triangulations of manifolds. *J. Indian Math. Soc. (N.S.)*, 40(1-4):1–52 (1977), 1976.

151. S. K. Donaldson. Irrationality and the h-cobordism conjecture. *J. Differential Geom.*, 26(1):141–168, 1987.

152. H. Donnelly. On L^2-Betti numbers for abelian groups. *Canad. Math. Bull.*, 24(1):91–95, 1981.

153. H. Donnelly. On the spectrum of towers. *Proc. Amer. Math. Soc.*, 87(2):322–329, 1983.

154. H. Donnelly and C. Fefferman. L^2-cohomology and index theorem for the Bergman metric. *Ann. of Math. (2)*, 118(3):593–618, 1983.

155. H. Donnelly and F. Xavier. On the differential form spectrum of negatively curved Riemannian manifolds. *Amer. J. Math.*, 106(1):169–185, 1984.

156. J. L. Dupont. *Curvature and characteristic classes*. Springer-Verlag, Berlin, 1978. Lecture Notes in Mathematics, Vol. 640.

157. P. B. Eberlein. *Geometry of nonpositively curved manifolds*. University of Chicago Press, Chicago, IL, 1996.

158. P. B. Eberlein and J. Heber. A differential geometric characterization of symmetric spaces of higher rank. *Inst. Hautes Études Sci. Publ. Math.*, (71):33–44, 1990.

159. B. Eckmann. Cyclic homology of groups and the Bass conjecture. *Comment. Math. Helv.*, 61(2):193–202, 1986.

160. B. Eckmann. Amenable groups and Euler characteristic. *Comment. Math. Helv.*, 67(3):383–393, 1992.

161. B. Eckmann. Projective and Hilbert modules over group algebras, and finitely dominated spaces. *Comment. Math. Helv.*, 71(3):453–462, 1996.

162. B. Eckmann. 4-manifolds, group invariants, and l_2-Betti numbers. *Enseign. Math. (2)*, 43(3-4):271–279, 1997.

163. B. Eckmann. Approximating l_2-Betti numbers of an amenable covering by ordinary Betti numbers. *Comment. Math. Helv.*, 74(1):150–155, 1999.

164. B. Eckmann. Introduction to l_2-methods in topology: reduced l_2-homology, harmonic chains, l_2-Betti numbers. *Israel J. Math.*, 117:183–219, 2000. Notes prepared by G. Mislin.

165. B. Eckmann. Idempotents in a complex group algebra, projective modules, and the von Neumann algebra. *Arch. Math. (Basel)*, 76(4):241–249, 2001.

166. E. G. Effros. Why the circle is connected: an introduction to quantized topology. *Math. Intelligencer*, 11(1):27–34, 1989.

167. A. Efremov. Combinatorial and analytic Novikov Shubin invariants. Preprint, 1991.

168. J. Eichhorn. L_2-cohomology. In *Geometric and algebraic topology*, pages 215–237. PWN, Warsaw, 1986.

169. G. Elek. Abelian coverings. *J. Funct. Anal.*, 141(2):365–373, 1996.

170. G. Elek. Coarse cohomology and l_p-cohomology. *K-Theory*, 13(1):1–22, 1998.

171. G. Elek. Amenable groups, topological entropy and Betti numbers. Preprint, Alfred Renyi Mathematical Institute of the Hungarian Academy of Sciences, to appear in *Israel J. Math.*, 1999.

172. G. Elek. On the analytic zero-divisor conjecture of Linnel. Preprint, 2001.

173. G. Elek. The rank of finitely generated modules over group algebras. Preprint, 2001.

174. K. D. Elworthy, X.-M. Li, and S. Rosenberg. Bounded and L^2 harmonic forms on universal covers. *Geom. Funct. Anal.*, 8(2):283–303, 1998.

175. I. Emmanouil. On a class of groups satisfying Bass' conjecture. *Invent. Math.*, 132(2):307–330, 1998.

176. I. Emmanouil. Solvable groups and Bass' conjecture. *C. R. Acad. Sci. Paris Sér. I Math.*, 326(3):283–287, 1998.

177. D. B. A. Epstein. Finite presentations of groups and 3-manifolds. *Quart. J. Math. Oxford Ser. (2)*, 12:205–212, 1961.

178. D. B. A. Epstein and K. Fujiwara. The second bounded cohomology of word-hyperbolic groups. *Topology*, 36(6):1275–1289, 1997.

179. T. Fack and P. de la Harpe. Sommes de commutateurs dans les algèbres de von Neumann finies continues. *Ann. Inst. Fourier (Grenoble)*, 30(3):49–73, 1980.

180. M. Falk and R. Randell. Pure braid groups and products of free groups. In *Braids (Santa Cruz, CA, 1986)*, pages 217–228. Amer. Math. Soc., Providence, RI, 1988.

181. M. Farber. Singularities of the analytic torsion. *J. Differential Geom.*, 41(3):528–572, 1995.

182. M. Farber. Homological algebra of Novikov-Shubin invariants and Morse inequalities. *Geom. Funct. Anal.*, 6(4):628–665, 1996.

183. M. Farber. Geometry of growth: approximation theorems for L^2 invariants. *Math. Ann.*, 311(2):335–375, 1998.

184. M. Farber. Novikov-Shubin signatures. I. *Ann. Global Anal. Geom.*, 18(5):477–515, 2000.

185. M. Farber. von Neumann Betti numbers and Novikov type inequalities. *Proc. Amer. Math. Soc.*, 128(9):2819–2827, 2000.

186. M. Farber. Novikov-Shubin signatures. II. *Ann. Global Anal. Geom.*, 19(3):259–291, 2001.

187. M. Farber and S. Weinberger. On the zero-in-the-spectrum conjecture. *Ann. of Math. (2)*, 154(1):139–154, 2001.

188. D. R. Farkas and P. A. Linnell. Zero divisors in group rings: something old, something new. In *Representation theory, group rings, and coding theory*, pages 155–166. Amer. Math. Soc., Providence, RI, 1989.

189. F. T. Farrell. The obstruction to fibering a manifold over a circle. *Indiana Univ. Math. J.*, 21:315–346, 1971/1972.

190. F. T. Farrell and L. E. Jones. K-theory and dynamics. I. *Ann. of Math. (2)*, 124(3):531–569, 1986.

191. F. T. Farrell and L. E. Jones. K-theory and dynamics. II. *Ann. of Math. (2)*, 126(3):451–493, 1987.

192. F. T. Farrell and L. E. Jones. Rigidity and other topological aspects of compact nonpositively curved manifolds. *Bull. Amer. Math. Soc. (N.S.)*, 22(1):59–64, 1990.

193. F. T. Farrell and L. E. Jones. Stable pseudoisotopy spaces of compact nonpositively curved manifolds. *J. Differential Geom.*, 34(3):769–834, 1991.

194. F. T. Farrell and L. E. Jones. Isomorphism conjectures in algebraic K-theory. *J. Amer. Math. Soc.*, 6(2):249–297, 1993.

195. F. T. Farrell and L. E. Jones. Topological rigidity for compact non-positively curved manifolds. In *Differential geometry: Riemannian geometry (Los Angeles, CA, 1990)*, pages 229–274. Amer. Math. Soc., Providence, RI, 1993.

196. F. T. Farrell and L. E. Jones. Rigidity for aspherical manifolds with $\pi_1 \subset GL_m(\mathbb{R})$. *Asian J. Math.*, 2(2):215–262, 1998.

197. F. T. Farrell and P. A. Linnell. Whitehead groups and the Bass conjecture. Preprint, 2000.

198. A. Fel'shtyn. Dynamical zeta functions, Nielsen theory and Reidemeister torsion. *Mem. Amer. Math. Soc.*, 147(699):xii+146, 2000.

199. S. C. Ferry, A. A. Ranicki, and J. Rosenberg. A history and survey of the Novikov conjecture. In *Novikov conjectures, index theorems and rigidity, Vol. 1 (Oberwolfach, 1993)*, pages 7–66. Cambridge Univ. Press, Cambridge, 1995.

200. R. H. Fox. Free differential calculus. I. Derivation in the free group ring. *Ann. of Math. (2)*, 57:547–560, 1953.

201. W. Franz. Über die Torsion einer Überdeckung. *J. Reine Angew. Math.*, 173:245–254, 1935.

202. M. H. Freedman. The topology of four-dimensional manifolds. *J. Differential Geom.*, 17(3):357–453, 1982.

203. M. H. Freedman. The disk theorem for four-dimensional manifolds. In *Proceedings of the International Congress of Mathematicians, Vol. 1, 2 (Warsaw, 1983)*, pages 647–663, Warsaw, 1984. PWN.

204. M. H. Freedman and P. Teichner. 4-manifold topology. I. Subexponential groups. *Invent. Math.*, 122(3):509–529, 1995.

205. P. Freyd. *Abelian categories. An introduction to the theory of functors.* Harper & Row Publishers, New York, 1964.

206. D. Fried. Analytic torsion and closed geodesics on hyperbolic manifolds. *Invent. Math.*, 84(3):523–540, 1986.

207. D. Fried. Torsion and closed geodesics on complex hyperbolic manifolds. *Invent. Math.*, 91(1):31–51, 1988.
208. B. Fuglede and R. V. Kadison. Determinant theory in finite factors. *Ann. of Math. (2)*, 55:520–530, 1952.
209. K. Fujiwara. The second bounded cohomology of a group acting on a Gromov-hyperbolic space. *Proc. London Math. Soc. (3)*, 76(1):70–94, 1998.
210. K. Fujiwara. The second bounded cohomology of an amalgamated free product of groups. *Trans. Amer. Math. Soc.*, 352(3):1113–1129, 2000.
211. A. Furman. Gromov's measure equivalence and rigidity of higher rank lattices. *Ann. of Math. (2)*, 150(3):1059–1081, 1999.
212. A. Furman. Orbit equivalence rigidity. *Ann. of Math. (2)*, 150(3):1083–1108, 1999.
213. D. Gabai. Convergence groups are Fuchsian groups. *Bull. Amer. Math. Soc. (N.S.)*, 25(2):395–402, 1991.
214. D. Gaboriau. Invariants l^2 de relation d'équivalence et de groupes. Preprint, Lyon, 2001.
215. D. Gaboriau. On orbit equivalence of measure preserving actions. Preprint, Lyon, 2001.
216. M. P. Gaffney. The harmonic operator for exterior differential forms. *Proc. Nat. Acad. Sci. U. S. A.*, 37:48–50, 1951.
217. M. P. Gaffney. A special Stokes's theorem for complete Riemannian manifolds. *Ann. of Math. (2)*, 60:140–145, 1954.
218. S. Gallot, D. Hulin, and J. Lafontaine. *Riemannian geometry*. Springer-Verlag, Berlin, 1987.
219. R. Geroch. Positive sectional curvatures does not imply positive Gauss-Bonnet integrand. *Proc. Amer. Math. Soc.*, 54:267–270, 1976.
220. P. B. Gilkey. *Invariance theory, the heat equation, and the Atiyah-Singer index theorem*. Publish or Perish Inc., Wilmington, DE, 1984.
221. H. Gillet and C. Soulé. Analytic torsion and the arithmetic Todd genus. *Topology*, 30(1):21–54, 1991. With an appendix by D. Zagier.
222. V. M. Gol'dshteĭn, V. I. Kuz'minov, and I. A. Shvedov. The de Rham isomorphism of the L_p-cohomology of noncompact Riemannian manifolds. *Sibirsk. Mat. Zh.*, 29(2):34–44, 216, 1988.
223. V. M. Gol'dshteĭn, V. I. Kuz'minov, and I. A. Shvedov. On Cheeger's theorem: extensions to L_p-cohomology of warped cylinders. *Siberian Adv. Math.*, 2(1):114–122, 1992.
224. K. R. Goodearl. *von Neumann regular rings*. Pitman (Advanced Publishing Program), Boston, Mass., 1979.
225. C. M. Gordon. Dehn surgery and satellite knots. *Trans. Amer. Math. Soc.*, 275(2):687–708, 1983.
226. M. J. Greenberg. *Lectures on algebraic topology*. W. A. Benjamin, Inc., New York-Amsterdam, 1967.
227. R. I. Grigorchuk. Some results on bounded cohomology. In *Combinatorial and geometric group theory (Edinburgh, 1993)*, pages 111–163. Cambridge Univ. Press, Cambridge, 1995.
228. R. I. Grigorchuk. An example of a finitely presented amenable group that does not belong to the class EG. *Mat. Sb.*, 189(1):79–100, 1998.
229. R. I. Grigorchuk, P. A. Linnell, T. Schick, and A. Żuk. On a question of Atiyah. *C. R. Acad. Sci. Paris Sér. I Math.*, 331(9):663–668, 2000.
230. R. I. Grigorchuk and A. Żuk. The Lamplighter group as a group generated by a 2-state automaton, and its spectrum. *Geom. Dedicata*, 87(1-3):209–244, 2001.

231. M. Gromov. Groups of polynomial growth and expanding maps. *Inst. Hautes Études Sci. Publ. Math.*, (53):53–73, 1981.

232. M. Gromov. Volume and bounded cohomology. *Inst. Hautes Études Sci. Publ. Math.*, (56):5–99 (1983), 1982.

233. M. Gromov. Large Riemannian manifolds. In *Curvature and topology of Riemannian manifolds (Katata, 1985)*, pages 108–121. Springer-Verlag, Berlin, 1986.

234. M. Gromov. Hyperbolic groups. In *Essays in group theory*, pages 75–263. Springer-Verlag, New York, 1987.

235. M. Gromov. Sur le groupe fondamental d'une variété kählérienne. *C. R. Acad. Sci. Paris Sér. I Math.*, 308(3):67–70, 1989.

236. M. Gromov. Kähler hyperbolicity and L_2-Hodge theory. *J. Differential Geom.*, 33(1):263–292, 1991.

237. M. Gromov. Asymptotic invariants of infinite groups. In *Geometric group theory, Vol. 2 (Sussex, 1991)*, pages 1–295. Cambridge Univ. Press, Cambridge, 1993.

238. M. Gromov. *Metric structures for Riemannian and non-Riemannian spaces*. Birkhäuser Boston Inc., Boston, MA, 1999. Based on the 1981 French original [MR 85e:53051], With appendices by M. Katz, P. Pansu and S. Semmes, Translated from the French by S. M. Bates.

239. M. Gromov and H. B. Lawson, Jr. Positive scalar curvature and the Dirac operator on complete Riemannian manifolds. *Inst. Hautes Études Sci. Publ. Math.*, (58):83–196 (1984), 1983.

240. M. Gromov and M. A. Shubin. von Neumann spectra near zero. *Geom. Funct. Anal.*, 1(4):375–404, 1991.

241. K. W. Gruenberg. Residual properties of infinite soluble groups. *Proc. London Math. Soc. (3)*, 7:29–62, 1957.

242. A. Guichardet. *Cohomologie des groupes topologiques et des algèbres de Lie*. CEDIC, Paris, 1980.

243. U. Haagerup and H. J. Munkholm. Simplices of maximal volume in hyperbolic n-space. *Acta Math.*, 147(1-2):1–11, 1981.

244. I. Hambleton and E. K. Pedersen. Non-linear similarity revisited. In *Prospects in topology (Princeton, NJ, 1994)*, pages 157–174. Princeton Univ. Press, Princeton, NJ, 1995.

245. I. Hambleton and E. K. Pedersen. Topological equivalence of linear representations for cyclic groups. Preprint, MPI Bonn, 1997.

246. I. Hambleton, L. R. Taylor, and E. B. Williams. On $G_n(RG)$ for G a finite nilpotent group. *J. Algebra*, 116(2):466–470, 1988.

247. D. Handelman. Rickart C^*-algebras. II. *Adv. in Math.*, 48(1):1–15, 1983.

248. Harish-Chandra. Harmonic analysis on real reductive groups. III. The Maass-Selberg relations and the Plancherel formula. *Ann. of Math. (2)*, 104(1):117–201, 1976.

249. R. Hartshorne. *Algebraic geometry*. Springer-Verlag, New York, 1977. Graduate Texts in Mathematics, No. 52.

250. J.-C. Hausmann and S. Weinberger. Caractéristiques d'Euler et groupes fondamentaux des variétés de dimension 4. *Comment. Math. Helv.*, 60(1):139–144, 1985.

251. S. Helgason. *Differential geometry, Lie groups and symmetric spaces*. Academic Press, 1978.

252. J. Hempel. *3-Manifolds*. Princeton University Press, Princeton, N. J., 1976. Ann. of Math. Studies, No. 86.

253. J. Hempel. Residual finiteness for 3-manifolds. In *Combinatorial group theory and topology (Alta, Utah, 1984)*, pages 379–396. Princeton Univ. Press, Princeton, NJ, 1987.

254. E. Hess and T. Schick. L^2-torsion of hyperbolic manifolds. *Manuscripta Math.*, 97(3):329–334, 1998.

255. G. Higman. A finitely generated infinite simple group. *J. London Math. Soc.*, 26:61–64, 1951.

256. N. Higson. The Baum-Connes conjecture. In *Proceedings of the International Congress of Mathematicians, Vol. II (Berlin, 1998)*, pages 637–646 (electronic), 1998.

257. N. Higson and G. Kasparov. E-theory and KK-theory for groups which act properly and isometrically on Hilbert space. *Invent. Math.*, 144(1):23–74, 2001.

258. N. Higson, J. Roe, and T. Schick. Spaces with vanishing l^2-homology and their fundamental groups (after Farber and Weinberger). *Geom. Dedicata*, 87(1-3):335–343, 2001.

259. J. A. Hillman. Deficiencies of lattices in connected Lie groups. Preprint, 1999.

260. J. A. Hillman and P. A. Linnell. Elementary amenable groups of finite Hirsch length are locally-finite by virtually-solvable. *J. Austral. Math. Soc. Ser. A*, 52(2):237–241, 1992.

261. M. W. Hirsch and W. P. Thurston. Foliated bundles, invariant measures and flat manifolds. *Ann. Math. (2)*, 101:369–390, 1975.

262. F. Hirzebruch. Automorphe Formen und der Satz von Riemann-Roch. In *Symposium internacional de topología algebraica – International symposium on algebraic topology*, pages 129–144. Universidad Nacional Autónoma de México and UNESCO, Mexico City, 1958.

263. N. Hitchin. Compact four-dimensional Einstein manifolds. *J. Differential Geometry*, 9:435–441, 1974.

264. C. Hog, M. Lustig, and W. Metzler. Presentation classes, 3-manifolds and free products. In *Geometry and topology (College Park, Md., 1983/84)*, pages 154–167. Springer-Verlag, Berlin, 1985.

265. W. C. Hsiang and W. L. Pardon. When are topologically equivalent orthogonal transformations linearly equivalent? *Invent. Math.*, 68(2):275–316, 1982.

266. W. C. Hsiang and V. Pati. L^2-cohomology of normal algebraic surfaces. I. *Invent. Math.*, 81(3):395–412, 1985.

267. I. Hughes. Division rings of fractions for group rings. *Comm. Pure Appl. Math.*, 23:181–188, 1970.

268. J. E. Humphreys. *Introduction to Lie algebras and representation theory*. Springer-Verlag, New York, 1972. Graduate Texts in Mathematics, Vol. 9.

269. D. Husemoller. *Elliptic curves*. Springer-Verlag, New York, 1987. With an appendix by R. Lawrence.

270. S. Illman. Whitehead torsion and group actions. *Ann. Acad. Sci. Fenn. Ser. A I Math.*, (588):45, 1974.

271. S. Illman. Smooth equivariant triangulations of G-manifolds for G a finite group. *Math. Ann.*, 233(3):199–220, 1978.

272. S. Illman. The equivariant triangulation theorem for actions of compact Lie groups. *Math. Ann.*, 262(4):487–501, 1983.

273. S. Illman. Existence and uniqueness of equivariant triangulations of smooth proper G-manifolds with some applications to equivariant Whitehead torsion. *J. Reine Angew. Math.*, 524:129–183, 2000.

274. H. Inoue and K. Yano. The Gromov invariant of negatively curved manifolds. *Topology*, 21(1):83–89, 1982.

275. N. Ivanov. Foundations of the theory of bounded cohomology. *J. Soviet Math.*, 37:1090–1114, 1987.

276. W. H. Jaco and P. B. Shalen. Seifert fibered spaces in 3-manifolds. *Mem. Amer. Math. Soc.*, 21(220):viii+192, 1979.

277. N. Jacobson. *Lectures in abstract algebra. Vol. II. Linear algebra*. D. Van Nostrand Co., Inc., Toronto-New York-London, 1953.

278. K. Johannson. *Homotopy equivalences of 3-manifolds with boundaries*. Springer-Verlag, Berlin, 1979.

279. F. E. A. Johnson and D. Kotschick. On the signature and Euler characteristic of certain four-manifolds. *Math. Proc. Cambridge Philos. Soc.*, 114(3):431–437, 1993.

280. J. Jost and Y. L. Xin. Vanishing theorems for L^2-cohomology groups. *J. Reine Angew. Math.*, 525:95–112, 2000.

281. J. Jost and K. Zuo. Vanishing theorems for L^2-cohomology on infinite coverings of compact Kähler manifolds and applications in algebraic geometry. *Comm. Anal. Geom.*, 8(1):1–30, 2000.

282. R. V. Kadison and J. R. Ringrose. *Fundamentals of the theory of operator algebras. Vol. I.* Academic Press Inc. [Harcourt Brace Jovanovich Publishers], New York, 1983. Elementary theory.

283. R. V. Kadison and J. R. Ringrose. *Fundamentals of the theory of operator algebras. Vol. II.* Academic Press Inc., Orlando, FL, 1986. Advanced theory.

284. D. A. Kajdan. On arithmetic varieties. In *Lie groups and their representations (Proc. Summer School, Bolyai János Math. Soc., Budapest, 1971)*, pages 151–217. Halsted, New York, 1975.

285. E. Kaniuth. Der Typ der regulären Darstellung diskreter Gruppen. *Math. Ann.*, 182:334–339, 1969.

286. R. M. Kashaev. The hyperbolic volume of knots from the quantum dilogarithm. *Lett. Math. Phys.*, 39(3):269–275, 1997.

287. T. Kato. *Pertubation theory for linear operators*, volume 132 of *Grundlehren der math. Wissenschaften*. Springer, 1984.

288. M. A. Kervaire. Le théorème de Barden-Mazur-Stallings. *Comment. Math. Helv.*, 40:31–42, 1965.

289. M. A. Kervaire. Smooth homology spheres and their fundamental groups. *Trans. Amer. Math. Soc.*, 144:67–72, 1969.

290. H. Kesten. Full Banach mean values on countable groups. *Math. Scand.*, 7:146–156, 1959.

291. N. Keswani. Von Neumann eta-invariants and C^*-algebra K-theory. *J. London Math. Soc. (2)*, 62(3):771–783, 2000.

292. R. C. Kirby and L. C. Siebenmann. *Foundational essays on topological manifolds, smoothings, and triangulations*. Princeton University Press, Princeton, N.J., 1977. With notes by J. Milnor and M. F. Atiyah, Annals of Mathematics Studies, No. 88.

293. A. W. Knapp. *Representation theory of semisimple groups*. Princeton University Press, Princeton, NJ, 1986. An overview based on examples.

294. K. Kodaira. On Kähler varieties of restricted type (an intrinsic characterization of algebraic varieties). *Ann. of Math. (2)*, 60:28–48, 1954.

295. K. Köhler. Holomorphic torsion on Hermitian symmetric spaces. *J. Reine Angew. Math.*, 460:93–116, 1995.

296. K. Köhler. Equivariant Reidemeister torsion on symmetric spaces. *Math. Ann.*, 307(1):57–69, 1997.

297. D. Kotschick. Four-manifold invariants of finitely presentable groups. In *Topology, geometry and field theory*, pages 89–99. World Sci. Publishing, River Edge, NJ, 1994.

298. M. Kreck, W. Lück, and P. Teichner. Counterexamples to the Kneser conjecture in dimension four. *Comment. Math. Helv.*, 70(3):423–433, 1995.

299. M. Kreck, W. Lück, and P. Teichner. Stable prime decompositions of four-manifolds. In *Prospects in topology (Princeton, NJ, 1994)*, pages 251–269. Princeton Univ. Press, Princeton, NJ, 1995.

300. P. H. Kropholler, P. A. Linnell, and J. A. Moody. Applications of a new K-theoretic theorem to soluble group rings. *Proc. Amer. Math. Soc.*, 104(3):675–684, 1988.

301. E. Laitinen and W. Lück. Equivariant Lefschetz classes. *Osaka J. Math.*, 26(3):491–525, 1989.

302. T. Y. Lam. *A first course in noncommutative rings*. Springer-Verlag, New York, 1991.

303. C. Lance. On nuclear C^*-algebras. *J. Functional Analysis*, 12:157–176, 1973.

304. B. Leeb. 3-manifolds with(out) metrics of nonpositive curvature. *Invent. Math.*, 122(2):277–289, 1995.

305. H. W. Lenstra, Jr. Grothendieck groups of abelian group rings. *J. Pure Appl. Algebra*, 20(2):173–193, 1981.

306. J. Lewin. Fields of fractions for group algebras of free groups. *Trans. Amer. Math. Soc.*, 192:339–346, 1974.

307. P. A. Linnell. Decomposition of augmentation ideals and relation modules. *Proc. London Math. Soc. (3)*, 47(1):83–127, 1983.

308. P. A. Linnell. Zero divisors and $L^2(G)$. *C. R. Acad. Sci. Paris Sér. I Math.*, 315(1):49–53, 1992.

309. P. A. Linnell. Division rings and group von Neumann algebras. *Forum Math.*, 5(6):561–576, 1993.

310. P. A. Linnell. Analytic versions of the zero divisor conjecture. In *Geometry and cohomology in group theory (Durham, 1994)*, pages 209–248. Cambridge Univ. Press, Cambridge, 1998.

311. P. A. Linnell, W. Lück, and T. Schick. The Ore condition, affiliated operators, and the lamplighter group. Preprintreihe SFB 478 — Geometrische Strukturen in der Mathematik, Heft 205 Münster, to appear in the Proceedings of the school/conference "High-dimensional manifold theory" in Trieste, May/June 2001, 2002.

312. P. A. Linnell and M. J. Puls. Zero divisors and $L^p(G)$. II. *New York J. Math.*, 7:49–58 (electronic), 2001.

313. P. A. Linnell and T. Schick. Finite group extensions and the Atiyah conjecture. Preliminary version.

314. N. Lohoue and S. Mehdi. The Novikov-Shubin invariants for locally symmetric spaces. *J. Math. Pures Appl. (9)*, 79(2):111–140, 2000.

315. E. Looijenga. L^2-cohomology of locally symmetric varieties. *Compositio Math.*, 67(1):3–20, 1988.

316. J. Lott. Heat kernels on covering spaces and topological invariants. *J. Differential Geom.*, 35(2):471–510, 1992.

317. J. Lott. Equivariant analytic torsion for compact Lie group actions. *J. Funct. Anal.*, 125(2):438–451, 1994.

318. J. Lott. The zero-in-the-spectrum question. *Enseign. Math. (2)*, 42(3-4):341–376, 1996.

319. J. Lott. L^2-cohomology of geometrically infinite hyperbolic 3-manifolds. *Geom. Funct. Anal.*, 7(1):81–119, 1997.

320. J. Lott. Deficiencies of lattice subgroups of Lie groups. *Bull. London Math. Soc.*, 31(2):191–195, 1999.

321. J. Lott. Delocalized L^2-invariants. *J. Funct. Anal.*, 169(1):1–31, 1999.

322. J. Lott and W. Lück. L^2-topological invariants of 3-manifolds. *Invent. Math.*, 120(1):15–60, 1995.

323. J. Lott and M. Rothenberg. Analytic torsion for group actions. *J. Differential Geom.*, 34(2):431–481, 1991.

324. A. Lubotzky. Group presentation, p-adic analytic groups and lattices in $SL_2(\mathbb{C})$. *Ann. of Math. (2)*, 118(1):115–130, 1983.

325. W. Lück. *Eine allgemeine algebraische Beschreibung des Transfers für Faserungen auf projektiven Klassengruppen und Whiteheadgruppen.* PhD thesis, Universität Göttingen, 1984.

326. W. Lück. *Transformation groups and algebraic K-theory.* Springer-Verlag, Berlin, 1989. Lecture Notes in Mathematics, Vol. 1408.

327. W. Lück. Analytic and topological torsion for manifolds with boundary and symmetry. *J. Differential Geom.*, 37(2):263–322, 1993.

328. W. Lück. Approximating L^2-invariants by their finite-dimensional analogues. *Geom. Funct. Anal.*, 4(4):455–481, 1994.

329. W. Lück. L^2-Betti numbers of mapping tori and groups. *Topology*, 33(2):203–214, 1994.

330. W. Lück. L^2-torsion and 3-manifolds. In *Low-dimensional topology (Knoxville, TN, 1992)*, pages 75–107. Internat. Press, Cambridge, MA, 1994.

331. W. Lück. Hilbert modules and modules over finite von Neumann algebras and applications to L^2-invariants. *Math. Ann.*, 309(2):247–285, 1997.

332. W. Lück. L^2-Invarianten von Mannigfaltigkeiten und Gruppen. *Jahresber. Deutsch. Math.-Verein.*, 99(3):101–109, 1997.

333. W. Lück. Dimension theory of arbitrary modules over finite von Neumann algebras and L^2-Betti numbers. I. Foundations. *J. Reine Angew. Math.*, 495:135–162, 1998.

334. W. Lück. Dimension theory of arbitrary modules over finite von Neumann algebras and L^2-Betti numbers. II. Applications to Grothendieck groups, L^2-Euler characteristics and Burnside groups. *J. Reine Angew. Math.*, 496:213–236, 1998.

335. W. Lück. The universal functorial Lefschetz invariant. *Fund. Math.*, 161(1-2):167–215, 1999. Algebraic topology (Kazimierz Dolny, 1997).

336. W. Lück. The type of the classifying space for a family of subgroups. *J. Pure Appl. Algebra*, 149(2):177–203, 2000.

337. W. Lück. A basic introduction to surgery theory. Preprintreihe SFB 478 — Geometrische Strukturen in der Mathematik, Heft 197, Münster, to appear in the Proceedings of the summer school "High dimensional manifold theory" in Trieste May/June 2001, 2001.

338. W. Lück. L^2-invariants and their applications to geometry, group theory and spectral theory. In *Mathematics Unlimited—2001 and Beyond*, pages 859–871. Springer-Verlag, Berlin, 2001.

339. W. Lück. The relation between the Baum-Connes conjecture and the trace conjecture. Preprintreihe SFB 478 — Geometrische Strukturen in der Mathematik, Heft 190, Münster, to appear in *Invent. Math.*, 2001.

340. W. Lück. Chern characters for proper equivariant homology theories and applications to K- and L-theory. *J. Reine Angew. Math.*, 543:193–234, 2002.

341. W. Lück. L^2-invariants of regular coverings of compact manifolds and CW-complexes. In *Handbook of geometric topology*, pages 735–817. Elsevier, Amsterdam, 2002.

342. W. Lück and D. Meintrup. On the universal space for group actions with compact isotropy. In *Geometry and topology: Aarhus (1998)*, pages 293–305. Amer. Math. Soc., Providence, RI, 2000.

343. W. Lück, H. Reich, and T. Schick. Novikov-Shubin invariants for arbitrary group actions and their positivity. In *Tel Aviv Topology Conference: Rothenberg Festschrift (1998)*, pages 159–176. Amer. Math. Soc., Providence, RI, 1999.

344. W. Lück and M. Rørdam. Algebraic K-theory of von Neumann algebras. *K-Theory*, 7(6):517–536, 1993.

345. W. Lück and M. Rothenberg. Reidemeister torsion and the K-theory of von Neumann algebras. *K-Theory*, 5(3):213–264, 1991.

346. W. Lück and T. Schick. L^2-torsion of hyperbolic manifolds of finite volume. *Geometric and Functional Analysis*, 9:518–567, 1999.

347. W. Lück and T. Schick. Approximating L^2-signatures by their finite-dimensional analogues. Preprintreihe SFB 478 — Geometrische Strukturen in der Mathematik, Heft 190, Münster, 2001.

348. W. Lück, T. Schick, and T. Thielmann. Torsion and fibrations. *J. Reine Angew. Math.*, 498:1–33, 1998.

349. M. Lustig. On the rank, the deficiency and the homological dimension of groups: the computation of a lower bound via Fox ideals. In *Topology and combinatorial group theory (Hanover, NH, 1986/1987; Enfield, NH, 1988)*, pages 164–174. Springer-Verlag, Berlin, 1990.

350. R. C. Lyndon and P. E. Schupp. *Combinatorial group theory*. Springer-Verlag, Berlin, 1977. Ergebnisse der Mathematik und ihrer Grenzgebiete, Band 89.

351. S. MacLane. *Categories for the working mathematician*. Springer-Verlag, New York, 1971. Graduate Texts in Mathematics, Vol. 5.

352. I. Madsen. Reidemeister torsion, surgery invariants and spherical space forms. *Proc. London Math. Soc. (3)*, 46(2):193–240, 1983.

353. I. Madsen and M. Rothenberg. On the classification of G-spheres. I. Equivariant transversality. *Acta Math.*, 160(1-2):65–104, 1988.

354. W. Magnus. Beziehungen zwischen Gruppen und Idealen in einem speziellen Ring. *Math. Annalen*, 111:259–280, 1935.

355. W. Magnus. Residually finite groups. *Bull. Amer. Math. Soc.*, 75:305–316, 1969.

356. A. Malcev. On isomorphic matrix representations of infinite groups. *Rec. Math. [Mat. Sbornik] N.S.*, 8 (50):405–422, 1940.

357. W. S. Massey. *A basic course in algebraic topology*. Springer-Verlag, New York, 1991.

358. V. Mathai. L^2-analytic torsion. *J. Funct. Anal.*, 107(2):369–386, 1992.

359. V. Mathai. Spectral flow, eta invariants, and von Neumann algebras. *J. Funct. Anal.*, 109(2):442–456, 1992.

360. V. Mathai. L^2 invariants of covering spaces. In *Geometric analysis and Lie theory in mathematics and physics*, pages 209–242. Cambridge Univ. Press, Cambridge, 1998.

361. V. Mathai. K-theory of twisted group C^*-algebras and positive scalar curvature. In *Tel Aviv Topology Conference: Rothenberg Festschrift (1998)*, pages 203–225. Amer. Math. Soc., Providence, RI, 1999.

362. V. Mathai and M. Rothenberg. On the homotopy invariance of L^2 torsion for covering spaces. *Proc. Amer. Math. Soc.*, 126(3):887–897, 1998.

363. V. Mathai and M. A. Shubin. Twisted L^2 invariants of non-simply connected manifolds and asymptotic L^2 Morse inequalities. *Russian J. Math. Phys.*, 4(4):499–526, 1996.

364. V. Mathai and S. G. Yates. Discrete Morse theory and extended L^2 homology. *J. Funct. Anal.*, 168(1):84–110, 1999.

365. T. Matumoto. On G-CW complexes and a theorem of J. H. C. Whitehead. *J. Fac. Sci. Univ. Tokyo Sect. IA Math.*, 18:363–374, 1971.

366. F. I. Mautner. The structure of the regular representation of certain discrete groups. *Duke Math. J.*, 17:437–441, 1950.
367. R. Mazzeo and R. S. Phillips. Hodge theory on hyperbolic manifolds. *Duke Math. J.*, 60(2):509–559, 1990.
368. C. T. McMullen. Iteration on Teichmüller space. *Invent. Math.*, 99(2):425–454, 1990.
369. C. T. McMullen. *Renormalization and 3-manifolds which fiber over the circle.* Princeton University Press, Princeton, NJ, 1996.
370. D. Meintrup. *On the Type of the Universal Space for a Family of Subgroups.* PhD thesis, Westfälische Wilhelms-Universität Münster, 2000.
371. R. J. Miatello. On the Plancherel measure for linear Lie groups of rank one. *Manuscripta Math.*, 29(2-4):249–276, 1979.
372. J. Milnor. On the existence of a connection with curvature zero. *Comment. Math. Helv.*, 32:215–223, 1958.
373. J. Milnor. Two complexes which are homeomorphic but combinatorially distinct. *Ann. of Math. (2)*, 74:575–590, 1961.
374. J. Milnor. A duality theorem for Reidemeister torsion. *Ann. of Math. (2)*, 76:137–147, 1962.
375. J. Milnor. Whitehead torsion. *Bull. Amer. Math. Soc.*, 72:358–426, 1966.
376. J. Milnor. Infinite cyclic coverings. In *Conference on the Topology of Manifolds (Michigan State Univ., E. Lansing, Mich., 1967)*, pages 115–133. Prindle, Weber & Schmidt, Boston, Mass., 1968.
377. J. Milnor. A note on curvature and fundamental group. *J. Differential Geometry*, 2:1–7, 1968.
378. J. Milnor. *Introduction to algebraic K-theory.* Princeton University Press, Princeton, N.J., 1971. Annals of Mathematics Studies, No. 72.
379. J. Milnor and J. D. Stasheff. *Characteristic classes.* Princeton University Press, Princeton, N. J., 1974. Annals of Mathematics Studies, No. 76.
380. A. S. Miščenko and A. T. Fomenko. The index of elliptic operators over C^*-algebras. *Izv. Akad. Nauk SSSR Ser. Mat.*, 43(4):831–859, 967, 1979. English translation in *Math. USSR-Izv.* 15 (1980), no. 1, 87–112.
381. E. E. Moise. *Geometric topology in dimensions 2 and 3.* Springer-Verlag, New York, 1977. Graduate Texts in Mathematics, Vol. 47.
382. B. Moishezon. On n-dimensional compact varieties with n algebraically independent meromorphic functions. *AMS Translations*, 63:51–177, 1967.
383. J. A. Moody. Induction theorems for infinite groups. *Bull. Amer. Math. Soc. (N.S.)*, 17(1):113–116, 1987.
384. J. A. Moody. Brauer induction for G_0 of certain infinite groups. *J. Algebra*, 122(1):1–14, 1989.
385. J. W. Morgan. On Thurston's uniformization theorem for three-dimensional manifolds. In *The Smith conjecture (New York, 1979)*, pages 37–125. Academic Press, Orlando, FL, 1984.
386. H. Moscovici. L^2-index of elliptic operators on locally symmetric spaces of finite volume. In *Operator algebras and K-theory (San Francisco, Calif., 1981)*, pages 129–137. Amer. Math. Soc., Providence, R.I., 1982.
387. H. Moscovici and R. J. Stanton. R-torsion and zeta functions for locally symmetric manifolds. *Invent. Math.*, 105(1):185–216, 1991.
388. G. D. Mostow. *Strong rigidity of locally symmetric spaces.* Princeton University Press, Princeton, N.J., 1973. Annals of Mathematics Studies, No. 78.
389. A. W. Mostowski. On the decidability of some problems in special classes of groups. *Fund. Math.*, 59:123–135, 1966.
390. W. Müller. Analytic torsion and R-torsion of Riemannian manifolds. *Adv. in Math.*, 28(3):233–305, 1978.

391. W. Müller. *Manifolds with cusps of rank one.* Springer-Verlag, Berlin, 1987. Spectral theory and L^2-index theorem.

392. W. Müller. L^2-index theory, eta invariants and values of L-functions. In *Geometric and topological invariants of elliptic operators (Brunswick, ME, 1988)*, pages 145–189. Amer. Math. Soc., Providence, RI, 1990.

393. W. Müller. Analytic torsion and R-torsion for unimodular representations. *J. Amer. Math. Soc.*, 6(3):721–753, 1993.

394. W. Müller. On the L^2-index of Dirac operators on manifolds with corners of codimension two. I. *J. Differential Geom.*, 44(1):97–177, 1996.

395. H. Murakami and J. Murakami. The colored Jones polynomials and the simplicial volume of a knot. *Acta Math.*, 186(1):85–104, 2001.

396. F. Murray and J. von Neumann. On rings of operators. *Annals of Math.*, 37:116–229, 1936.

397. M. Nagase. L^2-cohomology and intersection homology of stratified spaces. *Duke Math. J.*, 50(1):329–368, 1983.

398. M. Nagase. Remarks on the L^2-cohomology of singular algebraic surfaces. *J. Math. Soc. Japan*, 41(1):97–116, 1989.

399. H. Neumann. *Varieties of groups.* Springer-Verlag New York, Inc., New York, 1967.

400. S. P. Novikov and M. A. Shubin. Morse inequalities and von Neumann II_1-factors. *Dokl. Akad. Nauk SSSR*, 289(2):289–292, 1986.

401. S. P. Novikov and M. A. Shubin. Morse inequalities and von Neumann invariants of non-simply connected manifolds. *Uspekhi. Matem. Nauk*, 41(5):222–223, 1986. in Russian.

402. T. Ohsawa. On the L^2 cohomology of complex spaces. *Math. Z.*, 209(4):519–530, 1992.

403. T. Ohsawa. On the L^2 cohomology groups of isolated singularities. In *Progress in differential geometry*, pages 247–263. Math. Soc. Japan, Tokyo, 1993.

404. M. Olbrich. L^2-invariants of locally symmetric spaces. Preprint, Göttingen, 2000.

405. A. Ol′ shanskii. Groups of bounded period with subgroups of prime order. *Algebra und Logik*, 21:369–418, 1982.

406. R. Oliver. *Whitehead groups of finite groups.* Cambridge University Press, Cambridge, 1988.

407. A. Y. Ol′shanskii and M. V. Sapir. Non-amenable finitely presented torsion-by-cyclic groups. *Electron. Res. Announc. Amer. Math. Soc.*, 7:63–71 (electronic), 2001.

408. D. S. Ornstein and B. Weiss. Ergodic theory of amenable group actions. I. The Rohlin lemma. *Bull. Amer. Math. Soc. (N.S.)*, 2(1):161–164, 1980.

409. P. Pansu. Cohomologie L^p des variétés à courbure négative, cas du degré 1. Number Special Issue, pages 95–120 (1990), 1989. Conference on Partial Differential Equations and Geometry (Torino, 1988).

410. P. Pansu. Cohomologie L^p: Invariance sous quasiisometries. Preprint, Orsay, 1995.

411. P. Pansu. Introduction to L^2 Betti numbers. In *Riemannian geometry (Waterloo, ON, 1993)*, pages 53–86. Amer. Math. Soc., Providence, RI, 1996.

412. P. Pansu. Volume, courbure et entropie (d'après G. Besson, G. Courtois et S. Gallot). *Astérisque*, (245):Exp. No. 823, 3, 83–103, 1997. Séminaire Bourbaki, Vol. 1996/97.

413. P. Pansu. Cohomologie L^p, espaces homogènes et pinchement. Preprint, Orsay, 1999.

414. W. L. Pardon. The L_2-$\overline{\partial}$-cohomology of an algebraic surface. *Topology*, 28(2):171–195, 1989.

415. W. L. Pardon and M. Stern. L^2-$\bar{\partial}$-cohomology of complex projective varieties. *J. Amer. Math. Soc.*, 4(3):603–621, 1991.

416. W. L. Paschke. An invariant for finitely presented cG-modules. *Math. Ann.*, 301(2):325–337, 1995.

417. D. S. Passman. Semiprime and prime crossed products. *J. Algebra*, 83(1):158–178, 1983.

418. D. S. Passman. *Infinite crossed products*. Academic Press Inc., Boston, MA, 1989.

419. A. L. T. Paterson. *Amenability*. American Mathematical Society, Providence, RI, 1988.

420. G. K. Pedersen. C^*-*algebras and their automorphism groups*. Academic Press Inc. [Harcourt Brace Jovanovich Publishers], London, 1979.

421. G. K. Pedersen. *Analysis now*. Springer-Verlag, New York, 1989.

422. M. J. Puls. Zero divisors and $L^p(G)$. *Proc. Amer. Math. Soc.*, 126(3):721–728, 1998.

423. D. Quillen. Determinants of Cauchy-Riemann operators over a Riemann surface. *Funct. Anal. Appl.*, 19:31–35, 1986.

424. F. Quinn. Ends of maps. I. *Ann. of Math. (2)*, 110(2):275–331, 1979.

425. F. Quinn. Ends of maps. II. *Invent. Math.*, 68(3):353–424, 1982.

426. F. Quinn. Ends of maps. III. Dimensions 4 and 5. *J. Differential Geom.*, 17(3):503–521, 1982.

427. M. Ramachandran. von Neumann index theorems for manifolds with boundary. *J. Differential Geom.*, 38(2):315–349, 1993.

428. A. A. Ranicki. On the algebraic L-theory of semisimple rings. *J. Algebra*, 50(1):242–243, 1978.

429. A. A. Ranicki. The algebraic theory of surgery. I. Foundations. *Proc. London Math. Soc. (3)*, 40(1):87–192, 1980.

430. A. A. Ranicki. *Exact sequences in the algebraic theory of surgery*. Princeton University Press, Princeton, N.J., 1981.

431. A. A. Ranicki. On the Hauptvermutung. In *The Hauptvermutung book*, pages 3–31. Kluwer Acad. Publ., Dordrecht, 1996.

432. D. B. Ray and I. M. Singer. R-torsion and the Laplacian on Riemannian manifolds. *Advances in Math.*, 7:145–210, 1971.

433. D. B. Ray and I. M. Singer. Analytic torsion for complex manifolds. *Ann. of Math. (2)*, 98:154–177, 1973.

434. M. Reed and B. Simon. *Methods of modern mathematical physics. I*. Academic Press Inc. [Harcourt Brace Jovanovich Publishers], New York, second edition, 1980. Functional analysis.

435. H. Reich. *Group von Neumann algebras and related algebras*. PhD thesis, Universität Göttingen, 1999. http://www.math.uni-muenster.de/u/lueck/publ/diplome/reich.dvi.

436. K. Reidemeister. Homotopieringe und Linsenräume. Hamburger Abhandlungen 11, 1938.

437. J. Roe. An index theorem on open manifolds. I. *J. Differential Geom.*, 27(1):87–113, 1988.

438. J. Roe. An index theorem on open manifolds. II. *J. Differential Geom.*, 27(1):115–136, 1988.

439. J. Roe. Coarse cohomology and index theory on complete Riemannian manifolds. *Mem. Amer. Math. Soc.*, 104(497):x+90, 1993.

440. J. Roe. *Index theory, coarse geometry, and topology of manifolds*. Published for the Conference Board of the Mathematical Sciences, Washington, DC, 1996.

441. J. Rosenberg. *Algebraic K-theory and its applications*. Springer-Verlag, New York, 1994.
442. J. Rosenberg. Analytic Novikov for topologists. In *Novikov conjectures, index theorems and rigidity, Vol. 1 (Oberwolfach, 1993)*, pages 338–372. Cambridge Univ. Press, Cambridge, 1995.
443. M. Rothenberg. Torsion invariants and finite transformation groups. In *Algebraic and geometric topology (Proc. Sympos. Pure Math., Stanford Univ., Stanford, Calif., 1976), Part 1*, pages 267–311. Amer. Math. Soc., Providence, R.I., 1978.
444. J. J. Rotman. *An introduction to homological algebra*. Academic Press Inc. [Harcourt Brace Jovanovich Publishers], New York, 1979.
445. C. P. Rourke and B. J. Sanderson. *Introduction to piecewise-linear topology*. Springer-Verlag, Berlin, 1982. Reprint.
446. L. H. Rowen. *Ring theory. Vol. I*. Academic Press Inc., Boston, MA, 1988.
447. L. H. Rowen. *Ring theory. Vol. II*. Academic Press Inc., Boston, MA, 1988.
448. L. H. Rowen. *Ring theory*. Academic Press Inc., Boston, MA, student edition, 1991.
449. R. Roy. The trace conjecture—a counterexample. *K-Theory*, 17(3):209–213, 1999.
450. M. Rumin. Sub-Riemannian limit of the differential form spectrum of contact manifolds. *Geom. Funct. Anal.*, 10(2):407–452, 2000.
451. L. Saper. L_2-cohomology and intersection homology of certain algebraic varieties with isolated singularities. *Invent. Math.*, 82(2):207–255, 1985.
452. L. Saper. L_2-cohomology of algebraic varieties. In *Proceedings of the International Congress of Mathematicians, Vol. I, II (Kyoto, 1990)*, pages 735–746, Tokyo, 1991. Math. Soc. Japan.
453. L. Saper. L_2-cohomology of Kähler varieties with isolated singularities. *J. Differential Geom.*, 36(1):89–161, 1992.
454. L. Saper and M. Stern. L_2-cohomology of arithmetic varieties. *Ann. of Math. (2)*, 132(1):1–69, 1990.
455. L. Saper and S. Zucker. An introduction to L^2-cohomology. In *Several complex variables and complex geometry, Part 2 (Santa Cruz, CA, 1989)*, pages 519–534. Amer. Math. Soc., Providence, RI, 1991.
456. R. Sauer. Invariance properties of l^2-betti numbers and novikov-shubin invariants under orbit equivalence and quasi-isometry. PhD thesis, in preparation, 2002.
457. R. P. Savage, Jr. The space of positive definite matrices and Gromov's invariant. *Trans. Amer. Math. Soc.*, 274(1):239–263, 1982.
458. J. A. Schafer. The Bass conjecture and group von Neumann algebras. *K-Theory*, 19(3):211–217, 2000.
459. T. Schick. *Analysis on ∂-manifolds of bounded geometry, Hodge-de Rham isomorphism and L^2-index theorem*. Shaker Verlag, Aachen, 1996. PhD thesis, Johannes Gutenberg-Universität Mainz.
460. T. Schick. L^2-index theorem for boundary manifolds. Preprint 1998, to appear in *Pacific J. Math.*, 1998.
461. T. Schick. Integrality of L^2-Betti numbers. *Math. Ann.*, 317(4):727–750, 2000.
462. T. Schick. L^2-determinant class and approximation of L^2-Betti numbers. *Trans. Amer. Math. Soc.*, 353(8):3247–3265 (electronic), 2001.
463. T. Schick. Erratum: "Integrality of L^2-Betti numbers" [Math. Ann. **317** (2000), no. 4, 727–750; 1777117]. *Math. Ann.*, 322(2):421–422, 2002.
464. W. Schmid. L^2-cohomology and the discrete series. *Ann. of Math. (2)*, 103(2):375–394, 1976.

465. A. H. Schofield. *Representation of rings over skew fields.* Cambridge University Press, Cambridge, 1985.

466. P. Scott. The geometries of 3-manifolds. *Bull. London Math. Soc.*, 15(5):401–487, 1983.

467. R. T. Seeley. Extension of C^∞ functions defined in a half space. *Proc. Amer. Math. Soc.*, 15:625–626, 1964.

468. R. T. Seeley. Complex powers of an elliptic operator. In *Singular Integrals (Proc. Sympos. Pure Math., Chicago, Ill., 1966)*, pages 288–307. Amer. Math. Soc., Providence, R.I., 1967.

469. V. Sergiescu. A quick introduction to Burnside's problem. In *Group theory from a geometrical viewpoint (Trieste, 1990)*, pages 622–629. World Sci. Publishing, River Edge, NJ, 1991.

470. J.-P. Serre. *Linear representations of finite groups.* Springer-Verlag, New York, 1977. Translated from the second French edition by Leonard L. Scott, Graduate Texts in Mathematics, Vol. 42.

471. J.-P. Serre. *Trees.* Springer-Verlag, Berlin, 1980. Translated from the French by J. Stillwell.

472. T. N. Shorey and R. Tijdeman. *Exponential Diophantine equations.* Cambridge University Press, Cambridge, 1986.

473. M. Shubin. *Pseusodifferential operators and spectral theory.* Springer, 1987.

474. M. A. Shubin. A Lefschetz fixed point formula in reduced L^2-cohomologies on manifolds with cylinders. In *Symposium "Analysis on Manifolds with Singularities" (Breitenbrunn, 1990)*, pages 290–297. Teubner, Stuttgart, 1992.

475. M. A. Shubin. Semiclassical asymptotics on covering manifolds and Morse inequalities. *Geom. Funct. Anal.*, 6(2):370–409, 1996.

476. I. M. Singer. Infinitesimally homogeneous spaces. *Comm. Pure Appl. Math.*, 13:685–697, 1960.

477. I. M. Singer. Some remarks on operator theory and index theory. In *K-theory and operator algebras (Proc. Conf., Univ. Georgia, Athens, Ga., 1975)*, pages 128–138. Lecture Notes in Math., Vol. 575. Springer-Verlag, Berlin, 1977.

478. J. Smillie. An obstruction to the existence of affine structures. *Invent. Math.*, 64(3):411–415, 1981.

479. T. Soma. The Gromov invariant of links. *Invent. Math.*, 64(3):445–454, 1981.

480. T. Soma. Bounded cohomology of closed surfaces. *Topology*, 36(6):1221–1246, 1997.

481. J. Stallings. Whitehead torsion of free products. *Ann. of Math. (2)*, 82:354–363, 1965.

482. N. E. Steenrod. A convenient category of topological spaces. *Michigan Math. J.*, 14:133–152, 1967.

483. M. Steinberger and J. E. West. Equivariant h-cobordisms and finiteness obstructions. *Bull. Amer. Math. Soc. (N.S.)*, 12(2):217–220, 1985.

484. B. Stenström. *Rings of quotients.* Springer-Verlag, New York, 1975. Die Grundlehren der Mathematischen Wissenschaften, Band 217, An introduction to methods of ring theory.

485. M. Stern. L^2-index theorems on locally symmetric spaces. *Invent. Math.*, 96(2):231–282, 1989.

486. M. Stern. Index theory for certain complete Kähler manifolds. *J. Differential Geom.*, 37(3):467–503, 1993.

487. D. Sullivan. A generalization of Milnor's inequality concerning affine foliations and affine manifolds. *Comment. Math. Helv.*, 51(2):183–189, 1976.

488. R. M. Switzer. *Algebraic topology—homotopy and homology.* Springer-Verlag, New York, 1975. Die Grundlehren der mathematischen Wissenschaften, Band 212.

489. D. Tamari. A refined classification of semi-groups leading to generalized polynomial rings with a generalized degree concept. In J. Gerretsen and J. de Groot, editors, *Proceedings of the ICM, Amsterdam, 1954*, volume 3, pages 439–440, Groningen, 1957.

490. E. Thoma. Über unitäre Darstellungen abzählbarer, diskreter Gruppen. *Math. Ann.*, 153:111–138, 1964.

491. W. P. Thurston. Geometry and topology of 3-manifolds. Lecture notes, Princeton. http://www.msri.org/publications/books/gt3m, 1978.

492. W. P. Thurston. *Three-dimensional geometry and topology. Vol. 1.* Princeton University Press, Princeton, NJ, 1997. Edited by Silvio Levy.

493. T. tom Dieck. Orbittypen und äquivariante Homologie. I. *Arch. Math. (Basel)*, 23:307–317, 1972.

494. T. tom Dieck. *Transformation groups and representation theory.* Springer-Verlag, Berlin, 1979.

495. T. tom Dieck. *Transformation groups.* Walter de Gruyter & Co., Berlin, 1987.

496. H. Triebel. *Höhere Analysis.* VEB Deutscher Verlag der Wissenschaften, Berlin, 1972. Hochschulbücher für Mathematik, Band 76.

497. V. G. Turaev. Reidemeister torsion in knot theory. *Uspekhi Mat. Nauk*, 41(1(247)):97–147, 240, 1986. English translation in *Russian Math. Surveys* 41 (1986), no. 1, 119–182.

498. V. G. Turaev. Torsion invariants of spin^c-structures on 3-manifolds. *Math. Res. Lett.*, 4(5):679–695, 1997.

499. C. Vafa and E. Witten. Eigenvalue inequalities for fermions in gauge theories. *Comm. Math. Phys.*, 95(3):257–276, 1984.

500. B. Vaillant. Indextheorie für Überlagerungen. Diplomarbeit, Universität Bonn, http://styx.math.uni-bonn.de/boris/diplom.html, 1997.

501. A. Valette. Introduction to the Baum-Connes conjecture. To appear as ETHZ lecture notes, 2001.

502. N. T. Varopoulos. Random walks and Brownian motion on manifolds. In *Symposia Mathematica, Vol. XXIX (Cortona, 1984)*, pages 97–109. Academic Press, New York, 1987.

503. S. M. Vishik. Analytic torsion of boundary value problems. *Dokl. Akad. Nauk SSSR*, 295(6):1293–1298, 1987.

504. S. M. Vishik. Analytic torsion of finite-dimensional local systems. *Dokl. Akad. Nauk SSSR*, 300(6):1295–1299, 1988.

505. S. M. Vishik. Generalized Ray-Singer conjecture. I. A manifold with a smooth boundary. *Comm. Math. Phys.*, 167(1):1–102, 1995.

506. J. B. Wagoner. Diffeomorphisms, K_2, and analytic torsion. In *Algebraic and geometric topology (Proc. Sympos. Pure Math., Stanford Univ., Stanford, Calif., 1976), Part 1*, pages 23–33. Amer. Math. Soc., Providence, R.I., 1978.

507. F. Waldhausen. On irreducible 3-manifolds which are sufficiently large. *Ann. of Math. (2)*, 87:56–88, 1968.

508. F. Waldhausen. Algebraic K-theory of generalized free products. I, II. *Ann. of Math. (2)*, 108(1):135–204, 1978.

509. C. T. C. Wall. Rational Euler characteristics. *Proc. Cambridge Philos. Soc.*, 57:182–184, 1961.

510. C. T. C. Wall. *Surgery on compact manifolds.* Academic Press, London, 1970. London Mathematical Society Monographs, No. 1.

511. C. T. C. Wall. Norms of units in group rings. *Proc. London Math. Soc. (3)*, 29:593–632, 1974.

512. C. T. C. Wall. Geometric structures on compact complex analytic surfaces. *Topology*, 25(2):119–153, 1986.

513. D. L. Webb. Higher G-theory of nilpotent group rings. *J. Algebra*, 116(2):457–465, 1988.

514. N. E. Wegge-Olsen. *K-theory and C^*-algebras*. The Clarendon Press Oxford University Press, New York, 1993. A friendly approach.

515. C. Wegner. *L^2-invariants of finite aspherical CW-complexes with fundamental group containing a non-trivial elementary amenable normal subgroup*. PhD thesis, Westfälische Wilhelms-Universität Münster, 2000.

516. C. Wegner. *L^2-invariants of finite aspherical CW-complexes*. Preprintreihe SFB 478 — Geometrische Strukturen in der Mathematik, Heft 152, Münster, 2001.

517. B. A. F. Wehrfritz. *Infinite linear groups. An account of the group-theoretic properties of infinite groups of matrices*. Springer-Verlag, New York, 1973. Ergebnisse der Matematik und ihrer Grenzgebiete, Band 76.

518. C. A. Weibel. *An introduction to homological algebra*. Cambridge University Press, Cambridge, 1994.

519. S. Weinberger. *The topological classification of stratified spaces*. University of Chicago Press, Chicago, IL, 1994.

520. R. O. Wells, Jr. *Differential analysis on complex manifolds*. Springer-Verlag, New York, second edition, 1980.

521. G. W. Whitehead. *Elements of homotopy theory*. Springer-Verlag, New York, 1978.

522. H. Whitney. *Geometric integration theory*. Princeton University Press, Princeton, N. J., 1957.

523. K. Whyte. Amenability, bi-Lipschitz equivalence, and the von Neumann conjecture. *Duke Math. J.*, 99(1):93–112, 1999.

524. W. Woess. Random walks on infinite graphs and groups—a survey on selected topics. *Bull. London Math. Soc.*, 26(1):1–60, 1994.

525. J. A. Wolf. *Spaces of constant curvature*. Publish or Perish Inc., Houston, TX, fifth edition, 1984.

526. K. Yano. Gromov invariant and S^1-actions. *J. Fac. Sci. Univ. Tokyo Sect. IA Math.*, 29(3):493–501, 1982.

527. S.-K. Yeung. Betti numbers on a tower of coverings. *Duke Math. J.*, 73(1):201–226, 1994.

528. G. Yu. The Novikov conjecture for groups with finite asymptotic dimension. *Ann. of Math. (2)*, 147(2):325–355, 1998.

529. A. E. Zalesskii. On a problem of Kaplansky. *Soviet. Math.*, 13:449–452, 1972.

530. S. Zucker. L_2 cohomology of warped products and arithmetic groups. *Invent. Math.*, 70(2):169–218, 1982/83.

531. S. Zucker. L_2-cohomology and intersection homology of locally symmetric varieties. In *Singularities, Part 2 (Arcata, Calif., 1981)*, pages 675–680. Amer. Math. Soc., Providence, RI, 1983.

532. S. Zucker. L_2-cohomology and intersection homology of locally symmetric varieties. II. *Compositio Math.*, 59(3):339–398, 1986.

533. S. Zucker. L^2-cohomology and intersection homology of locally symmetric varieties. III. *Astérisque*, (179-180):11, 245–278, 1989. Actes du Colloque de Théorie de Hodge (Luminy, 1987).

534. S. Zucker. *L^2-cohomology of Shimura varieties*. In *Automorphic forms, Shimura varieties, and L-functions, Vol. II (Ann Arbor, MI, 1988)*, pages 377–391. Academic Press, Boston, MA, 1990.

535. S. Zucker. *L^p-cohomology: Banach spaces and homological methods on Riemannian manifolds*. In *Differential geometry: geometry in mathematical physics and related topics (Los Angeles, CA, 1990)*, pages 637–655. Amer. Math. Soc., Providence, RI, 1993.

Notation

st(σ), 61
T_f, 42
TM, 49
$T_{\min}$, 55
$\mathbf{T}M$, 237
$\mathbf{T}_{\dim}M$, 331
$\mathbf{T}_{\mathcal{U}}M$, 331
$T \subset S$, 54
$T = S$, 55
$T(G)$, 387
$T(R \subset S)$, 385
tr$_{\mathbb{C}G}(A)$, 194
tr$_{\mathbb{C}G}^u$, 358
tr$_{\mathcal{N}(G)}$, 15
v_n, 489
WH, 31
Wh(G), 121
$X[\infty]$, 265
$||y||_1$, 486
$\alpha_p(C_*)$, 81
$\alpha_p^{\Delta}(C_*)$, 81
$\alpha_p(X; \mathcal{N}(G))$, 96
$\alpha_p^{\Delta}(X; \mathcal{N}(G))$, 96
$\Delta(K)$, 218
Δ_p, 24, 50, 51
δ^p, 50
δ_r, 87
$\theta_F(t)$, 181
$\theta_p(M)(t)$, 178
$\Lambda(G, C_r^*)_{\text{fgp}}$, 411
$\Lambda(G, F)_{\text{all}}$, 372
$\Lambda(G, F)_{\text{fg}}$, 372
$\Lambda(G, F)_{\text{fgp}}$, 372
$\Lambda(G, F)_{\text{fp}}$, 372
$\rho(\alpha)$, 405
$\rho^{(2)}(f \colon G \to G)$, 305
$\rho_{\text{virt}}^{(2)}(f \colon G \to G)$, 310
$\overline{\rho}^{(2)}(f \colon G \to G)$, 311
$\rho_{\text{an}}(M; V)$, 126
$\rho_{\text{an}}(M, \partial_0 M; V)$, 190
$\rho_{\text{an}}^{(2)}(M)$, 179
$\rho_{\text{an}}^{(2)}(M, \partial_0 M)$, 190
$\rho_{\text{top}}(M; V)$, 125
$\rho_{\text{top}}(M, \partial_0 M; V)$, 190

$\rho_{\text{top}}^{(2)}(M)$, 176
$\rho_{\text{top}}^{(2)}(M, \partial_0 M)$, 190
$\rho(X; V)$, 123
$\rho^{(2)}(X; \mathcal{N}(G))$, 160
$\rho^{(2)}(\widetilde{X})$, 163
ΣC_*, 35
$\Sigma(G)$, 387
$\Sigma(R \subset S)$, 385
$\chi^G(X)$, 281
$\chi'(G)$, 378
$\chi^{(2)}(G)$, 277
$\chi^{(2)}(X; \mathcal{N}(G))$, 277
$\chi_{\text{virt}}(X)$, 280
$||\psi||_{\infty}$, 486
$\langle \psi, y \rangle$, 487
$\Omega^p(M)$, 49
$\Omega^{p,q}(M)$, 425
$\Omega_c^p(M)$, 51
$\Omega_d^p(M, \partial_0 M)$, 63
$\Omega_\delta^p(M, \partial_0 M)$, 63
$\Omega_1^p(M, \partial_0 M)$, 63
$\Omega_2^p(M, \partial_0 M)$, 63
$\omega^p \wedge \eta^q$, 49
$\mathcal{A}$, 336
$\mathcal{AM}$, 256
$\mathcal{B}_d$, 294
$\mathcal{B}(H)$, 15
$\mathcal{C}$, 378
$\mathcal{D}$, 379
$\mathcal{D}(G)$, 387
$\mathcal{D}(R \subset S)$, 385
$\mathcal{EAM}$, 256
$\mathcal{FIN}(G)$, 369
$\mathcal{G}$, 457
$\mathcal{H}_{(2)}^p(M)$, 51
$\mathcal{H}_{(2)}^{p,q}(M)$, 426
$\mathcal{K}(H)$, 392
$\mathcal{L}(f, \lambda)$, 73
$\mathcal{L}^0(H)$, 392
$\mathcal{L}^1(H)$, 392
$\mathcal{L}^p(H)$, 392
$\mathcal{N}(G)$, 15
$\{\mathcal{N}(G)^n\}$, 247
$\mathcal{R}(G)$, 387

Index

Printing: Strauss GmbH, Mörlenbach
Binding: Schäffer, Grünstadt

Printed in the USA
CPSIA information can be obtained
at www.ICGtesting.com
LVHW082356110124
768731LV00005B/148